THE BRYOPHYTA OF SOUTH AFRICA

COMPRISING

SPHAEROCARPALES, MARCHANTIALES, JUNGERMANNIALES,
ANTHOCEROTALES, SPHAGNALES, ANDREAEALES,
POLYTRICHALES, BRYALES

BY

T. R. SIM, D.Sc., F.L.S.
Maritzburg

CAPE TOWN

PUBLISHED BY THE ROYAL SOCIETY OF SOUTH AFRICA.

1926.

Reprint
by

 OTTO KOELTZ ANTIQUARIAT

Koenigstein - Ts./B.R.D.

1973

First published in: Transactions of the Royal Society of South Africa,
Volume XV, 1926.

Printed in Western Germany
ISBN 3-87429-053-0

PREFACE.

HEREBY another new world is opened up to the South African student. For 150 years specimens collected haphazard have been sent to Europe, named and published by experts there in many languages and in many books and pamphlets, then the types were locked away as in a fireproof safe, inaccessible to the South African public except by a long pilgrimage to many shrines ; the names have been added to the world's list, the descriptions are mostly out of print or unobtainable except second hand, and thus those here who wish to study the Bryophyta have found the difficulty increasing as time went on, new or reputedly new species being added, while past work remains accumulated in pigeon-holes which only the very determined European specialist is allowed to consult or can segregate for a local flora.

It is with a view to give easy access to what is known or to what is worth knowing that the present work has been prepared ; to give a base from which the local student, as well as the European expert, can start, and to which he can add or subtract as required, but which in any case acts as an index to the past, gathers together the scattered information, and conveys the views on many points of one who has worked on the subject for thirty-five years, after a long previous training in the same line in Europe and America. If the object desired be attained, and the Bryophyta become a favourite study, much pleasure is promised to all who participate ; an easily followed foundation is hereby given, but if no mistakes have been made the miracle will indeed be wonderful.

Thanks are accorded to all helpers in whatever direction, particularly to Mr. H. N. Dixon, M.A., F.L.S., who, from England, has been the ever-willing adviser and investigator there ; to Professor J. W. Bews, D.Sc., who has rendered invaluable aid here ; to Dr. I. B. Pole-Evans, Mrs. Bolus, and Dr. Schönland, who have lent Rehmann's specimens and assisted in other ways ; to Professor Petrie and Professor A. Reid in connection with literature ; to the many who have sent in specimens, and lastly to those who have provided the funds for publication, viz. the Government Research Grant Board and the Royal Society of South Africa.

T. R. SIM.

MARITZBURG,
30th August 1925.

ERRATA.

Page 13, line 37, *change* Genadeudal *to* Genadendal.
   ,,   39, ,, 28,   ,,   *Robustior* to *robustior*.
   ,,   39, ,, 30,   ,,   *Tenerior* to *tenerior*.
   ,,   82, ,,   4,   ,,   Form *to* form.
   ,, 146, ,,   9,   ,,   *Leptobryum* to *Leptotrichum*.
   ,, 146, ,,   21,   ,,   *Dryplodon* to *Dryptodon*.

THE BRYOPHYTA OF SOUTH AFRICA.

In early days the Bryophyta of South Africa attracted the attention of several noted botanical collectors, all of whom either took or sent their collections to Europe to be examined and reported upon by themselves or by other specialists there. On this account none of the earlier type specimens remain in South Africa, except a few which have been returned from Europe to the leading South African herbaria, and even these are mostly co-types and may be different.

Thunberg included several mosses as well as several liverworts in his Prod. Fl. Cap. (1794–1800) and in his Flora Capensis (1813), and these found their way into Schwaegrichen's Prodromus (1814) and Sprengel's Syst. Veget. (1827). Among Thunberg's were what he called *Hypnum pennaeforme*, *H. asplenioides*, *Jungermannia convexa*, and *J. podophylla*.

Burchell also collected and named several species, including his *Bryum umbraculum* and *Orthotrichum lycopodioides*.

Berg, Menzies, Mund and Maire, and Krauss each collected early in the nineteenth century, mostly near Cape Town, and Gueinzius in Natal.

European botanists were about that date laying the foundations of systematic bryology, and in doing so described many European species now known to be also South African, and whose original specific names still hold good.

Linneus, following Dillenius (1741), took the lead in this, others being Swartz, Sturm, Hedwig and Schwaegrichen, Dickson, Hornschuch, Mougeot and Nestler, Palisot Beauvois, Weber and Mohr, Hooker, etc.

In 1818–1820 Hooker's Musci Exotici appeared, in which are included several South African species of mosses, along with others from the Mascarenes, New Zealand, Australia, etc.

Ecklon, who arrived at the Cape in 1823, collected sets of South African plants for sale, including among them a comparatively large number of both Musci and Hepaticae, which specimens still exist in European herbaria, and some have found their way back to South Africa; unfortunately they are in some cases without locality, but they are mostly from the Western Province of Cape Colony, as other districts were not then opened up.

His Hepaticae were described by Lehmann as *Hepaticae Capensis Ecklonii* in Linnaea, 1829, 1831, and 1834; by Lindenberg in his Species Hepaticarum (1829); by Nees in Linnaea, 1831; and by Lehmann and Lindenberg in Lehm. Pugill. Pl. stirp. nov., 1828–1831, continued by Lehmann till 1857.

Drege, who arrived in 1826, collected plants throughout Cape Colony and Natal, and descriptions were published in his Zwei Pflanzen Geographische Documente (1843), in which are included 65 Musci and 21 Hepaticae in his lists.

In 1844 Hampe published his Icones Muscorum, a most important work.

Hornschuch also began a Muscorum Frondosum Novorum, in Linnaea, xv, 1841, dealing with the collections made in South Africa by Ecklon, Drege, and Mund and Maire; in the first part 45 species are dealt with—mostly new—but though intended to be continued, no further issue seems to have been made.

In 1838 Harvey's Genera of South African Plants (first edition) appeared, in which he dealt with the Musci in 38 genera and the Hepaticae in 5 genera, viz. *Riccia*, *Anthoceros*, *Marchantia*, *Fimbriaria*, and *Jungermannia*.

In the second edition, published in 1868, the Bryophyta are not included.

Hooker and Harvey also described *Wardia* in Comp. Bot. Mag., vol. ii, p. 183, and Harvey described and illustrated others in Thesaurus Capensis (pl. 100). About this date great activity was displayed by numerous European botanists in arranging and rearranging the Bryophyta of the world in accordance with specimens in hand and views held regarding them by these different botanists; these views varied considerably, and as in these early days each botanist acted more or less independently and without a common basis from which to start, names were produced in abundance, many of which had afterwards to be sunk as synonyms.

Sprengel's Syst. Veget. (1827), Bridel's Muscologia Recentiorum (1797–1801), Muscologia, Supplementum (1806–1817), and Bryologia Universa (1826, 1827), and Dumortier's Sylloge Jungermannidearum Europae (1831), however, stand out prominently as early practical endeavours to produce order. In the latter concise little work a curious sidelight is thrown upon the state of hepaticology up to that date in the following chronological table concerning European species, in which is given the principal works published, the date of publication, and the number of species described in each, his own work including the largest number, viz. 130 species from the whole of Europe :—

Linneus, Sp. Pl., 1764	28 species described.
Hoffmann, Fl. Germ., 1795	37 ,, ,,
Gmelin, Syst. Nat., 1796	42 ,, ,,
De Candole, Fl. Fr., 1805	32 ,, ,,
Weber and Mohr, Deutschl. Krypt., 1807	45 ,, ,,
Schwaegrichen, Prodr. Hist. Hep., 1814	57 ,, ,,
Hooker, Brit. Jung., 1813–1817	81 ,, ,,
Weber, Prodr. Hepat., 1816	86 ,, ,,
Martius, Fl. Crypt. Erlang, 1817	66 ,, ,,
Sprengel, Syst. Veget., 1827	87 ,, ,,
Lindenberg, Synops. Hepat., 1829	103 ,, ,,
Dumortier, Sylloge Jung. Eur., 1830	130 ,, ,,

1

During the next decade, however, collections from abroad came in freely, and in 1845–1847 Gottsche, Lindenberg, and Nees published their Synopsis Hepaticarum, describing the known species from all parts of the world, by that time a huge undertaking, but well done, and it is still a standard work.   During its production so much fresh material came in, especially from North America and from Hooker Jr. and Taylor's Hepaticae in Hooker's Flora Antarctica, vol. i, 1844 ; vol. ii, 1847, that a very large supplement was included in the last volume (1847).

The Synopsis Hepaticarum includes very many of the South African Hepaticae, and is indispensable to the South African bryologist.   About the same time Lindenberg and Gottsche began their illustrated Species Hepaticarum, which continued to appear in parts during a decade or more, many of the South African species being included.

Soon afterwards Dr. C. Müller's Synopsis Muscorum Frondosorum (1849–1851) brought the knowledge of the world's mosses up to date, describing 2303 species and 91 doubtful species, as compared with 937 species in Bridel's Bryologia Universa (1826–1827).   In this Synopsis Müller described 473 new species, including many from South Africa, and during subsequent years he continued to describe new species from all parts of the world—an enormous number altogether.

Dr. A. Rehmann, who resided in South Africa for several years, and travelled extensively in the Cape and Knysna districts and in Natal and Transvaal, collected Bryophyta as well as other plants, and also obtained the whole of MacLea's moss collection and all Wood's mosses, and on his return to Europe distributed sets of his Musci Austro-Africani (1875–1877), including up to nearly 1200 numbers or subnumbers, many of which were repetitions of plants sent out under the same name in earlier numbers, though in some such cases the plants were not identical, and confusion consequently arose.   He named a very large number as new species and some as new genera in this distribution, but they were not published and consequently do not hold.   He also included in this distribution many named as new by Müller, but not all published.

Rehmann did good work as a collector, but his further work can only be described as careless and unreliable, and practically all of his names are now synonyms.   No list of his Exsiccatae appeared till lately, and it has been most difficult to trace his names, as the three herbaria in South Africa which have each some of his specimens do not have a full set between them, and elsewhere, I understand, full sets are rarely if ever available, but British bryologists have now taken the matter up and a list appeared in the Kew Bulletin, No. 6, 1923, prepared by Messrs. H. N. Dixon and A. Gepp, which is of great assistance in tracing what he did among the mosses.

Breutel also visited the Cape and collected largely among the southern mountains, his collections being some time afterwards worked up by W. P. Schimper, but not published separately.

Macowan, MacLea, and Bolus all collected mosses in the Graaff-Reinet and Somerset East districts ; MacLea also collected round Pilgrims Rest, and Gerrard in Natal.   Macowan transmitted his specimens to Müller, who returned a schedule of the names, but afterwards published some of them under different names.

Müller, in addition to naming species in Rehmann's and Macowan's collections, described many South African mosses later in various publications, and often dealt again with his own species, using different names, and with his earlier name stated as a synonym ; and in his Contributiones ad Bryologiam Austro-afram in Hedwigia, 1899, in which he deals with 273 South African species, mostly described as new, he also appends a list of 82 further South African mosses described by him as new in various magazines from 1855 to that date, as follows :—

Botanische Zeitung, 1855	.	.	.	.	.	.	.	.	16 species.
,, ,, 1856	.	.	.	.	.	.	.	.	9 ,,
,, ,, 1858	.	.	.	.	.	.	.	.	20 ,,
,, ,, 1859	.	.	.	.	.	.	.	.	17 ,,
,, ,, 1862	.	.	.	.	.	.	.	.	1 ,,
,, ,, 1864	.	.	.	.	.	.	.	.	1 ,,
Botanisches Centralblat, 1881, No. 37	.	.	.	.	.	.	.	1 ,,
Flora, oder Regensburger Botanische Zeitung, 1887	.	.	.	.	7 ,,
,, ,, ,, 1888	.	.	.	.	5 ,,
Oesterreichische Botanische Zeitung, 1897	.	.	.	.	.	5 ,,

Müller worked with hair-splitting exactness, among herbarium specimens only, and sometimes on fragments picked out from among specimens of other kinds, and as a natural consequence and with very incomplete material, he made many species and varieties which have not stood the test of time and of further field investigation ; indeed in one species which varies considerably even on the same specimen according to surroundings, he published from Rehmann's barren specimens four species belonging to two genera, in one of which three subgenera are represented.

Müller, in his earlier days, did grand work for bryology, but the above tendency has to be remembered in working from his later descriptions, or even in comparing Rehmann's specimens named by him ; indeed many hundreds of his species were afterwards sunk as synonyms by himself or by others, and more remains to be done in that way.   This, however, was partly due to more extended and exact knowledge of groups allowing genera to be split up and new genera established ; Müller himself did much of this during his long and useful life, but he also had the habit of referring to earlier species by using the subgeneric instead of the generic name, and it is sometimes doubtful whether or not he then intended these to stand as genera or subgenera.   The net result, however, is that the list of synonyms is largely increased thereby.

C. F. Austin published one South African hepatic—*Jungermannia Kiaeri*, in Bull. Torrey Bot. Club, 1875—one of the very few South African endemic species published in America until recent times.

Mitten, in Hooker's Handbook of the New Zealand Flora (1867), introduced new genera and many southern species, and in Journ. Linn. Soc., vol. xvi, No. 91 (1877), he dealt with the Hepaticae collected round Table Mountain, Cape, in 1874 by Rev. A. E. Eaton, and described several new species.

In the Cape Monthly Magazine, vol. xvii (1878), appears a Catalogue of the Mosses of the Cape Colony on the basis of Müller's Synopsis, being a paper read by Dr. John Shaw before the Linnean Society, 7th May 1874, prepared from the collections of Macowan, Bolus, MacLea, and himself, in which some new species are shortly described in English, and though prepared by him after considerable work in Kew Herbarium, many species are included of which the South African identifications are doubtful, and these were probably included in error, while his new species cannot stand not being fully published.

In 1886 W. H. Pearson, in his Hepaticae Natalensis, dealt with a small collection made in 1882 at Umpumulo, Natal, by Mrs. Bertelsen, and in his Hepaticae Knysnanae (1887) he dealt with a collection made at Knysna by Mr. Hans Iversen, including as a new species *Lejeunea pluriplicata*, for which Dr. Spruce had formed a new subgenus *Anomalolejeunea*.

Meantime hepaticology had received an immense impetus and much fresh direction from the publication of Dr. Spruce's Hepaticae Amazonicae et Andinae (Trans. Bot. Soc. Edinburgh, vol. xv, 1885). Spruce lived for many years in South America, and did more towards advancing the classification of Hepaticae, and especially the subdivision of unwieldy genera, than anyone else of his time ; his genera and subgenera, though primarily intended for South America, cover the case of South Africa and of all other countries in which these occur. His On Cephalozia (1882) is also a standard monograph of that group.

Almost continuously since that date Stephani has been publishing new genera and species of Hepaticae, or revising old ones, and he has monographed many genera, including South African species along with others, in Hedwigia and in other publications. He critically dealt with the *Lejeuneae* in Hb. Lindenberg in Hedwigia, 1890 ; in Hedwigia, 1891, and onward he had several contributions on Hepaticae Species Novae and on Hepaticae Africanae, in which, so far as South African species are concerned, he dealt with the few Hepaticae found in Rehmann's collections, and with those collected by Bachmann in Pondoland and by MacLea and Dr. Wilms in Eastern Transvaal, while his most important work, Species Hepaticarum, which was in course of publication from 1898 to 1908 in the Bull. de l'Herb. Boissier, and since then in the Bull. de la Société Botanique de Genève, has now been completed and is being issued separately. It brings together all known species of each group, and adds an enormous number of new species, many of which are geographical, without specific differences, but much of it is already out-dated or was originally redundant, and I have had to reduce many of his species.

In comparatively recent years Brotherus has published much on mosses, including South African species, published in Hedwigia. An important contribution to South African bryology is his Musci Africani, in Engler's Botanische Jahrbucher, 1895.

The contributions of Brotherus, Ruhland, and Warnstorf (mosses), and Schiffner (Hepaticae), to Engler and Prantl's Die Naturlichen Pflanzenfamilien (vol. i, pt. 3, 1893 to 1909) constituted a key to all earlier work but soon became out-dated, and a new edition (vols. x and xi) has been published (1924 for the mosses, and 1926 is promised for the Hepaticae) in which Brotherus deals with the classification of mosses, Paul deals with the Sphagnaceae, and Ruhland with conformation, etc., of the mosses. Many changes have been found necessary, and much geographical reduction has still to be done.

Warnstorf, who dealt with the Sphagnaceae in the first edition of the above work, reviewed that group in Hedwigia in 1890, and again later in Engler's Pflanzenreich, and in all three works went beyond the present requirements of South Africa in subdivision, and Paul has, in this new edition, maintained an impracticable standard, giving South Africa fifteen species.

Fleischer's Neue Gattung und Arten of the mosses of the Indian Archipelago, which appeared in Hedwigia through several years after 1900, is exhaustive in regard to such new genera therein as have been made by subdivision, and mentions the South African and other species which fall into these genera, as well as those of the Indian Archipelago.

Dr. Marloth's Flora of South Africa, vol. i (1913), treats briefly on the Bryophyta, and gives a synopsis of the genera of the mosses by Brotherus and of the genera of the Hepaticae by Diels.

Professor Wager has distributed collections of about 170 South African species of mosses to several South African herbaria, including several new and named but still undescribed species, while in the Trans. R. Soc. S. Afr., vol. iv, pt. 1 (1914), he described some (reputedly) new South African species.

My paper on South African Hepaticae, read before the South African Association for the Advancement of Science in July 1915 (S.A. Journ. Sc., May 1916), deals with the local adaptations of the group, and includes a synoptical list of genera, indicating the species in each ; and in S.A. Journ. Sc., April 1918, I had a paper on The Geographical Distribution of the South African Bryophyta.

Three outstanding recent physiological and morphological works are also of highest importance in reference to the structure and relationships of the Bryophyta (including South African species), viz. :—

Mosses and Ferns, by Campbell (1895).

Organography of Plants, by Goebel (1905).

The Interrelationships of the Bryophyta, by Cavers, 1911.

Cavers also described an interesting new hepatic from Port Elizabeth—*Riella capensis*, Cav. in Revue bryologique, 1903, p. 81.

I have refrained from mentioning the many European or other highly important local works in which species which are also South African are dealt with, but considering how cosmopolitan or widely distributed many of the species are, it is difficult in the present chaotic condition of South African bryology to do satisfactory work without consulting these also. Especially useful are three recent well-illustrated British works, viz. :—

Student's Handbook of the British Mosses, by Dixon (1904), and of which a third edition has recently been published (1924) ; The Hepaticae of the British Isles, by Pearson (1899–1902) ; Student's Handbook of the British Hepaticae, by Macvicar, 1912 ; while Jaeger's Genera and Species Muscorum (1871–1879) and Schimper's Synopsis Muscorum Europaeorum (2nd edition, 1876) are also most helpful.

Indeed, in dealing with the Bryophyta of South Africa under present conditions, one must know a good deal about the Bryophyta of every other country, and especially those of comparatively near localities, such as Central Africa,

Madagascar, the Mascarenes, and the Canaries. Fortunately much systematic work has been done on these, but it is not easily available.

The collections made in Central Africa by Bishop Hannington and Mr. H. H. Johnson are dealt with by Mitten in Linn. Soc. Journ., xxii, 146 (1886); those made by Schweinfurth are treated by Müller in his Musci Schweinfurthiana (Linnaea, 1875); while those of Hepaticae made by Holst, Stuhlmann, Buttner, Mönckmeyer, Fischer, Schimper, Preuss, Volkens, Moller, Schrau, etc., have been dealt with by Stephani in Hepaticae Africanae in Engl. Bot. Jahrb. (1895).

P. Dusen published New and Little-known Mosses from the West Coast of Africa (1895, 1896).

Montagne, in Webb and Berthelot's Histoire Nat. des Iles Canar., deals with the Canary Islands; and Cardot published The Mosses of the Azores in 1897.

Müller also dealt with mosses from Central Africa, West Africa, and St. Thomé, in Flora, 1885, 1886, 1890.

For Madagascar and the Mascarenes the earlier works of Hampe, Bescherelle, Wright, Renauld, and others have been utilised in the preparation of Renauld's Prodrome de la Flore Bryologique de Madagascar, Des Mascareignes et Des Comores (1897), in which 746 species of mosses are dealt with, and to which is appended a list by Stephani of 229 species of Hepaticae; and also in the preparation of Renauld and Cardot's Histoire des Mousses de Madagascar (1899, 1914); and in Bescherelle's Florule Bryologique de la Reunion (1879) species are also described now known to be South African.

Pearson described some of the Hepaticae of Madagascar in Chr. Vid. Scl. Forh., 1892, pp. 8 and 14, and Bescherelle and Spruce dealt with others in Bull. de la Soc. Bot. de France, xxxvi (1889).

Australia and New Zealand have a considerable bryological literature of their own, partly condensed in Rodney's Tasmanian Bryophyta, and their species have some relation to those of South Africa, as has been shown by Dixon in his Studies in the Bryology of New Zealand (N.Z. Inst. Bull., iii, 1913, 1914).

Paris, Index Bryologicus (2nd edition, 1903), has come as an indispensable aid in locating synonymy and in indicating the geographical distribution of the species. It is a vast undertaking, carefully done, but unfortunately there are several important omissions of South African species (*i.e.* those of Shaw, Dr. Harvey, and some of Mitten's); also many are not mentioned as South African which have been so recorded, but to what extent this is intentional and how far unintentional is not clear, while, in the other direction, Rehmann's and other exsiccatae and manuscript names which have never been published are included.

He concludes that there are 476 endemic and 82 non-endemic species belonging to South Africa, a total of 558 species of mosses, as compared with 876 names of mosses and 169 of Hepaticae recorded as reported from South Africa in my check-list of the Bryophyta of South Africa (1915, published in manuscript, reproduced by plex process), but both in his list and in mine are many names now reduced to synonyms.

Another slightly rearranged check-list of the mosses of South Africa appeared under Professor Wager's name in 1917.

N. Bryhn published in 1911 Bryophyta nonnulla in Zululand collecta, in which Brotherus, Stephani, and Warnstorf are thanked for help, and several new species of mosses are described by Brotherus and Bryhn.

In 1913 Brunnthaler, who in 1909 had touched at various localities in South and East Africa, published in Vienna a paper in the Denkschriften der Mathematisch-naturwissenschaflichen Klasse der Kaiserlichen Akademie der Wissenchaften, in which, among other things, Stephani dealt with the Hepaticae and Brotherus with the Musci, including some new species and extended range for others.

One of the most active workers among mosses during recent times has been Mr. H. N. Dixon, who, besides his Handbook already mentioned and many papers on mosses from elsewhere, has published the following relating to Africa or having a bearing on African mosses, viz.:—

Miscellanea Bryologica, in Journal of Botany, 1912–1922 (8 contributions); New and Rare African Mosses from Mitten's Herbarium and other Sources (1916); Uganda Mosses collected by R. Dummer and Others (1918); The Mosses collected by the Smithsonian African Expedition, 1909–1910 (1918); New and Interesting South African Mosses (Trans. R. Soc. S. Afr., 1920); Reports upon Two Collections of Mosses from British East Africa (Dummer-Maclellan Exp. to Mount Elgon, 1918), 1920; Rhaphidosteguim caespitosum Sw. and its Affinities (1920); Dr. Stirton's New British Mosses, revised (1923); Studies in the Bryology of New Zealand (N.Z. Inst. Bull., No. 3, 1913, 1914); Some New Genera of Mosses, Journal of Botany, April 1922, p. 101; Bryophyta of Southern Rhodesia (S.A. Journ. Sc., June 1922: Mosses, H. N. Dixon; Hepaticae, T. R. Sim); Rehmann's South African Mosses, by H. N. Dixon and A. Gepp (Kew Bulletin, No. 6, 1923), and he has helped immensely as a correspondent and European referee in the preparation of the present work.

Foreign Bryophyta for comparison are not well represented by specimens in South African herbaria; indeed I believe my own herbarium contains much more of this nature than all others put together, including a nearly full set of the mosses and Hepaticae of Britain, collected by my father and myself about fifty years ago; Sphagnaceae Britannicae Exsiccatae (1877), by Dr. Braithwaite; Musci Appalachiani (1870), by C. F. Austin; Hepaticae Boreali-Americanae (1873), by C. F. Austin; North-American species collected by myself in 1879, 1880; and a set of Wright's Hepaticae Cubensis presented to me then by Dr. Asa Gray.

Though exceedingly useful as first aids, these are altogether insufficient for the critical work on exotic species which alone can correlate African with these exotic species and simplify the world's moss-flora, a duty which devolves on those who have abundant cosmopolitan material at hand.

The exceedingly scattered nature of the literature dealing with the South African Bryophyta and of the type specimens on which that is based, as revealed in the preceding pages, and the identity or close connection of many species with those of other countries, has so far rendered the subject a closed book for all except a few European experts who have added to our local difficulty by continuing to describe new species without giving easy access to the descriptions of those formerly described, or sinking those so deserving. It is with a view to relieve that difficulty that the present work has been prepared. It has been in preparation for years, but owing to the European War and connected causes its publication in the ordinary manner became impossible during the war period, so, to allow South African students an opportunity of becoming familiar with at least the genera, fifty-two copies of a manuscript volume describing the genera of the mosses (reproduced by plex process) were issued and distributed in 1917, and the same arrangement of material, with the same or extended descriptions of the genera, have been used herein, together with descriptions and illustrations of the species and of any added genera.

To assist in clearing the way I published papers on—

South African Hepaticae (S.A. Journ. Sc., May 1916) ; Geographical Distribution of the South African Bryophyta (S.A. Journ. Sc., April 1918) ; Bryophyta of Southern Rhodesia : Mosses by Mr. H. N. Dixon, Hepaticae by myself (S.A. Journ. Sc., June 1922) ; The Mosses of the South-West Portion of South Africa (S.A. Journ. Sc., 1924).

Table Mountain, with its daily bright sunshine and daily mist-cloud, has probably more species, forms, and conditions of Bryophyta, and especially of Hepaticae, than any other locality in South Africa, and it is still far from fully explored.

Unsuspected causes may be at work in connection with distribution, even locally, here and elsewhere. Shaw, in 1878, pointed out that several species of mosses are confined to trap rock, and that there is an almost total absence of alpine forms, but the latter statement is now disproved.

In the Journal of Ecology, November 1918 (vol. vi, Nos. 3 and 4), pp. 189–198, W. Watson deals with The Bryophytes and Lichens of Calcareous Soil, and gives British lists of calcifuge, indifferent and calcicole species, and says (p. 192), " The factors influencing the distribution of the higher plants are both chemical and physical, and there is some doubt as to the relative importance of these factors. In regard to many bryophytes and lichens there seems little room to doubt that the chemical factor is much more important," and then he gives instances.

This subject has received no attention in South Africa so far, though it is well-known that certain species are almost confined to walls or culverts where lime is used, or to lime rocks, or soil.

In the Annals of Botany, vol. xxxvi, No. cxlii, April 1922, p. 193, W. F. F. Ridler has a paper on a fungus present in *Pellia epiphylla*, Corda (a species of Phoma), in which he concludes that the relationship of the fungus to the gametophyte may be symbiotic, though it sometimes kills the tissues, but toward the sporophyte it is distinctly destructive. The paper contains valuable historical notes on records regarding Fungi on Mosses and Hepaticae and bibliography on the subject. Here the presence of *Nostoc* in *Anthoceros*, and the occasional destruction of capsules of Rhaphidosteguim by Fungi are about the only South African records taken so far.

EXPLANATION OF THE ILLUSTRATIONS.

The Thalloid Hepaticae have the figures explained in the text on each species ; thereafter the following common plan of explanation is followed in the Hepaticae, and that for the mosses precedes the classification of them.

A. Plant, natural size.	K. Male spikes or its bracts.
B. Upper surface of stem.	L. Androecia.
C. Under surface of stem.	M. Cell formation of leaf.
D. Leaf (moist), or part of leaf.	MM. Lower cells of leaf.
E. Leaf or plant (dry).	N. Leaf-apex.
F. Lobule.	O. Elaters and spores.
G. Foliole.	P. Archegonia.
H. Perichaetal leaves.	Q. Capsule.
I. Perianth.	S. and SS. See text for that species.
J. Section of perianth.	

Key to the Hepaticae.

(The key to each group or genus will be found in connection with it in its place.)

Cellular plants, thalloid or foliose, in which fertilisation is effected by antheridia and archegonia, producing spores from which fresh development starts.

From the mosses the foliose forms differ in that they have spiral elaters or sterile cells present among the spores in the capsule (except Ricciaceae), the rhizoids are unicellular, capsule seta hyaline and evanescent, peristome always absent, leaves frequently 2-lobed and bilateral, without midrib, and usually folioles are present.

Plant thalloid .	1.
Plant foliose (not thalloid)	12.
1. Archegonia and antheridia included singly in the prostrate thallus	RICCIACEAE.
1. Archegonia and antheridia included singly in the erect convolute thallus .	RIELLA.
1. Sporogonium an erect linear organ, splitting vertically	ANTHOCEROS.
1. Archegonia exserted in various other combinations from the thallus	2.
2. Thallus divided into layers of different tissue and usually has surface pores from air cavities (*Marchantiales* except *Riccia*)	3.
2. Thallus not divided into layers and with no pores or air cavities. Archegonia not terminal ; capsule destitute of lid, opening by four valves (*Thalloid Jungermanniales Anacrogynae*).	10.
3. Archegonial group terminal, almost immersed, enclosed in two scales	TARGIONIA.
3. Archegonial group on peduncled receptacles	4.
4. Capsule dehixing irregularly, or its lid becoming detached in one piece (*Operculatae*)	5.
4. Each involucre contains a group of archegonia ; capsule splitting into four to eight teeth (*Compositae*)	8.
5. Carpocephala dorsal, produced in succession on the thallus surface, one archegonium in each involucre. Androecia cushioned, surrounded by scales	PLAGIOCHASMA.
5. Androecia not cushioned nor surrounded by scales	EXORMOTHECA.
5. Carpocephala terminal ; involucre, with often more than one archegonium	6.
6. Perianth none	7.
6. Special basket-like perianth present, of many membranous segments	FIMBRIARIA.

7. Capsule lid falling away in fragments leaving a cup . . . . . REBOULIA.
7. Capsule lid hemispherical, separating bodily . . . . . GRIMALDIA.
8. Peduncle not grooved ; gemmae-cups crescent-shaped . . . . LUNULARIA.
8. Peduncle 2-grooved ; gemmae-cups round when present . . . . . 9.
9. Air-chambers and pores usually absent ; no gemmae-cups . . . . DUMORTIERA.
9. Air-chambers and pores conspicuous ; gemmae-cups round . . . . MARCHANTIA.
10. Sexual organs from short marginal branches . . . . . ANEURA.
10. Sexual organs from short ventral branches . . . . . METZGERIA.
10. Sexual organs from the upper surface (*Blyttiaceae*) . . . . . 11.
11. Perianth present ; calyptra thin . . . . . PALLAVICINIA.
11. Perianth absent ; calyptra succulent . . . . . SYMPHOGYNA.
12. Archegonia not terminal ; folioles absent, leaves waved or toothed ; rhizoids purple . FOSSOMBRONIA.
12. Archegonia at first terminal on stem or branch, the involucre representing true leaves (*Acrogynae*) . . 13.
13. Elaters one-spiral . . . . . 14.
13. Elaters with two or more spiral fibres in each (*Jungermanniae*) . . . . 15.
14. Branches intra-axillary ; lobule convex toward the upper lobe . . . . FRULLANIA.
14. Branches infra-axillary ; lobule concave toward the upper lobe . . . . LEJEUNEACEAE.
(For genera herein see Lejeuneaceae.)
15. Leaves incubous, closely conduplicate, 2-lobed, the lower lobe the smaller . . . . 16.
15. Leaves not as above . . . . . 17.
16. Folioles present . . . . . MADOTHECA.
16. Folioles absent . . . . . RADULA.
17. Leaves closely complicate, 2-lobed . . . . . 18.
17. Leaves 2-lobed, spreading or laxly conduplicate . . . . . 19.
17. Leaves deeply 2-fid, not conduplicate . . . . . 22.
17. Leaves undivided or 2-fid, but not deeply . . . . . 24.
17. Leaves various, not included above . . . . . 33.
18. Lobes equal ; central cells like a midrib . . . . . HERBERTA.
18. The upper lobe the smaller . . . . . SCHISTOCHILA.
19. Perianth not angled . . . . . 20.
19. Perianth 3-angled ; third angle postical ; leaves succubous . . . . 21.
20. Folioles present, like the leaves . . . . . ANTHELIA.
20. Folioles absent . . . . . MARSUPELLA.
21. Leaves equally bifid . . . . . CEPHALOZIA.
21. Leaf with saccate lobe on upper margin . . . . . NOWELLIA.
22. Lobes bifid and often ciliate . . . . . LEPICOLEA.
22. Lobes simple . . . . . 23.
23. Perianth 3-angled . . . . . Some species of LOPHOCOLEA.
23. Perianth laterally compressed, 2-labiate . . . . Some species of LEPTOSCYPHUS.
24. Leaves incubous . . . . . 25.
24. Leaves succubous . . . . . 26.
25. Marsupium descending . . . . . CALYPOGEIA.
25. Perianth erect . . . . . Some species of BAZZANIA.
26. Leaves not toothed beyond any bidenture . . . . . 27.
26. Leaves further toothed . . . . . 32.
27. Odd angle of perianth postical . . . . . 28.
27. Odd angle of perianth antical . . . . . 29.
28. Stem rhizomatous . . . . . ALOBIELLA.
28. Stem not rhizomatous . . . . . ODONTOSCHISMA.
29. Perianth free . . . . . 30.
29. Perianth an erect marsupium . . . . . 31.
30. Folioles usually absent . . . . . JAMESONIELLA.
30. Folioles bifid, leaves rounded . . . . . CHILOSCYPHUS.
30. Folioles bifid, leaves more or less 2-3-dentate . . . . LOPHOCOLEA.
31. Leaves alternate, rhizoids colourless . . . . . NOTOSCYPHUS.
31. Leaves alternate, rhizoids red . . . . . NARDIA.
31. Leaves opposite . . . . . GONGYLANTHUS.
32. Marsupium short ; lower side of leaf toothed . . . . . ADELANTHUS.
32. Perianth laterally compressed ; leaf toothed anywhere . . . . PLAGIOCHILA.
33. Leaves deeply 3-4-fid . . . . . 34.
33. Leaves 3-4-fid, not deeply . . . . . 35.
34. Leaves usually 3-fid, cut nearly to the base and ciliate . . . . CHANONDANTHUS.
34. Leaves cut to the base, segments setaceous . . . . . LEPIDOZIA.
34. Leaves usually 4-fid, cut deeply . . . . . LEPIDOZIA.
35. Leaves incubous ; apex 2-3-fid or almost entire . . . . Some species of BAZZANIA.
35. Leaves succubous . . . . . 36.
36. Perianth laterally compressed . . . . . Some species of LEPTOSCYPHUS.
36. Marsupium solid, descending . . . . . TYLIMANTHUS.

HEPATICAE.

(Names marked * are *nomina nuda, i.e.* unpublished.)

Order I. SPHAEROCARPALES.

Sexual organs superficial and almost sessile, but each separately enclosed in a special envelope developed from the thallus, and extending beyond the enclosed organ. Sterile cells are mixed among the spores, but no spiral elaters ; the capsule is single-layered, without fibrous thickenings, and bursts irregularly. Usually thalloid with no air-chambers in the thallus, but with or without leaves in addition. Rhizoids smooth.

Family 1. RIELLACEAE. Stem erect with one undulate one-layered dorsal thalloid expansion, and sometimes also provided with small leaves from each side of the stem. Archegonia in flask-like envelopes on the stem along the base of the wing ; antheridia on the free edge of the wing enclosed separately in specially thickened portions of the wing.

Genus 1. RIELLA Mont., Ann. Sc. Nat., 3, xviii, 11.

Plant unilaterally thalloid, thin, branched, usually small ; if growing in water, erect and rooting at the base only, but sometimes left dry, then more procumbent and rooting more freely. Wing unilateral on the rib, mostly one cell thick, much twisted and appearing spirally inserted but not so ; bracts (or leaves) not numerous, small, set on the rib, and on the female plant mixed with the archegonia. Archegonia mostly near the branch-points ; involucre bottle-shaped ; capsule almost sessile, included, globose. Sterile cells small, not spirally marked ; antheridia sunk in a nearly continuous marginal line, the lower ones being shed before the upper have all appeared. Tender water-weeds resembling Algae.

1. *Riella capensis* Cavers, in Revue Bryologique, No. 5, 1903, p. 81, with plate.

Dioecious, annual, 1–2 inches high, much-branched, growing in brak water or mud, resembling a sea-weed. Wing 1–3 mm. wide, much twisted, one-sided (dorsal) on the rib, and consisting mostly of one layer of cells, except the submarginal rim in which the antheridia are embedded. The wing continues to the base of the stem, but is wider upward. Bracts minute, cordate or reniform, placed on the rib in a single line, abundant at the growing points, more scattered lower, and in the female plant intermixed among the archegonia, which are also in a single line on the rib,

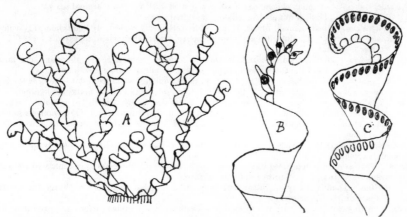

Riella capensis.

A, plant, nat. size ; B, female branch, ×5 ; C, male branch, ×5.

and most abundant near the growing apex. Involucre bottle-shaped, erect, as long as half the wideness of the wing. Capsule globose, included, shortly stalked. Antheridia sunk in a nearly marginal disc, continuous along the outer margin of the wing in the male plant.

Described in 1903 by Cavers from plants raised in England from submerged mud sent dry from Port Elizabeth by Mr. Hodgson, containing Crustaceae, which plants in three months (February to May) came to the fruiting stage. It has since been found in abundance in the more or less tidal pools on the Cape Flats by Mr. Garside and others. Probably often passed as Algae, for which reason also the known distribution of the genus is restricted to Europe and North Africa, with this one exception.

Order II. MARCHANTIALES.

Thalloid, the thallus usually differentiated into (1) a hyaline epidermis ; (2) an upper zone of green tissue, with or without air-chambers ; (3) a lower zone of large-celled hyaline tissue.   Pores from the air-chambers usually present in the upper surface ; lunate scales or ridges often present in rows on the under-surface, each usually bearing at first an appendage, which in its youngest stage folds over and protects the growing point, which consists of a group of initial cells.   Smooth and tuberculate rhizoids usually both present.   Sexual organs either embedded separately in the thallus or collected into groups (special receptacles), which are either sessile or more or less pedunculate.   No columella present.

Family 2. RICCIACEAE.   Antheridia and archegonia immersed singly in cavities of the upper surface of the thallus.   Sterile cells not present among the spores.

Genus 2. RICCIA Linn., Sp. Pl., p. 1138 ; Mich. Nov. Pl. Gen., p. 107.
= *Riccia* and *Ricciella*, Braun.
= *Riccia* and *Ricciocarpus*, Schiffner, Pfl., i, 3, 15.
= *Riccia, Ricciella, and Ricciocarpus,* Corda.

Small thalloid plants, growing on mud, or by immersion floating in water, usually more or less stellate or rosette-shaped in arrangement of the thalli, at least at first.   Thalli linear to obcordate, emarginate, usually succulent ; upper surface smooth, or warted, or sometimes bearing mamillate cells or cilia.   Lower surface usually protected by squamae crossing the thallus, but by rupture at the midline forming two rows of scales.   Cellular structure different in the sections, but always includes intercellular air-spaces, of various forms, which sometimes with age lose their epidermal layer of cells and appear as external pits.   Differentiated midrib not present (except under special conditions of *R. purpurascens*), but the thallus is usually channelled more or less along the middle of its surface.   Antheridia and archegonia immersed singly in cavities in the thallus ; antheridia opening to the upper surface by narrow osteoles, archegonia opening out by wider mouths, the capsule on maturity opening to either upper or lower surface in accordance with species.   Spores tetrahedral ; sterile cells not present among the spores.

Many species ; they occur in mud in most parts of the world.

Synopsis.

§ 1. LICHENOIDES.   Thallus without air-cavities further than perpendicular intercellular spaces between individual rows of cells.

  (*a*) Epidermal cells, or some of them, elongated or mamillate and free.
    R. coronata.   Marginal epidermal cells mamillate in one coronal row ; scales present.
    R. natalensis.   Marginal epidermal cells mamillate ; scales absent.
    R. albo-marginata.   All epidermal cells elongated, forming 5-celled pillars ; scales large, white.
  (β) Epidermal cells not much elongated ; thallus provided with ventral marginal scales.
    R. Pottsiana.   Thallus minute, flabellate ; marginal scales small.
    R. africana.   Thallus small, linear, 1 mm. wide, repeatedly forked, without channel ; scales pointed.
    R. atro-purpurea.   Thallus ultimately purple above, linear, channelled, repeatedly forked, 1 mm. wide ; scales not pointed.
    R. limbata.   Thallus medium-sized, irregularly forked ; ventral scales abundant and large, purple, extending beyond the thallus as a purple margin.   Cells in compact columns.
    R. concava.   Thallus large, subflabellate ; segments oblong, when fresh flat except at the boat-shaped apex, very concave when dry.   Cells in lax columns ; scales small, marginal, hyaline.
§ 2. FAVOIDES.   Thallus with only one layer of air-cavities ; these deeply cylindrical, hexagonal, separated by walls one cell thick, and when young furnished with roof and minute apical pore.   This resembles a honeycomb.
    R. bullosa.   Thallus large, succulent, bullate, channelled, convex both directions.   Spores with numerous minute areolae.
    R. Garsidei.   Thallus large, succulent, slab-like, hardly furrowed ; cavities eventually open ; spores with few large areolae.
§ 3. SPONGODES.   Thallus mostly composed of numerous large superimposed air-cavities, separated by walls one cell thick ; capsules eventually opening to the upper surface.
    R. crystallina.   Thallus firmly succulent and deep, with rounded margins ; with age the cavities mostly open as pits on the upper surface.
§ 4. RICCIELLA.   Thallus cavernose ; capsules formed in a swelling on the under-surface and eventually bursting to the under-surface.
    R. fluitans.   Thallus green, firm, usually forked or repeatedly forked ; segments linear, with hyaline margin.
      var. *limicola*, growing on mud.
      var. *aquatica*, floating in water.
    R. purpurascens.   Thallus stoloniferous, softly membranaceous, dichotomous ; laciniae ligulate, forked at the apex ; margin sometimes purple.
§ 5. RICCIOCARPUS.   Thallus composed mostly of superimposed air-cavities, with connecting and external pores ; sexual organs in several rows in a ridge immersed under the mid-furrow ; ventral scales numerous, independent, narrow.
    R. natans.
      var. *limicola*, growing on mud.
      var. *aquatica*, floating in water.

2. *Riccia coronata* Sim (new species).*

Thallus segments obcordate, 4 mm. long, 2 mm. wide, level on the upper surface, slightly convex on the under-surface, hardly grooved except at the sinus, and having a few hyaline cells along the under-surface, then 8–10-celled perpendicular lines of small round chlorophyllose cells, with one epidermal upper layer of similar but hyaline cells, compactly placed, but with quadrangular interspaces between the upright lines. Near the margin of the thallus one line of long white mamillate cells, erect or inflexed, gives a crown-like appearance. Scales fairly large, horizontal when moist. Archegonia scattered or in a double line. Spores flat on front, convex on back, evenly reticulated, and made roughly triangular by protruding hyaline border. Mooi River, Natal, on soil (Sim 8730).

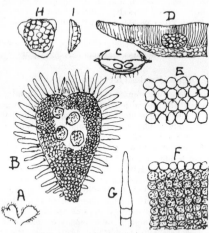

Riccia coronata.

A, thallus, nat. size; B, same, ×10; C, transverse section of thallus; D, longitudinal section of thallus; E, epidermal cells; F, chlorophyllose and epidermal cells; G, coronal cells; H, I, spores.

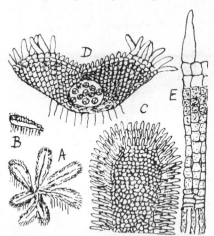

Riccia natalensis.

A, thalli, nat. size; B, longitudinal section; C, enlarged view of thallus; D, cross-section of same, showing mamillae and sporangium; E, cells in vertical section of thallus.

3. *Riccia natalensis* Sim (new species).†

Thallus 5–12 mm. long, 2–3 mm. wide, ligulate, wider upward, rounded at the apex, grooved along centre of upper surface, 1 mm. deep, rounded on the under-surface, without lamellae or scales on the under-surface; texture of the thallus dense, composed of minute round cells in lines of eight to ten cells in the section through the thallus, these lines rising obliquely forward, the surface of the thallus covered with about two layers of larger, much laxer cells, from which all along the outer portion of the thallus surface rise pellucid, single-celled mamillae, numerous and irregular, giving an appearance of white scales to the thallus when it is dry. Several thalli originate from the same point, but with age, by extension and the decay of old parts, the star form disappears, and they lose apparent connection. Sporangium embedded in the body of the thallus, globose, 1 mm. diameter, and containing about twenty to thirty opaque spores. Rhizoids abundant along the whole length of the thallus, occupying the central portion. Though the thallus is distinctly concave, and more so or channelled when dry, there is no nerve.

Scheeper's Nek, Natal (Sim 8228); Wellington, Rosetta, Natal (Sim).

4. *Riccia albo-marginata* G. L. N., Syn. Hep., p. 604; Steph., Sp. Hep., 1898, p. 329.

Thalli gregarious, at first somewhat stellate, soon scattered, simple or once forked, dark green, obcordate, 4–5 mm. long, 2–4 mm. wide, flat when moist, not furrowed, widely and persistently fringed with large, rounded, laxly celled hyaline white scales, which when dry are closely folded over the then inflexed thallus. Thallus solid, section 2 mm. deep, flat above, nearly semi-circular below, the lower third composed of lax empty round cells, the next third

---

* *Riccia* (*Lichenoides*) *coronata* Sim (sp. nov.).

Thallus obcordatus, cum 8–10 cellatis lineis rectis cellarum chlorophyllosarum parvarum ac rotundarum confertim collocatarum, spatiis autem quadrigonis inter lineas rectas interpositis. Stratum epidermale hyalinum, cum una linea marginali cellarum mamillatarum longarum albarumque. Squamae grandes adsunt.

† *Riccia* (*Lichenoides*) *natalensis* Sim (sp. nov.).

Thallus ligulatus; cellae densae in lineis rectis positae; earum, quae in stratis epidermalibus exsistunt maiores et laxiores, a quibus secundum aream marginalem oriuntur plurimae cellae singulae mamillatae pellucidae irregulares. Squamae desunt.

of pillars of short chlorophyllose cells about ten cells deep, in close proximity, which are continued above as pillars of long empty cells about five cells deep, gradually narrowed upward, especially the terminal cell ; these upper pillars quite free from one another and without epidermal covering, forming, with the white scales, excellent protection against drought and sunshine. Scales on the lower surface white throughout, closely placed, very large, rounded at the apex, giving a remarkable white margin to the thallus. Dioecious. Antheridia and archegonia each in a single central line, immersed to the base of the chlorophyllose tissue and without fissure, the osteole single ; archegonia only seen young ; spores not seen. Very distinct in its wide white border of scales (forming half the width of the surface) and in its pillars of empty cells narrowed upward. Suited for a dry climate or seasonal marshes.

Western Cape Province, Orange Free State, Natal, Transvaal, etc.

R. villosa Steph. (Brunnthaler, 1913, pp. 1–14), from Matjesfontein, C.P. (an arid locality), answers this description exactly.

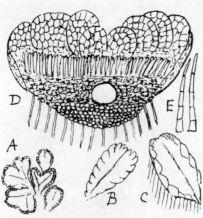

Riccia albo-marginata.

A, thallus, nat. size ; B, under-surface of same, ×3 ; C, upper surface, ×5 ; D, apex of thallus, including section, ×20 ; E, free pillars of cells, ×40.

5. *Riccia Pottsiana* Sim (new species).*

Gregarious, dioecious, minute ; thalli at first stellate, often permanently so, but the central portion dies out leaving large open holes ; lobes 1–2 mm. long, seldom more, simple or dichotomously or somewhat pinnately divided, the latter condition coming through the production of a young central bud in the sinus between two lobes, that bud afterwards growing out actively while the side lobes also fork. Lobes tumid, three times as wide as deep, furrowed at the apex ; under-surface widely lunate ; upper surface also widely lunate where not furrowed, the margins rounded. Section shows a few lines of rather large horizontal elliptical empty cells at the base, above which the chlorophyllose tissue about ten cells deep is made of closely placed perpendicular lines of rounded or somewhat compressed cells, all chlorophyllose except the epidermal layer which also is of globose cells, very tumid when moist, the surface showing these cells in parallel lines running straight to the margin of each lobe, except those nearest a sinus which curve rapidly inward toward it as if development were still proceeding there, and there are angular perpendicular interspaces between the chlorophyllose cells, but no larger

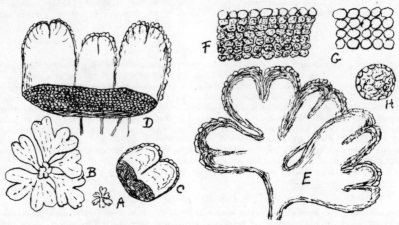

Riccia Pottsiana.

A, rosette, nat. size ; B, same, ×10 ; C, apex of thallus ; D, central bud developing ; E, thallus, ×15 ; F, chlorophyllose and epidermal cells (section) ; G, epidermal tissue, from above ; H, spore.

cavities. Ventral scales small, hyaline, or sometimes purplish, of few large cells, and extending only slightly beyond the margin, forming a slight white border. Archegonium central in the old thallus. Spores round, indistinctly reticulated ; antheridia not seen. Abundant among grass near Bloemfontein, usually mixed with *R. albo-marginata*,

---

* *Riccia (Lichenoides) Pottsiana* Sim (sp. nov.).

Thallus minutus, flabellatus ; lobi dichotome vel pinnatim divisi ; cellae epidermales haud elongatae, cellae inferiores densae in lineis rectis positae, spatiis angulatis rectis interiectis. Squamae ventrales parvae, marginem tenuem album efficientes.

from which its size, its smaller scales, and its cell-formation easily distinguish it. The smallest *Riccia* known to me, seldom more than 4 mm. across the whole plant, very succulent, and dark green, with a habit of dying off white at the centre of each rosette. Related to *R. concava*, but much smaller. (Potts, No. 5.)

6. *Riccia africana* Sim (new species).*

Dark green, growing in rosettes 8–12 mm. diameter with three to four primary divisions, each two to three times dichotomously forked. Thallus and segments linear throughout, 0·8–1·0 mm. wide, half that depth, without air-cavities or areolae, the section showing the upper surface horizontal or convex, or near the apex slightly channelled, the margins rounded, and the lower surface slightly convex. The epidermal cells are 0·04 mm. and hyaline, and several layers of cells along the lower surface are hyaline ; all the other cells are densely chlorophyllose and in close contact, not in perpendicular lines, globose, or through pressure polygonal. At the apex small hyaline, large-celled, ventral scales occur, mostly triangular-acute, forming a marginal border, but early caducous. Capsules numerous, irregularly placed, 0·4 mm. diameter, containing few mother-cells, the spores of which continue to adhere, and are papillose on the convex sides, without areolae.

Stellenbosch, Cape Prov., Garside No. 8, and Miss Duthie (Sim 8417).

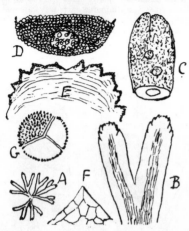

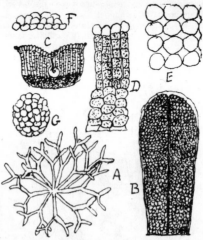

Riccia africana.

A, thallus, nat. size ; B, apex of segment, under-surface, ×7 ; C, apex, upper surface, with capsules visible, ×10 ; D, section of thallus, ×20 ; E, under-surface, ×30 ; F, apex of scale, ×100 ; G, spores, ×100.

Riccia atropurpurea.

A, thallus, nat. size ; B, apex of the thallus, upper surface, ×30 ; C, cross-section of thallus, ×20 ; D, same in part, ×50 ; E, epidermal layer, from above, ×100 ; F, scale, ×50 ; G, spore (immature).

7. *Riccia atropurpurea* Sim (new species).†

Dark brown or olive brown ; thalli numerous, at first spreading in an irregular whorl 1–2 cm. diameter, afterwards forming an overlapping mass several inches wide, each thallus linear, several times dichotomously branched, 0·5–1·0 mm. wide, channelled, rooting throughout, the terminal segments widest behind the apex and rounded or slightly emarginate at the apex. Section of segment 1·5–2 times as wide as deep, channelled, with high angular shoulders, perpendicular sides, and gently rounded base, which has about four horizontal layers of large cells, above which the smaller cells are in contiguous pillars with 4-angled intercellular spaces between the pillars and open at the top ; all the cells chlorophyllose except the flattened epidermal cells on the lower surface and the globose epidermal cells on the upper surface, which latter rupture early. Marginal scales evident at the growing apex, of two to three rows of hyaline cells, afterwards remaining as a narrow indistinct hyaline margin to the thallus, without points. Capsule on the median line ; mature capsule and spores not seen. Resembles *R. fluitans* in appearance but not in structure.

Edendale Falls, Natal (Sim) ; Wellington, Rosetta, Natal (Sim).

---

* *Riccia (Lichenoides) africana* Sim (sp. nov.).

Thallus parvus, linearis, bis terve dichotome furcatus, non canaliculatus. Cellae epidermales hyalinae ; cellae inferiores dense chlorophyllosae et confertae, non in lineis positae. Desunt cavernae. Squamae ventrales minutae, acutae, caducae.

† *Riccia (Lichenoides) atropurpurea* Sim (sp. nov.).

Thallus parvus, linearis, crebrius dichotome furcatus, canaliculatus ; cellae exiguae, in columnis contiguis dispositae, cum interstitiis quadrigonis. Desunt cavernae. Squamae marginales non acutae, tantum ad apicem apparentes, deinde marginem angustum thallo efficientes.

8. *Riccia limbata* G. L. N., Syn. Hep., p. 606 ; Steph., Sp. Hep., 1898, p. 328.

Dioecious, glaucous green, more or less rosulate when young, ultimately by decay of older parts scattered or matted in a mass closely resembling Fimbriaria ; thallus simple, or once or twice dichotomously forked, the primary divisions 1–2 cm. long, narrowed toward the base ; segments 2–4 mm. wide, flat or slightly convex above, with an indistinct central channel throughout except at the withered base, the channel pronounced near the apex. Small forms occur with divisions 2–3 mm. long, 2 mm. wide, in circles up to an inch or more in diameter, the inner parts having decayed away or been replaced by less regular circles. Mature segments two to five times as wide as deep, a section having widely semicircular outline on the lower surface ; thallus solid ; air spaces crowded out, the cells being in contact throughout and hexagonal in cross-section (except the epidermal layer), 0·04 mm. wide, 0·04 mm. deep near the surface, rather larger lower down, arranged in parallel lines and passing gradually into the hardly distinct though smaller basal layer; the epidermal cells when young globose or flatly mamillate, evanescent; the next layer roundly quadrate, hyaline, forming ultimately the epidermal layer. Ventral scales evident, incurved when dry, imbricated, purple, persistent, extending somewhat beyond the thallus margin, and giving it the appearance of having a purple margin. Capsules 0·4 mm. wide and deep, in one central line in some segments and in two lateral lines in other segments on the same plant, one to twelve in succession on a segment, ultimately bursting as black dots through the upper surface; the earlier capsules mature while successive capsules are still being formed near the apex. Where sporangia are numerous the tissue bursts as a channel, leaving the sporangia sitting free. Spores abundant, nearly black, very solid, without margin, 8–10μ diameter, reticulated with about six to eight areolae across the diameter each way, an end view showing the back to be abundantly papillate, these papillae being the highest points on the ridges. A cross-section near the apex of a thallus shows a deep channel, with its sides approaching, and the protecting purple scales at the apex are evident. Perennial, and endures drought as a xerophyte.

Cape, Natal, Transvaal, etc , frequent.

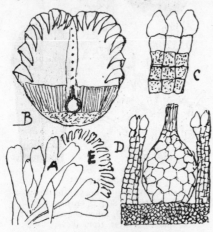

Riccia limbata.

A, thallus, nat. size ; B, apical cross-section, showing young portion of thallus and protective purple scales ; C, epidermal hyaline cells (two lines) and chlorophyllose cells below them, ×75 ; D, archegonium, in ruptured tissue, and surrounding cells ; E, small form.

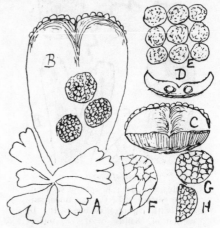

Riccia concava.

A, thallus, nat. size, young ; B, part of thallus with position of capsules indicated ; C, section and view of apex, ×7 ; D, section of thallus, ×7 ; E, cross-section of chlorophyllose tissue, ×100 ; F, ventral scale, ×10 ; G, H, spores, ×100.

9. *Riccia concava* G. L. N., Syn. Hep., p. 604 ; Steph., Sp. Hep., 1898, pp. 325, 378.

Dioecious, growing in open rosettes 1·5–3·5 cm. diameter, at first bright green, later scurfy grey ; branching palmate or dichotomous, segments 0·5–2 cm. long, 2–4 mm. wide, ligulate, flat on the upper surface except near the apex, which is acute and deeply channelled through the margins at first meeting face to face, expanding with age. Section of mature portion flat on the upper surface, forming a segment of a circle on the under-surface, the depth being one-third to one-half the width ; the basal small cells forming 0·3–0·4 of the depth, the upper stratum practically solid, composed of upright pillars of lax chlorophyllose cells oval lengthways, 0·07 mm. diameter, with a larger globose epidermal cell on each, these pillars standing in contact and having smaller 3–4-angled intercellular spaces between the adjoining pillars, but no larger cavities. Apex protected by tender hyaline scales, which in early stages form the lateral crest but soon disappear. Capsules several, in two lateral rows, alternate, immersed, opening to the upper surface, 0·8–1·0 mm. diameter ; the calyptra chlorophyllose but evanescent ; spores numerous, 8μ in diameter, somewhat angular, laxly reticulated, with about five areolae on the diameter each way, brown, and without hyaline margin. Antheridia single or few, oval, 1 mm. long, 0·04 mm. wide and deep.

Stellenbosch, etc., Western Cape Province.

10. *Riccia bullosa* Link., in Lindb. Syn. Hep., p. 119 ; G. L. N., Syn. Hep., p. 609 ; Steph., Sp. Hep., 1898, p. 377.

Dioecious ; thallus glaucous, terrestrial, very succulent, 7–15 mm. long, 5–7 mm. wide, usually once dichotomously forked, or through decay of the older portion the cordate branches stand singly, or two connected at the base only. Upper surface convex both longitudinally and laterally ; central channel very pronounced but narrow. Cross-section 1·5–2 times as broad as high, subquadrate, with the upper surface twice as wide as the base. Under-surface same colour ; ventral scales absent ; rhizoids abundant, all tubercled, and with somewhat inflated upper portion inside the thallus. Keel of the thallus composed of numerous small spherical cells 0·05–0·07 mm. diameter ; upper portion a honeycomb of deep hexagonal hyaline air-cavities in one layer, about 12 wide on each side of the central channel, separated by walls one cell thick, these cells hyaline, oblong-hexagonal, 0·15 × 0·05 mm. ; each cavity almost closed at the surface by a semispherical single layer of cells consisting of three to four circles surrounding a minute apical pore ; with age these cellulose coverings break down, leaving the air-cavities open at the top, also with age the green thallus becomes occasionally wrinkled on the upper surface longitudinally. When broken down by age and exposure the much-developed open air-cavities appear as a thick coat of irregular, lax, pellucid scales superlying a bright green thallus, coating it as if with raised white pouches.

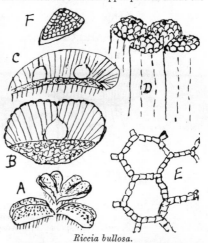

Riccia bullosa.

A, thalli, nat. size ; B, cross-section of thallus ; C, longitudinal section of same ; D, air-cavities ; E, section across same ; F, spore.

Antheridia and archegonia in separate plants, immersed under the mid-groove, singly or in one row in each segment. Calyptra at first 2-walled, the sporogonium single-walled, flatly spherical, 1·5 mm. diameter, containing numerous spores, mostly adherent in threes or fours, and no sterile cells ; the spores brownish, roundly tetrahedral, 17μ diameter, the reticulate outer surface having about ten areolae across the diameter each way, with raised margins. Antheridia not seen, but said by Mr. Garside to be confined to the middle line and to have very long osteoles. When young the superficial layer of cells covers the central channel, including undeveloped dorsal surface, but with age that layer ruptures and the surface develops in depth, still retaining, however, a deep channel. Mature plants apparently die off except the apex, which starts next season usually as a wedge-shaped bud or bulb.

Stellenbosch Flats and other Western Province localities. Ookiep, Namaqualand, on rocks (M. H. Giffen, July 1924) ; Premier Mine, Transvaal (Dr. Pole-Evans, 1925).

R. capensis Steph. (Brunnthaler, 1913, i, 14) appears to be a young sterile condition of this with " postical squamae crowded, dark purple," and is from Genadeudal, C.P.

11. *Riccia Garsidei* Sim (new species).*

Dioecious, glaucous, terrestrial ; thallus very succulent, simple or once dichotomously forked ; segments 1–1·5 cm. long, 6–8 mm. wide, nearly flat or toward the apex somewhat channelled, the upper surface at first areolate with a pore in each cavity, but with age the epidermal layer disappears leaving the air-cavities open, and sometimes several cavities then have no cell-wall between them, leaving the older portion of the thallus an open spongy mass. Tissues and transverse section as in *R. bullata*, and the same one-layered honeycombed arrangement of tubular cavities exists. Colour of under-surface same as that of upper surface ; ventral scales absent ; rhizoids abundant from the wide keel and swollen in the tissue. Sporogonium deeply embedded under the middle line, single or several in a line, opening eventually to the upper surface, 1·5 mm. diameter, flatly spherical, containing numerous separate spores. Spores 12μ diameter, black, globose, the reticulated under-surface having about six areolae across the diameter each way. Differs from *R. bullosa* in having larger, usually simple, thalli, flat and eventually pitted on the upper surface, and in the smaller and black spores with larger areolae, but nearly related.

Stellenbosch Flats (Garside 2).

12. *Riccia crystallina* Linn., Sp. Pl., p. 1605 ; Steph., Sp. Hep., 1898, p. 368.

Plant a rosette 10–20 mm. across, with about five firm cuneate-obovate primary divisions 5–10 mm. long, each once or twice dichotomously branched, the segments overlapping ; ultimate divisions obcordate. Cross-section shows width three times the depth, the upper and lower surfaces parallel, with rounded margins, the dense keel extending nearly the whole width, and the upper stratum composed of large air-cavities in one to three layers, separated by walls one cell thick and with similar epidermal walls in which no pores are naturally present, but with age the upper epidermal walls decay, leaving large pits into the cavities, some of which coalesce. Ventral scales not present. Capsules numerous in a central line, by decay of tissue opening to the upper surface, the spores mostly remaining three to four together. Spores 0·08 mm. diameter, echinate, the papillae mostly free but united to form indistinct areolation in the

---

Thallus grandis, sucidus, tabulatus, subsulcatus. Cavernae magnae, cylindricae, rectae, denique apertae. Sporae paucas areolas gerunt.

central part of the convex surface.   River-sides and vleys ; scattered over South Africa, usually as a few scattered plants, seldom gregarious.   Occurs also in Europe and in Rhodesia.

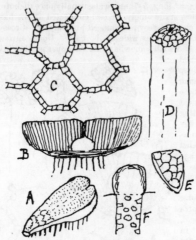

Riccia Garsidei.

A, thallus, nat. size ; B, cross-section of
thallus, ×5 ; C, section of cavities ;
D, cavities ; E, spore ; F, rhizoid.

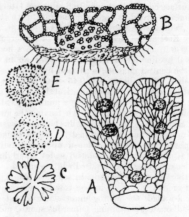

Riccia crystallina.

A, part of thallus, ×7 ; B, section of
same, ×20 ; C, plant, nat. size ; D,
four spores before separation, ×50 ;
E, spore, ×100.

13. *Riccia fluitans* Linn., Sp. Pl., p. 1606 ; Steph., Sp. Hep., 1898, p. 366.

Thallus branches linear, flattened, containing at least two layers of air-cavities, irregularly superimposed.   Capsules produced several in succession, each forming a protuberance from the under side of the median line of the thallus, and opening eventually to the lower surface.   Margin of thallus hyaline.   Widely distributed.

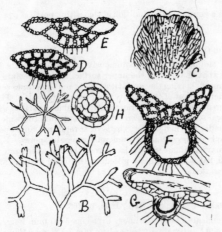

Riccia fluitans-limicola.

A, young plant, nat. size ; B, one division of mature
plant, nat. size ; C, apex of thallus, ×10 ; D,
section of thallus of dry growth, ×20 ; E, section
of more vigorous thallus, ×20 ; F, transverse
section of fertile thallus, ×20 ; G, segment, cut
longitudinally to show position of capsule, ×10 ;
H, spore, ×200.

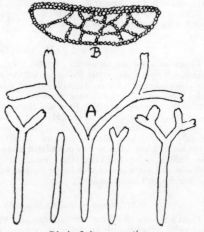

Riccia fluitans-aquatica.

A, thalli, nat. size ; B, transverse section.

Form, *limicola* ; growing on mud, brilliant green, at first in a rosette, the segments 0·5 mm. wide, but repeatedly dichotomous and soon passing into a mass of overlying segments, the individual then 3–5 cm. long and wide, five to six times forked, forming an openly flabellate arrangement, the linear segments varying with vigour from 1–2 mm. wide, and usually the more vigorous the less fertile. Thallus composed almost entirely of cavities in two to three irregular layers, separated and enclosed by walls one cell thick, the cells globose and 0·04 mm. diameter. In the vigorous forms the outer cavities are large and the central smaller, giving an appearance of midrib; in the less vigorous forms they are more equal in size.

Cross-section shows a flat or convex upper surface, a convex under-surface, and subacute margins, the wideness being two to three times the depth. Ventral scales are practically absent, but each outer cavity at first has the appearance and protective function of a short rounded scale, and through life the plant retains the crenate hyaline margin thus created. Capsules about 1 mm. wide and deep, in a swelling on the under-surface, while a distinct corresponding depression occurs on the upper side of the thallus. The tissue enclosing the capsule is abundantly rhizoid-producing. Spores 0·05 mm. wide, with a wide hyaline margin, and a few large areolae on the convex surface. Frequent on mud or moist soil throughout South Africa, enduring drought and standing the winter, and fruiting freely under rather dry conditions, but less so when moist and vigorous.

Form, *aquatica*; floating, perennial, densely packed in masses; plants light green, simple or one to three times forked, 3–5 cm. long, 1·5–2 mm. wide, rounded or emarginate at the apex, thin, without rhizoids, not rooting and never fertile, but forming adventitious ventral branches which detach and form new plants.

Common, sometimes forming a sud in cold pools.

14. *Riccia purpurascens* L. and L. Pugill., Pl. iv, 1832, p. 23; Steph., Sp. Hep., 1898, p. 363.

Light green in colour, softly herbaceous, not evidently rosulate, but more or less palmate; primary divisions 10–15 mm. long, once or twice dichotomously branched, 2–3 mm. wide, cuneate from a narrow base, rooting at the base only and with the basal and apical portions often horizontal with an arched portion between. Thalli overlapping in mass, undulate, deeply channelled throughout, at the apex the sides often plaited over, especially where dichotomy is in progress. Section of thallus 0·5 mm. deep at the centre, with an evident keel of small cells varying from two to ten cells deep; the rest of the thallus consists of air-chambers mostly about 0·15 mm. wide by 0·30 mm. long, two layers toward the centre, one layer toward the margins, separated by a wall having only one layer of cells, these 0·05 mm. diameter; the epidermal walls are of one layer of cells; the air-chambers are tumid when young, without pores, and showing on the surface as hexagonal areolae radiating from the centre, and each containing five to six hexagonal cells in each diameter. No ventral scales are present, but at the apex the margin forms semicircular hyaline scales, 1–2 cells long, 0·5 mm. wide, at first overlapping, and these form a permanent crenate margin to the whole thallus, and are rudimentary representations of the ventral scales more highly developed in other species.

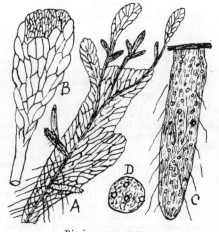

Growing among stream mosses in Knysna district, with a general appearance more like *Pallavicinia* than *Riccia*, but with decidedly Riccioid formation. Sexual organs not seen by me. L. and L. state concerning Ecklon's Table Mountain plant that it is dioecious and that they have only seen young spores, and have not seen the male plants; also that it is stoloniferous, the stolons starting from the middle of the lower side of the thallus, producing radicels below and a flat surface above. I have not seen the purplish colour they mention, from which it takes its name. Most of the specimens of this species sent me by Miss Duthie and by Mr. Garside were normal, as described above, but several were in a very abnormal condition which throws light on its propagation. These were occupying the ground surface and were dominated by others above them. Such an one is depicted in fig. A. The main thallus, 15 mm. long, 4 mm. wide, is irregular as to outline, crenate throughout, with distinct midrib on the lower surface and hardly channelled on the upper surface, but producing from the under-

Riccia purpurascens.

A, dominated thallus starting fresh colonies by tuberous rhizomes, ×3; B, apical portion becoming flattened again, ×15; C, one of the tuberous rhizomes descending into the soil, ×15; D, section of same, with oil-glands.

surface of the midrib several terete, fleshy, green tubers, 3 mm. long, simple or branched, descending into the soil, producing rhizoids, and containing numerous oil-glands which are absent from other parts of the plant.

The apex of the dichotomously forked midrib, however, extends beyond the lamina as three terete similar stems, also branched and containing oil-glands, the terminal branches again becoming flat, cavernose, and crenate, evidently the beginnings of a new series of thalli. These specimens answer the difficult question of how this species (at least sometimes) survives and multiplies by semi-detached thalli, each starting from a narrow terete stem.

Table Mountain (Ecklon); Stellenbosch (Garside); Knysna (Miss Duthie).

15. *Riccia natans* Linn., Syst. Veget., p. 956.

=*Ricciocarpus natans* Corda, in Opitz, Beiträge, p. 651, October 1898; Steph., Sp. Hep., 1898, p. 757.

Plant dimorphous; usually floating and 1–2-lobed, with long flat chlorophyllose scales pendent in the water, but when settled on mud flabellate or stellate with short scales. The section *Ricciocarpus*, containing only this species, is characterised "Thallus composed mostly of superimposed air-cavities with connecting and external pores. Sexual organs in several rows in a ridge immersed under the mid-furrow. Ventral scales numerous, independent, narrow."

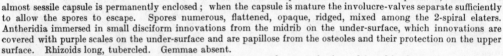

Riccia natans.

Form A, *limicola*. E, thallus, nat. size ; F, section, × 4.
Form B, *fluitans*. A, under-surface ; B, upper surface ; C, scales of thallus, × 4 ; D, point of scale, × 25.

The plant is dioecious and the thallus succulent and obcordate. The two extreme forms depending on the extent of immersion are :

Form A, *limicola* ; thallus flabellate or stellate, deeply forked, its branches 10 mm. long, obcordate or deeply forked, forks 2 mm. wide, rounded at the apex, succulent, deeply channelled, but the margins of the channel meeting ; cavities numerous, narrow ; scales usually absent except when quite young ; rhizoids present and capsules frequent. Grows on mud.

Form B, *fluitans* ; thallus branched, larger and widely obconic and less channelled. Squamae long, narrow, flat, serrate scales, pendent in the water. Rhizoids absent and capsules usually absent.

This species is very distinct, especially in its floating form, characterised by the numerous air-chambers and the long flat serrate scales. Found in Natal and Portuguese Maputa.

Family 3. TARGIONIACEAE. Archegonial group terminal, sessile, enclosed in two scales ; antheridia on special short branches. Elaters 2–3-spiral.

Genus 3. TARGIONIA Linn., Sp. Pl., p. 1604 ; Steph., Sp. Hep., 1898, p. 763.

Thallus ligulate, coriaceous, indistinctly areolate, innovating from the midrib on the under-surface, and consisting of a lower layer of large-celled colourless tissue, a cavernose layer containing branched columns of smaller chlorophyllose cells, and a hyaline epidermal layer, through which a few pores open into the cavities. Archegonia formed several in succession, in juxtaposition on the under-surface, immediately behind the growing point. After fertilisation of one or more archegonia the apical growth of the thallus ceases, but a large, sessile, purplish-black bivalve involucre is produced for each fertile perianth, in which the almost sessile capsule is permanently enclosed ; when the capsule is mature the involucre-valves separate sufficiently to allow the spores to escape. Spores numerous, flattened, opaque, ridged, mixed among the 2-spiral elaters. Antheridia immersed in small disciform innovations from the midrib on the under-surface, which innovations are covered with purple scales on the under-surface and are papillose from the osteoles and their protection on the upper surface. Rhizoids long, tubercled. Gemmae absent.

A small genus containing only two or three species, more or less xerophytic, and found in the tropics and warmer temperate regions.

16. *Targionia hyophylla* Linn., Sp. Pl., p. 1604 ; Steph., Sp. Hep., 1898, p. 764 ; Brunnthaler, 1913, pp. 1–14.

=*T. capensis* Hüben, Hep. Germ., p. 17 ; G. L. N., Syn. Hep., p. 574.

Monoecious. Thallus ligulate or cuneate-oblong, crenate throughout, green above, brownish-purple below, flat when moist, involute when dry, innovating freely from the midrib on the under-surface, the segments 1–1·5 cm. long, 2–3 mm. wide, with a distinct midrib in the older portion (sometimes reduced to a terete stem), and with about three lines of pores on each side of the midrib, those nearest the margin largest and most raised ; the crenate margin hyaline except the marginal line of oblong cells which are usually purplish ; other portions of the thallus chlorophyllose, with an epidermal layer of hyaline oblong-hexagonal cells 0·04 mm. long, occasionally containing oil-glands. Archegonial group subterminal under the apex of the main thallus or larger innovations ; usually only one archegonium becomes fertilised, and that produces a sessile, purplish-black, compressed, keeled bivalve involucre 2 mm. long, which makes the fertile plant easily recognisable, in which the globose subsessile capsule 1 mm. diameter is contained, the capsule-wall having annular brown thickenings. Involucre valves composed of one layer of irregularly quadrate or hexagonal purple cells 1 × 0·5 mm. diameter. Spores opaque, circular, somewhat flattened, distinctly crenulate-margined. Ventral scales purple, triangular-acute, arranged in one line on each side of the midrib, and reaching half-way to the margin, hardly overlapping except on the male innovations which barely reach the margin.

Mountain regions and interior, rare ; also Rhodesia, and Ookiep, Namaqualand (M. H. Giffen).

Family 4. MARCHANTIACEAE. Antheridia and archegonia in separate groups ; archegonial groups placed on peduncled receptacles ; sterile cells (usually spiral elaters) present among the spores.
Section 1. OPERCULATAE. Capsule dehiscing irregularly, or by its lid becoming detached in one piece.
Subsection 1. Carpocephala dorsal, produced in succession on the thallus surface, the peduncle not grooved or furrowed.

Genus 4. PLAGIOCHASMA, L. and L. in Lehm. Pug. Pl., iv, 13 ; G. L. N., Syn. Hep., p. 511 ; Steph., Sp. Hep., 1898, p. 775.

Thalloid, gregarious, monoecious or dioecious (sometimes in one species) ; thallus simple or forked, extending at the apex, or with innovations from the lower surface ; flat above, with cavernose air-chambers in two or more layers ;

stomata small ; chlorophyllose cells abundant ; epidermal cells hyaline ; midrib weak ; purple lunate scales abundant on the under-surface of the thallus, with simple or double folioles, or these mixed. Carpocephala from the upper surface of the thallus, several produced separately in succession on the mid-line, small, pedunculate, the peduncle

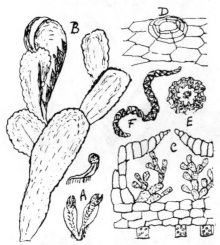

Targionia hyophylla.

A, thallus, nat. size ; B, plant, ×5 ; C, part of section of thallus, ×100 ; D, pore, ×200 ; E, spore, ×100 ; F, elater, ×200.

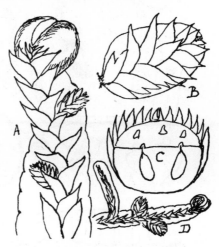

Targionia hyophylla.

A, thallus, showing under-surface and two male discs, ×10 ; B, under-surface of male disc, ×7 ; C, section and point of disc, showing osteoles, ×20 ; D, under-surface of thallus, ×2.

not grooved or furrowed, but barbed with filaments at the apex. Involucres one to four, with one archegonium in each, capsules spherical, eventually erect in South African species through absence of dorsal tissue on the receptacle, dehiscing irregularly. Spores large, reticulate, rough. Androecia sessile cushions in succession along the mid-line of the upper surface of the thallus, each surrounded by many scales, these scales remaining permanently after the antheridia have gone.

Many species, of which several belong to Africa.

Synopsis.

One fertile capsule produced . . . . 17. *P. capense.*
Usually two fertile capsules produced.
  Lunate scales bifid, segments triangular . . . . . . 18. *P. tenue.*
  Folioles with lanceolate hyaline points . . . . . . 19. *P. rupestre.*
  Horns of capitulum densely verrucose . . . . . . 20. *P. Dinteri.*

17. *Plagiochasma capense* Sim (new species).*

Thallus 10 mm. long, simple or forked, its branches oblong, 5 mm. wide, with an epidermal layer of large indurated hyaline cells papillose above, under which numerous air-cavities occur irregularly in two layers toward the centre of the thallus and in one layer further out, the intermediate tissue being chlorophyllose cells ; the lower stratum is of large hyaline cells, and the midrib absent. Scales not extending to the margin, except at the growing apex where evanescent lanceolate folioles incurve as protection, but the under-surface of the thallus has numerous purple lines (not scales) extending to the margin. Pores small, 6-sided. Peduncle 2–3 mm. long set in simple scales and barbed with numerous scales at the top, and bearing one white scarious globose calyptra sitting free on the top of the peduncle ; this splits on top and contains one almost sessile globose capsule included, which bursts irregularly and falls away in fragments. Spores granular ; elaters long and obtuse, hardly spiral. Androecium unknown ; tubercled rhizoids present.

Herschel, Cape Province, 5000 feet (Hepburn).

---

* *Plagiochasma capense* Sim (sp. nov.).
  Thallus ad 10 mm. longus, simplex vel furcatus ; cellae epidermales grandes, induratae, hyalinae, supra papillosae ; cavernae in 1–2 stratis cellis chlorophyllosis separatae ; squamae absunt praeterquam ad apicem, ibi evanescentes et lanceolatae. Una tantum capsula fertilis in alba calyptra ; haec dirumpitur irregulariter atque in fragmenta dilabitur.

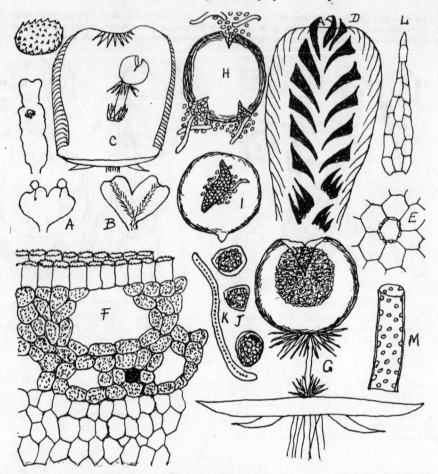

Plagiochasma capense.

A, thallus, nat. size, upper surface ; B, same, lower surface ; C, part of same, ×5 ; D, under-surface of same, ×5 ;
　E, pore ; F, section of mid thallus ; G, section of thallus, showing calyptra ; H, I, capsule bursting ; J, spores ;
　K, elater ; L, foliole, ×20 ; M, tubercled rhizoids.

18. *Plagiochasma tenue* Steph., Bull. de l'Herb. Boissier, vi (1898), 779.

Thallus 2 cm long, 2–4 mm. wide, ligulate, flat above, thin but firm, innovating from the apex ; the lower side purple ; midrib weak or absent. Pores small. Scales large, deeply bifid, the lobes triangular-acute, not constricted below. Peduncle ½ inch long, with many filaments at top. Receptacle 2-lobed, lobes bursting by an upward slit ; spores small, reticulated ; elaters long, 3-spiral.
　South and East Africa, rare.

19. *Plagiochasma rupestre* (Forst.), Steph., Bull. de l'Herb. Boissier, vi (1898), 779.

=*Aitonia rupestris* Forst., Comm. Soc. Gott., ix, 73.
=*Plagiochasma Aitonia* Nees, Natur. Eur. Leb., iv, 41.

Thallus ligulate, 5–15 mm. long, 2–4 mm. wide, firm, emarginate at the apex, flat, glaucous on the upper surface but not areolate, purple under, usually simple or sparingly and irregularly innovating from the midrib on the lower surface. Section of thallus shows one line of large hyaline epidermal cells, then a layer three or four cells deep of smaller chlorophyllose cells mixed with irregular air-cavities in two layers, then the lower layer several cells deep of empty and rather larger cells, but no evident midrib. Lower surface convex, midrib weak and wide or absent, pores small. Carpocephala usually 2-lobed, the cells each opening by a slit its whole length and having the appearance of a bivalve. Spores small ; elaters long, 3-spiral. Androecia described by Stephani as produced on small adventitious lobes from the under-surface, of which several may appear, but what I take to be the male of this species has the antheridophores produced in succession along the middle line of the upper surface of the thallus, each in a slight depression, one to six being sometimes present in various stages, the antheridophore about 1 mm. diameter containing six to nine antheridia embedded in it, and opening to papillose osteoles on its upper surface when mature, the antheridophore set among

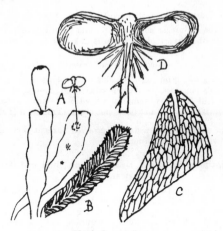

Plagiochasma tenue.

A, thallus, nat. size ; B, same, under-surface ;
C, scale, ×10 ; D, carpocephalum, ×10.

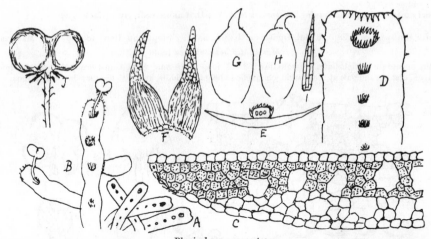

Plagiochasma rupestre.

A. thalli, nat. size ; B, large fertile thallus ; C and E, sections of thallus ; D, upper surface of thallus ;
F, folioles ; G, H, same ; I, fibril off peduncie ; J, carpocephalum.

numerous simple crimson scales which are persistent for years though the antheridophore itself disappears.   Usually or always dioecious.   When moist the thallus is flat, with the entire margins apparently ciliated from below by the folioles ; when dry the margins are much incurred and set with the numerous white lunate scales reaching nearly or quite to the margin, each ending in a bifid scarlet foliole, each portion ovate below, but suddenly contracted about half-way up, the lower portion scarlet; the upper portion lanceolate and hyaline, and it is these points that appear as cilia to the thallus, and that incurve to protect the growing point.

Natal, Transvaal, Orange Free State, etc.; not common.

20. *Plagiochasma Dinteri* Steph., Sp. Hep., vi, 7.

Not known to me ; described by Stephani as " Monoecious, frond thin, flat, red or green, gregarious on earth. Thallus to 15 mm. long, 6 mm. wide, with the margin lightly crenulate ; stomata large, convex, with six cells, 3-seriate.

Epidermal cells thin ; nerve low, wide ; scales purple, crowded, appendiculate, large, widely ovate, acute, hyaline, entire. Peduncle of the capitulum 7 mm. long, strong, paleaceous at the apex ; capitulum emarginate at the top, 2-horned ; horns short, obtuse, densely verrucose ; valves closely connate, conduplicate, concave, smooth. Other parts not seen. South-West Africa (Dinter)."

Genus 5. EXORMOTHECA Mitt., in Goodman, Hist. Azores, p. 325.

Much like *Plagiochasma*, but the androecia not disciform and not surrounded by scales. The species are from Africa and S. Europe ; the one described by Stephani below is unknown to me and described as follows :—

21. *Exormotheca africana* Steph., vi, 18.

Dioecious, small, nearly black, growing on soil. Thallus 5 mm. long, narrow, simple or lightly forked, strong, twice wider than thick ; surface highly papillose, papillae crowded, large, conical, obtuse ; mouth minutely perforated ; postical scales large. Seta bearded at the apex, sometimes 2-setose. Carpocephala small, subglobose (0·83 mm.), trilocular, truncate below, the mouth small. Other parts not seen. Transvaal (Pole).

Subsection 2. Carpocephalum terminal, its peduncle grooved (except *Lunularia*). Involucres horizontal or pendent, often containing more than one archegonium.

Genus 6. REBOULIA Raddi, Opusc. Sc. di Bol., ii, 357 ; Steph., Bull. de l'Herb. Boissier, vi (1898), 790 ; Macvicar, p. 35.

Dioecious, thalloid ; thallus thin ; air-chambers in one or two layers, chlorophyllose tissue abundant. Under-surface purple, scales few, long, and each ending in a single or double lanceolate foliole. Peduncle ½ inch long, rising from the upper surface behind the sinus, not quite terminal, 1-grooved. Pores minute ; midrib present but not pronounced. Carpocephalum conical, 4-lobed, the lobes descending, crenate at the margin. "Involucres arising from the ventral margin of the receptacle lobes, conchoidal and 2-valved, each enclosing a single sporogonium which does not fill the cavity. Pseudo-perianth absent. Capsule subglobose, shortly pedicillate, with a large foot, irregularly dehiscing at the apex, the lower portion being left behind as a hemispherical cup containing the spores and elaters " (Macvicar). Male inflorescence a cushion on the upper surface, behind the sinus, and surrounded by linear scales. Gemmae not present.

A widely distributed genus, having one species only, but that an exceedingly variable one.

22. *Reboulia hemispherica* (Linn.), Raddi, Opusc. Sc. di Bol., ii, 357 ; Steph., Sp. Hep., 1898, p. 790 ; Macvicar, p. 35.

= *Marchantia hemispherica* Linn.

Thallus obcordate or flabellate, emarginate, 1–3 cm. long, 5–10 mm. wide, thin toward the margin, innovating from the apex, crenate and pale at the margin, channelled along the middle, and with lighter side corrugations. Margins

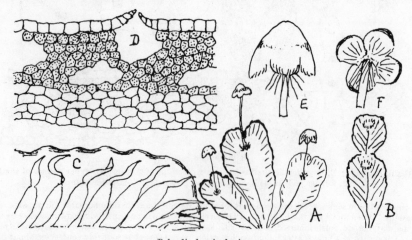

Reboulia hemispherica.

A, fertile thalli, nat. size ; B, male thallus, nat. size ; C, under-surface of thallus, ×4 ; D, section of thallus ; E, carpocephalum, upper surface, ×4 ; F, carpocephalum, under-surface, ×4.

ascending, undulated, surface smooth. Chlorophyllose tissue abundant, air-chambers numerous in it and in the midrib which merges gradually. Epidermal cells 4–6-angled with thickened angles. Peduncle 5–15 mm. long, not quite terminal on the thallus, surrounded by lanceolate scales at the base and barbed with linear scales at the apex. A rare species in South Africa.

Genus 7. GRIMALDIA, Raddi, Opusc. Sc. di Bol., ii, 356 ; G. L. N., Syn. Hep., p. 549 ; Steph., Sp. Hep., 1898, p. 792.

Thalloid, usually matted ; thallus dark green, the air-cavities densely packed with pillars of chlorophyllose cells, and opening by pores through the hyaline epidermal layer. Scales oblique on the under-surface, adherent their whole length, then produced into usually two segments, or with a detached basal one. Monoecious or dioecious. Female receptacle rising from the under-surface of the terminal sinus, peduncled, peltate, convex, and reticulated or papillose above, and containing on its under-surface two to four spherical capsules, each occupying a lobe, without perianth, opening downward, and bursting by dehiscence of a hemispherical lid. Peduncle one-grooved, the groove carrying down tuberculate rhizoids ; peduncle paleaceous at the base and apex, remaining short for a long time, then suddenly elongating on maturity. Spores large, very remarkable in the South African species, having a black or brown opaque cushion-like base, on which rests the larger hemispherical body of the spore, covered with transparent globes which make it appear to be bordered with transparent separate cells. Elaters highly developed, 3-spiral at the middle, 2-spiral at the ends. Antheridia embedded irregularly throughout the thallus, with papillose osteoles, or collected into variable cushions, which are not peduncled nor surrounded by scales.

A small xerophytic genus, found mostly in the Northern Hemisphere and in Africa.

23. *Grimaldia capensis* Steph., Bull. de l'Herb. Boissier, vi (1898), 793.

Thallus ½–2 inches long, strap-shaped or jointed, deep green, acutely angled at apex of segments or of thallus, simple or forked

Grimaldia capensis.

A, thallus (young plant), nat. size ; B, same, more mature ; C, side view, ×5 ; D, thallus, ×5 ; E, carpocephalum (young), ×15 ; F, same (mature), ×7 ; G, section of same, ×7 ; H, lid of capsule, ×10 ; I, capsule, ×10 ; J, spore, ×100 ; K, scale from base of peduncle, ×25 ; L, cells of scale ; M, cells of surface of thallus ; N, chlorophyllose scales ; O, tuberculate rhizoid ; P, section of peduncle, showing groove.

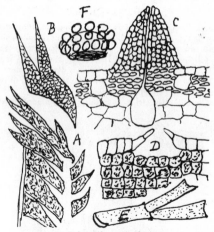

Grimaldia capensis.

A, scales of thallus, showing occasional double row ; B, one scale ; C, male osteole, ×30 ; D, pore, ×20 ; E, thallus, nat. size ; F, spore, ×200.

at the segments, or somewhat branched from the under-surface ; closely adherent to bare soil or moist rock ; surface flat and areolate when moist, with numerous minute raised pores ; margins closely inrolled when dry, and presenting a closed cylinder protected by purplish-black scales. Epidermal cells hyaline, hexagonal as seen from above and much thickened at the angles, but quadrate vertically. Pores small, simple, slightly raised, with several reduced surrounding cells. Air-chambers not evident, as they are full of chlorophyllose tissue, but there are occasional chambers in the lower layer. Lunate crimson scales on under-surface abundant, at first with two folioles, and often a basal one afterwards forming a free scale. Peduncle rises from among abundant simple, crimson, lanceolate scales, and has a few scarlet fimbrils at its apex. Receptacle 3–4 mm. diameter, hardly 2-3-4-lobed, from the under-surface of which the one to four globose, separate, black capsules protrude downward, without perianth, the capsule walls consisting of a single layer of honeycombed hyaline cells without markings, and the capsule ultimately opening by the dehiscence of nearly half its wall as a complete lid, the separation taking place at a line of smaller and differentiated cells. Spores as described for the genus, black, and making the capsule appear black. The capsules in a receptacle are normally four, but by abortion they are frequently one, two, or three. Gemmae-cups not present.

G. capensis is described by Stephani as monoecious, but I find the antheridia produced also on separate thalli intermixed among the fertile ones.

A South African species, xerophytic in habit, widely distributed, but not common.

Genus 8. FIMBRIARIA Nees, in Hor. Phys. Berol., 1820, p. 45 ; G. L. N., Syn. Hep., p. 555 ; Steph., vii, 84 (1899).
=*Hypenantron* Corda ; Schiffner, Pfl., i, 3, 33.

Thalloid, monoecious or dioecious, procumbent, rooting ; thallus ligulate, emarginate, usually simple or sparingly forked or innovated, flat and deep green above when moist ; the under-surface somewhat convex, with one double row of lunate purple scales, usually folioled at the apex and sometimes rooting below ; when dry the lower surface inrolls over the upper, showing mostly scales. Section of thallus has well-defined midrib and a layer of cavities in which free filaments are present in some species. Pores small, seldom prominent. Peduncle terminal, rising from among many scales in the sinus, 1-furrowed ; carpocephalum peltate, with a central area and normally four lobes, in the margin of each of which is an involucre, the perianth hanging, companulate, and divided into many parallel hyaline rib-bands coherent at the apex, basket-like, until mature. Calyptra delicate ; capsule included. Spores reticulated, yellow ; elaters short, 1–2-spiral. Frequent along streams or moist banks, often under very xerophilous conditions, and able through inrolling of the margins to endure much drought.

Synopsis.

Dioecious ; capitulum set with long narrow papillae   .   .   .   .   .   .   24. *F. Bachmannii.*
Monoecious.
   Capitulum papillose.   Thallus ½ inch long   .   .   .   .   .   25. *F. muscicola.*
   Capitulum with semi-globose stomata ; thallus 1–1½ inch long   .   .   26. *F. marginata.*
   Capitulum plane ; thallus often 2 inches long   .   .   .   .   27. *F. Wilmsii.*

24. *Fimbriaria Bachmannii*, Steph., Hedw., 1894, p. 7 ; Steph., Sp. Hep., 1899, p. 105.

Thallus 1–2 inches long, ¼ inch wide, ligulate, innovating from the apex or from the under side of the midrib ; pores prominent ; section of thallus shows midrib one-third of its wideness, the upper layer narrow, with shallow caverns and few free filaments. Scales sparse, purple, triangular-acute, often bearing delicate hyaline, oblong or lanceolate folioles, constricted at the base. Peduncle ½ inch long, with many linear scales at the apex. Capitulum hemispheric, set with long narrow papillae, deeply 4-lobed, lobes decurved, crenate, papillose. Perianth pendulous, hyaline. Spores small, rough. Elaters short, 2-spiral, obtuse. Plant dioecious. Androecia terminal, large, oblong, sessile convex, with large purple osteoles.

Throughout South Africa, especially along mountain streams.

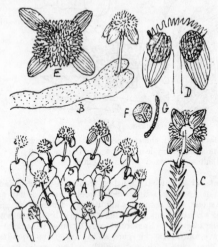 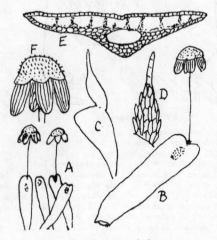

Fimbriaria Bachmannii.

A, thallus, nat. size ; B, same, ×2 ; C, same, under
   side, ×2 ; D, section of capitulum, ×4 ; E,
   capitulum, ×4 ; F, spores ; G, elater.

Fimbriaria muscicola.

A, thalli, nat. size ; B, same, ×2 ; C, scale
   and foliole, ×10 ; D, foliole, ×15 ; E,
   section of thallus ; F, capitulum, ×5.

25. *Fimbriaria muscicola*, Steph., Hedw., 1892, p. 121 ; Steph., Sp. Hep., 1899, p. 97.

Thallus ½ inch long, succulent, usually simple, channelled above, keeled below ; midrib small, elliptic in section, placed low, the upper layer more pronounced, cavernose, with few free filaments. Scales with a separated foliole, which is somewhat toothed below, and with subulate point of several superimposed cells. Peduncle $\frac{1}{2}$ inch long; capitulum small, hemispherical, papillose; lobes small, hyaline; perianth short, rounded. Plant monoecious; androecium near the peduncle.

Transvaal and Natal.

26. *Fimbriaria marginata*, Nees, in Hor. Phys. Berol., 1820, p. 44; Steph., Sp. Hep., 1899, p. 104; G. L. N., Syn. Hep., p. 559.

Thallus 1 inch or more long, $\frac{1}{4}$ inch wide, entire, forked or innovating from the apex, with small purple scales on the under-surface having long lanceolate folioles in pairs or single, also the under-margin of the thallus is often purple or hyaline streaked with purple lines; section of thallus succulent, the elliptic midrib only one-sixth its wideness and

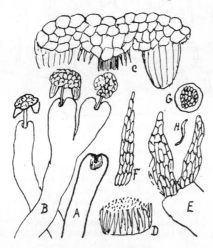

Fimbriaria marginata.

A, B, thalli, nat. size; C, capitulum, ×10; D, same, in youngest stage, ×10; E, folioles, in pairs, ×30;
F, same, single, ×30; G, spore; H, elater.

half its depth, the upper layer fully cavernose, with few free filaments. Peduncle terminal, $\frac{1}{4}$ inch long; capitulum large, hemispheric, its stomata semi-globose, lobes spreading and the perianth also spreading; spores small, narrowly margined; elaters 1-spiral. Plant monoecious; androecia in small innovations on under-surface of thallus.

South Africa (many localities), and up to Nababeep, Namaqualand (M. H. Giffen).

27. *Fimbriaria Wilmsii*, Steph., Hedw., 1892, p. 122; Steph., Sp. Hep., 1899, p. 103.

Thallus up to 2 inches long and $\frac{1}{2}$ inch wide, sometimes forked or innovated from the apex, emarginate, convex below, with large lunate scales bearing large purple oblong or lanceolate folioles, sometimes in pairs. Section of thallus shows strong midrib one-third the width, over which the upper very cavernose layer is shallow and the epidermis marked by large pores. Free filaments abundant. Peduncle terminal, up to 2 inches long, from among abundant scales in the sinus. Capitulum hemispherical, with long smooth bullate lobes tending downward, and each bearing an involucre. Perianth ovate-oblong, hyaline, conspicuous. Spores large, with crenulate margins. Plant monoecious. Androecia in small branchlets from the under-surface.

Cape, Natal, Transvaal, Mozambique, Madagascar, etc. This is the largest species here.

Section 2. COMPOSITAE. Each involucre contains a group of archegonia. Capsule splitting into 4–8 teeth.

Genus 9. LUNULARIA, Micheli, Nov. Gen., p. 4; G. L. N., Syn. Hep., p. 510; Steph., Sp. Hep., 1899, p. 216; Macvicar, p. 38.

Thalloid, dioecious, reproduced abundantly by gemmae formed in semi-circular gemmae-cups on upper surface of thallus. Thallus ligulate, prostrate, emarginate, areolate, warted with simple raised pores on the surface. Epidermal cells hyaline; air-cavity layer narrow, undivided, more or less filled with branched filaments composed of clavate chlorophyllose cells; lower layer of lax empty cells; midrib indistinct and merging. Under-surface of thallus with one double row of white scales and a central strand of tuberculate rhizoids.

One species only, widely distributed by gemmae, with its headquarters in the Mediterranean region where both sexes occur and stalked receptacles are produced. In North Europe only the female plant is known, and in South Africa neither male nor female inflorescence had been seen till this year (1924), when Mr. Giffen sent me plants bearing female inflorescence from Orangezicht, Cape Town. These had the subterminal sessile discs on the upper surface, enclosed among scales, and not unlike the gemmae-cups but not lunulate. The archegonia stood three to six on each of four sides of a very short receptacle, with some sterile hairs crowning the receptacle. As the male plant is not present

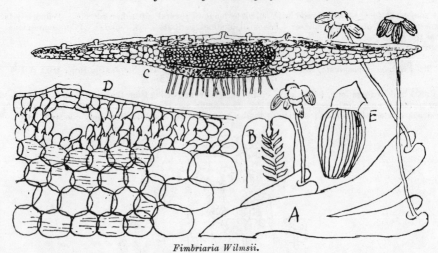

Fimbriaria Wilmsii.

A, thallus, nat. size ; B, lower surface of same, ×2 ; C, section of thallus, ×10 ;
D, cavity in same, ×30 ; E, perianth, ×10.

in South Africa (so far as is known) these archegonia could not be fertilised, and consequently no stalked receptacles or spores are ever produced here, but in South Europe the characters are found to be that a peduncle 2–3 cm. long from among scales in the terminal sinus of the thallus, hairy but not grooved, supports a receptacle of four horizontal tubular involucres containing one sporogonium each. " Pseudo-perianth absent, capsule rather longly pedicellate, exserted from the bilabiate involucre, dehiscing nearly to the base by four narrow valves. Cells of capsule wall without annular thickenings, elaters 2-spiral, threadlike ; male receptacle disciform, sessile, at the apex of a short branch, becoming, as with the female receptacle, apparently lateral, surrounded except in front by the elevated border of the thallus " (Macvicar).

28. *Lunularia cruciata* (Linn.), Dum., Comm. Bot., p. 116 ; Steph., Sp. Hep., 1899, p. 217 ; Macvicar, p. 39.

    = *Lunularia vulgaris* Micheli, Nov. Pl. Gen., p. 4 ; G. L. N., Syn. Hep., p. 511.
    = *Marchantia cruciata* Linn., Hort. Cliff., p. 477 ; Sp. Pl., p. 1137, and 2nd edition, p. 1604.

Thallus ½–3 inches long, 5–10 mm. wide, strap-shaped or cuneate, simple or dichotomously branched, bright green, with a terminal sinus and undulate margin, the older portions brownish yellow. Thallus many cells deep at

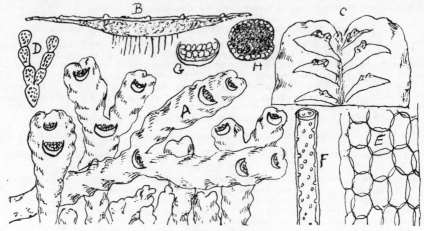

Lunularia cruciata.

A, nat. size ; B, section of thallus, ×10 ; C, under-side of same, ×5 ; D, filaments ; E, epidermal cells ;
F, rhizoids ; G, lunate gemmae-cup ; H, one of the gemmae, ×10.

the centre, tapering to one cell deep toward the margin, not furrowed, but having the lunar gemmae-cups on the central line sparingly placed. Epidermal cells hexagonal and hyaline, 0·04–0·02 mm., except those toward the margin which are quadrate and the marginal line in which the cells are oblong. Filaments 3–5 cells long, forked or single, seldom reaching the epidermis, the cells clavate, chlorophyllose. Scales white, appendiculate only when young.

Easily recognised by the lunate gemmae-cups and the filamentose tissue. Frequent in nurseries and town verandahs, and carried from these to country gardens ; seldom, if ever, indigenous in South Africa.

Genus 10. DUMORTIERA Reinw. Bl. and Nees, Nova Acta Leop. Carol., vii, 410 (1824) ; G. L. N., Syn. Hep., p. 542 ; Steph., Sp. Hep., 1899, p. 222 ; Macvicar, p. 40 ; Spruce, p. 565.

Thalloid, usually dioecious, submerged or wet, matted ; thallus large, herbaceous, dark green, translucent, emarginate, repeatedly forked ; air-chambers and pores absent. Areolation indistinct but ridges evident when young ; a wide midrib or thickened central portion sometimes evident externally, often obscure, but the continuation of this from the terminal sinus forms the peduncle of the receptacle, without basal scales, but the top of the peduncle has a few chaffy scales. Peduncle has two deep grooves, almost invisible externally, filled with tubercled rhizoids extending from the receptacle to the thallus above the younger portion of the midrib. Midrib on its under-surface set with large 1-celled descending root hairs, interspersed with numerous slender tubercled rhizoids 1–2 cm. long, adpressed to the thallus and grouped on radiating ridges pinnately arranged, and also to some extent scattered over the under-surface like long hairs. Peduncle remaining long undeveloped, but eventually growing rapidly, carrying up the receptacle, which is when subsessile a circular hairy disc set with fibrils, 6–8-lobed, each lobe containing about five sporogonia, successive in age, but on maturity the receptacle is 6–8-lobed, with a horizontal involucre under each, containing usually one, but occasionally two, developed sporogonia. Capsule enclosed in a membranous calyptra when young, shortly stalked inside the calyptra, not exceeding the involucre in length, globose, bursting to near the base into eight valves, and full of very numerous, slender, 2-spiral elaters and small triangular papillose spores mixed through its whole cavity. Male receptacle disciform, nearly sessile, placed at the apex of the thallus. Gemmae not produced.

A small genus, widely distributed in streams throughout the tropics and for some distance beyond them.

29. *Dumortiera hirsuta* (Sw.), R. Bl. and N., Nova Acta Leop. Carol., vii, 410 ; G. L. N., Syn. Hep., p. 542 ; Steph., Sp. Hep., 1899, p. 224 ; Macvicar, p. 41 ; Spruce, p. 566.

= *Marchantia hirsuta* Swartz, Prod. Fl. Ind. Occ., p. 145.

Thallus olive green, translucent, 4–6 inches long, 1–2 cm. wide, repeatedly dichotomously forked, elongating from the apex where the dichotomous division of the midrib occurs within the lamina ; the apex consequently is not emarginate except between the forks of a midrib, or where fertile. Adventitious innovations also occur. Margin undulate, occasionally somewhat crenate, sparingly set with long marginal hairs when young. Upper surface glabrous, areolae absent or indistinct, lower surface producing abundant smooth root hairs 0·025 mm. diameter from the midrib, which descend into the mud, and also abundant slender tuberculate rhizoids 0·005–0·010 mm. diameter and sometimes several centimetres long, which are crowded along the surface of the midrib and pinnately arranged ridges (which replace the squamae of other groups), and also scatter somewhat over the surface ; with age these disappear. Thallus 1–2 cells deep toward the margin, about 6–8 cells deep toward the centre, the inner cells large (0·075 mm.), loose, and almost hyaline, the epidermal layer of cells on both upper and lower surfaces much smaller, 0·05–0·025 mm. diameter, and containing abundant small, green, free chloroplasts. Fertile peduncle produced from a continuation of the midrib, and as the lamina continues to grow beyond it, the peduncle eventually rises from the base of a deep sinus, the margins of which overlap so closely that the seta often appears to rise from the surface and centre of a lobe. Peduncle when mature and vigorous 2 inches long, naked at the base, erect, firm, green, large-celled, and having two almost invisible grooves full of tuberculate rhizoids along its length. Receptacle when immature discoid, 3–6 mm. wide, nearly circular, 6–8-lobed on a 2-mm. peduncle, and abundantly hairy, having about five archegonia in each group, those nearest the peduncle being produced first, the others outward following in order ; in the above condition the receptacle remains for months till some are fertilised, then quickly the peduncle extends to full length, the receptacle develops a star-shaped form with six to eight rays hairy on the ridges and margins, and from under each ray, or by suppression only some of them, a horizontal, herbaceous, green, ovate involucre 3–4 mm. long, opening by a terminal slit. This contains usually one (occasionally two) capsules, each enveloped in a thinly membranaceous calyptra. Capsule on a short succulent green pedicel, globose or elliptical, bursting within the involucre into eight oblong valves which separate nearly to the base. Valves formed of a single layer of tubular cells each extending the whole length of the valve and about 0·035 mm. diameter, at first containing a red spiral band which eventually breaks into numerous red rings on the cell-wall. Elaters and spores produced in enormous abundance, filling the capsule, and without central columella. Elaters 2-spiral, short, and elliptical acuminate while contained within its cell-wall, afterwards up to 0·4 mm. long, widest at the middle, and tapering to a long slender point at each end. Spores triangular in section, twice as long as wide, papillose, 0·025 mm. long. Elaters and spores forming a purple brush for several days adherent to the mouth of the involucre. Under-surface of receptacle more or less set with tuberculate rhizoids.

This might be considered a distinct African species, differing from *D. hirsuta* in the greater wideness of its thallus, the length of its seta, the absence of areolation on the surface of the thallus, the form and hairiness of the receptacle, and the occasional presence of two sporogonia in the perianth, but when the range of variation within the genus and species in all parts of the world is fully known, as also the effect of stream friction or other external injury upon the surface hairs, it is probable that one species with several conditional forms will include all.

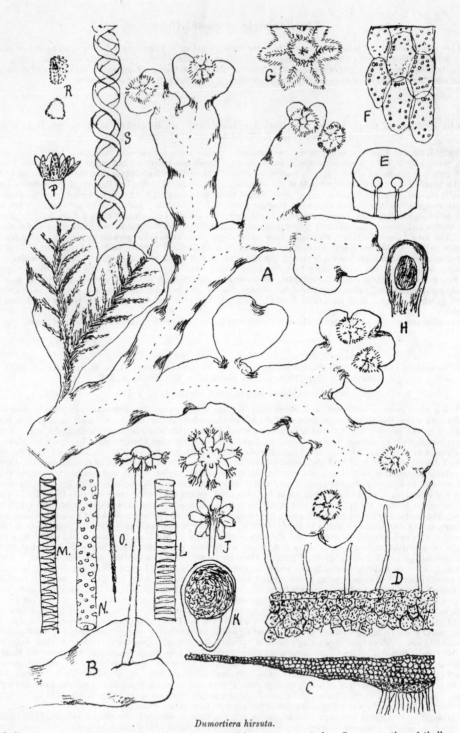

Dumortiera hirsuta.

A, thallus, nat. size, with young receptacles ; B, same, with mature receptacle ; C, cross-section of thallus, ×10 ;
  D, portion of margin of thallus, from above, with marginal hairs, ×50 ; E, section of peduncle, with two rhizoid
  channels ; F, cells of thallus, ×200 ; G, surface of young receptacle, with hairy ridges and margin ; H, section
  of perianth ; I, mature receptacle, from above, with perianths tipped with brushes of purple elaters and spores ;
  J, under-view of mature receptacle ; K, capsule in calyptra ; L, portion of cell of valve, mature ; M, same,
  younger ; N, portion of tuberculate rhizoid ; O, elater ; P, capsule, with open valves, ×6 ; R, spore, and
  section of same ; S, portion of elater, much enlarged.

Frequent in shaded forest streams upon stones subject to constant spray or capillary moisture, and to occasional inundation, throughout the eastern Cape Province, Transkei, Natal, Eastern Transvaal, and Portuguese East Africa.

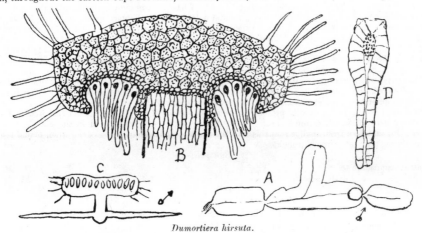

Dumortiera hirsuta.

A. thallus, with young receptacle ; B, section of young receptacle before fertilisation, ×30 ;
C, section of antherophore, ×5 ; D, archegonium, ×100.

Genus 11. MARCHANTIA, Linn., Sp. Pl., p. 1137 (1753) ; G. L. N., Syn. Hep., p. 521 ; Steph., Sp. Hep., 1899, p. 383 ;
Macvicar, p. 44 ; Spruce, p. 558 ; Schiffner, Pfl., i, 3, 36.

Thalloid, dioecious, matted ; thallus irregularly branched, strap-shaped, somewhat succulent, with flat upper surface dotted with pores, and with convex under-surface set with one permanent double row of scales, appendiculate

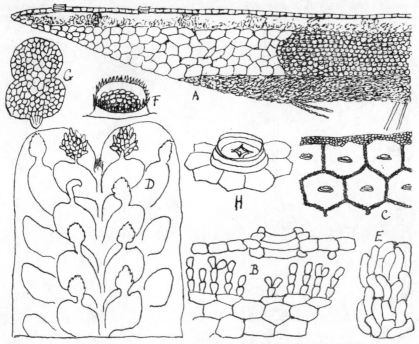

Marchantia.

A, section of thallus ; B, section of air-chamber, showing guard-cell and chlorophyllose tissue ; C, external appearance
of air-cavities and guard-cells ; D, scales on under-surface of thallus, with folioles ; E, twisted cells on same ;
F, gemmae-cup, ×5 ; G, one of the gemmae, second stage, ×20 ; H, guard-cell.

when young, and often with one or two outer double rows of quickly evanescent scales. Thallus consisting of an embedded midrib, a lower tissue of small cells, a mid layer of large loose hyaline cells, and an upper layer consisting of air-chambers containing many separate simple or branched lines of superimposed chlorophyllose cells, and each cavity opening through the epidermal layer by a compound pore or guard-cell, built up of four rows of cells and having a cruciate entrance. Rhizoids abundant, some tuberculate, from the lower surface. Peduncle stout, rising through the sinus from the under-surface of the thallus, set with abundant single or 2–3-fid scales at the base and having a few linear scales at the top under the capitulum. Capitulum peltate, ⅓–½ inch diameter; the female 8–10-rayed, 2-grooved, with the involucres under, embedded, lanceolate, 2-valved, fimbriated, extending inward from each sinus of the capitulum, and containing several sporangia, each surrounded by a pseudo-perianth; they consequently alternate with the rays, which have one rhizoid groove along the under-surface of each; the male peltate but somewhat one-sided, 5–7-rayed, the rays narrow, having the margins incurved toward the central papillose cushion in which are the antheridophores, the under-surface set with protective lunate scales.

Saucer-shaped gemmae-cups usually present on the upper surface of the thallus, these cups circular, fringed, and each containing many flattened gemmae, each at first discoid, afterwards constricted in the middle, and after distribution growing on into a fresh thallus; in this way the plant spreads along streams even when sexual reproduction is absent, or during times of extreme drought.

Synopsis.

Rays of female capitulum terete . . . . 30. *M. tabularis.*
　　,,　　　　,,　　　flatly cuneate . . . . 31. *M. Wilmsii.*

30. *Marchantia tabularis* Nees, Hep. Eur., iv, 71.

=*M. Berteroana* L. and L., in Lehm. Pug., vi, 21; Steph., Sp. Hep., 1899, p. 393; G. L. N., Syn. Hep., p. 525.

Thalloid, dioecious, often matted, growing on moist banks along streams, or occasionally flooded. Thallus 1–3 inches long, ½ inch wide or more or less, regularly strap-shaped, simple or irregularly forked, flat, often not showing mid-line, dark green, and usually showing areolae indicating internal air-cavities, each opening to the surface by a minute pore. Lower surface of thallus convex, having a double central line of rounded permanent scales, together with one or two lines of outer smaller scales which are more or less evanescent, younger scales appendiculate, the foliole ovate or cordate, dentate, its cells large except the marginal rows which are smaller. Midrib not prominent, half the depth of the thallus and a quarter its width. Peduncle 1–2 inches long, slightly paleaceous at the top, and abundantly paleaceous at the base, these scales narrowly lanceolate or linear. Female capitulum large, peltate, symmetrical, 8–10-lobed, the lobes clavate, terete, smooth. Male capitulum smaller, somewhat one-sided, about 6-rayed.

Closely related to the widely distributed *M. polymorpha* Linn., which has the female capitulum smaller and its lobes papillate above. A South African species found on Table Mountain and other moist southern localities.

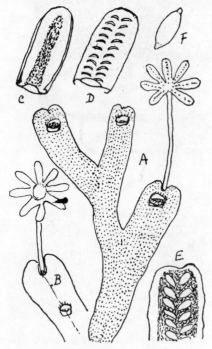

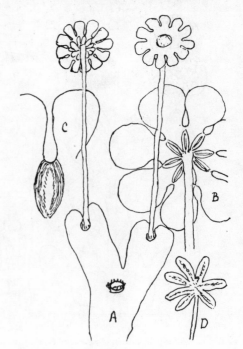

Marchantia tabularis.

A, male thallus, nat. size; B, female thallus, nat. size; C, D, upper and under surfaces of male rays; E, section of same; F, antheridium.

Marchantia Wilmsii.

A, female thallus, nat. size; B, female capitulum, under-surface, showing sporangia, ×3; C, same, ×6; D, male capitulum.

31. *Marchantia Wilmsii*, Steph., Hedw., 1892, p. 126 ; Steph., Sp. Hep., 1899, p. 398.

This corresponds in vegetative characters with *M. tabularis* except that the pores are often more pronounced and the folioles on the scales have the marginal cells about equal to the others. Involucral scales numerous, widely lanceolate or 2–3-fid, the few scales at the top of the peduncle filiform. Female capitulum large, convex at the centre, 8–10-rayed, the rays symmetrical, flat, smooth, widely clavate from a narrow base, and having a rounded sinus between each pair.

A South African species common in Natal, Transvaal, and northward.

Order III. JUNGERMANNIALES.

Plant foliose in most cases, thalloid in some ; when thalloid the thallus is not differentiated into layers of different tissue and is without pores. Tubercular rhizoids not present. Sexual organs usually in groups, but not on special pedunculate receptacles, and seldom immersed. Capsule, which is usually on a long seta, is destitute of lid, opens by four valves, and contains spiral elaters as well as spores. Apical growth of stem or thallus proceeds from a single apical cell.

Suborder 1. ANACROGYNAE. Archegonia not terminal ; involucre of sexual organs consequently not representing leaves.

Family 5. ANEURACEAE. Plant thalloid. Sexual organs from marginal or ventral branchlets. Elaters uni-spiral. Elaterophores remain as tufts on the apex of the capsule valves.

Genus 12. ANEURA Dum., Comm. Bot., p. 115 ; G. L. N., Syn. Hep., p. 493 ; Steph., Sp. Hep., 1899, p. 662 ; Macvicar, p. 49 ; Spruce, p. 540.

=*Riccardia*, S. F. Gray, Nat. Arr. Brit. Pl., i, 683 (in part) ; Schiffner, Pfl., i, 3. 52.

Thalloid, submerged or constantly wet, dark green ; thallus exceedingly variable in accordance with water-supply, usually having a prostrate parent-stem with an indefinite midrib, from which rise irregularly pinnatifid or pectinate stems or branches, upright where crowded, more or less prostrate where space is available, and usually biconvex in section. The sexual organs are from short, lateral, marginal branches which are dwarfed by the process, so the archegonia often appear afterwards to be from the under-surface. Archegonia two to eight in two rows, covered by the margin of the thallus, and usually on a main stem. Pseudo-perianth absent ; calyptra large and fleshy, tubular, papillate at the apex, seldom seen. Capsule oblong-ovoid or shortly cylindrical, 4-valved, on a 10–20 mm. pedicel. Spores smooth or papillose ; elaters 1-spiral, remaining attached in a tuft at the end of the valves. Male branches scattered more over the plants, short, crenulate, with three to five pairs of antheridia. Gemmae sometimes produced in the tissues of the thallus. Exceedingly variable ; species not easily separable and usually intermixed.

Synopsis.

32. *A. fastigiata.* Thallus often erect, irregularly pinnate, an inch or more high, its branches flattened.
33. *A. compacta.* Thallus prostrate with short erect simple pectinate or subpinnate branches, with terete segments.

32. *Aneura fastigiata* L. and L., Syn. Hep., p. 500 ; Steph., Sp. Hep., 1899, p. 732.

Thalloid, dioecious, the stems usually more or less erect, 1–2 inches high, 2 mm. wide, irregularly 2-pinnatifid or nearly 2-pinnate ; the stem and segments flattened and narrowly winged. Sporangia on the main stem, and male inflorescence more scattered. Midrib indistinct. Calyptra large, tubular, fleshy, erect ; capsule on pedicel about 1 inch high, its valves tufted with elaters. Present in nearly every stream from Cape Town to the Zambesi, and also often on wet ground in marshes, round waterfalls, etc., where it is more vigorous and has room for development.

A. multifida (Linn.), Dum. Comm., p. 115, is larger, more pinnate, thicker, with less-winged narrower branches.

A. pinnatifida Dum., Rec. d'Observ., 1835, p. 26 (=*A. major* (Lindb.), K. Mull.), is less pinnate and not winged.

These two names, which belong to European monoecious species, were used for *A. fastigiata*, which is South African only and dioecious, before the differences were understood.

33. *Aneura compacta*, Steph., Hedw., 1893, p. 19 ; Steph., Sp. Hep., 1889, p. 755.

Thalloid, dioecious, ½ inch high, caespitose, matted over submerged stones ; main thallus prostrate, with erect, simple, pectinate or subpectinate slender branches with slender filiform segments, terete or elliptical in section, not winged, and seldom even flattened. Smaller in all parts than *A. fastigiata*, but often intermixed with it, forming a lower layer in running streams under forest conditions.

Frequent in forest districts throughout South Africa, less common in exposed streams.

Genus 13. METZGERIA Raddi, in Att. soc. scient. Mod., xviii, 34 ; G. L. N., Syn. Hep., p. 501 ; Schiffner, Pfl., i, 3, 53 ; Steph., Sp. Hep., 1899, p. 927 ; Macvicar, p. 58 ; Spruce, p. 551.

Thalloid, thallus linear, thin, pale yellow, branched either irregularly or dichotomously ; midrib well defined and the lateral lamina consists of one layer of pellucid cells. Long 1-celled hairs are often present, singly or two-ranked, on the margins, nerve, and under-surface, and sometimes on the upper surface also. These hairs are not difficult to rub off ; the thallus then resembles *Riccia fluitans* in general appearance. Sexual organs on very short hairy branch-lets from the under-surface of the midrib. Calyptra very hairy ; capsule spherical or shortly oval, on a slender seta,

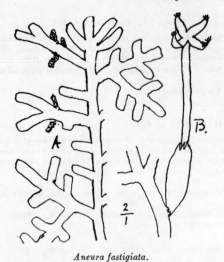

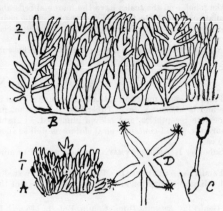

Aneura fastigiata.

A, thallus, ×2 ; B, fruiting branch, ×2.

Aneura compacta.

A, thallus, nat. size; B, same, ×2; C, young
capsule ; D, open capsule.

4-valved, often with terminal tufts of hairs on the valves. Elaters long, slender, 1-spiral. Male branchlet very short, incurved, usually not hairy. Usually this genus occurs intermixed among corticolous mosses or on rocks in the wettest parts of the forests ; seldom does it occur on the ground, or under less than 35 inches of rainfall per annum.

The species are very similar in general appearance, straggling and flattened in habit (one figure describes all), and possibly they are conditional. Nearly similar species occur under similar conditions in all countries, especially in the tropics, but in colder regions they are more often on decayed bark or humus. Gemmae occur occasionally on the upper surface of the thallus.

Synopsis.

Marginal hairs single ; plant dioecious.
    Section of nerve has two cells above and two below .   .   .   .   .   34. *M. Perrotana.*
        „   „   two cells above and four below .   .   .   .   .   35. *M. furcata.*
Marginal hairs in pairs.
    Nerve section has two cells above and four below. Plant monoecious   .   .   36. *M. conjugata.*
        „   „   four cells above and six below. Plant dioecious .   .   .   37. *M. muscicola*

34. *Metzgeria Perrotana*, Steph., Bull. de l'Herb. Boiss., 1899, p. 938, and Sp. Hep., 1899, p. 938.

Thallus linear, small, matted, up to 1 inch long, ½ mm. wide, repeatedly forked ; the midrib with two cells above and two below, hairy below, hairs slender ; marginal hairs in single series, and occasional hairs are present on the under-surface. Dioecious ; male branch naked, very small ; female larger, hairy.

Stephani cites it from Montagu Pass (Dr. Rehmann) and Madagascar.

35. *Metzgeria furcata* (Linn.), Lindb., Monog. Metz., p. 35 ; Dum., Rec. d'Obs., p. 26 ; Steph., Sp. Hep., 1899. p. 941.

          = *Jungermannia furcata*, Linn., Sp. Pl., p. 1002.
          = *Metzgeria planiuscula*, Spruce, Rev. Bryol., 1888, pp. 34–35.
          = *Metzgeria flavovirons*, Col., Trans. N.Z. Inst., 1888, vol. xxi.

Thallus linear, strap-shaped, up to 2 inches long and 1 mm. wide, repeatedly and irregularly forked, with flat, naked upper surface and rounded but emarginate apex. Midrib distinct, with two cells above and four smaller below ; hairs abundant on the under-surface of the midrib, occasional and separate on the lamina and to the margin. Cells of the lamina large, horizontal, often pellucid. Dioecious ; calyptra very hairy. Gemmae occasionally present on upper surface of the thallus, at first 1-celled, gradually growing to a many-celled, flat, elliptical body from which several or many simple hairs protrude like rhizoids.

Common in forests throughout South Africa from Table Mountain to and beyond Zimbabwe and Victoria Falls ; also in Europe, Asia, North Africa, South America, and Australia.

M. tabularis Steph. (Brunnthaler, 1913, i, 15, and Steph., Sp. Hep., vi, 61), from Table Mountain, agrees in everything except that the wings are said to be nude, a common condition in *M. furcata*.

36. *Metzgeria conjugata* Lindb., Monog. Metz., p. 29 ; Steph., Sp. Hep., 1899, p. 951.

Thallus up to 1 inch long, 2 mm. wide, repeatedly forked ; branches flat, naked above, and with midrib protruding below, with two cells above and four below, the latter with many single hairs, but the margin with short, stout, straight, 2-geminate hairs and the calyptra very hairy. Monoecious. Frequent in forest, often in masses on overhanging rocks, throughout eastern South Africa ; also Europe, Asia, Africa and its islands, N. and S. America, and New Zealand.

37. *Metzgeria muscicola* Steph., Bull. de l'herb. Boissier, 1899, p. 955.

= *M. nudifrons*, Steph., Hedw., 1892 (which he sank in the above since it represented specimens from which the hairs had fallen, as often happens).

Thallus 1–2 inches long, vigorous, linear, 1 mm. wide, naked on the upper surface. A section of the wide midrib has four cells above and six to eight cells below, with many long, simple, straight hairs on its under-surface and occasional simple hairs on the lamina, and numerous 2-geminate long straight hairs on the margin. Dioecious ; male branches comparatively large, often nude ; female branches and calyptra hairy.

Not uncommon in the forest country of eastern South Africa from Cape Town to Transvaal ; also in Cameroons ; usually prostrate among mosses.

M. Brunnthaleri Steph., Brunnthaler, 1813, i, 15, and Steph.,

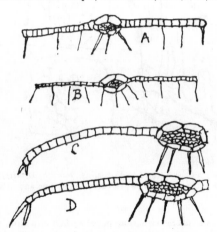

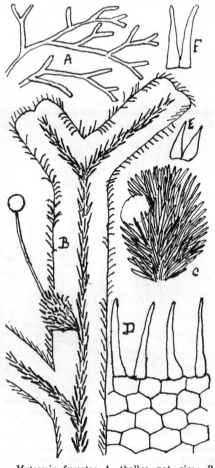

Metzgeria Perrotana : B, section of thallus.
Metzgeria furcata : A, section of thallus.
Metzgeria conjugata : C, section of thallus.
Metzgeria muscicola : D, section of thallus.

Metzgeria furcata : A, thallus, nat. size ; B, thallus, ×10 ; D, margin. *Metzgeria conjugata* : C, calyptra ; and E, marginal hairs. *Metzgeria muscicola* : F, marginal hairs.

Sp. Hep., vi, 48, seems to be this with vigorous thallus 2·33 mm. wide and the hairs all rubbed off. It is from van Reenen, Natal.

Family 6. BLYTTIACEAE. Plant thalloid ; sexual organs from the upper surface, not marginal ; capsule ovoid ; elaters 2-spiral ; elaterophores absent. Ring-fibres absent from cells of capsule wall ; dehiscence of capsule valves incomplete.

Genus 14. PALLAVICINIA Schiffner, Pfl., i, 3, 55 ; Macvicar, Student's Handbook of British Hepaticae, p. 64.

= *Pallavicinius*, S. F. Gray, Nat. Arr. Brit. Pl., i, 775 ; Steph., Sp. Hep., 1900, p. 5.
= *Dillaena*, Dum., Comm. Bot., p. 114.
= *Blyttia*, Endl., Gen. Pl., p. 1339 ; Sim, S.A. Hep., p. 14 ; G. L. N., Syn. Hep., p. 474.
= *Steetzia*, Lehm., Pl. Preissiana, ii, 129 ; G. L. N., Syn. Hep., p. 785.

Thalloid, usually prostrate or the point ascending ; thallus simple or branched or with innovations from the under-surface ; ligulate, the margin undulate, incurved when dry ; midrib distinct, solid ; lamina one-cell thick, without scales on the under-surface. Dioecious ; inflorescence on the midrib on the upper surface, the female as groups of archegonia surrounded by toothed or ciliate involucral scales, from among which rises the pseudoperianth, enclosing a fleshy calyptra with mamillate terminal segments. Capsule oblong-cylindrical, 4-valved, inner wall without ring-thickenings ; valves adherent at the apex. Spores large, papillose ; elaters wide, 2–3 spiral. Antheridia in several rows along the midrib of male plants, each at first protected by a toothed scale on the lower side ; arranged as a long central cushion along the thallus.

38. *Pallavicinia Lyellii* (Hook.), Gray, Nat. Arr. Br. Pl., i, 775 ; Schiffner, Pfl., i, 3, 55 ; Macvicar, p. 65.

> =*Pallavicinius Lyellii*, Gray (as above) ; Steph., Sp. Hep., 1900, p. 13.
> =*Jungermannia Lyellii*, Hook., Brit. Jung., t. 77.
> =*Dilaena Lyellii*, Dum., Comm. Bot., p. 114.
> =*Blyllia Lyellii*, G. L. N., Syn. Hep., p. 475.
> =*Steetzia Lyellii*, Lehm., Pl. Pr., pp. 1–129.

Thalloid, usually prostrate, adherent to soil by numerous rhizoids ; simple or branched or with innovations from the under-surface ; occasionally more or less erect, or horizontal from a perpendicular bank. Thallus ligulate, glabrous, 1–2 inches long, 2–4 mm. wide ; margin usually undulate or crenulate ; midrib strong, flat above, rounded below,

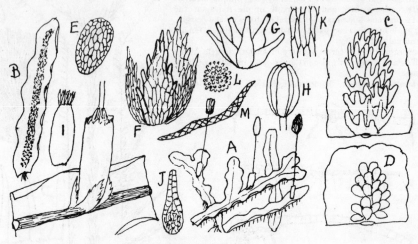

Pallavicinia Lyellii.

A, fertile thallus, nat. size ; B, male thallus, slightly enlarged ; C, surface of male thallus, young ; D, same, old ; E, antheridium ; F, involucral scales ; G, group of archegonia ; H, capsule ; I, calyptra ; J, one archegonium ; K, cells of lamina ; L, spore.

with central woody tissue ; under-surface without scales ; male inflorescence from upper surface of midrib, its involucral scales short, more or less united, toothed or ciliated ; pseudo-perianth ¼ inch long, with ciliated mouth ; calyptra fleshy below ; seta up to 1 inch long ; capsule oblong-cylindrical, 4-valved. Spores 25μ diameter, papillose. Male thallus smaller ; antheridia abundant along the midrib, at first protected by fingered scales, ultimately nude, cushioned, and golden-yellow.

Frequent throughout South Africa from Cape to Transvaal, on stream banks, mud flats, etc., and often among other moisture-loving mosses ; even along dry-country streams it is not uncommon on moist, shady banks, usually fertile. (For *P. Stephani* Jack, see *Symphyogyna Lehmanniana*.)

Genus 15. SYMPHYOGYNA Mont. et Nees, in Ann. des. Sc. Nat., 2nd series, p. 36 ; G. L. N., Syn. Hep., p. 479 ; Spruce, p. 533 ; Schiffner, Pfl., i, 3, 55 ; Steph., Sp. Hep., 1900, p. 23.

Thalloid, dioecious in South African species ; thallus rising from a wiry rhizome, procumbent and rooting, or erect and more or less palmate on a wiry stipe, with linear segments and beautifully areolated lamina. Female inflorescence on the midrib on upper surface near the frond-base ; involucre of several laciniate scales ; perianth absent ; calyptra, fleshy, crowned with barren archegonia, the fertile archegonium included in it. Seta slender, ¼–½ inch long ; capsule subcylindric, 4-valved, valves seldom free ; elaters 2-spiral ; spores small, minutely papillose ; antheridia in a continuous line along each side of the upper surface of the midrib, at first protected by scales, afterwards nude, usually on simple prostrate thalli.

A mist-belt and forest-stream genus, growing on moist banks, or on sandstone, usually in shade, but sometimes on rocks in exposed streams.

39. *S. Lehmanniana.* Usually prostrate and rooting, simple or nearly so : teeth usually absent.
40. *S. podophylla.* Erect ; simple or irregularly or palmately 1–6-branched ; teeth 1–3-celled.
41. *S. spinosa.* Erect ; palmately many-branched ; teeth several-celled.

39. *Symphyogyna Lehmanniana* M. and N., Syn. Hep., p. 483 ; Steph., Sp. Hep., 1920, p. 28.

Thallus 1–1½ inch long, 2–4 mm. wide, procumbent or arched, simple or irregularly branched, often attenuated and rooting at the apex. Margin entire or nearly so, midrib narrow, firm. Involucre small, of laciniated scales : barren archegonia numerous, crowning the calyptra. Male thallus simple, strap-shaped, prostrate.

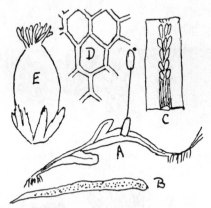

Symphyogyna Lehmanniana.

A, thallus, nat. size, fertile ; B, same, male ; C, same, male, enlarged ; D, areolation ;
E, involucre and calyptra.

Table Mountain and Western Mountains.
Very similar to *Pallavicinia* ; and *P. Stephani*, Jack (Hedw., 1892) ; Steph., Sp. Hep., 1900, p. 30, from Natal, appears, from description, to belong here.

40. *Symphyogyna podophylla* (Thunb.), M. and N., Syn. Hep., p. 481 ; Steph., Sp. Hep., 1920, p. 44 ; Brunnthaler, 1913, i, 15.

=*Jungermannia podophylla* Thunb., Prodr. Fl. Cap., ii, 174.
=*Symphyogyna Harveyana*, Tayl., Journ. of Bot., 1846, p. 408 ; G. L. N., Syn. Hep., p. 786.

Thallus usually erect and stiped, ½–1½ inch long, very irregular in form, from simple to palmately 3–4–6-branched, the branches linear, with distinct rather wide midrib containing central ligneous strand, and lamina 2–3 cells deep, or toward the margin 1-cell deep. Margin distantly toothed, the teeth 1–3-celled, straight. Involucres near the base of the palmate portion, short and much divided, ultimately absent or near the lower forks, fimbriate. Frequent along forest and alpine streams, usually in masses, in which simple and variously flabellate thalli are intermixed. Common on Table Mountain and occurs in all eastern forest country, and often along exposed mountain streams and in deep clay holes.
S. valida, Steph., Sp. Hep., vi, 69, is this species, unbranched as it often is, and is from Zululand (Bryhn).

41. *Symphyogyna spinosa* L. and L., Syn. Hep., p. 786 ; Steph., Sp. Hep., 1900, p. 44.

More vigorous than *S. podophylla*, but closely allied to it ; the fronds 1–2 inches high, palmately many-branched, the lower branches short but incurved, the upper ½–1 inch long, 1–2 mm. wide, expanded. Margin when young closely spinose, the spines 3–4 cells long, the upper cells smaller ; the spines near the growing-point of the segment bending over that point. Involucral scales cut into several segments ; calyptra crowned with many barren archegonia.
S. spinosa L. and L. occurs in Mascarenes, Central Africa, and eastern forest districts of South Africa.
Jungermannia rhizoloba Schw. (=*S. rhizoloba*, Nees) appears in Drege's list (iii, A, e, 9), as well as *S. podophylla* Thunb. *S. rhizoloba* is a Mascarene species, and Drege's record presumably applies to *S. spinosa* L. and L., which hardly differs.

Family 7. CODONIACEAE. Plant with stem and leaves. Capsule globose.

Genus 16. FOSSOMBRONIA Raddi, in Atti Soc. Ital. Sc. Mod., xviii, 40 ; G. L. N., Syn. Hep., p. 467 ; Schiffner, Pfl., i, 3, 59 ; Macvicar, p. 79 ; Spruce, p. 526.

Stem subthalloid, simple, or little branched, rooting along its whole length (except the apex) by purple rhizoids, and cut to the stem or midrib into more or less regular, lateral, leaf-like, quadrate lobes, or more properly obliquely

3

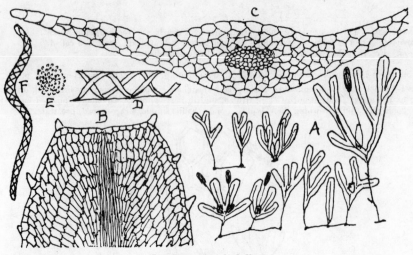

Symphyogyna podophylla.

A, thallus, nat. size ; B, point of thallus ; C, section of same ; D, part of elater ; E, spore ;
F, elater.

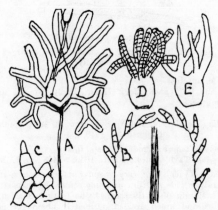

Symphyogyna spinosa.

A, thallus, nat. size ; B, apex of thallus ; C, marginal spines ; D, calyptra, with barren archegonia ;
E, involucral scale.

inserted leaves having succubous connection, decurrent on the upper surface, and either entire or variously cut along the margin. Leaves without nerve, usually wider than long ; cells large and lax. Archegonia on the upper surface of the stem, the fertile one near the apex ; at first nude, ultimately enclosed in a laciniated leafy pseudo-perianth in which the calyptra is included. Capsule globose, irregularly 4-valved ; seta short, delicate ; spores large, papillate or lamellate or reticulate ; elaters short, 2-spiral. Antheridia on upper surface of the stem, yellow, bracted or nude. Amphigastria absent.

A widely distributed genus in which the vegetative characters are often unreliable and the spore characters difficult to see.

Synopsis.

Spores lamellate.
   Leaves entire   .   .   .   .   42. *F. tumida.*
   Leaves lobed   .   .   .   .   43. *F. pusilla.*
Spores reticulated ; leaves almost entire   .   .   .   .   .   44. *F. Zeyheri.*
Spores papillose.
   Leaves bluntly lobed   .   .   .   .   45. *F. crispa.*
   Leaves with few large sharp teeth   .   .   .   .   .   46. *F. spinifolia.*
   Leaves abundantly dentate-serrate and ciliate   .   .   .   47. *F. leucoxantha.*

42. *Fossombronia tumida* Mitt.

Plant 2–4 mm. long, prostrate, simple, with purple rhizoids ; leaves light green or nearly white, few, very concave, suberect, and with the apices from opposite sides of the stem habitually incurved and folded over one another, forming an inflated, concave, subglobular ball, the earlier leaves 1 mm. long and wide, the apical leaves 2 mm. long, 1 mm. wide, all roundly subquadrate, entire, the apex rounded and always incurved, not crisped even when dry. Cells lax, oblong-hexagonal, thin-walled, pellucid ; occasional cells brown, containing oil or other stores. Perianth turbinate, rather shorter than the leaves, with undulate-crenate margin ; seta 3 mm. long ; capsule large, globose, the outer wall hyaline, the inner wall with many dark brown elongated thickenings. Spores dark brown, 40μ diameter, somewhat flattened, the convex face with deep lamellae radiating from a few central areolae, and showing as twenty-four to thirty spines on the margin. Elaters 2–3-spiral. Sent by Mr. Garside from Stellenbosch Flats ; also received from other S.W. localities, but not seen eastward. Apparently annual or biennial, maturing during the western wet season, June, July.

Stephani (Sp. Hep., 1900, p. 39) mentions this species as widely spread in South Africa, but badly described by Mitten, and he is in doubt in which section to place it, knowing it only by the tumid, almost colourless, leaves, and in this doubt he mentions it near *F. leucoxantha*, which is not its place.

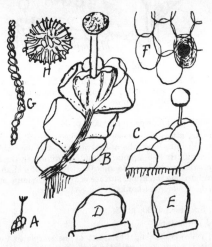

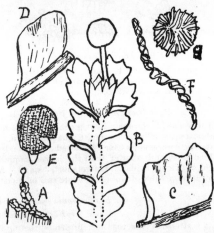

Fossombronia tumida.

A, plant, nat. size ; C, same, ×5 ; B, same, pulled open, ×7 ; D, E, leaves ; F, cells, ×100 ; G, elater, ×300 ; H, spore, ×300.

Fossombronia pusilla.

A, plant, nat. size ; B, same, ×5 ; C, upper leaf, ×10 ; D, lower leaf, ×10 ; E, calyptra, ×10 ; F, elater, ×200 ; G, spore, ×200.

43. *Fossombronia pusilla* (Linn.), Dum., Rec. d'Obs., p. 11 ; Steph., Sp. Hep., 1900, **p. 24.**

=*Jungermannia pusilla* Linn., Sp. Pl., p. 1603.

Gregarious ; paroecious ; stems 10–15 mm. long, simple or once forked, prostrate ; leaves spreading or suberect toward the apex ; lower leaves smaller and rounded or emarginate ; mid leaves quadrate, 2 mm. wide, 1·5 mm. long, crowded, undulate-crisped, bluntly 3–4-lobed, with margins entire. Upper leaves deeply 3–5-lobed, lobes cut, usually incurved. Upper cells 40μ, lower cells 50–70μ. Pseudo-perianth 1–1·5 mm. long, widely campanulate, plaited and crisped, many-lobed ; capsule-wall with dark semi-annular thickenings ; spores 40μ diameter, brown, with high ridges from near the centre outward, showing as about twenty spines on the margin. Elaters wide, 2–3-spiral.

Frequent from Cape Town to Rhodesia ; occurs also in Europe and North America.

44. *Fossombronia Zeyheri* Steph., Sp. Hep., 1900, p. 20.

Plant 2–4 cm. long, matted, dioecious, glaucous, often once forked, prostrate on moist mud, and having abundant long violet rhizoids. Stem flat on the upper surface, rounded below, twice as wide as deep ; leaves horizontal when moist, ascending when dry, firm, imbricate, roundly quadrate, 2 mm. wide, 1·5 mm. long, nearly flat, usually neither lobed nor undulate, with entire margin ; cells small and quadrate at the margin, 45μ, much larger (60–80μ) and of irregular hexagonal form toward the centre. Pseudo-perianth as long as the leaves, deeply cut into a few large segments. Spores 35μ, dark, with reticulated surface, the areolae large and few, the walls showing as teeth at the margin. The largest, most common, and most glaucous South African species, perennial, firm in texture, usually found on stream-banks or wet stones and expanded by moisture ; seldom fertile.

Table Mountain to the Transvaal, along the eastern aspect. Kloofnek, Brunnthaler (1913, i, 15).

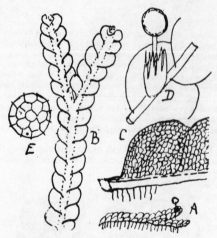

Fossombronia Zeyheri.

A, plant, nat. size ; B, same, ×5 ; C, two
leaves, on half-stem ; D, part of fertile
stem ; E, spore, ×200.

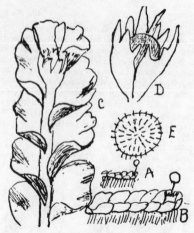

Fossombronia crispa.

A, plant, nat. size ; B, same, ×5 ; C, point
of stem, ×10 ; D, pseudo-perianth and
capsule ; E, spore, ×200.

45. *Fossombronia crispa* Nees, Syn. Hep., p. 469 ; Steph., Sp. Hep., 1900, p. 35.

Plant dioecious, 2 cm. long, light green ; stems at first simple, afterwards repeatedly forked ; usually matted ; leaves imbricate, spreading when moist, quadrate, 1 mm. wide, 1·5 mm. long, the apex truncate, undulate-crisped, not dentate, the lower margin usually inflexed.   Cells laxly and irregularly 5–6-angular, about 55μ diameter, the lower larger.   Pseudo-perianth shorter than the leaves, campanulate, with wide lacerate mouth.   Spores 45μ diameter, dark, hispid, with long papillae.   Elaters short, 2-spiral.   Antheridia along the upper surface of stem, with large bracts.
Frequent from Table Mountain to Transvaal on moist river-banks or road-banks.

46. *Fossombronia spinifolia* Steph., Sp. Hep., 1900, p. 35.

=*F. capensis* Sim (ined.) in letters.

Plants scattered or few together, prostrate, rooting freely, yellowish green, dioecious ; stems simple or forked, 5–10 mm. long, the leaves spreading, quadrate, 0·50–1·00 mm. wide, 0·8 mm. long, with the margins suberect, undulate-crisped, bluntly triangular-dentate, the three to six teeth pointing alternately inward and outward, and occasionally having additional smaller teeth.   Cells variable in size.   Archegonia at first naked on the upper surface of the stem ; pseudo-perianth much larger than the leaves, campanulate, plicate, the wide mouth dentate-lobate, lobes triangular-acute.   Capsule wall with incomplete brown thickenings.   Spores 45μ diameter, opaque, densely papillose, without ridges ; papillae long, slender, truncate.   Elaters large, 2–3-spiral.   Antheridia not seen.
Jonker's Hoek, Stellenbosch, Gnadenthal, and other south-western localities.

47. *Fossombronia leucoxantha* L. and Lg., Syn. Hep., p. 469 ; Steph., Sp. Hep., 1900, p. 39.

Dioecious ; stems 7–12 mm. long, simple or forked, prostrate, with long purple rhizoids ; leaves deep green, uni-stratose, suberect or sometimes spreading, numerous, quadrate, with rounded corners and undulate or incurved rounded apex, acutely dentate along the whole upper margin, but with several teeth larger than the others, the appearance under a lens being crisped and hairy.   Cells oblong-hexagonal, 0·1×0·05μ, those toward the base of the leaf larger. Perianth campanulate, plaited, toothed, and ciliated.   Seta 3–4 mm. long ; capsule large, globose.   Spores with numerous short papillae.
Table Mountain, Stellenbosch Flats, and other western localities.

Suborder 2. ACROGYNAE.   Archegonia at first terminal on stem or branch, the involucre (perianth) representing true leaves, perianth often lateral later through innovation.   Antheridia borne in axils of more or less modified leaves. Stem always producing two lateral rows of leaves, and an additional row of small leaves (=folioles or amphigastria) is often present on the lower surface.
Tribe 1. JUBULOIDEAE.   Elaters few, with only one spiral fibre, and fixed by one end to the capsule-wall, and pendent, extending to the base of the capsule-cavity.   Archegonia usually one to four.   Leaves incubous, usually complicate 2-lobed, the lower lobe (*lobule*) small, sometimes absent through abortion or breakage.   Folioles usually present ; perianth usually ridged, and contracted above to a narrow mouth till burst by the capsule.   Seta short ; capsule globose, the lower part solid.
Family 8. FRULLANIACEAE.   Branches intra-axillary ; innovations rarely present.   Archegonia usually two or few in a group.   Lobule convex toward the upper lobe, and usually pouched, galeate, lunate, cylindrical or crested, with a short attachment to the larger lobe, and separate from the stem, a small subulate process being usually present between.

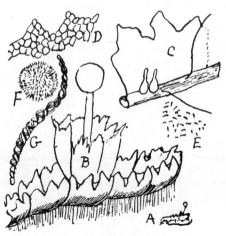

Fossombronia spinifolia.

A, plant, nat. size ; B, same, ×10 ; C, leaf,
  ×20 ; D, apical teeth, ×25 ; E, markings
  on capsule-wall, ×50 ; F, spore, ×200 ;
  G, elater, ×200.

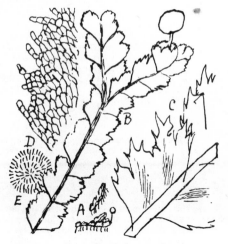

Fossombronia leucoxantha.

A, plants, nat. size ; B, same, ×5 ; C, leaves,
  ×10 ; D, leaf-margin, ×20 ; E, spore,
  ×200.

Genus 17. FRULLANIA Raddi, in Atti Soc. Ital. Sc. Mod. ; Syn. Hep., p. 408 ; Schiffner, Pfl., i, 132 ;
Spruce, p. 3 ; Macvicar, p. 434.

Characters of the family : usually epiphytic in forest surroundings, either on bark, stones, leaves, or mosses, seldom on soil ; sometimes submerged. Stems 1–3 inches long, pinnately or irregularly branched, varying from bright red or brown-red to green in the same patch in accordance with light, usually red, but green if very wet or submerged or shaded. Leaves incubous, shortly inserted, the upper lobe very regular, entire, and almost round, the lobule shortly attached to a sinus at its base, reflexed under its surface, but convex toward that, *i.e.* whatever its form (and it has many forms even on the same plant) the lobule is either inflated opening downward, or it is more or less concave or boat-shaped opening away from the upper lobe, but with its back adpressed to it. Foliole always present at first, 2-lobed, often absent from dried specimens through adherence to its site by adventitious rhizoids produced on central warts. Perianths are at first terminal on stem and branches ; true innovations from immediately below the perianth are rarely present, but new branches from lower down (*i.e.* normal branches) get rapidly ahead and soon leave the perianth-branch as apparently lateral. Basal and apical leaves, lobules, and folioles are seldom normal ; mid-stem specimens show the mature characters better, but all parts should be examined. Cells and trigones regular when young, very variable with age. Monoecious or dioecious ; floral leaves larger than others, usually with long, open, entire, or serrate lobules and enlarged folioles ; perianth usually free, pear-shaped, 3–4 or many-winged, the wings with straight, waved, or toothed margins. Archegonia usually two. Male inflorescence flatly cone-shaped, with short leaves and concave folioles containing an androecium in each axil. The genus is easily distinguished from all others ; the species in it are more difficult to separate.

The European species mostly have a central line of hyaline cells like a midrib ; in South African species this has not been seen. Gemmae sometimes occur on leaf or perianth.

Synopsis.

Perianth 4-winged (*Chanonthelia*).
    48. *F. Ecklonii.* Lower half of lobule expanded, upper half twisted-galeate ; all parts large.
    49. *F. caffraria.* Lobule variable from expanded to shortly galeate, with deflected point ; foliole usually
        shouldered.
    50. *F. sylvestris.* Lobule often deeply galeate, foliole not shouldered.
    51. *F. trinervis.* Densely congested, small.
Perianth 3–5 winged ; wings tuberculate, waved, or straight. Lobules seldom much longer than wide (*Galeiloba*).
  A. Floral leaves free.
    52. *F. squarrosa.* Perianth wings papillate or mamillate.
    53. *F. diptera.* Perianth wing-margin usually waved, seldom straight ; style small.
    54. *F. socotrana.* Perianth 8-plicate ; lobule twice longer than wide.
    55. *F. affinis.* Perianth wing-margin straight.
  B. Floral leaves, lobules, and folioles united.
    56. *F. natalensis.* Lax ; all parts large.
  C. 57. *F. bursicula* Steph. Perianth not known ; style very prominent.

Perianth 3-winged, smooth, on small lateral branches ; leaves concave ; lobules longer than wide (*Thyopsiella*).
  58. *F. serrata.*   Wiry ; leaves acute, incurved ; floral leaves serrate.
  59. *F. Rehmannii.*   Leaves obovate, fairly large ; floral lobules and folioles dentate.
  60. *F. capensis.*   Minute, much branched, all leaves entire, rounded.
Perianth 3-winged, smooth, terminal on main branches ; lobules obliquely patent, clavate, or cylindrical (*Diastaloba*).
  61. *F. brunnea.*   Leaves mucronate, style large, folioles dentate-laciniate ; floral parts serrate.
  62. *F. cuculliloba.*   Leaves obovate, rounded ; lobule cucullate ; floral leaves toothed at the base, other parts
        entire.
  63. *F. Lindenbergii.*   Small ; leaves rounded ; folioles bifid, entire ; lobule cuneate, long ; floral parts entire.

  48. *Frullania Ecklonii* Spreng., Syst. Veget., iv, 2, 324 ; Syn. Hep., p. 772 (in part) ; Spruce, p. 10 ;
          Schiffner, Pfl., i, 133, fig. 70 ; Steph., Sp. Hep., iv, 323.

              =*F. Mundiana*, L. and L., Syn. Hep., p. 772.
              =*Jungermannia Ecklonii* Spreng.
              =*F. emergens* Mitt. (Cameroons).

   Monoecious ; stem pinnate or often irregularly branched, 2–3 inches long, suberect or prostrate, matted or more usually separately epiphytic on mosses to which they adhere.   Leaves vertical, imbricate, cordate, circular in outline and very convex, the rounded apex usually undulate-cucullate, the base suddenly cut away by the large lobe being inflexed.   Leaves on the main stem 0·75–2·5 mm. diameter ; those on the branches considerably smaller, the margin

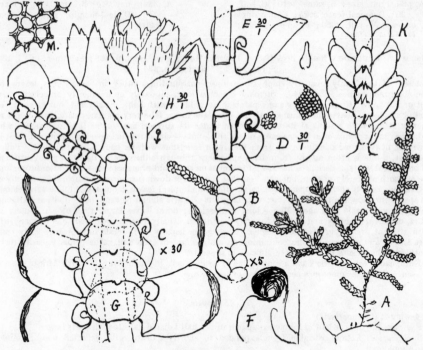

Frullania Ecklonii.

entire] and often undulate, lobe on old stems standing erect, expanded at the base, then suddenly twisted to the opposite plane and curved over away from the stem to form a tubular crest, below which the flat lobe has a small projecting point on its surface, the whole lobe and crest being somewhat ear-shaped, but on young stems the lobe is very variable.
   Folioles at the bases of the leaves on one side of the stem only, each foliole half the diameter of the leaf, rounded-cordate, usually emarginate, on some branches bluntly angled, margin entire.   Fruit terminal on the stem and branches until outgrown by branches from below.   Involucral leaves four or six, large, 2-fid, each keeled, the upper lobe bluntly ovate, more or less toothed or entire, the lower lobe inflexed, as long but much narrower (fig., p. 39), acute, the margins reflexed and often several-toothed, those of the inner leaves lanceolate and more or less spinose.   Pistilidia four.   Perianth

oblong, standing higher than the spreading perichaetal leaves, 1·5–2 mm. long, 0·50–0·75 mm. wide, suddenly contracted to a short tubular point ; two prominent ridges on the back (making it 4-sided) and three slight ridges in front ; the keels not papillose. Male spike short, dense, almost globose ·or ovate-globose, with 2-lobed imbricate keeled leaves and acutely emarginate folioles. Leaf-cells roundly hexagonal, 0·025 mm. diameter, with numerous trigones, marginal cells rather larger and more hyaline than the others, forming a distinct marginal row. Leaves when dry have the apex reflexed. It often happens that all the folioles along a branch develop a wart on the back of each from which rhizomes are emitted.

As pointed out by Spruce in 1884 (p. 10), some confusion exists in the Syn. Hep., where the name *F. Ecklonii* of p. 413 is *F. Arecae* (Spreng.), with South African localities wrongly attached, while on p. 771 two varieties of South African *F. Ecklonii* are mentioned together with one of the South American forms of *F. Arecae*, while *F. Mundiana* L. and G. is added and described, which is a synonym of the true South African *F. Ecklonii* Spreng.

They all belong to the section separated by Spruce as *Chonanthelia*, in which the back of the perianth has two prominent ridges, thus forming a 4-sided figure, while in all the other sections the mid-front ridge is prominent, thus giving a more or less triangular section. Stephani refers to this confusion in Hep. Afr. (Engler's Bot. Jahrb., 1895, p. 314) and states that in any case *F. Arecae* occurs in Central Africa, African islands, and Natal. I am not aware that *F. Arecae* Spreng., with "cucullate slightly toothed bracteoles" as described by Spruce, occurs in our area, but the two are near relations.

The varieties mentioned in the Syn. Hep., p. 771, are

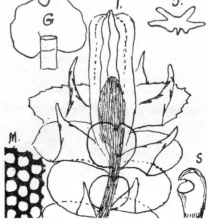

Frullania Ecklonii.

S, calyptra, showing mode of opening.

(*a*) *Robustior.* Leaves wide ; involucral leaves deeply dentate ; auricles large.

(β) *Tenerior.* Leaves and auricles small, involucral leaves with six to seven subserrate teeth.

F. Ecklonii occurs in all forest districts of South Africa.

49. *Frullania caffraria* Steph., Hep. Sp. Nov., vi ; Hedw., 1894, p. 141 ; Steph., Sp. Hep., iv, 382.

Dioecious ; stems 2 inches long, many-branched, the branches short and smaller-leaved than the stem. Leaves nearly circular, somewhat horizontal ; lobule close to the stem or on it, and style not evident. Lobule often open and lanceolate, in length more than half the width of the leaf, and with incurved margins, but when closed or nearly

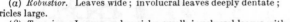

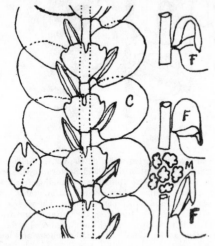

Frullania caffraria.

Frullania caffraria.

closed it is shortly but very widely hooded, the decurved outer side forming a long hook or point to the lobule, and these pointed lobules are sometimes very abundant and regular. Foliole half as wide as the leaf, rotund-cuneate, shortly and narrowly emarginate, and with angled or shouldered outline, or on the larger folioles with one tooth on each shoulder. Cells at first quadrate, soon with very irregular outline·; central cells of leaf large, outer rather smaller.

Perianth at length well exserted, 4-winged, the outer wings with straight edges, the inner often waved or lobed. Floral leaves about two pairs, oval-oblong, entire, rounded, with large, open, lanceolate, entire or once-toothed lobule, and inner folioles longer and more acute than usual. Below the floral leaves an innovation with pointed lobules usually appears, and pointed lobules occur in abundance on the main stem below that.

Stephani's original description, which is from a Van Reenen plant, is very incomplete and unsatisfactory. I have it from many Eastern localities.

50. *Frullania sylvestris* Sim (new species).*

Dioecious; densely matted on stones and trees, usually dark purplish brown. Stems 2–3 inches long, pinnately branched, vigorous. Leaves nearly circular and convex, but through the rounded apex being reflexed a squarrose appearance is presented. Margin entire; lobule large, rather widely separated from the stem, and with a short linear style between, helmet-shaped, oblong or widely cylindrical, rounded above, twisted at the base, not extending below the junction with the leaf, standing slightly divergent from the stem and extending half across the leaf. On the branches open lobules are common. Foliole ovate, roundly emarginate to $\frac{1}{4}$–$\frac{1}{3}$ its length, entire and without

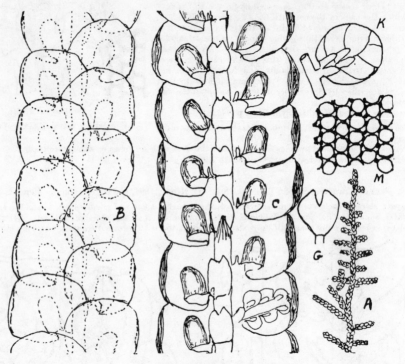

Frullania sylvestris.

angled shoulders; usually inflexed round the stem, and often producing rhizoid warts on the back. Cells rounded, with six most distinct trigones to each cell. A most vigorous grower, frequent in Natal and Eastern forests, but very sparingly fertile; perianth obovate-elliptical, with straight or waved edges to the wings; standing free from the perichaetal leaves, and with only about two pairs of perichaetal leaves, below which an innovation starts. Section of perianth shows two opposite wings with a slight ridge on the back, and two separate wings in front (*i.e.* on the under side), having wings waved or nearly straight and the folioles and lobes free from the perianth and from one another. Perichaetal leaves oblong with rounded apex, entire, and with only a short undulation as lobule. Male cone short, almost circular. *F. caffraria* differs in having pointed lobule, usually shouldered foliole, and long and pointed perichaetal lobule.

---

* *Frullania (Chanonthelia) sylvestris* Sim (sp. nov.).

Dioica, vigens, 2″–3″ longa, pinnatim ramosa. Folia orbicularia, convexa, squarrosa, apice deflexo, rotundata, integra; lobuli magni, galeati, ad basin torti, in ramis saepe aperti. Foliola ovata, emarginata, sine umeris angulatis. Perianthium obovato-ellipticum, quadrigonum.

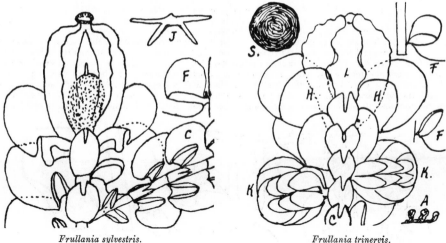

Frullania sylvestris.   *Frullania trinervis.*

51. *Frullania trinervis*, L. and Ldbg., Syn. Hep., p. 427 ; Steph., Sp. Hep., iv, 325.

A minute, densely congested, monoecious species, usually green, in which the fertile branch is usually subtended by one or two male cones ; apart from this the general appearance is much like that of *F. diptera* except that the perianth is short and wide-mouthed and has two undulate wings on the lower surface besides the two undulate lateral wings. Capsule very large and green, opening irregularly. Leaves rounded, imbricate, squarrose ; lobule open or shortly galeate ; foliole ovate, rather deeply cut, its margins entire. Involucral leaves rounded, with large round lobule ; the involucral leaves, lobules, and folioles more or less connate but not enough to be a good section character. Perianth with two wings and two folds.

In the Synopsis Hepaticarum five varieties are described, four of which are African, one South American. Variation through surroundings accounts for the South African forms. It occurs from Table Mountain to Natal.

F. Hildebrandtii Steph., from Madagascar, Natal, and Cape Town (Brunnthaler, 1913, i, 19), does not seem to differ.

52. *Frullania squarrosa* N. ab. E. (Syn. Hep., p. 416) ; Steph., Sp. Hep., iv, 388.

=*Jungermannia tuberculata* L. and Lg., in Linn., ix, 434 (South African specimens).

Dioecious, usually red ; stems matted or rambling separately among mosses, 3–4 cm. long, pinnate or somewhat 2-pinnate ; leaves subvertical, overlapping, 1 mm. diameter, convex, cordate-circular with rounded flattened apex,

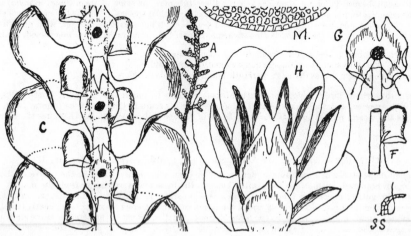

Frullania squarrosa.
SS, style.

the lower shoulder usually folded inward so that there appears to be a deep basal lobe which, however, is only part of the circular lamina. Lobe attached close to the stem, elsewhere free, exceedingly variable even on the same plant; when closed, obovate helmet-shaped, seldom longer than wide or not much longer, truncate, and somewhat pointed at the base, which does not extend below the point of junction; often not inflated (*i.e.* convex toward the outer lobe, keeled along its margin, and the two surfaces closely adpressed), but the lobe is more frequently either expanded or partly expanded. Style rather large and arching toward the lobe. Foliole cordate at the base, roundish acuminate, emarginate, its segments more or less acute, not distant, the margin entire and not lobed, but standing out from the stem over all and much reflexed toward the base, so that the foliole is deeply concave outwardly and often provided with a central rhizoid-bearing wart. On young branches the lobes are arched lunate, and as at that stage they stand out from the stem and are seen more or less sideways, they often appear club-shaped though not really so. Cells

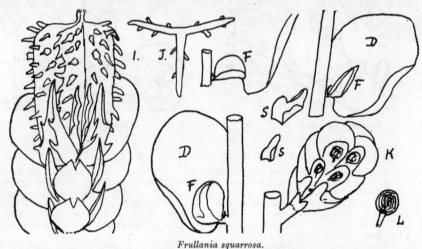

Frullania squarrosa.

S, perianth tubercles developed into lobules.

0·025 mm., irregularly shaped, except the marginal cells which are quadrate and often coloured; no other distinctive cells are present. Fructification on short lateral branches on the main stem; involucral leaves about three pairs, roundish oblong, the lobe of each larger than usual, expanded, ovate-lanceolate, the reflexed margins of the inner lobe sometimes several-toothed. Involucral folioles larger than usual, more deeply cut and their lobes more pointed, the inner foliole sometimes several-toothed. Perianth terminal on short lateral branchlets, free, 3-winged, the back flat or channelled, the wings more or less tuberculate along the margins, or sometimes exceedingly tuberculate, in which case the surfaces of the perianth are also tuberculate, and some of these tubercles occasionally develop into small 3-lobed folioles. Macvicar regards these tubercles, or some of them (in *F. dilatata*) as gemmae, since similar gemmae sometimes occur on the leaves of that (and this) species. Male plant more slender, the cone terminal on very short side branches, usually ovoid and 6–10-leaved, but sometimes longer, with growth afterwards proceeding from its apex. In the Synopsis Hepaticarum, Suppl. 772, three varieties are mentioned, viz.:

(α) *vulgaris.* Robust; keel and angles distinctly tuberculate.
(β) *laevior.* Perianth smooth, with occasional crenatures or papillae.
(γ) *laxa.* Stem long; leaves distant; perianth minutely papillose.

Frequent throughout forest regions of South Africa. Said to be found also in India, East Indies, Australia, South and Mid America, and the nearly related *F. usagara* Mitt. occurs in Central Africa (Journ. Linn. Soc., 1886, p. 326). It is also closely related to *F. dilatata* (L.) Dum. of Europe and North Africa, which has the foliole lobed at the sides.

53. *Frullania diptera* L. and Ldbg., Syn. Hep., p. 420; Steph., Sp. Hep., iv, 374.

Matted, on moist stones or bark, usually near water or upon wet rocks; stems in the centre of the mat short and dense, much branched, with closely imbricated leaves; the loose stems extending from the margin of the mat, elongated, lax, irregularly pinnate or bipinnate, with leaves separate, so imbricate and lax leaves occur on the same plant. Leaves horizontal (*i.e.* set across the plane of the stem), so standing erect from the stem, nearly circular, convex, entire, the largest 1 mm. diameter, those on the branches much smaller, lobule at right angles to the stem and tending toward the apex of the leaf, varying exceedingly in form on one stem, from an expanded, somewhat concave, ovate-lanceolate, or deeply channelled lobe to a lunate hood, which standing more or less erect and compressed often appears under the microscope shorter and more hooded than it actually is. Style usually present but small. Folioles ovate to roundish cordate, more or less deeply emarginate or bifid, the smaller ovate entire, the larger often bluntly angled.

Fructification on very short terminal, but eventually lateral, branchlets; in this way many perianths frequently occur at short intervals on one stem which is still elongating by innovation. Perianth subtended by two to three pairs of free involucral leaves, on which the lower lobe is longer than usual, extending across the width of the larger lobe, is convex outward, and toward the apex has its margin inflexed and usually somewhat dentate. Perianth folioles deeply bifid, the lobes lanceolate-acute. Perianth at first obovate and shorter than the perianth leaves; afterwards oblong with a long beak and much longer than the perianth leaves; upper face flat, lower face rounded when young but

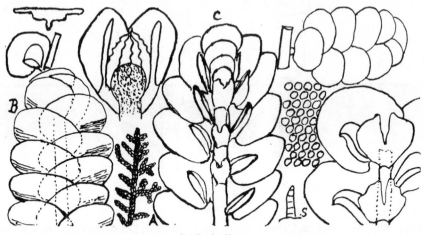

Frullania diptera.

S, style.

eventually somewhat 3-angled, the middle angle the larger; angles smooth or usually waved. Male inflorescence on a separate branch or separate plant, lateral, shortly cone-shaped or almost globose, almost sessile. Leaf-cells irregular, usually not in contact, and without trigones except in youngest state. The most common and the most variable species in South Africa, usually dense and green, often submerged, and with very diverse but usually wide and shallow lobules. It occurs from Table Mountain to Rhodesia.

54. *Frullania socotrana* Mitt., Trans. Roy. Soc. Edin., 1888, p. 335; Steph., Sp. Hep., iv, 368 (of Socotra), is recorded by Brunnthaler (1913, i, 19) from Zwartkops, near Port Elizabeth. Stephani includes it in § Galeiloba, but the description hardly answers that, viz.:

Monoecious, small; leaves obovate, 0·7 mm. long, imbricate; cells 18μ, basal cells 36μ; trigones small. Lobule widely cylindrical, twice longer than wide, erect near the stem, style large, narrowly ligulate; folioles small, narrowly obcuneate, one-third bilobed. Perianth shortly compressed, on the front 2-plicate, on the back 4-plicate (altogether 8-plicate), plicae sinuate undulate. Floral leaves larger, lobes widely lanceolate; inner floral lobule ovate, 2-dentate; androecium small, rounded. I doubt if the Zwartkops identification is correct.

55. *Frullania affinis*, Nees, Syn. Hep., p. 430; Steph., Sp. Hep., iv, 374.

A very small dioecious plant, usually in closely pressed mats or scattered as single stems 1–3 cm. long, adhering so tightly to the bark that on removal many folioles break off and remain fixed to the bark by the rhizoid group, and the plant then on examination presents the appearance of being a *Frullania* without rhizoids and often without folioles, but the youngest tips escape this since rhizoids have not yet developed on them. Stems mostly simply pinnate, with short branches. Leaves imbricate, closely adpressed, roundly quadrate, wider than long, very convex, and with rounded apex. Lobule attached near the stem by its nearest part but leaning away from it, often expanded and concave; when saccate short and wide, deeply lunate or shortly galeate, and free except for the short attachment. Style large, sometimes very large, many celled, and occasionally even lobed. Foliole cuneate, 2-lobed, the lobes rather acute, rhizoid-tubercle often present on every foliole, but sometimes absent; rhizoids short, much branched, and discoid. Perianth terminal, free, well exserted, heart-shaped, mucronate, with two wide wings, two small ridges on the lower surface, and one central short dorsal ridge, cross-section consequently 5-angled or 5-winged; the wings not tuberculate, but all the ridges more or less undulate. Floral leaves larger than the others, pointing forward, and with rather long, concave, open, entire lobes; floral foliole long-cuneate, deeply divided into two acute lobes, occasionally shouldered. Innovation present with the second pair of leaves. Leaf-cells round, 0·025 mm., with large trigones, and each cell containing three to five distinct chloroplasts.   Near *F. diptera*, but the perianth is distinctive. Coast and midlands of Natal and Zululand and said to be in East Africa and East African islands.

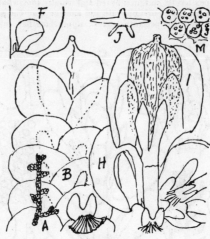

Frullania affinis.

F. tatanarivensis Steph., in Hedw., 1894, p. 148, and in Sp. Hep., iv, 373, is described from a sterile specimen and cannot be placed.   In Brunnthaler (1913, i, 19) it is ascribed to Van Reenen's Pass, which identification is likely wrong ; formerly it was known from Madagascar.

56. *Frullania natalensis* Sim (new species).*

Matted and closely adherent to bark.   Stems irregularly branched, 2–3 cm. long, 1·5 mm. wide, the leaves over-lapping, very convex, nearly vertical, circular except the base, 0·75 mm. diameter ; the margin entire, the rounded apex

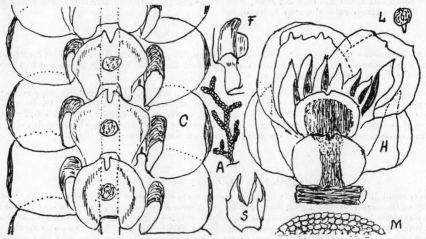

Frullania natalensis.

somewhat reflexed-hooded, the lobule obovate-obtuse, connected far from the stem, the connection extending to the base of the leaf-angle ; the upper end of the lobule widest, plane, and somewhat reflexed toward the other lobe, but

---

* *Frullania (Trachycolea) natalensis* Sim (sp. nov.).

Monoica, vigens, adhaerens, irregulariter ramosa, 2–3 cm. longa ; folia imbricata, convexa, rotundata ; lobulus altus, longe a caule connexus, ita ut connexio usque ad basin folii anguli excurrat ; foliola reniformia, maxima ; folia involucralia interiora dentata, ibique lobuli et foliola acuta ac magna, floralis quaeque pars plus minus connata. Perianthium oblongum, tri-alatum, alae leves ac rectae.   Partes universae magnae laxaeque.

with a deep central cavity below this. The lobule is hooded only in the upper portion, but the hood is not inflated, its sides being so closely applied to one another that the lobule appears to be fully expanded, and only the closest examination reveals the presence of the hood. No style has been seen. Foliole reniform, very large, deeply cordate at the base, emarginate, its segments connivent and acute, the margin always entire, but usually reflexed toward the base ; foliole often saccate in the centre and bearing rhizoids from a wart. Monoecious, the fertile branch 6-leaved, with two rounded, lightly emarginate folioles. Perichaetal leaves very convex, incurved, and enclosing the young inflorescence, more or less apiculate, the inner pair smaller and toothed, all the lobules large, convex, and lanceolate-acute, and several small inner bracts are present, bifid and toothed ; all these floral parts more or less connate. Perianth oblong, free, almost all exserted, 3-lobed, wings neither tuberculate nor undulate. Innovation with deeply galeate lobules and roundly emarginate folioles. Male cone sessile, elliptical or globose, of about twelve closely imbricating circular leaves, each with an open, concave, ovate-lanceolate lobule not larger than usual, and folioles smaller and less emarginate than usual, and with large pedicellate antheridia, two together, in the axils of the lobules. Marginal cells quadrate, all others rounded or elliptical, 0·025 mm. diameter, the lower cells 0·035 × 0·025 mm.

Wellington, Rosetta, Natal (Sim 8007) and elsewhere in the Natal forest region or near it.

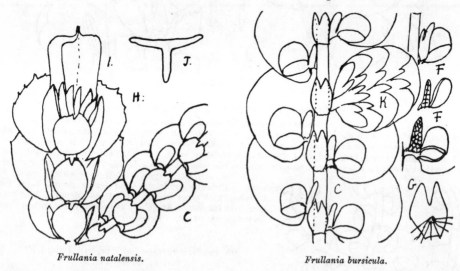

Frullania natalensis.　　　　　　*Frullania bursicula.*

57. *Frullania bursicula* Steph., Sp. Hep., iv, 678.

Small, slender, pale green, usually on rocks ; only known sterile. Stems 3–4 cm. long, shortly pinnate ; stem leaves imbricate, nearly round, spreading, concave, 0·6 mm. long and wide ; style large or very large, simple or forked, sometimes as long as the lobule; lobule large, round, hooded, and inflated, with straight open mouth. Folioles not larger, narrowly cuneate, widest at the middle, one-third split. Fruit not known.

Natal and Mauritius.

(*F. exigua* Steph., Sp. Hep., iv, 367, is stated there to be from the Congo, though in the list on p. 359 it is stated to be from the Transvaal. The latter is an evident mistake.)

58. *Frullania serrata* Gottsche, Syn. Hep., p. 453 ; Steph., Sp. Hep., iv, 478.

A slender but vigorous and wiry monoecious species ; stems 2–3 inches long, irregularly but closely and shortly 2-pinnate, purplish in colour, and more or less matted or epiphytic. Leaves overlapping, concave on the under-surface, oblong, with cordate base and incurved acute apex. Lobule cylindric, tubular, or with inflexed margins, divergent from the stem and exceedingly regular, those on the young branches usually with inflexed margins, but appearing tubular. Style present. Foliole ovate on young branches, widely ovate on old branches, opposite the base of leaves on one side of the stem only (as is usual among *Frullaniae*) spurred, and the older ones margined, all emarginate one-third depth, the segments acute. Perichaetia terminal on short lateral branches, which are 10–12-leaved ; perianth free, 3-angled, smooth, inner per. leaves and lobules laciniate, lower ones less so ; perichaetal folioles more acute than others. Male catkin short, nearly globular, terminal on branchlets. When dry the foliage incurves more and closely envelops the stem. Not matted and not frequent, except in forests in Umtali and elsewhere in North-Eastern Rhodesia.

It occurs also in East Africa and in the East Indies (Eyles 2626, etc.). This is evidently what Pearson called *F. apiculata* Nees, from Knysna, in Hep. Knysnanae, p. 5, pl. 1, figs. 2–13. Stephani now limits *F. apiculata* to Asia and Oceania tropica.

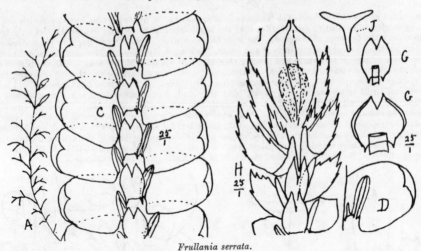

Frullania serrata.

59. *Frullania Rehmannii* Steph., Sp. Hep., iv, 485.

Not known to me ; described by Stephani as autoecious, small, red, up to 4 cm. long, regularly pinnate ; pinnae to 5 mm. long, with few or no pinnules. Leaves widely obovate, 0·8 mm. long, middle 0·6 mm. wide ; apex widely rounded, leaf crossing the stem widely. Cells, upper 14μ without trigones, basal $18 \times 27\mu$ with large trigones ; walls everywhere straight. Lobule large, elliptical, distant from the stem, erect, more than twice as long as wide, the top obtuse, mouth attenuated ; folioles large, twice as wide as the stem, obcuneate below, angled or subdentate at the middle on each side ; apex 2-lobed to one-third depth, lobes acute. Perianth pyriform, floral leaves entire, floral lobule narrower, with outer margin crispate and dentate, floral folioles large with the margin irregularly and roughly dentate and spinose. Androecia numerous. Montagu Pass, C.P.

60. *Frullania capensis* Gottsche, Syn. Hep., p. 449 ; Steph., Sp. Hep., iv, 482.

Frullania capensis.

Dioecious, minute, abundantly branched and innovated, the branches smaller leaved than the main stem. Leaves distant, almost round, very fragile, often mostly absent, the lobule deeply helmet-shaped or slightly contracted at the mouth, connected close to the stem, more than half as long as the leaf ; style small, usually present ; foliole comparatively large and conspicuous, divided half-way, the segments acute and often slightly shouldered, more permanent than the leaves. Floral leaves about two pairs, twice as long as others, oblong-rotund, with long, open, almost entire, free lobules, and enlarged but almost normal folioles. Perianth and male cone not seen. Frequent on Table Mountain, Stellenbosch, Paarl, etc. ; absent eastward, but Stephani records it (Engl. Bot. Jahrb., 1895, p. 461) from Mascarenes and Central Africa, and I have it from Matopas (Sim 9099B) where the foliole is regularly 4-fid. Gottsche and Stephani both include it in the group now known as *Thyopsiella* ; the lobule is hardly clavate, and in the absence of perianth I doubt its position in that group.

Gemmae are often present on the leaves ; apparently reproduction is effected mostly by that means.

61. *Frullania brunnea*, Syn. Hep., p. 441 ; Steph., Sp. Hep., p. 621.

=*Jungermannia brunnea* Sprengel, Syst. Veget. Cur. post., p. 325.

Matted among other mosses, $\frac{1}{2}$–$1\frac{1}{2}$ inch long, prostrate, simple or irregularly branched ; foliage very regular, the branch leaves about half the size of those on the main stem, but similar, lying along the stem, imbricate, rotund, with a mucro of three or four cells ; lobule far distant from the stem, attached only along the outer edge of the cut-away leaf-base ; the mouth small and attached there, the lobule itself balloon-shaped and lying away from the stem. An inflexed edge, carrying a leafy style, separates the lobule from the stem. Foliole large, cut half-way into two, each deeply cut along the outer edge into three to four segments. Small branches are similar in all these parts, but there are many undeveloped small branches producing only two to four lobules without leaves. Cells in regular lines but separate, and large and turgid, making the leaf appear papillose. Perianth and androecium not seen, but perichaetal branches have the upper leaves pointed and ciliate-serrate at the apex (which is regularly recurved over the

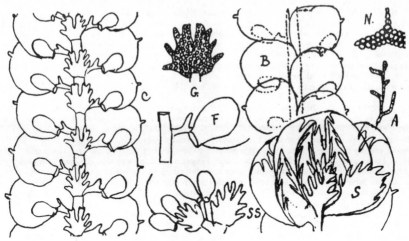

Frullania brunnea.

per. lobes and folioles), long, lanceolate, serrate lobules, and large ciliated folioles. Very distinct from all other South African species. Top of Table Mountain, Cape Town (many specimens), and Steenbras Valley, Sir Lowry's Pass Mountains (Miss Stephens). Not seen in eastern forests, Natal, Transvaal, or Rhodesia.

62. *Frullania cuculliloba* Steph., Sp. Hep., iv, 622.

Not known to me; described by Stephani as dioecious, minute; stem up to 2 cm. long, longer branches floriferous, but extending by innovation. Leaves obovate, 0·46 mm. long, middle 0·33 wide, apex rounded, base crossing far over the stem. Cells, upper 18μ, basal 27μ, with large trigones and thin walls. Lobule large, cucullate, scarcely longer than wide, distant from the stem, inclined, somewhat constricted behind the mouth. Foliole small, 2-lobed half-way, and angled at the middle on each side. Floral leaves very large, entire. Floral lobules partly united, toothed at the base. Floral folioles oblong, bifid half-way, and angled to the middle on each side.

Table Mountain, C.P.

63. *Frullania Lindenbergii* Gottsche, Syn. Hep., p. 447; Pearson, Hep. Knysnanae, p. 4, pl. 1, fig. 1.

A small monoecious species growing among mosses or hepatics, seldom matted alone, usually purplish brown in colour; stems firm, 2–3 cm. long, closely pinnate or 2-pinnate, the pinnae short; the leaves not overlapping on the

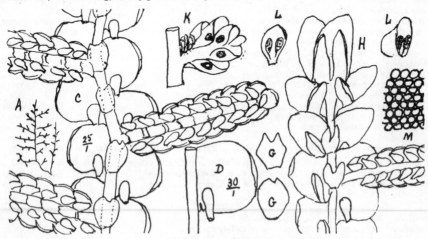

Frullania Lindenbergii.

main stems, but closely overlapping and much smaller on the branches. Leaves small, 0·2–0·4 mm. diameter, nearly circular or rather longer than wide, very convex, the rounded apex reflexed ; lobule with a style between but near the stem, cylindrical club-shaped, one-third to one-half as long as the width of the leaf on the main stem, as long as the width of the leaf on the branches, the upper part of the lobule approaching the stem, the narrow mouth extending below the junction with the leaf, the lobules on the branches closely placed and parallel-divergent from the stem. No lobes are expanded except three or four pairs below the involucre, and sometimes two occur without lamina as in *F. brunnea.* Foliole ovate-emarginate with entire margins, occasionally repandly shouldered, the sinus slight or to one-third of the foliole, with obtuse segments, except those on fertile branches which are more deeply cut and have more acute segments. Involucral leaves rather larger, ovate, obtusely pointed, with more or less expanded lobules, those of the upper leaves lanceolate-concave, sometimes toothed. Perianth terminal, 1 mm. long, compressed, obovate, keeled in front. Male cone lateral, sessile, globose or oblong, its bracts hooded and closely imbricated, with two antheridia in each axil. Cells roundly hexagonal, thick-walled, 0·015–0·017 diameter. Frequent in forest country.

Family 9. Lejeuneaceae. Branches infra-axillary ; innovations usually present immediately below the female inflorescence, which is monogynous. Lobule usually present, concave toward the upper lobe, or flat, usually with a long attachment, and often attached to the stem also ; sometimes the lobule is absent or only the lower margin inflexed. Capsule wall of two layers, the inner thickened irregularly. Usually epiphytic.

The group of species forming Lejeuneaceae is a very large one, widely distributed over the world. These species were formerly, so far as then known, included in the genus Jungermannia, as was also the case with many other groups now separated as genera, hence all earlier species were first described as species of that genus. Although some sub-division took place in which almost all these species of this group became Lejeunea, it was not till Spruce's Hepaticae Amazonicae et Andinae appeared in 1884 that the classification now adopted came into use. Spruce recognised the relationship of the group, but tried to reduce the unwieldy conglomeration by separating on very artificial and often vegetative characters into subgenera, each of which retained the name Lejeunea, but with a prefix founded on character. His subgenera were afterwards made genera, mostly by Schiffner, and are likely to endure in cosmopolitan work, though for a local flora fewer genera would be better and more equal in value to other genera. The difficulty is increased by most characters being more or less unstable, and by some characters varying at different parts of the plant.

In Hedwigia, 1890 (three issues), Stephani reviews the species of Lejeunea represented in the Herbarium Lindenberg. So far as South Africa is concerned nineteen species are listed, but at a distance, and not easily accessible ; this helps us little, except that they are placed in Spruce's subgenera, since raised to generic rank. Stephani, Sp. Hep., vol. v, uses these and many others.

The following synopsis indicates definite characters which separate the generic types, but these are apt to approach, or vary, or overlap in places.

Synopsis of Genera of Lejeuneaceae.

Folioles absent .   .   .   .   18. Cololejeunea.
Folioles entire, rounded ; colour yellowish brown or green.
 Perianth compressed, with ciliate margins .   .   .   .   .   19. Thysanolejeunea.
 Perianth 3-plicate .   .   .   .   20. Mastigolejeunea.
 Perianth 4–5-keeled.
  Yellow. Innovations usually one at each perianth (or dichotomous at apex only) ; leaves all entire.
                                                                         21. Archilejeunea.
  Green. Innovations usually dichotomous. Floral leaves sometimes toothed   .   22. Brachiolejeunea.
Folioles not 2-fid, but with several-toothed apex .   .   .   .   23. Ptycholejeunea.
Folioles 2-fid.
 Folioles one to each leaf .   .   .   .   .   .   24. Diplasiolejeunea.
 Folioles one to each second leaf.
  Perianth ultimately 7–12-wrinkled almost the whole length .   .   .   25. Anomalolejeunea.
  Perianth 3–5-ridged, or compressed and plane ; lobule along the incurved leaf-base (or absent).
   Upper lobe lanceolate and toothed .   .   .   .   .   26. Drepanolejeunea.
   Leaves entire, colourless or pale yellow, with rounded apex.
    Plants very minute ; leaves 6–10 cells across .   .   .   .   27. Microlejeunea.
    Plants fairly large ; leaves 10–30 cells across .   .   .   .   28. Eulejeunea.
   Leaves entire, with acute apex, colour various .   .   .   .   29. Taxilejeunea.
   Leaves cellulose-crenulate, colour various.
    Lobule obliquely ascending, arched, in slight sinus .   .   .   30. Hygrolejeunea.
    Lobule linear, straight, without sinus .   .   .   .   31. Cheilolejeunea.
  Perianth unknown ; lobule an erect style .   .   .   .   .   32. Stylolejeunea.

Genus 18. Cololejeunea (Spruce as subgenus, p. 291), Schiffner, Pfl., i, 3, 121 ; Macvicar, p. 408.

Plants small or very small, usually among or upon other hepatics or mosses. Leaves various, 2-lobed, plane or papillose, and with a stylus present adjoining the stem, often caducous. Folioles absent. Perianth more or less inflated, obtusely 5-angled above ; innovations on one side only, floriferous.

Synopsis.

64. *C. Rossettiana.* Minute, every cell mamillate.
65. *C. minutissima.* Not mamillate ; foliole nearly as large as leaf.
66. *C. oleana.* Not mamillate ; folioles one-fourth length of leaf, or less.

64. *Cololejeunea Rossettiana* (Mass.), Schiffner, Pfl., i, 3, 122 ; Macvicar, p. 410.

=*Lejeunea Rossettiana*, Mass., in Nouve Giorn. Ital., p. 487 ; Pears., Journ. Bot., p. 353, pl. 292.
=*Physocolea Rossettiana*, Steph., Sp. Hep., v, 915.

The most weird and wonderful of all the Micro-hepaticae. Stems 1–3 mm. long, irregularly much branched, abundantly rhizoid-producing. Leaves ovate, subacute, imbricate or scattered, convex, one layer of cells thick, the upper surface densely covered with mamillate papillae, one on each cell, the under-surface not papillate, but fringed with the outer mamillae of the upper surface. Stem not mamillate, composed of fragile thin-walled cells. Leaf connected by a narrow base, and when dry rhomboid and spreading. Lobule one-third the length of the leaf, flat, ovate, acute, its surface mamillate. Foliole absent ; stylus minute, caducous. Cells 15μ. Fertile inflorescence terminal on a short branch, soon rendered lateral by one innovation. Involucral leaves similar to others ; perianth elliptical, inflated, with five obtuse ridges above, the whole surface mamillate. Androecium on a short branch, few-bracted, the bracts equally 2-lobed, swollen, mamillate. Macvicar mentions stalked discoid gemmae on the leaf-surface ; these I have not seen.

Perie Forest, Kaffraria (Sim 9779), rare, and other moist forests. It occurs also in Britain. Stephani locates it from Algeria, Britain, and Italy (not South Africa).

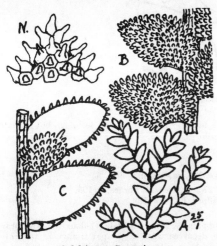

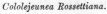

Cololejeunea Rossettiana.

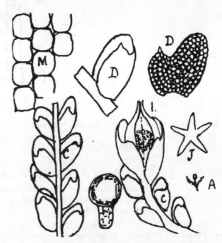

Cololejeunea minutissima.

65. *Cololejeunea minutissima* (Smith), Schiffner, Pfl., i, 3, 122 ; Macvicar, p. 411 ; Spruce, Hep. Am. And., 1884, p. 293.

=*Jungermannia minutissima* Smith, Eng. Bot., pl. 1633.
=*Lejeunea minutissima* Dum., Comm. Bot., p. 111.
=*Physocolea minutissima* (Smith), Steph., Sp. Hep., v, 914.
=*Cololejeunea parvula* Steph., Bot. Gaz., 1892, p. 171.

Minute, 1–3 mm. long, monoecious, dichotomously branched, or sometimes with only one innovation, in which case the perianth soon becomes lateral. Leaves ovate, erect, convex, rounded at the apex, almost in contact, and with the nearly as large, convex, and acute lobule forming a rounded tumid moisture container. Folioles absent. Cells 15μ diameter, roundly quadrate, tumid, with small but distinct trigones. Perianth pyriform, terminal on a branch, with large floral leaves, five acute smooth ridges near its apex, and tumid cells. Androecium terminal on a branch, few bracted, the bracts concave, nearly equally 2-lobed, lobule not toothed. Found among other hepatics under moist eastern forest conditions.

Victoria Falls (Sim 9776), Nels Rust, Natal (Sim 9762), etc. Much like *Microlejeunea gracillima*, but is rather larger, has no folioles, has the leaves more closely placed, the cells smaller, and the two perianth leaves remain permanently after the perianth has gone.

66. *Cololejeunea oleana* Sim (new species).*

Stems closely adherent on leaves of Olea, 5–12 mm. long, slightly branched, dark green, chlorophyllose. Leaves lingulate, spreading, rounded at the apex, with inflated inflexed lobule one-fourth the length of the leaf, quadrate

---

* *Cololejeunea oleana* Sim (sp. nov.).

Adhaerens in foliis Oleae ; caules 5–12 mm. longi ; folia lingulata, rotundata ; lobulus inflatus, inflexus, ¼ folii longus, quadratus, 2-dentatus supra ; foliolum abest ; stylus ordo cellarum, caducus ; folio-cellae leves ; perianthium dorsis 5 acutis, leve.

and 2-toothed above, and rounded on lower margin.   Folioles absent ; stylus a line of cells, soon gone ; rhizoid tufts abundant on the stem.   Perianth with five acute ridges, smooth.   Perianth leaves like others, but with larger

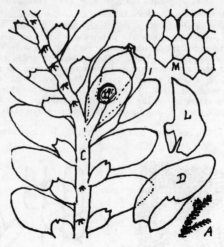

Cololejeunea oleana.

lobules.   Androecium on short lateral branches, with few acute bracts.   Perianth at first terminal, but left lateral by one innovation.   Leaf cells 20μ hexagonal, thin walled, smooth.

Kentani, Transkei, 1914 (Miss Pegler, 1952).

Genus 19. Thysanolejeunea (Spruce, 105, as subgenus) Sim.

=*Lejeunea*, subgenus *Mastigolejeunea*, section *Thysanolejeunea*, Spruce, p. 105.
=*Thysanthus*, Ldbg., Syn. Hep., p. 286 ; Schiffner, Pfl., i, 3, 129.
=*Bryopteris*, Nees, p.p.

Stems adherent, 1-2 inches long, irregularly branched or subpinnate, often flagellate ; colour yellowish brown ; leaves imbricate, lingulate, spreading alternately, convex, entire, decurved when dry ; apex triangular, subacute or blunt, the greater portion of the lower margin inflated-incurved, with or without lobule ; cells thick-walled ; foliole oblong-cuneate, rounded or retuse, entire (in our species).   Inflorescence monoecious or dioecious, usually lateral through single innovation, or occasionally terminal with dichotomous innovation.   Floral leaves like the others or more acute, entire or serrulate.   Perianth compressed or triquetrous, wings ciliate-spinose.

67. *Thysanolejeunea africana* Sim (new species).*

Dioecious, yellowish brown ; stems 1 inch long, irregularly branched, flagellate below ; leaves crowded, regularly ovate-lingulate, 1 mm. long, imbricate, convex, entire, with rounded upper margin, triangular subacute apex and incurved-inflated lower margin, especially in the lower half ; foliole from a narrow base, rounded or cuneate-oblong, entire.   Central leaf-cells fairly large, outer much smaller, all yellowish brown, pellucid, and thick-walled, with distinct trigones.   Perianth usually lateral through innovation, compressed, with ciliate-spinose margins.

Perie Forest, Kaffraria, among *Frullania* (Sim 346), etc.

Caudalejeunea africana Steph., Hedw., 1895, p. 233 ; and Sp. Hep., v, 10=*Thysanthus africanus* Steph., Engl. Bot. Jahrb., pp. 8, 93, from tropical West Africa, might be confused with this, but has emarginate folioles and hidden but present lobules.

Genus 20. Mastigolejeunea Spruce.

Brown or dark plants, caespitose on bark.   Stems 3-6 cm. long, repeatedly branched ; leaves crowded, concave, ovate, obtuse, or rounded at the apex, entire, crossing the stem above, and with the lower margin incurved, with very large, oblong, inflated lobule ; foliole large, rounded, entire ; perianth pyriform, triplicate.

---

* *Thysanolejeunea africana* Sim (sp. nov.).

Dioica, flavo-fusca ; caules 1″ longi, irregulariter ramosi, flagellati.   Folia imbricata, ovato-lingulata, convexa, integra, apice trigono subacuto, margine autem inferiore incurvato-inflato, praecipue basin versus.   Foliola cuneato-oblonga, integra.   Perianthium fere laterale compressum, marginibus ciliatis spinosis.

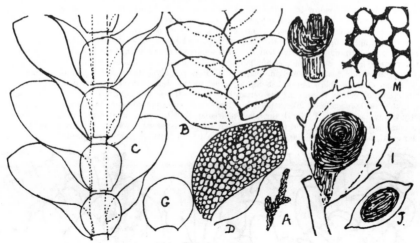

Thysanolejeunea africana.

68. *Mastigolejeunea trigona* Steph., Engl. Bot. Jahrb., 1895, p. 319 ; Sp. Hep., iv, 761 ; Brunnthaler, 1913, i, 22.

Known from East Africa, and recorded by Brunnthaler from Palm Grove, Victoria Falls.

Monoecious, yellowish ; stem 2 cm. long, branched ; leaves spreading, widely ovate, lower margin incurved ; lobule small, narrowly oblong, keel inflated ; foliole three times wider than the stem, subrotund from obcuneate base. Perianth obovate triplicate, folds low, with other rudimentary folds between. Floral leaves small, narrow ; lobule ligulate ; foliole as long as the leaves. Androecium median, bracts to 7-jugate. Not known to me.

Genus 21. ARCHILEJEUNEA (Spruce, 88, as subgenus), Schiffner, Pfl., i, 3, 130.

Plants fairly large, caespitose on bark, stones, or soil, brown or dark as seen in growth, much more regularly yellowish brown as seen under the microscope, irregularly branched, innovations usually one below each perianth, or dichotomous at apex only, and usually without rhizoids. Lower branches often flagellate. Leaves large (1–1·5 mm.), often imbricate, spreading, usually rounded at apex or only bluntly pointed, entire, and with the lower margin more or less inflexed and with an inflexed lobule ; foliole large, rounded, entire ; cells small, thick-walled ; inflorescence usually dioecious, terminal on stems or innovations ; floral leaves entire, similar to other leaves ; perianth pear-shaped, 3–4–5-angled, keels smooth or waved or toothed ; androecea usually terminal bracts, often numerous, rather small, almost equally 2-lobed, 2-androus.

Synopsis.

Perianth wings waved or toothed (*Archilejeunea*).

　　69. *A. Pappeana.* Lower leaves rounded ; upper leaves often triangular, subacute. Innovations often dichotomous.

Perianth wings smooth (*Leucolejeunea*).

　　70. *A. xanthocarpa*, whole of lower and apical margin usually inflexed.

Only lower half of lower margin inflexed.

　　71. *A. rotundistipula*, leaves rounded at the apex.

　　72. *A. chrysophylla*, leaf-apex pointed, lobule evident.

　　73. *A. palustris*, leaf-apex pointed, decurved ; lobule absent or closely adpressed ; perianth unknown.

69. *Archilejeunea Pappeana* (Nees), Sim.

=*Phragmicoma Pappeana* Nees ab E. ; Syn. Hep., p. 296 ; Mitten, in Linn. Soc. Journ., xxii, No. 146, p. 323 ; Schiffner, Hedw., 1894. Revision.

=*Acrolejeunea Pappeana*, Steph., Hedw., 1890, p. 8 ; Steph., in Engl. Bot. Jahrb., 1895, p. 317 ; Steph., in Hedw., 1890, pp. 3, 133.

=*Archilejeunea erronea* Steph., Hedw., 1888, p. 13 ; and Sp. Hep., iv, 707.

=*Ptychocoleus Pappeanus* (Nees), Steph., Sp. Hep., v, 28 (which is said to be eplicate in a genus described as pluriplicate).

Usually golden-brown under the microscope, sometimes nearly green. Monoecious, at least sometimes, the perianth terminal on main or fairly long side branches, and the androecium median or terminal on long basal branches. Stem ½–2 inches long, in some cases closely adherent to bark, with the main stem straight but having many irregular long or short lateral branches and flagellae, and with perianth terminal on larger main and side branches, the innovation there often dichotomous ; in other cases it is caespitose on wet stones along streams, irregularly branched, the branches abundant below, but some of these extend as long, simple, beautifully regular stems ½ inch or more long, with branching either simple or dichotomous below the perianth. Leaves numerous, imbricate, 1 mm. long, ½ mm. wide, expanded when moist, twisted round the stem when dry, the upper margin very rounded, the apex rounded on the leaves on lower half of the stem, but triangular subacute on terminal innovations; the lower margin laxly inflexed for half its length—often increasingly inflexed toward the stem, and with a large rounded decurve in the lower leaves and a large rounded incurve in the terminal innovation leaves, which renders them very unlike. Lobule usually present but often invisible, fairly large, closely adpressed, either pointed and truncate, or rounded or crenulate-emarginate, and the inflexion of the leaf-margin ceases where this begins. Folioles usually imbricate, rounded or repand or subreniform at the apex, rather cuneate below through being attached to the stem and having the margins reflexed when dry; upper folioles larger, especially those near the perianth, and sometimes that next the perianth is bifid, with two acute segments. Involucral leaves almost like others, but larger and more pointed, or 3–5-toothed, with the lobule larger and distinctly triangular-acute. Perianth terminal, pear-shaped, with compressed wings and one to three compressed ridges; the margins of these wings and ridges are either irregularly crenate-waved or sharply dentate. Male bracts few when terminal, many when median, smaller than the leaves, with rather large, pointed lobules. Cells of the

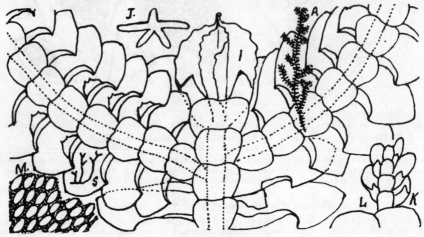

Archilejeunea Pappeana.

leaf rather small near the margin, much larger toward the centre, thick-walled, obliquely elliptical, chlorophyllose, the cells of the folioles and infolded leaf-bases more clearly and early thick-walled and transparent than the chlorophyllose leaf lamina.

This species varies in habitat more than most, and characters consequently change somewhat, but these changes occur even on one plant and are not specific. Not common, but scattered through all the forest districts of South Africa, also extends to Victoria Falls, Malenji, Central Africa, Madagascar, and Mascarenes. (Sim 7964, 9094, 9774, 9092, etc.)

The *Acrolejeunea pulopenangensis* Steph., recorded from Victoria Falls, S. Rhodesia, no doubt belongs to *A. Pappeana*, the said species being an Asiatic one having deeply bifid foliole. See Steph., Sp. Hep., v, 51.

Stephani (Sp. Hep., v, 38) introduces near this, under the name *Ptychocoleus aulacophorus* (Mont.) Steph. =*Phragmicoma aulacophora* Mont., Ann. Sc. Nat., 1843, p. 259, a species which he allocates to "Asia, Africa, Australia, tropical and subtropical, very common," but which I am not aware is South African. Apart from being pluriplicate it might belong to *Archilejeunea*, but the mamillate leaves indicate no South African plant. His description is :—

"Autoecious, of medium size, brownish green, very flaccid, caespitose on soft bark. Stem up to 3 cm. long, much and irregularly branched; stem leaves crowded, squarrose, subrotund when spread out, 1 mm. long, 0·86 mm. wide, inserted on a wide base and with apex rounded. Upper cells $18 \times 27\mu$, trigones fairly large; basal cells $27 \times 45\mu$, trigones large. Lobules triangular when *in situ*, subrotund when spread out; keel almost straight, apex and upper margin regularly mamillate, mamillae 7–8. Stem folioles rather large, three times wider than the stem, obcuneate, wider than long, apex widely rounded. Perianth pyriform, pluriplicate (plicae to fifteen), beak small. Inner floral leaves obovate-oblong, obtuse, the lobule sub-equal in size, truncate, with obtuse angle; inner floral foliole with lobes of equal length, widely spathulate, apex widely truncate-rounded. Androecium small, near to the perianth, long spiked; bracts 8–10-jugate."

70. *Archilejeunea xanthocarpa* (Lehm. and Ldbg.), Steph., Hedw. (1890), i, 20; and iii, 134.

= *Jungermannia xanthocarpa* L. and L., in Lehm., Pug. Pl., v (1833), p. 8, No. 8.
= *Lejeunea (Archilejeunea) xanthocarpa* (L. and L.), Pearson, Hep. Knysnanae, p. 4, pl. 1.
= *Lejeunea xanthocarpa* (L. and L.), Syn. Hep., p. 330.
= *Leucolejeunea xanthocarpa*, Evans, Torreya, 1907, p. 229; Steph., Sp. Hep., iv, 739; Brunnthaler, p. 22.

Plants matted or scattered, light green or pale, or under the microscope yellowish, ½–1½ inch long, 2 mm. wide; stems simple or irregularly branched, the fertile branch very short and lateral, and male cones are frequently placed lower on the same stems (autoecious). Leaves closely imbricated, numerous, 1–1·5 mm. long, oblong-rotund, very convex, rounded at the apex, the lower and apical margins infolded, the lower half of the lower margin being some-what inflated and having a long narrow truncate or bluntly pointed lobe so closely adpressed inside the leaf as often to be difficult to detect. Folioles numerous, large, circular or reniform, somewhat cordate at the base, very regular, usually entire at the apex, but occasionally slightly emarginate, and often bearing a rhizoid wart, lower margins often

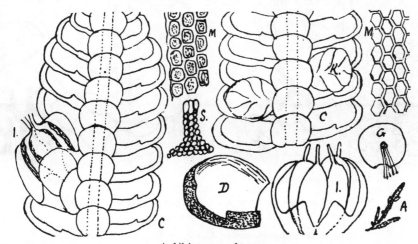

Archilejeunea xanthocarpa.

S, apex of peristome valve.

reflexed when dry. Cells at first regularly hexagonal, 0·025 mm. diameter, adjacent but separate, afterwards sub-globose or roundly quadrate, very turgid when the chlorophyll dries into a central mass ; trigones not evident. Perianth lateral, or terminal on a short lateral branch ; obcordate, inflated, with two ridges on lower surface, and eventually dividing into four long-pointed valves with smooth roundish surfaces. Per. leaves large, oblong, entire, with large lobule or subequally 2-lobed, the margin either hooded or not hooded, and sometimes with an emarginate or shortly 2-lobed foliole having a tooth on each side besides the terminal teeth. Male cone shorter than the leaves and included under them, about 6–8-bracted, with one androecium in each axil.

Zwartkop, Natal (Sim 8413) ; Knysna (B. D. 17,003) ; Grahamstown (Hepburn 4) ; and many other forest localities in South Africa and East Africa, as also in Brazil, etc., unless the latter all belong to *A. Bongardii* St. (see Hedw., 1890, pp. 1, 20).

On young branches the leaves are often less hooded, and the lower margin sometimes less inflexed.

71. *Archilejeunea rotundistipula* (Ldbg.), Steph., in Hedw., 1890, pp. 1, 21.

= *Leucolejeunea rotundistipula* (Ldbg.), Steph., Sp. Hep., iv, 738.
= *Jungermannia rotundistipula* Ldbg., in Linnaea, iv (1829), 360.
= *Lejeunea rotundistipula* Ldbg., Syn. Hep., p. 331.

Dioecious, laxly caespitose, prostrate or the upper part suberect ; leaves crowded, closely imbricate, subvertical, convex, shortly rounded, 0·75 mm. long, entire, the apex rounded, the lower margin with a rounded inflated not adpressed quadrate lobule, having one acute point, and gradually connecting with the margin upward. Foliole imbricate, rounded, from a small base, entire ; floral leaves larger and with erect ovate-lanceolate lobule, the inner foliole largest and sometimes emarginate ; perianth not seen, but described in Syn. Hep., p. 331, as lateral, oblong, on the back 1-keeled, on the front 2-keeled, with margins smooth. In Syn. Hep. the folioles are described as small and remote, but I find them fairly large and imbricate, and the perianth is evidently terminal at first though soon lateral by innovation. Male cone terminal, few-bracted, the bracts shorter and rounder than the leaves, more convex, and with more pointed lobe.

Top of Table Mountain, Cape Town (Ecklon, Pappe, and Sim 9772) ; Dohne Hill, Cape Province, 1898 (Sim 9777), etc.

I found some examples of an unusual form, with some fasciated terminal branches in which the folioles were present in great abundance, so close as to be standing pectinate at right angles to the stem, but leaves and folioles quite absent.

The Syn. Hep., p. 331, also records from the Cape (Ecklon) the South American *Lejeunea unciloba* Ldbg., which differs from *A. rotundistipula* in having the foliole point often uncinate and the perianth shorter, and with the margins not much compressed. I doubt if it exists here, and think the record refers to either *A. rotundistipula* or *A. xanthocarpa.*

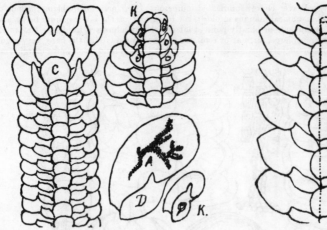
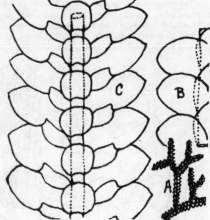

Archilejeunea rotundistipula. *Archilejeunea chrysophylla.*

72. *Archilejeunea chrysophylla* (Lehm. and Ldbg.), Steph., in Hedw., 1890, pp. 1, 20.

= *Lejeunea chrysophylla* L. and L., Syn. Hep., p. 330.
= *Jungermannia chrysophylla*, Lehm., in Linnaea, ix, 4 (1834), p. 423.
= *Marchesinia chrysophylla* Steph., Sp. Hep., v, 144.

Caespitose, prostrate, fuscous; stems ½–1 inch long, dichotomously branched. Leaves numerous, imbricate, 1 mm. long, roundly ovate, convex, all more or less acute, entire, the apex decurved so that seen from above the leaves appear rounded. Lower leaf-margin roundly decurved opposite the lobule, inflated, the lobule adpressed but usually visible, quadrate, truncate, usually with an acute point. Folioles hardly in contact, entire, round; cells large, fuscous, thin-walled. Perianth usually terminal, but sometimes eventually lateral, obovate, the lower surface keeled, the wings and keel compressed, entire. Floral leaves rather larger than others and more pointed, but entire.

Giant's Castle, Natal, 8000 feet, 1915, R. E. Symons (Sim 9775), Phillipstown, Cape Province (Ecklon), East Pondoland, etc.

73. *Archilejeunea palustris* Sim (new species).*

Dioecious; stems ½–1½ inch long, sparingly and irregularly branched, neither dichotomous nor pinnate, but with numerous small-leaved flagellae; dark brown in colour, and growing suberect in loose patches, either alone, or with

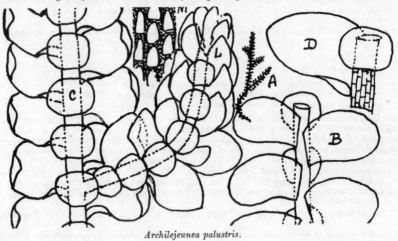

Archilejeunea palustris.

* *Archilejeunea palustris* Sim (sp. nov.).
Dioica, mascula tantum nota; caules ½″–1½″ longi, irregulariter ramosi, cum pluribus foliorum exiguorum flagellis. Colore fusca, in locis palustribus laxe proveniens. Folia secundum plantam varia ad magnitudinem, speciem, propinquitatem, admodum convexa, apice decurvato et obtuse acuto; lobulus pars exigua involuta vel absens vel arte adpressus; foliola rotundata vel breviora, plana, libera a caule, integra; quae omnes partes minores exsistunt in ramis flagellisque.

other Hepaticae or mosses, on swampy soil, or in wet bushland near streams. Leaves exceedingly irregular on different parts of the same plant as to size, form, and proximity ; leaves on old stems are usually large, closely imbricated, very convex, and with apex decurved and more or less bluntly pointed ; the lobule a small involute portion, and the foliole wider than long ; foliole rounded at apex and base, placed half-way between two leaves, three times as wide as the stem, which is rather brittle and lax-celled. On smaller branches the leaves are distant and smaller, and the upper aspect of the whole plant shows the leaves as very convex, with rounded apex, because deflexed. Male cone terminating short lateral branches, 8–12-bracted, the bracts shortly boat-shaped, hardly imbricate. Cells fairly large, thick-walled, 6-sided except the outer line, translucent and brown, but when old the cells become nearly triangular, with very thick walls.

Unique among South African Lejuneae in selecting exposed perennial swamps as its habitat, where it occurs along with Radula and Sphagnum ; there are acres of it between Rosehaugh and Graskop in the scattered small swampy patches of grassland, especially near Pilgrim's Rest, but all barren, and it is not known from elsewhere (Sim 9656, 8025, 7494, etc.).

Apparently near *A. Pappeana*, but the inflorescence is dioecious, the leaf-apex decurved, foliole round, flat, and free from the stem, small-leaved flagellae more abundant, cells larger, colour brown, and it has no dichotomous branching.

Genus 22. BRACHIOLEJEUNEA (Spruce, 75 and 129, as subgenus), Schiffner, Pfl., i, 3, 128.

=*Phragmicomae* species in Syn. Hep., p. 292.

Corticolous, green, monoecious, or, as Spruce puts it, " often also by abortion dioecious." Stems ½–2 inches long, dichotomously branched below the terminal perianth ; the male plant often irregularly branched. Leaves rounded or pointed, entire ; lobule large, adpressed, one or several-toothed or crenulate ; foliole large, rounded, entire. Perianth compressed, 2-10-keeled or plaited, keels with entire margins. Perichaetal leaves larger, entire or toothed. Stephani, in Sp. Hep., v, 110, gives ten African species, none of which are recorded there as from South Africa.

Synopsis.

74. *B. natalensis.* Leaves more or less acute, floral leaves toothed, lobule entire ; perianth 4–5-ridged.
75. *B. crenata.* Leaves rounded at apex, floral leaves entire, lobule crenate or toothed ; perianth 8–10-ridged.

74. *Brachiolejeunea natalensis* Sim (new species).*

Corticolous, adherent, ¼–1 inch long, green, with terminal perianth and dichotomous innovations. Leaves numerous, imbricate, spreading, oval-oblong, subacute, with decurrent inflated base, and small, inflexed, usually one-pointed lobule. Folioles round, with rounded apex, imbricate or nearly so, entire. Cells large, rounded-quadrate,

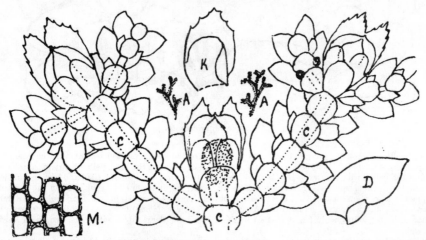

Brachiolejeunea natalensis.

hardly in contact. Per. leaves larger than others, crenate or toothed at the acute apex, the lobule large, rounded, entire. Per. foliole large, rounded, entire, at first hiding the perianth. Perianth pear-shaped, compressed at first, afterwards with two front folds ; margins entire. Capsule full of spores, but with only about six disc-based elaters and one central columella which escapes first. Spores 3-winged, papillose.

Wellington, Rosetta, Natal (Sim 7696, 8058), and elsewhere in forest regions of Natal.

---

* *Brachiolejeunea natalensis* Sim (sp. nov.).

Corticolosa, adhaerens, ¼″–1″ longa, viridis, perianthio terminali atque innovationibus dichotomis. Folia imbricata, ovo-oblongata, subacuta, integra, basi decurrente inflata, lobulo parvo inflexo, plerumque simpliciter acuto ; foliola rotunda, integra. Folia involucralia acuta, grandia, crenata vel dentata ; lobulus magnus, rotundatus, integer. Perianthium primo compressum, deinde duplex ; margines integri.

75. *Brachiolejeunea crenata* Sim (new species).*

= *B. africana* Sim, MSS.

Corticolous and adherent, or epiphytic singly on mosses, vigorous, green or yellowish, with terminal perianth and dichotomous branching there, but lower branches irregular or on one side mostly. Apparently dioecious, no male inflorescence seen on the many specimens examined. Stems ½–1 inch long, much branched; leaves vary considerably in density, but usually leaves and also folioles imbricate freely, though sometimes they are scattered. Leaves up to 1 mm. long, ½ mm. wide, rounded at the apex or the upper sometimes subacute; convex, the upper margin much rounded, the lower margin roundly decurved and inflated in its lower half, the lobule large and rounded, but usually so closely adpressed as to be difficult to detect, its margin crenated or about 8–10-toothed, the teeth of three to four

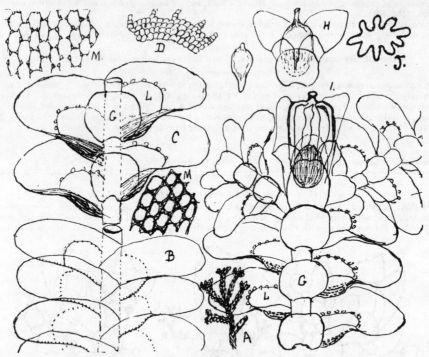

Brachiolejeunea crenata.

superimposed cells. Foliole very large, four times the width of the stem, rounded, entire, with narrow connection, and often producing rhizoids. Cells 25μ toward the apex, 35 × 25μ toward the base, pellucid, marginal cells quadrate, others at first hexagonal with thin walls and distinct trigones, ultimately very thick-walled and having a short straight base and a wider rounded upper end; the trigones distinct when in direct line of vision, but on slight displacement they disappear. Perichaetal leaves two, erect, entire, bluntly pointed, with large rounded lobule; per. foliole one, longer than others, and with apex rounded or emarginate. Perianth oblong-obovate, abruptly mucronate, equalling the perichaetal leaves, pluriplicate (eight to ten) in the upper half, these ridges rather inflated, with hardly acute smooth margins.

When young the pistil is far exserted. Capsule eventually bursting to near the base into four valves, to which a few elaters adhere by a swollen base. The bead-like teeth on the closely adpressed lobule easily distinguish this species in South Africa.

Elandskop and Hilton Road, Natal (Sim 9770, 9773, 9771, 8298); Spelonken, Transvaal, Rev. H. A. Junod (Sim 7965); and many other eastern localities.

---

* *Brachiolejeunea crenata* Sim (sp. nov.).

Corticolosa, adhaerens, vigens, ½″–1″ longa, viridis vel subflava, perianthio terminali atque ibi dichotome ramosa, infra autem irregulariter. Folia rotundata, convexa, margo inferior rotunde decurvatus et inflatus; lobulus grandis, rotundatus, arte adpressus, margine 8–10-dentato. Foliola magna, rotundata, integra. Folia involucralia magna, erecta, integra, lobulo magno rotundato. Perianthium pluriplicatum (8–10), dorsa subinflata, marginibus levibus subacutis.

Genus 23. PTYCHOLEJEUNEA (Spruce, 97, as subgenus) Sim.

=*Ptychanthus* Nees, Hep. Eur., iii, 211 ; Syn. Hep., p. 289 ; Schiffner, Pfl., i, 3, 130.

Very vigorous, loosely tufted on trees and stones, green, monoecious or dioecious, pinnately much-branched, the male and female inflorescence usually on separate branches in monoecious species. Leaves acute, entire or serrate, with slightly inflexed lower margin and small lobule, foliole attached to the stem in the lower half, with reflexed margins and crenate or several-toothed rounded apex—not 2-fid. Perianth oblong, eventually bursting and bell-shaped, in our species about 10-striate or plicate, the plicae rounded, without keel or cilia. Male spike many bracted, median (*i.e.* included as part of a continuous branch).

Synopsis.

76. *P. striata.* Leaves serrate ; plant monoecious.
77. *P. dioica.* Leaves entire ; plant dioecious.

76. *Ptycholejeunea striata* (Nees), Steph., Hedw., 1890, p. 140 ; Hedw., 1892, p. 120.

=*Ptychanthus striatus* N. ab E., Hep. Eur., iii, 212 p.p. ; Syn. Hep., p. 289 ; Schiffner, Pfl., i, 3, 130 ; Steph., Hedw., 1890, p. 5 ; and Sp. Hep., iv, 753.

=*Jungermannia striata* L. and L., in Lehm., Pug. Pl., iv, 1832, p. 16, No. 3.

Abundant in all forest districts of South Africa as loose cushions on stones or tree stumps in forest. Stems 1–5 inches long, more or less regularly pinnate, the pinnae 1–2 inches long, sometimes simple, more frequently irregularly branched, some pinnae bearing the perianths laterally or soon lateral, other pinnae having male spicate portions which are at first terminal but by leafy prolongation of the axis become median, and occasionally several occur on the same branch with leafy portions between, but in all cases male and female are on separate branches. Leaves distichous,

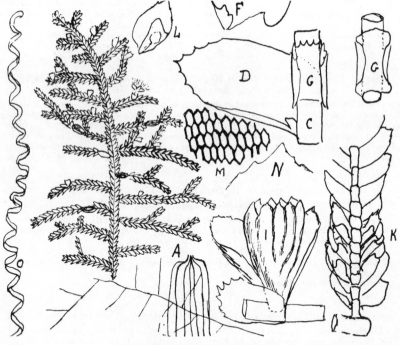

Ptycholejeunea striata.

spreading, imbricate, convex, cordate-ovate, usually acute, dentate or crenate-dentate toward the apex, the lower margin inflexed, and with a very small infolded lobule usually present but often difficult to see. Upper base produced into a rounded auricle. Foliole cordate-quadrate, attached to the stem for most of its length, the margins always reflexed on mature parts, giving it an oblong outline ; apex rounded (not bifid), more or less sharply serrate when mature, especially those near the perianth, but on young shoots the folioles are often flat, with the apex rounded-entire, on the same plant as bears serrate folioles lower, and the leaves are also usually almost entire on the young parts while acutely dentate on the older portions of the stem. Cells regular, hexagonal, thick-walled, pellucid, without trigones. Perianth rather longer than the leaves, sessile, oblong, shortly pointed, ultimately bursting into five valves and about ten incurved points, the valves rounded, without ridge or roughness. Perichaetal leaves two, similar to the others but more toothed. Male spike median, composed of four to six pairs of small, convex, imbricate, unequally 2-lobed, keeled bracts, each bract containing one pedicillate antheridium between its lobes, the lobule one-half to two-thirds the length and width of the other. Prostrate old stems give rise to adventitious fresh stems which develop and mature.

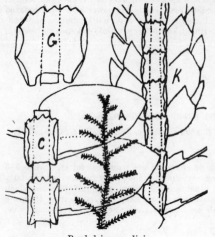

Ptycholejeunea dioica.

Stephani (Hedw., 1890, p. 5) makes *P. squarrosus* Lehm. (Syn. Hep., p. 290)=*P. striatus*, which appears to me to be the case, *P. squarrosus* being the subentire young condition; and he there separates from *P. striatus* an East Indian form as *P. Perrottetii* St., so maintained in his Sp. Hep., iv, 750, where *P. squarrosus* is separated as from tropical Asia, Japan, and East Africa, while he gives wide African, Asiatic, and Oceanic distribution to *P. striatus*.

Common in every South African forest; Perie, Cape (Sim 7272, etc.); Natal (Sim 7532, etc.); Rosehaugh, Transvaal (Sim 7493, etc.); Umtali, Rhodesia (Eyles 1733); Victoria Falls (Sim 9095, etc.).

77. *Ptycholejeunea dioica* Sim (new species).*

A fairly vigorous dioecious species, 2–3 inches high, growing in loose tufts under forest conditions, simply pinnate or somewhat irregularly branched, dark green, and when dry very much curled up. The general habit and characters correspond with those of *P. striata* except that the plant is smaller and dioecious, the branching is simpler, the leaves are bluntly pointed but without teeth, and the male cone which consists of about three pairs of bracts, near the base of a branch, has the bracts longer and more equally 2-lobed. Fertile inflorescence and perianth unknown.

The leaves are normal above and below the cone.
Perie Forest, Kaffraria, 3000 feet (Sim 9780).

Genus 24. DIPLASIOLEJEUNEA (Spruce, 80 and 301, as subgenus) Schiffner, Pfl., i, 3, 121 ; Steph., Engl. Bot. Jahrb., xx (1895), 318.

Unknown to me, but described by Spruce as having plants small, adherent, fragile, often slender, with large, flat, pale red, oblong-rotund leaves, having very narrow insertion and the lower half of the lower margin complicate, with the lobule 1–2-dentate. *Folioles one for each leaf*; cuneate, 2-fid or 2-parted, the segments acute and spreading. Floral folioles bifid, segments erect, male bracts small, complanate, subequal-lobed. Perianth oblong, somewhat compressed, 5-keeled, margins smooth. Monoecious or dioecious; innovations simple; androecia on small branches; two to eight pairs of small bracts.

78. *Diplasiolejeunea Kraussiana* (Ldbg.), Steph., Engl. Bot. Jahrb., xx (1895), 318 ; Steph., Sp. Hep., v, 919.

=*Lejeunea Kraussiana* Ldbg., Syn. Hep., p. 393.

Unknown to me, but described in Syn. Hep. as having adherent stems ⅓–½ inch long, slightly branched, prostrate, leaves large, distant, ovate-oblong, the apex rounded and ascending; lobule large, inflexed, 2-dentate, the marginal tooth the larger; foliole double the usual number (*i.e.* one to each leaf), 2-fid, rhizoid-bearing, with segments spreading widely. Fruit lateral, sessile, at the base of a branch; perianth 5-sided; mouth truncate, with one high keel on the back and two keels in front. Involucral leaves like the others, 2-lobed; lobule large; involucral folioles obovate, 2-fid, segments acute.

Outeniqua, Cape Province (Dr. Krauss).

Genus 25. ANOMALOLEJEUNEA (Spruce, as subgenus in Pearson, Hep. Knysnanae, p. 5), Schiffner, Pfl., i, 3, 127.

Stems small, dichotomously or irregularly branched. Leaves spreading widely, imbricate; lobule small, pointed; folioles large, nearly circular, with sinus one-fourth depth. Monoecious; female inflorescence terminal, with usually two innovations, floral leaves like the others, per. lobules narrow; perianth oval-acute, irregularly 7–10-ridged almost to the base; ridges smooth. Male cones lower, nearly circular, sessile. The only pluriplicate genus having bifid folioles.

---

* *Ptycholejeunea dioica* Sim (sp. nov.).
    Species vigens dioica, 2″–3″ alta, simpliciter pinnata, *P. striatam* referens, nisi quod haec de qua agimus minor est et dioica, rami simpliciores, folia obtuse acuta sed non dentata, conus autem masculus, qui ex paribus fere tribus bractearum constat, prope basin rami, bracteas longiores itemque aequius bi-lobatas habet. Inflorescentia fertilis et perianthium ignota.

79. *Anomalolejeunea pluriplicata* (Pearson), Schiffner, Pfl., i, 3, 127 ; Steph., Sp. Hep., v, 297.

= *Lejeunea pluriplicata*, Pears., Hep. Knys. (1887), p. 5 and pl. 2.

A small, dark green, monoecious species, growing in masses. Stems up to $\frac{1}{2}$ inch long, dichotomously branched by innovations, and having the perianth terminal and almost sessile and the male cone lower, sessile, and circular, with almost equally 2-lobed bracts. Leaves imbricate, spreading, convex, ovate, with a deeply decurved base facing the small, acute, inflexed, inflated lobule, and with the margin entire and the apex deflexed and rounded or subacute. Folioles large, almost imbricate, nearly circular, with sinus one-fourth depth and segments subacute. Cells separate, pellucid, small, or closer, and with trigones. Perianth oval, compressed, mucronate, with seven to ten blunt unarmed ridges extending nearly to the base. Distinct among the Lejeuneae with 2-fid folioles in having a 7–10-ridged perianth. Forest regions from Knysna eastward, rare.

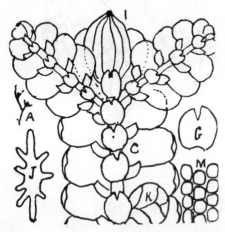

Anomalolejeunea pluriplicata.

Drepanolejeunea hamatifolia.

Genus 26. DREPANOLEJEUNEA (Spruce, 76 and 186, as subgenus) Schiffner, Pfl., i, 3, 126 ; Macvicar, p. 426.

= *Lejeunea*, subgenus *Drepanolejeunea*, Spruce, 186.

Minute plants, straggling parasitically over mosses or hepatics in forest, not forming a mat of themselves. Leaves not overlapping, often distant on young branches, the upper lobe with acute incurved apex and toothed upper margin ; the lower lobe half as large, inflated, usually entire. Folioles cut nearly to the base into two widely spreading acute segments, with lunate sinus. Perianth pyriform, 5-winged, the wings ciliate-dentate ; male cones small, terminal on branches, 6–8-bracted ; bracts 2-lobed, lobes inflated, nearly equal, almost entire, with one antheridium in each axil.

80. *Drepanolejeunea hamatifolia* (Hk.), Spruce, Hep. Am. And., p. 187 ; Schiffner, Pfl., i, 3, 126 ; Macvicar, p. 426 ; Pearson, Hep. Natalensis, pl. 1 ; Steph., Hedw., 1890, pp. 3, 136 ; Steph., Sp. Hep., v, 360.

= *Jungermannia hamatifolia* Hook., Br. Jung., pl. 51.

= *Lejeunea hamatifolia*, Dum., Comm. Bot., vol. iii ; Syn. Hep., p. 344 ; Mitten, Journ. Linn. Soc., xvi, 91, 192 (1877) ; Pears., Hep. Knys., p. 4 ; and Hep. Nat., p. 5.

A minute monoecious species, parasitic or epiphytic, usually found as scattered, simple or slightly branched plants, with separate or distant leaves, subtended by folioles with widely spreading acute segments. Leaves ovate-acute, the apex incurved, the upper margin acute and more or less ciliate toothed ; lower lobe half as large, pointed, entire ; foliole lunate, with acute segments. Cells irregularly rounded-quadrate, 20μ diameter, tumid, especially on the leaf-keel. Perianth at first terminal, soon lateral to an innovation, with two larger perichaetal 2-lobed ciliate leaves, and a lunate ciliate foliole. Frequent on forest mosses, but never in masses or in quantity. (Sim 8201, etc.)

D. hamatifolia Schiffner var. *gracillima*, Syn. Hep., p. 344 = *D. capensis* Steph., Sp. Hep., v, 322.

Smaller in all its parts than *D. hamatifolia* ; stems simple or sparsely branched, 2–4 mm. long, pellucid ; leaves distant, ventricose, spreading, ovate, with a long, subulate, ascending point ; the upper margin of leaf with about two 2-celled teeth ; the lower lobe almost as large as the upper, pointed by one protruding cell, ventricose, the cells of the keel tumid. Foliole spreading widely, lunate ; cells rounded-oblong, thick-walled, without trigones. Male cone oval or nearly round ; perianth terminal till an innovation grows, then acutely 5-angled, the angles crested with long cilia, the terminal one being forked.   Extreme forms of the species and of this variety are widely dissimilar, but intermediate grades occur and I find no specific difference.   Frequent among forest mosses, more common than the species.

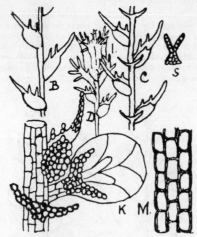

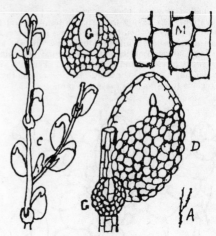

Drepanolejeunea hamatifolia var. *gracillima.*   *Microlejeunea gracillima.*
S, forked summit of valve.

Genus 27. MICROLEJEUNEA (Spruce), Jack et Steph., Bot. Centralblad, p. 60 (1894).

= *Lejeunea*, subgenus *Microlejeunea*, Spruce, p. 286.
= *Eulejeunea*, subgenus *Microlejeunea* Schiffner, Pfl., i, 3, 124.

Minute dioecious plants, usually scattered, seldom matted, and not much branched.   Leaves very small, colourless, separate or distant, erect, 2-lobed, entire, six to ten cells across ; the apex rounded ; the lobule very large ; folioles bifid.   Perianth 5-keeled, keels acute, smooth.   Distinguished by size more than anything else from *Eulejeunea*, and by presence of folioles from *Cololejeunea*.

Synopsis.

81. *M. gracillima.*   Lobule half as large as leaf, or more.
82. *M. Helenae.*   Lobule one-fourth size of leaf.

81. *Microlejeunea gracillima* (Mitt.), Pears., Hep. Nat., p. 7, pl. 4 ; Pears., Hep. Knys., p. 7 ; Steph., Sp. Hep., v, 830.
= *Lejeunea gracillima* Mitt., Linn. Soc. Proc., v, No. 191, p. 115 (1861).

Minute, dioecious, light green, 2–5 mm. long, sparingly branched ; leaves distant, the larger 0·15 mm. long and wide, about ten cells across, with the upper lobe rounded and the lower lobe nearly as large but usually pointed, these lobes clasping the stem, very convex, and keeled, the upper leaves smaller, more distant, more tumid, and more pointed ; some branches flagellate and small-leaved ; folioles small, wide from a narrower base, the acute lobes converging round a lunate sinus ; cells hexagonal, 0·02 mm. diameter, with one small chloroplast in each.   Perianth not seen. Greytown (Sim 8433) ; Nels Rust, Natal ; Houtbosch, Transvaal, etc.   Very near the British and Madeiran *Lejeunea ulicina* Tayl. (Syn. Hep., p. 387).   Gepp includes it among the plants of Milanji, Nyasaland (Trans. Linn. Soc. (1894), iv, pt. 1, p. 63).   *Microlejeunea cucullata* Nees (Syn. Hep., p. 389) is shown by Stephani (Hedw., 1890, ii, 89, and iii, 138) to have been made up in the Lindenberg herbarium of various specimens from East and West Indies, South America, and Cape (No. 983, Dr. Krausse, 6720), and being unable to decide which was the type he dismissed the name and described two new species instead, but allocated the South African specimen to *M. gracillima* Carr. and Pears.

82. *Microlejeunea Helenae* Pears., Hep. Nat., vi, Tab. III, 1.
= *Lejeunea Helenae* (Pears.) Steph., Sp. Hep., v, 714.

A minute, dioecious, glaucous plant, 1–5 mm. in length, prostrate but not adherent, usually on bark and more or less gregarious.   Stems simple or with several branches from the base, the leaves on the older stems overlapping, roundish-ovate, slightly convex, bluntly rounded at the apex, saccate at the base, eight to ten cells wide, twelve to eighteen cells long ; the lobule one-fourth the length of the leaf, very tumid, oblong, with one-toothed apex, often absent or so closely incumbent as to be hid.   Foliole oval, wider than the stem, divided half-way into two widely lanceolate blunt segments. It varies much in one specimen, leaves scattered and ascending or more or less imbricate and more spreading. Stem, leaves, folioles, and perianths pellucid. Younger branches more slender, and with leaves separate and more pocket-like. Perianth 5-angled, sessile near the base of a young shoot or on small basal branches, having two large, obliquely rounded-cuneate, bractlike perichaetal leaves twice as long as the ordinary leaves, each having a small lobule. Cells large, lax, the outer line smaller and quadrate, the central rather larger and hexagonal; margin cellulose-crenulate. Androecium not seen.

Hilton Road, Natal (Sim 8201, 8198), etc.

It hardly exceeds in size *M. gracillima* and *D. hamatifolia* with which it is usually associated, but has the general appearance of a minute *Eulejeunea*; indeed, Stephani, in Sp. Hep., v, 714, has placed it in *Lejeunea*, but he describes a larger plant than Pearson did.

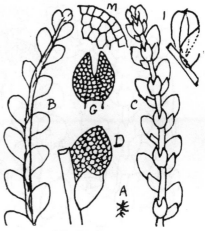

Microlejeunea Helenae.

Genus 28. EULEJEUNEA (Spruce, 79 and 260, as subgenus), Schiffner, Pfl., i, 3, 122.

= *Lejeunea*, Libert, Ann. Gen. Sc. Phys., pp. 6, 372; Macvicar, p. 414.

Plants usually monoecious, medium sized or fairly large, usually branched at the base or irregularly branched above, glaucous green, colourless or pale yellow under the microscope, and rather brittle when dry. Leaves usually imbricate, spreading, rounded at the apex, entire, ten to thirty cells across; lobules small, inflated, pointed, in some species absent. Folioles roundish, small, 2-fid. Leaves pellucid, thin-walled. Perianth at first terminal, soon lateral, 3–5-valved, the valves either rounded or acute, unarmed. Androecium usually sessile and lateral. Innovation either single or dichotomous.

Synopsis.

Monoecious (Stephani, in Sp. Hep., v, 708, 709, described *E. capensis* and *E. caespitosa* as dioecious; I do not find them so).

    84. *E. capensis.* Perianth 3-angled. Lobule present.
    85. *E. elobulata.* Perianth 3-angled. Lobule absent.
Perianth acutely 5-angled.
    86. *E. tabularis.* Stems erect, long; branches mostly basal. Lobule fairly large.
    87. *E. Wilmsii.* Stems widely branched, lobule very small. Involucral leaves free.
    88. *E. Eckloniana.* Stems branched, lobule small or absent; involucral leaves connate with folioles.
Perianth bluntly 5-angled, or inflated.
    89. *E. flava.* Branches mostly basal. Lobule small.
    90. *E. isomorpha.* Branches mostly basal. Lobule very large.
    91. *E. caespitosa.* Branches continuous and irregular.
Dioecious.
    92. *E. Breutelii.* Floral parts more or less connate.

84. *Eulejeunea capensis* (Gott.), Steph., Hedw., 1890, ii, 83, and iii, 136.

= *Lejeunea capensis*, Gottsche, Syn. Hep., p. 374; Steph., Sp. Hep., v, 709.

Monoecious, prostrate, densely chlorophyllose, with dichotomously branched stems; leaves short, with a round apex, almost circular except for the deeply arched base; lobule not pointed, but it and the leaf together form a decidedly ventricose vessel, the leaf being convex throughout and the base and lobule inflated and not adpressed. Leaves very regular, erect, and convex, often in long, straight, exactly regular-leaved branches. Folioles distant, circular or longer, small, 2-fid, one-third to one-half their depth, with short rounded sinus and acute lobes. Cells small, separate, and chlorophyllose, at first with large trigones. Inflorescence terminal, with usually two innovations; perianth oval-turbinate or pyriform, at first hid under the larger perichaetal leaves, which are unequally lobed, the lobule being much smaller and oval-lanceolate, the large lobe like the leaves but larger. Perianth flat on the back, 1-ridged on the lower side, giving a 3-angled section. Male cones occur lower on the same stem, and are either short, circular, and about 6-bracted, or elliptical and 12–18-bracted, the bracts nearly equally 2-lobed. On and near Table Mountain, Cape (Sim), and Newlands (Brunnthaler, pp. 1, 20). *Eulejeunea serpyllifolia* Libert (Syn. Hep., p. 374), there recorded as from Cape, Europe, America, India, etc., is dealt with by Stephani (Hedw., 1890, pp. 2, 83), where he places Lindenberg's specimens 841, 848, 849, 852 in *L. capensis* and others elsewhere, and so maintains the species from Europe and North America, but not from Africa or its islands, South America, India, or Java. It is dealt with as a British species in Macvicar, p. 418, under the name *Lejeunea cavifolia* (Ehrh.) Lindb., which has the perianth sharply 5-angled at the apex, and Steph., Sp. Hep., p. 802, keeps it there. *Eulejeunea laeta* (L. and L.), Steph., recorded from India,

East Indies, Brazil, and Cape, is stated by Stephani (Hedw., 1890, pp. 2, 85) to be broken fragments only. and he sinks the name, but in Sp. Hep., v, 742, he uses it again for the American plant only.

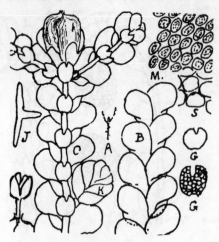

Eulejeunea capensis.

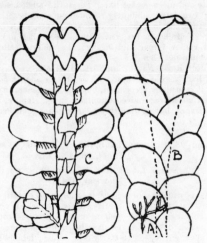

Eulejeunea elobulata.

85. *Eulejeunea elobulata* Sim (new species).*

Monoecious; stems 1–2 cm. long, irregularly branched, the longer branches straight and simple, with exceedingly regular, imbricate, alternate ovate leaves, shortly inserted, entire, rounded at the apex, convex on the ovate portion, but with an expanded and almost reflexed rounded lobe on which the lobule would naturally be placed if present, but there is no lobule there. Leaf usually undulated just above this lobe, the undulation giving it its characteristic expanded position. Cells small, pellucid, separate, round. Folioles almost in contact, slightly wider than the stem, wide below, ovate-bifid to the middle, the segments acute with lunate sinus. Female inflorescence terminal, the perichaetal leaves much larger than the stem leaves, and with the lobe inflexed and enlarged. Floral folioles much enlarged, deeply emarginate, with rounded lobes. Perianth 3-valved, widely cylindrical or pyriform, shortly pointed, valves with acute junction but without keel; section triangular, with flat postical surface. Androecium a short sessile lateral cone, about 8-bracted.

Table Mountain only, and not common there (Sim 9783, etc.).

86. *Eulejeunea tabularis* (Spr.), Steph., Hedw., 1890, ii, 83, and iii, 137.

= *Lejeunea tabularis* Spreng., Syn. Hep., p. 374 ; Steph., Sp. Hep., v, 721.

Monoecious ; loosely tufted, with erect, mostly barren stems ½ inch or more long, and numerous short basal branches, some of which are fertile, with terminal perianth and one or two innovations, the branching of these being irregularly subpinnate, and others bear the terminal or lateral male cones. Leaves numerous, spreading, pellucid, much longer than wide, narrower toward the leaf-apex, sometimes bluntly triangular at apex, more usually rounded, entire ; the lower margin somewhat arched downward and inflated at the short, oblong, truncate, pointed lobule ; folioles meeting or imbricate, longer than wide, convex, 2-fid to the middle, the lobes nearly meeting at the apex. Fruit terminal on short basal branches, or eventually lateral through innovation, then sessile and cuneate below, but when mature pyriform, shortly mucronate, acutely 5-angled, angles smooth. Small basal branches have much smaller leaves than the main stem.

Table Mountain, common.

In Hedw., 1900, ii, 83, in dealing with Lindenberg's specimens of *E. tabularis*, which were partly South African and partly South American, Stephani places one South African and two South American numbers in *E. flava*, one South American in *L. pulvinata*, and the other South American as a broken sterile fragment, but leaves Nos. 800, 801,

---

* *Eulejeunea elobulata* Sim (sp. nov.).

Monoica ; caules 1–2 cmm. longi, irregulariter ramosi. Folia regularia, convexa, ad apicem rotundata, integra, nullo lobulo, lobo autem basali expanso subreflexo, cui lobulus, si adesset, adiungeretur. Cellae parvae, pellucidae, separatae, rotundae. Foliola ovata, infra lata, bifida ad medium, segmentis acutis sinuque lunato. Folia perichaetalia maiora, lobo inflexo ; foliolum perich. amplificatum et tantum emarginatum. Perianthium 3-gonum, pyriforme, inerme. Androecium laterale, sessile, parvum, rotundum.

802, and 805 (all South African) in *E. tabularis.* This shows its close relationship with *E. flava.* In Sp. Hep., v, 721, Stephani ascribes *L. tabularis* to New Calabar, West Africa, which seems another slip.

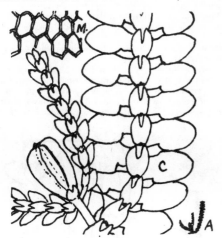

Eulejeunea tabularis.   *Eulejeunea Wilmsii* (after Stephani).

87. *Eulejeunea Wilmsii* Steph., Hedw., 1892, pp. 124 and 120, and Taf., vi, figs. 7–9.

= *Lejeunea Wilmsii* Steph., Sp. Hep., v, 722.

Unknown to me ; described by Stephani as :—Monoecious, small, hyaline, laxly caespitose ; stem widely branched ; leaves contiguous, suberect, spreading, widely ovate or almost ligulate, flat ; lobule infolded or absent. Cells : marginal 0·017 mm., median 0·025 mm., basal scarcely longer ; trigones minute, hyaline. Folioles more than twice as wide as the stem, almost circular, incised to one-third, lobes blunt. Perianth terminal, with one lateral innovation, widely ovate, very muticous ; beak small ; 5-plicate, the plicae on the lower side near the margin, low, extending far down. Floral leaves three times shorter than the perianth, ovate, their lobules shorter by half, lanceolate. Floral foliole widely ovate, bifid to one-third, with obtuse segments. Androecium on the stem, small ; bracts 2-jugate.

Greytown, Natal (Dr. Wilms).

Adult perianth elongated or semi-stipitate. Nearest to *E. Eckloniana* Ldbg., which differs in having leaves narrower, with larger lobules, folioles cut to the middle, subreniform ; involucral leaves connate with the folioles ; from all other related species *E. Wilmsii* differs in having the perianth muticous, its apex rounded-truncate, and its beak very short.

88. *Eulejeunea Eckloniana* (Ldbg.), Steph., Hedw., 1890, ii, 86, and iii, 136 ; Hedw., 1892, p. 124 ;
Mitten, Journ. Linn. Soc., xvi, 91, 192 (1877).

= *Lejeunea Eckloniana* Ldbg., Syn. Hep., p. 381.
= *Lejeunea Ecklonii* Steph., Sp. Hep., v, 711.

Unknown to me : described in Syn. Hep. as :—Stem prostrate, branched ; leaves imbricate, semi-vertical, distichous, ovate-orbiculate, obtuse or somewhat acute, the base decurrent, not or not much complicate. Folioles distant, quarter size of leaf, ovate, acutely bifid, the segments acute. Fruit on lateral branches ; involucral leaves small, oblong-obtuse ; its lobule lanceolate, its foliole bifid at the apex. Perianth elongate, pyriform, acutely 5-sided at the apex. = *Jungermannia serpyllifolia* L. and L., in Linnaea, iv, 361. North and east sides of Table Mountain (Ecklon).

Stephani credits it also to West tropical Africa.

89. *Eulejeunea flava* (Swartz), Steph., Hedw., 1890, iii, 137 ; Steph., Engl. Bot. Jahrb., xx (1895), **317.**

= *Jungermannia flava* Swartz, Prod. Fl. Ind. Occ., p. 144.
= *Lejeunea flava* Nees, Leb., iii, 277 ; Syn. Hep., p. 373 ; Macvicar, p. 415 ; Steph., Sp. Hep., **v, 755.**
= *Lejeunea* (subgenus *Eulejeunea*), Spruce, p. 268.

Monoecious (dioecious, *fide* Stephani), ¼–1 inch long, usually simple ; perianths at first terminal on stem and branches ; innovations usually simple, occasionally dichotomous. Leaves imbricate, up to 1 mm. long, spreading, varying from pellucid to light yellowish, widest at the middle, narrowed toward the narrow base, also narrowed toward the rounded apex, more or less convex, the upper margin not crossing the stem, the lower margin nearly straight, the lobule small, truncate, pointed, inflated along the keel, inflexed ; folioles small, half the wideness of the leaf, nearly circular, one-third to one-half 2-lobed ; floral leaves longer and narrower than others, obtuse, with acute triangular lobule and long foliole with acute segments ; innovation usually single. Perianth pyriform, bluntly 5-angled. ultimately inflated with five rounded valves without keels. Frequent in all forestal districts of South Africa, and widely dispersed elsewhere including tropical Africa. Mitten records it from Usagara (Journ. Linn. Soc., 1886, xxii, 146, 326).

Spruce enters in great detail into the variations of this species, and into changes which occur when the perianth passes maturity, and also into the occasional variations accompanying sterile flowers

Pearson, Hep. Nat., p. 5, and pl. ii, fig. 1, describes and figures a Natal specimen under the name *L. flava* Sw. variety *convexiuscula*, which does not appear to me to differ except that the foliole segments are very blunt, but he figures 3-valved perianths (fig. 13) and 5-angled section (fig. 14). In the Syn. Hep. some similar discrepancy occurs, where *L. flava* is described (p. 373) as having perianth acutely pentagonous, while on p. 374 it is " Angles 3, obtuse." The close relationship with *E. tabularis* is mentioned under that species. In Sp. Hep., v, 755, Stephani confines *E. flava* to the tropics.

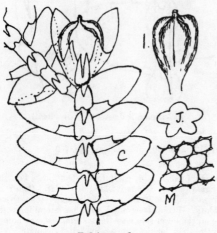

Eulejeunea flava.

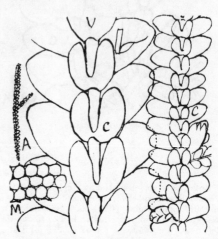

Eulejeunea isomorpha.

90. *Eulejeunea isomorpha* Gott., Abh. Ver. Brem., vii, 355 ; Steph., in Hedw., 1892, p. 120.

=*Lejeunea isomorpha* (G.), Steph., Sp. Hep., v, 715.

Monoecious ; stem 2–3 cm. long, simple or slightly branched. Leaves and lobules as in *E. flava*, but folioles much larger, usually imbricate, cordate at the base, cut almost or quite to the base into two ligulate (or subacute) segments frequently unequal in size and occasionally even free from one another, and one placed lower than the other. Cells round, pellucid, with evident trigones. Perianth sessile, lateral, 5-valved, with or without two reduced floral leaves. Androecium lateral, sessile, about as long as the leaf protecting it, about 10-bracted.

Among mosses in forest in eastern districts and Natal (Sim 275, etc.).

Lejeunea hepaticola Steph., Hedw., 1888, Heft 2 ; Sp. Hep., v, 714, a Central African species, is included in Bryhn's list as from Eshowe, Zululand. It is likely that *Eulejeunea isomorpha* is referred to, as they have much in common.

91. *Eulejeunea caespitosa* (Ldbg.), Steph., Hedw., 1890, ii, 86, and iii, 136 ; Steph., Sp. Hep., v, 708.

=*Lejeunea caespitosa* Ldbg., Syn. Hep., p. 382.

A small, yellowish green or glaucous, monoecious species (dioecious, *fide* Stephani), 5–15 mm. long, usually epiphytic on mosses and irregularly branched, but sometimes growing as pendent masses under rocks or on bark, and then usually less branched or nearly simple. Leaves usually imbricate, oblong-rotund, very convex, entire, suddenly arched downward from where the small inflated acute lobule joins ; lobule variable in size, but regular on one plant. Foliole almost circular, with one-third sinus and rather blunt entire segments. Perianth at first terminal, soon apparently lateral through innovation ; usually without floral leaves, or sometimes with two rather large perichaetal leaves, the perianth growing more inflated and more pedunculate with age. Perianth shortly pear-shaped, shortly pointed, with five rounded valves. Androecium a short, sessile, lateral cone, or occasionally longer and included in the stem. Leaf-cells large, pellucid, separate, roundly quadrate, the outer cells smaller.

Table Mountain (Sim 9778) ; Storms River (B. D. 17,023). One of the most common hepatics in forest country, but so small as to be usually overlooked.

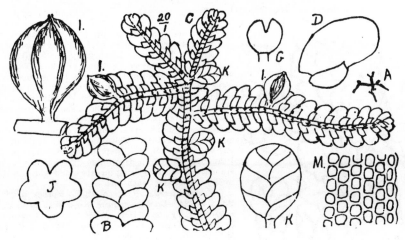

Eulejeunea caespitosa.

92. *Eulejeunea Breutelii* Steph., Hedw., 1896, xxxv, 85.

=*Lejeunea Breutelii* St., Sp. Hep., v, 708.

Unknown to me, and evidently described from very incomplete material. Stephani's description is:—" Dioecious, very small, pale yellow, caespitose on bark ; stem 10–15 mm. long, widely branched ; leaves somewhat imbricate (0·5 mm. long, at the middle 0·33 mm. wide), spreading obliquely, ovate, obtuse or rotundate, entire, very pellucid. Cells large for the size of the plant. Cells at the apex 12μ, middle 20μ, base 17×35μ, with large acute trigones. Lobules ovate, inflated, the keel very arched, constricted under the apex, the apex itself much narrowed, shortly dentate, with narrow lobe. Folioles large, distant, transversely inserted, cordate, 2-lobed half depth, the margin narrow, lobes obtuse. Perianth absent. Female inflorescence pseudo-lateral, floral leaves smaller than stem leaves, similar ; lobule large, partly connate, oblong, obtuse, much smaller than others, and infolded. Floral foliole much larger than its leaves, equal to stem leaves, and united on both sides with corresponding lobule, widely obovate, bifid to half-depth, lobes acute, sinus acute. Androecium unknown." Kookebosch, Cape, (Breutel) under the name *L. caespitosa* Ldbg., which differs in having the lobule three times longer than wide, acuminate, and the cells of the leaves not thickened, etc., etc. It is included in Bryhn's list from Entumeni, Zululand (1911, p. 5).

In Hedw., 1890, iii, 134, Stephani includes *Ceratolejeunea Breutelii* G., which, however, appears to be a different plant, and is not maintained under that name in Steph., Sp. Hep., vol. v, or as a *Ceratolejeunea*.

Genus 29. TAXILEJEUNEA (Spruce, as subgenus) Schiffner, Pfl., i, 3, 125.

=*Lejeunea*, subgenus *Taxilejeunea*, Spruce, p. 212.

Plants small or fairly large, monoecious or dioecious, yellow-brown or pale green or pellucid ; leaves convex, ovate, apex triangular-acute, entire ; lobule small or absent ; foliole almost circular, entire, shortly bidentate. Perianth pyriform, in our species inflated and without keel. Male cone lateral.

Synopsis.

93. *T. krakakammae.* Monoecious ; yellow-brown. Lobule difficult to see.
94. *T. elobulata.* Pellucid ; lobule absent.
95. *T vallis-gratiae.* Pale green, dioecious. Lobule evident.

93. *Taxilejeunea krakakammae* (Ldbg.), Sim.

=*Lejeunea krakakammae*, Ldbg., Syn. Hep., p. 353.

=*Strepsilejeunea krakakammae* (Ldbg.), Steph., Hedw., 1890, ii, 74, and iii, 141 ; and Sp. Hep., v, 276.

Monoecious, though in the absence of perianths it may appear dioecious, slender, yellowish brown, living scattered on mosses ; stem $\frac{1}{4}$–1$\frac{1}{2}$ inch long, adherent, straggling, irregularly branched, the branches mostly weak and small-leaved at first, ultimately like the main stem. Leaves widely ovate, squarrose, 0·5–1 mm. long, convex, the subacute apex regularly decurved, the upper margin rounded, the lower margin much undulated, the base having a keel deeply arched downward and containing a much-inflated cup-like adjunct to the leaf ; the lobule either absent or so inflexed as to be not visible ; folioles separate, abundantly rhizoid-producing, small, nearly round, or somewhat narrowed below, the sinus one-half to one-third deep, with acute base, and the segments acutely pointed. Branch leaves much smaller but of similar form, though occasionally showing a shallow one-pointed lobule. Cells elliptical 20×40μ,

5

thick-walled, trigones indefinite.   Perianth lateral, usually near the base of a branch, hardly longer than the leaves, usually if not always without perichaetal leaves, cordate-pyriform, mucronate, 3–5-valved, not compressed, the valves inflated-rounded, smooth, and without keel.   Androecium lateral, near the base, short, with about six pairs of bracts.

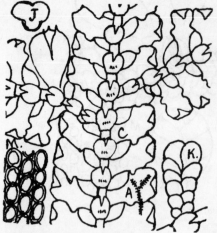

Taxilejeunea krakakammae.

Eastern Transvaal, Rosehaugh (Sim 9769); Krakakamma, Eastern Cape Province (Ecklon); Eveleyn Valley, Perie (Sim 7061), etc. A large form of this, having leaves 1 mm. long but otherways similar, has been found sparingly in the Eastern Transvaal, having adherent stems 1½ inch long, irregularly branched, bearing male but not female inflorescence, which may indicate a dioecious plant or only an incomplete monoecious plant, but the material is too sparse to decide upon.   Macmac (MacLea) intermixed among mosses (Sim 9768).   This may be *T. conformis* (=*Lejeunea conformis* N. and M., Syn. Hep., p. 355) St., Sp. Hep., p. 454, which I cannot separate from description, except that in it the leaf-points are not mentioned as being incurved, and which is there credited to Mascarenes and South Africa.

Euosmolejeunea trifaria (N. ab E.), Spruce, p. 242; Schiffner, Pfl., i, 3, 124=*Lejeunea trifaria* Nees, Syn. Hep., p. 361, and *L. rufescens* Ldbg., Syn. Hep., p. 366, a tropical species, is recorded by Mitten, Journ. Linn. Soc., xvi, 91, 192, from the base of Table Mountain, Cape Town (Rev. A. E. Eaton), who adds, "The specimens agree with those from Magellan, and are fertile."   I have seen nothing which belongs to it and think *T. krakakammae,* which closely resembles it, has been mistaken for it.   It has the same brown colour, lateral unarmed perianth, pointed deflexed leaves, divided folioles, and habit of growth, but differs in having leaves shorter and rounder, involucral leaves elongate, acuminate, with flat oblong-ovate lobule, perianth short, obovate or sub-pyriform, with five compressed angles.   I see no generic distinction, and they are not included in *Euosmolejeunea* by Steph., in Sp. Hep.

94. *Taxilejeunea elobulata* Sim (new species).*

Stems 2 cm. long, simple or branched, not rooting, tender, the stem of about six rows of transparent cells.   Leaves not in contact, shortly and almost transversely inserted, ovate acute, or with some obtuse or even repand leaves intermixed, pellucid, the upper side much rounded and free, the lower side slightly decurrent, not inflexed, and with no lobule.   Cells 25×30μ, hexagonal, but deep and thick-walled, and so appearing circular, without trigones.   Folioles free, distant, nearly circular but 2-fid, with sinus one-third to one-half deep.   Monoecious; perianth lateral, sessile, axillary, 3-lobed and shortly 3-pointed, with two concave floral leaves; androecium a short sessile cone lower.   Maritzburg, Natal (Sim 8302, fertile; 8257, etc.).   So like a juvenile *Calypogeia* that I was inclined to pass it as such till found fertile.

95. *Taxilejeunea vallis-gratiae* Steph., Hedw., 1896, xxxv, 136.

A pale green, little, gregarious, dioecious species, growing on soil.   Stems 1–2 cm. long, branched below, with the perianths terminal on short basal branches, but with sterile flowers along the longer branches, at first terminal, soon lateral from the production of one innovation.   Leaves pellucid, 0·5 mm. long, nearly as wide, spreading, flat, the apex triangular-acute, not deflexed, imbricate, the upper margin very much rounded as the leaf has a narrow connection, the lower margin slightly arched downward at the base and incurved there, forming a small oblong lobule, pointed at its truncate bend, regularly opposite a foliole and often apparently attached to it though not really so.   Foliole transversely attached, cordate at the base, rounded above, bifid one-third to one-half depth, with acute sinus and acute points.   Cells hexagonal, 25×50μ, with very thick walls and no trigones.   Perianth oblong, three times as long as wide, with three inflated rounded valves (Stephani says five), especially the under one, all rounded at the apex.   Floral leaves larger than others, the upper pair acuminate, acute, with two to three teeth on their upper margins, the lobules one-third the length, triangular acute, the floral folioles larger than others, one-third to one-half bifid.   Floral leaves of sterile flowers like those of fertile flowers but smaller.   Androecium small, circular, about 6-bracted.   Capsule oval; elaters numerous.

Port Shepstone, Natal (Sim 7448).   Stephani's specimen is from Genadendal, Cape (Breutel), and he adds, "Well distinguished by the inflated-plicate perianth."

---

* *Taxilejeunea elobulata* Sim (sp. nov.).
Monoica; caulis tener, haud radicans, simplex ramosusve, 2 cmm. longus.   Folia distantia, ovata, acuta, pellucida, supra rotundata libera, infra vix decurrentia, non inflexa lobuloque carentia.   Cellae 25–30 mm. diam. hexagonae.   Foliolum distans, paene orbiculare, bifidum ad tertiam partem altitudinis.   Perianthium sessile, laterale, binis foliis floralibus reductis.   Androecium conus brevis sessilis, inferior.

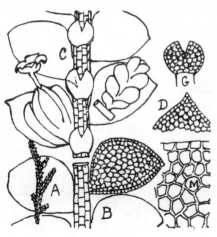

Taxilejeunea elobulata.

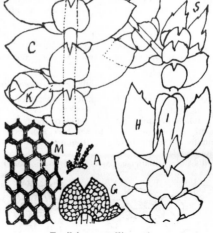

Taxilejeunea vallis-gratiae.
S, sterile inflorescence.

Genus 30. HYGROLEJEUNEA Sp.

Large, leaf-apex decurved, leaves rounded or subacute, cellulose-crenulate, when young pale, when old brown; lobule ascending; foliole large, reniform, shortly bidentulate or bifid. Plant dioecious; perianth 5-keeled. Usually on tree-bark.

96. *Hygrolejeunea Breuteliana* Steph., Sp. Hep., p. 522.

Unknown to me; described by Stephani from Genadendal as, "Dioecious, small, reddish brown, caespitose on bark. Stems up to 15 mm. long, many-branched; leaves crowded, obliquely spreading, very concave, when flattened ovate-triangular, 0·83 mm. long, base 0·67 mm. wide, apex rounded, front base truncate. Cells 27µ, everywhere equal, trigones minute, margin scarcely crenulate. Lobules obliquely ascending, rectangular, slightly longer than wide; keel slightly arched in a slight sinus, apex a little narrower than the base, truncate at a right angle, its angle obtuse. Foliole large, subrotund (0·5 mm. long and wide), somewhat sinuately inserted, 2-lobed to one-third depth, sinus narrowly obtuse, lobes widely triangular, obtuse. Perianth with innovation on one side, pyriform, 1 mm. long, middle 0·67 mm. wide, apex rounded, beak short, strong; postical folds narrow, widely divergent, decurrent to the middle. Floral leaves small, 0·5 mm. long, narrowly lanceolate, obtuse; lobule half as long, narrow, united to the middle, obtuse; foliole also half as long, elliptical, 2-lobed to one-third depth, sinus narrow, obtuse, lobes narrowly triangular, obtuse; androecia unknown."

Genus 31. CHEILOLEJEUNEA Spruce.

Plants small; leaves imbricate but not crowded, usually subfalcate and spreading, semicordate oblong, chloro phyllose, colourless, green or red; folioles large, subrotund, 2-lobed to the middle; perianth compressed, 5-plicate.

97. *Cheilolejeunea latistipula* Steph., in Brunnthaler, 1913, i, 21, and in Sp. Hep., v, 647.

Unknown to me, but described from Victoria Falls, Rhodesia, as dioecious, small, pale green, growing on bark stems to 10 mm. long, leaves spreading, concave, widely ovate when flattened, 0·7 mm. long, middle 0·6 mm. wide, symmetrical, the apex rounded, front base rounded covering the stem. Cells 18 × 27µ, basal 27 × 36µ, with thin walls. Lobule small (often obsolete), one-sixth the length of the leaf, linear, twice longer than wide, apex hardly narrower than base, obliquely truncate, keel spreading at a right angle, straight, without sinus; foliole large, three times wider than the stem, rather wider than long, 2-lobed to the middle. Female flower terminal on branches, innovations none, floral leaves small, as long as the stem-leaves, narrowly ligulate, apex rounded; lobule shorter by half, linear, obtuse, slightly united. Floral folioles as long as the floral leaves; lobes connate, obovate-oblong, more than twice as long as wide; apex 2-lobed, one-quarter depth. Other parts unknown.

Genus 32. STYLOLEJEUNEA Sim (new genus).*

Minute, light green plants growing on bark in small masses resembling small *Radulae*. Stems branched; leaves pellucid-chlorophyllose, nearly circular from a narrow base, the lowest corner slightly inflexed and bearing an erect

---

* *Stylolejeunea* Sim (genus novum).

Minutissima, foliis alternantibus ovatis rotundatisve integris, parte inferiore folii marginis inferioris tenuiter inflexa, stylum erectum 2–6 cellarum alius super aliam gerentibus. Foliolum evidens, bifidum, cellae hexagonae. Inflorescentia lateralis, sessilis.

style-like lobule of about four cells on end, near to and parallel with the stem. Foliole variable, deeply cut, and with acute spreading segments of one to three rows of cells.

Synopsis.

98. *S. Duncanii.* Lobule four to five cells long ; foliole segments spreading.
99. *S. rhodesiae.* Lobule one to two cells long ; foliole rounded, segments not spreading.

98. *Stylolejeunea Duncanii* Sim (new species).*

Very minute, much branched, tender, brittle ; the stems 1–2 mm. long, geniculate ; the leaves 0·1 mm. long, near or imbricate, in form and lobule as described for the genus ; leaf-cells large, pellucid, regularly hexagonal, rather tumid ; foliole variable, but very regularly present, and often remaining with the lobule after the larger lobe has broken off ; foliole cut almost to the base, segments acute and spreading ; the sinus usually acute but sometimes even lunate, the segments varying from one to three lines of cells wide. Sexual parts not seen. Texture much like a minute *Radula*.

Hell's Gate, Uitenhage, Cape (Mrs. Duncan, 1922).

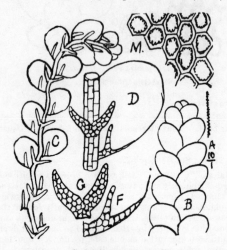

Stylolejeunea Duncanii.

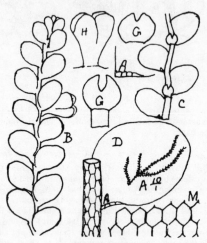

Stylolejeunea rhodesiae.

99. *Stylolejeunea rhodesiae* Sim (new species).†

A very minute, dioecious species, with stems up to 2 mm. in length ; branches mostly at the base, but a sterile inflorescence subtended by an innovation occurs anywhere on the side of the longer stems. Stems geniculate, the larger leaves up to 0·1 mm. long, more than half as wide, shortly elliptical or rounded, flat, pellucid, usually separate, with a narrow base ; the lower margin thickened or slightly inverted at the base, and carrying an erect style two cells long. Foliole short and wide, round or shorter, standing at right angles from the stem, the sinus about half-depth or less, the points more or less acute and not spreading, and the segments four to five cells wide. Cells large, hexagonal, thin-walled. Floral leaves of sterile inflorescence two, larger than others, with equal lobes and rounded apex, or sometimes the lobule about half the size of the leaf. Other parts not seen.

Victoria Falls, Rhodesia, on bark—abundant in small masses where present (Sim 9088) ; Matopas (Sim 9091) ; Zimbabwe (Sim 9090).

Tribe II. JUNGERMANNIAE. Elaters many, variously arranged but never as in *Jubuloideae.* Elaters with two or more spiral fibres in each. Archegonia usually numerous in a group, always more than four. Leaves various ; perianth often not ridged ; seta long ; capsule 4-valved, bursting to the base.

Family 10. PORELLACEAE. Leaves incubous, complicate 2-lobed, the lower lobe the smaller. Capsule wall of two layers, the inner thickened irregularly. Folioles present. Perianths terminal on short lateral branches, free. Branches all axillary, *i.e.* not infra-axillary = *Bellincinioideae*, Schiffner, Pfl., i, 3, 115.

---

* *Stylolejeunea Duncanii* Sim (sp. nov.).
  Caules 2–4 mm. longi, tenues, infirmi, foliis distantibus vel subimbricatis ovatis rotundatisve, stylum erectum 4–5 cellarum gerentibus. Foliolum magnum, bifidum ; segmenta patula, lanceolata.

† *Stylolejeunea rhodesiae* Sim (sp. nov.).
  Caules 3–6 mm. longi, inter muscos. Folia elliptica, stylum fere binarum cellarum in basi tenuiter inflexa gerentia. Foliola parva, distantia, prope rotunda, apice ad medium bifido, segmentis acutis, convergentibus.

Genus 33. MADOTHECA Dum., Comm. Bot., p. iii ; Syn. Hep., p. 262 ; Macvicar, p. 397 ; Steph., Sp. Hep., iv, 241.

> =*Porella* Dill., Hist. Musc. ; Spruce, p. 326.
> =*Bellincinia* Raddi, emend., O. Kuntze, 1891 ; Schiffner, Pfl., i, 3, 115.

Dioecious ; perianths lateral on side branches, sessile, ovate 2-lipped ; pistils many ; calyptra globose ; seta short ; capsule globose, opening in four valves to the middle or thereby, full of long 2-spiral elaters and few large roundish spores. Male inflorescence spicate ; spikes almost sessile, lateral on side branches, many-bracted, each bract having one globose shortly pedicillate antheridium enclosed between its two lobes. Plants large, 2–8 inches long, regularly pinnate or 2–3-pinnate ; branches and leaves distichous ; leaves complicate-bipartite, the lower lobe much smaller than the other and parallel to the stem, usually ligulate, foliole present, widely ligulate, simple, sometimes ciliate ; floral leaves ciliate.

˙100. *Madotheca capensis* Gottsche, Syn. Hep., p. 270.

> =*Porella capensis* Mitt., Linn. Soc. Journ., xxii, 146, 513 (1886).

Stems ascending or pendulous, frond-like, 3–9 inches long, regularly pinnate or slightly bipinnate, complanate, rising from the prostrate axis of a former stem, growing in loose cushions in forest or pendent from tree stems. Branches almost regularly alternate, 1–2 inches long, spreading or arching, usually simple, but occasionally with irregular branches rising from them, especially from old lower branches. Colour brownish green where exposed, bright green and the plant much more vigorous when hanging around shaded forest streams, but not otherways different. Leaves 2–3 mm. long, 1·5 mm. wide, overlapping somewhat or nearly separate, spreading, bluntly ovate-triangular, entire, convex or almost flat, the upper margin undulate, and sometimes with the upper shoulder separated by an undulation or slight sinus ; the lower margin usually somewhat reflexed ; lobule oblong-ligulate, adpressed to the stem, nearly parallel with it, rounded at the apex, about 3-spurred at the base (but these spurs often difficult to see), half as long as the leaf, usually prominent but occasionally smaller or absent on parts of the plant which has them elsewhere. Foliole as long as the lobule, widely oblong, entire, rounded or very slightly emarginate at the apex, the sides folded round the stem, decurrent, and about 3-toothed on each side at the base, these teeth usually closely adpressed to the stem and difficult to see. Cells irregularly 5–6-angular, about 0·03 mm. diameter with small trigones. When dry the leaves clasp the stem and have the margin more wavy-crenulate. Fructification lateral on the branches, erect, one or several on a branch. Pistils ten to twelve, several often enlarge considerably as if fertilised, but only one matures. Perianth almost sessile, 3 mm. long, ovate, flattened, 2-lipped, the lips adpressed but separate half-way down, and ciliate at the apex, the seta emerging where the lips join. Seta 4 mm. long, stout ; capsule 2 mm. long, globose, brown, shortly exserted, the valves splitting apart more than half-way. Perichaetal leaves smaller than the others, 2-lobed, toothed or ciliated. Male inflorescence spicate, spikes shortly stalked, 4 mm. long, flattened, the upper lobes on one side and the enlarged lower lobes on the other ; one shortly stalked antheridium between the lobes of each bract.

One of the prettiest and most widely distributed forest hepatics in South Africa. Mitten records it from Usagara Mountains, and remarks, " Very different in appearance from the South African specimens, being deep green and luxuriant," but the South African plant occurs so also.

Stephani (Sp. Hep., iv, 260, 261) separates this into three species without distinctions, viz. :—

1. *M. Höhneliana* St.=*Porella Höhnellii* Steph. in Hedw., 1891, p. 266=*M. Steudnera* St., MSS., slightly smaller, and with slightly smaller cells, recorded from North and East Africa, Madagascar, and Natal ; included in Bryhn's list from Ekombe, Zululand.

2. *M. capensis* G., Syn Hep., p. 270, as above, from South Africa.

3. *M. vallis-gratiae* (G., MSS.), St., small, 2 cm. long, branches 1 cm. long, perianth (young) pilose, from Genadendal, C.P.

I find only one slightly variable species, variable in size in every tuft, and variable in other respects mostly in accordance with how far it is expanded through soaking. It is represented by close, if not identical, relations in most other tropical forests.

Family 11. RADULACEAE. Leaves incubous, complicate 2-lobed, the lower lobe the smaller. Capsule-wall of two layers, the inner thickened irregularly. Folioles absent. Perianth at first usually terminal on the stem, compressed, free. Rhizoids often produced from a wart on the lobule near its base. Branches all lateral and infrafoliar in origin=*Stephaninoidea*, Schiffner, Pfl., i, 3, 113.

Genus 34. RADULA Dum., Comm. Bot., 1822, p. 112, emend. ; Spruce, p. 314 ; Macvicar, p. 385 ; Steph., Sp. Hep., iv, 151.

> =*Martinelius* S. F. Gray, p.p. Nat. Arr. Brit. Pl., i, 691.
> =*Stephanina* O. Kuntze, Rev. Gen. Pl., p. 839 ; Schiffner, Pfl., i, 3, 113.

Stems ½–3 cm. long, growing in flat cushions, or straggling epiphytically on mosses and ferns, or on bark ; usually irregularly pinnate or dichotomous, and much branched, all branches originating below the leaves. Leaves incubous, complicate-bilobed, entire, or through gemmae slightly erose ; lobule usually adpressed to the leaf and to the stem, except the keel, which is usually somewhat inflated ; the near side of the lobule is attached to the stem, and sometimes has a lobe which crosses it. A rhizoid-producing wart often occurs upon the outer face of the lobule near the keel. Folioles absent. Archegonia five to sixteen. Inflorescence at first terminal ; through innovation soon lateral. Perianth campanulate-compressed, truncate or 2-lipped. Capsule oval, 4-valved to the base, shortly exserted when mature, the valve-walls of two layers of cells. Elaters long, slender, 2-spiral. Involucral leaves two, like others but rather smaller, unequally 2-lobed. Inflorescence paroecious or usually dioecious, in which latter case the androecia are spicate, terminal on the branches or included in them, the bracts concave, with usually one, occasionally two to three, antheridia. Gemmae are sometimes present on the leaf-margins. A small but distinct cosmopolitan genus.

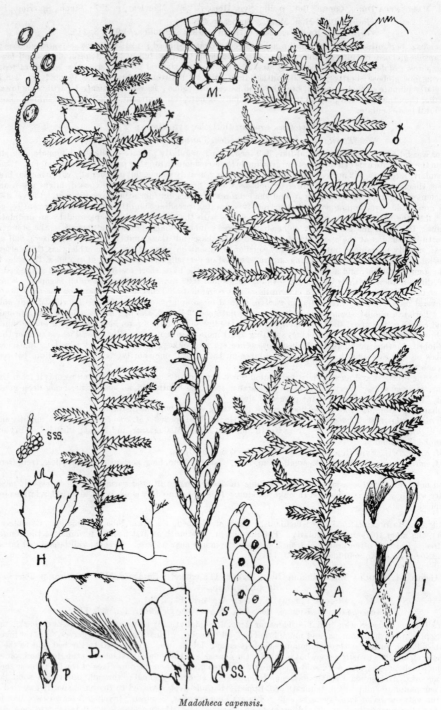

Madotheca capensis.

S, ciliate base of foliole; SS, ciliate base of lobule; SSS, cilia on perichaetal leaves.

Synopsis of Radula.

101. *R. Boryana.* Lobule of old leaves far crossing the stem.
102. *R. vaginata.* Perianth very long and narrow, cylindrical, narrower below.
103. *R. capensis.* Lobule not crossing the stem. Perianth not twice as long as wide.
   (? *R. complanata*, paroecious.)

101. *Radula Boryana* (Weber), Nees, Syn. Hep., p. 254 ; Steph., Sp. Hep., iv, 157.

=*Jungermannia Boryana* Weber, Prodr., p. 58.

Very vigorous, yellowish brown or brown, growing in wet, shaded mountain forests. Stem 2–5 cm. long, with few primary branches, but many alternate, pinnately arranged, seldom dichotomous branches on which the leaves are much smaller than on the main stem but otherwise similar. Stem-leaves plano-distichous, 1·5–2 mm. long, branch-leaves 0·5–1 mm. long ; the keel short or very short, ventricose, erecto-patent ; lamina spreading at right angles to the stem, oval, rounded at the apex and very bluntly subacute, crossing the stem at the base, when dry often much inflexed ; lobule with upper point bluntly triangular, crossing the stem far on old stems, and often with several undulations and sometimes distinctly pointed or even 2–3-pointed at the base, and only shortly attached to the stem ; on young stems the lobule hardly crosses the stem, but the keel is very ventricose. Cells hexagonal, 25μ, basal cells hardly larger, all very thick-walled, with small oval centre. Only known sterile, but occurs also in tropical Africa and East African islands.

Cape, E., Perie Forest (Sim 7264) ; Natal, Buccleuch (J. M. Sim), Polela, Maritzburg, Zwartkop, Blinkwater, etc. (Sim), Eshowe (Bryhn) ; Transvaal, Houtbosh.

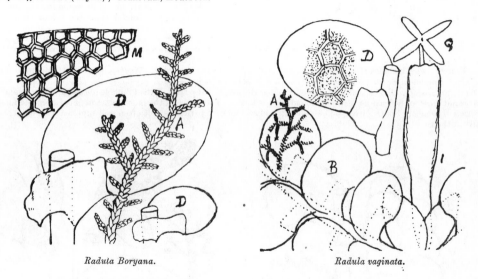

Radula Boryana.   *Radula vaginata.*

102. *Radula vaginata* Steph., Sp. Hep., iv, 167.

Dioecious ; light green ; stems 2–3 cm. long on fern leaves, etc., 2-pinnate or many-branched, the branches alternate, ending in a female inflorescence, with simple or dichotomous innovation from there. Older stem-leaves distant, 1 mm. long, 0·7 mm. wide, roundly ovate, spreading at right angles from the stem ; younger leaves gradually smaller but of similar form. Lobule 0·3 mm. long and wide, quadrate, connected far up the stem, acute, slightly rounded out at the base ; leaf cells 25μ ; basal cells hardly larger, all densely chlorophyllose with pellucid walls, or eventually the cells pellucid with thick walls in many layers. Floral leaves two, small, clasping the base of the long, narrowly cylindric perianth, 2·5 mm. long, of which the lower half is narrower than the upper half. Capsule cruciate, orange yellow, composed of long spirals. Perianth pendulous, at right angles to the stem. Not common. Perie (C.P.), Karkloof (Natal), etc. I do not see where the West African *R. tubaeflora* Steph. (Sp. Hep., iv, 193) differs.

103. *Radula capensis* Steph., Sp. Hep., iv, 193.

(Too near *R. Lindbergii* Gott., in Hartm. Sk. Fl., 2nd ed., ii, 98 ; Macvicar, p. 387 ; now restricted to the Northern Hemisphere=*R. commutata* Gott., Flora, Nos. 23 and 25, 1881.)

Dioecious. Plant in small cushions or straggling among Hepaticae and mosses, or on bark or stones, pale green, prostrate, ½–3 cm. long, abundantly branched, branches complanate, with a stem-bend at each branch ; where fertile often dichotomous ; the stem distinctly reduced in size as each branch is given off, and the branching is either irregularly pinnate or somewhat palmate. Leaves regularly imbricate, semi-horizontal, the upper lobe elliptical, deflexed outward from the keel, nearly crossing the stem, and shortly attached, slightly convex, rounded at the apex, 0·5–1·0

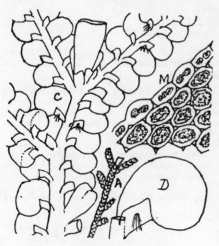

mm. long, 0·5–0·75 mm. wide, very variable as to size on different parts of the plant ; lobule quadrate, about 0·2 mm. long, adpressed to the large lobe, with a usually acute apex and somewhat inflated keel, the inner margin connected with the stem, closely pressed against it and seldom rounded out. Cells 0·025–0·035 mm. oblong-hexagonal, tumid, chlorophyllose, trigones small or absent ; marginal line of cells irregular in size giving an erose outline. Inflorescence terminal, involucral leaves two, almost similar to the others but erect and less rounded, and with the lobule more rounded. Perianth shortly cuneate-oblong (said by Stephani to be 2 mm. long), flat, with wide truncate mouth. Capsule oval, elliptical, brown, shortly exserted when mature. Rhizoid warts present on many of the lobules on the older parts of the stem. Androecia terminal as short cones of small bracts containing one antheridium each, or through extension of the vegetative growth included in the stems, with ordinary leaves below and above.

Mitten records *R. physoloba* Mont. from base of Table Mountain (Journ. Linn. Soc., 1877, xvi, 91, 191), which species was sunk as a synonym of *R. aquilegia*, Syn. Hep., p. 260 (Syn. Hep., p. 724) (though restored by Stephani as from Australasia), and this again differs from *R. Lindbergii* mostly in colour, being brown, which to my mind is an insufficient character since the colour varies with exposure.

Though Stephani places *R. Lindbergii* Gott. in § *Ampliatae*, and *R. capensis* in § *Communes*, the difference is exceedingly slight and mostly geographical.

Radula capensis.

Pearson, who includes (Hep. Nat., p. 7) *R. commutata* (a synonym of *R. Lindbergii*), from Umpumulo, Natal, says he " can detect no character of sufficient value to separate it from the European form. *Radula capensis* Steph. is no doubt a nearly allied species, to judge from description, but the apex of the lobule is obtuse, not acute, and the leaves not erose, or at least no mention is made of the fact." This is one of the most common hepatics in the forestal districts throughout South Africa, seldom absent, and seldom with perianths, but often well supplied with floral leaves without them.

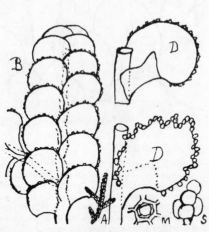

Radula capensis, forma *gemmifera.*

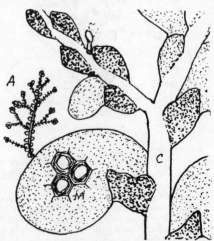

Radula capensis, forma *fragilis.*

Forma *gemmifera.*   Leaf-margin frequently set more or less closely with gemmae, giving it an erose appearance. In other respects identical with *R. capensis.* Table Mountain, Newlands Ravine, Window Gorge Waterfall, Skeleton Gorge, Knysna Forests, Alexandra Forests, Perie Forest, Kokstad, Giants Castle, etc.

(*Radula complanata* Dum., Comm. Bot., p. 112 ; Syn. Hep., p. 257, is very similar in general appearance and in glaucous-green colour to *R. capensis* but is paroecious, the antheridia occupying the ventricose bracts immediately under the perianth, usually one in each bract ; the perianth also is wider and more oblong, and the cells, though hexagonal, are nearly round and shorter. It is recorded in Syn. Hep. from the Cape, collected by Ecklon, Pappe, and Krauss ; and by Mitten (Linn. Soc. Journ., xiv, 91, 191) from the base of Table Mountain, near Cape Town, who adds, " the specimens are in that form which has the margin of the leaf fringed with gemmae." As *R. capensis* often *appears* to have paroecious inflorescence through the presence of air-bubbles in the ventricose axils I consider these records made in error, at least I have seen no paroecious inflorescence, but lest it should be present one figure shows the plant from a European specimen. It is widely distributed in the Northern Hemisphere, where it often bears marginal gemmae, but those I have found on Table Mountain and elsewhere in South Africa bearing gemmae belong to forma *gemmifera* of *R. capensis*.)

Forma *fragilis*. Leaves as in the type, but the lamina often broken off, leaving the stem clad in ventricose bases and lobules ; the terminal as well as the basal leaves usually complete. Probably frost or snow is the cause of breakage, but the plants have the appearance of having male bracts all over the stem, which is not the case.

Perie, C.P. ; Maritzburg, Giants Castle, Karkloof, Natal, The Downs, Pietersburg, Transvaal, etc.

Stephani (Hedw., 1884, p. 131, and Sp. Hep., iv, 195) describes as *R. tabularis* St. a sterile form from Table Mountain without difference from *R. capensis*, except that the leaf-margins are incurved and the lobule narrowly fusiform *in situ*, rectangular when spread out. He says it has Antarctic relationship, but I fail to find it or its distinctions.

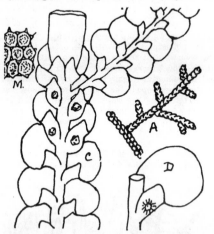

Radula complanata, from European specimen.

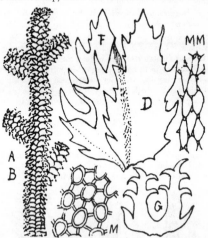

Schistochila alata.

Family 12. Scapaniaceae. Leaves transverse or succubous, complicate 2-lobed, the upper lobe the smaller, or the lobes nearly equal, the margins often dentate or ciliate. Folioles usually absent, but present in the South African plant. A large family in the Northern Hemisphere, represented in South Africa by one genus and only one species, which, however, is a conspicuous plant.

Genus 35. Schistochila Dum., 1835 ; Schiffner, Pfl., i, 3, 110.

=*Gottschea* Nees, Syn. Hep.

Very vigorous, erect or suberect, simple or forked or pinnately branched mosses, with terminal fructification. Perianth connate with the calyptra into an erect fleshy marsupium, with ciliate parts. Leaves distichous, alternate, 2-lobed, the lobes almost separate, the upper lobe the smaller, both lobes serrate, and the lower lobe usually deflexed upon itself. Foliole bifid, toothed, very large. Leaf-cells nearly circular, distant, with very large angular trigones separating them. Rhizoids red. A curious genus confined to the Southern Hemisphere, and in South Africa to the southern portion.

104. *Schistochila alata* (Lehm.), Schiffner, Pfl., i, 3, 111 ; Steph., Sp. Hep., iv, 71.

=*Gottschea alata* Nees, Syn. Hep., p. 16.

=*Jungermannia alata* Lehm., in Linnaea, iv, 359 ; Pug., pl. iii, p. 44.

A vigorous erect or suberect plant, 1–4 inches long, usually in Sphagnum clumps or forming dense masses in swamp. Stems simple or forked or pinnately branched, the branches few and short. Leaves distichous, the upper lobe rather the smaller, 2 mm. long, 1 mm. wide, ovate-acute from a narrow base, strongly toothed on both margins and with small teeth intermixed ; the lower 2–3 mm. long, widely ovate from a narrow base by which it is united to the upper lobe either flat against the upper lobe, or more usually with its upper half deflexed upon itself. Folioles 1 mm. long and wide, bifid to near the base and artistically and strongly 2–3–4-toothed on each margin. Leaf-cells pellucid, those in the upper portion of the leaf very large, nearly circular, thick-walled, distant, with pellucid angular trigones nearly as large separating them. In the lower half of the leaf the cells are longer, more hexagonal, and with small trigones. Fructification terminal, ciliate, connate with the calyptra into a small erect fleshy marsupium. Rhizoids red, abundant in the lower portion of the stem, growing out of the stem itself.

Table Mountain, abundant. The other S.-W. mountains have not been closely examined, but it is not reported and is too large to have easily escaped observation ; it is absent from all eastern localities.

Family 13. Ptilidiaceae. Leaves either incubous or almost transversely inserted, deeply cut into two or more segments, not conduplicate ; folioles present, similar and almost as large. Involucral leaves polyphyllous. Perianth plaited at the mouth, often concrescent with the involucre.

Genus 36. CHANONDANTHUS Mitt., in Hook. Handb. New Zealand Fl., ii, 750 (1867) ;
Schiffner, Pfl., i, 3, 105 ; Macvicar, p. 332.

Inflorescence terminal. Leaves deeply several-lobed, dentate at base ; folioles similar and about as large, but 2-lobed. Plants 2–5 cm. long, stout, rigid, stems mostly simple, prostrate or suberect ; leaves imbricate on the upper surface, transversely inserted, deeply 3–4-lobed, the folioles imbricate on the lower surface, deeply 2-lobed, the lobes spinulose at the base or throughout. "Perianth terminal, with one innovation, deeply multiplicate, the mouth slightly contracted, ciliate or dentate. Calyptra thin, free, surrounded at the base by sterile archegonia " (Macvicar). Androecia in the middle of the main stem, with bracts in several pairs, resembling the leaves but less deeply lobed, and antheridia two to three, large, mixed with paraphyses.

105. *Chanondanthus hirtellus* (Web.), Mitt., Journ. Linn. Soc., xxii, No. 146, p. 321 ; Steph., Sp. Hep., iii, 643.
=*Jungermannia hirtella* Web., Prod., p. 50, No. 43 ; Syn. Hep., p. 130.
=*Jungermannia fimbriata* Rich., in Hook., Musc. Exot., tab. 79 ; Spreng., Syst. Veget., iv, 1, 224, No. 89.

Plants brown, 2–5 cm. long, simple or nearly so, suberect or prostrate on swamp mosses, with few rhizoids except at the base. Leaves 1 mm. long and wide, densely imbricate but not adpressed, divided almost to the base into three to four (usually three) widely lanceolate lobes overlapping somewhat at the base, the two central lobes with few nearly basal teeth and long spinose points, the lateral lobes smaller, much more spinose-dentate or ciliate, especially on the outer margin, with the outer basal teeth long and reflexed and sometimes again toothed on the lower margin. Folioles deeply 2-lobed, lobes often as large as the outer leaf lobes, similar in shape to them and more spinulose-dentate, those near the apex of the stem large and prominent, those toward the base often smaller and less toothed or more weather-worn. The spinulose teeth give the plant a somewhat hairy appearance under a lens. Cells 0·03 mm. long, 0·01 mm. wide, the contents of each congested into a very irregular central mass, having numerous papillae in an irregularly stellate arrangement on each cell-surface ; lowest cells much larger and hyaline. Perianth not seen in South Africa.

Macmac, Transvaal, sterile, 8000 feet, Maclea (mixed in Rehmann M.A.A. 456 (Sim 8332)). Occurs also at Usagara, Usambara, and Kilimanjaro, East Africa, and in Mauritius, Java, Australia, and India, and is very (if not too) closely allied to *C. setiformis* (Ehrh.), Lindb., of Europe.

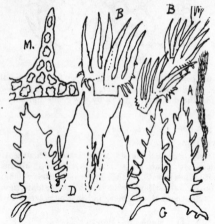

Chanondanthus hirtellus.

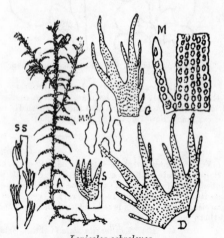

Lepicolea ochroleuca.
S and SS, leaves on flagellate branches.

Genus 37. LEPICOLEA Dum., 1835 ; Schiffner, Pfl., i, 3, 108.

Stems erect or suberect, in loose mats, several inches high, with few main branches but many flagellate side-branches. Inflorescence cladogynous, *i.e.* on very short lateral branchlets on the main stem. Leaves much cut, but usually more or less irregularly twice bifid, with narrow and somewhat ciliate segments. Folioles usually bifid with ciliate narrow segments. Cells in lines, irregular in outline, the lower larger than the upper, but no differentiated nerve-cells.

106. *Lepicolea ochroleuca* (Spreng.), S. O. Lindb. ; Schiffner, Pfl., i, 3, 108 ; Spruce, Hep. Am. And., p. 345 ;
Steph., Sp. Hep., iv, 32.

=*Sendtnera* (§ *Schisma*) *ochroleuca* (Spr.), Nees ab. E., Syn. Hep., p. 240.
=*Jungermannia ochroleuca* Spreng., Syst. Veget., iv, 2, 325, No. 210 ; Lehm., Hep. Cap., in Linn., iv, 3, 365, No. 26 ; Nees., Hep. Jav., p. 17, No. 14.
=*Jungermannia hirsuta* Tayl., Hep. Ant., in Lond. Journ. of Bot., 1844, p. 389, No. 50, and p. 475, No. 49.
=*Leperoma ochroleuca* (Spreng.), Mitten, Linn. Soc. Journ., xvi, 91, 191, and Handb. of N.Z., Flora, p. 754.

Stems 1–3 inches high, erect or suberect, growing in large loose mats, the main stem simple or 2–3-branched, but from it and from these main branches many ¼–½ inch flagellate, simple, somewhat pendulous branches are produced, pinnately arranged, each having nearly full-sized leaves in the lower half, but the leaf-size is reduced gradually toward the apex where the leaves are very small, 3–4-fid, and distant. Leaves abundant, almost transversely inserted, 1 mm. long and wide, deeply cut into two divergent channelled lobes, each of which is again deeply cut into two lobes, all the obes lanceolate subulate, occasionally entire, but much more frequently the outer lobe has several large cilia toward the base, while all the lobes are subject to having occasional irregular cilia two to eight cells long, one cell wide, along the margin. Folioles deeply bifid, the segments usually shortly toothed or ciliated canaliculate. On the flagellate branchlets the leaves and folioles gradually reduce in size and in number of cilia, and the leaves usually have four simple segments. Cells distant, large, irregularly oblong, in rows, the lower cells being larger than those above. When dry the leaf dries up so that the cells seem like lines of round beads, and the margins contract round these beads, but when moist the margin is straight. Inflorescence on very short lateral branches in the upper part of the main stem, lost in the dense tuft of cilia produced by the much-divided numerous leaves and floral folioles ; perianth campanulate.

Abundant on the upper part and top of Table Mountain, Cape. I have no specimens from elsewhere and have not seen it on the Drakensberg or other eastern mountains. Recorded also from South America.

Genus 38. HERBERTA S. F. Gray, Nat. Arr. Brit. Pl., i, 705 (as *Herbertus*) ; Macvicar, p. 8.

=*Sendtnera*, Sect. 1. *Schisma*, Syn. Hep., p. 239.
=*Schisma* Dum., Syll. Jung., p. 77 (1831) ; Steph., Engl. Bot. Jahrb., 1895, p. 307 ; and Sp. Hep., iv, 1.

Inflorescence terminal ; leaves secund, deeply 2-lobed ; lobes long and narrow with a central band of long cells resembling a midrib. Perianth free but nearly concealed among bracts. Dioecious. Plants several inches high, simple or slightly branched, and with flagellate branches, rigid, green, suberect, from a slender rhizomatous base. Leaves transversely inserted, with the base narrow, the lower portion spreading, from which the two long, narrow lobes stand erect, face to face, nearly parallel with the stem, or more or less secund, the hyaline cells forming a midrib-like band along the centre of the base and the centre of each lobe for about half its length. Folioles like the leaves and nearly as large. Cells quadrate with irregular margins. Perianth terminal, 3-angled, subulate, with plicate and laciniate mouth. Androecia at first terminal, but by continued growth of the stem ultimately included, with few long bracts, less deeply cut and more saccate than ordinary leaves and with recurved margins ; floral folioles also bearing antheridia.

107. *Herberta capensis* (Steph.) Sim.

=*Schisma capensis* Steph., Sp. Hep., iv, 6.
=*Chalubinskia africana* Rehm., in Rehm., Musci Austr.-Afr. Exsicc., p. 595.
(=*Herberta juniperina* Sw., Dixon, The Bryologist, vol. xxi, No. 5, Sept. 1918 ; Kew Bull., No. 6, 1923, p. 229, but Stephani now restricts that name to the West Indian plant having basal teeth on the leaves.)

Stems erect, wiry, 2–3 inches long, simple or with 2–3 strong branches, and often several or many shortly flagellate small-leaved branches. Leaves 2–4 mm. long, sometimes secund, imbricate, deeply 2-lobed, the lobes erecto-patent lanceolate, face to face and not widely spread, but with a rounded connection so there is no keel though usually the connate basal portion, which has a narrow connection with the stem and no basal teeth, spreads from the stem at a right angle, while the lobes themselves stand erect or suberect at a right angle to that. The centre of the connate base has a wide nerve-like band in its lower half, which then forks, one fork being central to each leaf-lobe for half its length ; this band is of thickened hyaline cells much larger than the chlorophyllose cells forming the rest

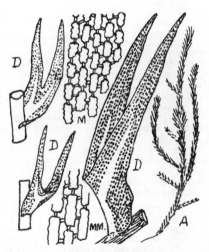

Herberta capensis.

of the leaf. All the cells are quadrate-oblong, with irregularly waved outline, distant, and surrounded by large shapeless and irregular trigones. Androecium at first terminal, ultimately central on a main stem, with bracts similar to the leaves but rather larger. Perianth not seen in South Africa ; Spruce describes that of his *H. juniperina* Sw., which is closely allied, as : " With the apex exserted, subulate-fusiform, five times longer than wide, highly 3-plicate, the ridges canaliculate (hence the perianth appears to be 6-plicate), the upper third part 6-fid, segments channelled-acuminate, cut at the base." Other African species are known only sterile or with young fruit. On the flagellate branches though the leaf is much reduced in size the nerve-like mark is still present. Folioles similar to the leaves and as large.

Distributed by Dr. Rehmann as *Chalubinskia africana* Rehm., Exsicc. A. Afr., p. 595, from tree-trunks in the forests at Lechaba, Mount Snellkop, Transvaal ; I have seen no other South African specimen, which is surprising since it is a large and conspicuous plant. It is recorded also from Madagascar.

Genus 39. ANTHELIA (S. O. Lindb.), Spruce, p. 1885 ; Schiffner, Pfl., i, 3, 106 ; Macvicar, p. 335.

Dioecious. Inflorescence terminal on stem and branches, leaves imbricate, keeled, with two equal acute lobes ; folioles similar. Involucral bracts adnate to the base of the perianth, or forming instead a pseudo-perianth. Plants small, in some species julaceous ; stems firm, somewhat branched, without flagellae.   Leaves transversely or somewhat obliquely inserted.   Antheridia large, solitary.

108. *Anthelia africana* Steph., Hedw., 1892, p. 123, and Tab. v.

This species is unknown to me, so Stephani's figure is reproduced, and his description is : " Dioecious, small, hyaline, on soil among mosses, simple, innovating under the flower.   Leaves transversely inserted from a narrow base, two-thirds bifid, complicate-carinate, keel arched (concave), at right angles to the stem, segments spreading,

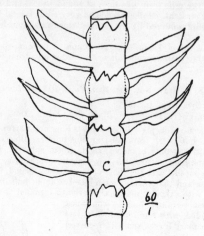

Anthelia africana (after Stephani).

incurved-erect, the lower margin on the under-side recurved, the rest equally subtriangular, acuminate acute.   Cells 0·010–0·020 mm., everywhere almost equal, with the walls subequally thickened.   Folioles much smaller than the leaves, standing at right angles from the stem, as wide as the stem, almost twice as wide as long, the apex 3–4-dentate, with acute irregular teeth.   Perianth terminal, innovating.   Androecia unknown.

" Near Lydenburg, Transvaal, Dr. Wilms."

In a German footnote he discusses whether this plant should be placed in *Anthelia*, but decides in favour, and in Sp. Hep., iii, 647, he retains it there without additional information.

Family 14. Cephaloziaceae.   Leaves not complicate, usually cut into two or occasionally several segments, in some species entire ; folioles usually present.   Perianth sometimes cylindrical, more usually triangular in cross section with one side of the triangle on the upper surface.   Female inflorescence usually on short branches from the lower surface.   Distinguished from *Lophoziaceae* by the position of the third angle of the perianth, but when that is absent it is often difficult to separate them.

Synopsis of Genera of Cephaloziaceae.

§ 1. Leaves rounded, succubous, margin entire or toothed.

40. *Odontoschisma*.   Leaves entire, cells lax, with thickened walls.   Folioles small or absent.   Stem not rhizomatous, but flagellae present.

41. *Alobiella*.   Stem rhizomatous ; leaves entire or emarginate ; folioles simple ; cells very lax, thin-walled.

42. *Adelanthus*.   Leaves secundly bent downward, lower margin dentate.

§ 2. Leaves incubous, entire, or 2–3-dentate.

43. *Calypogeia*.   Leaves entire or bidentate ; marsupium pendulous.

44. *Bazzania*.   Leaves oblique, usually 3-fid or subentire.   Perianth erect.

§ 3. Leaves 2-lobed, succubous.

45. *Cephalozia*.   Leaves equally 2-fid.

46. *Nowellia*.   Leaves 2-fid, the margin on under-side inflexed and saccate, forming a lobe.

§ 4. Leaves cut half-way or to near the base into narrow segments.

47. *Lepidozia*.   Leaves incubous.

    (*Psiloclada*.   Leaves succubous.)

Genus 40. Odontoschisma Dum., Rec. d'Obs., p. 19 (1835) ; Schiffner, Pfl., i, 3, 99 ; Macvicar, p. 295.

        =*Sphagnocetes*, Syn. Hep., p. 148.
        =*Cephalozia*, subgenus *Odontoschisma* Spruce, On Cephalozia, p. 59.

Stems prostrate on the ground or straggling among mosses, not rhizomatous, but with rhizoids and sometimes flagellae present, the latter postical.   Leaves succubous, somewhat succulent, obliquely inserted, often sparse, entire or almost so, rounded ; cells lax, roundly hexagonal, with thickened walls. Folioles small or absent, or among floral leaves bifid. Inflorescence cladogenous, lateral on short postical branches ; perianth subcylindrical, well exserted, the mouth crenulate and shortly ciliate. Androecium lateral with shortly 2-lobed complicate bracts, monandrous.

Stephani does not record this genus from South Africa, but Pearson described a South African form, and I find the genus common, both on the ground and among mosses, but am not aware that it has been found fertile in South Africa.

Synopsis.

109. *O. africanum.* Straggling among mosses, vigorous ; folioles present, simple ; leaves entire.
110. *O. variabile.* Folioles usually present and small ; leaves rounded or subemarginate, variable.
111. *O. sphagni.* Prostrate on or near the ground, rooting ; folioles absent except minute deeply bifid ones among floral leaves.

109. *Odontoschisma africanum* (Pears.) Sim.

=*Cephalozia (Odontoschisma) denudata* Mart. var. *africana* Pears., Hep. Knys. (1887), p. 10, and pl. iv.

Plant 1–2 inches long, straggling among mosses, never a mass by itself but seldom absent in forest collections. Stem simple or few-branched, stems and branches large-leaved about the middle, tapering to small-leaved at each end, the branches rising at right angles from the under-side of the stem, as also do leafless flagellae near the base, but there is no rhizomatous basal stem. Rhizoids abundant below, occasional above, colourless. Leaves obliquely or almost horizontally inserted, slightly concave, longer than wide, rounded at the apex, ovate from a wide base, sometimes somewhat acuminate, succulent, seldom imbricate, often scattered, with smooth surface ; cells roundly elliptical, separate, 0·04 mm., thick-walled, irregular except the marginal cells which form a straight margin. Folioles regularly present, at right angles from the stem, simple, five to six cells wide at the base tapering to one cell wide at the apex. Sexual parts not seen in South Africa, but probably agree with those of *O. denudatum* Dum., to which species Pearson attached this as an African variety.

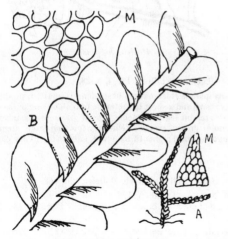
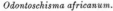

Odontoschisma africanum.

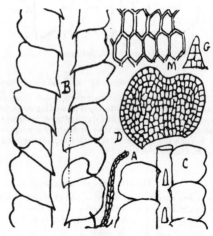

Odontoschisma variabile.

110. *Odontoschisma variabile* Sim (new species).*

Stems ½–1½ inch long, straggling, rooting by few long rootlets, usually simple, but with leafless flagellae near the base and no rhizomatous stem. Leaves usually imbricate, obliquely inserted, wider than long, the apex varying from entire and rounded to emarginate or wavy several-pointed, these points rounded and seldom alike, the lower junction slightly decurrent, and the lower margin sometimes incurved as a rounded lobe. Cells numerous, hexagonal, 0·03 mm. long, thick-walled, but without trigones. Folioles present on younger parts, small, simple, conical, and few-celled. Much like *O. africanum* except in the irregular leaf-form and small folioles. Common.

---

* *Odontoschisma variabile* Sim (sp. nov.).

Caules 1–1½″ longi, vagi, radicantes, plerumque simplices, infra flagellati sed non rhizomatosi. Folia plerumque imbricata, oblique inserta, latiora quam longiora, apice variante ab integro rotundatoque ad emarginatum undulatumve pluri-acutum ; hi apices rotundati et raro consimiles. Foliola adsunt in partibus recentioribus, parva, simplicia, conica, pauci-cellata.

111. *Odontoschisma sphagni* (Dicks.) Dum., Rec. d'Obs., p. 19 ; Macvicar, p. 296.

Stems ½–1½ inch long, usually simple, prostrate or straggling, rhizoid-bearing, and rooting by few strong rootlets, the point of the stem usually small-leaved or flagellate ; leaves succulent, suberect, seldom imbricate, slightly concave or almost flat, spreading, round from a wide base or occasionally longer (as in fig. D), the lower junction slightly decurrent ; cells hexagonal except the marginal quadrate line, thick-walled, 0·03 mm. ; without trigones. Folioles absent except in the youngest tips, those among young floral leaves deeply 2-fid and very tender (as in fig. G). Frequent everywhere, but sexual parts not seen. Occurs also in Europe, where it is dioecious.

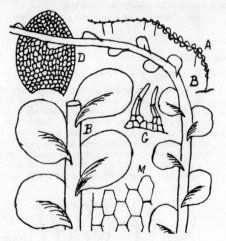

Odontoschisma sphagni.   *Alobiella heteromorpha.*

Genus 41 ALOBIELLA (Spruce, 1882, subgenus) Schiffner, Pfl., i, 3, 28.

Plant minute, somewhat succulent, much branched at the leafless rhizomatous base, densely matted, producing succulent leafless flagellae with short ascending or arching leafy stems and a short basal fertile branch having paroecious inflorescence ; colourless radicels produced interruptedly ; leaves on leafy branches almost horizontally inserted, succubous, nearly round, entire, or slightly emarginate ; folioles present, small ; cells very few and large, angular ; fertile branch few-leaved. Perianth free, oblong-oval, trigonous upward ; mouth incurved, subentire. Very closely related to *Odontoschisma*.

112. *Alobiella heteromorpha* (Lehm.), Steph., Sp. Hep.; Bull. de l'Herb. Boissier, viii, 568.

=*Cephalozia* (*Lembidium*) *heteromorpha* (Lehm.), Pears., Hep. Nat., p. 11, Tab. VII.
=*Jungermannia heteromorpha* Lehm., Hep. Cap., in Linn., iv, 362 (1829) ; L. and L., in Lehm., Pug., iii, 48 ; Syn. Hep., p. 131 ; Spruce, On Cephalozia, p. 59.

Plant 3–5 mm. long, rhizomatous, flagellate, and with ascending or arching leafy stems and short basal paroecious fertile stems. Leaves on sterile branches round or often submarginate, suberect and thus face to face, wide based ; folioles regularly present, small, three to four cells long and one cell wide, or mostly so, incurved toward the stem. Cells few and large, irregularly 5–6-sided, about 75μ long, thin-walled, without trigones ; fertile branch basal, postical, erect, fleshy, short, the upper leaves much the larger, truncately 2-lobed ; then several large bracts, also lobed, containing two androecia each, then a few small ordinary leaves.

Pearson says, " The postical branches and trigonous perianth induce me to refer this species to the genus *Cephalozia*, and its habit with its arcuate branches and thick texture to the subgenus *Lembidium*." Its resemblance in growth and habit to *Notoscyphus* and *Nardia* is very great, but the erect free perianth and large cells and colourless rhizoids differ. It occupies similar ground (decayed granite or dolerite) from Table Mountain to Natal, frequent.

Hilton Road, Natal (Sim 8252, 8238, etc.).

Genus 42. ADELANTHUS Mitt., Journ. Linn. Soc. (1864), p. 264 ; Schiffner, Pfl., i, 3, 99 ; Macvicar, p. 301 ; Steph., Sp. Hep., viii, 2nd ser., 595.

=*Adelocolea* Mitt., Challenger Exp., Bot. i, 2, 106.

Leaves succubous, transversely inserted, more or less secundly bent downward, the upper margin entire and incurved, the lower margin dentate. Folioles absent or rudimentary. Stem ascending, not rooting. Marsupium short.

113. *Adelanthus unciformis* (Tayl.), Mitt., Linn. Soc., vii, 243 ; Schiffner, Pfl., i, 3, 99 ; Steph., Hedw., 1892, p. 120 ; Steph., Sp. Hep., viii, 2nd ser., 600.

=*Jungermannia unciformis* Tayl., Journ. of Bot., 1844, p. 457.
=*Plagiochila unciformis* Tayl. and Hook., Syn. Hep., p. 653.
=*Plagiochila Lindenbergiana* Lehm., Pug., iii, 53.
=*Jungermannia haliotiphylla*, De Not. Acad. Turin, 1885.

Stems about 1–2 inches high, rigid and erect or suberect from a rhizomatous base, brown, either caespitose or scattered among mosses, simple or slightly branched, branches from the back ; apex regularly recurved. Leaves succubous, transversely inserted on a comparatively narrow base, all homomallous, bent downward and more or less meeting below, obliquely ovate, the upper margin much incurved, and rounded and entire ; apex acute ; lower margin rounded and strongly toothed, the teeth increasing in size outward, the apical tooth the largest. Cells rounded and separate, 20μ long, lower cells rather larger, but the central basal portion of the leaf has an area with very large hexagonal cells, immediately adjoining smaller cells. Folioles absent. Inflorescence dioecious, marsupium terminal ; perianth oblong-elliptical, the mouth set with many spinose lobes ; floral leaves and folioles acute and entire ; apex acute or emarginate. Androecia in short branches on the upper side, bracts crowded, monandrous.

Abundant on exposed rock-cliffs on upper part of Table Mountain, Cape. I have no specimens from elsewhere, and it is absent from the eastern mountains. It occurs also at the Straits of Magellan and on the Mascarene islands (Pillans 4079, etc.).

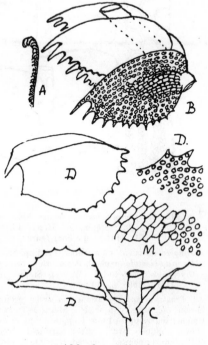

Adelanthus unciformis.

Genus 43. CALYPOGEIA Raddi, Mem. Soc. Ital. d. Sc. Mod., xviii, 42.

=*Kantia* S. F. Gray, Nat. Arr. Brit. Pl., i, 706.
=*Cincinnulus*, Dum.

Leaves entire or bidentate ; folioles present ; often bifid or emarginate ; marsupium tubular, fleshy, pendulous, investing the mature sporogonium.

Stems prostrate, flaccid when dry, usually simple or slightly branched, more or less matted on open or shaded ground such as river banks or under trees. Branches from under-surface leafy ; rhizoids colourless, in tufts from under-surface of folioles. Leaves incubous, entire, or 2–3-dentate, alternate, flat or sligthy convex, variable as to form. Cells large, pellucid, with thin walls and usually without trigones. Folioles present, large, usually 2-dentate. Inflorescence ($\male$ and $\female$) cladogenous, *i.e.* on very short branches axillary to the folioles, the fertile plant developing a pendulous cylindrical fleshy rhizoid marsupium which descends into the soil. Androecium cone-like with few 2-lobed saccate bracts. Much confusion has existed as to what constituted *Calypogeia* of Raddi. Spruce claims it for plants with *opposite succubous* leaves, and even Stephani, in Hedw., 1892, p. 120, included therein two such, viz. *C. renifolia* (Mitt.) Spr. and *C. scariosa* (Lehm.) Spr.—now placed in *Gongylanthus*, but in Sp. Hep. he describes the genus as having *alternate* incubous leaves, which is followed by Macvicar, and agrees with our plants. Stephani deals with *Kantia* Gray as (*Calypogeia* Nees, non Raddi) including this group, in Hedw., 1895, p. 55.

Synopsis.

114. *C. fusca.* Leaves entire or slightly bifid, not decurrent ; folioles rounded, emarginate half-way with basal tooth.
115. *C. bidentula.* Leaves entire or bidentulate ; folioles deeply bifid, not toothed.
116. *C. trichomanis.* Leaves entire, ascending, decurrent ; foliole almost round, slightly emarginate.

114. *Calypogeia fusca* (Lehm.), Steph., Sp. Hep., viii, 2nd ser., 666.

=*Kantia fusca* (Lehm.), Steph., Hedw., 1895, p. 55.
=*Jungermannia fusca* L. and L.
=*Lejeunea fusca* Lehm., Linnaea, iv 360 (1829), (insufficiently described) ; Syn. Hep., p. 408 (among doubtful species) ; Drege's Zwei-Pflan. Doc. (1843).

Dioecious ; stem prostrate, straggling, 1–2 inches long, equally wide the whole length, from deep green to brown, simple or nearly so, rooting, usually growing on ditch-banks under trees, and often having several small basal branches having similar but smaller scattered leaves. Leaves complanate, succubous, 0·7–1 mm. long, ovate, obliquely attached, or the upper side almost transversely attached, convex or nearly flat, spreading, rounded at the apex, entire, crenulate or more usually slightly emarginate with round lobes, the upper side much rounded and shortly inserted, the lower side not or only slightly decurrent.　Cells lax, large; elliptical-hexagonal, tumid, $30 \times 35\mu$, the lower much longer than the upper; trigones absent, surface rough.　Folioles distant, roundly quadrate, usually bifid half-way with or without shoulders on the segments; rhizoids abundant from foliole warts.　Floral leaves conduplicate.

Not uncommon in mountain and forest-outskirt localities from Cape Town eastward.　Hilton Road, Natal (Sim 8204); Vryheid, Natal (Sim 8222), etc.

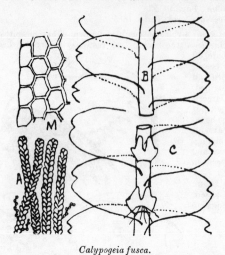

Calypogeia fusca.

115.　*Calypogeia bidentula* (Weber) Nees, Syn. Hep., p. 199; Steph., Sp. Hep., viii, 2nd ser., 671; Mitt., Linn. Soc. Journ., xvi, 91, 191.

=*Jungermannia bidentula* Weber, Prod., p. 38.
=*Kantia bidentula* (Weber), Pears., Hep. Nat., p. 13.
=*Cincinnulus bidentulus* (Weber), Steph., in Engl. Bot. Jahrb., 1895, p. 314.

Prostrate and gregarious, on moist ground or among mosses, etc., usually in shade.　Stems 1–3 cm. long, usually simple; leaves up to 1 mm. long, incubous, crossing the stem above, obliquely ovate, rather narrowed toward the apex, either bluntly pointed, rounded, truncate or somewhat emarginate all on one stem, without regularity in this respect, and occasionally some leaves are rather deeply 2-lobed, the lower margin not decurrent.　Folioles twice as wide as the stem, and of similar length, deeply cordate-decurrent at the base, bifid two-thirds, standing out from the stem, and with the basal margins also often erect, consequently the foliole is deeply concave outwardly.　Rhizoids produced from the stem in the lower sinus of the foliole.　Cells mostly hexagonal and pellucid, 40–50μ diameter.　Marsupium fleshy,

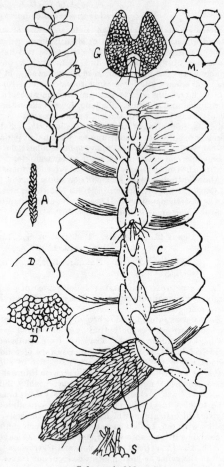

Calypogeia bidentula.
S, apex of marsupium.

cylindrical, terete, descending from the axil of a foliole, 1·5 mm. long, 0·5 mm. diameter, producing long rhizoids irregularly from its surface, and with the point ended by about five short several-celled hairs or bracts.　Androecium not seen.

Cape Town to Natal, in forest country or moist mountain localities (Sim 8309, 8204, etc.).　*Kantia arguta* (N.), Pears., in Hep. Nat., p. 13, and Tab. VIII, probably belongs here, representing a more deeply 2-lobed leaf form.

116.　*Calypogeia trichomanis* (Linn.), Corda, in Opiz. Beitr., p. 653; Macvicar, p. 307.

=*Kantia trichomanis* Lindb., Hep., in Hib. Lect., p. 508; Steph., Hedw., 1895, p. 55.
=*Mnium trichomanis* Linn., p.p., Sp. Plant, p. 1114.
=*Cincinnulus trichomanis* var. *communis*, Boulay.
=*Calypogeia fissa* var. *integrifolia* Raddi.

Stems 2–4 cm. long, straggling among mosses, nearly simple, innovating in continuation of apex and with colourless rhizoids from below the foliole along its length.　Leaves nearly separate or more or less overlapping, incubous, 1 mm. long, spreading widely but ascending, obliquely ovoid, the apex rounded or truncate, the upper margin nearly at right angles to the stem and with rounded shoulder, the lower margin descending and so long-decurrent as to give the stem a winged appearance. Folioles very concave, nearly round, slightly emarginate, and the cells drawn toward the sinus. Leaf-cells 50μ diameter, irregularly hexagonal, pellucid, the walls thin.

Table Mountain (Skeleton Ravine, Mrs. Bolus; Sim 8414), etc.

C. sphagnicola W. and L. (*Kantia sphagnicola* Arn. and Pers.) introduced into my list in error is closely related to this, but has the folioles half-bifid, and other differences.

Genus 44. Bazzania, S. F. Gray, Nat. Arr. Br. Pl., i, 704 (as *Bazzanius*); Macvicar, p. 315; Carrington, Trans. Bot. Soc. Edin, 1870, p. 309.

=*Pleuroschisma* Dum., Recueil. d'Obs., p.p.; Steph., Engl. Bot. Jahrb., 1895, p. 308.

=*Mastigobryum*, Syn. Hep., p. 214; Schiffner, Pfl., i, 3, 101; Steph., Sp. Hep., viii, 2nd ser., 681.

Dioecious. Stems erect, somewhat branched, and with small-leaved flagellae from the under-side. Leaves incubous, oblique, narrowed to a rounded or 3-toothed apex from a wide base. Folioles present, abundant, sometimes connate with the leaf above. Cells small, rounded, but in some species several lines of large pellucid cells occur, and in some the lower margin of the leaf is reflexed and large-celled. Male and female inflorescence on short branches axillary to the folioles, the perianth without marsupium.

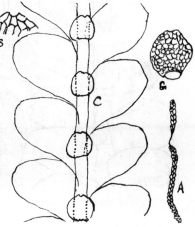

Calypogeia trichomanis.
S, apex of foliole.

Synopsis.

117. *B. convexa.* Leaves 3-fid, folioles deeply 4-fid, three lines of large pellucid cells in most leaves.
118. *B. adnexa.* Leaves 3-fid, folioles crenate; no large cells.
119. *B. radicans.* Leaves entire or crenate, decurved; folioles crenate; strongly anchored by flagellae.
120. *B. pellucida.* Leaves entire, lower margin reflexed, large-celled.

117. *Bazzania convexa* (Thunb.), Mitt., in Hook. f. Fl. N. Zel., ii, 147; Mitt., Journ. Linn. Soc., xxii, 146, 322.

=*Jungermannia convexa* Thunb., Prod. Fl. Cap., p. 173; Schw. Prod., ii, No. 22.
=*Jungermannia Thunbergii* Meisn. in litt.
=*Jungermannia nitida* Web., Prod., p. 43; Lehm., Hep. Cap. Ecklon., in Linn., v, 364.
=*Mastigobryum convexum*, Ldbg., Syn. Hep., p. 215; Steph., Sp. Hep., p. 134.
=*Mastigobryum exile* Ldbg., Syn. Hep., p. 217.
=*Pleuroschisma convexa* Steph., Engl. Bot. Jahrb., 1895, p. 308.

Stem ½–1 inch long, erect from a creeping base, pinnately or dichotomously branched above, often attenuated at the apex, and with flagellate small-leaved branches in the lower part, but seldom on erect branches. Leaves incubous, 1 mm. long, nearly longitudinally inserted but subvertical in position, covering only a small portion of the stem; ovate-oblong, somewhat concave, the upper margin rounded, the lower almost straight, both entire, the apex truncate, shortly 3-toothed or occasionally 2-toothed, or on some plants all almost entire, upper tooth usually the larger; cells in lines, quadrate, tumid, irregular as to size, but all small except three adjoining lines near the lower margin which are much larger and pellucid, forming a nerve-like band extending from the leaf-base nearly to its apex. Folioles standing out from the stem, wider than the stem, usually free from the leaf, roundly quadrate or somewhat cuneate, cut about half-way into four equal segments, each about four lines of cells wide below, the cells pellucid, all similar in size to the larger leaf-cells but hexagonal. Leaves on young branches often more distant, narrower, entire or 1-2-denticulate. Perianth ovate, incurved, plicate at the apex and with the mouth denticulate.

Table Mountain (Sim 9791), Eastern forests, etc., and Stephani (as above) records it also in the Mascarenes, and in tropical Africa (Sim 9792) and Rhodesia (Sim 9797), etc., Asia and Chile, but not in Australasia. Stephani (Sp. Hep., p. 534) introduces into his group *Vittata* (having a nerve) both *M. convexa* Ldbg., Syn. Hep., p. 215, and *M. exile* Ldbg., Syn. Hep., p. 217, the former from tropical and South Africa and elsewhere, and the latter from Cape Town only (Ecklon). He separates thus :—

M. convexum. Leaf ovate-ligulate, shortly 3-toothed; foliole acutely 4-fid.

M. exile. Leaf ovate, entire; foliole subquadrate, with obtuse angles.

I have not found a vittate specimen with crenulate foliole, all are quadrifid, so cannot maintain *M. exile.* The crenate foliole in South Africa belongs to all the non-vittate species.

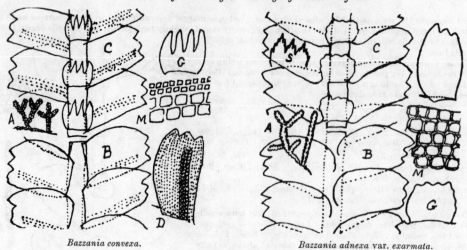

Bazzania convexa.   *Bazzania adnexa* var. *exarmata.*

S, apex of leaf of type.

118. *Bazzania adnexa* (L. and L.) Sim.

=*Mastigobryum adnexum* (L. and L.), Sp. Hep., viii, 2nd ser., 855.
=*Jungermannia adnexa* L. and L., Pugill., iv, 56.
Form *exarmata*, Steph., Sp. Hep., viii, 2nd ser., 855.

Plant usually erect, stems 2–3 cm. long, yellowish, regularly narrow, somewhat branched above or with large simple branches from the base, and with flagellae below from axils of the folioles. Leaves incubous, imbricate, sub-opposite, 1–1·5 mm. long, of almost the same shape as in *B. convexa*, but attached lower on front of the stem, more horizontal, and without any different cells. Leaf-apex 1–3-toothed with entire lobes, or sometimes the leaf is almost entire. Cells all fairly large, 20–30μ, pellucid, quadrate-hexagonal, and showing no band of different cells. Folioles standing out from the stem, roundly quadrate, wider than the stem, the apex varying from almost entire to erosely 4–6-crenate, usually rounded and free at the base, but sometimes connate to the base of the adjoining leaf on one or both sides on the upper part of stems on which they are free lower. Perianth large, funnel-shaped, mouth ciliate. Floral leaves crowded, the inner 6-fid half-way, with narrow segments; floral folioles similar, 4–6-fid. Capsule oval on a long seta; elaters short, 1–2-spiral. This variable species is also Australasian, where the teeth of the leaves often have denticulate margins (which is the type of the species), but here they have usually been seen entire, which is var. *exarmata*.

Table Mountain to Natal (Sim 9798) and Rhodesia (Eyles 2631, 2595β), etc., except Sim 10,277 from Inyanga, South Rhodesia (Coll., I. S. Henkel), which has them denticulate, as in fig. S.

119. *Bazzania radicans* Sim (new species).*

Stems 1–1½ inch long, usually simple, straggling over mosses in forest, and anchored by strong, leafless, root-like flagellae from the under-surface of the upper part of the stem. Branches axillary to folioles, laterally divergent. Leaves oblique, very uniform, 1 mm. long, widely ligulate, convex, with rounded, entire, or subcrenate apex, usually decurved. Folioles quadrate, with irregularly crenate apex, the base of a foliole being regularly opposite the base of the adjoining leaf, and sometimes but not regularly connate. Cells thin-walled, all alike except the marginal line on the reflexed lower margin, which are rather larger; trigones distinct.

Perie Forest, Kaffraria (Sim 78), etc.

120. *Bazzania pellucida* Sim (new species).†

Sterile, growing in dense masses; known from Table Mountain only. Stem 2–3 cm. long, with a few widely divergent lateral branches from foliole axils, and toward the base some flagellae. Leaves hardly imbricate, very

---

* *Bazzania radicans* Sim (sp. nov.).

Caulis 2–3 cmm. longus, inter muscos vagans, flagellibus paucis validis rigidulis radicans; caules simplices, vel nonnumquam ramis posticalibus. Folia obliqua, late ligulata, apice rotundato integro vel subcrenato, plerumque decurvata. Foliola quadrata, apice irregulariter crenato. Cellae leptodermes, omnes consimiles praeter marginales, praecipue in margine inferiore reflexo, quae paullo grandiores sunt. In silvis orientalibus.

† *Bazzania pellucida* Sim (sp. nov.).

Sterilis, cumulis densis proveniens. Caulis 2–3 cmm. longus, ramis paucis lateralibus divergentibus. Folia vix imbricata, admodum regularia, maxima in medio caule, in caule principali ramisque ad apicem versus sensim multum attenuata. Folia pellucida, oblonga, apice rotundato vel nonnumquam retuso, margine inferiore multum ac regulariter reflexo, cellis paucis maximis; alibi cellae parvae multaeque, ordinibus positae. Foliola quadrata, apice crenato. In Monte Tabulari.

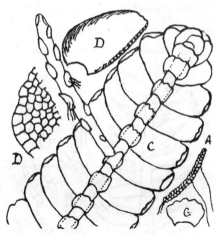

Bazzania radicans.

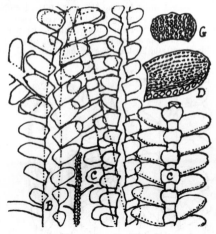

Bazzania pellucida.

regular, shortly inserted, not decurrent, largest in mid-stem, gradually much reduced in size on main stem and branches toward the apices. Leaves pellucid, oblong, with rounded or sometimes retuse apex, the lower margin much and regularly reflexed with a line of few and very large cells along that reflexed portion, giving a characteristic appearance to the plant ; elsewhere cells are small and many, in straight rows. Folioles quadrate, with crenate apex. A remarkable plant, not found fertile (Sim 9784, 8494, etc.).

Genus 45. Cephalozia Dum., Rec. d'Obs., p. 18 (emend.) ; Steph., Sp. Hep., viii, 2nd ser., 271.

Included in *Jungermannia*, Syn. Hep.

Includes *Cephalozia* and other genera of Macvicar, p. 250, etc.

Plants usually minute, branched or branched at the base only, the branches postical, with pellucid cells, and flagellae often present. Leaves usually bifid, keeled, and laxly conduplicate, sometimes flat, obliquely inserted, succubous ; cells large ; folioles usually absent except among the floral leaves, which are larger than the stem leaves, and usually somewhat dentate. Perianth terete or 3-sided, with the third angle postical ; female inflorescence usually cladogenous, and androecium either hypogynous or cone-like. Small tender plants, on soil, or among mosses. Several species formerly included in *Cephalozia* are now placed in *Odontoschisma*.

Synopsis.

Leaf segments connivent.

121. *C. connivens.* Fairly large, sinus small, rounded, the segments connivent.

Leaves bicuspidate, not connivent.

122. *C. bicuspidata.* Leaves comparatively large, with few but large cells.
123. *C. atro-viridis.* Dioecious, leaves nearly flat, longitudinally inserted ; cells deeply bifid.
124. *C. radicans.* Dioecious, plant rooting freely, leaves conduplicate.
125. *C. Kiaeri.* Monoecious, plant erect, leaves conduplicate.
126. *C. lycopodioides.* Leaves imbricate, julaceous, minute.
127. *C. natalensis.* Dioecious, plant erect, leaves conduplicate.
128. *C. capensis.* Leaf with lunate sinus and spreading segments.
129. *C. tenuissima.* Leaves denticulate along the margins.

Leaves somewhat 2-lobed, lobes rounded.

130. *C. Pillansii.* Leaves intermediate with *Odontoschisma*.

121. *Cephalozia connivens* (Dicks.), Lind., Proc. Linn. Soc., xiii, 190 ; Spruce, On Cephalozia, p. 46 ; Pears., Hep. Brit. Isles, p. 157, pl. 60 ; Macvicar, p. 259 ; Steph., Sp. Hep., viii, 2nd ser., 278.

=*C. connivens* var. *flagellifera* Pears., Hep. Nat., p. 10, pl. 6.

=*Jungermannia connivens*, Dicks., Pl. Crypt. Fasc., iv, 19, pl. 11 ; Hook., Brit. Jung. Pl., p. 15 ; Syn. Hep., p. 124, with many varieties where var. *diversifolia* is recorded from Ecklon's Cape collection.

=*Cephalozia multiflora*, Lindb., Musc. Scand., p. 4 ; Lindb., Act. Soc. Sc. Fenn, x, 501 (not Spruce, On Cephalozia, p. 37).

Monoecious ; scattered among mosses ; stems ½–1½ cm. long, stems prostrate, with numerous rhizoids, simple or slightly branched, without flagellae, the stem delicate, of few pellucid cells. Leaves to 40μ long, largest at mid-stem, almost longitudinally inserted, oblique, roundly quadrate, bifid at the apex, lower margin decurrent, apex rounded but with a round sinus, leaving the acute lobes connivent. Cells large, elliptical, $35 \times 50\mu$, the upper cells smaller. Folioles absent except among the floral leaves where they are mostly 4-fid. Perichaetium a short postical branch, with about three pairs of floral leaves, and ovate perianth contracted at the 3-angled ciliate apex. Androecium in mid-branch, bracts small, monandrous, acute, unequal.

Table Mountain, Eastern forests, and Transvaal, rare. Occurs also in Europe, Asia, and North America.

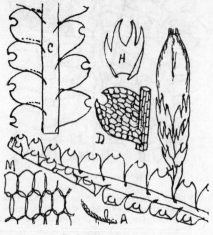

Cephalozia connivens.

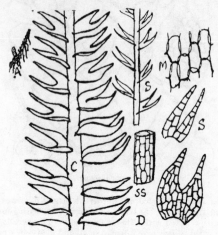

Cephalozia bicuspidata.

S, leaf of attenuated stem ; SS, cell-formation of stem.

122. *Cephalozia bicuspidata* (Linn.) Dum., Rec. d'Obs., p. 18 ; Syn. Hep., p. 138, p.p. ; Macvicar, p. 252.

　　　=*Cephalozia vallis-gratiae* Steph., Sp. Hep., viii, 2nd ser., 431.
　　　=*Jungermannia biscuspidata* Linn., Sp. Pl., p. 1132 ; Hook., Brit. Jung., pl. 11, p.p.

Monoecious, scattered among mosses. Stems 1–2 cm. long, or less, of few pellucid cells, usually simple, rooting freely, comparatively large-leaved below, the leaves decreasing in size up the stem. Folioles absent. Leaves separate, very shortly and almost horizontally attached, ovate, bifid half-way, somewhat concave in transverse position, not keeled nor conduplicate, but the points ascending and acute. Leaves on attenuated branches much smaller but still patent and deeply 2-fid. Cells very large, few, hexagonal, $40 \times 50\mu$, the terminal part of leaf-segments often of 3–5-superimposed cells. Perichaetium on a short postical branch, with few but large bifid floral leaves, nearly entire at the margins, and similar floral folioles. Perianth elliptic or cylindrical, trigonous above, with denticulate mouth. Androecium spicate in mid-stem, the bracts resembling the leaves.

Table Mountain (Sim 9806), Perie Forest, Kaffraria (Sim 268), etc., rare. This is the *C. vallis-gratiae*, Steph., Sp. Hep., 2nd ser., p. 431, a local form of *C. bicuspidata*, which he records from Genadendal. The *C. bicuspidata* of the Syn. Hep., p. 138, is a very composite group, in which the one form assigned to South Africa (Ecklon), *i.e.* var. *concinna* form δ *Brauniana*, is a julaceous form unknown to me and not corresponding with that described above from my own specimens. In Britain *C. bicuspidata* also varies greatly.

Stephani (Sp Hep., viii, 2nd ser., 431) describes another South African species of this group, *C. robusta* St., which I cannot by his description separate from that described above, except that the leaf-segments are said to be " very unequal." He gives as localities Claremont, Montagu Pass, Genadendal (Breutel). I have not seen any specimen of *Cephalozia* having the segments very unequal, though sometimes when some segments are seen sideways they appear to be so.

123. *Cephalozia atro-viridis* Sim (new species).*

Dioecious, minute, very dark green, common on decayed tree-trunks and on humus below them ; stems 2–6 mm. long, spreading in flat mats, very much branched, with flagellae below ; cells few, pellucid ; adult leaves seldom imbricate, usually separate, spreading, nearly flat, convex but not conduplicate, equally 2-fid half-way, longitudinally inserted, with subacute entire segments ; leaves on younger branches smaller, shortly inserted, more conduplicate and more crowded. Folioles absent. Cells roundly quadrate, small, separate. Perichaetium nearly sessile on a small postical branch, shortly elliptical, 5-valved, with incurved entire points ; when young, shortly cup-shaped with five rounded

---

* *Cephalozia atro-viridis* Sim (sp. nov.).

Dioica, minuta, caespitosa, atro-viridis ; caules 2–6 mm. longi, multum ramosi, flagellati infra ; folia adulta separata, patula, convexa (non conduplicata), aequaliter 2-fida ad dimidium, in longitudinem inserta ; segmenta sub-acuta, integra. Folia in ramis recentioribus minora, breviter inserta, conduplicata, conferta. Perich. clodogenum, ellipticum. Folia floralia maiora, inaequaliter 2-fida, conduplicata, carinata, prope integra. Androecium terminale.

crenations. Floral leaves much larger than stem-leaves, about three pairs, unequally 2-fid, conduplicate, keeled, the inner bluntly toothed. Androecium terminal. Common as a pioneer in all forest and humid districts ; forming a dark green perennial surface on new ground.

Hilton Road, Natal (Sim 8983, etc.).

The name *C. diraricata* (Sm.) Spr. was introduced into my list in error in connection with this.

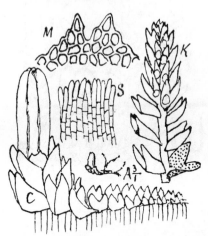

Cephalozia atro-viridis.   *Cephalozia radicans.*

124. *Cephalozia radicans* Sim (new species).*

Dioecious, minute, usually on soil ; stems solid, up to 5 mm. long, prostrate, rooting freely, especially under the perichaetium, by colourless rhizoids ; leaves spreading, flat below, conduplicate above, keeled where connected, two-thirds bifid, the segments triangular-acute, entire. Folioles absent. Perianth terminal, cylindric, 5-valved to the base, the apex incurved, bordered with long but not free cells. Floral leaves much larger than stem-leaves, keeled and equally 2-fid, segments acute, entire or only slightly crenate from cell to cell. Enlarged bifid folioles present among floral leaves. Androecium terminal, about 10-bracted, bracts smaller than the leaves, with inflexed lobes.

Vryheid, Natal, and other upland semi-exposed localities (Sim 8205, etc.).

125 *Cephalozia Kiaeri* (Aust.), Pears., Hep. Knys., p. 8, pl. 3 ; Steph., Sp. Hep., viii, 2nd ser., 499.

=*Jungermannia Kiaeri* C. F. Austin, Torrey Bot. Club, vi, 3, 18 (1875).

Monoecious, the perianth on a short postical nearly basal branch, and the androecium at first terminal on stem, but through continued growth median on stem, and several occur in succession. Stems 5 mm. to 2 cm. long, usually among mosses or caespitose, solid, slender, irregularly branched, without flagellae. Leaves not quite imbricate, sometimes scattered, shortly attached, clasping, laxly conduplicate, very small, widely ovate, 2-fid to the middle, spreading and keeled, the segments ascending and subacute ; cells minute, 15μ, hexagonal, thick-walled, without trigones. Folioles absent (Pearson illustrates folioles, which I have not seen). Perianth cladogenous, cylindrical, 5-valved, with constricted denticulate mouth. Floral leaves few, larger than the stem-leaves, 2-fid, segments sub-acute, the margin of the inner leaves subdenticulate. Bracts of androecium, about eight pairs, closely imbricate, small, bifid, monandrous.

In the forest districts, rare.

126. *Cephalozia lycopodioides* Sim (new species).†

Very minute ; stems prostrate, 2–4 mm. long, with short erect branches, somewhat succulent, many-celled, solid ; leaves closely imbricate, almost julaceous, hardly wider than the stem, ovoid, two-thirds bifid, entire,

---

* *Cephalozia radicans* Sim (sp. nov.).

Dioica, minuta ; caulis ad 5 mm. longus, prostratus, libere radicans ; folia patula, carinata, ⅔ bifida ; segmenta trigono-acuta, integra. Perianthium terminale, cylindricum, 5-valvatum, apex incurvatus, cellis longis fimbriatus. Folia floralia multo maiora, carinata, 2-fida, acuta, paene integra. Androecium terminale, circiter 10-bracteatum ; bracteae minores quam folia, lobis inflexis.

† *Cephalozia lycopodioides* Sim (sp. nov.).

Minutissima, prostrata, ramis brevibus erectis ; caules subsucidi, multi-cellati ; folia dense imbricata, paene iulacea, caule vix latiora, ovoidea, ⅔ bifida, integra. Inflorescentia lateralis ; perianthium ovoido-oblongum, sub-tentum non nisi 2–4 foliis floralibus bifidis amplificatis, segmentis subaequis integris. Androecium non vidi.

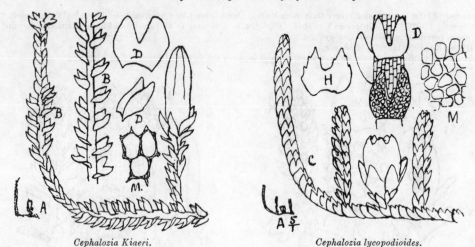

Cephalozia Kiaeri.   *Cephalozia lycopodioides.*

the segments subacute.   Folioles absent.   Inflorescence lateral from a short postical branch, having one or two pairs of large bifid floral leaves with subequal almost entire segments, except the inner pair which are dentate.   Perianth ovoid-oblong, apex 5-fid, segments entire, incurved.   Leaf-cells minute, rounded.

Natal, rare.

127. *Cephalozia natalensis* Sim (new species).*

Dioecious ; stem 3–6 mm. long, erect, simple or almost so, slender, wiry, rooting near the base only, and with long, erect, slender sterile branches from the base, on which branches leaves are sometimes reduced to two segments of three to four cells each.   Leaves distant, very small, shortly attached, spreading, conduplicate, keeled, with entire triangular segments.   Folioles absent.   Cells very small, 4–5–6-sided, with thin walls.   Inflorescence terminal, perianth ovoid, shortly pointed, with apical lobes inflexed, entire.   Floral leaves much enlarged, subequally 2-fid, strongly keeled, conduplicate, the inner two usually slightly toothed.   Androecium terminal ; bracts much larger than the leaves, usually red, loosely imbricate, 2-fid, monandrous.   Scattered among mosses in forest districts, never in quantity, but occurs from Grahamstown to Natal (mixed in Sim 8313, 9799, etc.).

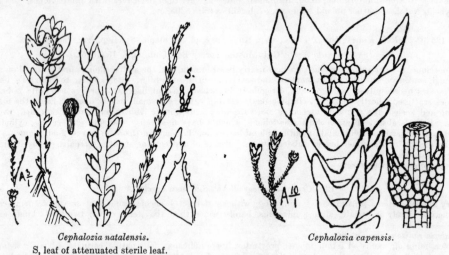

Cephalozia natalensis.   *Cephalozia capensis.*
S, leaf of attenuated sterile leaf.

* *Cephalozia natalensis* Sim (sp. nov.).
   Caulis 3–6 mm. longus, erectus, fere simplex, tenuis, ramis longis erectis gracilibus sterilibus a basi radicanti. Folia distantia, minima, breviter adfixa, patula, conduplicata, carinata ; segmenta integra, trigona, obtusa.   Inflor. terminalis, perianthium ovoideum, breviter apicatum ; lobi apicales integri, inflexi.   Folia floralia multum amplificata, sub-aequaliter bifida, carinata, conduplicata ; bina interiora leviter dentata.   Androecium terminale, bracteae multo maiores quam folia, sub-imbricatae, bifidae, monandrae.

128. *Cephalozia capensis* Sim (new species).*

Minute, dark green ; stems 2–3 mm. long, ten to twelve cells round, pellucid, solid, prostrate below, apex suberect, with few branches. Leaves deeply bifid, closely imbricate, regularly placed, horizontally inserted, sinus lunate, segments divaricate, lanceolate, subacute, entire, about three cells wide below. Floral leaves much larger, in two to three pairs, long-keeled, spreading, segments triangular-acute, conduplicate, entire ; folioles absent elsewhere but present among floral leaves, shortly bifid, very delicate, few-celled, the segments each of one to two turgid cells only, with one to two connecting lines of cells below. Other parts not seen.

Below the viaducts, Constantia Nek, Table Mountain (Sim 9787) ; not seen elsewhere.

129. *Cephalozia tenuissima* (L. and L.), Steph., in Sp. Hep., viii, 2nd ser., 514.

=*Jungermannia tenuissima* Lehm., in Hep. Cap., in Linn., iv, 367 (1829) ; Lehm., Pug., iii, 53 (1831) ;
L. and L., Syn. Hep., p. 143.

A minute plant described from Ecklon's Cape Town specimen in Herb. L. and L., which I have not seen. The description is " Folioles none ; stem prostrate, simple ; leaves distant, vertical, subcomplicate, bifid ; segments erect, lanceolate-subulate, acute ; margin subrepand, denticulate. Fruit terminal ; perianth ovate, plicate, mouth denticulate." Stephani (as above) gives some further particulars including " Adult stem-leaves 0·4 mm. long, squarrose-spreading, to two-thirds bifid, keeled ; keel sinuate, lobes very divergent, spreading widely, unequal, the upper ovate-acuminate, bluntly toothed, the lower subentire or repandly angulate. . . . Folioles small, spreading, to one-half bifid, margin minutely dentate."

Mitten (Journ. Linn. Soc., xvi, 91, 187 (1877)) records it from " Base of the Lion's Head, on damp soil, above the Kloof Road, Cape Town " (Rev. A. E. Eaton), and adds : " Scattered stems with perianths are observable ; it is a very minute species, nearly allied to *C. dentata*, Raddi."

Stephani (Sp. Hep., viii, 2nd ser., 274) includes it in the subgenus *Prionolobus* of *Cephalozia*, which Spruce (p. 508) decided could stand as a genus, having only lateral and no postical branches, and having dentate leaves. Macvicar also adopts this genus. I include *C. tenuissima* in *Cephalozia* since the only material available is in Europe. Collected a hundred years ago, and then scarcely visible ; all our species of *Cephalozia* have some of the floral leaves dentate, but not the stem-leaves, and it is desirable to see more of this before bringing in another genus on it.

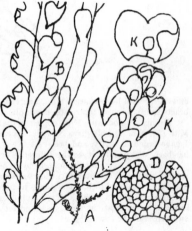

130. *Cephalozia Pillansii* Sim (new species).†

Very vigorous for *Cephalozia* ; stems 2–3 cm. long, erect, stout, succulent, twelve to twenty cells round, pellucid, with cells as large as the leaf-cells ; simple or branched, branches postical, the lower branches with terminal male inflorescence, other branches in all stages of small-leaved conditions. Leaves scattered, succubous, with short transverse attachment, the upper leaves the smaller, loosely clasping, ovate to ovate-truncate, the apex in mature leaves varying from emarginate to one-third 2-lobed, the lobes loosely conduplicate, somewhat concave, with the rounded segment-points usually incurved. Folioles absent. Flagellate branches stout but small-leaved, the leaves varying from cuneate-entire to obovate and emarginate Androecium terminal on lower short branches, bracts spreading, imbricate, in four to five pairs, much larger than the stem-leaves, monandrous, bluntly 2-lobed, the lower lobe usually the smaller and incurved. All cells irregularly 4–5–6-angled, thin-walled, large, pellucid. Other parts not seen. Matted above *Sphagnum*, in Pillans 3588, from wet loam. mountain between Muizenberg and St James, 15/6/1912 (N. S. Pillans). More vigorous than *Odontoschisma variabile*, and distinctly lobed and clasping.

Cephalozia Pillansii.

Genus 46. Nowellia Mitt., in Godman's Nat. Hist. Azores, p. 321 ; Schiffner, Pfl., i, 3, 97 ; Steph., Sp. Hep., iii, 347 ;
Macvicar, p. 271.

=*Cephalozia* Dum., Rec. d'Obs., p. 18.

A small genus formerly included in *Cephalozia* and corresponding with it in every respect except that the upper leaf-margin is inflexed and saccate, forming an incurved lobule in an unusual position. The leaves are narrowly and transversely attached ; they are concave (not conduplicate), 2-lobed, with incurved capillary segments.

---

* *Cephalozia capensis* Sim (sp. nov.).
Minuta ; caules 2–3 mm. longi, 10–12 cellas rotundi, pellucidi, solidi, infra prostrati ; apex suberectus. Folia alte bifida, sinus lunatus, segmenta divaricata, lanceolata, subacuta, integra, cellas fere tres lata infra. Folia floralia multo maiora, in 2–3 paribus, longe-carinata, segmenta trigono-acuta, conduplicata, integra ; foliola desunt alibi, adsunt inter folia floralia, breviter bifida, delicatissima, pauci-cellata, segmentum utrumque tantum 1–2 cellarum turgidarum. Cetera parum cognovi.

† *Cephalozia Pillansii* Sim (sp. nov.).
Caules 2–3 cmm. longi, leviter ramosi, folia superiora minora. Folia sparsa, connexa, breviter adfixa, variantia a prope integris ad bifida ; segmenta rotunda, foliola absunt. Androecium terminale in ramis brevibus basalibus ; bracteae grandes, breviter cordatae, convexae, uno lobo incurvato. Monandra. Refert *Odontoschisma*.

131. *Nowellia curvifolia* (Dicks.), Mitt. (as above), p. 321.
=*Jungermannia curvifolia* Dicks., Plant. Crypt. Fasc., ii, 15, pl. 5 ; Syn. Hep., p. 142.
=*Cephalozia curvifolia* Dum., Rec. d'Obs., p. 18 ; Spruce, On Cephalozia, p. 47.
=*Jungermannia Baueri* Mart., Fl. Crypt. Erlang.

Scattered among mosses or on decayed tree trunks, prostrate, rooting ; stems 1–2 cm. long, pellucid, sparingly branched ; leaves spreading from the stem but ascending further out, narrowly attached, concave, 2-fid, the segments ending in incurved capillary points, but a third lobe without point is formed by the inflexed saccate upper leaf-margin. Folioles absent. Cells quadrate, 20μ diameter. Inflorescence cladogenous, perianth oblong, trigonous, with ciliate

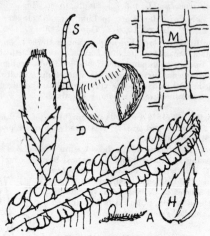

Nowellia curvifolia.
S, apex of leaf-segment.

mouth. Floral leaves large, bifid, acute, with reflexed dentate margins. Androecium terminal on short branches, bracts small, 2-lobed, imbricate, conduplicate. Forest districts ; rare.

(Genus Psiloclada Mitt. comes in here, and Stephani, in his Sp. Hep., iii, 549, in dealing with *P. cladestina* Mitt. (Fl. N.Z., ii, 143), gives New Guinea and Amboina as localities and also " Cape Town, Wilms," which latter record I have failed to verify, and believe to be a mistake. I believe that a species of *Lepidozia* (having incubous leaves) has been mistaken for it, as the difference between incubous and succubous is sometimes not very evident there.

Psiloclada has short, pinnately branched stems having small, distant, quadrate, *succubous* 4–5-cleft leaves and similar folioles, short lateral fruiting branches, and crowded, large, almost spinose, involucral leaf and foliole segments and cylindrical perianth. The genus has another species in Australia and one in Brazil. *P. cladestina* is described as having slender 2-pinnate stems 3 cm. long, distant spreading leaves 4–6-fid to mid-leaf, with spreading setiform laciniae, incurved at the apex, incurved folioles as wide as the stem and hardly less than the leaves, and 4-lobed for two-thirds of the leaf.)

Genus 47. Lepidozia Dum., Rec. d'Obs., p. 19 ; Syn. Hep., p. 200 ; Spruce, p. 357 ; Schiffner, Pfl., i, 3, 102 ;
Macvicar, p. 321.

Leaves incubous, oblique, deeply cut into two to six long narrow segments ; folioles similar but smaller.

Plants varying from minute to several inches high, the larger usually pinnately branched, the branches lateral, and often attenuated and small-leaved toward the apex, and small-leaved postical flagellae also occur. Leaves usually cut half-way or more, or to the base, transversely inserted. Inflorescence on short postical branches, perianth surrounded by crowded, much-cut bracts. Androecia various.

Synopsis of Lepidozia.

Leaves cut less than half-way—cells few.
 132. *L. reptans* var. *tenera.*
Leaves asymmetrical, cut half-way or more, segments incurved.
 133. *L. cupressina.* Segments four, bluntly triangular, entire.
 134. *L. truncatella.* Segments four, subulate, often dentate.
 135. *L. natalensis.* Small, segments four to five, expanded ; folioles 3-fid.
Leaves cut to the base into capillary segments (*Blepharostoma*).
 136. *L. succulenta.* Leaves 3-fid, narrower than the succulent stem.
 137. *L. capillaris.* Segments three, hardly capillary, strongly incurved.
 138. *L. nematodes.* Segments capillary, spreading.
 139. *L. bicruris.* Tender, segments usually two, capillary.

132. *Lepidozia reptans* (Linn.) Dum., Rec. d'Obs., p. 19; Macvicar, p. 323; Syn. Hep., p. 205.

=*Jungermannia reptans* Linn., Sp. Pl., p. 1133.
=*Pleuroschisma reptans* Dum., Syll. Jung., p. 69.
=*Herpetium reptans* Nees, Eur. Leb., iii, 31.

Monoecious; plant small, stems prostrate, strong, 2–3 cm. long, tender, pinnately branched, the branches often flagellate and small-leaved toward the apex. Leaves seldom in contact, obliquely inserted, nearly flat, 3–4-cut at the apex, the teeth triangular, about four cells wide at the base, subacute at the apex, the cells quadrate, separate, in straight lines; folioles erect from the stem, 4-fid to one-third depth, the teeth one to two cells wide, those on branches often 3-fid. Perichaetium cladogenous, floral leaves crowded, large, rounded, but 3–6-dentate, foliole oval, dentate. Androecium on a short postical branch.

The South African plant is var. *tenera*, which is more slender than the type.

Frequent among mosses in the forests of Natal and Kaffraria (Sim 248).

Stephani confines *L. reptans* to the Northern Hemisphere and makes no provision for this Cape form, though his Congo *L. ubangiensis* comes near it (Sp. Hep., iii, 561 (1909)).

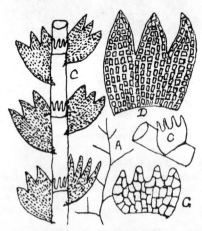

Lepidozia reptans var. *tenera*.

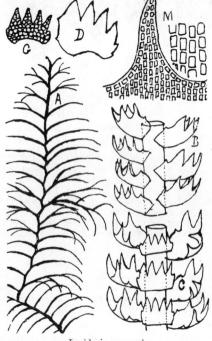

Lepidozia cupressina.

133. *Lepidozia cupressina* (Sw.), Ldbg., Syn. Hep., p. 207; Steph., Sp. Hep., iii, 576.

=*Jung-cupressina*, Swartz, Prod. Fl. Ind. Occ., p. 144.

Plant several inches high, pinnately or sub-bipinnately much-branched, when prostrate an occasional new stem rises from it and repeats the growth, but usually the growth is suberect or erect. Branches distichously arranged, similar to the main stem where they leave it, but each attenuated and smaller-leaved toward its apex where segments get down to two cells wide. Leaves transversely inserted, approximate, oblique, 4-fid to or near the middle, the upper segments the larger, the segments triangular, bluntly acute, entire, 6–7-celled across the base; the leaves convex, spreading directly from the stem but incurved further out, and with all segments incurved, and often they have one or two small lateral teeth near the base. Folioles patent, usually 4-fid half-way, segments mostly two cells wide below, 1-celled at the bluntly pointed apex. Leaf-cells large, quadrate, in rows; cells of folioles smaller and rounder. Perianth cylindrical, mouth dentate.

L. cupressina occurs in tropical and South America, tropical Africa, and is common on the top of Table Mountain, Cape, and present on other mountain sites eastward. Too near it are *L. laevifolia*, Syn. Hep., p. 208, of Australia, and *L. asperifolia* Steph. of New Zealand.

134. *Lepidozia truncatella* Nees, Syn. Hep., pp. 209, 716; Steph., Sp. Hep., iii, 563.

=*Jungermannia cupressina, β capensis* L. and L., in Lehm., Pug., pl. iv, p. 39.

Stems 1–3 inches long, suberect, pinnately many-branched, the branches flagellate at the apex. Leaves concave, 4-fid to the middle, obliquely trapezoid, the segments triangular below but subulate toward the apex, longer and narrower than in *L. laevifolia*, and the segments are often 1–2-toothed per segment and several-toothed at the rounded base of the upper segment. Foliole much more ciliate than in *L. laevifolia*, or than its own leaves, especially at the base of the upper segment. Cilia more abundant toward the apex of main stem ; everywhere the leaves are more incurved than in that species.

The plants and leaves in *L. laevifolia* and *L. truncatella* are much larger than in any of the other species.

Common on Table Mountain and occurs on eastern mountains. Nees, Syn. Hep., p. 209, separated a Cape plant as var. *minor*, characterised as having the branches not attenuated and the leaves and folioles more denticulate, but I find no plant answering that description.

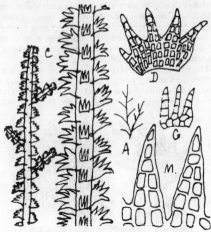

Lepidozia truncatella.   *Lepidozia natalensis.*

135. *Lepidozia natalensis* Steph., Sp. Hep., iii, 562.

Stems 1–3 cm. long, firm and straight, branched freely at nearly right angles, stems and branches being very regular in narrow linear form, caused by the small short leaves being of equal length, fully expanded, palmately 4–5-fid half-way, the segments 2-celled below, 1-celled above ; folioles narrower than the stem, usually deeply 3-fid, the segments of about five cells on end. A much smaller plant than the previous two species, more regular, the leaves nearly flat, and flagellae absent.

Table Mountain, Cape Town, Pillans 4900, p.p. (*Leptotheca*) ; Eastern Transvaal (Sim 8345) ; Perie Forest, Kaffraria, 1888 (Sim 8620) ; South Africa (Ecklon, Rehmann).

136. *Lepidozia succulenta* Sim (new species).*

Stems 2–4 cm. long, twelve to fifteen cells round, growing in dense masses and branching pinnately. Leaves not so wide as the stem, 3-fid to or nearly to the base ; segments capillary, of about five cells on end, incurved, and consequently hardly showing on the thick stem. Folioles usually present, 2–3-fid : segments of three to four cells on end.

May be a juvenile form ; if so, of what ? Slongoli and elsewhere on Table Mountain (Sim).

137. *Lepidozia capillaris* (Sw.), Ldbg., Syn. Hep., p. 212.

=*Jungermannia capillaris* Swartz, Prod. Fl. Ind. Occ., p. 144 ; Lehm., Hep. Capens., in Linnaea, iv, 364.
=*Jungermannia hippurioides* Tayl., Hep. Antarctic, in Lond. Journ. of Bot., 1844, p. 387.

Stem 1–2 cm. long, prostrate on soil, irregularly branched, the branches divergent. Leaves approximate, 3–4-fid almost to the base, the segments inflexed, subulate, two cells wide at the base, one cell wide upward, the folioles much smaller, 2–3-fid, the segments capillary, about five cells on end. Perianth with ciliate mouth and floral leaves ciliate with the margins denticulate.

Table Mountain (Sim 8496, 9716, 9717, 9815, etc.). *L. setacea* (Web.) Mitt. was included in my list in error for this species.

Stephani (Sp. Hep., iii (1909), 560) separates the South African plant under the name *L. tabularis* St. (=*L. capillaris* Lindb. p.p.) on geographical grounds only, and also records *L. Stephani*, Ren. Bot. Gaz., 1890, p. 287, from Cape Town (Ecklon) as well as from the Mascarene Islands, with leaves 3–4-trifid, which I cannot separate in Cape specimens.

---

* *Lepidozia succulenta* Sim (sp. nov.).

Caules 2–4 cmm. longi, 12–15 cellas rotundi, densis cumulis provenientes et pinnatim ramosi. Folia non tam lata quam caulis, 3-fida usque vel prope ad basin ; segmenta cellarum fere quinque alius super aliam, incurvata, atque inde vix apparentia in caule crasso. Foliola plerumque adsunt, 2–3-fida, segmenta 3–4 cellarum alius super aliam. Fertilem haud cognovi, fortasse species iuvenilis, sed cuius incertum.

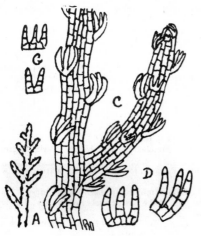

Lepidozia succulenta.

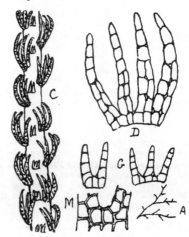

Lepidozia capillaris.

138. *Lepidozia nematodes* (Aust.), Howe, N.Y. Bot. Gard., 1902, p. 284 ; Steph., Sp. Hep., iii (1909), p. 559.

> =*Cephalozia nematodes* Aust., Torr. Bot. Cl., 1879, p. 302.
> =*Jungermannia nematodes* Gottsche, in Wright's Hep. Cub.
> =*Lepidozia chaetophylla* Spruce, Trans. Bot. Soc. Edin., p. 365 (1885).

Stems 1–3 cm. long, very tender and fragile, slightly branched, the branches lateral, divaricate. Leaves distant, patent, usually 3-fid to the base, branch-leaves mostly 2-fid ; the segments setaceous, of six to eight cells on end, the upper cells the smaller and acute ; branch-leaves smaller. Folioles much smaller, 3-fid to the base on stems, 2-fid on branches, the segments capillary, of three to four cells ; cells elliptical. Perichaetium cladogenous ; floral leaves and folioles crowded, the segments numerous, setaceous, nearly as long as the cylindrical perianth. Monoecious ; androecium a short branch with few but crowded bracts, ciliate and monandrous. Pearson (Hep. Nat., p. 8 and Tab. V) describes the Natal plant as *L. chaetophylla* Spr. var. *tenuis* Pears., and in finding a character for the variety, says : " In the type the stems are thicker, often bipinnate, with shorter leaves closer together, perianth sometimes terminal on chief stem." I find no sufficient difference.

Among mosses, Table Mountain and eastward, often on soil, confervoid in apparance, and never matted.

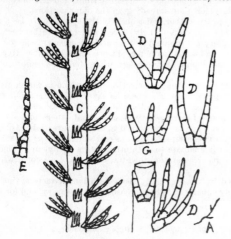

Lepidozia nematodes.

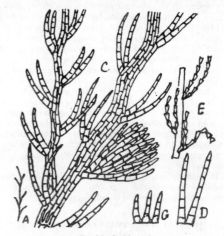

Lepidozia bicruris.
E, appearance of dry leaves.

139. *Lepidozia bicruris* Steph., Hedw., 1885.

Minute, confervoid, to 1 cm. long, irregularly branched, the stems about eight cells round of quadrate pellucid cells. Leaves patent or somewhat incurved, 2-fid to the base, segments capillary, six to ten cells long, the upper cells the smaller, and all cells contracted and bead-like when dry, regaining substance after some exposure in water.

Folioles much smaller, often absent, 2–3-fid to the base, segments three to five cells long. Dioecious. Perianth cylindrical, enclosed by crowded segments of 5–7-fid floral leaves.

Frequent, usually on soil among grass, often on soil which is very dry for several months (Sim 8169, 8205, etc.). Why Stephani omits this from his Sp. Hep. is not evident.

Family 15. LOPHOZIACEAE. Leaves succubous or transversely inserted (not incubous), entire or dentate or 2-lobed. Folioles often absent or rudimentary. Perianth when present either laterally compressed, cylindrical or ovate, or if triangular in cross-section having one angle of the triangle on the lower surface. Inflorescence usually terminal on stem or larger branches, except in *Chiloscyphus.*

Synopsis.

§ 1. Leaves entire or slightly emarginate, succubous, alternate except in *Gongylanthus.*
    48. *Jamesoniella.* Leaves decurrent, usually face to face. Folioles usually absent. Perianth free, terete plicate, contracted at the mouth.
    49. *Notoscyphus.* Leaves not decurrent ; folioles present. Perianth forming an erect marsupium, sometimes bulbous at the base. Rhizoids colourless.
    50. *Nardia.* As in *Notoscyphus,* except : rhizoids purple, folioles minute or absent, when present bifid.
    51. *Gongylanthus.* Leaves entire, opposite, somewhat connate at the base, closely imbricated ; folioles absent. Marsupium succulent, long, descending into the soil.
    52. *Chiloscyphus.* Larger ; leaves entire or slightly emarginate. Folioles bifid ; inflorescence on short lateral branches. Perianth campanulate, with wide 3-lobed mouth.
§ 2. Leaves succubous, more or less longitudinally inserted, simple, not concave, usually more or less toothed or spinose.
    53. *Lophocolea* Perianth 3-angled ; leaves with two to three large tapering teeth.
    54. *Tylimanthus.* Marsupium solid, descending. Leaves various, truncate to 2–3-toothed. Folioles present, small, subulate.
    55. *Leptoscyphus.* Leaves subopposite, entire, or with two to three teeth. Foliole attached to the adjoining leaf, bifid and toothed or ciliated. Perianth terminal, laterally compressed.
    56. *Plagiochila.* Leaves decurrent Folioles absent, perianth terminal, laterally compressed.
§ 3. Leaves concave, 2-lobed, transversely inserted.
    57. *Marsupella.* Leaves complicate-concave, folioles absent, marsupium erect.

The genera which compose the first group, *i.e. Jamesoniella, Notoscyphus, Nardia,* and *Gongylanthus,* together with *Alobiella* and *Kantia,* comprise, so far as South Africa is concerned, a series of very similar plants forming pioneers on mountain soil, mostly gregarious and prostrate in habit though seldom densely caespitose, the larger species connecting through *Chiloscyphus* with the bifid-leaved genus *Lophocolea.* The tendency of the prostrate stems to arch for a portion, then become prostrate and rooting by long strong radicels, then arch again, is common to most of these small pioneers, as are the simple, rounded, succubous leaves, usually somewhat decurrent in front, and generally erect and meeting on the younger portions, and either similarly placed or more spreading on the older portions.

The limits of these genera have been changed by almost every author, consequently the species have been repeatedly transferred from one genus to another, not because they had been wrongly placed earlier, but rather because the limits of some genus have changed. That is still going on, and no final and natural arrangement can yet be claimed. *Lindigina* and *Notoscyphus* were formerly included in *Acolea* and in the recent work—Stephani's Sp. Hep.— the South African species are placed in *Notoscyphus, Jamesoniella,* and *Gongylanthus,* while *Acrobolbus* approaches our borders, and the generic names *Nardia* and *Lindigina* sink.

The presence of a marsupium in some of these and other genera forms a connecting link, which has been given high place by some, but Spruce deals well with this in On Cephalozia, p. 92, where he says : "*It is probable that there is not in Nature any separate tribe of pouch-fruited Jungermanniaceae* (=*Marsupiocarpeae*=*Geocalyceae*=*Saccogyneae*), but that almost every tribe may have a genus (or genera) of marsupial species, and that, where none such is known to exist, it is either because it has hitherto eluded our search, or has succumbed to other plants in the struggle for place, or has not yet been evolved. The transitional stage, between supraterraneous and subterraneous perianths, is to be found in those genera whose floral whorls are more or less adnate to each other into a fleshy cup, which is apt to become turgid and gibbous at the rooting base. A further extension downward results in a pouch, which buries itself in the matrix."

It must be noted that even the generic distinctions used by me, especially the vegetative characters, though convenient for separation of genera in a local flora, do not apply equally or fully in a cosmopolitan survey, and for the latter further investigation is still desirable.

Genus 48. JAMESONIELLA Schiffner, Pfl., p. 82 ; Steph., Sp. Hep., 1901, p. 1024 ; Macvicar, p. 149.

=*Jungermannia,* subgenus *Jamesoniella,* Spruce, Journ. of Bot., pp. 26–29, 1876.

Dioecious ; stems simple or slightly branched, ½–2 inches long, suberect, rising from an underground stem, growing in mats on soil or on or among fully exposed mosses, often bright red, wiry, incurved at the apex, with numerous strong uncoloured radicels in the lower portion, fewer upward, and occasional stolons from the under side. Innovations occasional, from below the perianth. Leaves alternate, entire, round or nearly round, succubous, obliquely inserted, so somewhat decurrent, those on the younger portions or often on the whole plant standing erect, face to face, almost flat, those on the older portions often spreading ; folioles usually absent except in the involucre ; cells round, with distinct trigones or thickened angles. Inner floral leaves small, much cut, outer as large as or rather larger than the stem-leaves, entire, or with one to several teeth ; floral folioles present, large, lacerate. Perianth oblong, free, erect, not forming a marsupium, several or many-ribbed upward. Androecia at first terminal, then by continued growth they occupy the mid-stem ; bracts ventricose, with an inflated lobule.

Synopsis.

140. *J. colorata.* Perianth pluriplicate, its mouth subentire.
141. *J. Rehmannii.* Perianth 3-plicate, its mouth lobed, the segments acute, incurved.

140. *Jamesoniella colorata* (Lehm.), Spruce, Journ. of Bot., 1876, p. 30 ; Steph., Sp. Hep., 1901, p. 1025 ; Schiffner, Pfl., i, 3, 83.

=*Jungermannia colorata* Lehm., Linn., iv, 366 ; Syn. Hep., p. 86.

Stem usually suberect, 1–2 inches long, straight, with the apex incurved, growing in loose cushions or among mosses and rooting by strong radicels below, few above ; occasionally ½ inch long, prostrate, and rooting more. Stems wiry, usually simple or once innovated below an inflorescence, with also occasional small-leaved succulent basal flagellae ; leaves alternate, crowded, imbricate, succubous, somewhat obliquely inserted, or subtransverse, semi-amplexicaul, almost round, entire, somewhat decurrent, those in the younger parts erect and meeting face to face, those in older parts sometimes so but often spreading ; leaf-surface rough, cells rounded, 30μ diameter, with thickened walls and evident trigones ; the marginal cells smaller, more quadrate, and often more highly coloured ; the basal cells larger and more elliptical. Folioles absent except in the inflorescence. Upper leaves rather larger than others, entire or few-toothed. Floral leaves smaller, included, deeply 5–6-laciniate with lanceolate ciliate segments ; inner folioles laciniate and somewhat connate to the leaves. Perianth obovate, erect, almost exserted, the upper half many

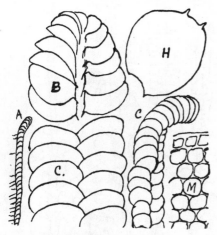

Jamesoniella colorata.

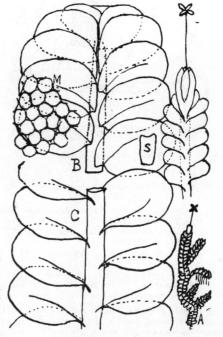

Jamesoniella Rehmannii.

S, involucral foliole.

ribbed (nine to ten), with small entire mouth. Androecium at first terminal, then median, with few, obovate, adpressed, concave, monandrous bracts.

Very common on Table Mountain and other western mountains (rarely fertile), and absent from eastern localities. Widely distributed in South America, Australasia, etc.

141. *Jamesoniella Rehmannii* Steph., Sp. Hep., 1901, p. 1035.

=*Jungermannia Rehmannii* Steph., Hedw., 1892, p. 123, Tab. 5.

Dioecious ; a straggling, prostrate, slender plant, 2 inches long, branched, the branches rising from the upper axils of leaves, the branches arched and rooting at the tips, the lower leaves spreading, the upper leaves spreading or more usually approaching face to face on the upper side. Leaves alternate, imbricate, obliquely inserted, slightly decurrent in front, shortly inserted on the back, the leaves there overlapping in one curved line. Cells round, 20–30μ diameter, hyaline, the basal cells largest, the marginal cells smaller, the leaf-surface smooth. Folioles absent except in the involucre where one or two cuneate-truncate or 2–3-toothed folioles appear. Floral leaves larger, erect, adpressed. Perianth cylindric, 3-plicate ; mouth lobed, lobes incurved, entire, acute, the mature perianth dividing into about ten teeth. Androecia median on the stem, bracts 2-lobed, the front lobe not connected, its apex 2-spinose.

Eastern Transvaal. Originally described from specimens from Lydenburg District, mixed in one of Rehmann's moss-numbers, where Stephani mentions the foliole as minute, 3-fid to the base ; segments of about three cells on end, but I have failed to find any stem foliole.

J. Rehmannii was introduced into my list as *Aplozia Rehmannii*, but I now follow Stephani in placing it under *Jamesoniella*, and as *Aplozia caespiticia* Dum. was included there in error the genus *Aplozia* now drops out of the South African list.

Genus 49. Notoscyphus Mitten, 1862 ; Steph., Sp. Hep., 1901, p. 171, p.p.

Leaves entire, not decurrent, the cells containing oil bodies. Folioles present, lanceolate, subulate or bifid. Perianth included in and concrete with the involucral bracts and the hollowed-out stem-apex, forming an erect, succulent marsupium, sometimes having a succulent bulb at the base. Rhizoids white. Plants small, prostrate, greenish to red.

Synopsis.

142. *N. vermicularis.* Leaves truncate ; folioles bifid and toothed.
143. *N. lutescens.* Leaves rounded ; folioles of two setaceous segments, or these united.
144. *N. natalensis.* Leaves rounded ; folioles linear, setaceous.

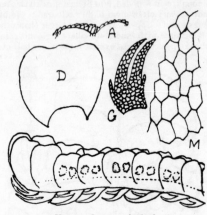

Notoscyphus vermicularis.

142. *Notoscyphus vermicularis* (Lehm.), Steph., Sp. Hep.,
1901, p. 172

=*Jungermannia vermicularis* Lehm., Linnaea, iv, 361 (1829) ;
     Pug. Pl., iii (1831), 47.
=*Jungermannia flexuosa* Lehm., Linnaea, iv, 361 (1829) :
     Pug. Pl., iii, 47 (1831).
=*Alicularia vermicularis* Lehm., Syn. Hep., p. 11.
=*Alicularia flexuosa* Nees, Syn. Hep., p. 11.
=*Notoscyphus variifolius* Mitt., Linn. Soc. Journ., xvi, 91,
     188, and pl. iv ; Pears., Hep. Nat., p. 18 and pl. 12.
=*Notoscyphus flexuosus* (Nees), Sim, S.A. Hepaticae, p. 20.

Dioecious, prostrate, arching and rooting at the tips, with colourless rhizoids ; stems ½–1½ inch long, slightly branched ; younger leaves ascending and nearly meeting, the older ones lying more flat, rather scattered, varying in form, the lower leaves oblong, longer than wide, with rounded or truncate apex, the upper leaves wider than long, with apex 2–3-waved. Upper cells round, 25µ, lower cells elliptical-hexagonal, irregularly 30µ long, thin-walled, without trigones. Folioles nearly as long as the leaves, arched toward the stem, bifid to the middle, sometimes 1-toothed at each side, many-celled, the main segments each about four cells wide where they unite. Inflorescence terminal ; floral leaves longer than others, and more deeply waved and emarginate ; floral folioles large and toothed. Other features not seen. Common from **Table Mountain** to Natal on deep red soil, usually decomposed dolerite. The large many-celled folioles are very characteristic.

     Edendale, Natal (Sim 9801), etc.

Notoscyphus lutescens.
(S, SS, and SSS, see text page 95.)

143. *Notoscyphus lutescens* (L. and L.), Mitt., Fl. Viti., 1862, p. 407 ; Pears., Hep. Nat., p. 16 and pl. xi.

=*Jungermannia lutescens* L. and L., Pug., iv, 16.
=*Gymnomitrium lutescens* G., Syn. Hep., p. 4.
=*Acolea, fide* Stephani, Sp. Hep., 1901, p. 141.

Dioecious, prostrate in dense patches on sour wet mountain ground, green and with the leaves spreading when grown in shade, brown or purplish and with the leaves erect and meeting when fully exposed. Stems mostly simple, producing numerous long white rhizoids from the stem at the base of the foliole or attached to the foliole ; barren stems and male stems up to 1 cm. long, fertile stems shorter, prostrate or arching, with a solid terminal process on the ground from which the inflorescence rises erect like a very short stem in which the parts are more or less united at the base. Leaves succubous, imbricate, 0·5–0·8 mm. long, when fully exposed roundly quadrate or sometimes retuse, when shaded roundly oval, slightly decurrent at the base, very equal and similar on one plant except the floral leaves and bracts. Cells almost circular, 35μ, with very large trigones sometimes meeting. Several oil vessels are present and persistent in almost every cell when fully exposed, these appearing without order in young parts, but with age approaching in pairs from neighbouring cells with the cell-wall between. On the female plant the older leaves are usually small, but rapidly increase in size toward the inflorescence, the two inner leaves are larger, oval, and bifid, inside these are several small, wide-ciliated, hyaline bracts (fig. S). Inner foliole large with lanceolate segments. Pistilidia numerous (forty to sixty), perianth immersed ; seta 5 mm. long, capsule bursting to the base into four roundly oblong valves which have delicate ephemeral hyaline epidermis and two inner layers of larger cells beautifully marked by red annular bands (fig. SS). After maturity these bands on the outer surface burst and stand up as projections (fig. SSS). Elaters numerous throughout the capsule but inclined to adhere to its surface, 2-spiral, 0·15 mm. long ; spores globose, 15μ diameter, papillose. On the male plant the upper and lower portions have usual leaves, but those in the mid-portion are smaller, separate, hyaline, very concave, somewhat 2-lobed, monandrous. The nearer leaves above and below are usual size and nearly plane, but have an oblong lobe infolded from the base, while the first two leaves above the hyaline antheriferous leaves are usually 3–4-lobed. Folioles present on the younger parts, small, usually deeply bifid into narrow spreading segments of one row of cells each, but also frequently these segments adhere and form an ovate acuminate undivided foliole. Stems continue to extend at the apex, or sometimes produce a slender innovation at the apex, or one or more flagellae from the under-surface near the base of the stem.

This species is much affected in colour, growth, size, position of leaves, etc., in accordance with light and shade, in the same patch ; certainly the same species, but quite fit to make two species from dry herbarium specimens if not known in growth.

The tumid base of the perianth also varies, sometimes evident and producing rhizoids, while on other adjacent stems it is absent ; probably it is only produced when the part is in direct contact with the soil.

Hilton Road, Natal (Sim 8313), etc.

144. *Notoscyphus natalensis* Sim (new species).*

Dioecious, pale green to red, growing in extensive dense mats. Stems ¼–1 inch long, usually simple or sparingly branched, or with a few slender, small-leaved flagellae near the base ; prostrate or repeatedly arched, producing numerous long colourless radicels where in contact with the ground. Leaves imbricate, suberect so as to approach face to face, obliquely inserted, oblong-rotund, 0·8–1 mm. long, longer than wide, rounded or often somewhat truncate at the apex, somewhat decurrent in front, shortly attached on the back, somewhat concave, the leaves beginning a new arch rather smaller than those above and below. Cells round, 25–30μ diameter, hyaline, but with chlorophyllose granules, all alike ; folioles abundant and permanent, simple, lanceolate, composed of about six cells on end, free, standing out from the stem, and thus easily seen, usually present on old as well as on young parts. Androecia at first terminal, afterwards left in mid-stem ; often occupying short basal branches ; bracts similar to other leaves but with a small infolded lobe covering the antheridia which are usually two together. Female inflorescence and perianth not seen, but the general appearance of the plant suggests a more or less developed marsupium.

Found in great abundance on the roadside above English's nursery, Town Bush, Maritzburg, in patches yards across, but all male (Sim 9799), etc.

Genus 50. NARDIA, S. F. Gray, Nat. Arr. Brit. Pl., i, 694.

As in *Notoscyphus* except : rhizoids purple ; folioles minute or absent, when present bifid.

Synopsis.

145. *N. Jackii.* Perianth a fleshy marsupium on or in the soil.
146. *N. stolonifera.* Perianth erect, free.

---

* *Notoscyphus natalensis* Sim (sp. nov.).
Dioica, dense caespitosa ; caules ¼–1″ longi, simplices vel pauci-ramosi supra, flagellati ad basin, identidem arcuati, radicati cumulis rhizoidorum longorum et colore carentium. Folia imbricata, sursum curvata, oblongo-rotunda ; foliola abundantia et permanentia, simplicia, lanceolata, cellarum fere sex alius super aliam, libera, exposita. Androecia terminalia, postea in medio caule, bracteae ceteris foliis similes, plerumque autem lobo implicato ; antheridia 2. Perianthium nusquam vidi.

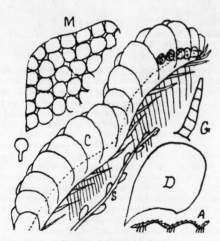

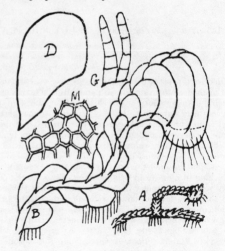

Notoscyphus natalensis.

S, small-leaved flagellae.

Nardia Jackii.

145. *Nardia Jackii* Steph., Hedw., 1892, p. 127, figs. 10, 11, 12.

=*Notoscyphus Jackii* Steph., Sp. Hep., 1901, p. 175.

Dioecious, prostrate, caespitose; stems ½–1 inch long, branched, arching, rooting by long red radicels where it touches the ground, especially under the swollen base of the perichaetium   Leaves roundly ovate, widely attached, somewhat decurrent, spreading, or the younger erect and meeting. Cells large, roundly hexagonal; folioles 2-fid to the base, the segments setaceous. Floral leaves and folioles large, connate, serrate. Perianth included, its bracts small and lacerate. Androecium at first terminal, becoming mid-stem by continued growth, the bracts few, 2-lobed. Monandrous.   Described by Stephani from Spitzkop, Lydenburg, Transvaal (Wilms); found also in Natal.

146. *Nardia stolonifera* Steph., Hedw., 1892, p. 128, figs. 13, 14.

=*Jungermannia stolonifera* Steph., Sp. Hep., 1901, p. 509.

Dioecious, prostrate, caespitose, abundantly stoloniferous; stem to ½ inch long, rooting freely by red rhizoids, the fertile branch short, erect, few-leaved, and with free terminal perianth. Leaves small, spreading, nearly round, slightly decurrent; folioles absent; cells round, 25µ, with small trigones. Floral leaves much larger than others, but similar, somewhat connate to the base of the perianth. Perianth pear-shaped, 3–4-plicate, the lobes incurved, crenulate. Androecia terminal, few-bracted, bracts monandrous.

Eastern Transvaal and Natal.

Genus 51. GONGYLANTHUS Nees, 1836; Steph., Sp. Hep., iii, 383.   Includes *Lindigina* Gottsche, Ann. des Sc. Nat., 4th ser., p. 1, 1864.

Leaves entire, opposite, somewhat connate at the base, closely imbricated; folioles absent.   Marsupium succulent, long, descending into the soil.

147. *Gongylanthus renifolius* Steph., Sp. Hep., iii, 2nd ser., 387.

=*Lindigina renifolia* Mitt., Journ. Linn. Soc., iv, 91, 190.
=*Gongylanthus scariosus* Steph., Sp. Hep., vi, 2nd ser., 389.
=*Jungermannia scariosa* Lehm., Linnaea, 1829, p. 365.
=*Gymnomitrium scariosum* Nees, Syn. Hep., p. 3.
=*Lindigina scariosa* Mitt., Journ. Linn. Soc., pp. 16, 91, 189.

Dioecious, prostrate, very small, gregarious; stems simple or nearly so, ½ inch long or less, closely adherent to the ground by numerous rhizoids, the leaves gradually increasing in size upward and largest above the deeply descending cylindrical fleshy marsupium. Leaves opposite, succubous, reniform or almost round, concave, slightly connate on both sides of the stem, somewhat decurrent in front, closely imbricate and standing so erect on each side that the two lines of leaves meet with only a slight groove between. Cells variable, the lower hexagonal and long, the upper roundly hexagonal, trigones often distinct. Mitten, in Linn. Soc. Journ., pp. 16, 91, 189, describes this under the above name, and redescribes *Lindigina scariosa* Lehm., claiming that they are different, and ends in regard to *L. renifolia*, " Very different from *L. scariosa* in its greater stature and in the imbrication of its leaves (which have no

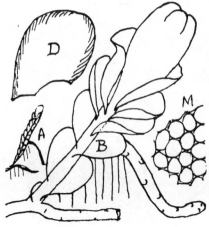

Nardia stolonifera.

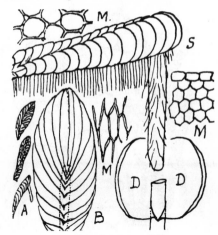

Gongylanthus renifolius

scarious margin) but approaching in the outline of the leaf." I fail to find two species, but I find specimens answering both descriptions in each patch, and so conclude they are identical. The name *Jungermannia scariosa* is the older and has priority, but as Stephani points out in regard to *Gongylanthus scariosus* that name is altogether unsuitable since the leaves are not naturally scarious but only become so by decay, and I prefer to sink it and use the name *G. renifolius* to cover the single species. Mitten (p. 190) refers to fragments of another which may be *L. prostrata* G., and on this it was included in my list, but the identification is too incomplete to allow it to stand.

Gongylanthus renifolius is a pioneer in dry country and is found on Slongoli and Signal Hill, Cape Town (Sim 9798), Stellenbosch Flats, etc., but not eastward.

Genus 52. CHILOSCYPHUS Corda, in Opitz, Beitr., i, 651 (as *Cheiloscyphos*); name corrected, Dum., 1831; emend. Schiffner, Oest. bot. Zeit., No. 5 (1910); Stephani, Sp. Hep., iii, 2nd ser., 686; Macvicar, p. 239.

Plants usually larger than the previous genera, and with larger cells; simple or branched, branches lateral and rhizoids colourless. Leaves succubous, usually alternate, rounded or slightly emarginate; margins entire or toothed, folioles bifid, free or connate to the leaves, segments often with a lateral tooth. Inflorescence on very short lateral branches; perianth campanulate with wide 3-lobed mouth. Not well distinguished from *Lophocolea*, in which the leaf-apex has two to three large teeth, but that is sometimes the case with the lower leaves in *Chiloscyphus*, where the upper leaves are all rounded, and the same condition occurs in *Lophocolea* and in *Leptoscyphus*.

Synopsis.

(See young condition of *Leptoscyphus Bewsii*.)

148. *C. expansus.* Leaf round, margin entire.
149. *C. Lindenbergii.* Leaf rounded, longer than wide, margin entire.
150. *C. dubius.* Leaf oblong, margin entire.
151. *C. lucidus.* Leaf much narrowed to the apex, margin almost entire.
152. *C. Rabenhorstii.* Leaf round, margin angled or toothed.
153. *C. fasciculatus.* Leaf round, margin ciliate-dentate.

148. *Chiloscyphus expansus* Nees., Syn. Hep., p. 179; Steph., Sp. Hep., iii, 2nd ser., 850.

=*Jungermannia expansa* Lehm., Linnaea, 1829, p. 361.
=*Jungermannia congesta* Lehm., Pug. Pl., iii, 51.

Dioecious; usually prostrate on soil or on wet rocks, or in masses on rocks in running water under trees. Stems 1–2 inches long, firm, irregularly branched, rooting from rhizoid branches just below the folioles or occasionally from the lower leaf-surface. Leaves spreading, succubous, slightly convex, imbricate, 1–1·5 mm. diameter, round or roundly quadrate or occasionally some leaves emarginate, both upper and lower base rounded to the stem, each base half round the stem, almost longitudinally inserted. Cells irregularly 5–6-angled, 20μ diameter, thin-walled, usually without trigones, the marginal line narrow and oblong, the next line quadrate. Folioles present on the younger branches, usually gone from older parts, small, free, somewhat tumid, deeply bifid, with lanceolate inflexed lobes entire or 1–2-dentate, the segments usually setaceous. Androecia in mid-stem, bracts about five pairs, smaller than the leaves, 2-lobed. The upper leaves on a stem are often more or less erect and meeting.

C. polyanthus Corda was introduced into my list in error on this.

Frequent from Table Mountain eastward.

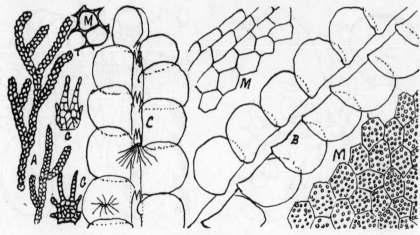

Chiloscyphus expansus.

149. *Chiloscyphus Lindenbergii* Nees, Syn. Hep., p. 187; Steph., Sp. Hep., iii, 2nd ser., 851.

Stems 1–2 inches, prostrate, rooting; the whole plant very similar to *C. expansus* but usually in water, more slender, with leaves longer than wide, more deeply 4-fid folioles and larger cells. The leaf-apex is often more or less obliquely 1-pointed.

It is not known fertile and I doubt if it is distinct from *C. expansus*. In pools, top of Table Mountain, Cape (Sim 9810).

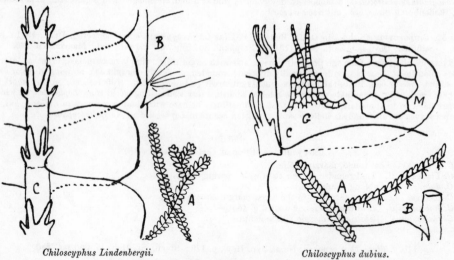

Chiloscyphus Lindenbergii.   *Chiloscyphus dubius.*

150. *Chiloscyphus dubius* Gottsche, Reliq. Rabenh. Abh. Nat. Ver. Bremen, vii, 346; Steph., Hedw., 1893, p. 319; Steph., Sp. Hep., iii, 49 (=682).

=*C. oblongifolius* Mitten, Linn. Soc. Journ., xxiii (1860), 57, Tab. V (not Taylor); Steph., Engl. Bot. Jahrb., 1895, p. 308.

Monoecious, prostrate, on soil or stones; stem several inches long, seldom branched, brown if exposed. Leaves subopposite, 1·5 mm. long, laxly imbricate, spreading oblong-ligulate with rounded entire apex or occasionally with one to two teeth, the upper margin rounded to the stem and connate with the foliole, the lower margin slightly decurrent. Upper cells small, lower large, the marginal line smaller and quadrate. Foliole small, bifid to below the middle and 1-dentate below that; the segments setaceous. Inflorescence on a short branch; inner floral leaves and folioles small, enclosed, more or less cut; perianth campanulate, 3-lobed, the margins ciliate. Androecia spicate, few-bracted; bracts bifid, segments acute. Table Mountain, eastern forests, and Natal.

According to Stephani this species is widely distributed in Africa and its islands.

151. *Chiloscyphus lucidus* (L. and L.) Nees, Syn. Hep., p. 182 ; Steph., Sp. Hep., iii, 2nd ser., 51 (=689).

=*Jungermannia lucida* L. and L., in Lehm., Pug., pl. v, p. 2 (1833).

Dioecious, prostrate, usually on soil, brown when exposed ; stems several inches long, much-branched ; leaves subopposite, very large, 2 mm. long, spreading, imbricate, flat, wide at the base but narrow at the apex which is truncate or emarginate, or sometimes obliquely 1-pointed. Foliole large, 4-fid, connate to the top of the adjoining leaf ; cells large, the lower longer. Androecium spicate, lateral, few-bracted, 2-fid, monandrous, the bract lobule acute.

Table Mountain and other western localities.

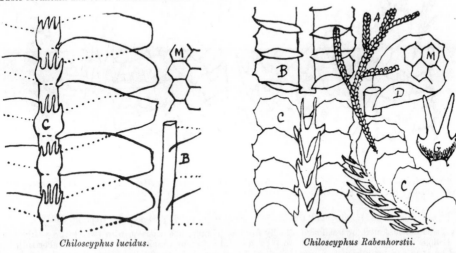

Chiloscyphus lucidus.   *Chiloscyphus Rabenhorstii.*

152. *Chiloscyphus Rabenhorstii* Steph., Sp. Hep., iii, 2nd ser., 51.

Stems several inches long, suberect, branched, usually rising from a prostrate basal stem on which the leaves are more or less bifid ; leaves large, 2 mm. diameter, round, several-angled, the angles hardly visible on the lower leaves, but more evident or even toothed toward the upper part of the stem. Leaves subopposite, imbricate, clasping, convex ; cells roundly hexagonal, thin-walled, 20μ diameter, basal cells much longer ; folioles very concave, shovel-formed, with two long lanceolate segments and two short lateral teeth, the segments parallel to the stem, but the base of the foliole rising at a right angle from the stem.

Sim 8790, 8500, 9530, 9521, and Pillans 3538, found floating on water on Table Mountain, has the leaves on prostrate stems deeply bifid and the folioles bifid, whereas on the upper parts of the same plant the leaves are circular and the folioles toothed as well as bifid.

Abundant on Table Mountain but not seen fertile ; the remarkable folioles are very large and distinct, and the bifid lower leaves resemble those of *Leptoscyphus*.

153. *Chiloscyphus fasciculatus* Nees, Syn. Hep., p. 190 ; Steph., Sp. Hep., iii, 2nd ser., 51.

=*Jungermannia fasciculata* Nees, Hor. Phys. Berol., p. 46, t. 5.
=*Jungermannia Bergiana* L. and L., ined.

Plant erect or suberect, 2–4 cm. high, simple or irregularly branched, or fasciculate above ; branches mostly by terminal innovations ; stems thick and solid ; leaves transversely inserted, very convex, standing out from the stem when moist, much more incurved and somewhat succubous in arrangement and closely imbricate when dry ; numerous and closely placed, circular, connected half-way round the stem above and below, set round the margin with ten to fourteen ciliate teeth of three to six cells length ; folioles very concave, rising out from the stem when moist, imbricate when dry, wider than the stem, 2-fid half-way, with shoulders or tooth-like side lobes, the two main segments bluntly lanceolate or sometimes roundly ovate. Leaf-cells large, irregularly hexagonal, rising some distance in the centre in corresponding form like a crystal ; marginal and teeth cells more quadrate-rounded. Male cone elliptical, closely 8–10-bracted, lateral, sessile, the bract margins entire or slightly ciliated. Floral leaves and perianths not seen. The plant is unique in bending backward toward the under-surface so that the toothed leaf-margins—which approach—almost meet over the upper surface, while on the under-surface (which often stands uppermost) the leaves imbricate away and the folioles stand out or imbricate as a raised central ridge. A very remarkable plant.

Window-gorge Waterfall, Table Mountain, Cape (Sim 9782) ; also present mixed in *Adelanthus unciformis* in Pillans 4079, from south slopes, upper plateau, Table Mountain ; Table Mountain, Brunnthaler (1913, i, 17). This plant varies immensely, and it appears to me that there is no permanent distinction between the three varieties—

α, ramosior, β, Dregeana, and *γ, exarida,* given in the Syn. Hep., p. 190, and that *C. semiteres* L. and L. (Syn. Hep. = *Jungermannia semiteres* Lehm., Hep. Cap., in Linnaea, 1829, iv, 363 ; Pug., iii, 49, 190) belongs to the same group of

Chiloscyphus fasciculatus.

forms or conditions. It is, however, included separately by Stephani, Sp. Hep., iii, 304, in *Lophocolea,* section A, *integrifolia,* for which I see no reason. See my remarks under *Lophocolea Rehmannii.*

Genus 53. LOPHOCOLEA Dum., 1835 ; Syn. Hep., p. 151 ; Steph., Sp. Hep., iii, 536.

Leaves succubous, more or less longitudinally inserted, simple, not concave, with two or three large tapering teeth. Folioles bifid, with a lateral tooth on each segment ; free or connate with the leaves. Perianth usually terminal on stem and main branches, tubular below, 3-angled and shortly 3-lobed above, the angles often winged. The plants are mostly large, scattered among mosses or laxly caespitose on the ground. There is no good difference between *Chiloscyphus* and *Lophocolea,* except that in the former the leaves are rounded at the apex, while in *Lophocolea* they are usually deeply 2-fid, or at least truncate. But as has been shown, some species of *Chiloscyphus* answer *Lophocolea* in the lower portion and *Chiloscyphus* in the upper, and it is also the case in *Leptoscyphus Bewsii* that the older leaves are those of *Lophocolea* or *Leptoscyphus* while the younger growth is that of *Chiloscyphus.*

L. *Rehmannii* St. and *L. cambouena* Steph. appear from description to belong to *Chiloscyphus.* With sterile specimens a difficulty also arises in regard to *Leptoscyphus,* which though distinguished by its laterally compressed perianth has no distinguishing character when the perianth is absent, as is often the case. The foliole is connate with the leaf in *Leptoscyphus,* and only sometimes so in *Lophocolea,* and is sometimes so in *Chiloscyphus* also.

Concerning *L. Elliotii* Steph., recorded from Newlands, C.P., in Brunnthaler, 1913, i, 17, I have no further information.

Synopsis.

154. *L. bidentata.* Leaves 2-fid, entire, points short.
155. *L. setacea.* Leaves 2-fid, entire, points setaceous.
156. *L. muricata.* Leaves 2-fid, ciliate and muricate.
157. *L. diversifolia.* Leaves 3-fid ; younger leaves 2-fid.
158. *L. Rehmannii.* Leaves quadrate-rotund, emarginate.
159. *L. Cambouéna.* Leaves ovate, apex obtuse or repand, on branches bidentate.

154. *Lophocolea bidentata* Nees, Hep. Eur., ii, 327 : Syn. Hep., p. 159 ; Macvicar, p. 233.

=*Jungermannia bidentata* Linn., Sp. Pl., ed. 2, ii, 1598.

Dioecious. Plant 1–2 cm. long, prostrate, simple or slightly branched, light green, flaccid, pellucid ; rhizoids few upward, rising from the back or base of the foliole. Leaves about 1 mm. long, succubous, with long insertion ; slightly imbricate, ovate-oblong, asymmetrical, the lower base descending and somewhat decurrent, the upper margin shorter and more rounded, the apex having two long triangular-acuminate lobes with large rounded or bluntly angular sinus between ; margins entire ; cells roundly hexagonal, 25–30μ diameter, somewhat irregular with thin walls, which sometimes appear thick through being coated with chlorophyll. Folioles free, wider than the stem, cut two-thirds into two subulate lobes, each having a short subulate tooth on its outer margin giving the foliole a 4-lobed appearance, the lobes one cell wide at the apex, two to four cells wide below, the teeth one cell wide throughout ; perianth lateral, sessile, 3-angled, the angles without wings, mouth laciniate ; involucral leaves bifid and toothed, large ; folioles bifid. Androecium terminal, spicate, few-bracted ; bracts imbricate, 2–3-lobed.

Frequent on soil throughout eastern South Africa.

Nees describes several varieties, among which he calls a Cape plant var. δ *capensis* (Syn. Hep., p. 159), having involucral leaves like the others, and leaves not deeply emarginate, which describes the usual condition here.

Stephani includes in Brunnthaler's list (1913, i, 17) *L. difformis* Nees, from Port Elizabeth. It is described (sterile) from Abyssinia, and has no point of difference.

Stephani, Sp. Hep., iii, 2nd ser., 309, makes a new species *L. Macleana* from the Eastern Transvaal which appears to agree, and it is included under that name in Bryhn's list, 1911, p. 5, and in Brunnthaler, 1913, i, 17, from Table Mountain and Newlands.

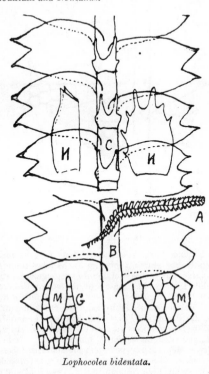

Lophocolea bidentata.

Lophocolea setacea.

155. *Lophocolea setacea* Steph., Hedw., 1892, p. 125; Steph., Sp. Hep., iii, 2nd ser., 308.

Monoecious, straggling on the ground; stems 1–2 inches long, not much branched; leaves alternate, spreading, flat, flaccid when dry, oblong, slightly decurrent, the apex with two long setaceous teeth separated by a large round sinus, the teeth sometimes spreading, sometimes straight or connivent. Cells circular hexagonal, 40μ diameter, basal cells twice as long. Folioles large, 4-fid, the segments setaceous, and the two central segments longest. Perianth terminal, 3-angled and 3-lobed, the lobes bifid and setaceous. Floral leaves like the others. Androecium spicate, bracts small, emarginate, monandrous.

Common on Table Mountain and in eastern districts.

156. *Lophocolea muricata* Nees, Syn. Hep., p. 169.

=*Jungermannia muricata* Lehm., in Linnaea, iv, 363.

Dioecious. Stems 3–10 mm. long, straggling separately among mosses, simple or branched; leaves imbricate, almost flat, sublongitudinally inserted, ovate, bifid with triangular points and acute sinus, ciliated and muricate on the upper surface of the upper half of each leaf, the under-surface without these; the cilia of about five cells on end, gradually less upward. Cells large, 35μ diameter, the basal cells twice as long: Folioles deeply bifid, each segment with a lateral tooth near the base. Perianth puriform, terminal, 3-angled, floral leaves larger, deeply bifid. Androecium terminal, spicate, monandrous.

Table Mountain, Perie Forest, etc.; rare.

157. *Lophocolea diversifolia* Gottsche, Syn. Hep., p. 166; Steph., Sp. Hep., iii, 2nd ser., 482.

Monoecious. Plant prostrate, rooting; stem firm, dark, $\frac{1}{4}$–$1\frac{1}{2}$ inch long, simple or with one or two branches. Leaves flat, spreading or rising to face one another, longitudinally inserted, slightly decurrent, irregularly oblong, usually wider than long, irregularly 2–3-toothed at the apex, those on the lower part of the stem imbricate and with mostly 3-toothed leaves intermixed, those on the branches more scattered, more decurrent, more narrowed to the point and usually 2-fid, and conduplicate with a wide lunate sinus; the teeth triangular, many-celled, the points with about two to three cells on end; cells very thick-walled, roundly hexagonal, large; folioles free, prominently incurved on the branches, deeply bifid, segments linear or lanceolate, entire Rhizoids abundant, colourless. Fructification terminal, the leaves increasing in size up to it and floral leaves usually bifid and toothed. Floral folioles also large and rhizoids abundant in tufts below them.　Perianth ovate, its mouth 3-lobed, laciniate, wings ciliate.　Androecium terminal, spicate, many-bracted ; bracts bifid.

Table Mountain (Sim 8046, 8003, 9523) ; Vryheid (Sim 9526).

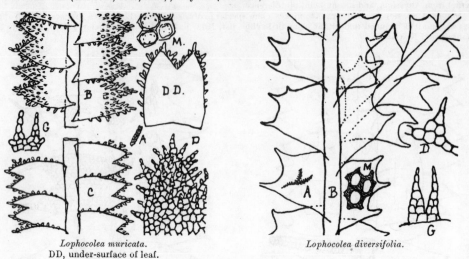

Lophocolea muricata.
DD, under-surface of leaf.

Lophocolea diversifolia.

158. *Lophocolea Rehmannii* Steph., Hedw., 1892, p. 124 ; Steph., Sp. Hep., iii, 2nd ser., 304.

Stephani describes this species, which is unknown to me, as being monoecious, hypogynous ; stems 1–1½ cm. long, slightly branched, with large, crowded, quadrate-rotund leaves, slightly emarginate, 2-lobed, with the lobes rounded and recurved, base shortly decurrent.　Cells very small.　Folioles large, cut to the middle into two segments with lunate sinus ; connate with the leaf.　Perianth terminal, deeply 3-lobed, lobes bifid, margin lobate-incised, laciniate below ; androecium few-bracted, bracts monandrous, the incurved lobule somewhat spinose.　Floral foliole like those of the stem but much larger, highly connate with the leaves.　Molmonspruit, Transvaal, from Dr. Rehmann's intermixed fragments.　He compares it with *L. semiteres* (L. and L.) Steph.=(*Chiloscyphus semiteres* L. and L.) which, he says, is not hypogynous but dioecious.　This group altogether requires further investigation.　In Sp. Hep., p. 302, he includes both *L. Rehmannii* St. and *L. semiteres* (L. and L.) in his section A, *integrifoliae.*

Lehmann and Lindberg's description of *C. semiteres* L. and L. (Syn. Hep., p. 190) is " Stem procumbent, ascending, branched ; leaves semivertical, subconnivent, roundly ovate, entire, obtuse, or slightly emarginate ; folioles free, large, adpressed, bifid, segments lacinulate.　Fruit. . . . ?　' Teufelsberg,' Cape (Ecklon, in Hb. L. and L.)," and he compares it with *Chiloscyphus polyanthus* as having " leaves more imbricate, subrotund, rarely emarginate, repand, cell-formation much smaller and denser."

He also records it from Usambara, East Africa (Brunnthaler, 1913, i, 17).　See my remarks on this under *Chiloscyphus fasciculatus.*

159. *Lophocolea Cambouéna* Steph., Sp. Hep., iii, 2nd ser., 478.

This is another species unknown to me, which, though placed by Stephani in his group C, *heterophyllae* of *Lophocolea*, appears to me to fit better into *Chiloscyphus* as used herein.　His description is " Monoecious, hypogynous, small, densely caespitose on bark.　Stem to 10 mm. long, slender, brown, rigid, slightly branched.　Leaves 1·5 mm. long, densely imbricate, spreading, alternate, flat, hardly decurrent, ovate or ovate-triangular, apex widely obtuse or repand, sometimes repand 2-lobed, the branch-leaves often emarginate-bidentate.　Apical cells 18μ, basal cells 36μ, trigones none.　Folioles small, as wide as the stem, free, transversely inserted, 4-lobed half-way, lobes narrow, acute, the inner the longer.　Inner floral leaves somewhat larger than stem-leaves, shortly inciso-bilobed.　Male bracts conduplicate concave, apex shortly 2-lobed, sinus narrow, lobes obtuse or acute ; floral lobule large, as long as its leaf but half the width, connate to the middle, oblong-acute, undulate ; inner floral lobule shorter than its leaf, free, emarginate 2-lobed half-way, lobes lanceolate-acute, straight, with a strong tooth on each outer margin.　Perianth oblong-obcuneate, deeply 3-lobed, lobes lacerate, without wings.　Molmonspruit, Transvaal (Dr. Rehmann) ; also Madagascar (Camboué)."

He reports it from Van Reenen's Pass, Natal (Brunnthaler, 1913, p. 17).

Genus 54. TYLIMANTHUS Mitten, 1867 ; Steph., Sp. Hep., iii, 2nd ser. (1905), 1129.

Leaves succubous, truncate, or with two or more large teeth at the apex.　Folioles present, small, subulate. Stem suberect, with rhizoids in lower part.　Marsupium solid, descending.　The marsupium gives this genus its character in this group, and has not yet been seen in our species.

T. Wilmsii Steph., Sp. Hep., vi, 252, from a sterile plant, with obtuse leaves and no mention of folioles, cannot belong to this genus and is unrecognisable as described.　It is from Cape Town.

160. *Tylimanthus africanus*, Pears., Hep. Knys., p. 14 and pl. vi ; Steph., Sp. Hep., v, 2nd ser. (1905), 1139.

Dioecious ; rhizomatous stems prostrate, wiry, branched, ordinary stems erect, ½–1 inch long, branched, leafless below ; leaves largest in mid-stem and reduced toward the apex, succubous, not decurrent, longitudinally inserted, imbricate, quadrate or oblong, with rounded margins, and exceedingly variable at the apex which is widely truncate with 2–3–4-serrate points or crenations, which are usually short and acute, the upper the larger, but as there is a tendency to be bifid the lower is also large sometimes, or it may be rounded or absent ; the reduced upper leaves usually emarginate. Cells small, irregular. Folioles small, subulate, usually connate with the adjoining leaf ; androecium terminal or left in mid-stem, bracts with incurved saccate basal lobule, serrate at the apex and diandrous.

Table Mountain (Sim) ; Knysna (Iversen) ; also Rehmann, *fide* Stephani.

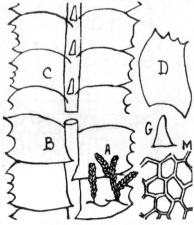

Tylimanthus africanus.

Genus 55. LEPTOSCYPHUS Mitten, Journ. Bot., p. 358 (1851) ; Macvicar, p. 228.

=*Leioscyphus* Mitt., in Hook., Handb. New Zealand Fl., ii, 134 ; Steph., Sp. Hep., iii, 13 (1906), and vi, 253 (1922).
=*Mylius* S. F. Gray, Nat. Arr. Brit. Pl., i, 693 ; Carr., Trans. Bot. Soc. Ed., x, 305 ; and Brit. Hep., p. 66.
=*Coleochila* Dum., Hep. Eur., p. 105.

Leaves succubous, subopposite, imbricate, entire or with two to three teeth ; the older leaves flat, the younger often folding up and meeting ; foliole attached to the adjoining leaf, bifid and toothed or ciliated ; cells large ; floral leaves and folioles usually like others. Perianth lateral or terminal on main stem, erect, inflated below, laterally compressed above, 2-labiate, lips toothed. Stem procumbent, vigorous, with rhizoids from the base of the foliole. Androecia median or terminal, bracts many.

As mentioned under *Lophocolea*, fructification is necessary to separate satisfactorily species of that genus and this, especially as the juvenile and mature leaves sometimes differ considerably.

Synopsis.

161. *L. Bewsii.* Lower leaves wide, irregularly 3-crenate at the apex, younger leaves entire, narrower ; folioles toothed.
162. *L. Iverseni.* Leaves variable from entire to 2–3-crenate ; folioles ciliate.
163. *L. Gottscheanus.* Leaves oblong, truncate-emarginate ; foliole 4-fid ; teeth setaceous.
164. *L. Leightonii.* Leaves oblong, truncate with acute or toothed angles ; foliole small, bifid.
165. *L. Stephensii.* Leaves ovate to oblong, with two long setaceous teeth ; foliole of four setaceous segments.
166. *L. Henkelii.* Leaves oblong, with three to four strong apical teeth ; foliole bifid with many long setaceous segments.

161. *Leptoscyphus Bewsii* Sim (new species).*

Caespitose, 1–2 inches long, prostrate below where it branches freely and has adult leaves, upper stem longer, erect, and with juvenile leaves   Adult leaves quadrate, wide-based, shortly inserted, those on one side of the stem connate with the foliole above, all rounded on the upper and lower sides ; lower base slightly decurrent ; apex varying exceedingly from entire or waved to 3–4-crenate ; cells small, 5–6-sided, thin-walled ; folioles in the lower portion shortly 2-fid, bluntly lanceolate, not toothed, but further up the stem they are as long as the wideness of the leaf, deeply 2-fid, with segments linear-lanceolate and 2-dentate near the outer base.   Leaves on the branches altogether different from those below, oblong with narrowed and rounded entire apex, and the foliole 2-fid with or without a lateral tooth or two.   Androecium terminal on short basal branches, cone-like ; bracts clasping, ventricose, 3–4-crenate at the apex.   The old and the young condition are so dissimilar that I had not thought of connecting them till I found that both were stages of one plant.

Table Mountain, Dr. Bews (Sim 9807), etc.

---

* *Leptoscyphus Bewsii* Sim (sp. nov.).
Caespitosa ; caules 1–2″ longi, prostrati infra, ubi folia adulta ramosque libere praebet. Caulis superior longior, erectus, foliis iuvenilibus, folia adulta quadrata, late basata, breviter inserta, in altero latere caulis connata cum foliolo supra, omnia rotundata supra et infra, basis inferior leviter decurrens, apex admodum varians ab integro vel undulato ad 3–4-crenatum ; cellae parvae, 5–6-gonae, leptodermes ; foliola in parte inferiore breviter bifida, obtuse lanceolata, haud dentata, superius autem in caule tam longa sunt quam latitudo folii, alte bifida, segmentis lineo-lanceolatis ac 2-dentatis prope basin exteriorem   Folia in ramis oblonga, apice attenuato rotundatoque integro ; foliolum bifidum, cui adest vel deest unus et alter dens lateralis.   Androecium terminale in ramis brevibus basalibus. coniformis ; bracteae inter se complectentes, ventricosae, 3–4-crenatae ad apicem.

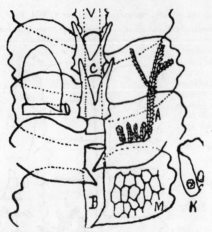

Leptoscyphus Bewsii   *Leptoscyphus Bewsii* (form).

162. *Leptoscyphus Iverseni* (Pearson) Sim.

=*Leioscyphus Iverseni*, Pears., Hep. Knys., 1887, p. 12 and pl. v ; Steph., Sp. Hep., iii, 2nd ser., 220.

Dioecious, at first prostrate, then suberect ; stems slightly branched, 1–2 inches long, rooting below from the base of the folioles, rhizoids in bunches, brown. Leaves imbricate, subopposite, oblong-quadrate, usually rounded and entire but occasionally on the same stems slightly emarginate or 1–2–3-crenate, or with one or two basal cilia. Cells round, 40μ diameter, thin-walled. Folioles very prominent, connate with the leaf below, deeply bifid, the segments setaceous at the apex and ciliate on both margins in the upper part of the stem, bifid without cilia in the lower part. Inflorescence terminal on stem and branches ; floral leaves entire, floral foliole bifid above with acute unidentate segments. Perianth oblong, laterally compressed, with truncate serrate mouth and entire wing. Androecia at first terminal, few-bracted ; bracts rounded, monandrous, unequally 2-lobed, the small lobe undulate. The deeply bifid ciliated folioles are very characteristic.

Knysna (Iversen) ; Uitenhage (Sim 9011) ; Table Mountain (Sim 9519, 9520, 9525) ; Piet Retief, Transvaal (Sim 9795) ; Ngoya, Zululand (Sim 9793) ; Inyanga, Rhodesia, 9796.

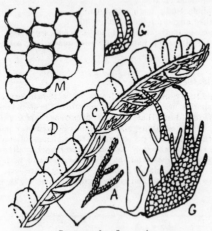

Leptoscyphus Iverseni.

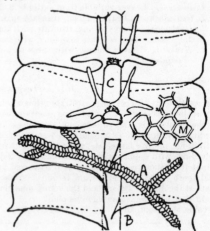

Leptoscyphus Gottscheanus.

163. *Leptoscyphus Gottscheanus* (Lind.) Sim.

=*Leioscyphus Gottscheanus* (Lind.), Steph., Sp. Hep., iii, 2nd ser., 227.
=*Plagiochila Gottscheana* Lindenb., Lehm., Pug., viii, 2 ; Syn. Hep., p. 646.
=*Jungermannia repanda* Schw. (of Sprengel, not *Plagiochila repanda* (Schw.), which still stands as Mauritian).

Dioecious ; stems straggling, 1–2 inches long, green to brown, rooting somewhat, sparingly branched, the branches spreading. Leaves longitudinally inserted, subopposite, spreading, semivertical, imbricate, oblong, the sides parallel, the apex truncate-emarginate with rounded lobes, the base hardly decurrent; cells roundly hexagonal, 40μ diameter, the lower cells longer. Foliole 4-fid, of two erect linear central segments with wide sinus, and two horizontal lateral linear segments, with sometimes a basal tooth each, one side connate with the leaf below. Fruit on a short lateral branch, perianth laterally compressed, with crenulate truncate mouth, and not winged.

Mont aux Sources (Sim 9523); Table Mountain (Ecklon; Sim 9524), etc.; Mascarenes, *fide* Stephani.

164. *Leptoscyphus Leightonii* Sim (new species).*

Stems 1–2 inches long, strong, many-celled; leaves 1 mm. long, very equal, regularly oblong, slightly narrowed to the apex with a small terminal tooth of one to three cells at each corner, and wide shallow sinus or occasionally a third tooth or rounded crenation between. Marginal cells small, square; inner cells large, rounded, thick-walled. Folioles small, shallow, lunately 2-fid with short blunt segments sometimes unidentate; connate with the top of the leaf below.

Buccleuch, Natal, W. Leighton (Sim 9805).

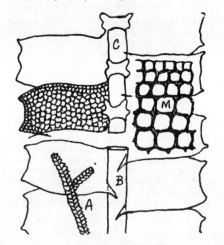

Leptoscyphus Leightonii.

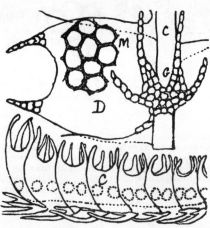

Leptoscyphus Stephensii.

165. *Leptoscyphus Stephensii* Sim (new species).†

Stems prostrate but not rooting, caespitose, 1–2 inches long; leaves on the lower part expanded, and in the upper part facing each other upward. Adult leaves 1·5 mm. long, oval, deeply bifid, with large lunate sinus and almost setaceous points of about six cells on end above what is gradually wider below. Cells large, round, thick-walled, without trigones. Foliole as long as the wideness of the leaf, attached at one side to the leaf below, deeply cut into two setaceous segments, each having a long setaceous tooth near its base. Juvenile leaves and folioles smaller than adult but similar. Androecium terminal on prostrate stem, many-bracted, the bracts diandrous, obliquely oblong, with two setaceous points separated by deep lunate sinus; many leaves produce antheridia, and folioles as above are present in abundance.

Table Mountain. Miss Stephens (Sim 9527, 9528); Stellenbosch (Sim 9529).

---

* *Leptoscyphus Leightonii* Sim (sp. nov.).

Caules 1–2″ longi, validi, pluri-cellati; folia 1 mm. longa, aequissima, regulariter oblonga, tenuiter attenuata ad apicem, dente parvo terminali 1–3 cellarum in utroque angulo, sinu lato brevique, nonnumquam tertio medio dente vel crenatione rotundata. Cellae marginales parvae, quadratae; cellae interiores maiores. Foliola parva, brevia, lunate bifida segmentis brevibus obtusis, interdum unidentatis; alterum latus cum summo folio infra coniunctum.

† *Leptoscyphus Stephensii* Sim (sp. nov.).

Caules prostrati sed non radicantes, caespitosi, 1–2″ longi; folia in parte inferiore expansa, in superiore alterum alteri ex adversa sursum. Folia adulta 1·5 mm. longa, ovata, alte bifida, sinu magno lunato et apicibus paene setaceis cellarum circiter sex alius super aliam, supra quod infra sensim in latitudinem crescit. Cellae grandes, rotundae, pachydermes, sine trigonis. Foliolum tam longum quam folii latitudo, altero latere folio infra adiunctum, alte sectum in bina segmenta setacea, quorum utrique adest dens setaceus prope basin. Folia iuvenilia itemque foliola minora quam adulta, similia tamen. Androecium terminale in caule prostrato, multi-bracteatum, bracteae diandrae, oblique oblongae, apicibus binis setaceis sinu alto lunato separatis. Folia multa ferunt antheridia, foliola autem ut supra descripta plurima adsunt.

166. *Leptoscyphus Henkelii* Sim (new species).*

Sterile ; stems simple, straggling among forest mosses in Rhodesia, rooting by numerous rhizoid bunches.　Leaves 2 mm. long, ovate, apex usually 3-toothed, the teeth strong, the central tooth largest, with rounded sinus.　Foliole connate with the leaf below on one side, very large, rounded in outline but palmately cut, deeply 2-fid at the apex and each side having five to six long segments or cilia, each of one row of cells. Cells large, thick-walled.

Umtali, Rhodesia, Henkel (Eyles 2633).

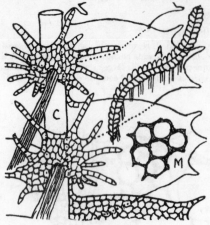

Leptoscyphus Henkelii.

Genus 56. PLAGIOCHILA Dum., Rec. d'Obs., p. 14 ; Ldbg., Sp. Hep., pp. 22 and 625 ; Spruce, p. 449 ; Schiffner, Pfl., i, 3, 87 ; Steph., Sp. Hep., iii, 2nd ser., 657 ; Macvicar, p. 215.

　=*Radula*, sect. 3, *Plagiochila*, Dum., Syll. Jung., p. 42.

Stems erect, almost without rhizoids, rising from a prostrate rhizome-like structure.　Leaves decurrent ; lower margin straight, apex and upper margin rounded and dentate, or with long spinous teeth.　Folioles absent.　Perianth terminal, laterally compressed, the mouth dentate or ciliate.　Epiphytes or subepiphytes.

This genus contains the largest, most attractive, and most common hepatics in South Africa, abundant wherever humidity is present.

Stephani states (Sp. Hep., p. 657), " Folioles in a few well developed," and in several South African species he mentions or describes folioles ; and Pearson (Hep. Nat., p. 15) says *re P. natalensis*, " Under-leaves only present on some branches, bifid, segments with a marginal tooth."　But after having examined thousands of specimens I have failed to find any foliole.

All the species have the leaves on young branches different from the mature leaves—sometimes very different— and usually where a fresh branch starts its lower leaves begin as bracts and go through all the stages of form up to the branch-leaves, including many which have the appearance of folioles, but they have the position of leaves and I cannot consider them to be folioles.　In no other genus is there so great variation in leaf-form upon almost every stem, but in probably no other genus does the general similarity so clearly place a plant in its genus.　In no other genus, however, is the difficulty of separating the species so great.

Synopsis.

167. *P. sarmentosa.*　Leaves round, often serrate all round.
168. *P. heterostipa.*　Leaves oblong, serrate at apex only.
169. *P. corymbulosa.*　Leaves oblong, strongly dentate at the apex, often serrate elsewhere.
170. *P. natalensis.*　Leaves oblong from an ovate base, entire and plane at both bases, which are short.
171. *P. Henriquesii.*　Leaves large, rounded from a wide base ; upper base toothed, lower base short, entire.
172. *P. crispulo-caudata.*　Leaves oblong from a wide decurrent, very undulate base both above and below.

167. *Plagiochila sarmentosa* Lehm., Syn. Hep., p. 57 ; Lindb., Spec. Hep., p. 86, t. 15 ; Steph., Sp. Hep., iii, 158.

　=*Jungermannia sarmentosa* Lehm., in Linnaea, ix, 4, 427.
　=*Plagiochila subquadrata* Steph., in Brunnthaler, 1913, i, 16.

Dioecious, growing in loose masses several inches across, with prostrate or subterranean leafless wiry stems 1–2 inches long, from which rise erect or suberect stems 1–2 inches long, bare below, leafy above, branched at nearly right angles, the leaves spreading when moist, closely imbricate and closely connivent when dry, the apex of the stem then curved round and facing down.　Flagellae are also present and vary much.　Adult stem-leaves 2 mm. diameter, obliquely circular, very shortly and transversely inserted, not decurrent, concave, serrate all round except a short distance at the lower base, the teeth triangular-acute, mostly equal, but two or three of the terminal ones are often larger, with smaller interspersed.　The upper margin is usually reflexed and dentate where parallel to the stem.　Cells round, marginal cells square, the upper small ($18 \times 25\mu$), the lower elliptical and much longer (70μ).　Upper leaves near the perianth gradually larger and more distinctly round, lower sometimes somewhat truncate or bilobed, young branches usually small-leaved at the base, these leaves very diverse in form and the upper sharply 2–3-fid (see page 107, where two teeth are prominent).　Perianth nearly triangular, laterally compressed, well-exserted, fringe with long narrow cilia or spines.　Innovation frequent.　Androecium terminal, few-bracted ; bracts rounded, denticulate.

Abundant on and round Table Mountain (Sim 9809, 8295, 8290, etc.) ; not seen by me from elsewhere, though Ecklon's original specimen was from the Winterbergen (Herb. L. Lg. N.).

---

* *Leptoscyphus Henkelii* Sim (sp. nov.).
Sterilis ; caules simplices inter muscos silvestres in Rhodesia, cumulis pluribus rhizoideis radicantes.　Folia 2 mm. longa, ovata ; apex plerumque 3-dentatus, dens medius maximus, sinu rotundato.　Foliolum connatum cum folio infra altero latere, maximum, forma rotundatum sed palmate sectum, alte bifidum ad apicem et utrinque 5–6 longa segmenta vel cilia praebens, quorum quodque ex uno ordine cellarum constat, cellae grandes, pachydermes.

I have Table Mountain specimens (Sim 9813, etc.) on which the lower leaves are entire ; the condition shown in the lower figure, if taken by itself, might well be regarded as a distinct species or variety (Sim 8628), and has been

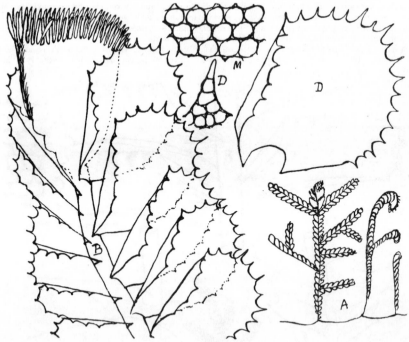

Plagiochila sarmentosa.

described as *P. Wilmsiana* St., Sp. Hep., vi, 240, from Cape Town, and I have also a form with long etiolated branches (Sim 9812) from intake of the Woodhead Tunnel, Table Mountain, but all these grade into one species. There

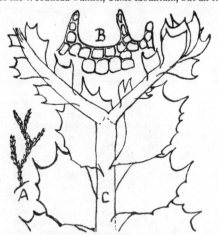

Plagiochila sarmentosa (form).

seems to be a misprint in Sp. Hep., vi, 240, where *P. Wilmsiana*, described as small, is afterwards stated to have stems 15 cm. long ; probably 15 mm. is intended.

Mitten, Linn. Soc. Journ., xx, 146, 319, credits tropical and West Africa and Mascarenes with *P. sarmentosa*, which he compares with the widely distributed *P. asplenioides* M. and N., which species was included in error in my list upon specimens of *P. sarmentosa*.

168. *Plagiochila heterostipa* Steph., in Hedw., 1892, p. 129 and pl. vii ; Steph., Sp. Hep., iii, 2nd ser., 126.

Dioecious ; stems to 2 inches long, firm, slightly branched, branches spreading widely.   Stem-leaves 2–5 mm. long, hardly in contact, spreading, oblong, the sides subparallel, the apex squarely rounded and strongly serrate, the lower margin reflexed, entire, the upper margin shortly serrate, undulate, its base reflexed, entire ; branch-leaves much smaller, various in form, and with fewer teeth.   Cells small, round, with distinct trigones, lower cells longer.   Perianth lateral, strongly spinose, floral leaves larger, rounder, and more spinose.

Knysna, C.P., Burtt-Davy (Sim 9811) ; Stephani described it from scraps of Rehmann's from Blanco and Tow River, and describes folioles as " numerous, very irregular, linear or lanceolate, entire or deeply bifid, sometimes

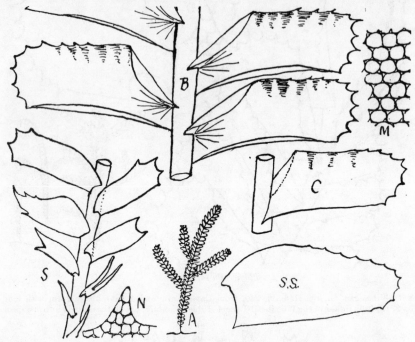

Plagiochila heterostipa.

shortly many-fid, or large with the margin remotely and irregularly laciniate," and he illustrates these in positions on the stem which seem to me doubtful.   I can only think of these as basal leaves on young stems (fig. S), not as folioles. The oldest leaves are wider near the base, with the margin crenate instead of serrate at or near the apex.   Stephani divides *Plagiochila* into two groups, viz. :—

(A) *Patulae* (Folio vulgo angusta, in paucis latiora, margine postico stricte patula).

(B) *Ampliatae* (Folia ventre ampliata), which includes all the South African species except as follows, viz. :—

In *Patulae* he has § II *Vastifoliae* (adult leaves widely ligulate), which includes *P. heterostipa* Steph. (as above, and it is also included in Bryhn's list from Eshowe), and § III *Angustifoliae*, which includes *P. laxifolia* Gott. from Madagascar, but which occurs in Bryhn's list from Ekombe, Zululand (1911), and of which I know nothing more.

169. *Plagiochila corymbulosa* Pears., Hep. Nat., p. 14 and pl. 9 ; Steph., Hedw., 1892, p. 120 ; Steph., Sp. Hep., iii, 2nd ser., 353.

Dioecious ; loosely caespitose or straggling separately among mosses.   Stem 1–3 inches long, leafless at the base, firm, wiry ; the leaves rather laxly placed, 2–3 mm. long, oblong or ovate-oblong, not much widened toward the base, the whole leaf folding back from both sides when dry and being then cuneate, and giving the plant a thinner and more pectinate appearance than the other species.   Stems rising from prostrate former stems, irregularly but frequently branched, so as to be somewhat corymbose, and with occasional long slender flagellae.   Leaves strongly dentate at the rounded or truncate apex, the teeth few but very large and irregularly triangular, the upper margin entire or with one or several smaller teeth, the lower margin apparently entire through being usually reflexed its whole length but often bearing several teeth.   First leaves on innovations small and nearly entire, the size of leaf and the number and size of teeth increasing up to the fructification, the upper leaves being strongly spinulose.   Perianth terminal until outgrown by one or two lateral innovations rising from below it, but remaining permanent laterally compressed, longer and wider than the neighbouring leaves, strongly triangular-spinulose at the mouth, triangular pear-shaped ;

old lateral perianths are smaller and with much shorter and more triangular teeth than fresh terminal perianths. Calyptra pear-shaped, bursting at the top, the capsule shortly exserted, 4-valved, the valves when old externally papillose through the rupture of the tissue of the cell-walls. Androecium of about six to eight pairs of closely placed

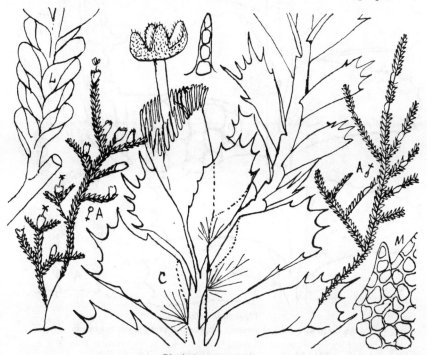

Plagiochila corymbulosa.

short somewhat dentate bracts, clasping and concave at the base, each having two stalked antheridia in its axil, the androecia at first terminal but by continued growth soon median, and several may occur in succession on the same branch.

Common in the Amatola forests, Cape (Sim 7010, etc.), less common elsewhere, but described by Pearson from Umpumulu, Natal (Bertelsen).

Rehmann's 504β from Houtbosch has small, round, entire leaves on the prostrate portion of the stem.

170. *Plagiochila natalensis* Pears., in Christ. Vid. Sels. Fornh., 1886, p. 15 ; Steph., Sp. Hep., iii, 166 ; Brunnthaler, 1913, p. 16.

Dioecious, loosely caespitose, often in large masses. Stems 2–3 inches long, suberect or pendent, wiry, simple or pinnate or 2-pinnate ; leaves 2–3 mm. long, somewhat imbricate, oblong from an ovate base, *i.e.* the lower portion is much wider than from the middle upward, due to the upper margin forming a rounded arch there, the portion near the stem being recurved and entire, with short connection there, not decurrent ; the lower margin descending and somewhat decurrent but not waved. The apex is strongly 3–4-dentate, the lower margin usually entire, and the upper margin toothed or subentire from the apex to the widened part. Cells large, round, with distinct trigones ; lower cells twice as long. Floral leaves rather larger than others and more spinose. Perianth terminal or ultimately lateral, winged, mouth closely spinose (fig. SS, page 110). New branches usually small-leaved at the base, these leaves being rounded, entire, cuneate, lanceolate, bifid, trifid, etc., and more like ordinary leaves further up the stem, but the small leaves might be mistaken for folioles ; rhizoids are also frequent among these small leaves.

Abundant in all eastern forests, present also in Rhodesia and on Table Mountain (Sim 8295, etc.).

From description I cannot separate *P. capensis* Steph., Sp. Hep., v, 2nd ser., 350, collected by Macowan on the Boschberg. *Plagiochila spinulosa* Nees and Mont, a very variable and widely distributed species, is credited to the Cape in Syn. Hep., p. 25. No doubt this is the plant referred to, though the more inclusive aspect of *P. spinulosa* might include all the South African species except *P. sarmentosa*.

Pearson in publishing it says, " Differs from *P. javanica* N. in its ramification, its leaves being more undulate, perianth shorter and broader and winged ; from *P. hypnoides* Lindb. in the less decurrent leaves which are entire on the postical margin near stem and more undulate. Very near the Mexican *P. crispata* G., which has leaves still more crisped and longer perianth cilia."

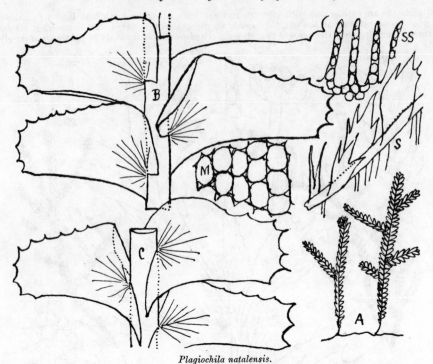

Plagiochila natalensis.

SS, teeth at mouth of perianth; S, lower leaves on young branch.

171. *Plagiochila Henriquesii* Steph., Sp. Hep., iii, 348.

Stems 3–5 cm. long, erect, branched, the branches spreading; leaves 3 mm. long, comparatively wide, ovate-oblong, with widely rounded, crenate-dentate apex, the crenatures extending partly along each margin, and strongly evident on the reflexed upper base which is shortly connected and not decurrent; the lower base is shortly decurrent, entire, and not undulate. The teeth throughout are shortly and widely triangular. First leaves on new branches ligulate or cuneate, one to several-toothed at the apex but not at the sides. Cells small.

Frequent in the Transvaal and Rhodesia; rare elsewhere (Sim 9659, 8033, etc.).

Probably this represents the *P. javanica* N. and M. recorded in Syn. Hep., p. 29, as collected at the Cape by Ecklon, as our plant answers the description there given and slender shoots or innovations answer the var. *β spinulosa* there described as a Cape plant. I do not find *P. javanica* in South Africa, as distinct from this, though Mitten records it (Linn. Soc. Journ., pp. 22, 146, 320, and also in Linn. Soc. Journ., pp. 16, 91, 187).

172. *Plagiochila crispulo-caudata* Gottsche, Abh. Nat. Ver. Bremen, vii (Reliq. Rabenh.), 240; Steph., Sp. Hep., iii, 347; Steph., Engl. Bot. Jahrb., 1895, p. 309; Steph. (Brunnthaler, 1913, i, 16).

Dioecious, robust, laxly caespitose; stem 2–4 inches high, strong, wiry, more or less branched; leaves 2–4 mm. long, oblong, twice as wide at the base as at the apex, strongly toothed in the upper half, the reflexed upper base entire but strongly undulate and extending below its connection, the lower base also long and undulate, so the two form a continuous undulate line down each side of the stem. Branch-leaves similar but smaller and less decurrent and less undulate. Leaves near the perianth largest, strongly and irregularly dentate, the inner pair with long, serrate, simple, acute apices. Perianth laterally compressed, mouth dentate-spinose. Androecium at first terminal, afterwards outgrown and median, few-bracted, the bracts small, entire, clasping, ventricose. Cells small toward the margin and teeth, much larger lower.

Frequent in the Eastern Transvaal, Natal, and Rhodesia, not seen further south (Sim 8291, etc.); found also in Madagascar and tropical Africa.

I cannot by description separate from this *P. Sprengeri* Steph. (Sp. Hep., iii, 345) from the Transvaal, except that the perianth is said to be connate with adjoining leaves. *P. fusco-virens* Steph. (Sp. Hep., vi, 158) from Ekombe, Zululand (Titlestad), does not appear to differ. *Plagiochila mascarena* (Gott.), Steph., Sp. Hep., iii, 2nd ser., 154 and 161, is included by Stephani in Hedw., 1892, p. 121, among species from South Africa and Mascarenes, but in the above (p. 154) it is only cited from Bourbon, though in Bryhn's list it is included from Entumeni, Zululand. He places it in his group, "(H) *Ampliatae* II, stem-leaves ovate," which has no other South African representative, and I have no further record of it.

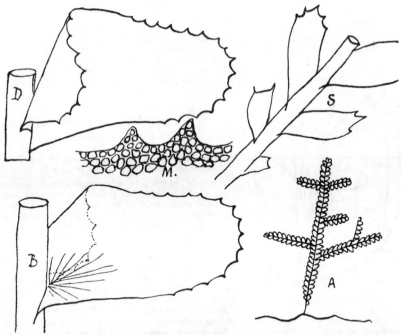

Plagiochila Henriquesii.

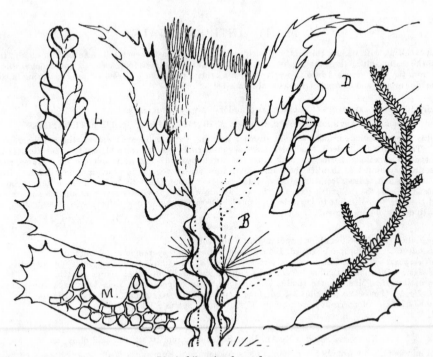

Plagiochila crispula-caudata.

Genus 57. MARSUPELLA Dum., Comm. Bot., p. 114 ; Macvicar, p. 105 ; Stephani, Sp. Hep., i, 2nd ser., 151.

Plants large or small, very diverse ; stems suberect, with rhizoids near the base only ; leaves complicate-concave, equally 2-lobed, transversely inserted, succubous, the base decurrent. Folioles absent. Inflorescence terminal, involucral leaves large, connate below and also with the perianth which is immersed below, free above, making an erect marsupium. Unfortunately the one South African species has not been seen fertile, so its place can only be assumed from its leaves.

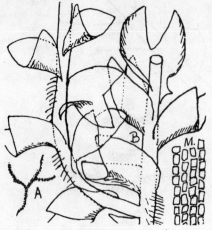

Marsupella aurita.

173. *Marsupella aurita* (Lehm.) Sim.

=*Anastrophyllum auritum* (Lehm.), Steph., Sp. Hep., i, 2nd ser., 1137.
=*Jungermannia aurita* Lehm., Linnaea, 1829, p. 368.
=*Sarcoscyphus auritus* Nees, Syn. Hep., p. 9.

Sterile ; stems ½–1½ inch long, slender, simple, or with two to three spreading branches ; growing in open situations on soil, and with numerous colourless rhizoids from the stem. Leaves separate, spreading, amplexicaul, 2-lobed, the lobes folded openly toward one another, but with a rounded connection and no keel, distichous ; leaf lobes ovate-triangular, subacute, equal on the older parts, but the upper lobe is set somewhat sideways on the young branches and so produces an appearance of unequal lobes (the upper the smaller) which is much modified when these leaves are pressed out flat. Connection with the stem short, upper lobe attached near the centre of the stem, lower lobe somewhat decurrent and sometimes 1-toothed (from which the name *auritum*). Folioles absent ; cells small, quadrate, in lines. No sexual parts seen.

This is described and drawn from a specimen collected on the top of the Giant's Castle, Natal, Drakensberg, about 8000 feet altitude, by Mr. R. E. Symons (Sim 9781), but seems to agree with the sterile specimen sent to Europe from near Cape Town on which the synonyms already cited have been founded, though Stephani (Sp. Hep., i, 154) excludes this plant from *Marsupella*.

ORDER IV. ANTHOCEROTALES.

Plants thalloid, with smooth rhizoids, and with one large chloroplast in each cell. Antheridia sunk in the upper surface of the thallus, ultimately bursting free. Sporogonium with a bulbous foot, a sheath, and a long sessile capsule, bursting from the top downward into two valves, with a central columella between. Spores maturing in succession from apex downward, and having sterile cells intermixed.

Family 16. ANTHOCEROTACEAE. Characters of order. Always on moist soil.

Genus 58. ANTHOCEROS Mich., Gen. Pl., p. 10 (1729) ; Linn., Sp. Pl., p. 1139 (1753).

Thallus often flabellate, usually without distinct midrib. Capsule linear, 2-valved, the valve-surface having stomata. This genus and, indeed, this order have a considerable general resemblance to other thalloid hepatics, and still there is something distinctive both in its outward appearance and in its microscopic features, and only a little practice is required to identify the genus, apart from the sporogonium, which is widely different from other thalloid hepatics and, indeed, from all other plants. It is a group distinctly by itself, hence its inclusion at this point instead of near the thalloid hepatics. The species are apt to grow intermixed, as all like similar saturated sites, and they are then difficult to separate, the specific characters being unsatisfactory in classification, and they change with degree of development.

Synopsis.

174. *A. minutus.* Thallus 5 mm. long ; sporangium 5 mm. long.
175. *A. natalensis.* Thallus much crisped, 1–4 cm. long, autoecious ; sporangium 3–4 cm. long.
176. *A. dimorphus.* Dioecious, dimorphous, female crisped ; sporangium 3–5 cm. long, male not crisped.
177. *A. Bolusii.* Dioecious ; thallus 2–5 cm. long ; sporangium 8–10 cm. long ; spores smooth.
178. *A. usambarensis.* Monoecious ; thallus 1–2 cm. long, entire ; sporangium 3 cm. long ; spores pale, rough.
179. *A. Eckloni.* Monoecious ; thallus 2–5 cm. long ; sporangium 8–10 cm. long ; spores nearly smooth.
180. *A. sambesianus.* Monoecious ; sporangium 35 mm. long ; spores echinate.

174. *Anthoceros minutus* Mitten, Linn. Soc. Journ., pp. 16, 91, 195, and pl. iv, fig. 6.

Plant minute ; thallus usually 5 mm. long, or less, seldom more, simple or forked ; lobes obovate, entire, smooth, with a small midrib ; involucre large, cup-shaped ; sporangium cylindrical, 5 mm. long, laxly areolated, fleshy ;

when dry linear, split to the base, showing the columella, numerous smooth round spores, and scattered cylindrical elaters six cells long, without spinal bands ; sporangium cells long-hexagonal, lax, with stomata intermixed.

Mitten describes it from " Base of Table Mountain, in wet places near the Kloof Road, above the Round House," and I find it frequent in moist ploughed ground on the eastern slope and in Natal, so evidently it is annual.   Antheridia not seen (Sim 9531, etc.).

Stephani (Sp. Hep., v, 997) gives a similar description under this name, " Mitt. Venus Exp.," and adds, " plant dioecious, thallus apex tuberiferous, tubers long-stalked ; capsule 1 cm. long, spores brown, 45μ "; but he also introduces (p. 1021) *Notothylas minuta* (Mitt.), Steph., on Mitten's Table Mountain plant from Proc. Linn. Soc., 1877, p. 195, quoted above, with nearly similar description but " irregularly many-branched, spores 90μ, brown, densely muriculate, elaters spiral, 700μ ; androecia on the main branches few," which items are not in Mitten's description.

I have collected many specimens and find only one plant, and see no reason for the *Notothylas.*

That genus is always monoecious, more succulent in its short sporogonium than is *Anthoceros,* and has the elaters more developed, but I find them present and more or less developed in all South African species of *Anthoceros.*

175. *Anthoceros natalensis* Steph., in Brunnthaler, 1913, p. 22, and Sp. Hep., v, 979.

Thallus congested, nearly circular, much undulated and lamellate, depressed in the centre, raised toward the margin, 1–3 cm. across, without midrib, many-lobed, the lobes irregular, some deeply multifid, the whole central depression being sometimes full of plates like a double rose, as if due to monstrosity, but some among these bear the antheridia so it is autoecious.   Involucres large, more or less central, entire ; sporogonium 3–4 cm. long, slender, stomatose, eventually splitting to the base ; columella full length of sporogonium ; elaters of three to four long cells, sometimes branched ; spores brown, 40–50μ, the convex surface echinate with darker, mostly 3–4-pointed plates which appear like simple spines at the margin.

A similar crisped form is known in Britain as *A. crispulus* (Mont.) Douin (Rev. Bryol., p. 25 ; Macvicar, p. 449) =*A. punctatus* Linn. var. *crispulus* Mont. in Webb. and Berth., p. 64. Pearson, Hep. Nat., p. 19, records *A. punctatus* Linn. from Umpumulo, Natal, but he had no spores, and I have seen no species here having the strongly spinose spores of that species.

A. natalensis is frequent wherever *Anthoceros* occurs in South Africa, usually mixed among other species.

Anthoceros minutus.

A, plant, nat. size ; B, same, magnified ; C, section of thallus ; D, cells of sporogonium ; E, capsule, open (young) ; F, section of same ;   G, spores ; H, elaters.

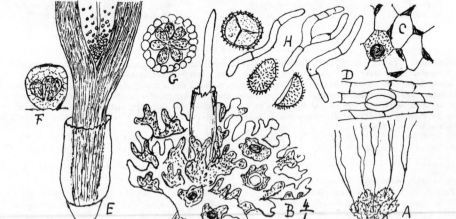

Anthoceros natalensis.

A. nat. size ; B, plant, $\frac{4}{1}$; C, cells ; D, cells of sporogonium ; E, base of open sporogonium ; F, G, androecia ; H, spores and elaters.

176. *Anthoceros dimorphus* Sim (new species).*

Dioecious, dimorphous, the male plant with dichotomous, flat, lorate branches, forming a stellate group 2–3 cm. across, and not crisped ; the female plant a similar or smaller group of nearly circular, concave thalli each about 4 mm. diameter, having irregular crisped margin and muricate surface, the outgrowths sometimes irregularly scale-like. Involucres large, 2–3 mm. long, one or more on a thallus, sporangium 3–4 cm. long, spores dark, 50µ wide, echinate, elaters about 3-celled.

A common plant on damp grasslands in Eastern Province and Natal ; and extending to Victoria Falls, S. Rhodesia.

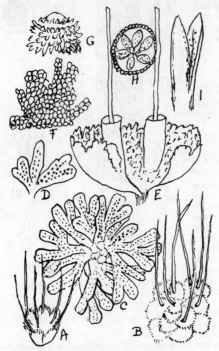 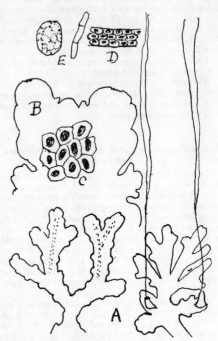

Anthoceros dimorphus.

A, B, female plant ; E, same, magnified ; C, male plant ; D, same, magnified ; F, scale ; G, spore ; ·H, androecia ; I, tip of young sporangium.

Anthoceros Bolusii.

A, male and fertile thalli, nat. size ; B, same, magnified ; C, cells ; D, section of thallus ; E, spore and elater.

177. *Anthoceros Bolusii* Sim (new species).†

Dioecious. Thallus 2–4 cm. long, dichotomously and irregularly forked into separate ligulate segments 2–3 mm. wide, somewhat lobed and crenate along the margin and showing frequent indications of suppressed furcation. Thallus thin, about three cells deep at the centre, without midrib, pellucid when young, dark green on the older parts. No tubers seen. Involucres single, central on a lobe, 2 mm. long, cylindrical, the mouth not contracted. Sporogonium 3–4 inches long, slender, brownish ; stomata not seen ; spores 35–40µ diameter, yellowish, somewhat quadrate, the surface indistinctly areolate by numerous minute points. Elaters brownish, 60–80µ long. Distinct in habit and in spores, and in being dioecious.

Skeleton Ravine, Table Mountain, Mrs. Bolus (Bolus Herb. 14,823) (Sim 8425, etc.).

As this genus has not been largely collected it is quite likely that this species occurs elsewhere.

---

178. *Anthoceros usambarensis* Steph., in Brunnthaler, 1913, p. 23, and in Sp. Hep., v, 977.

Autoecious, small, caespitose ; thallus 1–2 cm. long, slightly forked or lobed, entire ; involucre solitary, cylindrical, 3 mm. long ; sporangium 3 cm. long, spores 36μ, pale, granular.   Androecium near the involucre.   Described from Usambara, East Africa, included also by Brunnthaler from Van Reenen's Pass, Natal, December 1909.

179. *Anthoceros Eckloni* Steph., Sp. Hep., v, 978.

Monoecious.   Thallus simple, 1–3 inches long, ½ inch wide, or radiating into several simple and cuneate or pinnately 3-lobed divisions each 1–2 cm. long, 3–6 mm. wide, flat, dark green, without midrib, but with the margins crenate by abortive furcation.   Section about six cells deep, without lacunae ; central cells large and lax, surface

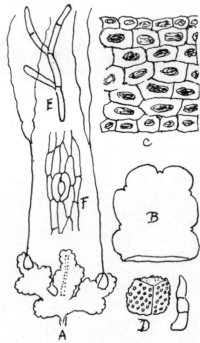

Anthoceros Eckloni.

A, plant, nat. size ; B, point of thallus, magnified ; C, section of same ; D, spore and
elater ; E, occasional branched elater ; F, cells and stomata of sporogonium.

cells smaller, about 70μ long, 30μ deep, all containing one chloroplast each.   Involucres one or two on a segment, 2–3 mm. long, narrowed upward.   Sporogonium 10 cm. long ; stomata numerous toward the base ; spores usually angular, 45μ diameter, yellowish, papillate but not echinate ; elaters two to three cells long, sometimes branched. Antheridia few in a group, in lines on upper surface of thallus.

Paarl, Cape (Sim 8423) ; Greytown, Natal (Sim 8424), and almost every other *Anthoceros* site in South Africa. Near *A. laevis* Linn. of the Northern Hemisphere.

180. *Anthoceros sambesianus*, Steph., Sp. Hep., v, 996.

" Monoecious, small, yellowish-green, thin, growing on soil.   Thallus up to 10 mm. long, widely obconic, cavernose, furcate, irregularly lobed, somewhat lacerated, lobes spathulate, often crispulate.   Involucre solitary, cylindric, 3 mm. long, cavern inflated, rough.   Capsule 35 mm. long, spores 36μ, black, strongly echinate.   Androecia near the involucre, few.   Antheridia not seen.   Boroma, Zambesia " (Steph.) ; also Usambara, Brunnthaler, p. 22.   Not known to me.

MOSSES.

EXPLANATION OF THE ILLUSTRATIONS—MOSSES.

A. Plant, natural size.
B. Leaf.
BB. Dry leaf.
C. Capsule.
CC. Branch leaves where these differ.
D. Leaf-apex.
E. Cell-formation.
EE. Cell-formation of lamellae.
F. Cell-formation toward leaf-base.
G. Section of leaf.
H. Leaf-margin.
K. Branch-tip, enlarged.
L. Lid.

M. Male inflorescence.
N. Perichaetal leaves.
NN. Perigonial leaves.
O Peristome, or its teeth.
P. Plant, or part of it, enlarged.
Q. Plant, dry condition.
R. Calyptra.
S. Special features.   See text of that species.
T. Alar cells.
TT. Marginal cells and those within them.
V. Nerve.
W. Leaf-base.

Explanation of the illustrations of the Hepaticae will be found preceding the descriptions of that group.

Specific names marked by an asterisk are unpublished so far as known, but as these occur with extraordinary frequency among South African mosses they have been dealt with throughout, or else included among synonyms, in order to clear them off.

MUSCI.

Cellular plants having a leaf-bearing stem or axis, with many-celled rhizoids, the leaves undivided, and usually arranged round the stem in several or many rows (in only a few cases distichous), with a midrib (nerve) or one or two shorter nerves present in the leaf in most but not in all species. Differentiated folioles are present in only a few cases, but dimorphous leaves are more frequent. The capsule, which bursts through its calyptra at an early stage (except in Sphagnales and Andreaeales), matures slowly in full exposure to a coriaceous condition, in which it remains permanently, even after the spores are shed ; the seta also is firm and persistent, and there are no special elaters or sterile cells present among the spores in the capsule.

The mosses are cellular plants having distinct alternation of generations, *i.e.* (1) the protenema, usually confervoid or thalloid, produced direct by the germinating spore ; (2) the mature plant or gametophore, starting as a bud from the protenema, and having a distinct stem, variously clad with sessile undivided leaves, and producing in characteristic positions antheridia (male organs) and archegonia (female organs). The antheridia contain numerous 2-ciliate antherozoids, some of which, on dispersal, reach and enter the archegonia, fertilising them ; each fertilised archegonium produces a 1-celled capsule, which at an early stage breaks through the membrane enclosing it ; the capsule is either sessile or pedicellate, and contains the numerous spores by which the plant is reproduced. These usually rise from a central columella and do not have sterile cells or elaters mixed among them.

Asexual reproduction by division of the plant, by gemmae or by other means, also occurs in some species.

The characters differentiating the Musci from the Hepaticae have been mentioned above. The general texture and appearance is also distinct in most cases, though not in all.

Mosses occur in all terrestrial parts of the world except the highest latitudes ; they occur on rocks, tree-trunks, humus, and soil, and in swamps and waterfalls, and even submerged in fresh water ; each species has its own nature of habitat, climate, and nidus ; some occur only on lime, others cannot endure lime ; some enjoy and require sunshine, others shade ; some are always epiphytic or saprophytic, others never are so ; the majority enjoy cool humid conditions ; some are among the first colonists on bare and perpendicular rocks, or in rock crevices or caves ; a few belong to the coast only, a good many belong to the highest mountain localities, many are denizens of the forest, while a few extend into the Karroo ; some live on more or less saline soil, but none endure salt water.

In domestic economy their uses are few and unimportant, but in the economy of nature many play an important part, forming a first clothing on bare soil or rock, from which humus fit to carry higher plants is produced, and thereafter in absorbing available moisture and retaining it till again required to maintain the humidity of the local atmosphere.

The mosses form a very large group, of which almost all are possessed of a uniform general plan, while the smaller groups forming the orders Andreaeales, Sphagnales, and Polytrichales are very decidedly different in plan, though sufficiently near to be also included among the Musci.

117

Order V. SPHAGNALES.

=Order III. *Stegocarpae*, Schimper, Syn. Musc. Eur.
=Classis III. *Stegocarpi*, Tribe IX. *Sphagnaceae*, C. Mull., Syn. Musc.
=Underklasse I. *Sphagnales* ; *Sphagnaceae*, Warns., in Pflanzenfamilien, i, 1.

Semiaquatic plants, having main stem usually simple, with numerous fasciculate lateral branches of two kinds, some strong and spreading, others slender and pendent ; the production of both proceeds from the apex of the main stem. Leaves all concave, sometimes cucullate, frequently 3–5-pointed, spreading or imbricate ; stem-leaves composed of hyaline cells only ; branch-leaves often of different shape from those on the stem, and composed of one layer of long narrow chlorophyllose separate cells or tubes, alternating with and almost enclosed between much larger, usually elliptic hyaline cells, which in almost all cases are curved and contain spiral thickenings of the cell-wall, and often have numerous intercellular pores through the cell-walls. Leaf-margin incurved sufficiently to make the leaf deeply concave, but not involute beyond that. Inflorescence in most cases dioecious, produced on the youngest lateral branches, the male branches being catkin-like where antheriferous, but prolonged in ordinary form beyond that, the antheridia are spherical and stalked, axillary to short branch-leaves, which are often highly coloured red or brown. Archegonia terminal on short bud-like branches, surrounded by long green clasping leaves. The capsule, which is enclosed in the calyptra almost till maturity, is globose, almost sessile on a short pseudopodium, without peristome, and opening by the sudden detachment of a lid. Spores produced in abundance in the upper part of the rather solid capsule, round a central columella. After fertilisation the upper part of the stem bearing the capsule develops into a very short, stout, leafless stem (the pseudopodium) in the apex of which the absorbing foot of the capsule stands, separated from the body of the capsule only by a constriction. The spores on germination produce a flat thallus from which the young plant arises, without any further fertilisation.

The main stems are of peculiar structure, containing an inner tissue of long slender cells, surrounded by one or more layers of much larger rectangular or hexagonal hyaline epidermal cells, some of which in certain species are much larger than the others, flask-shaped, with a curved open mouth. These doubtless take a part in the hygroscopic action for which the Sphagna are noted. Young plants having only stem-leaves often differ in appearance from mature plants, and in mature plants the pendent branches have more narrow and acute leaves than the spreading branches, the apical leaves being subulate and acute.

The Sphagna differ so completely from all other mosses in their general morphology that they cannot be mistaken ; and as the reproductive organs are more or less alike in all, it is on vegetative characters that classification has to be based, and opinions vary much as to the value of the respective characters in that respect, and in the consequent arrangement and number of species.

They occur more or less in all parts of the world, always in moist or swampy places where lime is absent, and some species are present but rare throughout South Africa, from the coast to the mountain-tops.

The peculiar relation of Sphagna with acidity is still under investigation

Family 17. SPHAGNACEAE. Characters of the order.

Genus 59. SPHAGNUM Linn. Characters of the order and amily.

Order VI. ANDREAEALES.

Simple or dichotomously branched, perennial, rock-grown mosses, of tufted habit, firm texture, and dark colour, with small, firm, usually papillose leaves, with or without a midrib, and resembling the leaves of the Bryales, having thick-walled, roundish, obscure, small cells toward the apex and more rectangular cells toward the base, sometimes sinuose. Inflorescence terminal, the sexes on different branches, and both sexes surrounded by numerous perichaetal leaves, and with numerous filamentous paraphyses intermixed in each.

After fertilisation the apex of the branch grows into a slender elongated pseudopodium with a swollen apex in which the absorbing foot of the capsule rests, separated from the rest of the capsule only by a constriction as in Sphagnum.

The calyptra is several layers thick, and continues to enclose the capsule till both are almost mature, when it ruptures round more or less irregularly, and the apex remains on the top of the capsule. The capsule remains closed at the apex, but opens by four (or sometimes eight) longitudinal slits along its middle portion. The columella does not penetrate the spore-bearing layer, but is persistent, and the spores smooth or somewhat papillose. The spores germinate by producing filamentous and somewhat thalloid protenema mixed, from any of which plants may arise, and the rhizoids, which are confined to the base of the stem, also vary from cylindrical form to adherent plates.

In the manner of its capsule-production this order resembles the Sphagnales, in the manner of the capsule bursting it resembles some of the Hepaticae, while in general habit and foliage it more closely resembles the Bryales. From all these it differs sufficiently to form a separate order.

Family 18. ANDREAEACEAE. Characters of the order.

Genus 60. ANDREAEA, Ehrh. Characters of the order and family. The only genus, and represented in South Africa as elsewhere only upon the mountains ; very rare.

Order VII. POLYTRICHALES.

Stems erect, leafy, simple or branched, often tall, rising from an underground rhizome. Leaves serrate, lamellose on the upper surface of the midrib, which in some genera occupies almost the whole leaf-surface except at the base. Cells without spiral thickenings. Capsule cylindrical or angular on a long seta ; having calyptra and lid, and a distinctive columella expanded and connected with the peristome teeth. Peristome teeth short and solid, not transversely barred. No pseudopodium is present.

This group is included as a family in Bryales by Brotherus in Pflanzenfamilien, and as an order in group Nematodonteae of Bryales by Dixon and Jameson, which latter arrangement is followed in my check-list, but it is dealt with as a separate order by Lorch, Fleisher, Cavers, and others, and deserves to be so dealt with here.

Family 19. POLYTRICHACEAE.

Plants erect, often large ; stems usually simple or slightly branched, rising from an underground rhizome, firm, and having a special central strand. Leaves numerous, usually toothed, placed all round the stem, usually long and narrow, firm, with a strong nerve sometimes occupying nearly the whole leaf-surface, on the upper surface of which several or many (up to forty) longitudinal plates (lamellae) are present, which are several cells high and one cell in thickness, but often have a different form of cell as the terminal row. Leaf-cells more or less rounded-hexagonal, with thin walls, those toward the base longer. Inflorescence terminal, usually dioecious, the male inflorescence large and discoid, often innovating from its centre. Capsule cylindrical or terete or several-angled, erect or inclined, with a long seta, a narrow cucullate calyptra, glabrous, scabrid or thatched with reversed hairs, a beaked lid, a peristome of thirty-two or sixty-four short, solid, ligulate teeth, triangular in transverse section and without bars, and an expansion of the columella (the epiphragm) covering the mouth of the capsule, and united at its edges with the teeth of the peristome. All are pioneers on clay banks or loam.

Synopsis of Genera of the Polytrichaceae.

Calyptra almost glabrous, but rough toward the apex. Leaves with few lamellae.
Leaves not vaginate and not narrowed above the base, but strongly bordered. Capsule cylindric, often curved,
and without stomata . . . . 61. ATRICHUM.
(Leaves vaginate, suddenly subulate ; capsule ovoid-oblong . POLYTRICHADELPHUS.)
Leaves vaginate, not narrowed above, not bordered. Capsule with stomata . . 62. PSILOPILUM.
Calyptra densely thatched with deflexed hairs, leaves with many lamellae but not bordered.
Plants small, with rosulate, basal sheathing foliage. Capsule not angled . . . 63. POGONATUM.
Stems elongated, wiry, with firm leaves along their lengths. Capsule angled . . 64. POLYTRICHUM.

Order VIII. BRYALES.

Habit varied ; stems and leaves of many forms, but leaf-cells without thickenings, and the capsule, which bursts its calyptra at an early stage and is usually raised on a seta, opens by the detachment of a lid, or in a few cases by bursting irregularly, never splitting centrally into perpendicular valves, but often having a toothed peristome protecting the mouth. Columella penetrating the spore-bearing layer (absent only in *Archidium*). There is no pseudopodium, the foot of the capsule being at the base of the seta, and remaining where the inflorescence was produced.

These characters distinguish the Bryales from the Sphagnales and the Andreaeales, while from the Polytrichales they are distinguished by the nature of the peristome teeth, which when present, as is usually the case, are thin and membranous, transversely barred. The peristome may be in a single series, or a double series, or it may be absent, and the columella is usually not expanded, nor connected with the peristome teeth.

The Bryales includes all other mosses than the three orders already described.

This is Bryales, suborder B. Arthodonteae of my check-list, and now becomes a separate order when the Polytrichales are separated from the Bryales.

Tribe I. APLOLEPIDEAE

(or Haplolepideae of many Authors).

Peristome (absent in some cases) consisting of a *single* series of teeth, each tooth formed of an outer membrane of which each bar extends across the tooth, and an inner membrane of which each bar has a longitudinal division down its centre. The teeth are normally sixteen, but in many cases they are forked above, and in some they are forked to the base, when the above arrangement is difficult to trace. In addition to the above, numerous other markings or striae are present on the peristome teeth in some groups.

Usually erect and acrocarpous, *i.e.* having the capsules normally terminal on the stem or ordinary branches, but in some cases the seta ultimately appears to be lateral through development of subterminal innovations subsequent to fertilisation.

Families of the Aplolepideae.

Spores few, ten to twenty, very large. Capsule sessile, bursting irregularly, and without columella, lid, or peristome,
20. ARCHIDIACEAE.

Spores small and numerous; capsule various, supplied with columella, and usually with lid and peristome.
  Leaves distichous, vertically placed, and having a sheath-lobe, a ventral lamina, and a dorsal lamina,
23. FISSIDENTACEAE.
    Leaves placed round the stem (or in *Distichium* distichous and horizontally placed).
      Leaves of two or more layers of large hyaline cells, between which are separate narrow tubular chlorophyllose
        cells . . . . 22. LEUCOBRYACEAE.
      Leaves having only chlorophyllose tissue apart from the base, and usually no larger differentiated hyaline epidermal
        cells on the upper half of the leaf. Plant usually erect, often cushioned or matted.
          Areolation all rather lax and papillose, sometimes having pores in the cell-walls. Peristome of sixteen short,
            entire, or rudimentary teeth, or absent.
              Capsule short, oval, or pear-shaped, with cucullate calyptra, and peristome of sixteen short, widely lanceolate,
                entire teeth . . . . 26. SYRRHOPODONTACEAE.
              Capsule cylindrical, with cylindric calyptra covering the whole capsule, and fringed at the base. Peristome
                of sixteen short, lanceolate, entire, inflexed teeth, or rudimentary, or absent . 27. ENCALYPTACEAE.
          Cells small in the upper half of the leaf. Peristome teeth usually long, entire, forked, or bifid.
            Basal membrane usually present.
              Leaves from narrow to spathulate, often with nerve excurrent as a hair-point. Cells toward the apex
                small, obscure, and usually papillose. Capsule erect or almost so; peristome (in some cases absent or
                imperfect) short and straight, or more often spirally twisted, of sixteen slender papillose teeth, sometimes
                connate below, and often divided deeply or to the base, then appearing as thirty-two teeth,
25. TORTULACEAE.

  This has two sections :—
    Leaves wide, cells seldom minute . . . . . . § 1. POTTIEAE.
    Leaves narrow, cells toward the apex often minute . . . . § 2. TRICHOSTOMEAE.
            Basal membrane absent.
              Leaves usually narrow, usually smooth, often with special alar cells, and having the nerve often excurrent
                as a hair-point. Calyptra (when present) smooth, usually cucullate; capsule globose to narrow, often
                cernuous or gibbous; peristome (sometimes absent) of sixteen long straight teeth, usually deeply cleft
                into lanceolate divisions having numerous transverse bars, and with fine vertical striae between these,
21. DICRANACEAE.
              Leaves lanceolate, concave, often with hair-points which are in continuation of the lamina rather than of
                the nerve; the cells often opaque and several deep in the upper half of the leaf; the lower cells often
                sinuose. Calyptra usually mitriform. Capsule oval, usually erect, peristome of sixteen wide teeth,
                entire or somewhat forked, without vertical striae. Outer layer thicker than the inner. Tufted rock
                plants . . . . 24. GRIMMIACEAE.

Tribe II. DIPLOLEPIDEAE.

Peristome (absent in a few cases) consisting normally of two series of teeth, though in some only one is present.
In the outer peristome each tooth is usually transversely barred, and consists of an outer membrane having a longitudinal division down its centre, and an inner membrane without such a division. The inner peristome when present
is usually of more delicate tissue, variously divided or plaited, its segments usually alternating with the teeth of the
outer peristome. A very large tribe, represented everywhere. Dixon and Jameson say (p. 254)," Philibert has shown
that the single peristome of the Aplolepideae is the homologue, not of the outer but of the inner peristome of the Diplolepideae. It is, therefore, misleading to speak of the outer layer of teeth in the latter as the peristome, and the inner
as the endostome, a term occasionally used; it would, indeed, be more in accordance with the actual facts to term the
latter the peristome and the outer layer the exostome."

Synopsis of Subtribes.

Subtribe A. Diplolepideae Acrocarpae.

The fruit normally *terminal* on the stem or branches (except in a few cases), though by innovation subsequent to
fertilisation it sometimes appears to be lateral. Stems usually erect, often pinnate.

Subtribe B. Diplolepideae Pleurocarpae.

Fruit produced from a *lateral* bud on the side of the stem or branches, not from the apex. Stems usually prostrate,
and often more or less pinnate.

Subtribe A. Diplolepideae Acrocarpae.

Characters as above.

Families of the Diplolepideae Acrocarpae.

Leaves distichous, horizontally attached . . . . . 29. EUSTICHIACEAE.
Leaves in many rows, placed all round the stem.
  Leaf-cells small.
    Leaf-cells dense, rounded-hexagonal, seldom transparent, often papillose. Capsule erect, terminal or lateral,
      symmetrical, often striate . . . . . . 28. ORTHOTRICHACEAE.
    Leaf-cells usually papillose, often narrow, the papillae often at one or each end of cell. Capsule usually globose,
      ultimately striate or ridged, terminal or lateral on a long seta, erect or cernuous; peristome, when present
      and developed, of sixteen outer teeth and sixteen inner cleft processes . 35. BARTRAMIACEAE.
    Fertile branch basal or nearly so, barren branch much larger. Leaf-cells smooth . 38. RHIZOGONIACEAE.
  Leaf-cells larger, or very large, usually smooth.
    (Capsule with distinct apophyses; leaf-cells smooth . . . . SPLACHNACEAE.)
    Capsule without apophyses; leaf-cells smooth.
      Capsule erect or almost erect, cylindrical, regular, ribbed, furrowed when dry. Leptotheca in,
                                                                      34. AULACOMNIACEAE.
      Capsule erect or cernuous, globose, oval or pyriform, terminal, usually with a distinct neck but no apophyses.
        Peristome sometimes absent, usually single or double and well developed, with the inner teeth opposite to
        the outer. Leaf-cells parenchymatous, lax . . . . . 32. FUNARIACEAE.
      (Capsule curved or inclined, clavate or ovate-oblong. Peristome double, outer of sixteen teeth, inner of sixteen
        processes alternating to these teeth, and with basal membrane. Probably included in error,
                                                                      33. MEESIACEAE).
      Capsule sessile or nearly so, globose, eperistomate; plant rhizomatous . 30. GIGASPERMACEAE.
      Capsule sessile, globose, cleistocarpous; protonema abundant . . 31. EPHEMERACEAE.
      Capsule pendulous or inclined, pyriform, oval, cylindrical or clavate, not striated and usually with a neck;
        peristome, when present and developed, of sixteen outer teeth and the inner of sixteen perforated processes,
        more or less connate below, and alternating with the outer teeth; leaf-cells of very different form in different
        genera, but smooth and usually long-hexagonal. Intermediate cilia present or absent from the inner peristome,
                                                                      36. BRYACEAE.
      Capsule without a neck, sterile stems prostrate, cells roundly hexagonal . . 37. MNIACEAE.

Subtribe B. Diplolepideae Pleurocarpae.

Usually more or less prostrate or arching perennial plants, continuing to grow from the apex of the stem and
branches, often matted, the branches often pinnately or subpinnately arranged, seldom dichotomous, but in some
cases the branches are erect from a prostrate, nearly leafless rhizome or stolon. Fructification always lateral, *i.e.*
the inflorescence from which it grows is not terminal, but produced as a lateral bud, though in some cases this bud
develops, so that the seta appears to be terminal on a short lateral branch. Very seldom do the male and female
organs occur in the same inflorescence. Peristome usually double, or more or less perfect.

A few genera hold a somewhat indefinite position, or depart from their type, and have been placed alternately
in Acrocarpae and Pleurocarpae, according to the views of the various authors. Thus, Hedwigiaceae has terminal
fructification, and is included by some in Grimmiaceae, while *Anoectangium* in Orthotrichaceae has lateral inflorescence,
and *Macromitrium* and various other related genera have pleurocarpous habit of growth.

Families of the Diplolepideae Pleurocarpae.

Peristome absent; leaves without nerve; capsule immersed or almost so.
  Small prostrate bark-epiphytes . . . . 39. ERPODIACEAE.
  Robust, suberect, straggling, or hanging plants . . . . . 40. HEDWIGIACEAE.
Peristome absent, rudimentary, single or double; leaves nerveless in South African genera. Aquatics,
                                                                      41. FONTINALACEAE.

Outer peristome present; inner usually present.
  Leaves all round the stem, not complanate.
    Capsule immersed; calyptra laciniate at the base, rough above; cells parenchymatous; sometimes papillose;
      nerve single. Small bark epiphytes . . . . . . 42. CRYPHAEAE.
    Capsule just emergent; calyptra cucullate; cells parenchymatous, papillose; marginal cells at leaf-base
      modified. Stems stoloniferous or rhizome-like . . . . 43. PRIONODONTACEAE.
    Capsule exserted; calyptra cucullate; cells smooth.
      Cells parenchymatous; marginal cells at leaf-base modified. Stems stoloniferous 44. LEUCODONTACEAE.
      Cells prosenchymatous, usually smooth. Outer cells of stem not thickened; basal leaf-cells quadrate but no
        differentiated alar cells. Inner peristome absent or of thread-like processes without basal membrane. Nerve
        single, seldom absent. Capsule erect and regular. Small bark parasites or epiphytes,
                                                                      45. FABRONIACEAE.
      Outer cells of stem thickened. Nerve single; capsule inclined, horizontal or pendulous,
                                                                      46. BRACHYTHECIACEAE.

Leaves either round the stem or in flattened arrangement. Dorsal leaves not different from the others.
    Cells parenchymatous, papillose ; no differentiated alar cells. Nerve single in South African species. Paraphyllia
        usually numerous. Capsule exserted, erect, or inclined. Calyptra cucullate   .   47. LESKEACEAE.
    Cells small, prosenchymatous or parenchymatous, usually smooth ; alar cells not much modified ; capsule
        usually immersed, erect, symmetrical, occasionally inclined. Leaves thin and scarious, usually complanate,
                                                                      48. NECKERACEAE.
    Cells prosenchymatous.
        Capsule erect and regular ; leaf-cells sometimes papillose ; alar cells differentiated, numerous, square or
            broader than long. Nerve usually single and long ; sometimes short and double or absent.
                                                            49. ENTODONTACEAE.
        Capsule usually cernuous or horizontal, not pendulous ; leaves often falcate ; nerve very short and weak, single,
            double, or absent. Cells usually smooth ; alar cells all alike   .   .   . 50. HYPNACEAE.
        Capsule inclined, irregular ; lid sharply beaked ; cells smooth or papillose, one row of the alar cells inflated,
            elongated, thin-walled ; midrib short and double, or absent   .   .   51. SEMATOPHYLLACEAE.
Leaves (in South African genera) distinctly complanate, delicate ; cells large, parenchymatous, smooth (except
    Callicostella).
    Sparsely branched, prostrate, with spreading side-leaves, and adpressed, often different back and front leaves.
        Nerve double, single, or absent. No differentiated alar cells. Cells of stem not thickened. Peristome
        double   .   .   .   .   .   .   52. HOOKERIACEAE.
    Rhizomatous, with erect frondose branches. Leaves distichous, with a row of small leaves on under-surface of
        stem. Marginal cells of stem not thickened. Peristome double, or the outer absent. No differentiated
        alar cells. Capsule pendulous   .   .   .   .   53. HYPOPTERYGIACEAE.
    Pinnately branched ; leaves distichous, with two rows of small leaves on under-surface of stem. Marginal cells
        of stem thickened ; lumen small   .   .   .   .   .   54. RHACOPILACEAE.

GENERA OF THE BRYALES.

Tribe I. APLOLEPIDEAE.

Family 20. ARCHIDIACEAE.

Characters of the family   .   .   .   .   65. ARCHIDIUM.

Family 21. DICRANACEAE.

Capsule without peristome ; plants small.
    Calyptra cucullate ; capsule almost included, and without special neck   .   66. PLEURIDIUM.
    Calyptra mitriform ; capsule immersed   .   .   .   .   67. CLADOPHASCUM.
    Calyptra campanulate, fringed at the base ; capsule often with wide neck and short seta   .   74. BRUCHIA.
Capsule provided with peristome.
    (Leaves distichous, horizontally placed, with sheathing base, subulate. Plants small   .   DISTICHIUM.)
    Leaves placed all round the stem.
        Capsule tapering at the base into a long neck ; peristome teeth perforated or split   .   75. TREMATODON.
        Neck absent or short.
        Peristome teeth very short, undivided. Minute plants with erect capsule, striate when dry.
            Leaves subulate, nerve excurrent ; peristome teeth perforated   .   70. BRACHYDONTIUM.
           Leaves ligulate, nerve ceasing below the apex , peristome teeth not perforated   .   73. RHABDOWEISIA.
        Peristome teeth longer, usually forked.
           Lid conical.
               Plants minute ; capsule minute, inclined, leaves subulate   .   .   72. NANOBRYUM.
               Capsule narrow, inclined, strumose, striate, furrowed when dry ; leaves often widely lanceolate, with
                    small quadrate cells in the upper portion   .   .   .   71. CERATODON.
               Capsule not furrowed, erect, not strumose ; basal membrane sometimes more or less present ; leaves
                    subulate ; slender little green plants   .   .   .   68. DITRICHUM.
           Like *Ditrichum* but glaucous   .   .   .   .   .   69. SAELANIA.
        Lid subulate or subulate-rostrate. Peristome without basal membrane.
           Alar cells not modified.
               Capsule erect, regular. Calyptra minute. Male inflorescence terminal and discoid. Leaves closely
                    imbricated   .   .   .   .   .   .   76 AONGSTROEMIA.
              Capsule usually cernuous, sometimes strumose ; calyptra large ; male inflorescence bud-like ; leaves
                    spreading, or falcate-secund   .   .   .   77. DICRANELLA.
           Alar cells larger, sometimes coloured.
              Lobes of peristome teeth usually cohering   .   .   .   78. HOLOMITRIUM.
              Lobes of peristome teeth free above.
                 Nerve wide ; seta flexuose, usually cygneous   .   .   .   81. CAMPYLOPUS.
                 Nerve rather narrow ; calyptra not fringed at base. Capsule arcuate or erect.
                    (Leaf not margined by narrow hyaline cells. (Not South African)   .   DICRANUM.
                 Leaf margined by narrow hyaline cells.
                    Capsule arcuate, unequal   .   .   .   .   79. DICRANOLOMA.
                    Capsule erect, regular   .   .   .   .   80. LEUCOLOMA.

Family 22. LEUCOBRYACEAE.

Peristome teeth regularly sixteen. Chlorophyllose cells of leaf 4-angled in section.
Capsule unequal, often strumose ; calyptra cucullate . . . . . 82 LEUCOBRYUM.
Capsule regular ; calyptra campanulate, cylindrical . . . . . 83. SCHISTOMITRIUM.
Peristome teeth usually eight, variable to sixteen. Chlorophyllose cells 3-angled in section 84. OCTOBLEPHARUM.

Family 23. FISSIDENTACEAE.

Capsule without stomata, very small, on short axillary setae. Peristome of sixteen short truncate teeth ; calyptra very small, conical, undivided. Aquatic plants ; stems much branched, without central strand,
86. CONOMITRIUM.

Capsule larger, and having stomata, terminal or terminal on short lateral or basal branches. Peristome of sixteen subulate teeth ; calyptra cucullate or mitriform ; stems simple or little branched, terrestrial or amphibious, with central strand . . . . 85. FISSIDENS.

Family 24. GRIMMIACEAE.

Calyptra plicate, covering most of the capsule. Leaf-cell walls not sinuate . . . 89. PTYCHOMITRIUM.
Calyptra not plicate, covering only the top of the capsule ; leaf-cell walls sinuate.
Peristome teeth cut to the base into filiform segments . . . . . 88 RHACOMITRIUM.
Peristome teeth entire, or perforated, or partly split . . . . . 87. GRIMMIA.

Family 25. TORTULACEAE.

Section 1. POTTIEAE. Leaves rather wide, ovate to spathulate-lanceolate. Cells seldom minute.

Capsule cleistocarpous (*i.e.* bursting irregularly).
Capsule globose, hardly pointed, included. Calyptra conical, with torn base . . 90. SPHAERANGIUM.
Capsule ovate or ovate-oblong, mucronate, included or exserted. Calyptra cucullate . 91. PHASCUM.
Capsule usually either gymnostomous or peristomate.
Peristome teeth when present straight, rather imperfect.
Leaf-margins more or less flat or reflexed. Areolation very dense. . . 92. POTTIA.
Leaf-margins ascending or involute, especially when dry ; areolation rather lax.
Nerve lamellate . . . . 93. PTERYGONEURUM.
Nerve not lamellate.
Capsule thick-walled, without peristome . . . . . 94. HYOPHILA.
Capsule thin-walled ; peristome of sixteen teeth . . . . 95. WEISIOPSIS.
Peristome teeth usually twisted, and of thirty-two equal linear segments.
Peristome teeth more or less connate below, sometimes half-way . . . 96. TORTULA.
Peristome teeth free, on a very short basal membrane, nerve of leaf bearing granulose threads on its upper surface,
97. ALOINA.

Section 2. TRICHOSTOMEAE. Leaves narrow (lanceolate or narrower) ; upper leaf-cells minute.

Leaves 3-seriate. Capsule peristomate . . . . . . 106. TRIQUETRELLA.
Leaves many-seriate.
Leaves serrate or dentate, capsule peristomate . . . . . 103. LEPTODONTIUM.
Leaves minutely serrulate by prominent acute hyaline cells ; capsule peristomate . 104. OREOWEISIA.
Leaves entire or almost so.
Peristome present.
Margin of leaf distinctly recurved, especially at the base.
Peristome teeth usually short, erect, or almost so . . . 102. DIDYMODON.
Peristome teeth long and much twisted spirally . . . 98. BARBULA.
Margin of leaf flat or slightly incurved.
Peristome teeth much twisted, long and slender . . . 99. TORTELLA.
Peristome teeth erect and straight, often imperfect . . . 100. TRICHOSTOMUM.
Peristome teeth perforated . . . . . 101. EUCLADIUM.
Margin of leaf usually incurved or involute ; outer layer of peristome thicker than the inner one, with conspicuous transverse ridges . . . . 107. WEISIA.
Peristome teeth minute or often absent ; ring dehiscing . . . 105. GYROWEISIA.
Peristome absent.
Lid more or less persistently attached or cleistocarpous . . . 108. TETRAPTERUM.
Lid separating.
Mouth of the capsule closed by a membrane . . . . 109. HYMENOSTOMUM.
Mouth not closed by a membrane.
Nerve not reaching the apex of the leaf, lid falling with columella attached 110. HYMENOSTYLIUM.
Leaves nerved to or beyond the apex, the margins not involute ; columella permanent, lid remaining attached. Ring permanent . . . . . . 111. GYMNOSTUMUM.
(Leaf margin involute. See . . . . . . 25. HYOPHILA.)

Family 26. SYRRHOPODONTACEAE.

Calyptra cucullate, permanent . . . . 113. SYRRHOPODON.
Calyptra bell-shaped, deciduous . . . . 112. CALYMPERES.

Family 27. ENCALYPTACEAE.

Only genus . . . . 114. ENCALYPTA.

Tribe II. DIPLOLEPIDEAE.
Sub-Tribe A. Diplolepideae Acrocarpae.
Family 28. ORTHOTRICHACEAE.

Plants erect, slender, cushioned ; inflorescence truly lateral . . . . 115. ANOECTANGIUM.
Plants erect, slender, cushioned ; inflorescence terminal, though the seta is often apparently lateral through subsequent
    innovation. Calyptra smooth, cucullate.
  Leaves narrow, with plane margins.
    Capsule on a very short seta, almost immersed, without peristome . . 116. AMPHIDIUM.
    Capsule on a longer seta, well exserted, usually peristomate . . . 117. ZYGODON.
  Leaves wider . . . . 118. RHACHITHECIUM.
Plants usually erect, short, and rather stout, in dense tufts, but occasionally with short prostrate stems from which
    the erect ones rise. Calyptra conical bell-shaped, lobed at the base.
  Capsule well exserted ; calyptra covered with erect yellow hairs. Leaves with hyaline border, and when dry
    curled . . . . 119. ULOTA.
  Capsule often immersed. Calyptra naked or sparsely hairy ; leaf-border usually not hyaline,
                                  120. ORTHOTRICHUM.
Plants with long prostrate stems from which short fertile branches rise, the whole forming flat carpets on bark or
    stones, with pinnated flat stems extending outward.
  (Calyptra inflated, cucullate ; capsule subglobose, smooth ; peristome single. (Probably included in error)
                                   DRUMMONDIA.)
  Calyptra subulate-campanulate or conical-campanulate ; capsule pear-shaped or cylindrical, often striated ; peri-
    stome often double, or the inner rather imperfect.
    Capsule pear-shaped.
      (Leaf with hyaline border below ; cells papillose . . . DASYMITRIUM.)
      (No hyaline border . . . . COLEOCHAETIUM.)
    Capsule cylindrical.
      Calyptra plicate or cut to the beak into spreading segments ; leaves seldom spirally twisted,
                                121. MACROMITRIUM.
      Calyptra not plicate, the basal lobes incurved ; leaves often spirally twisted . . 122. SCHLOTHEIMIA.

Family 29. EUSTICHIACEAE.

The only genus . . . . 123. EUSTICHIA.

(Family SPLACHNACEAE.

Peristome single . . . . SPLACHNUM.)

Family 30. GIGASPERMACEAE.

Leaf with nerve, capsule sessile . . . . 124. OEDIPODIELLA.
Leaf with nerve, capsule exserted . . . . 125. CHAMAEBRYUM.
Leaf without nerve . . . . 126. GIGASPERMUM.

Family 31. EPHEMERACEAE.

Calyptra conical, campanulate, and torn at the base, or cucullate . . . 127. EPHEMERUM.

Family 32. FUNARIACEAE

Capsule cleistocarpous . . . . 128 PHYSCOMITRELLOPSIS.
Capsule gymnostomous.
  Capsule on a well-developed seta . . . . 131. PHYSCOMITRIUM.
  Capsule almost sessile.
    Calyptra shorter than the lid . . . . 129. MICROPOMA.
    Calyptra covering the capsule . . . . 130. GONIOMITRIUM.
Peristome absent, or rudimentary or single . . . . . . 132. ENTOSTHODON.
Peristome double and well-developed . . . . 133. FUNARIA.

Family 33. MEESIACEAE (probably included in error).

Outer peristome short, inner longer ; leaf-cells smooth . . . . . 134. AMBLYODON.
The two peristomes of equal length ; leaf-cells papillose . . . . 135. PALUDELLA.

Family 34. AULACOMNIACEAE.

The only South African genus . . . . 136. LEPTOTHECA.

Family 35. BARTRAMIACEAE.

Branches irregular, not verticillate ; leaves narrow, male inflorescence bud-like.
  Peristome present . . . . 137. BARTRAMIA.
  Peristome absent . . . . (137) GLYPHOCARPUS.
Branches when present verticillate below the inflorescence ; leaves usually wide and short ; male inflorescence usually
  discoid in the dioecious species.
  Leaves not furrowed.
    Peristome present, single or double . . . . . 139. PHILONOTIS.
    Peristome absent . . . . 138. BARTRAMIDULA.
  Leaves furrowed . . . . 140. BREUTELIA.

Family 36. BRYACEAE.

Peristome very imperfect, consisting usually of basal membrane only . . . . (LEPTOSTOMUM).
Peristome usually single ; inflorescence on short lateral branches near the root.
  Outer peristome absent . . . . 141. MIELICHHOFERIA.
  Inner peristome absent . . . . 142. HAPLODONTIUM.
Peristome double ; male and female inflorescences usually near together at the top of the main stem.
  No basal membrane present . . . . 143. ORTHODONTIUM.
  Basal membrane present.
    Outer peristome longer than the inner . . . . 144. BRACHYMENIUM.
    Outer and inner peristome about equal.
      Upper leaf-cells narrow.
        Leaves narrow from a rather wider base ; spreading.
          Segments not ciliated . . . . . . 145. WEBERA.
          Segments ciliated . . . . . . 146. LEPTOBRYUM.
        Leaves more or less ovate, closely adpressed and imbricated . . . 147. ANOMOBRYUM.
      Leaf-cells all lax, rhomboid, or hexagonal.
        Plants stoloniferous . . . . . . 150. RHODOBRYUM.
        Plants not stoloniferous.
          Leaves lanceolate, acute. Ring absent . . . 148. MNIOBRYUM
          Leaves ovate. Ring present . . . . . 149. BRYUM.

Family 37. MNIACEAE.

Cells very large . . . . 151. MNIUM.

Family 38. RHIZOGONIACEAE.

Cells small . . . . 152. RHIZOGONIUM.

Sub-Tribe B. Diplolepideae Pleurocarpae.

Family 39. ERPODIACEAE.

Calyptra cylindrical, twisted, covering the whole capsule ; prostrate stems somewhat flattened, with leaves dimorphous.
  153. AULACOPILUM.
Calyptra bell-shaped, lobed, striate, not twisted, covering only the upper half of the capsule. Leaves all round the
  stem, not dimorphous . . . . 154. ERPODIUM.

Family 40. HEDWIGIACEAE.

Capsule immersed. Papillae 2- or more-pointed, small.
  Leaves with a spinosely denticulate hyaline point . . . . 155. HEDWIGIA.
  Leaves not having a hyaline point . . . . . 156. HEDWIGIDIUM.
Capsule more or less emergent ; papillae mostly 1-pointed or nearly smooth.
  Leaves not bordered ; seta long or short ; alar cells not differentiated . . . 157. BRAUNIA.
  Leaves bordered, seta long ; alar cells prominent . . . . 158. RHACOCARPUS.

Family 41. FONTINALACEAE.

Stems rigid, submerged ; seta produced, peristome rudimentary . . . . 159. WARDIA.
Stems long, floating ; capsule sessile, peristome double, imperfectly developed . . 160. FONTINALIS.

Family 42. CRYPHAEAE.

The only South African genus . . . . 161. CRYPHAEA.

Family 43. PRIONODONTACEAE

The only South African genus . . . . 162. PRIONODON.

Family 44. LEUCODONTACEAE.

Nerve absent ; leaf-cells smooth . . . . 163. LEUCODON.
Nerve single ; leaf-cells smooth . . . . 164. FORSSTROEMIA.
(Nerve single, or one strong nerve and one or two small side nerves at its base . . ANTITRICHIA.)
Nerve double ; leaf-cells papillose . . . . 165. PTEROGONIUM.

Family 45. FABRONIACEAE.

Inner peristome absent ; outer usually present.
  Silky-shining epiphytes.
    Slender little plants ; teeth of the outer peristome wide and blunt, or absent . . 166. FABRONIA.
    Stronger ; teeth of the outer peristome lanceolate, long and narrow . . . 167. ISCHYRODON.
  Slender plants, not silky shining ; teeth of the outer peristome widely lanceolate, in pairs 168. DIMERODONTIUM.
Peristome double.
  Teeth of the outer peristome separate, lanceolate, not striate ; basal membrane of inner peristome very narrow or
    absent . . . . 169. SCHWETSCHKEA.

Family 46. BRACHYTHECIACEAE.

Capsule erect, regular ; basal membrane very narrow ; leaves with many deep furrows.
  (Inner peristome somewhat attached to the outer . . . . . . HOMALOTHECIUM.)
  Inner peristome free . . . . 170. PLEUROPUS.
Capsule inclined or horizontal, irregular ; basal membrane of inner peristome wide ; leaves smooth or slightly furrowed.
  Lid conical, shortly pointed ; leaves somewhat furrowed ; alar cells differentiated . 171. BRACHYTHECIUM.
  Lid long-beaked ; alar cells not numerous or not much differentiated, not forming a group at the leaf-margin.
    Stem- and branch-leaves little different ; cells long and smooth, not thickened ; leaves not furrowed.
      Leaves widely ovate ; nerve often protruding as a mucro on the back of the leaf ; branches mostly flattened ;
        seta usually rough . . . . 172. OXYRRHYNCHIUM.
      Leaves narrowly ovate to lanceolate ; nerve not protruding ; branches somewhat flattened ; seta smooth ;
        plants robust . . . . 173. RHYNCHOSTEGIUM.
      Leaves narrowly lanceolate, sometimes pinnately arranged ; seta rough ; plants slender,
                                            174. RHYNCHOSTEGIELLA.
  (Stem- and branch-leaves often unlike, evidently furrowed ; leaves widely ovate ; nerve often protruding as a
    mucro. Cells long, smooth, not thickened. Seta rough . . . . EURHYNCHIUM.)

Family 47. LESKEACEAE.

§ ANOMODONTEAE. Main stems stoloniform ; seta rising from the branches. Paraphyllia absent ; leaves uniform ;
  processes of inner peristome thread-like, rudimentary, or absent.
Slender plants, the nerve not passing the middle of the leaf, often shorter. Inner peristome processes absent,
                                            175. HAPLOHYMENIUM.
Robust plants, the nerve ending at or near the leaf-apex.
  Nerve twisted above ; cells very small, quadrate, smooth ; processes rudimentary 176. HERPETINEURON
Seta rising from the main stem, which is not stoloniform.
§ LESKEAE. Leaves uniform.
  Capsule erect, regular, usually straight ; basal membrane of inner peristome narrow. Processes linear or absent.
    Dioecious ; teeth without lamellae ; processes thread-like . . . . 177. LESKEELLA.
    (Teeth of outer peristome with well-developed lamellae ; processes narrowly linear ; autoecious . LESKEA.)
    Teeth of outer peristome with narrow lamellae ; processes absent ; autoecious . . 178. LINDBERGIA.
  Capsule inclined or horizontal, more or less irregular ; basal membrane of inner peristome wide, processes wide,
                                            179. PSEUDOLESKEA.

§ Thuideae. Leaves not uniform (except *Haplocladium*).

Autoecious slender plants ; stems pinnate ; leaves dimorphous ; cells with numerous papillae on both sides ; end cell pointed or blunt. Lid shortly beaked from a conical base . . . . . 180. Rauia.

Stems pinnate ; leaves uniform, almost smooth or with one papilla on the lumen, or with projecting papillose angles ; end cell pointed. Lid shortly conical . . . . . 181. Haplocladium.

Dioecious ; stems 1–2–3-pinnate ; leaves heteromorphous ; cells with numerous low papillae ; end cell 2–4-pointed, 182. Thuidium.

Stems 1–2 pinnate, end cell blunt, 2-pointed ; lid with long narrow beak . . 182a. Thuidiella.

Stems 2–3·pinnate ; end cell of the branch-leaves (except *T. tamariscinum*) blunt, 2–4 pointed ; lid long, with thicker beak . . . . 182β. Euthuidium.

Family 48. Neckeraceae.

§ Pterobryeae Leaves not flattened, closely symmetrical ; alar cells more or less differentiated. Peristome usually double, with properistome ; outer teeth smooth, irregularly tumid ; inner peristome with very rudimentary processes and without cilia.

Nerve absent (or occasionally very short and double) . . . 183. Renauldia.

Nerve single (in the South African species). Alar cells differentiated . 184. Pterobryopsis.

§ Meteoreae. Usually epiphytal and hanging. Leaves not or hardly complanate, symmetrical, often auricled ; cells longish ; alar cells usually lax or distinct ; peristome double ; outer teeth not tumid, inner well developed. Seta smooth.

Nerve single.

Leaves not at all complanate.

Cells smooth ; alar cells quadrate ; distinct . . . . . 185. Squamidium.

Cells with several papillae on each ; alar cells hardly differentiated . . 186. Papillaria.

Leaves or some of them somewhat complanate.

Cells linear, several papillae on each. Seta short and smooth. Outer peristome striate, papillose only toward the apex . . . . 187. Floribundaria.

Cells oblong, one papilla on each. Seta long and rather rough. Outer peristome papillose, not striate, 188. Aerobryopsis.

Nerve absent, leaves in rows ; cells smooth, alar cells distinct . . . 189. Pilotrichella.

§ Trachypodeae. As in Meteoreae, but seta papillose.

Only South African genus . . . . 190. Trachypodopsis.

§ Neckereae. Secondary stems distinctly complanate, usually pinnate, seldom dendroid. Leaves unsymmetrical, in eight rows, the side leaves spreading, the back and front leaves adpressed, and arranged alternately right and left. Cells smooth, the upper rhomboid, the lower linear. Peristome double ; outer teeth not tumid, inner peristome well developed.

Branches closely inrolled when dry, paraphylliate ; nerve single ; properistome not present 191. Leptodon.

Branches not closely inrolled when dry ; usually not paraphylliate.

Nerve single ; leaves auricled ; properistome present . . . 192. Calyptothecium.

Nerve very short and weak, forked or absent ; leaves not auricled. Properistome not present 193. Neckera.

§ Thamnieae. Secondary stems erect and dendroid or frondose, usually more or less complanate. Leaves symmetrical or almost so, serrate, with single nerve not reaching the apex. Cells smooth or nearly so. Peristome double, the outer teeth not tumid ; the inner well developed.

(Leaves papillose . . . . Pinnatella.)

Leaves smooth.

(Outer peristome teeth papillose above, striate at the base only . . Porotrichum.)

Outer peristome teeth striate . . . . . . 194. Porothamnium.

Family 49. Entodontaceae.

Cells, or some of them, papillose ; nerve double, short, or absent . . . 195. Trachyphyllum.

Cells all smooth.

Processes much shorter than the teeth, thread-like ; nerve double, often short or absent 196. Erythrodontium.

Processes as long as the peristome teeth, or nearly so.

Nerve absent ; inner peristome without basal membrane ; branches not flattened, leaves not narrowed to the base . . . . 197. Platygyrium.

Nerve single ; inner peristome absent . . . . . 198. Levierella.

Nerve single, usually present ; leaves narrowed to the base ; peristome double . 199. Stereophyllum.

Nerve double, very short or absent.

Alar cells segregated ; branch-leaves usually flattened ; inner peristome without basal membrane, 200. Entodon.

(Alar cells not sharply segregated ; branches not flattened, often incurved. Basal membrane of inner peristome wide . . . . 148. Pylaisia.)

Family 50. HYPNACEAE.

A. Subfamily HYLOCOMIEAE. Leaves usually dimorphous, symmetrical, transversely inserted ; nerve double or absent ; stem-leaves usually from a wide base, drawn to a long or short point.

    Leaves more or less spreading, pointed ; seta smooth. Alar cells differentiated. Lid conical or shortly beaked.
      (Capsule erect, regular ; paraphyllia absent . . . . . . LEPTOHYMENIUM.)
      Capsule inclined or hanging, more or less irregular ; leaves dimorphous or heteromorphous ; stem-leaves lanceolate subulate . . . . 205. MICROTHAMNIUM.
    (Stem-leaves imbricate, rounded, or with short bluntish points. Alar cells numerous, forming a concave group,
      [Restricted definition not strictly adhered to here.]   HYPNUM.)

B. Stem-leaves and branch-leaves little different.

Subfamily AMBLYSTEGIEAE. Leaves transversely inserted, symmetrical ; nerve single, more or less long, seldom double. Lid never beaked.

    Leaves bordered . . . . 204. SCIAROMIUM.
    Leaves not bordered.
      Cells hexagonal, 2–4–6 times as long as wide (in the South African species) ; leaves secund,
              201. HYGROAMBLYSTEGIUM.
      Cells laxly quadrate or oblong-hexagonal, seldom linear ; leaves spreading widely . 202. AMBLYSTEGIUM.
      Cells long and linear, usually very narrow ; leaves usually falcate-secund . . 203. DREPANOCLADUS.
      (Cells long and linear ; leaves not falcate-secund.
      See *Catagonium* . . . . ACROCLADIUM.)

Subfamily STEREODONTEAE. Leaves transversely inserted and symmetrical, or somewhat obliquely inserted and more or less unsymmetrical. Nerve double or absent. Lid sometimes beaked.

    Alar cells evidently differentiated.
      Basal alar cells similar to the others . . . . . 206. STEREODON.
      Alar cells in a small group ; leaves denticulate . . . 207. ACANTHOCLADIELLA.
      Basal alar cells large, long ; others small, parenchymatous . . 208. ACANTHOCLADIUM.
    Alar cells not evidently differentiated, or few and small . . . 209. ECTROPOTHECIUM.

Subfamily PLAGIOTHECIEAE. Branches mostly flattened-leaved, side-leaves obliquely inserted, spreading complanately, usually unsymmetrical. Nerve double or absent. Lid conical or short—seldom long-beaked.

    Stem with indurated outer cells.
      Leaves in many rows, little differentiated. Leaf-cells prosenchymatous. Lid shortly pointed or conical.
        Cells very long ; the alar cells not differentiated ; leaves slightly concave . . 210. ISOPTERYGIUM.
        Cells rhomboid or linear ; the alar cells wider ; hyaline and thin-walled . . 211. PLAGIOTHECIUM.
      Leaves in two rows, very concave . . . . . 212. CATAGONIUM.
    Outer cells of stem not indurated ; leaf-cells parenchymatous ; outer line longer toward the base, making a border ;
      alar cells not differentiated . . . . . . 213. VESICULARIA.

Family 51. SEMATOPHYLLACEAE.

Inner peristome absent ; leaf-cells smooth . . . . . . 214. MEIOTHECIUM.
Peristome double ; leaf-cells smooth or ends projecting.
    Mid-line of teeth furrowed . . . . . . 215. SEMATOPHYLLUM.
    Mid-line of teeth zig-zag . . . . . . 216. RHAPHIDOSTEGIUM.
Peristome double ; leaf-cells with papillae on the lumen . . 217. TRICHOSTELIUM.

Family 52. HOOKERIACEAE.

(Nerve absent ; leaves bordered ; cells very lax ; seta smooth . . . HOOKERIA.)
Nerve single, undivided . . . . . . 218. DISTICHOPHYLLUM.
Nerve single at the base ; forked above . . . . . 219. ERIOPUS.
Nerve double at the base, divergent ; sometimes toothed along the ridge.
    Leaves margined . . . . 220. CYCLODICTYON.
    Leaves not margined.
      Cells usually smooth . . . . . . 221. HOOKERIOPSIS.
      Cells usually with one papilla per cell . . . . 222. CALLICOSTELLA.

Family 53. HYPOPTERYGIACEAE.

(Secondary stem simple ; under-leaves in one line ; seta very short, axillary . . CYATHOPHORUM.)
Secondary stem pinnate, fan-like or dendroid ; under-leaves in two lines ; seta long   223. HYPOPTERYGIUM.

Family 54. RHACOPILACEAE . . . . . . 224. RHACOPILUM.
           Genus discredited in its South African record.

(PTYCHOMNIACEAE=the *Achyrophyllum* (Mitt.) *aciculare* Brid. of Shaw's list . . PTYCHOMNION)

ORDER V. SPHAGNALES.

Family 17. SPHAGNACEAE.

Genus 59. SPHAGNUM Linn.

The only genus. Sixteen species have been recorded from South Africa, but there is still great force in Dr. Harvey's dictum (Genera of South African Plants, 1st ed.): " After all the labours of botanists on this genus, perhaps the so-called species may be reduced to one (*S. palustre* Linn.), which appears to be common to all parts of the world."

The Sphagna of South Africa have been described mostly from a few specimens sent to Europe by Drege and Dr. Pappe nearly a hundred years ago, further lots collected about fifty years ago by Dr. Rehmann, MacLea, and Breutel, and a specimen sent from Pondoland about thirty-five years ago by Bachmann ; the whole lot represents probably fifty tufts of Sphagnum, in which sixteen so-called species are claimed to have been detected, and either distributed in Rehmann's Exsiccatae or described, all except two being claimed as endemic to South Africa, and in these two exceptions one occurs also in Central Africa and the other in East African islands.

C. Müller, who named most of them, afterwards found that some of his species were synonyms, and his descriptions do not satisfactorily define or distinguish those left ; Warnstorf, in Hedwigia, 1891, dealt with what he placed as twelve South African species, and in 1901 he again had twelve, though not all the same as previously.

Taking into consideration the huge list of so-called species and varieties belonging to the Northern Hemisphere which have been grouped by most systematists into a few species (or groups of forms), and that even in some of these species the range of variation is considerable, it seems to me that to adopt Warnstorf's classification, wherein nearly every known tuft of Sphagnum is a distinct species, is inflicting immense trouble on those many botanists who wish in the first case to know the mosses of South Africa before they begin to search for slight varietal or form modifications, and wider grouping has consequently been here adopted. Fifty or a hundred years hence, when systematists have no further problems to solve, they may settle down to further separate the Sphagna, but meantime that task is a century ahead of the present stage of bryological research in South Africa.

Compared with the Sphagna of the Northern Hemisphere, *S. capense* Horns. (or *S. capense* group) is distinct and endemic, but the truncate species or forms can be closely matched elsewhere. No South African species has the leaves subsecund, but Warnstorf's classification puts most of them in the *subsecundum* group of the Sphagna of the world.

On the other hand, most of the northern groups are absent from South Africa, and there is thus an indication that long segregation, under generally drier and warmer conditions, has cut out many old types and produced some new ones.

The distribution of the genus in South Africa is peculiar, but in accordance with the local conditions. There are here practically no perennial swamps, moors, bogs, or wet stream-banks of the nature occupied by them in Europe or North America ; either water stands in pools or lakes constantly varying from being either too deep for Sphagna or quite dry, or else it flows in rapid streams of varying depth but always with dry banks, consequently the swamp and pool type of Sphagnum is absent or rare, and peat is absent. But on Table Mountain and other south-western mountains where rock-layers outcrop on a slope, and bring constant moisture as drainage, tufts of Sphagnum are frequent in rock-clefts and shelves, wherever these clefts are within the mist-belt and not subject to prolonged drying-out. Below the mist-belt Sphagna are absent or occur only in permanent rock-pools. Throughout the Karroo and north-west Sphagna are absent ; in the eastern forest country they are absent from inside the forest and from the thorn-veld below, but occur occasionally on moist rocks in the mist-belt above the forest ; on the Drakensberg they are rarely present ; in the eastern Transvaal they occur mostly in grassy springs ; in the Pondoland and Natal coast-belt they occur occasionally in small pools in small slow perennial streams, always where fully exposed to light and usually where constantly saturated but seldom flooded for many days ; and from Rhodesia and Portuguese East Africa they are not recorded, and I can say they are not present coastward there. In most of these localities *S. capense* Horns. is present ; the other types occur more rarely and under more special conditions. But the total quantity of Sphagnum in South Africa is extremely small, and the probabilities are all against specific differentiation beyond the stage of conditional forms.

Although male and female amenta occur, capsules are not recorded as having been collected in South Africa, except that Warnstorf mentions having found a capsule on one of Rehmann's specimens in Berlin. The illustration of a capsule on page 131, P, is consequently from a European specimen, and not connected with the species therein depicted, though there is little variation throughout the genus in respect to the capsule.

Synopsis.

Branch-leaves oval, deeply concave throughout.

181. *S. capense.* Stem-leaves short, rounded ; branch-leaves with erect, narrow, truncate, crenate apex. Hyaline cells all short, and very variable as to pores.

182. *S. pycnocladulum.* Stem-leaves short, acute ; branch-leaves with erect, acute, entire apex ; outer surface with very many pores in rows ; lower cells of stem-leaves without pores.

183. *S. panduraefolium.* Stem- and branch-leaves with bent back and incurved, cucullate, narrow, but crenate, rounded apex. Pores few.

Branch-leaves longer than oval, most concave at the base.

184. *S. Pappeanum.* Stem-leaves short, rounded ; branch-leaves spreading, boat-shaped ; pores absent ; hyaline cells very wide.

9

185. *S. Rehmanni.*   Stem-leaves large, concave below, with wide, flattened, rounded apex ; branches tapering, with closely imbricate leaves, the lower leaves with wide apex, the upper leaves gradually narrower, the apex varying from 5–3-toothed to acute.   Pores present in varying quantity, often abundant.

186. *S. truncatum.*   Stem- and branch-leaves spreading, with widely truncate, toothed apex, usually flat ; pores absent.

181. *Sphagnum capense* Horns., in Linnaea, xv, 113 (1841) ; C. M., Syn., i, 103 ; Warnst., in Hedw., 1891, p. 30 ; Warnst., in Pfl., i, 3, 262 ; Marloth's Flora of S. Afr., pl. 6, fig. A ; Paul, Pfl., x, 123.

S. mollissimum C. M., in Rehm. M.A.A. 434*b* and *c* ; C. M., in Flora, 1887, p. 418.

S. mollissimum C. M., var. *elongatum*, Rehm. M.A.A. 18.

S. austro-molle C. M., in Rehm. M.A.A. 16*b*, 433*a*, *b*, and *c* ; C. M., in Flora, 1887, p. 419.

S. Beyrichianum Warnst., in Hedw., 1897, p. 157 ; and in Pfl., i, 3, 262, from Pondoland, seems also to belong here.

Plant more or less slender, dense, 2–4 inches high, with numerous fasciculate, slender, spreading branches 6–10 mm. long, or occasionally longer, and many more slender deflexed branches.   Branch stems with occasional retort cells.   Stem-leaves 1 mm. long, triangular-concave from a very wide base, and with rounded entire apex.   Branch-leaves 1–1·5 mm. long, ovate-acute, very concave or tumid, closely imbricated, but the margins incurved till they almost meet, especially toward the erect or somewhat reflexed apex, which is narrowly truncate and crenate.   Margin bordered by one to three rows of long narrow cells.   Hyaline cells twisted, shortly oblong toward the leaf-apex, twice as long toward the base, having few to many pores, and having the chlorophyllose cells small, oval, and central in leaf-section.   Perichaetal leaves few, very large, oval.   C. Müller made too many species out of inseparable forms of this species, but reunited them and admitted that he had confused *S. panduraeforme* with *S. capense* Horns. under the name *S. austromollis* in Rehm. M.A.A. 16 and 16*c*.

Cape, S.W.   Table Mountain, Sim 9120, Rehm. 433*a*, *b*, 434*b*, *c* ; Window Gorge, Kirstenbosch, 1500–2000 feet altitude, Sim 9486, 9465, and Pillans 3545 ; Newlands Ravine, 2000 feet altitude, Sim 9417 ; Blinkwater Ravine, 1700 feet altitude, Sim 9260, and Dr. Bews (Sim 8609, 8610) ; Disa Ravine, Sim 9139, and Dr. Bews (Sim 8491) ; Platteklip Ravine, 3000 feet altitude, Sim 9276 ; Stinkwater, Rehm. 18 and 433*c*.

Cape, E.   Egossa, Pondoland, 1000 feet altitude, 1899, Sim 7440, and see *S. Beyrichianum* above.

Natal.   Van Reenens Pass, mixed in Rehm. 355*f* ; Little Berg, Cathkin, Drakensberg, 6000 feet altitude, July 1922, Miss Owen 30.

Transvaal.   Macmac, MacLea (Rehm. 434), and same locality, mixed in Rehm. 456 ; Rietfontein, Wager.

Rehmann's label 334*c* is named *S. mollissimum* on some packets in error.

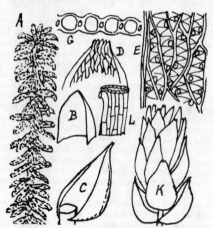

Sphagnum capense.

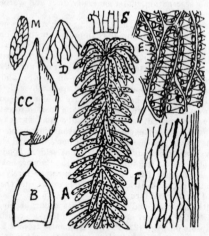
Sphagnum pycnocladulum.

182. *Sphagnum pycnocladulum* C. Müll., in Flora, 1887, p. 420 ; Warnst., in Hedw., 1891, p. 130 ; and in Pfl., i, 3, 259, fig. 159*c* ; and Paul, Pfl., x, 117.

Stem robust, rigid, 4–6 inches high, many-branched, the larger branches arched, the smaller pendent, and altogether giving a dense columnar aspect.   Cortical cells in several layers, the outer with occasional retort cells.   Stem-leaves small, oblong-acute from a wide base, concave and with much-incurved margins, the narrow marginal cells forming a distinct border, and toward the leaf-base the hyaline cells are longer than those above and without pores.   Branch-leaves imbricate, very concave, ovate-acute from a cordate or ovate-saccate base, the apex either acute and simple, or formed of two converging cells which occasionally separate somewhat ; leaf-margins incurved, and along the grooves of the outer surface the strings of pearl-like pores are continuous over the whole surface ; the hyaline cell-form varies little, and there is a narrow border of long cells.   Perigonial cone dense, its leaves short, ovate-acute, and the apex often 2–3 split.

Known only from Rehm. M.A.A. 13 and 17, both from Montagu Pass; the latter distributed by Rehmann as *S. mollissimum* C. M., but not corresponding with his other specimens distributed under that name.

This species is included by Warnstorf (Pfl., p. 259) in section *Mucronata*, a small group confined to South Africa and East African islands, distinguished by its short leaf, with acute, entire or almost entire apex, but the species is too closely related to *S. capense* Horns. to allow them to belong to separate groups.

Cape, S.W., Montagu Pass, Rehm. 13 and 17.

183. *Sphagnum panduraefolium* C. M., in Flora, 1887, p. 418; Warnst., in Hedw., 1891, p. 26; Warnst., in Pfl., i, 3, 260.

Stems slender, several inches high, with very few branches, mostly spreading. Cortical layer one cell deep, with few retort cells. Stem-leaves fairly large, spreading, oval, very concave, very tumid in the lower half, and less so above the middle, with hooded apex, and consequently often appearing as if spreading or bent outward from the middle. Apex much incurved, rounded, slightly several-toothed; the leaf narrowly bordered by two to three rows of long narrow cells; hyaline cells much longer than wide, many-banded; the upper cells shorter. Branch-leaves similar to the stem-leaves but smaller, more numerous, and imbricated at the branch-apex, and with cells as in the stem-leaves; the chlorophyllose cells wedge-shaped in section, with the short base of the triangle near the inner leaf-surface. Pores very few or absent in the young plant; more frequent but still few in the mature plant.

Known first from Rehm. 15, a slender little plant 1 inch high, simple or sparingly branched below—not above—concerning which Warnstorf says the whole appearance is that of a plant not yet fully developed, in which I concur; but under Rehm. 433 Müller gives the observation that Rehm. 16 and 16c, distributed under the name *S. austro-molle* C. M., belong to *S. panduraefolium* C. M., and having seen Rehm. 16 I agree that it does so, though more branched toward the apex than is Rehm. 15. Thus *S. austro-molle* C. M., as distributed by Rehmann, belongs in part here and in part to *S. capense* Horns. But my 9670 and other specimens are a still more developed state of the same species, and from these the above description is drawn. This species is closely related to *S. capense*, but with cucullate leaves, consequently the back of the leaf is bent and the tip incurved in this, while the tip is erect or reflexed in *S. capense*.

Cape, S.W. Table Mountain, Rehm. 15, 16c (as *S. austro-molle* C. M.); Lower Plateau, Table Mountain, Sim 9670; Stinkwater, Rehm. 16 (as *S. austro-molle* C. M.); Blinkwater Ravine, 2000 feet, Dr. Bews (Sim 9673); French Hoek Pass, 2000 feet, Miss Stephens (Sim 9672).

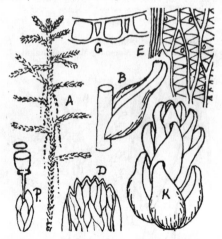

Sphagnum panduraefolium.

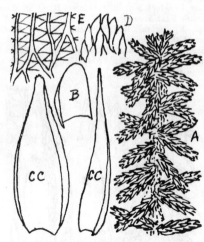

Sphagnum Pappeanum.

184. *Sphagnum Pappeanum* C. Müll., Syn., i, 101 (1848); Warnst., in Hedw., 1890, p. 248; Warnst., in Pfl., i, 3, 254; and Paul, Pfl., x, 116, where it is included as the only South African species in his group *Rigida microphylla*.

Stem robust, 3–4 inches long, abundantly branched all up the stem, with robust, leafy, spreading branches ½–¾ inch long, and also with pendent slender branches along the stem. Cortex one to two cells deep, without retort cells. Stem-leaves small, oval or oblong, with very wide base, rounded apex, and narrowly bordered margin. Branch-leaves large, clasping below, spreading above, oblong to lanceolate, from a concave-ovate base, the larger narrowly truncate and toothed, the smaller subulate-acute or narrowly 2–3-toothed at the apex, all widely bordered below, hardly bordered above. Cells long, crooked, without pores. Perichaetal cone large, its leaves few, very large, widely ovate, with 5–7-toothed apex, and widely bordered by long narrow cells from apex to base; inner cells short, twisted, banded, poreless, those toward the leaf-apex shorter.

The type of this species is Dr. Pappe's specimen from Zwellendam (C. M., Syn., i, 101), and the description agrees with Rehm. M.A.A. 12, from Montagu Pass, from which my figure and description are taken. It is the most squarrose South African species.

Stuhlmann's 2385 from Central Africa has also been placed in this species by Brotherus.

Cape, S.W. Zwellendam, Dr. Pappe; Montagu Pass, Rehm. 12.

185. *Sphagnum Rehmanni* Warnst., in Hedw., 1891, p. 16, and includes *S. Rehmanni* Warnst., in Pfl., i, 3, 261, and p. 260, fig. B, and Paul, Pfl., x, 121 ; and includes also *S. eschowense* Warnst., Pfl., x, 121 : *S. oligodon*, Rehm. 14 ; C. M., in Flora, 1887, p. 412 ; Warnst., in Hedw., 1891, p. 39 ; and in Pfl., i, 3, 261 ; and Rehm. 431 : *S. oxycladum* Warnst., 1869 ; and in Hedw., 1891, p. 15 ; and Pfl., i, 3, 261 = *S. coronatum* C. M. var. *cuspidatum*, Rehm. 10 : *S. Bordasii* Besch., as claimed for South Africa in Rehm. 432 by Warnst. (Hedw., 1891, p. 25, and Pfl., i, 3, 261), distributed by Rehmann as *S. coronatum* C. M., and as equal Rehm. 9, which Warnstorf says is different, but in this I do not concur. *S. coronatum* C. M. is described in Flora, 1887, p. 412 : * *S. transvaalense* C. M., in litt., and Warnst., Hedw., 1891, p. 32, and Pfl., i, 3, 262, seems to belong here, but to have up to 10-toothed branch-leaves ; I have seen no specimen.

Plant usually vigorous, 3–4 inches high, light coloured, well clad with strong branches, and with a crown of strong spreading branches ¾–1 inch long, these branches wide at the base but tapering to an acute point ; it also occurs several inches high, unbranched or almost unbranched, and then with all its leaves as very large, entire, rounded stem-leaves. Stem-leaves very large, very concave from widely ovate base, three times as long as wide, with widely rounded, entire apex and narrow margin of long cylindrical cells and rather short hyaline cells. Branch-leaves on lower branches and at the base of all branches similar to the stem-leaves but rather shorter, and sometimes with three to five protruding apical cells ; further along the branch all leaves are gradually narrower, closely imbricating, 5–3-toothed, or, in the subulate apical leaves, acute and untoothed. Margin composed of long narrow cells. Pores few, inconspicuous.

Warnstorf separated Rehm. 14 (*S. oligodon* Rehm.) from this by its having the stem-leaves more widely bordered, and *S. oxycladum* Warnst. and his *S. Bordasii* by number and position of the pores on the branch-leaves, which characters appear to me insufficient.

S. convolutum Warnst., in Hedw., 1890, p. 220, Tab. VIII, pp. 10–12, and x, 6 ; and *S. fluctuans* C. M., in Flora, 1887, p. 414 ; Warnst., in Bot. Centralblat, 1900, p. 20, both from Cape Town, are included by Warnstorf, in Pfl., i, 3, 257, in the lanceolate-leaved section of group *S. cuspidata* and along with the cosmopolitan *S. cuspidatum* R. and W., but seem to me to belong to *S. Rehmanni* Warnst., whose branches produce leaves very varied in form, but mostly long, somewhat wavy, often several-toothed at the apex, and imbricate or sheathing.

Cape, S.W. Table Mountain, 1917, Dr. Bews ; Top of Table Mountain, floating in pools, 1919, Sim 9299 ; Paarl Mountain, 1500 feet, Sim 9628 ; Mossel River, in Water-courses, Jan. 1913, Professor Potts 22 (Sim 9675) ; Montagu Pass, Rehm. 9 (as *S. coronatum* C. M.) ; Brandvley, Worcester, Rehm. 432 (as *S. coronatum* C. M.).

Cape, E. Egossa, E Pondoland, 100 feet altitude, Sim 9671 (not typical) ; Eveleyn Valley, Perie Mountains, 4500 feet, 1893, Sim 7166.

Natal. Inanda, Rehm. 14 (as *S. oligodon* Rehm.) ; Ngoya Forest, Zululand, Sim.

Transvaal. Sabie, Sim 8014, 8580 ; Rustenburg, J. Playford (Sim 8334) ; Macmac, MacLea (Rehm. 431 as *S. oligodon* Rehm.).

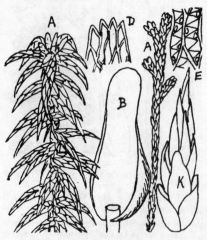

Sphagnum Rehmanni.   *Sphagnum truncatum.*

186. *Sphagnum truncatum* Horns., in Linnaea, xv, 114 ; C. M., Syn., i, 103 ; Warnst., in Hedw., 1891, p. 28, Tab. 21*a*, *b*, and IV*p* ; and in Pfl., i, 3, 261 ; Paul, Pfl., x, 122.

A long and very lax floating plant ; stems 6–8 inches long, with few branches ; the spreading branches ¾–1 inch long, sparsely leaved except at the point, the pendent branches loosely appressed, very slender, and with the long scattered terminal leaves narrowly lanceolate, acute or 2–3-toothed at the apex. Branch-leaves very large, ovate-oblong, widely truncate, nearly flat except at the concave sheathing base, the margin not involute, narrowly margined in the upper half, rather widely margined below ; the apex nearly as wide as the upper part of the leaf, flat, 7–10-toothed, the teeth often in pairs and, like all the upper tissue, dense and chlorophyllose, consisting of the massed ends of the chlorophyllose tubes which are abundant and large throughout the leaf, elliptical and central in leaf-section ; the hyaline cells shortly vermicular toward the leaf-base, shorter upward, all with abundant spiral bands ; pores very small, mostly in the upper half of the leaf on the inner surface only. The above description is from a specimen collected by Dr. Marloth in the pool of a spring shaded by rocks, Muisenberg (Marloth 147).

The type was collected by Drege in swamps at Dutoits-Kloofberg, at about 2700 feet altitude, and is in Herb. Laurer, Berlin. Sim 9261, from Blinkwater Ravine, Table Mountain, is a much smaller plant, and was not floating, but agrees in formation.

From Warnstorf's description (Hedw., 1891, p. 28) * *S. marginatum* Schimp. in Hb. Kew (Warnst., in Pfl., i, 3, 261) depends on a single specimen collected by Breutel (Hb. Zichendrath) at " Cap Sonderend," and appears to agree with *S. truncatum* in everything except that the stem-leaf borders are up to eight cells wide and the branch-leaf borders up to four to six cells wide. Further proof is required concerning it.

All the above are placed by Paul (Pfl., x, 113) in his Section I. *Lithophloea*, defined " stem- and branch-cells without spiral fibres," but he also has a Section II. *Inophloea*, " stem- and branch-cells, or only the latter, with spiral fibres," in which, besides *S. palustre* L. (=*S. cymbifolium* Ehrh.) already mentioned (and which he locates as almost cosmopolitan, except Africa), he includes *S. Balfourianum* Warnst., from East African islands and South Africa, and *S. Marlothii* Warnst., from the Cape, both in one group, having the smaller leaves without fibres and the larger with many.

He maintains altogether fourteen South African species of Sphagnum, viz. : *S. Pappeanum* C. M. ; *S. pycnocladulum* C. M. ; *S. Bordasii* Besch. ; *S. coronatum*, C. M. ; *S. oxycladum*, Warnst. ; *S. transvaalense*, C. M. ; *S. marginatum*, Schpr. ; *S. eschowense*, Warnst. ; *S. oligodon* Rehm. ; *S. Beyrichianum*, Warnst. ; *S. truncatum*, Horns. ; *S. capense*, Horns. ; *S. Balfourianum* Warnst. ; and *S. Marlothii*, Warnst.

Order VI. ANDREAEALES.

Family 18. ANDREAEACEAE.

Genus 60. ANDREAEA Ehrh.

The only genus. Tufted, firm, dark-coloured rock-mosses, with numerous small firm leaves, thick-walled small cells in straight lines, terminal inflorescence, and the capsule opening by four longitudinal slits in its middle portion but closed at the apex, or ultimately opening by four spreading valves. The capsule resembles that of an hepatic and the leaves those of a moss.

Synopsis.

187. *A. petrophila.* Leaves ovate-oblong, obtuse, erect.
188. *A. subulata.* Leaves subulate, subfalcate.

187. *Andreaea petrophila* Ehrh., in Hann. Mag., 1784, p. 140 ; and in Beitr., i, 192 ; Pfl., x, 129 ; Dixon, 3rd ed., p. 24.

=*A. rupestris* Hedw. (not Roth), var. *capensis*, Rehm. 430.
=*A. amblyophylla* (C. M.), Broth., an Australian name.

Plant ½–¾ inch long, somewhat branched, slender but firm and brittle, julaceous when dry, clothed below with the remains of very many leaves ; leaves dark brown, ovate-oblong, obtuse, cucullate, entire, nerveless, concave but

Andreaea petrophila.

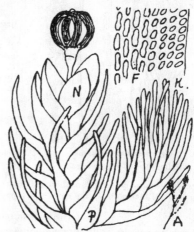

Andreaea subulata.

with margins not incurved beyond that except near the apex when dry. Upper and outer cells roundly quadrate, so swollen on the outer surface as to appear papillose, especially when dry ; lower central cells gradually more oblong, irregular in outline. Perichaetal leaves much larger, sheathing, more pointed, nerveless.

Cape, E.   Dohne Hill, 5000 feet, 1897, Sim 7205 ; Rhenosterberg, 6800 feet, MacLea (Rehm. 430).
O.F.S.   Nelson's Kop, Wager (as *A. amblyophila*).
Natal.   Top of Giant's Castle, 8000 feet, Symonds.

188. *Andreaea subulata* Harv., in Hook. Icon. Pl. Rar., iii, t. 201 ; C. M., Syn. Musc., i, 10 ;
Broth., Pfl., i, 3, 268, and x, 130 ; Marloth's Flora of S. Afr., pl. 6, fig. B.

Plant ½–1 inch high, irregularly branched, tufted. Capsule shortly exserted ; leaves entire, subulate, from a short, wide sheathing base, widely but indistinctly nerved below by oblong cells up to the subula ; subula terete, solid, obtuse, often falcate ; cells between nerve and margin nearly round. Perichaetal leaves larger, sheathing, elliptical, apiculate but not subulate.

Cape, S.W., top of Table Mountain on rocks, Sim 9223 ; Rehm. 8. Harvey (Genera of S. Afr. Plants, 1st ed.) says common there ; I found it present, but not common.

Order VII. POLYTRICHALES.

Family 19. POLYTRICHACEAE.

Genus 61. ATRICHUM P. B., Prod., p. 40 (1805).

Plants 1–2 inches high, gregarious, having simple or forked stems from an underground rhizome. Leaves numerous, spreading when moist, crisped when dry, clasping but not sheathing, strapshaped-acute, strongly bordered, serrate, undulate, with comparatively narrow central nerve, having three to seven lamellae on the upper surface, and often spinulose at the back near the apex. Capsule smooth, cylindrical, terete, frequently curved, with a long seta, a glabrous or scabrid calyptra, a long rostrate beak, and a peristome of thirty-two ligulate rigid teeth, with a narrow basal membrane. Columella round.

Terrestrial forest mosses, growing in extensive patches and fruiting freely.

Syn. *Catharinea* Ehrh., in Hannov. Mag. (1780), p. 913.

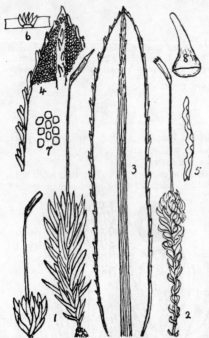

Atrichum androgynum.
6, lamellae on front of leaf.

189. *Atrichum androgynum* (C. M.), Jaeg. Ad., i, 703 ; Paris, i, 67 ; *Catharinea androgyna* C. M., Syn., i, 193 ; Broth., in Pfl., i, 3, 672 ; Dixon, S. Afr. Journ. Sc., June 1922, p. 323 ; Brunnthaler, 1913, i, 28.

Monoecious, the male and female flowers near together. Plants gregarious but seldom crowded, erect, 1–3 cm. high, dark green, usually simple, clothed laxly with leaves except at the base, but the leaves gradually larger upward, the upper 8–12 mm. long, oblong-lanceolate or ligulate-lanceolate, 2–2·5 mm. wide, the lowest leaves scale-like. Leaves rather laxly placed, plane and erecto-patent when moist, much undulated at the margins and twisted inward when dry. Nerve wide, pronounced, terminating just below the apex, and having three to five raised longitudinal lamellae on the upper surface. Margin strongly bordered by long narrow cells, the ends of which protrude as sharp teeth, usually in pairs ; the teeth produced along the whole length of the leaf, but small toward the base and larger upward, and similar teeth are often produced on the back of the nerve in its upper portion. Areolation very fine ; cells small, varying from square to hexagonal, in regular longitudinal lines. Seta 2–3 cm. long, yellowish ; capsule inclined, 5–6 mm. long, cylindrical, often somewhat curved, sometimes erect ; lid half as long, rostrate.

Frequent near the forest or on moist banks in all forest districts from Montagu Pass, Blanco, and Knysna eastward through the Eastern Province, Natal, Zululand, Transvaal, and Rhodesia, but no specimens are on hand from Table Mountain or that neighbourhood. It is common also in moist, shady eastern mountain localities.

Included in this are Rehmann's 266, distributed as *A. polyphyllum* Rehm., and Macowan's Somerset East plant (*C. synoica* C. M., in sched.). Endemic. *A. angustatum* (Brid.), Br. Eur., which is dioecious, has been ascribed in error to South Africa.

(Genus POLYTRICHADELPHUS (C. Müll.) Mitt.)

Shaw includes in his South African list *Polytrichum dendroides* Hedw., as in the Kew Herbarium, with locality and collector not named, and mentions that it is also found at the Straits of Magellan, New Zealand, and Peru. I am not aware of its having been recorded again from South Africa, and Shaw's record seems doubtful. *Polytrichum dendroides* Hedw. was in C. Müller's Syn. Musc. the only species in his § *Dendroligotrichum* of *Catharinea* having these characters : " Leaves long-sheathing at the base, twisted ; capsule cylindrical, urn-shaped or oval, not incurved, more or less erect. Stem stalked at the base, branched above like a tree," while his next section *Polytrichadelphus* is described as having the capsule and leaves of the preceding section, except that the leaves are juniperoid, not twisted, but densely imbricated. Brotherus (Nat. Pfl., p. 680) illustrates the plant as *Dendroligotrichum dendroides* Hedw., and mentions these other countries but not South Africa, and the plant shown is one I have not seen.

Mitten made *Polytrichadelphus* a genus and included *dendroides* (Hedw.) Mitt. in it, and this is still upheld by Paris, who places fifty-three species in the genus. The above characters will identify this plant, if found, but I have doubts about its being South African.)

Genus 62. Psilopilum Brid., Bryol. univ., ii, 95 ; *Catharinea* § II *Psilopilum*, C. M., Syn. Musc., i, 194.

Small, erect, simple mosses, growing gregariously in lax mats from underground rhizomes ; leaves vaginate, not narrowed above, not bordered, slightly toothed, the leaf lamina consisting of a single layer, the wide nerve ceasing below the leaf-apex and having about six leafy plates rising from the whole length of its upper surface ; cells fairly large, quadrate, lax ; outer and lower cells not different ; plants dioecious, male inflorescence discoid, terminal. Fertile inflorescence terminal, with hardly differentiated leaves. Seta usually erect, smooth ; capsule erect or somewhat inclined, laterally compressed, ovoid, with or without peristome ; lid beaked or conical · calyptra cucullate, glabrous.

The species belong mostly to the Southern Hemisphere and the Arctic Zone.

Synopsis.

190. *P. afro-laevigatum.* Peristome absent.
191. *P. Wageri.* Peristome present.

190. *Psilopilum afro-laevigatum* Dixon, Trans. Roy. Soc. S. Afr., viii, 3, 207, and pl. xi, fig. 10 (1920).

Plant erect, 5 mm. high, with six to ten oblong lamellate leaves, which are erecto-patent and slightly concave when moist, but incurved and with the margins much inflexed and the apex hooded when dry. Leaves somewhat sheathing at the base, roundly acute at the apex, the margins not bordered, bluntly toothed or almost entire, but with an occasional tooth-like projecting cell. Nerve wide, ceasing just below the somewhat incurved leaf-apex, and with about six green, wavy, leafy lamellae on its inner surface, each many cells deep, one cell thick, and with crenate margin ; the leaf lamina a single layer of large clear lax cells somewhat tumid at the back, $16–20\mu$ in regular lines ; involucral leaves like others but more toothed. Seta ½ inch high, straight ; capsule oval-elliptical, erect or somewhat inclined, with wide mouth and no peristome. Lid calyptra and male inflorescence not seen.

Moist mountain banks, not under 4000 feet altitude, rare.

Natal, Rosetta (Sim 6068) ; Gildats, Boston (Sim 9682) ; Little Berg, Cathkin, 6000 feet, Miss Owen No. 20 (fertile), etc.

191. *Psilopilum Wageri* (Broth., M.S.), Dixon, Trans. Roy. Soc. S. Afr., viii, 3, 207, and pl. xii, fig. 12 (1920).

Unknown to me, but Dixon describes it as being about the same size as *P. afro-laevigatum*, with leaves more acuminate and more cucullate, concave and involute ; cells rather smaller, lamellae about five, with slightly crenate margins ; nerve percurrent, glabrous on the back. Seta 5–7 mm. high, red. Capsule small, oval, curved, compressed with age ; peristome $60–70\mu$ high, of thirty-two remote, triangular, unequal; acute, yellow teeth.

Rydal Mount, O.F.S. (Wager 45).

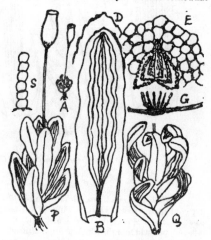

Psilopilum afro-laevigatum.

Genus 63. Pogonatum P. B., Prod.

Plants gregarious, usually small, simple, and rosulate, the female innovating from the base or rhizome, the male from the centre of the male inflorescence. Leaves sheathing at the base, spreading, suddenly narrowed above the membranous and narrowly nerved sheath into a firm, many-lamellated, wide nerve with almost no lamina, except the serrate margin. Calyptra thatched with deflexed hairs, and covering most of the capsule ; capsule shortly cylindrical, erect or inclined, regular or slightly incurved, in transverse section round without angles, and not or hardly apophysate. Peristome of thirty-two short teeth.

Sometimes included as a section of *Polytrichum.* Pioneers on clay soil, cut banks, road-sides, etc.

Synopsis.

192. *P. simense.* Leaves entire, upper leaves bristle-pointed.
193. *P. capense.* Leaves bluntly toothed, obtuse ; capsule elliptical.
194. *P. subrotundum.* Leaves bluntly toothed, subobtuse ; capsule shortly cup-shaped.

192. *Pogonatum simense* (B. and S.), Jaeg. Ad., i, 719 ; Broth., Pfl., i, 3, 692 ; *Polytrichum simense*, B. and S., Pl. Abyss. Schimper ; C. M., Syn., i, 206 ; *Cephalotrichum simense*, Br. and Sch., Br. Eur., vol. iv.

A rigid little rosulate moss, laxly gregarious, usually among other mosses.   Plant up to ¼ inch high, with ten to twenty rigid leaves, the lower small and oblong-acute, without lamellae, the ordinary leaves widely lanceolate from a short pellucid sheathing base, margin entire and inflexed, apex acute or rigidly brown-pointed, incurved when dry ; nerve occupying the whole wideness except at the base, and having twelve to eighteen low lamellae ; inner and perichaetal leaves longer, with longer pellucid sheath and longer and narrower lamina, having six to twelve lamellae and ending in a long, subulate, horny, brown, entire point.   Cells large and lax in the pellucid base, much smaller and chlorophyllose in the lamellae and lamellose portion.   Paroecious ; seta ¾–1 inch long, smooth ; capsule erect, without apophyses, terete, elliptical, when young tapering below and wide-mouthed, somewhat tuberculate ; lid convex with an erect short blunt point ; calyptra rough at the point, hairy below, the deflexed hairs covering the capsule.   Capsule faintly ribbed ; more pronounced when dry.

This species takes its name from the Abyssinian Mountains where W. P. Schimper first found it ; it also occurs in the Cameroons, and on Mount Elgon, Uganda.

Natal.   In mountain districts only : Van Reenens Pass, Rehm. 574 (distributed unnamed), also H. A. Wager 52 ; Bulwer, 4000 feet, 1914, W. J. Haygarth ; Garden Castle, 6000 feet, Sim 9693.

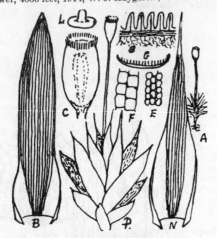

Pogonatum simense.

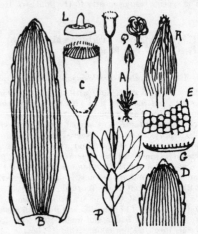

Pogonatum capense.

193. *Pogonatum capense* (Hpe.), Jaeg. Ad., i, 711 ; Broth., Pfl., i, 3, 687.

= *Polytrichum capense*, Hampe, Ic. Musc., t. 21 ; C. M., Syn., i, 203.
= *Polytrichum* *yuccoides, Rehm. 575.

A rigid little moss like a miniature aloe.   Rosette 3–4 mm. high, 6–12-leaved ; leaves usually erecto-patent, when dry contorted inward, when growing on wet banks sometimes flattened out.   Leaves from a short, sheathing, pellucid base, ovate-lanceolate, bluntly acute, with about twelve low lamellae occupying the whole surface except a narrow, inflexed, pellucid margin which is often almost entire, but sometimes distinctly inciso-dentate toward the apex, especially in the perichaetal leaves.   Nerve not toothed on the keel.   Dioecious, the male inflorescence discoid.   Seta 5–7 mm. long, smooth ; capsule erect, terete, ovate-cylindric, when young tapering below and wide-mouthed, without apophysis, surface somewhat warty toward the base, smooth above ; peristome teeth short ; columella not winged ; lid convex, with short blunt central teat ; calyptra rough at the point, hairy below.   The pellucid teeth on the leaf-margin are more evident when dry than when moist.

Specimens distributed by Wager, named in error *Pog. nanum* (No. 72) and *Pol. aloides*, both belong to *P. capense* ; the former is distinguished herein, the latter has the leaves much more sharply serrated, and though the perichaetal leaves of *P. capense* are often more or less pellucidly dentate they are less so than *P. aloides*, which also has the leaf narrower and has horned teeth on the keel.

Abundant in mist-belt, mountain, and forest localities in Eastern Cape, Natal, and Transvaal ; present also in Rhodesia at Victoria Falls, Sim 8932 ; but I have no specimen or record from Western Province, Cape, except Sim 9743 from Kalk Bay, 1000 feet.

194. *Pogonatum subrotundum* Huds., Fl. Angl., 1st ed., p. 400 ; Broth., Pfl., i, 3, 686.

= *P. nanum* (Schreb.) P. B., Prod., p. 84 ; and many British works.
= *Polytrichum nanum*, Schreb., Spicil. Fl. Ups., p. 74, No. 735 ; C. M., Syn., i, 204 ; Dixon, 3rd ed., p. 39.

Very similar to *P. capense* Jaeg., but with leafy stem 3–7 mm. high, seta 6–10 mm. high, capsule shortly cup-shaped or almost globose, and lid conical-beaked.   Leaves shortly lanceolate from a sheathing base, tapering to the apex which is bluntly acute and composed of several semi-detached cells ; the pellucid margin crenate or bluntly toothed, the teeth usually of single cells. Nerve comparatively narrow along the whole length of the leaf, but the lamellae rising from the lamina as well as from the nerve.

A common moss in the Northern Hemisphere, but I have only seen one South African specimen, viz. Rehm. 573, from Mahaliesberg, Transvaal (MacLea), distributed by Dr. Rehmann as *P. capense*, Hpe.

(*Pogonatum Borgeni* Jaeg. =*Polytrichum Borgeni* Hampe, in Bot. Zeit., 1870, p. 35, from Natal, is placed by Brotherus, in Pfl., i, 3, 689 (without his having seen it), as probably belonging to a group having stem up to 10 cm. long, single or forked ; nerve not protruding, leaves and keel sharply toothed upward, lamellae three to four rows deep, with marginal cells larger and rounded ; capsule inclined, cylindrical ; calyptra hairy. I have seen no such plant, but the name is recorded from Eshowe, Zululand, by Bryhn., Videns. Forh., 1911, iv, 17. *Polytrichum transvaalense* C. M., in Hedw., 1899, p. 63, is described as near this but much stronger.)

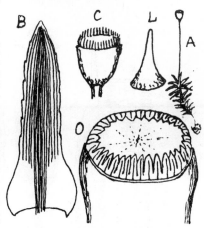

Pogonatum subrotundum.

(*Pogonatum convolutum* (Linn.) Brid., Brit. Univ., ii, 110, from Bourbon and Java, is placed by Müller, Syn., i, 213, in his section *Catharinella*, having crisped, curled secund leaves, and terete or subbicarinate capsule, and resembling *Atrichum*, but distinguished from it by its villose calyptra. The species is described as being dioecious, several inches high, with lanceolate-lamellate leaves having short sheathing pellucid base, the margin bluntly spinose, the keel with horny spines upward. Perichaetal leaves very minute, the inner reduced to ligulate form. Capsule cylindric ; calyptra small, very hairy.

Its variety β *cirrhatum* C. M. is smaller, more slender, with leaves narrower, and not secund, and is said to have been sent from " Prom. bona spei " by Thunberg.

This describes a distinct species which has not been seen here, and I suspect the country is wrongly stated.

Brotherus (Pfl., i, 3, 690) omits South Africa from its distribution, but includes the species in a group 7–30 cm. high, the var. *cirrhatum* having leaves toothed to the base, lamellae one to two rows of cells deep, with forked marginal cells. The terete capsule separates this from *Polytrichum*.

In Shaw's list *P. convolutum* is included from Graaf-Reinet as collected by MacLea.)

Genus 64. POLYTRICHUM Dill.

Prominent gregarious plants, usually having erect, wiry, leafy stems 1–12 inches long, simple or sparingly branched, and innovating from underground rhizomes or from the centre of the male inflorescence. Leaves sheathing, similar to those of *Pogonatum*, but set all along the stem, and usually longer and more rigid and opaque, but equally lamellose along the nerve, which occupies the whole width of the leaf, and often serrate along the margin, and sometimes at the back also, near the apex. Seta long, capsule usually inclined or horizontal, prismatically 4–6-sided, seated on a discoid or subglobose apophyses, and having a convex and pointed lid, and the calyptra thatched with deflexed hairs. Peristome teeth short, usually sixty-four.

A remarkable genus, having an erect rigid habit which cannot be mistaken when once known.

In North Europe the larger species are used for making brooms. Some authors still include *Pogonatum* in *Polytrichum*, and without capsules there are no sufficient distinctions.

Synopsis.

195. *P. piliferum.* Leaves entire, awned.
196. *P. natalense.* Leaves shortly lanceolate, serrate, without awn.
197. *P. commune.* Leaves linear-lanceolate, serrate, with or without awn.

195. *Polytrichum piliferum*, Schreb. Spic. fl. lips., p. 74, No. 1031 ; C. M., Syn., i, 217 ; Broth., Pfl., i, 3, 696 ; Dixon, 3rd ed., p. 43.

Plant 1–3 inches high, laxly gregarious, leafless below, the upper half closely clad with imbricate entire leaves, closely appressed when dry. Leaves firm, two to four lines long, lanceolate from a short, wide, sheathing base, quickly tapering at the apex to a long, hyaline, denticulate awn ; twenty to thirty lamellae, occupying nearly the whole lamina, each about five cells deep with terminal cell larger and cruciform ; leaf-margin entire, narrow, inflexed, and composed of transversely elliptical cells ; sheath hyaline except the narrowed dark nerve ; cells, the nerve rectangular-oblong, those toward the leaf-margin longer and very narrow, those separating the oblong cells from the chlorophyllose lamina shortly transverse-oblong. Upper leaves with longer arista and inner perichaetal leaves sometimes with long and wide sheathing hyaline lamina to or near the long awn. Seta 1½–2 inches long ; capsule small, erect when young only, sharply 4-angled, with separate apophysis and short teeth ; lid convex with an oblique beak ; calyptra covering the capsule, with numerous deflexed hairs.

Almost cosmopolitan, and having many varieties or forms, but all with the hyaline denticulate awn, joining the coloured nerve where the leaf widens.

Cape, W. Paarl Mountain, near the rock, 1500 feet, Jan. 1919, Sim 9610 ; Langeberg, Oudtshoorn, Dr. Bews, 1925 ; Table Mountain, Rehm. 269. It occurs also on Mount Elgon, Uganda, and in the Canaries, but I have seen no other specimens from South Africa, though it is likely to be present.

P. juniperinum Willd., which has been reported from South Africa, but evidently in error, has almost similar *entire* leaves with a *red arista* ; probably forms of *P. commune* with red arista have been mistaken for it.

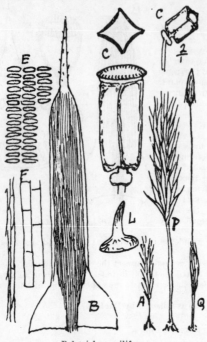

Polytrichum piliferum.

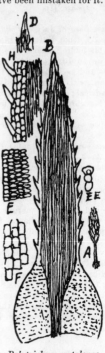

Polytrichum natalense.

196. *Polytrichum natalense*, Sim (new species).*

Plants ½–1 inch high, gregarious, leafless below, and with a coma of very solid, shortly and widely lanceolate, awnless leaves from short, wide, hyaline sheaths, and tapering gradually to the acute apex. Nerve almost as wide as the leaf, with many lamellae three to four cells high with rounded terminal cells. Leaf-apex acute, horny. Margin not inflexed, sharply serrate, the horny teeth each composed of a single cell ; the clear lamina of about three rows of oblong cells. Nerve narrow in the sheath, the lower sheath-cells shortly oblong or quadrate, passing gradually outward and upward into small transversely elliptical cells. Inflorescence and capsule not seen, so this may possibly be a *Pogonatum* of the *P. aloides* group, but it differs much from European specimens of *P. aloides* Hedw. in being a firmer, more horny plant, the leaf-teeth composed of single cells, the lamellae lower, and in the cell-formation ; altogether its appearance is that of a *Polytrichum* rather than that of a *Pogonatum*. It is evidently quite distinct, not a juvenile form of another species.

Natal. Home Rule, Polela, 4000 feet, Sim 9694 ; Upper Bushman's River, 7000 feet, Sim 8671 ; Upper Tugela Valley, 7000 feet, Sim 9695 ; Mont-aux-Sources, 10,000 feet, Miss Edwards (Sim 9696) ; Giant's Castle, 8000 feet, R. E. Symons (Sim 9697).

197. *Polytrichum commune* Linn., Sp. Pl., ii, 1109, No. 1 ; C. M., Syn., i, 220 ; Dixon, 3rd ed., p. 46.

An exceedingly variable species, even in one patch varying from 1–18 inches in height, and varying also in other characters ; laxly gregarious when short, matted when long, and then usually tomentose at the base ; stems erect,

---

* *Polytrichum natalense* Sim (sp. nov.).

Plantae ½–1″ altae, gregariae, infra foliis carentes, cum coma foliorum solidissimorum itemque breviter lateque lanceolatorum asubulatorum e brevibus latisque vaginis hyalinis, ad apicem acutum sensim decrescentium. Nervus prope tam latus quam folium, pluribus lamellis 3–4 cellas altis, cella terminali rotundata. Folii apex acutus, corneus. Margo non inflexus, acute serratus, lamina hyalina trium fere ordinum cellarum oblongarum, dentes marginales e singulis quaeque cellis constantes, nervus angustus in vagina, inferiores vaginae cellae breviter oblongae vel quadratae, sursum et extrorsum sensim in parvas transverse ellipticas cellas desinentes. Inflorescentiam et capsulam nondum vidi.

wiry, firm, usually simple, equally clad in leaves throughout except at the base.   Leaves ¼–¾ inch long, spreading when moist, twisted up to or round the stem when dry, lanceolate or linear-lanceolate, tapering regularly from the sheath to the apex, clasping by a much wider hyaline sheathed base, and somewhat attached along the inner surface of this base. Leaves occupied along the inner surface of almost the entire lanceolate portion by very many parallel raised lamellae (fifty to sixty) about six cells deep, the terminal cell depressed along its centre. Margin hyaline, of one to two rows of cells, and sharply serrate from the sheath upward, the teeth usually equal and each composed of a single cell, but sometimes with smaller intermediate teeth. Apex either a sharply serrate acute point, or extended into a long, subulate, denticulate hair-point of red colour except at its apex. Back of leaf often toothed on keel toward the point. Sheath-cells oblong, the outer longer and very narrow; those above the oblong cells roundly quadrate, forming a distinct zone.   Perichaetal leaves more awned and more sheathing, and often with long scarious margin.   Male inflorescence terminal, discoid, with very short, wide, acute, scale-like leaves, but the stem continues later to grow from the centre again, becomes leafy, and again terminates with a male inflorescence; this is repeated annually, and each old whorl remains.   Seta 1–4 inches long, wiry; apophysis distinct; capsule at first erect, finally horizontal, 4-angled; lid convex, with conical beak; calyptra large, brown, enclosing the capsule.

Extreme forms are very different, and this has led to many species having been described from apparently distinct herbarium specimens of this almost cosmopolitan species, but in growth it is found that these forms grade from one into another, sometimes separately, sometimes in juxtaposition, so that the small and the large, the long-leaved and the short-leaved, the awned and the awnless, the serrate and the dentate, and the simple and branched forms cannot be separated as permanent species or varieties, and can only be regarded as conditional forms subject to further variation under changed environment.

The following names have been applied to South African forms :—

(a) P. commune Linn., type = P. *Buchanani Rehm. 580.   Leaves ½ inch long or more, laxly serrate.   Plants usually 4–12 inches long.

A form of this has the leaves with long narrow points as in form trichodes.

(b) P. commune var. minus Weis., Pl. Crypt. Gött., p. 171; C. M., Syn., p. 221.
= P. commune var. *africanum C. M., in Rehm. 276 (from Montagu Pass).
= P. commune var. *Macleanum, Rehm. 577 (from Macmac).
= P. radulifolium C. M., in Hedw., 1899, p. 62.   Sterile (from Transvaal).
= P. *Rehmanni C. M., in Rehm. 270 (Montagu Pass) and 270b (Table Mountain).
= P. *Rehmanni C. M. var. julaceum Rehm. 576 (Houtbosh) and 576b (Macmac).

Leaves ¼ inch long, or less, sharply and closely serrate; plant usually 2–4 inches long.   Leaves acute but not awned.

Polytrichum commune.

S.1, type ; S.2, male plant ; S.3, form *minus* ; S.4, form *fastigiatum* ; S.5, form *trichodes.*

(c) *P. trichodes* C. M., in Rehm. 277, 578, and in Hedw., 1899, p. 63, is described as a small form having the nerve extended as a long, strong, entire awn, the leaf-tip ferruginous, and the margin sparsely serrated upward. This awned condition is frequent both in the large and small forms of the species, the awn being usually red in its lower half, while the leaf-margin varies from laxly to closely serrated. Tyson's 1265 from Griqualand East, named *P. Ecklonianum*, Hpe., belongs here.

(d) *P. Höhnelii* C. M., in Flora, 1890, p. 471, described from Central Africa, is supposed to differ from var. *minus* in having the teeth very large and numerous, and with smaller intermediate teeth present. This occurs among var. *minus* through South Africa.

　　P. elatum P. B. is included in Dr. Shaw's list as collected at Boschberg, Somerset East, by Macowan, and *P. remotifolium* Schw. is shown in Drege's Zwei Pfl. Doc., 1843; both belong to this group and are Mascarene, but not known from South Africa.

(e) *P. commune* var. *fastigiatum* Wils., Bry. Brit., p. 212, is forked or repeatedly branched, which appears to me to be the result of the tips being damaged by wagon traffic, as it usually grows on roadsides. It occurs at Gillitts, Natal, and also in Europe.

(f) *P. armatum* Broth., in Engler's Bot. Jahrb., 1897, p. 253, and its var. *minor*, from Zambesia and East Africa, have leaves ciliate at the base, closely and sharply toothed upward. Known only sterile.

(g) *P. *flexicaule* C. M., in Rehm. 275, 275*b*, and Hedw., 1899, p. 62; *P. *atrichoides* C. M., Rehm. 273; and *P. flaccido-gracile* C. M., in Rehm. 274, and Hedw., 1899, p. 62, are all lax, small, immature conditions from Knysna.

　　P. commune occurs in all mist-belt, forest, and mountain regions from Table Mountain to the Zambesi (and elsewhere), and from Inanda on the Natal coast to the Drakensberg and Transvaal Houtbosch, and the Matopos, Zimbabwe, and Victoria Falls.

Order VIII. BRYALES.

Tribe I. APLOLEPIDEAE.

Family 20. Archidiaceae.

　　Plants small, slender, branching somewhat by innovations, usually from below the perichaetium. Leaves lanceolate or ovate-lanceolate, with hexagonal cells. Capsule globose, not pointed, sessile, formed of a single layer of cells, and bursting irregularly. Columella absent, the few (ten to twenty) large spores rising from a single basal cell. Calyptra thin, membranous, tearing irregularly from near the base of the capsule. Male flower gemmiform; sterile plant often flagellate. Slender plantlets, usually less than 1 cm. long, and growing in patches on rock or soil or stream-drift. The sessile capsule usually appears axillary through innovation, and flagellae are often an inch long and exceedingly variable in accordance with surroundings and conditions.

　　This family differs from all other mosses in its few and very large spores, and in the absence of columella and the nature of its capsule and calyptra.

　　It is included in Dicranaceae by Dixon and Jameson, but is sufficiently distinct to deserve equal rank.

Genus 65. Archidium Brid.

　　The only genus. Characters are those of the family; all our species are monoecious.

Synopsis.

A. Leaves spreading (*Euarchidium*).
　198. *A. alternifolium.* Areolation very lax; cells rhomboid.
　199. *A. capense.* Areolation dense; cells quadrate.
(? B. Leaves closely appressed, julaceous (*Schlerarchidium*), *A. julicaule*.)

　　198. *Archidium alternifolium* Schimp., Syn. Musc. Eur., i, 26, and ii, 810; Dixon, 3rd ed., p. 53.

=*A. Rehmanni* Mitt., in Journ. Linn. Soc., pp. 146, 300, and pl. 15, figs. 5–8; C. M., in Rehm. M.A.A. 427; Broth., Pfl., x, 155.
=*A. *falcatulum* C. M., in Rehm. 429 and *b, d.*
=*A. laterale* Bruch, in Flora, 1846, p. 132; C. M., Syn., i, 14.
　A. campylopodium C. M., in Hedw., 1899, p. 52=*A. *compactum* C. M., in Rehm. Herb., forma *tenerior*, is described as having more cuspidate leaves than *A. Rehmanni* Mitt., and *A. africanum* Mitt. from Central Africa hardly differs (Journ. Linn. Soc., pp. 146, 299, and pl. 15, figs. 1–4).

　　Plants minute, 2–6 mm. high, at first erect with terminal inflorescence, but afterwards with prostrate innovations from within or below the perichaetium, and these innovations again produce terminal inflorescence and further innovations. Primary stem almost leafless except for the perichaetal leaves (about six) which are spreading, lanceolate-subulate, and hyaline, the subula long and narrow and almost entire; nerve rather thin, not evident in the subula; cells large and lax, rhomboid, long-hexagonal, and somewhat twisted. Leaves of innovations nearly as large, closely erecto-patent, lanceolate, with wide nerve into the point, and lax rhomboid chlorophyllose cells. Capsule sessile, large, entirely included, with twelve to twenty large pale spores and convex-mucronate lid.

A frequent pioneer species in Europe ; in South Africa my specimens are from—
Cape, S.W. Devil's Peak, Rehm. 427 (as *A. Rehmanni* C. M.) ; Rondebosch, Rehm. 429*d* ; Lion's Head, Rehm. 429 ; and Greenpoint, Rehm. 429*b* (all as *A. falcatulum* C. M.).

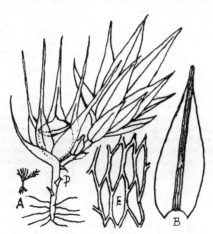

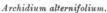

Archidium alternifolium.

Archidium alternifolium (form, Sim 8186).

Natal, Jan. 1908, H. A. Wager ; River Umslutie, Krauss (as *A. laterale*, a name which is quite excusable after innovation).

I cannot separate Sim 8186 from moist, ploughed ground, Scheepers Nek, Natal, in which, however, the perichaetal leaves are more lanceolate and denticulate, not subulate.

199. *Archidium capense* Horns., in Linnaea, 1841, p. 135 ; C. M., Syn., i, 13 ; Marloth's Flora of S. Afr., p. 1, pl. 6*c* ; Broth., Pfl., x, 155.

=*A. Ecklonianum* Hpe., in Sched.=*A. Ecklonii* C. M., in Hedw., 1889, p. 53.

(**A. synoicum* C. M., in list of Macowan's mosses, No. 25, probably belongs here.)

Primary plant erect, 2–5 mm. high, but innovating from inside the perichaetium or from lower on the stem, with similar short fertile branches, or from the base with abundant sterile small-leaved flagellae, up to an inch or more in length. Per. leaves much longer than the others, few, lanceolate-subulate, entire ; branch-leaves much smaller, ovate-lanceolate to lanceolate-acuminate, concave, clasping, erecto-patent, sometimes numerous ; flagellae distantly leaved, the leaves triangular to ovate-acute, keeled, reflexed at the apex, and somewhat squarrose ; all leaves with wide excurrent nerve, and the cells in all leaves small and quadrate. Capsule included, globose, with convex-mucronate lid and a few large pale spores (twelve to twenty).

This species varies much in the development of its branches and flagellae, and can only be satisfactorily separated from *A. alternifolium* by the much smaller quadrate areolation ; indeed, Schimper's description, which has no reference to areolation, is wide enough to include all forms of *A. alternifolium* and *A. capense* as here described. It is a pioneer on flat rocks, frequent in the Western Cape Province, scarce elsewhere, and often the flagellae form mats with few fertile branches.

Cape, S.W. Miller's Point, Pillans 4073 ; Newlands, Pillans 3984 ; Hout Bay, Rehm. 425 ; Kloof Nek, Dr. Bews (Sim 8421, 9566) ; Constantia, Sim 9341 ; Blinkwater Ravine, Sim 9259 ; Paarl Rock, Sim 9624, 9654 ; Robertson, Sim 9600.

Natal. Van Reenen, Jan. 1911, H. A. Wager.

A. julicaule C. M., in Hedw., 1899, p. 52 =*A.* *chrysosporum (Schpr. herb.) (from Saldanha Bay, *fide* C. M.), distributed as Rehm. 426, from hills above Cape Town, is exactly *Pleuridium nervosum* H. and W. (see fig. on page 143), with few spores (twelve to twenty) instead of many ; I cannot distinguish it by any other character in Rehmann's specimen, which, however, certainly has the capsule and spores of *Archidium*. Mitten's illustration of *A. Rehmanni* Mitt. shows small-leaved flagellae, which look very different in foliage from the fertile stem and are almost julaceous (though

Archidium capense.

he describes the leaves as "patent"); in Hedw., 1899, p. 53, C. M. mentions *A. Rehmanni* as having already been recognised as *A. *falcatulum* C. M., MSS., and in the unstable condition of the vegetative characters in the genus I look on the cells as the only permanent character, even though Brotherus maintains several on leaf-form and position.

A. subulatum C. M., in Flora, 1888, p. 8, is unknown to me, but is described in Broth., Pfl., x, 155, as belonging to § *Euarchidium*, monoecious, nerve percurrent, and leaves not subulate-pointed.

In a packet collected by Professor Pearson at Erongo Mountains, S.W. Africa (9849), the bulk of the collection was *Brachymenium Borgenianum* Hampe, but Mr. Dixon picked out one stem with one immature capsule of what in letter (14/10/1920) he calls **Archidium pellucidum* Dixon sp. nov., and says it is a very distinct new species in the lax, empty, pellucid cells. Unfortunately, neither his packet nor mine gave any further specimen of this plant, so it must wait further finding.

Family 21. DICRANACEAE.

Plants varying from minute up to several inches in height, simple or dichotomously branched. Leaves usually subulate or lanceolate, occasionally round or ovate-acute, in almost all cases nerved to the apex or near to it; lower cells rectangular, upper cells rounded, quadrate, rectangular or linear; alar cells often specialised. Seta usually present and terminal. Capsule erect, inclined or cernuous, globose or oval to cylindrical, sometimes gibbous with (except in some cleistocarpous members of § Ditricheae) narrow, usually cucullate calyptra, and peristome having no basal membrane, but consisting of sixteen straight teeth, each usually more or less forked into two lanceolate or subulate lobes, the outer layer thinner than the inner layer and having numerous transverse bars, and with fine vertical striae between these; inner layer with more or less distinct transverse bars. This is what is known as a Dicranoid peristome tooth-form, but appears also elsewhere than in Dicranaceae.

This family, though fairly natural, has no definite distinctive character, and includes several abnormal forms, especially in the distichous *Distichium* and in the cleistocarpous forms, which latter, however, appear in all groups as the lowest grade.

The following arrangement into sections is that used by Dixon and Jameson in The Student's Handbook of British Mosses, who, however, include in their Dicranaceae, as separate further sections, the groups here raised to family rank as Archidiaceae and Leucobryaceae.

§ I. *Ditricheae*. Leaves lanceolate or wider at the base, rarely short and wide, seldom papillose, and without special alar cells. Capsule erect and symmetrical, or somewhat inclined and unequal, in some cases rounded and cleistocarpous, in others oval to cylindrical; lid conical; peristome when present of sixteen narrow teeth, each cleft more or less forked or perforated; genera eight to fifteen.

Genus 66. PLEURIDIUM Brid.

Plants small or very small, simple or slightly branched, with or without innovations, and sometimes with flagellate growth. Cells rather large, not papillose, the lower rhomboid or hexagonal, the upper shorter, except those in the nerve; perichaetal leaves often setaceous; leaves lanceolate or subulate or rounded, not curled when dry. Capsule erect on a very short seta, ovate-globose, mucronate, cleistocarpous, almost included among the perichaetal leaves. Calyptra usually cucullate. Spores small, rough. Plants usually only one or two lines long, growing gregariously on soil, but occasionally much longer as flagellae.

This genus includes various plants which were placed in *Astomum* by Hampe. Paris sinks *Astomum*, and places our species under *Pleuridium*. Brotherus, in Marloth's Flora of South Africa, includes *Pleuridium*, with three species in Dicranaceae, and also *Astomum* with one species in Pottiaceae. In his illustration (Plate 6, *g*) it is *Astomum tetragonum* Harv. which is shown, which is *Tetrapterum capense* (Harv.), Paris, herein.

Synopsis.

200. *P. nervosum.* Leaves julaceous, acute, with excurrent nerve.
201. *P. Pappeanum.* Leaves spreading.

200. *Pleuridium nervosum* (Hook.), Brotherus, Pfl., i, 3, 294, and x, 157;
Dixon, Bull. N.Z. Inst., No. 3, pt. 2, p. 39 (1914).

=*Phascum nervosum*, Hook., Musci Exot., vol. ii, Tab. 105; Hk. f. and Wils., Fl. of N.Z., xi, 58.
=*Phascum *capense*, Sprengel, MS.
=*Phascum *curvatum* Taylor, in M. Drumm.
=*Astomum nervosum* C. M., Bot. Zeit., 1847, p. 98; and Syn. Musc., i, 15 (except Chile and except syn. *Phascum Robinsoni* Mont.).

Caespitose, erect, 4 mm. high, light green; stems mostly simple with innovations from the base, but occasionally carrying fertile innovations from below an old capsule; leaves densely imbricated, numerous, julaceous, ovate-acute, somewhat concave, with a very wide excurrent nerve, and crenate-dentate margin or almost entire. Per. leaves much larger, concave, lanceolate, enclosing the capsule, with the wide nerve excurrent as a long point. Venation beautifully reticulated, the cells toward the base hyaline and oblong-hexagonal, 0·05 mm. long, 0·02 mm. wide, with narrower cells along the margin; those in the upper part of the leaf 0·03–0·04 mm. long, tumid, irregularly elliptical-hexagonal or quadrate, the outer forming slight crenations (fig. D, page 143) of the margin. Capsule reticulated, brown, nearly sessile, quite included among the long per. leaves (which meet when dry), ovate with a short mucro, very tough, splitting irregularly, and containing very numerous small spores 0·03 mm. diameter. Calyptra dimidiate, with long slender point. Male flower disciform in erect, closed, bud-like, short, basal, and axillary branchlets, with four to six small, ovate-acute, slightly toothed bracts enclosing the few ovoid club-shaped antheridia and longer, slender, jointed paraphyses (fig. M).

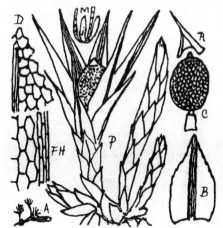

Pleuridium nervosum.

Very similar to *Archidium alternifolium* in general appearance but larger, and has numerous small spores instead of the few large ones found in that species.

For its resemblance to *Archidium Rehmanni* Mitt., see under that species.

Shaw includes *Astomum *subulatum* Hpe. (=*P. subulatum* (Linn.) Sch.) in his South African list, but I find no proof of its presence distinct from *P. nervosum* ; in it the nerve is narrow and does not reach the leaf-apex, and the antheridia are naked in the axils of the per. leaves.

P. alternifolium Rabenh. (of Europe) has similar male inflorescence in the axils of the upper leaves, but differs in the per. leaves being suddenly narrowed from a wide base to a long setaceous point.

Cape, S.W. Lions Mountain, Rehm. 462*b* ; Stellenbosch Flats, Miss Duthie 22. Müller (Syn., i, 15) cites Menzies, Ecklon, Mund, and Pappe as collectors.

Cape, E. Dohne Hill, 4000 feet, 1897, Sim 7195 ; Perie Mountains, 3000 feet, 1892, Sim 7333, 7336 ; Graaf-Reinet, MacLea, Rehm. 462.

Natal. Arcadia, Scheepers Nek, 4000 feet, Sim 9708.

General distribution : South Africa, Australia, and New Zealand.

201. *Pleuridium Pappeanum* (C. M.), Jaeg. (M. Cleist., p. 31 (1869), et Ad., i, 22 (1871)) ; Broth., Pfl., x, 157.

=*Astomum Pappeanum* C. M., Syn., i, 15.

This is described by Müller as having stems erect, subsimple ; stem-leaves spreading, loosely imbricate, linear-subulate from a widely oval base, entire, with wide excurrent nerve. Capsule globose, shortly pointed, almost sessile, brown, shining. Calyptra dimidiate, base entire, membranaceous, subcompressed. Male flowers gemmaceous, sessile, small, in the axils of the lower leaves ; perigonial leaves small, spreading. He adds, " Easily distinguished from *A. nervosum* by the stem-leaves being spreading—not julaceous, and by the male flowers."

Cape, S.W. Zwellendam, among *A. nervosum* (Dr. Pappe) ; also in woods in same locality (Ecklon).

I have not seen this, and the description seems to me to refer to forest-grown specimens of *P. nervosum*.

P. Breutelianum (Hampe), Jaeg., M. Cleist., p. 32 ; Broth., Pfl., x, 157.

=*Astomum Breutelianum* (Hpe. MSS.), C. M., in Bot. Zeit., 1859, p. 97, is unknown to me. Brotherus, in Pfl., i, 3, 295, placed it in § *Sclerastomum*, having sterile branches julaceous, but in Pfl., x, 157, he moves it into § *Eupleuridium* (along with *P. Pappeanum*), having spreading leaves, with quadrate upper cells and autoecious inflorescence.

Genus 67. CLADOPHASCUM Dixon (new genus).*

The plant on which this genus is founded was formerly, through incomplete material, described as *Aongstroemia gymnomitrioides* Dixon, in S.A. Journ. Sc., xviii, 301 (1922). Further material having been obtained, Mr. Dixon has now kindly favoured me with the following description of the new genus, together with additional description of the one species known, for insertion here, viz. :—

" Plants small, very slender, in form and structure of leaves recalling some species of *Illecebraria* Hampe, in the immersed capsule much like *Eccremidium*, but with the fruit often lateral, though sometimes terminal. Perhaps allied to the genus *Eccremidium* of Dicranaceae. Calyptra mitriform ; ring none ; spores large."

202. *Cladophascum gymnomitrioides* Dixon.

Mr. Dixon's additional description is :—

" *Cladophascum gymnomitrioides* Dixon, comb. nov. Syn. *Aongstroemia gymnomitrioides* Dixon, in S.A. Journ. Sc., xviii, 301 (1922) ; Broth., Pfl., x, 179.

" Sporophyte (undescribed) very conspicuous ; inflorescence usually at first terminal, occasionally lateral ; always rapidly becoming lateral by innovation ; synoecious ; male flower gemmiform, minute, perigonial leaves three to four,

---

* *Cladophascum* Dixon (gen. nov.).

Stirps humilis, gracillimus, foliorum forma et structura species nonnullas *Illecebrariae* Hampe referens, theca immersa *Eccremidio* sat similis, fructu tamen saepe laterali, nunc autem terminali. Generi *Eccremidio* Dicranacearum forsan affine. Calyptra mitriformis ; annulus nullus ; spori magni.

broadly oval, shortly and obtusely pointed ; antheridia four to five ; perichaetia large, bracts falcate and homomallous when immature, straw coloured, membranaceous, 1–2 mm. long, from a broad, concave, almost sheathing, ovate base abruptly narrowed to a falcate subula ; areolation lax, pellucid, basal rectangular, upper linear-rhomboid. Capsule almost sessile, quite immersed, subglobose or slightly urceolate, orange-brown, when ripe coriaceous and tough ; lid rather large, convex, with a blunt apiculus ; annulus none, stomata at base of capsule superficial, normal ; exothecium cells lax ; calyptra mitriform or bell-shaped, large, covering about half the capsule, of rather inflated cells, the upper overlapping the lower to some extent, like the tiles of a roof, irregularly protuberant at the basal rim. Peristome none. Spores large, 50–55μ, sometimes 60μ, orange-yellow ; almost smooth ; very finely reticulated on the surface by what a high magnification shows to be exceedingly minute superficial pits. Each spore contains a highly refractive region, which I do not recollect having seen before in a moss spore ; it may be an oil globule, but I do not think so ; at any rate it is always single, not like the oil globules in the spores of *Archidium*. It conveys more the impression of a vacuole or air-bubble.

"*Hab.*—To the localities given in the publication cited above may be added Kimberley Flats, 1918 (Moran), Sim 9706 ; Kimberley, Alexanderfontein, 1922 (Sim 9707) ; Burttholm, Vereeniging, Transvaal, 4000 feet, 1918 (J. Burtt-Davy 17,740).

"I published this plant from immature specimens, as an *Aongstroemia*, on account of the very close resemblance in the gametophyte to *A.* (*Illecebraria*) *julacea* Hook. Recently, however, Dr. Sim has detected it in fruit among earlier

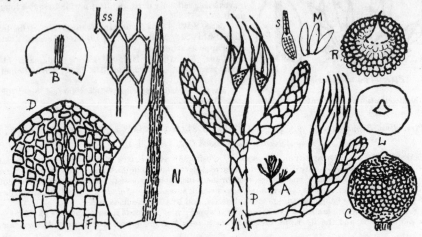

Cladophascum gymnomitrioides.

S, archegonium ; SS, cells of perichaetal leaves.

gatherings from Kimberley, and an examination of the fruiting characters shows at once that it has no close relationship with *Aongstroemia*, nor indeed with any other known genus. The description of the sporophyte in the published diagnosis is incorrect ; what was then taken for the male flower being really a very immature female flower, with a minute male flower, containing the antheridia at its base, so close indeed that it is easily removed as part of the perichaetium, and the inflorescence appears at first sight as synoecious. Dr. Sim, indeed, concluded that it was truly synoecious, but if so it is exceptional, as all the sporophytes I have dissected showed the autoecious inflorescence.

"The flowers are, no doubt, normally terminal, or at least principally so, becoming lateral at once by the stem producing a single innovation below the flower, in which case the cladocarpous position is rather apparent than real. That it is, however, at times actually cladocarpous seems clear from the fact that quite young perichaetia are frequently found laterally on a stem the upper part of which is of considerable age, while now and then a lateral perichaetium is distinctly younger and more recent than a more nearly terminal one on the same stem. Whatever the origin the very conspicuous perichaetia, two or three on a stem, are for the most part finally lateral, and the resultant plant bears a close resemblance to *Gigaspermum*, a resemblance carried even to most of the fruiting characters, which are very similar. The affinity is, however, not really close, as the stomata on the capsule of *Gigaspermum* are Funarioid in type, which is not the case here ; while *Gigaspermum* has the leaves nerveless and the calyptra very minute. *Eccremidium* has a very similar sporophyte, but the seta there is arcuate and the capsule not actually immersed, the fruit does not become lateral, and the calyptra is minute. *Astomiopsis* C. M. differs in much the same way, and the capsule moreover is annulate.　　　　　　　　　　　　　　　　　　　　　　　　　　　　　　　　H. N. D."

My own description of the plant is :—

Stems erect, closely caespitose, autoecious, 4–6 mm. high, with inflorescence lateral or at first terminal, ultimately lateral from growth of innovations from below the perichaetium ; abundantly fertile ; the perichaetal leaves conspicuous by their light green colour. Stem and sterile branches julaceous, the leaves numerous, closely imbricated, very concave, entire, rounded from a wide base, and as wide as long, or wider. Cells in nerve and lamina all in one layer ; nerve of two lines of cells darker and more regular than the others, vanishing below the apex ; upper cells lax, separate, and irregularly rhomboid ; lower cells 4–sided, in regular lines. Per. leaves much longer, entire, ovate-acute, with a long subula, the nerve and subula consisting of three to six lines of long narrow cells ; the cells of the lamina long-hexagonal, pellucid, much larger than those of the stem-leaves. Capsule immersed, hard, globose or shortly oval on a very short seta, erect, the lid convex-acuminate, marked by a dark line before dehiscence and separating at that line ; ring absent ; calyptra mitriform, convex-rostellate, not dimidiate, composed of numerous large loose overlapping cells, the lower cells giving a crenate margin.

A dry-country moss, abundant where present, but only known from the localities mentioned by Mr. Dixon, viz. on flat rocks at Zimbabwe, Matopos, and Khami in Rhodesia, and Kimberley (Cape Province), and Vereeniging, Transvaal. In the Rhodesian localities it is the most common moss on bare rock, densely caespitose on granite, but absent from soil or decomposed rock ; in the Kimberley sites it is on the shallow soil overlying lime-shales, and at Vereeniging it is on hard, stiff clay loam, dry for many months every year.

Genus 68. DITRICHUM, Timm.

Plants small, tufted ; leaves lanceolate-subulate, placed all round the stem, usually smooth, with excurrent nerve and with rectangular or rhomboid cells, the upper cells smaller and sometimes rounded. Capsule smooth, erect or nearly so, regular or nearly so, on a long seta ; from oval to cylindrical in form, without an elongated neck, and with a ring. Calyptra cucullate ; peristome teeth sixteen, usually inserted at, not below, the mouth, red or purplish, erect, papillose, each cleft to the base into two long *filiform* segments, or these segments occasionally united more or less above ; basal membrane short or almost absent ; no properistome. Spores very small, smooth. Lid conical.

The name *Leptotrichum* which has been in use for this genus is superseded, having been earlier used as a generic name among Fungi. This genus has by recent reviewers been removed from Dicranaceae, and as a family Ditrichaceae, along with Trematodontaceae, Aongstroemiaceae, and Trichostomaceae assembled under the name Ditrichocranoideae.

The South African species belong to section *Euditrichum*, having capsule usually slightly irregular and the peristome teeth divided to the base, and our plants are monoecious.

Synopsis.

Plants 2–10 mm. high, usually simple.
203. *D. brachypodium.*   Plant minute ; leaves crowded in a comal tuft ; perichaetal leaves hardly sheathing.
204. *D. flexifolium.*   Leaves lax, scattered ; perichaetal leaves long-sheathing. Peristome teeth well developed, filiform, papillose.
205. *D. amoenum.*   Peristome teeth not well developed.
Plants 8–12 mm. high, branched, cells rhomboid or rounded except those at the base.
206. *D. strictum* (as above).
Plants 2–3 inches high, simple or branched.
207. *D. spirale.*   Awn twisted when dry. Apical cells rounded or elliptical, mid-leaf cells rhomboid, basal cells long-oblong, with narrow-celled hyaline leaf-border.

203. *Ditrichum brachypodium* (C. M.), Broth., Pfl., i, 3, 300, and x, 162.

= *Leptotrichum brachypodium* C. M., Hedw., xxxviii, 89 (1899).
= *Leptotrichum *dolichopodon* Rehm. M.A.A. 86.
= *Ditrichum *dolichopodon* (Rehm.), Paris.

Very minute, gregarious or caespitose, autoecious soil mosses ; stem 1–2 mm. high, leafless below, with a dense comal tuft ; outer leaves very short and blunt, and almost nerveless ; other leaves spreading, and so involute above

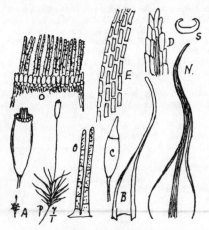
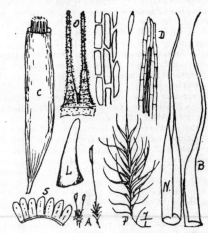

Ditrichum brachypodium.
S, section of subula.

Ditrichum flexifolium.
S, ring.

10

as to seem subulate from a wide base. Nerve strong almost to the apex, forming a keel, not wide, but occupying all the subula except the incurved margins. Cells oblong, small, nearly all alike. Per. leaves longer, flexuose, hardly sheathing, but with comparatively wide, nearly ovate base, the inner without nerve and with rhomboid cells. Seta erect, 4–5 mm. long; capsule small, erect, oblong-cylindrical, crowned by the dehiscent ring; peristome from below the mouth, well developed; teeth separate, split to the base, segments slender, papillose, red; lid somewhat obliquely conical-rostrate.

Much smaller in all parts than *D. flexifolium*; leaves more crowded, per. leaves with wider base and less sheathing; cells shorter, capsule regular.

Leptobryum brevifolium C. M. (Hedw., 1899, p. 88), described from Dr. Wilms' specimen in Herb. Jack, collected at the gold mines near Spitzkop, Transvaal, in 1887, and showing only a few immature fruits, does not appear to me to be different and I am not aware that it has been gathered again. Müller distinguishes it by its short firm leaves and narrowly cylindrical capsule. There is now a different *D. brevifolium* (Kindb.) Paris found in North America.

O.F.S. Kadziberg, Rehm. 86.

Natal. Van Reenen, Wager 88.

204. *Ditrichum flexifolium* (Hook.), Hampe, in Flora, 1867, p. 182; Dixon, Journ. Bot., Nov. 1913, p. 324; Dixon, N.Z. Inst. Bull., iii, pt. 2, p. 45; Broth., Pfl., x, 162.

    =*Dicranum flexifolium* Hook, Musc. Exot., Tab. 144 (1820).
    =*Ditrichum capense* (C M.), Par., 2nd ed., p. 392; Broth., Nat. Pfl., i, 3, 300.
    =*Leptotrichum capense* C. M., Syn., i, 453; Rehm. 84, 464, and 464b.
    =*Ditrichum *vallis-gratiae* Hpe., in Musc. Breutel.
    =*Dryplodon crispus*, Brid., p. 206=*Dicranum crispum* Thunb., *fide* C. M., Syn., p. 810.
    =*Ditrichum *Macleanum* (Rehm. 465 as *Leptotrichum*), Paris.
    =*Aongstroemia abruptifolia* C. M., in Hedw., xxxviii, 89.
    =*Dicranella abruptifolia* (C. M.), Paris,
and many other synonyms given by Dixon in papers cited above.

An erect little gregarious plant, having stems 2–10 mm. long, slender, simple or forked, or with innovations; with terminal slender seta 5–15 mm. long, and almost erect but slightly irregular elliptic-cylindrical capsule, often narrowed to the mouth, and when dry sometimes somewhat 4-ridged. Leaves very long, not very numerous, rather scattered, spreading widely from the sheathing base, but flexuose, and the apex ascending or somewhat falcate-secund; the margin strongly involute for the whole length of the subula, making the leaf appear linear-subulate from a wide base though really lanceolate. Nerve strong, wide, disappearing in the rather obtuse apex; leaf-margin entire; cells at the apex and base oblong, those in the mid-leaf rather larger, hexagonal-elongate, varying into oblong-hexagonal. Peristome of erect teeth split to the base into two linear, densely papillose segments, red below, hyaline above; ring prominent, lid obliquely conical. Stem usually short but often nearly leafless below; lower leaves short, upper longer; per. leaves very long and slender, high-sheathing, but otherwise similar to others. This species varies somewhat in size, length of seta, form of capsule, etc., but not enough to separate into species, and these variations occur together. Sometimes stems are an inch long and matted.

Hooker's figure and description in Musc. Exot. agree entirely with what C. Müller afterwards named *Leptotrichum capense*, except that the leaves are most abundant at the base of the stem, so I see no reason to maintain Müller's name though it is maintained by Brotherus and also by Paris.

Paris places *Dicranum flexifolium* Hook. (which was described from specimens collected at the Cape by Menzies in 1791) under the Australian *Ditrichum laxifolium* (Hk. and Wils.) Mitt., described by Hk. f. and Wils. in Flora of New Zealand, ii, 72, in 1855; the two are found by Dixon to be identical, but in that case the name *D. flexifolium* (Hk.) Hpe. has priority. Dixon gives its distribution as South Africa, Madagascar, East African islands, India, Java, New Caledonia, Australia, Tasmania, New Zealand, Patagonia, and Chile.

I see no difference in Rehm. 465 from Pilgrims Rest, named **Leptotrichum Macleanum* Rehm., except that the fragile peristome teeth have been broken off; it is mixed among *Mielichhoferia*, also fertile, which is usually present along with this species.

The stronger forms may be mistaken for species of *Dicranella*. A small and more delicate form with narrow capsule was described by Müller as var. *tenellum* (Hedw., xxxviii, 89 (1899)), for which he cites Lydenburg, Transvaal, and Greytown, Natal, as localities.

He also cites many localities for var. *vallis-gratiae* C. M., founded on Rehm. 85 (wrongly printed 185), which I cannot separate; among these are Genadentdal, Boschberg, Blanco, Esternek, Portland, Montagu Pass, and Rondebosch.

Frequent in all forest-margin and mist-belt situations throughout South Africa; also in mountain shady banks, etc., from the coast to the top of the Drakensberg; always a pioneer.

Cape, W. Peak, Sim 9711, 9713, 9718; Kirstenbosch, Sim 9488; Platteklip Valley, Sim 9387; top of Table Mountain, Sim 9116, 9296; Tokai, Sim 9509; Paarl Rock, Sim 9625; Rondebosch, Rehm. 84; Montagu Pass, Rehm. 85; Knysna, Rehm. 84c; Camp's Bay, Ecklon; Genadendal, Breutel; Kalk Bay, Potts (Sim 8541); Esternek, Knysna, Rehm. 25 (as *Dicranella abruptifolia* Rehm., n. sp.).

Cape, E. Boschberg, Somerset East, Macowan; Perie, Sim; Dohne Hill, 5000 feet, 1897, Sim 7192; Springs, Uitenhage, Sim 9048; Evelelyn Valley, Dr. v. d. Bijl 497.

Natal. Natal, Gueinzius (*fide* C. M.); Inanda, J. M. Wood (Rehm. 464b); Benvie, York, 4000 feet, Sim; Van Reenen, Wager 90; Maritzburg, Sim 9748; Lidgetton, Sim 9712; St. Ives, Sim 9714; Little Berg, Cathkin, 6000 feet, July 1922, Miss Owen 21; Giant's Castle, 8000 feet, R. E. Symons (Sim 9716); Gildats, Boston, 4000 feet,

Sim 9719 ; Hilton Road, Sim 8712 ; Nels Rust, Sim 9749 ; Elandskop, 5000 feet, Sim 9750 ; Mont Aux Sources, 7000 feet, Sim 9751.

Transvaal. Pilgrims Rest, Rehm. 464 ; Johannesburg, 6000 feet, Miss Edwards (Sim 9717).

205. *Ditrichum amoenum* (Thw. and Mitt.), Par., Ind., p. 391 (1895), and 2nd ed., ii, 90 ; Broth., x, 162.

=*Cynodontium amoenum* Thw. and Mitt., Journ. Linn. Soc , 1872, p. 296 ; Jaeg. Ad., ii, 634.

Plant 4–10 mm. high, erect, matted, mostly simple, few-leaved and red at the base, many-leaved above ; the leaves erecto-patent, long-subulate from a short oval-elliptical base, the per. leaves longer-clasping and more spreading ; cells of lamina long and narrow ; from the shoulder upward one to three rows are oblong-rhomboid ; the nerve, which extends almost to the apex, has long narrow opaque cells ; margin almost entire. Capsule cylindrical, red, on a 2 cm. red seta ; capsule hardly regular at the base. Peristome more or less rudimentary, the teeth short, irregular in development, densely papillose.

Dixon (Trans. Roy. Soc. S. Afr., viii, 3, 181), after citing localities as under, states : " These all agree with the Indian plant, which differs from *D. flexifolium* (Hook.) solely in having the peristome very short and rudimentary. Since *D. amoenum* is found only in collocation with *D. flexifolium*, in India and South Africa, and since the two agree in every other respect, I think there can be little doubt that this plant is but a variety of *D. flexifolium* with depauperate peristome."

Cape, W. George, Wager 514, 565.
Cape, E. Hogsback, 4000–6000 feet, Rev. Jas. Henderson.
Transvaal. Kaapsche Hoop, Wager 302, 330.

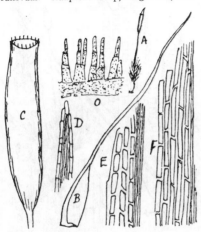

Ditrichum amoenum.

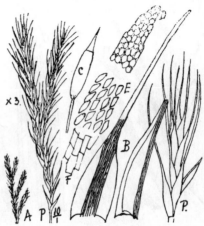

Ditrichum strictum.

206. *Ditrichum strictum* (Hook. f. and Wils.), Hampe, in Linnaea, 1867, pp. 181, 182 ; Dixon, Trans. R. Soc. S. Afr., viii, 3, 181 (1920) ; Dixon, N.Z. Inst. Bull., iii, 2, 41 (1914), pl. v, fig. 4 (with list of synonyms) ; Broth., Pfl., x, 162.

=*Ditrichum australe* (Hook. f. and Wils.), Mitt. ; Broth., Pfl., i, 3, 229.
=*Lophiodon strictus* Hook. f. and Wils., Lond. Journ. Bot., 1844, p. 543, and Fl. Antarct., i, 130, Tab. 59, fig. 2 ; C. M., Syn., i, 455, and ii, 613 ; Par., Ind., p. 195.

Plant 2–3 cm. high, branched somewhat, with fairly compact lax foliage, the leaves about 5-ranked, sheathing widely for about one-third their length in oval-elliptical form, then contracted easily into a long narrow subula which points more or less upward ; the keeled nerve wide and composed of long narrow cells, and extending to the base of the subula, but the subula itself composed of short round cells ; it is subterete below and smooth, but flattened at the apex ; its margin entire except at the apex where there are often three or four blunt crenations on each side. Basal cells of lamina oblong and hyaline, those in the shoulder of the leaf rather smaller and quadrate-rhomboid. Inflorescence and capsule not yet seen in South African specimens (my illustration of capsule is from Dixon's figure). *Lophiodon* in C. M., Syn., contains only *D. strictus* and is described as having wider peristome teeth than *Leptotrichum* (*Ditrichum*) in which there is a more or less high basal membrane, and the stronger dicranoid peristome teeth in *Dicranella* form the best distinction there also from a general resemblance to *Ditrichum* and to this species. Dixon (N.Z. Inst. Bull., 3, ii, 41–44) shows that this species has been much confused in N.Z. literature with *D. punctulatum* Mitt., and also with *Distichium capillaceum*, which latter, however, has distichous foliage, usually papillose leaf-apex, short lid, and paroecious inflorescence. Müller describes *Lophiodon strictus* as dioecious, but Dixon (as above, p. 43) places it as autoecious and *D. punctulatum* Mitt. as dioecious. Dixon describes the capsule as small, 1–1·5 mm. long (without lid) on a stout seta, pachydermatous, widely elliptical, with a narrow mouth ; the lid is " longly and finely subulate, the straight erect beak being about as long as the capsule itself."

D. strictum occurs in southmost America, Antarctic Islands, Tasmania, New Zealand, and now South Africa, though its distribution here, so far as is known, is confined to the tops of the Drakensberg where it has been collected in several Giant's Castle (Natal) localities by Mr. R. E. Symons, always in company with *Aongstroemia julacea*, with which also it is associated in South America. I also found it below the cliffs on Giant's Castle, Natal. Giant's Castle, 8000 feet, R. E. Symons (Sim 8538); Upper Bushman's River, 7000 feet, Sept. 1905, T. R. Sim 8685.

207. *Ditrichum spirale*, Dixon, Trans. R. Soc. S. Afr., viii, 3, 181, pl. xi, fig. 1, 1920.

A vigorous species, 2–3 inches high, growing in compact, dark green tufts; stems simple or sparingly branched; leaves erecto-patent or somewhat secund, up to 5 mm. long, concave but hardly sheathing at the oval-subulate base; the long awn solid and strap-shaped, spirally twisted when dry; awn twice as long as the lamina, the margin entire except near the apex where two or three bluntly toothed crenations appear. Cells at the leaf-base long-oblong with oblique divisions, the marginal ones there long, narrow, and hyaline; cells in the shoulder more or less elliptic, in the subula minute, rounded-quadrate, obscure. Nerve wide and ill-defined within the lamina, and occupying almost all the subula. Dioecious. Capsule unknown. The plant somewhat resembles *Leucoloma* in appearance. Dixon says, " Near *D. flexicaule* Schl. . . . The regular corkscrew-like twisting of the subula when dry, the leaves being otherwise unaltered in position, is a marked character best observed by holding the plant up to the light. . . . The rather coarse toothing of the apex and the elongated basal cells, with a very narrow border of very delicate hyaline ones, are also distinguishing characters."

Cape, E. Gaikas Kop, Tyumie, C.P., 6000 feet, 1916, D. B. and M. Henderson 232.

Ditrichum spirale.

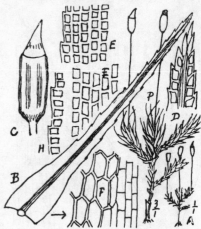

Saelania glaucescens.

Genus 69. Saelania Lindb., Utkast, p. 35 (1878); Dixon, Handb., 2nd ed., p. 71; Broth., Pfl., i, 3, 300.

A monotypic genus closely allied to *Ditrichum*, but having the leaves glaucous, smooth, more widely lanceolate, and less regularly subulate. Capsule oblong-cylindric, erect, regular, somewhat angular when dry; lid shortly conical; peristome as in *Ditrichum*, inserted in the capsule-mouth, with low basal membrane, and the segments usually connected above.

Placed in Cynodonteae by Dixon, but the leaves are not papillose, the capsule is regular, and it fits into Ditricheae better.

208. *Saelania glaucescens* (Hedw.), Broth., Pfl., i, 3, 300, and x, 163; Dixon, N.Z. Inst. Bull., iii, 2, 50.

=*Saelania caesia* Lindb., Utkast, p. 35; Dixon, Handb., 2nd ed., p. 71; 3rd ed., p. 68.
=*Leptotrichum glaucescens* Hampe, in Linn., 1847, p. 74; W. P. Sch., Syn., p. 146.
=*Trichostomum glaucescens* Hedw., Descr., iii, 91, T. 37, β; Bry. Eur., p. 2, T. 184; C. M., Syn., p. 569.
=*Ditrichum glaucescens* Hampe, in Flora, 1867, p. 182; Par., Ind., 2nd ed., p. 93.

Plants 5–10 mm. long, slender and simple below, much branched above, the several branches bearing terminal setae, innovations occur lower. Autoecious; male flowers bud-like, on short branches below the perichaetia. Lower portion of the stem almost leafless or with small deltoid leaves; upper portion densely leaved, the leaves longer, lanceolate from an ovate base; inner and per. leaves still longer and bearing a linear-subulate awn; all leaves erecto-patent, more erect when dry. Nerve not very wide, extending to the leaf-apex, and much thicker than the lamina. Cells of the lamina at the base irregular as to size and form, long-hexagonal, empty, with two to three rows of smaller oblong cells along the margin; mid-leaf cells oblong, in straight lines, the marginal line transversely oblong; further up the cells are oblong but smaller, and at the apex they are cylindrical and often with free points. In some but not all of the per. leaves long cells occupy the subula. Margin plane, bluntly serrate on the smaller leaves, but on larger leaves usually sharply serrate from where the leaf narrows upward, though on some leaves the serrations are only near the apex. Capsule erect, regular, on a half-inch seta, oblong-cylindric, somewhat angular or furrowed when empty; lid sharply conical, regular; teeth purple, cleft to the base into two filiform papillose segments, much united above.

A very rare moss, found only at high altitudes here, but found also in mountain habitats in Europe, Asia, and New Zealand, and easily detected by its colour.

Natal. Giant's Castle, in several localities over 8000 feet altitude, R. E. Symons (Sim 8681, 9720, 9721).

O.F.S. Rensburg Kop, Jan. 1911, H. A. Wager 86.

(Genus DISTICHIUM Br. and Sch. (1846)).

Plants slender, tufted, shining, with more or less distichous, subulate, reflexed leaves, from a wider white sheathing base; the nerve papillose, the cells at the base large, oblong, and hyaline, changing gradually to small, roundly quadrate, papillose cells in the subula. Capsule erect or almost so, oval-oblong or cylindric, with conical lid and cucullate calyptra. Peristome of sixteen free linear-lanceolate, smooth teeth, inserted much below the mouth; they are irregularly cleft into unequal lobes, or perforated, or entire. The plants form shining masses, often on mountain rocks, and vary a good deal in accordance with the presence or absence of water.

The name *Distichium* is used by Bruch and Schimper, and in Schimper's Synopsis and most other works, and is maintained in Paris' Index and by Brotherus, Pfl., i, 3, 302; but Dixon and Jameson raise a doubt as to its validity; the name *Distichia* having been used by Nees for a genus of Juncaceae in 1843, they use the name *Swartzia* Ehrh. instead, and *Distichium capillaceum* B. and S.=*Swartzia montana* Lindb. It is almost cosmopolitan and is reputedly South African, but I have seen no specimens to bear that out, and even the records are doubtful. In New Zealand, though it does occur, what was originally recorded as this species proved to be *Ditrichum punctulatum* Mitt. (near *D. strictum* Hampe), so the genus is described here so as to be recognised if found, though its presence is very doubtful. Concerning *S. montana*, Lindb., Dixon says (Handbook, 3rd ed., p. 62), " A very pretty plant, easily recognised by its distichous, setaceous leaves with very conspicuous, white, glossy, sheathing bases, so that the stems appear white and shining," which description does not apply to *Fissidens* or *Eustichia* or any of our distichous pleurocarpous mosses.)

§ II. *Seligerieae.* Minute, nearly stemless, annual or biennial, gregarious, rock-plants having lanceolate-subulate leaves, with excurrent nerve and without papillae or special alar cells. Seta short; capsule erect or nearly so, regular, oblong or roundish pyriform, smooth or slightly striate; peristome of sixteen short lanceolate obtuse teeth, free or confluent at the base, usually entire, sometimes cleft, perforated, or absent. Calyptra small, conical, and lobed; lid shortly rostrate.

Genus 70. BRACHYDONTIUM Fürnr.

Characters are those of the section; also the capsule is erect and oblong, and longitudinally striate when dry; the calyptra conical and 5-lobed, and the peristome teeth confluent at the base, very short, broad and truncate, dotted and thin.

A small genus of microscopic mosses, of which the one species recorded from South Africa is said by Paris to inhabit granite and volcanic rocks but not calcareous rocks, in Europe and elsewhere.

209. *Brachydontium trichodes* Fürnr., in Flora, 1827, ii, 1, 37.
=*Brachyodus trichodes*, Br. Germ., ii, 2, 5.

Shaw, in his Catalogue of the Mosses of Cape Colony (1878), includes *Brachyodus trichodes* Fürnr., without local locality. His records are so unreliable that unsupported no credence can be placed on this, and I doubt its presence. It is a minute rock-plant, similar to *Dicranella minuta*, but has mitriform calyptra, lobed at the base; the dry capsule longitudinally striated, and short wide peristome teeth hardly cleft (see Broth., Pfl., i, 3, 305, fig. 178 E, F, G).

§ III. *Cynodontieae.* Leaves linear to widely lanceolate, nerved to or beyond the apex, often more or less papillose, the upper cells small, quadrate or hexagonal-quadrate, and chlorophyllose, the lower larger, oblong, and pellucid (in *Nanobryum* all linear), but there are no special alar cells. Capsule erect or inclined, oblong or oblong-cylindrical, usually striate, sometimes strumose. Peristome teeth sixteen, entire or divided.

Genus 71. CERATODON Brid.

Small terrestrial or wall plants, often crowded, with erect habit, erect seta, slightly inclined, oblong-cylindrical, rather unequal striate or furrowed capsule, with a small struma at the base, the lid conical, ring broad, and the peristome of sixteen lanceolate, deeply forked teeth, outwardly papillose, the lobes subulate-filiform, the teeth having close transverse lines below, sparse above, deep red at the base; the lobes red-margined below by the inner membrane. Leaves lanceolate, or wider or narrower; margin more or less revolute; lower cells large, oblong, and pellucid; upper cells small, quadrate, and chlorophyllose.

A cosmopolitan genus, mostly through one widely distributed and variable species, *C. purpureus* (Linn.) Brid. Almost every bryologist finds a different place for this genus in his systematic arrangement, as it has relationships with or resemblances to many other mosses.

210. *Ceratodon purpureus* (Linn.), Brid., Bry. Univ., i, 480 ; C. M., Syn., ii, 633 ; W. P. Sch., Syn., 2nd ed., i, 132 ; Par., Ind., 2nd ed., i, 339 ; Broth., Pfl., i, 3, 301 ; Dixon, Handb., 3rd ed., p. 69 ; Dixon, N.Z. Inst. Bull., iii, 2, 50.

     = *Mnium purpureum* Linn., Sp. Pl., 1st ed., ii, 1111.
     = *Ceratodon *capensis*, W. P. Sch., in Breutel, Musc. Cap. ; Shaw, p. 377.
     = *Ceratodon *condensatus*, W. P. Sch., in Breutel, Musc. Cap.
     = *Ceratodon corsicus*, Br. Eur. ; C. M., Syn., ii, 633.
     = *Ceratodon streptocarpus*, Br. Eur. Mon., p. 4 ; C. M., Syn., i, 647 ; Shaw, p. 377.
     = *Ceratodon chloropus*, of Horns., Linnaea, 1841, p. 124 ; Drege, iii, *a, e*, 31.

Dioecious ; stems caespitose, simple or somewhat branched ; 2–10 or sometimes 15 mm. high, usually crowded, and having terminal, but through innovation apparently lateral setae. Leaves numerous, sometimes spread regularly along the stem, sometimes mostly in a comal tuft, erecto-patent when moist, incurved and somewhat twisted round the plant when dry ; lower innovations often somewhat etiolated with distant and small leaves, upper innovations abundantly clad with normal leaves. Leaves keeled, lanceolate, tapering gradually to the point from near the base, not papillose ; the margin recurved somewhat along its whole length except near the acute apex where a few blunt teeth often appear ; nerve fairly wide from base to apex, forming the protruding keel, and attached down the stem ; cells small, quadrate, chlorophyllose, set in lines, those nearest the nerve at the base larger, oblong, and pellucid, but no alar cells. Per. leaves hardly different, except more sheathing. Seta ½–1½ inch long, pale red ; capsule oblong-cylindrical, erect, thick-walled, regular when young, often inclined or curved and irregular when old ; the short neck usually regular when young and often strumose when old ; capsule bluntly 4–8-angled or striate when young, more deeply furrowed and somewhat twisted when old ; ring pronounced and caducous ; peristome of sixteen teeth on a basal membrane of about three lines of horizontal cells, each tooth cleft to the base into two subulate, nodose, papillose segments, red and closely jointed in the lower half, nearly hyaline in the filiform apex ; lid shortly conical, overlapping the capsule like the roof of a hut ; calyptra conical. On the dry capsule the teeth are often absent but the membrane present. In exposed and alpine situations the whole plant is usually of a clear purple colour, elsewhere it is green, but purple shades occur occasionally in the stem, nerve, and leaf-base. When dry the peristome teeth are inflexed.

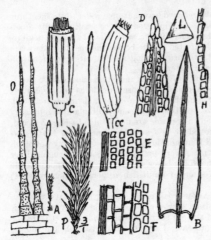

Ceratodon purpureus.
CC, dry capsule.

Male plants slender, male flower bud-like.

This species is cosmopolitan, very abundant in some localities, rare in most parts of South Africa except the Cape Peninsula, but occurs also high on Kilimanjaro and elsewhere in Africa. In the Northern Hemisphere it varies considerably and sometimes has the nerve well excurrent, the seta purple, and the peristome teeth more or less bordered, and it varies with age, hence the enormous number of subspecies, varieties, forms, and synonyms ; here only one form has been seen (var. *xanthopus* Sull., Musc. Bor. Am., 2nd ed., p. 29) in which the nerve ends just below the leaf-apex and the seta is pale.

Cape, S.W. Table Mountain, Rehm. 122 (as *C. corsicus*) ; Slongoli, Sim 9190, 9170, 9187, 9219 ; Woodhead Tunnel, Sim 9125 ; The Saddle, Sim 2200 ; Disa Ravine, Sim 9152 ; Rondebosch, Rehm. 122*b* (as *C. corsicus*) ; Paarl Rock, Sim 9621 ; Stellenbosch, Sim 9577, 9583, 9585, 9590 ; Montagu Pass, Rehm. 122*c* (as *C. corsicus*).

Natal. Van Reenen, Wager 420 (as *C. capense*) ; Maritzburg, Wager 450 (as *C. corsicus*) ; Little Berg, Cathkin, July 1922, Miss Owen 16 ; Polela, J. W. Haygarth 5.

Genus 72. NANOBRYUM Dixon, in Journ. Bot., Apr. 1922, p. 101.

Minute dioecious plants having annual stems on perennial protenema. Fertile plant larger than the male, few-leaved, the leaves comal, subulate from a clasping base ; the nerve forming most of the filiform subula. Cells linear, alar cells none. Capsule minute, inclined, mouth wide ; peristome as in *Fissidens*, lid conical.

211. *Nanobryum Dummeri* Dixon, in Journ. Bot., Apr. 1922, p. 101.

Male plants minute, leaves few, concave, clasping, short-pointed. Fertile plant with seta 4–6 mm. high, with about three scale-like lower leaves and about three much larger, suberect or subsecund, shining comal leaves 1–2 mm. long, with widely clasping, ovate base and long, flexuous, entire, filiform subula, mostly occupied by the weak excurrent nerve ; cells narrow, twisted, rhomboid-linear, 8–13μ wide, chlorophyllose ; basal cells shorter and wider (10–16μ wide) ; alar cells none. Seta 3–4 mm. long, vaginula large, red capsule minute, horizontal, when dry pendulous ; peristome teeth spreading when dry, inflexed into the capsule when moist, trabeculate, high-crested inside, with filiform, spirally thickened segments. Spores large, lid pale, conical-rostellate, large. Described and figured from specimens kindly lent by Mr. Dixon.

First found in Uganda by Dummer ; afterwards at Port St. Johns, Cape Province, 1921, H. A. Wager (955).

(By an evident slip *Nanobryum* is included in the key of Funariaceae in Journ. Bot., Apr. 1822, p. 108, instead of *Nanomitrium*.)

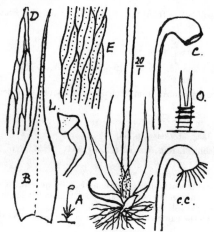

Nanobryum Dummeri.

C, capsule, moist ; CC, capsule, dry.

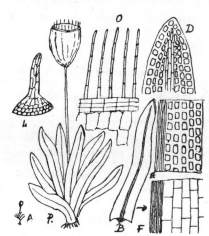

Rhabdoweisia fugax.

Genus 73. RHABDOWEISIA Br. and Sch.

Small, tufted, perennial rock-plants, having short, ligulate, channelled, level-margined, chlorophyllose leaves, nerved nearly to the apex, the lower cells oblong and pellucid, the upper cells small, roundly quadrate or irregular, dark green. Seta short ; capsule small, erect, regular, oval, urn-shaped when dry, 8-striate, and without struma. Peristome teeth sixteen, undivided, short, narrow, subulate and fugacious, consequently often apparently absent. A small but widely distributed genus of little alpine mosses.

212. *Rhabdoweisia fugax* (Hedw.), Br. and Sch., Bry. Eur. Mon., p. 4, t. 41 ; W. P. Sch., Syn., 2nd ed., p. 58 ; Par., Ind., p. 152 ; Broth., Pfl., i, 3 (not x, 194) ; Dixon, Handb., 2nd ed., p. 74, t. xiv A, and 3rd ed., p. 71, t. ix A (both ?).

= *Oncophorus striatus* Lindb., M. Scand. ; Brathw., M. Flora.

= *Weisia fugax*, Hedw. Sp. M., p. 64, t. 13, fig. 5.

Stemless ; simple and rosulate or sometimes branched at the base ; leaves 2 mm. high, six to ten, ligulate, hardly narrowed till near the obtuse apex, keeled ; margin plane, entire ; nerve strong below, disappearing just below the apex ; basal cells oblong, hyaline ; cells of lamina much smaller but oblong, separate, and chlorophyllose in a pellucid leaf with about five to seven rows of cells on each side of nerve. Seta 3 mm. long ; capsule erect, oval, regular, small, longitudinally striate when dry, with a permanent rim, and with sixteen narrow, linear, undivided teeth, which are not papillose ; lid convex, with conical beak as long as the capsule ; calyptra conical entire. Leaves incurved or curled when dry but still plane.

Not unlike *Weisia gracilis* D. and W., but that has no teeth and has cells alike to the base of the leaf.

R. fugax occurs in North Europe, North Asia, North America, and Juan Fernandez, but our plant has wider, more ligulate leaves.

Natal, Van Reenen, Jan. 1911, H. A. Wager 87 (mixed with *Weisia viridula*).

Brotherus, Pfl. x, 193, brings *Amphidium cyathicarpum* (Mont.) Broth. next to this, differing in having no peristome ; there forming his subfamily Rhabdoweisioideae.

§ IV. *Trematodonteae.* Plants small, gregarious, erect ; leaves lanceolate-subulate, smooth, with nerve the whole length or excurrent ; cells lax, pellucid, and oblong at the leaf-base, small, chlorophyllose, and obscurely quadrate upward, and with no special alar cells. Seta short or long ; capsule erect or cernuous, with a long and more or less inflated neck ; peristome sometimes absent, when present consisting of sixteen lanceolate papillose teeth, which are simple, and usually more or less cleft or perforated.

Genus 74. BRUCHIA Schw.

Plants very small, simple or slightly branched ; leaves lanceolate-subulate, with nerve the whole length or excurrent ; seta short ; capsule oval, globose or pyriform, tapering at the base into a neck ; cleistocarpous, almost included among the perichaetal leaves. Calyptra central, campanulate, fringed at the base. Inflorescence dioecious or monoecious.

I have seen no specimen and am not aware that either species has been collected other than as stated below, which happened long ago, but these minute soil-plantlets are so like *Pleuridium nervosum* that they are easily overlooked, and Müller, in describing them, refers to this. They all differ most evidently in having companulate calyptra, and *Eubruchia* in also having the capsule continued downward somewhat, as an apophysis, below the spore sac.

From descriptions the sections and species are :—

§ 1 (*Sporledera*). Protenema abundant. Capsule erect, pointed, without apophysis. Paris and also Brotherus retains *Sporledera* as a genus, Pfl., x, 158.

213. *Bruchia Rehmanni* C. M., in Flora, 1888, No. 1.

=*Sporledera Rehmanni* (C. M.), Par., Ind., p. 1229, and 2nd ed., iv, 24, 319 ; Broth., Pfl., x, 158 ; Rondebosch, 1875, Rehm. 463.

§ 2 (*Eubruchia*). Caulescent perennial. Seta short ; capsule pyriform from a considerable apophysis, apiculate ; calyptra campanulate, fringed or lobed at the base.

214. *Bruchia elegans* (Hsch.), Jaeg. M. Cleist., p. 35.

=*B. brevipes* Hook. (in part, excluding America), in Hook. Icon. Pl. Rar., vol. iii, t. 231 (1844) ; C. M., Syn., i, 18 ; Broth., Pfl., i, 3, 292, and x, 173.
=*Phascum elegans* Hsch., in Linnaea, 1841, xv, 114.

Autoecious ; leaves subulate from a wide base, obscurely denticulate, shortly acuminate, solid nerved. Capsule globose-oval-pyriform, apiculate, sub-exserted ; calyptra with many deeply cut crenate lobes. Male flowers in basal branches, perigonial leaves suddenly subulate from a wide base, nerved, irregularly crenate.

Cape, S.W. Clay soil, north side of Devil's Mountain above the gardens (Ecklon), and roadside near Newlands, Aug.

Harvey, in Genera of South African Plants, 1st ed. (1838), shortly described the genus and says, " We have one species *B. brevipes*, Mihi." As his name does not seem to be connected now with *B. brevipes*, I think it likely that he sent specimens to Europe under that name, but that it was published in Hook. Icon. Pl. Rar. under the name he gave it, but without his name.

215. *Bruchia Eckloniana*, C. M., Syn., i, 19 ; Broth., Pfl., x, 173.

Dioecious ; leaves lanceolate-subulate from a rather wide base, minutely areolate, entire ; nerve excurrent, wide, acute. Capsule elliptical (apex and base attenuate), long and acutely apiculate, erect, almost included. Calyptra small, covering one-third of the capsule, cut into segments at the base. Male flowers terminal on stem and branches.

Cape, S.W. Prom. bona spei (Ecklon).

Müller adds : " Near *B. elegans* in habit but the dioecious inflorescence, the form of the capsule and calyptra, and the entire leaves make it quite distinct."

Genus 75. TREMATODON Rich.

Plants with short stems, usually subulate, setaceous smooth leaves and long setae. The leaves are channelled and have a nerve often reaching to the apex ; the cells lax and pellucid below, but smaller and denser along the nerve toward the apex. Perichaetal leaves larger. Capsule straight or curved, cylindrical above, tapering to a very long twisted and often arched neck, which in some species or conditions ends abruptly on the inner side like a struma, but this is unreliable as a specific character. Calyptra cucullate-rostrate ; lid convex and rostrate ; ring broad ; peristome teeth confluent at the base on a short membrane, irregularly perforated and divided, or simple or cleft to the base, or absent. Dioecious or monoecious.

The usually subulate leaves and long-necked capsule make this genus resemble somewhat some species of *Webera*, which genus belongs to tribe Diplolepideae and so is distinguished from *Trematodon* by having both outer and inner peristome.

Hornschuch, in describing *T. paradoxus* (Linnaea, 1841, p. 122) as apparently the first eperistomate species of *Trematodon* known to him, excuses the assembling of peristomate and eperistomate species in one genus, as bringing together natural relations, as had already happened in *Gymnostomum* (*Weisia*), *Orthotrichum, Macromitrium, Encalypta*, etc. Indeed it is now known that in some species of *Trematodon*, as well as in *Weisia*, etc., developed, undeveloped. and absent teeth may occur not only in the same species but in the same capsule.

Synopsis.

Peristome absent (*Gymnotrematodon* C. M., in Hedw., 1895, p. 118).
216. *T. paradoxus.* Leaf entire ; spores round, rough.
217. *T. Pillansii.* Leaf denticulate-serrulate ; calyptra mitriform ; spores hemispherical, smooth, very large.
Peristome present, teeth perforated, segments more or less united.
218. *T. mayottensis.* Leaf wide up to the rounded apex.
219. *T. africanus.* Leaf narrow, acute, short.
220. *T. pallidens.* Leaf very narrow, long and flexuose, acute, with margins sometimes recurved.
Peristome present, teeth well developed, segments free.
221. *T. flexifolius.* Nerve wide, strong ; neck longer than capsule.
222. *T. aequicollis.* Nerve narrow and indistinct ; neck about equal in length to the capsule.

216. *Trematodon paradoxus* Horns., Linnaea, 1841, p. 122 ; C. M., Syn., i, 456 (1849) ;
Besch., Fl. R. (from Reunion and Cape Peninsula).

= *T. intermedius* Welw. and Duby, in Mem. Genève, 1870, p. 12, Tab. 11, fig. 6 ; Broth., Pfl., x, 175.
= *T. *Macleai*, Rehm. 439.

Monoecious ; male flowers discoid on short basal branches. Stem 3–6 mm. high, slender ; leaves few and small in the lower portion, those scattered up the stem or in a comal tuft longer ; per. leaves longest. Leaves oblong-lanceolate, 2 mm. long, expanded and nearly flat in the lower half, tapering gradually to a channelled subulate-acute apex ; margin entire ; nerve strong, usually extending to the apex ; cells laxly oblong above, chlorophyllose, and in the upper part forming one to three lines bordering the cylindrical-celled nerve ; those in the lower part of the leaf much longer and wider and pellucid ; perigonial leaves short, wider below, and more quickly subulate. Seta pale, 3–6 mm. long, erect ; capsule ovate-oblong, regular, on an erect or, with age, cernuous obconic neck one and a half times its length ; peristome absent, ring conspicuous, double, the inner cells rounded, the outer twice as long and oblong. Lid convex, with conical somewhat oblique beak. Calyptra very oblique, inflated below, cucullate.

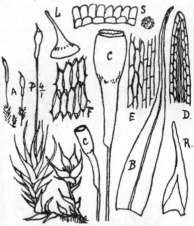

Trematodon paradoxus.
S, part of ring, and a spore.

Var. *nanus* = *T. intermedius* W. and D., var. *nanus* W. and D. ; Dixon (S.A. Journ. Sc., June 1922, p. 301) = *T. Pechuelii* C. M., in Flora, 1886, p. 508. Smaller ; seta shorter ; nerve often excurrent.

Dixon (Bry. of S. Rhodesia, S.A. Journ. of Sc., June 1922, p. 300) uses the name *T. intermedius* W. and D. for some of the specimens cited below, but unless that differ from *T. paradoxus* Horns., the latter name has precedence. Hornschuch's description of *T. paradoxus*, from Ecklon's Table Mountain specimen, gives leaves linear-lingulate from a wide base, entire, nerve disappearing above the middle, lower leaves lanceolate, with shorter nerve ; neck equal the capsule or a little longer. As the length of nerve and of neck varies, and as C. Müller's description in C. M., Syn., p. 456, fits our plant and also is earlier than the publication of *T. intermedius* W. and D., I give the name *paradoxus* preference.

Concerning Rhodesian specimens Dixon says :—

" *T. intermedius* may be quite gymnostomous, or it may have a very rudimentary peristome consisting of fragmentary bases of teeth. This latter condition occurs in the specimen of Eyles 2169. Number 2286 does not show any trace of peristome."

T. platybasis (C. M., in litt.), Renauld, Prod. Fl. Bry. Madag., p. 58, is included as South African in Wager's check-list. Renauld described it as eperistomate, with neck equal to the capsule or scarcely longer ; very near *T. paradoxus* Horns., but with more lax areolation. If distinct I have seen no proof of it.

Wager also includes *T. reticulatus* C. M. (in Abhandl. Bremen, vii, 205). Renauld (Prod. Fl. Bry. Madag., p. 59) says C. Müller is not sure of the presence or absence of peristome in it, but it certainly differs from *T. pallidens* in the more open areolation. *T. pallidens* is peristomate ; Brotherus places *T. reticulatus* as eperistomate ; I have no satisfactory description, and no proof that it exists in South Africa.

Brotherus (Pfl., x, 175) divides thus :—

Leaves pointed, upper cells 7–10μ wide : *T. intermedius* W. and D., and *T. platybasis* C. M.

Leaves lingulate with rounded point, upper cells 15–18μ wide : *T. paradoxus* Horns.

Cape, S.W. Table Mountain, Ecklon (type).

Cape, E. Fern Kloof, Grahamstown, 1867, Macowan and Shaw (*fide* Shaw's list) ; Dohne Hill, 4500 feet, 1898, Sim 7179.

Natal. Town Bush, Maritzburg, Sim 8450 ; Nottingham Road, Sim 9722 ; Sweetwaters, Sim 9723 ; Zwartkop Valley, Sim 9724 ; York, Sim 9725 ; Knoll, Hilton Road, Sim 9725 ; Bush below Cathkin Peak, 7000 feet, Sim 9726 ; Little Berg, Cathkin, 6000 feet, Miss Owen 27 ; Zwartkop, 5000 feet, Professor Bews (Sim 9727) ; Greenfield, Mooi River, Sim 9729 ; Entumeni, Bryhn.

O.F.S. Wolhuters Kop, Wager 93.

Transvaal. Pilgrims Rest, MacLea (as *T. Macleai*, Rehm. 439).

Rhodesia. Zimbabwe, Sim 8835, 8836, 8826, corticolous ; Salisbury, 5000 feet, Eyles 2286 ; Salisbury, Eyles 2169 (var. *nanus*).

I also have specimens from Uganda, Dummer 971, 2963.

217. *Trematodon Pillansii* (Dixon, MSS.), Dixon (new species).*

Autoecious, small, densely caespitose on wet banks. Stems 2–3 mm. long, simple or forked or branched, and with small basal male branches ; seta and capsule ½ inch high, erect when young, neck arched when old. Leaves few, short, 1–1·5 mm. long, narrowly lingulate from a wide base, erect or homomallous, the lower being smallest, and others

* *Trematodon Pillansii* (Dixon, MSS.), Dixon (sp. nov.).
Autoica ; caules 2–3 mm. longi, parvis basalibus ramis masculis. Seta et capsula ½″ alta, erecta iuvenescens, senescenti collum arcuatum. Folia pauca, brevia, 1–1·5 mm. longa, anguste lingulata e basi latiore, erecta vel homo-

gradually longer, the inner per. leaves longest, the wide base extending nearly half-way, and the upper portion strap-shaped, or sometimes widest at the apex; nerve wide, rather indistinct, but present to near the leaf-apex. Lamina in widened base composed of large, lax, oblong, pellucid cells; from where the leaf narrows to the apex, the cells are rounded or roundly quadrate, very tumid, the marginal ones protruding, others rising at the ends to meet one another, leaving a depression between, thus giving a papillose appearance to each surface of the leaf-subula; the leaf-apex varying from long and narrowly acute to short, wide, and obtuse, but always with projecting hyaline cells, and the margin, down to where the leaf widens, also irregular or denticulate-serrate by similar cells. Seta pale; capsule red; the neck one and a half times as long as the ovate-oblong body of the capsule; ring and peristome absent; lid convex and rostrate; calyptra usually mitriform with five equal lobes, occasionally one of these more deeply cut than the others; spores very large, smooth, with flat base and rounded apex. Male flower bud-like on a short basal branch; the perigonial leaves few, short, lanceolate, or wider, very concave, seldom with subula, but always with protruding apical cells. Androecia and paraphyses few.

Cape, S.W. Platteklip Ravine, Sim 9282; top of Table Mountain, Sim 9297; slopes above Miller's Point, Cape Peninsula, Oct. 1919, N. S. Pillans 4058, 4060.

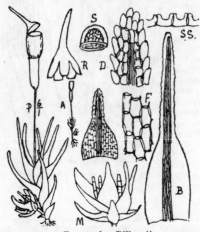

Trematodon Pillansii.
SS, raised cell-walls.

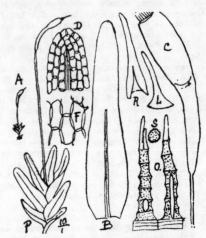

Trematodon mayottensis.

218. *Trematodon mayottensis* Besch., in Ann. Sc. Nat., vii, ser. 2, p. 84; Dixon, S.A. Journ. Sc., June, 1922, p. 301; Broth., Pfl., x, 176.

=*T. *ligulatus* Rehm.

Plant yellowish, gregarious, simple or nearly so, 3–4 mm. high, with terminal ½-inch straight seta. Leaves eight to ten, the lower small, the upper about 2 mm. long; all comparatively short and wide, lingulate from a wider base, with a wide rounded apex; margin entire except that some turgid cells protrude; nerve weak, fairly wide, extending almost to the leaf-apex, or half-way in the longer leaves; cells quadrate, very lax, mostly pellucid; basal cells much larger and very irregular in size and form among themselves. Body of capsule oblong-cylindric, its neck arched and one to one and a half times as long and tapering gradually, but pale and empty; peristome fairly well developed, of sixteen teeth on a membrane three cells deep; teeth not divided throughout but usually with one or more perforations in the lower portion, longitudinally striated with papillose lines. Spores very large, 22–25µ, papillose. Lid convex, with long slender beak; calyptra cucullate. Quite distinct in its rounded leaf-apex and short undivided teeth. Known from Mayotte Island and Madagascar, not previously from continental Africa.

Natal. Oakford, Rehm. 22 (as *T. *ligulatus* Rehm., n. sp.).

Rhodesia. Acropolis, Zimbabwe, Sim 8819; Matopos, 4600 feet, Eyles 935; Rua River, near Salisbury, 5000 feet, Eyles 1343.

219. *Trematodon africanus* Wager, in Trans. R. Soc. S. Afr., iv, 1, 4 (1914), pl. 1G; Dixon, S.A. Journ. Sc., June 1922, p. 301.

=*T. *obtusifolius* Mitt. MS. in Hb. (from Umbilo, Natal, Nov. 1853, Gueinzius).

Plant light green, gregarious, usually simple, 4–5 mm. high, with terminal straight seta ¾–1 inch long. Lower leaves small, upper mostly comal, the inner per. leaves rather longer than others. Leaves 2 mm. long, lanceolate

mala; folia involucralia interiora longissima, basi lata ad dimidiam fere partem patente, cetera ligulata. Nervus latus, subobscurus, prope tamen ad folii apicem apparens. Cellae marginales tumidissimae, prominentes, aliae ad extrema ita conniventes ut cavi aliquid intersit, quo fit ut superficies utroque folii subulae speciem papillosam praebeat: folii vero apex semper cellis prominentibus hyalinis ornatur. Capsulae collum altero tanto et dimidio longius quam capsula ipsa ovato-oblonga. Annulus et peristomium desunt, calyptra mitriformis lobis subaequis quinque. Sporae leves, basi plana apiceque rotundato. Flores masculi gemmiformes ramo basali brevi impositi.

from a wide, rather clasping base, subulate at the apex, and the lower part so incumbent that it is difficult to flatten out a leaf to see its form or to see the nerve. Nerve extending to the apex, forming a keel below ; margin entire, or at the apex the upper ends of some cells protrude slightly. Cells quadrate, lax, fairly large ; basal cells longer, pellucid, laxly quadrate-hexagonal. Body of capsule ovoid-cylindric, slightly bent, at first erect, soon inclined ; neck one and a half to two times as long, somewhat arcuate ; ring evident ; lid convex with slender beak three-fourths the length of the capsule-body ; calyptra cucullate. Body of capsule of long narrow cells ; basal membrane three to six cells deep. Peristome teeth sixteen, well developed, seldom cleft, but usually with several perforations, and set with longitudinal striae or lines of papillae ; the upper sections pale ; the teeth inflexed and interlocked when dry. Spores very large, somewhat papillose, 20μ diameter. Wager says, " This moss appears to be closely allied to *T. tortile*, but it is more straw-coloured, of a much greater stature, the struma is more pronounced, with a longer neck, and the peristome is longer." Wager illustrates the two lines of marginal cells smaller than those inside ; I do not find them to be so.

Natal. Maritzburg, 9730 ; New Hanover, 9733 ; Umbilo, Gueinzius (as above).

O.F.S. Wolhuters Kop, Wager 92 (wrongly distributed as *T. ligulatus* Rehm.).

Transvaal. Tzaneen, Dec. 1911, Wager 91 ; Melville, Johannesburg, Sim 8559, 9731, 9732.

S. Rhodesia. Rua River, Salisbury, Eyles 1342 (in part), 1341, 2288.

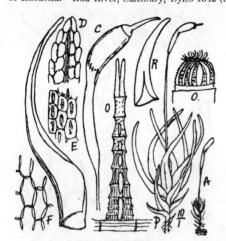

Trematodon africanus.

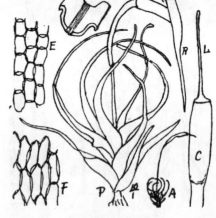

Trematodon pallidens.

220. *Trematodon pallidens* C. M., in Linnaea, xl, 242 (1876) ; Broth., Pfl., x, 176.

Very similar in habit, leaf-form, areolation, and lid to *T. africanus*, but more robust, lid as long as the capsule, neck longer, leaves much longer (*i.e.* 4–5 mm. long), more flexuose, and with the margin more or less recurved. Dixon says concerning Wager's specimen 273 from Woodbush, Transvaal (Trans. R. Soc. S. Afr., viii, 3, p. 181 (1920)), " Sent to me for description as ' *Trem. transvaaliensis* Wag. and Broth., n. sp., Broth. M.S. in litt.' I cannot separate it, however, in any way from *T. pallidens*. The teeth are not particularly pale, but that is not a character insisted on in the description of the original plant. It agrees in all other respects. The leaf-margin is frequently or usually very narrowly recurved above—a character which, I think, being unusual in the genus, goes far to confirm the identification with *T. pallidens*."

T. pallidens was formerly known from East African islands, whence another and more vigorous species *T. lacunosus* R. and C. has been sent which Renauld in comparing it with *T. pallidens* (Prod. Fl. Bryol. Madag., p. 59) describes as more robust, with neck 5–7 mm. long and with teeth perforated and outwardly striated. In *T. pallidens* the neck is 3 mm. long and the capsule 1·5–2 mm. long.

When all are better known *T. africanus* Wag. may prove to be only a smaller form of *T. pallidens*, and *T. lacunosus* a larger form, in which case the name *T. pallidens* has precedence.

Natal. Edgehill, Mooi River, 4000 feet, June 1921, Sim 9735.

Transvaal. Woodbush, Wager 273.

I also have it from Uganda (Dummer 1212) with young capsules, in which the peristome is immature.

221. *Trematodon flexifolius* C. M., in Flora, 1886, p. 278 ; Broth., Pfl., x, 176.

Plant 3–4 mm. high, with numerous strong almost basal leaves ; seta and capsule 1 inch high, pale. Lower half of the leaf oblong, incumbent, laxly areolated, then narrowed gradually to an acute point, the margins approaching inward, or sometimes incurved. Nerve strong, almost reaching the apex, acting as keel in the lower half. Cells very lax in the lower half, oblong, pellucid ; much smaller and quadrate in the upper half, those of apex margin elliptical. Margin entire. Capsule-body ovoid, neck curved, nearly twice as long ; lid nearly as long as capsule-body, longitudinally striate ; calyptra cucullate. Peristome on a high basal membrane ; teeth mostly cleft to the base, well-developed, outwardly striate, red below, pale near the apex, papillose when young, later apparently not papillose. Ring evident, composed of a single line of large oblong hyaline cells. Male flowers discoid on very short stems, intermixed with but apparently distinct from the fertile plants, at least sometimes ; perigonial leaves short, very wide and concave at the base, shortly and bluntly subulate.

Concerning the Victoria Falls specimens Dixon says :—

" Agrees well with the description of *T. flexifolius.* Notable for the extremely long collum, and closely allied to the northern *T. longicollis* Rich. New to continental Africa " (S.A. Journ. Sc., June 1922, p. 301).

This is, no doubt, *T. divaricatus* Br. and Sch. (in Bry. Eur. Trematodon, p. 4, note) concerning which Müller, who speaks of it as known by name only, says :—

" Near *T. ambiguus* Horns. but slender, leaves narrower, and teeth perfectly bifid and remotely articulated. Cape, Krauss " (C. M., Syn., i, 460).

Natal. Blackridge, Sim 9734.

S. Rhodesia. Victoria Falls, Sim 8936, 8940.

Bryhn records *T. divaricatus* from Eshowe (p. 6).

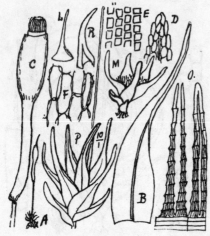

Trematodon flexifolius.

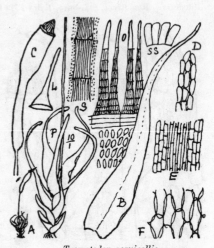

Trematodon aequicollis.

SS, part of ring ; S, part of tooth of peristome, much enlarged.

222. *Trematodon aequicollis*, Ren. and Card. ; Dixon, S.A. Journ. Sc., June 1922, p. 332.

Plants 7-10 mm. high, with seta and capsule five lines long ; stem usually simple ; lower leaves few, lax, and small ; upper leaves long and flexuous, 3-6 mm. long, 1 mm. wide at the base, somewhat clasping, but flat above the sheath or inflexed ; the oblong sheath openly and laxly areolated, and with nerve narrow and indistinct or almost absent ; the lamina above the sheath still wide but gradually tapering to the long acute point, and in all that portion the marginal cells are short and quadrate, often exserted at the upper end, giving a denticulate margin, but the inner cells are all rather longer and narrower, and the central cells narrowest, but through the whole leaf it is often difficult to say if a nerve is present. Sometimes several rows of marginal cells oblong, central cells cylindrical. Capsule ovoid-cylindric, with neck of about equal length, slightly cernuous. Ring distinct, of large oblong cells ; peristome fully developed, with deep basal membrane ; the teeth cleft to the base, slender above, longitudinally striate but not papillose, but with a hyaline papillose membrane as border. Capsule areolation very irregular, the cells short and pale, the wide walls red ; lid as long as the body of the capsule ; calyptra and male flowers not seen. Apparently perennial. Dixon remarks (S.A. Journ. Sc., June 1922, p. 332), concerning Junod's specimens, " These agree exactly with the description and figures of *T. aequicollis* as given by Roth., differing from *T. divaricatus* B. and S. in the narrower nerve and denticulate subula of the leaves ; from the other peristomate African species in the peristome teeth, mostly split through their whole length, not united above." Brotherus maintains both, but only *T. divaricatus* from South Africa.

Natal. Ngoya, Zululand, 1000 feet, abundant 1915, Sim 9736 ; near Durban, Apr. 1898, Dr. J. M. Wood (as *T. natalensis* n. sp.) ; Sea View, Dr. van der Bijl 27.

Portuguese East Africa. Antioka, Rev. H. A. Junod 322 ; Chicumbane, Junod 333.

§ V. *Dicranelleae.* Plants small, with lanceolate-subulate, smooth, channelled, and nerved leaves, often falcate-secund, the cells oblong or elongated-rectangular at the base, smaller and from shortly rectangular to narrow-linear above, without alar cells. Capsule erect or inclined, short, oval-oblong, with or without struma ; sometimes striate. Lid subulate-rostrate ; peristome large, red or purple, of sixteen teeth, each wide at the base, and forked to or beyond the middle into subulate segments.

Genus 76. AONGSTROEMIA Br. Eur.

Plants slender, erect, loosely caespitose, leaves small or minute, shining, densely imbricated or julaceous, crowded at the apex of the stem, usually acute from a wide base, with lax areolation and the thin nerve either disappearing below the apex or excurrent. Inflorescence dioecious, the male terminal and discoid, with large antheridia and long paraphyses, both abundant. Calyptra minute, cucullate ; capsule erect, oval, regular ; lid widely conical, more or less rostrate ; peristome teeth lanceolate, entire or bifid, purple, punctate or longitudinally striate. Spores large, smooth. Habit terrestrial.

This genus, as used by C. Müller, included *Dicranella* before the separation of the two genera took place, and as now used here it includes Hampe's genus *Illecebraria*, having no peristome, a distinct ring, and nerve ceasing below the leaf-apex.

Synopsis.

223. *A. vulcanica.* Nerve excurrent ; leaves scale-like, appressed when dry ; leaf-margin entire.
224. *A. julacea.* Nerve ending below the leaf-apex ; leaves rounded, appressed ; leaf-margins denticulate.

223. *Aongstroemia vulcanica* (Brid.), C. M., Syn., i, 427 ; Besch., Fl. Reunion ; Renauld's Exsicc., p. 153 ;
    Dixon, Trans. R. Soc. S. Afr., 1920, p. 182 ; Broth., Pfl., x, 179.

=*Weisia vulcanica* Brid., Sp. Musc., i, 124 (1806).
=*Dicranum vulcanicum* Brid., Bryol. Univ., i, 466.
=*Dicranum filiforme*, Schwaegr., Suppl. II, p. 72, t. 122.
=*Aongstroemia transvaalensis* C. M., Hedw., xxxviii, 89 ; Broth., Pfl., x, 179.
=*Aongstroemia *Macleana* Rehm., M.A.A. 440.

A slender little perennial moss ; stems ½–1 inch long, simple or forked, the leaves julaceous, imbricate, and closely appressed when dry, spreading slightly when moist, incumbent and keeled ; the lowest leaves round and small, crenated by protruding vermicular cells, next leaves small and oblong-mucronate, increasing in size up the stem, the upper and perichaetal leaves having a solid, well-developed, entire subula terminating the excurrent nerve, and an oblong-lanceolate lamina. Nerve very strong, protruding as a keel, widest where the lamina terminates, decreasing in width to the base, but connected down the stem ; composed of a single layer three to five cells wide of pellucid short columnar cells ; no differentiated alar cells are present, but the leaf-base cells are all irregularly oblong, not much longer than wide ; then longer and narrower oblong cells occupy the lamina, but upward it is held by irregular and often twisted shortly vermicular cells. Male inflorescence terminal, or by innovation ultimately lateral, its leaves distinctly subulate and the subula channelled. I am not aware that capsules have been seen in South Africa, but C. Müller, who describes the plant from Bourbon specimens, says, " Capsule included, cylindrical, small, erect, on short peduncle ; lid conical-rostrate, short, straight ; peristome teeth short, transversely sulcate, incurved, apex slightly bifid, brownish-yellow." Concerning *A. *falcicaulis* C. M. in Herb. ; Renauld, Exsicc., p. 202, Renauld says (Prod. Fl. Bryol. Madag., p. 57), " from the similar *A. vulcanica* it differs in the smaller size and falcate stem," which differences seem to me insufficient.
    Natal. Maritzburg, 1908, H. A. Wager (wrongly distributed as *A. subcompressa*).
    Transvaal. Pilgrim's Rest, MacLea (Rehm. 440) ; Spitzkop, near Lydenburg (Dr. Wilms, 1884, Herb. Jack).
    Found also in the Mascarenes.

224. *Aongstroemia julacea* (Hook.) Mitt., M. Austr. Am., p. 27 (1869) ; Dixon, Trans. R. Soc. S. Afr., 1920, p. 182.

=*A. andicola*, C. M., in Bot. Zeit., 1847, p. 189, and Syn., i, 428.
=*Gymnostomum julaceum*, Hook., Musc. Exot., t. 42 (1820) ; Brid., Bry. Univ., i, 92.
=*Illecebraria julacea*, Hamp., Prod. Fl. Nov. Granat., p. 345.

Plant ¾–1 inch high, simple or forked by innovations, closely julaceous, dioecious, and the male plant only is known so far in South Africa. When dry the leaves are closely imbricate, as in *Anomobryum* (which it closely resembles), and so appressed that the curious leaf-margins cannot be traced ; when moist they expand somewhat. It is then found that the leaves are shortly oblong, with round apex, or often almost orbicular, concave, strongly nerved to near the apex ; unique in its marginal formation along the upper portion of the leaf where, instead of serrations or denticulations, two marginal cells from different directions project together, back to back, with short interval between them and the next ; the lower margin is almost entire. Basal cells hyaline, irregularly oblong-hexagonal, from the nerve to the margin, and without differentiated alar cells ; cells of upper part of leaf irregularly elliptical ; marginal cells not different except those indicated ; leaf-apex rounded or mucronate, almost entire. Male inflorescence terminal, or through innovations ultimately lateral, discoid, enclosed in imbricated leaves similar to the others but rather longer. Capsules not seen, but Hampe formed his genus *Illecebraria* on this and another species, then both known only from America, in which the peristome is absent and the ring differentiated, all other known species of *Aongstroemia* having a peristome present, but ring not differentiated.
    Dixon (Trans. R. Soc. S. Afr., 1820, p. 182) says, " With *Ditrichum strictum*, together with which it was growing (on Giant's Castle), this constitutes a most interesting addition to the African moss-flora. The two form a further link in the curious chain of mosses, having a very definite while very restricted geographical distribution, viz. southern (or high alpine) South America, South Africa, and New Zealand. They are found in either markedly high southern latitudes or at high altitudes, and seem to imply an early distribution from a common antarctic or subantarctic centre.

" The distribution of the two as hitherto recorded is :—

" *Ditrichum strictum,* Chimborazo, Fuegia, Falkland Islands, Kerguelen, Marion Islands, Auckland Island, Campbell Island, New Zealand, Tasmania.

" *Aongstroemia julacea,* Andes of New Grenada and Quito.

" *A. julacea* is a very striking plant in appearance ; the almost orbicular leaves, which are curiously cristate-papillose

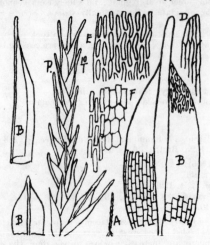

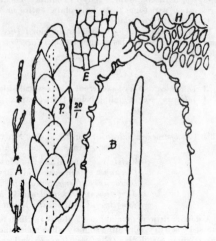

Aongstroemia vulcanica.   *Aongstroemia julacea.*

round the upper margin, are very closely appressed to the stems, which are quite terete and julaceous, and the plant may easily be mistaken for an *Anomobryum.*"

Natal. Giant's Castle, 8000 feet, 1915, Sim 8537*b*, and also R. E. Symons (Sim 9737) ; Upper Bushman's River, 7000 feet, Sept. 1905, Sim 8686 ; Mont aux Sources, 10,000 feet, Sept. 1922, Miss Edwards (Sim 9738) ; Little Berg-Cathkin, 6000 feet, Miss Owen 19, p.p., with *Mielichhoferia.*

(*Aongstroemia orthocarpa* Mitt. is mentioned in Shaw's list as from the Cape, but I have no information concerning it, or if it has been published.)

Genus 77. DICRANELLA W. P. Sch.

Plants usually small and caespitose ; leaves smooth, sparingly chlorophyllose, lanceolate or subulate, spreading, frequently secund-falcate and arched, usually keeled, with a distinct nerve ceasing below the apex, and with lax, usually rectangular areolation at the base and smaller elongate-hexagonal or narrow areolation upward, but no distinct alar cells. Inflorescence dioecious or rarely monoecious ; the antheridia aggregated in a bud-like body. Capsule erect or cernuous, usually oval or oblong, sometimes strumose, and occasionally striate ; peristome large and regular, or sometimes more or less undeveloped, teeth deeply bifid, with filiform granulose or punctate segments.

Not uncommon, growing on soil or on decayed tree-trunks ; very similar to *Ditrichum* but lid subulate-rostrate.

Synopsis.

Leaves erecto-patent. Capsule not furrowed when dry (*Microdus* Schimp., in Besch., Prod. bryol. Mexic., p. 17, as genus).

225. *D. minuta.* Leaves short, wide, and bluntly lanceolate ; capsule shortly oval. Peristome teeth irregular.
226. *D. subcompressa.* Leaves lanceolate ; capsule oval-oblong. Peristome teeth well developed.

Leaves sheathing, then patent, then ascending. Capsule often furrowed when dry (*Eudicranella*).

227. *D. subsubulata.* Leaves numerous ; shoulder and margin crenulate.
228. *D. rigida.* Leaves numerous ; shoulder and margin not crenulate.
229. *D. Symonsii.* Leaves scattered, shortly sheathing ; margin rugose-tuberculate.

Dicranella *pervaginata Broth. is included in Bryhn's list as from Eshowe, Zululand (1911), but is not mentioned in the much more recent Broth., Pfl., x, 181, under *Dicranella* nor under *Aongstroemia* (p. 179), where the Australasian *D. vaginata* Hook. is placed, and it is probably a manuscript name now sunk.

225. *Dicranella minuta* (Hampe), Jaeg. ; Broth., Pfl., i, 3, 309.

=*Microdus minutus* (Hampe), Besch., Fl. Bry. Reunion, p. 17 ; Paris, 2nd ed., p. 238 ; Dixon, Trans. R. Soc. S. Afr., 1920, p. 182 ; Broth., Pfl., x, 181.

=*Aongstroemia minuta* Hampe, in Linn., 1874, p. 269.

Plant usually minute, 2–6 mm. high, simple or forked, dioecious, gregarious or matted ; leaves numerous, erecto-patent, when dry erect, often slightly homomallous, bluntly lanceolate, keeled or somewhat incumbent, but with flat entire margins; nerve wide, strong almost to the apex; the inner (per. leaves) not longer but more sheathing and somewhat ligulate. Cells mostly oblong, with a few elliptical at the leaf-apex; lowest cells on per. leaves long-hexagonal and hyaline. Capsule minute, shortly oval, erect, regular, pale, with red band at the ring; seta 2–3 mm. long; lid convex and obliquely rostrate, as long as the capsule, the lid having long cells in the beak, smaller rounded cells on the convex portion, and the ring of large hyaline cells often attached to the base thereof; calyptra cucullate, obliquely fitting the lid, white. Peristome teeth wide and striate below, narrow and hyaline-papillose above, very irregular and badly developed, often truncate at one or two bars up, possibly through the segments breaking off. Leaves shorter, wider, and more open than in any related plant. The minute oval capsule and oblique cucullate calyptra are very distinctive. The leaves, which are shorter and wider than in *D. subcompressa* Jaeg., somewhat resemble those of *Ceratodon* in form, but the areolation is different.

Natal. Knoll, Hilton Road, 4500 feet, Sim 9744, 9752; Town Bush, Maritzburg, 2500 feet, Sim 7634, 8711, 9753; Chase Valley, 1907, Sim 7468; hill above Vryheid, 4000 feet, Sim 8210.

Transvaal. Houghton, Sim 9746, and Zoo Stream, Johannesburg, 5000 feet, Sim 9745. Dixon cites it also from Moordrift, Waterberg Dist., 1916, H. A. Wager (402 and 410), as *Microdus*, and adds, " No. 402 represents a small reduced form, which, however, intergrades with No 410; the latter corresponds with *M. limosus*, Besch., which, however, from a study of Bescherelle's specimens I should certainly refer to *M. minutus*. I think in all probability all these will have to be reduced ultimately to *M. pallidisetus* (C. M.), of which, unfortunately, specimens have not been available.''

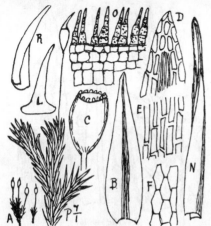

Dicranella minuta.

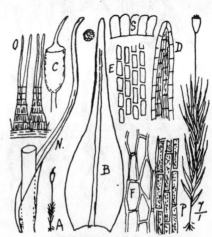

Dicranella subcompressa.
S, portion of ring.

226. *Dicranella subcompressa* (Hampe), Jaeg. Ad., i, 91.

=*Aongstroemia subcompressa*, Hampe, MSS.; C. M., in Bot. Zeit., 1859, p. 215; Broth., Pfl., x, 179.
=*Dicranum *sclerocarpum* W. P. Sch., in Breut. Musc. Cap.

Caespitose, ¼ inch high; stems simple or slightly branched, or with basal or sub-basal innovations, all stems slender, usually red, with short and wide scattered leaves toward the base, gradually longer and more subulate upward; all directed upward and not much spread, and all ordinary leaves with the lower half ovate-lanceolate and openly clasping or nearly flat and the upper half subulate, flattened below and almost or quite to the apex. Nerve strong, composed of dense oblong-cylindrical cells, and disappearing just below the leaf-apex; cells of lower lamina large, irregularly oblong or latticed, shorter near the shoulder, and above that and at the apex shortly oblong or quadrate. Margin entire. Per. leaves high clasping, and with much longer subula. Seta ¼ inch long, pale; capsule erect, red, regular, oval-oblong, with a very short, wide neck; hardly narrowed at the mouth; ring evident; peristome of sixteen short teeth, the upper three-quarters composed of two slender, papillose, and separate segments, united below into a wide, closely barred, and densely striate base. Lid obliquely rostrate. Male inflorescence not seen.

Cape, W. Montagu Pass, Rehm. 24.

Natal. Hilton Road, Sim 9754, 9757; Maritzburg, Sim 8529, 9755, 9756, and Rehm. 441 (as *D. graciliramea* Rehm.).

Transvaal. Lydenburg, MacLea (Rehm. 442, as *D. Macleana* Rehm.).

D. graciliramea, Rehm. 441, has numerous slender red innovations.

D. Macleana, Rehm. 442, differs only in having the lamina less expanded toward the leaf-base. MacLea's specimens, being over-mature, are not in condition to show the peristome.

227. *Dicranella subsubulata* (Hampe), Jaeg. Ad., i, 79; Dixon, Trans. R. Soc. S. Afr., 1920, p. 183; Broth., Pfl., x, 182.
=*Aongstroemia subsubulata* Hampe, MSS.; C. M., in Bot. Zeit., 1858, p. 162.

Caespitose, ½–1 inch high, simple or slightly branched; leaves few, small, and scattered below, but crowded and imbricated in a coma above, and with innovations again small-leaved below and many-leaved above, the inner per. leaves being largest and often homomallous.   Stem often red, rather slender.   Leaves sheathing and appressed at the base, then spreading, but the point usually erect.   Sheath widely oblong, clasping; its cells empty and irregularly oblong, the distinct nerve of longer, narrower, and denser cells; leaf quickly reduced to a long lanceolate-subulate subula, boat-shaped with erect margins, and nerved in the lower half, but more solid and nearly terete, and without evident nerve toward the apex, the cells there being shorter and more chlorophyllose.   From the shoulder of the sheath upward the margin is irregularly crenulate, through projecting cells.   Per. leaves have sheath longer and high-clasping and the subula longer.   Capsule on a ½-inch yellowish seta, oval-elliptical when young, with an obliquely rostrate beak nearly as long; with age the capsule becomes 4-striate or bluntly 4-angled, and the peristome then is usually gone.   Peristome placed well inside the mouth, of sixteen teeth cleft to the base into slender papillose segments, red below, pale above; the base wider and somewhat striate, the peristome often disappearing early, and often rather rudimentary.   Ring evident, of one row of large cells; upper three or four rows of cells of capsule smaller, rounded, and thick-walled, then others below are irregularly latticed and twisted, but short.   The oval-elliptical capsule, angled when dry, and crenulate margin of the leaf-shoulder distinguish this species.

Natal.   Knoll, Hilton Road, Sim 8710, 8712; Maritzburg, Sim 9758; Mistyhome, Hemu-hemu, 5000 feet, Sim 9759; Blinkwater, J. M. Sim.

Rhodesia.   Inyanga, Dr. E. A. Nobbs (Eyles 1361).

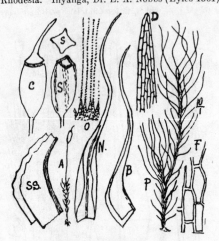

Dicranella subsubulata.

S, dry capsule and section of same.

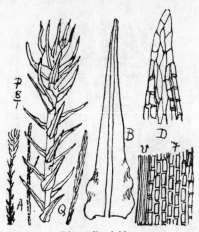

Dicranella rigida.

228. *Dicranella rigida* (Dixon, MSS.), Sim (new species).*

Plant slender, rigid, brown, up to 1 inch long; leaves acuminate from wide base, cuspidate (but not subulate nor terete at the apex), 1–2 mm. long, crisped toward the leaf-base, the lower scattered and subpatent, the upper closer and rather longer, curved upward, all appressed when dry; nerve strong, percurrent; apex acute, flat; margin entire from base to apex; cells quadrate, the lower shortly oblong, the nerve cells longer.   Not seen fertile.

Cape, W. Paarl Rock, 1500 feet, Sim 9632β and 9633.

This is closely related to D. *subsubulata*, but larger, more rigid in habit, with the leaves crisped below, the leaf-margin entire, the apex flat and margined by lamina, and the cells shortly oblong to the leaf-base.

229. *Dicranella Symonsii* Dixon, Trans. R. Soc. S. Afr., 1920, p. 183, and pl. xi, fig. 2.

=*Anisothecium Symonsii* (Dixon), Broth., Pfl., x, 177.

=*Bryum leptotricheaceum*, Rehm. 256.

=*Bryum bartramioides*, Rehm. 257.

" A lax, slender, caespitose, yellowish-green alpine moss, 1–2 inches high, simple or somewhat branched, usually crowded (or mixed with *Bryum*) and radiculose at the base.   Leaves scattered, 2 mm. long, with short, wide, sheathing, appressed base, suddenly contracted into a suddenly horizontal ligulate-lanceolate, channelled subula; in the upper leaves it is again bent upward.   The upper leaves are the larger and more crowded; nerve wide and distinct in the sheath, less distinct and more chlorophyllose in the subula, disappearing below the terete subacute apex.   Margin

---

* *Dicranella rigida* (Dixon, MSS.) Sim (sp. nov.).

Planta gracilis, rigida, ad 1″ longa; folia acuminata a basi lata, cuspidata, 1–2 mm. longa, folio-basin versus crispata, inferiora sparsa ac sub-patentia, superiora artiora et paulo longiora, sursum curvata, omnia adpressa ubi sicca; nervus fortis, percurrens; apex acutus planus; margo omnis integer; cellae quadratae, inferiores breviter oblongae, nervi cellae longiores.   Fertilem n.v.

entire ; that in the sheath straight, the cells there being large, oblong, empty, somewhat irregular, wider than the denser oblong nerve-cells, but throughout the subula the margin is erect or incurved, and the marginal cells there are shortly quadrate and have thickened transverse walls which project outward and give the whole subula the appearance of being rugose-tuberculate, the inner cells being oblong and chlorophyllose. Inflorescence not seen, apparently dioecious " (Dixon). Dixon adds, " A very distinct species from those hitherto recorded from Africa, and showing a connection with a certain Australian and sub-Antarctic group, including *D. clathrata* H. f. and W., *D. vaginata* (Hook.), etc., which for the most part have smooth cells. A New Zealand species, *D. wairarapensis*, Dixon, is still nearer it in the papillose subula, but is very different in habit and in other respects."

Natal. Tugela Valley, Mont aux Sources, 7000 feet, July 1921, Sim 10,145 ; Giant's Castle, 7000–8000 feet, on wet stones, 1915, R. E. Symons (Sim 8665, 9739, 9740) ; Upper Bushman River, 7000 feet, Sept. 1905, Sim 8670, and R. E. Symons (Sim 9741).

O.F.S. Kadziberg, Rehm. 256 (distributed as *Bryum *leptotrichaceum* Rehm.) and Rehm. 257 (distributed as *Bryum *bartramioides* Rehm.).

Dicranella Borgeniana (Hampe), Jaeg. Ad., i, 88 ; Broth., Pfl., x, 182.

=*Aongstroemia Borgeniana* Hampe, in Bot. Zeit., 1870, p. 33.

From Natal ; is unknown to me. Brotherus, Pfl., i, 3, 310, places it with *D. subsubulata* and *D. abruptifolia* (C. M.), Par., in section *Dicranella* (*sens. strict.*). *D. abruptifolia* is *Ditrichum flexifolium* (Hk.) Hampe with its lower leaves burned and the capsule starved through that.

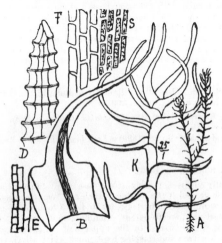

Dicranella Symonsii.

E, marginal quadrate cells and longer inner cells ; S, nerve-cells.

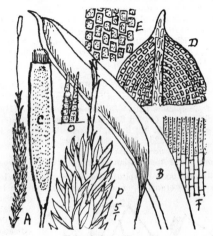

Holomitrium affine, var.

§ VI. *Dicraneae.* Plants larger than in the last section, in some cases several inches high, often branched, caespitose. Leaves lanceolate or narrowly lanceolate, usually long, often falcate-secund. Alar cells usually enlarged, often inflated, pellucid, or coloured. Capsule erect, cernuous or cygneous, usually oval or oblong, with long subulate lid, cucullate and sometimes fringed calyptra, and peristome of sixteen more or less forked teeth.

Genus 78. HOLOMITRIUM Brid.

Densely caespitose plants, suberect and more or less branched ; leaves convolute-lanceolate from a clasping base, closely imbricated, short, firm ; nerve slightly excurrent and reflexed ; lower cells elongated ; upper small, obscurely quadrate ; capsule erect on a long seta, oval-cylindrical, narrowed to the mouth ; lid rostrate ; peristome rising below the mouth ; teeth short, sixteen, recurved when dry, lanceolate-subulate, sometimes split but the lobes usually more or less coherent. Inflorescence monoecious ; perichaetium shortly exserted, with subulate leaves from a long sheathing base.

Holomitrium is maintained by Paris for many species, which belong mostly to all parts of the Southern Hemisphere ; none are European or continental North American.

230. *Holomitrium affine* Card. and Ther., var. *cucullatum* (Besch.) Ther. (see Theriot, in Bull. Soc. bot. Genève, 2nd ser., iii, 245 (1911)) ; Broth., Pfl., x, 201.

=*H. *capense* C. M., in Rehm. 73.

Formerly included in *H. vaginatum*, Brid., i, 227 ; C. M., Syn., i, 351 ; Broth., Pfl., i, 3, 320.

A vigorous moss growing in large tufts, green above, brown below ; monoecious, caespitose ; stems ½–1½ inch long, erect, usually unbranched, densely leafy, the leaves imbricate, 2–3 mm. long, boat-shaped, spreading, and with

11

inflexed apex when moist, but each leaf involute and spirally twisted when dry and then only appressed at the base, the upper twisted portion spreading, but with inflexed apex.   Leaves vaginate from a wide base, wider above that, then narrowed into concave boat-shaped form ; the nerve not covered by lamina cells, straight and smooth as a keel along the back but curving up toward the apex, then exserted as a sharp reflexed point ; margins entire, much raised, producing boat-shaped lamina when moist, incurved when dry ; nerve cells all long and translucent, lamina cells quadrate, 15–20μ diameter, exceedingly irregular as to size and form, smooth and tumid, chlorophyllose, but set in pellucid, pale green lamina so that the cells are obscure and the leaf darkly opaque when dry, but laxly cellular when moist ; lower cells long and cylindrical, merging into uninterrupted chlorophyllose straight lines, but at the base there are a few rectangular hyaline cells up to 80×20μ, the outer of these often chlorophyllose to the base. Alar cells not differentiated.   Perichaetal leaves (about three) large, vaginate, embracing the seta far up, with only the point free. Seta about 1 inch long, yellowish ; capsule erect, cylindrical, 2–3 mm. long, yellow ; peristome set within the mouth, of sixteen teeth, more or less split, and the segments often unequal. Teeth short, erect, red, reflexed when dry, trabeculate-papillose ; setaceous and pale at the apex.   Lid long-beaked, straight.

Cape, W.   Rehmann 73 from Knysna was distributed as " *H. *capense* C. M.''
Cape, E.   Eastern forests, Perie, etc., scarce, T. R. Sim.
Natal. Weenen County, Sim 7978 ; Little Berg, Cathkin, 6000 feet, July 1922, Miss Owen 8 ; Van Reenen, Jan. 1910, H. A. Wager 143 (distributed as *Syrrhopodon Dregei* C. M. wrongly).

Genus 79. Dicranoloma Renauld (Prod. Fl. bry. de Madag., etc., as a subgenus of *Leucoloma*, afterwards as a genus).

Plants robust, loosely tufted ; leaves long, somewhat falcate-secund, with a clasping base, and narrowed gradually and concave upward to an acute point ; nerve narrow, extending to the point, ridged on the back upward, these ridges and the upper part of the leaf-margin more or less serrate ; lower cells lax and cylindrical ; the cells near the nerve not different from the others, the upper cells linear-oblong and the marginal cells narrow, forming an indistinct hyaline border.   Cell-walls porose. Capsule irregular, cernuous, arcuate, more or less strumose. Lid with a curved beak. Calyptra cucullate.   Very near *Dicranum*, but this usually has a border of hyaline cells along the leaf-border, which is not the case in *Dicranum*.

231. *Dicranoloma Billardieri* Schwaegr., Par. ; Brid., Musci Rec., ii, 181 (1798) ; Dixon, N.Z. Bull., iii, 23 (1913) ; Dixon, Journ. Bot., Dec. 1916, p. 356 ; Broth., Pfl., x, 209.

=*D. *commutatum* Paris =*Dicranum commutatum* (Hpe., MS.), Knysna, Rehm. 31 =*Leucoloma* (Broth., p. 323).
=*D. dichotomum* (P. B.) Ren. =*Dicranum dichotomum* P. Beauv.
=*D. nitidulum* (C. M.) Broth. =*Dicranum nitidulum* C. M., Hedw., xxxviii, 88 (1899) =*Leucoloma nitidulum* Broth., Pfl., i, 3, 323 ; Paris, Suppl. Ind., p. 233.
=*Dicranum tabulare*, Rehm. 32 ; and
=a long list of Australasian synonyms given by Dixon in N.Z. Bull., iii, 23 (1913).

Robust, laxly caespitose, the stems 1–4 inches long, usually suberect, rooting at the base only, simple or with two to three branches, or sometimes branching freely near the ground if the leading branch has been damaged or dwarfed.   Leaves numerous, imbricate, up to 4 mm. long and 1 mm. wide at the widest part, erecto-patent, often homomallous or falcate, in which case they are twisted in the upper half.   Leaves ovate-lanceolate from a narrow but clasping base, much wider above that, then tapering gradually to a subulate point, margin entire except toward the point where the margin of the leaf and the back of the nerve are toothed by strong protruding cells ; nerve narrow, of about three lines of linear cells ; alar cells 40μ diameter, roundly square, very incrusted bright yellow and the base of leaf between the alar groups is yellow though with usual cells. Cells very lax, separate, 40–50μ long, oblong-cylindrical, connected by pores, the upper cells hardly different from the lower. Margin indistinctly bordered by long, narrow, hyaline cells. Inner perichaetal leaves short, blunt. Seta 1 inch long, red ; capsule red, curved, strumose, irregular ; peristome red, the teeth rather wide, of two equal segments connected to near the apex. Common on exposed localities. Table Mountain (Sim 9291, etc.) ; Knysna (Rehm. 31) ; Natal (Sim 1854), etc.

Forma, *aquatica.*   A very dissimilar plant to the type, having leaves few, all spreading, not falcate or homomallous and not known fertile, occurs floating in pools on Table Mountain, but does not differ specifically (Sim 9300).

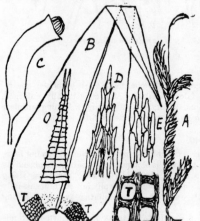

Dicranoloma Billardieri.

(Dicranum, Hedw.

Plants usually robust, in some species several inches long, simple or branched, caespitose ; leaves usually more or less falcate-secund, long, lanceolate or lanceolate-subulate, bright green, smooth or papillose, entire or serrate ; nerve usually narrow, excurrent, often ridged along the back ; the alar-cells large, sometimes swollen, and often quadrate and coloured, those between them and the nerve often narrow ; the cells in the rest of the leaf usually linear-rectangular, but shorter, smaller, and more obscure toward the apex. Cell-walls porous. Seta usually long, straight ;

capsule erect or cernuous, in some cases strumose, often curved and unequal, smooth or only faintly striate ; lid rostrate ; calyptra not fringed ; peristome teeth sixteen, straight, red, slightly confluent at the base, each divided about half-way into rather unequal segments, minutely vertically striate below, transversely barred within.

Dicranum was at one time a very comprehensive genus, from which many other genera have now been separated out, consequently the name now appears as a generic synonym in regard to many species, though many species still remain in *Dicranum*, some of which are cosmopolitan or nearly so, usually growing on decayed stumps. So far as South Africa is concerned, the whole group requires revision to definitely place the species (see note under *Campylopus*), but it is believed that the genus as now defined is not South African, *D. tabulare* *Rehm. 32 being *Dicranoloma Billardieri* Brid. It was a sterile specimen, and as the plant grows easily from detached buds or shoots it was the subject of Professor Wager's paper to the S.A. Assn. Adv. Sc. on the propagation of *Dicranum tabulare*.)

Genus 80. LEUCOLOMA Brid., Bryol. Univ., ii, 218 ; Broth., Pfl., i, 3, 323.

=*Dicranum*, § *Leucoloma* C. M., Syn. Musc., p. 352.

Plants short, robust, more or less fragile, softly caespitose ; leaves lanceolate or subulate, concave ; the nerve thin, yellowish, not extending to the apex ; the cells in the upper part of the leaf small, quadrate, thick-walled, papillose on the back and chlorophyllose ; usually extending down along the nerve as longer and more oblong cells ; those in the lower part of the leaf narrowly linear, thickened, smooth ; the leaf distinctly margined by long, narrow, hyaline cells, and the alar cells rectangular, lax, plane, or swollen ; capsule erect, cylindrical-oval, small, regular, the lid with erect beak ; calyptra torn at the side or many-lobed at the base, often scabrid at the apex. Renauld (Prod. Fl. Br. Madag., p. 61) describes forty-three species of *Leucoloma*, including *Dicranoloma* ; only two of these have been recorded from the opposite African coast forests, but more may be expected.

In Wager's list *Dicranoloma Billardieri* Schw. and its synonym *D. nitidulum* (C. M.) are included here (as well as *L. rugescens* C. M., unknown to me).

Synopsis of Leucoloma.

232. *L. syrrhopodontioides.* Separation of cell groups not sharply defined ; plants very small.
233. *L. Zeyheri.* Leaves lanceolate or the upper subulate ; margin a few lines of narrow cells.
234. *L. Sprengelianum.* Leaves strongly convolute-subulate, slightly papillose on the back. Margin a few lines of narrow cells.
235. *L. Rehmanni.* Leaves convolute-subulate, strongly papillose on back. Margin a few lines of cells wide.
236. *L. chryso-basilare.* Hyaline margin one-fourth of the lamina on each side ; leaves acuminate or subulate, slightly papillose.

232. *Leucoloma syrrhopodontioides* Broth., in Engl. Bot. Jahrb., 1896, p. 236, where it is said to be sterile on bark, in Pondoland ; Broth., Pfl., x, 211.

I have not seen the description or a verified specimen of this ; Brotherus (Pfl., i, 3, 325) put it in § 3, *Transmutantia*, §§ *Pseudocaespitulosa*, which Renauld, who defined these groups (Prod. Fl. Br. Madag., etc., pp. 72–75), thus describes, " §, Inner cells of the lamina filling the whole of the upper part of the leaf, downward gradually longer and smoother, not sharply defined from the thin-walled lower cells ; auricle more or less concave with subhexagonal or rectangular twisted and indurated cells. Seta long. Capsule narrow, oblong or cylindrical. §§, Very small plants in wide-spreading tufts. Leaves yellowish-green, when dry flexuose, sometimes falcate, hardly crisped, delicate, narrow, scabrid on the back. Upper papillae often uncinate." In Pfl., x, 211, the groups are changed, and he places it along with the following three species in § *Albescentia*, without specific description.

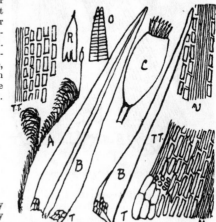

233. *Leucoloma Zeyheri* (C. M.), Jaeg. Ad., i, 115 ;
Broth., Pfl., x, 211.

=*Dicranum Zeyheri* C. M., Syn. Musc., p. 353.
=*Dicranum palescens* Wils., in Breutel, Musc. Cap.
=*Leucoloma palescens*, Jaeg. Ad., i, 118.
=*Weisia* n. sp., Zeyher, Pl. Cap., p. 496.
=*Poecilophyllum* C. M.

Leucoloma Zeyheri.

Dioecious, densely caespitose, ½–1 inch long, erect, sparingly branched, usually green ; leaves spreading when moist, strongly secund when dry, lanceolate, 2 mm. long, the upper more subulate and longer, entire ; marginal cells long, narrow, and hyaline, about six lines forming a border ; next them about two lines are of shorter quadrate cells than the lax oblong cells between them and the nerve, which are larger in the lower half of the leaf than in the upper, and are more or less papillose in the upper half. Alar cells yellow, about six on each side, forming a small corner group ; nerve-cells all very long, narrow, and cylindrical ; inflorescence ultimately lateral ; seta ¾ inch long, red, straight ; capsule oval, erect, regular ; teeth short, two subacute segments connate ; inflexed when dry ; lid conical ; calyptra lobed at the base, not fringed.. Perichaetal leaves few, shortly sheathing, suddenly subulate.

Very near *L. Sprengelianum* and possibly grading into it under changed conditions.   Table Mountain, abundant (Sim 9425), and in all forest districts.

Var. *compactum C. M., Rehm. 28.   More dwarf, dense, and compact ; leaves shorter and more crisped.   Frequent in exposed sites, Table Mountain, Paarl Rock (Sim 9642), etc.

234. *Leucoloma Sprengelianum* (C. M.), Jaeg. Ad., i, 115 ; Broth., Pfl., x, 211.

=*Dicranum Sprengelianum* C. M., Syn. Musc., p. 353.
=*Leucoloma Ecklonii* (Ltz.), Jaeg. Ad., i, 115 ; Broth., Pfl., x, 211.
=*Dicranum Ecklonii* Ltz., Moost., p. 158.
=*D. fasciatum* Spreng., Syst. Veg., iv, 2.

Dioecious, densely caespitose, often tomentose below ; stems ½–2 inches long, erect, simple or sparingly branched ; leaves subulate-flexuose from rather a wide base, 4 mm. long, entire ; margin closely involute, usually twisted, the lower leaves short, the upper longer and strongly falcate, forming an incurved coma ; usually very secund, firm, bright yellow at the top of the stem.   Nerve narrow, about six cells wide, distinct to near the apex.   Yellowish ; marginal cells long, narrow, and hyaline, four to six lines forming a usually distinct border ; cells of the lamina oblong, lax, slightly papillose at the back, the lower longer than the upper, those near the nerve not differentiated ; alar cells swollen, forming a distinct little group of nine to twelve large red cells on each side with a few small round or quadrate cells above them.   Inflorescence terminal but by innovation the capsule is soon lateral.   Seta ¾ inch long, erect ; capsule small, oval-oblong or elliptical erect, regular, with short erect lid and several-lobed mitriform calyptra.   Peristome teeth rather short, the segments connate and shortly acute.   Perichaetal leaves few, long-sheathing. suddenly subulate.   Abundant on Table Mountain and neighbouring localities, also in the eastern forests, Natal, and Transvaal.

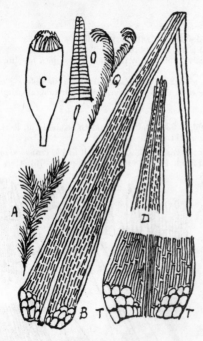

Leucoloma Sprengelianum.

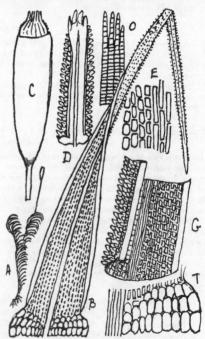

Leucoloma Rehmanni.

235. *Leucoloma Rehmanni* (C. M.) Rehm., Exsicc. M.A.A., p. 29 ; Wager, in Trans. R. Soc. S. Afr., iv, 1, 2 ; Dixon, Bry. of S. Rhod., S.A. Journ. Sc., xviii, 302 ; Broth., Pfl., x, 211.

=*Dicranum Rehmanni* C. M., in Hedw. (1899), xxxviii, 87.
=*Leucoloma Haakoni*, Broth. and Bryhn, 1911, p. 6 ; Broth., Pfl., x, 211.

Laxly caespitose, pale yellow, 1–2 inches high, firm, erect, slightly branched, but with the upper leaves secund and forming a decurved coma.   Leaves numerous, the upper 4 mm. long, subulate, entire, except the denticulate apex, strongly convolute and keeled, the upper part of each being tubular, strongly papillose on the back, but still margined with hyaline smooth cells and the keel is smooth.   Cells oblong, chlorophyllose, lax, all strongly papillose on the back ; about twelve lines of these occupy from the nerve to near the margin ; then three to nine lines of long narrow hyaline cells form a distinct border, wider below ; the nerve has five to eight lines of long, narrow, translucent cells, not papillose, and the alar cells form distinct auricles containing about twenty crowded, roundish-elliptical, pale cells each. Perichaetal leaves somewhat sheathing, widely margined, and with yellow base. Seta erect, $\frac{3}{4}$ inch long, red, twisted ; capsule cylindrical, widest near the mouth, erect, regular, red ; the teeth more or less connate in various combinations, some not well developed. Lid, calyptra, and male inflorescence not seen.

Clermont (Rehm. 29) ; Knysna (Rehm. 29*b*) ; Pilgrim's Rest, Transvaal (Rehm. 443) ; Inyanga, Rhodesia (Henkel, in Eyles, 2623*b*, 2635), and in many eastern forest localities.

236. *Leucoloma chrysobasilare* (C. M.), Jaeg. Ad., ii, 643 ; Renauld, Prod. Fl. Bry. Madag., p. 65 ; Dixon, S.A. Journ. Sc., xviii, 302 ; Broth., Pfl., x, 210.

Leucoloma *Woodii Rehm. and Maco, Rehm. 444.
Dicranum chrysobasilare C. M., in Linnaea, 1876, p. 238 (from Anjouan).
Leucoloma Zuluense, Broth. and Bryhn, 1911, p. 6 ; Broth., Pfl., x, 210.

Stems erect, 1–2 inches high, somewhat branched, the lower part sometimes leafless ; leaves numerous, very long (4 mm.), imbricate at the base, straight, spreading, or at the apex crowded all upward, sides somewhat inflexed, especially upward ; subulate, entire, except that the tips have protruding cells ; nerve narrow, about five lines of narrow cells wide, extending almost to the apex ; alongside the nerve are many lines of chlorophyllose cells, oblong in the lower half, more quadrate and smaller upward and somewhat papillose on the back ; these occupy about half the lamina, the rest being occupied by long, narrow, cylindrical, hyaline cells, of which the outer three or four lines are narrower than the others and form a border. These combine to give the leaf an appearance of being 3-striate on the back, under a low power. Alar cells twenty to thirty, small, oblong, crowded, yellow. Capsule axillary ; seta $\frac{1}{2}$ inch long, red ; capsule 2 mm. long, cylindrical, tapering downward. Male inflorescence and surroundings yellow, axillary, bud-like, with about five antheridia and as many paraphyses, enclosed in small, clasping, subulate, nerveless floral leaves. The calyptra has not been seen, and may possibly be fringed.

Woodbush, Transvaal (Sim 7728).

Leucoloma chrysobasilare.
S, var. *gracilicaulon.*

Var. *gracilicaulon* Ren., Prod. Fl. Br. Madag., p. 65.

=*L. Woodii* R. and M., of Wager, Trans. R. Soc. S. Afr., iv, 1, 2, and pl. 1*c.*

Stem more slender ; leaves shorter, sparse, somewhat crisped, slightly falcate ; capsule axillary, erect, very small, 1 mm. long, regular, shortly cylindrical ; peristome from inside the mouth, segments separate, fugaceous. Seta 2 mm. long, several in succession up a stem. Wager (as above) describes and illustrates the hardly exserted capsule as having very short teeth, but in Sim 10,143 the teeth are beautifully developed and 35μ long, the points being long and slender.

Inyanga, S. Rhodesia, 6000 feet, 1920, J. S. Henkel (Sim 10,143). *L. scabricuspes* Broth. does not differ (Eyles 2623*b* and 2624, and Eyles 2635β).

Genus 81. CAMPYLOPUS Brid.

Plants usually rather large, simple or forked, erect, caespitose, dioecious ; leaves more or less lanceolate, often setaceous or hair-pointed, usually straight, firm or rigid, not papillose and not margined. Nerve very broad, extending to the apex, often furrowed or lamellose along the back, and with one or more layers of larger hyaline cells either on the upper surface or included in the nerve ; the lower leaf-cells large, narrowly rectangular-oblong, and hyaline ; upper leaf-cells smaller, more or less elliptical and chlorophyllose, and the alar-cells often though not regularly (even on the same plant) forming distinct concave auricles, which are red, brown, or pellucid. Capsule small, elliptical, regular, and striate, on a flexuose, usually cygneous, short seta (which becomes erect when old and dry). Peristome of sixteen long forked teeth, divided about half-way down. Calyptra cucullate, usually fringed at the base. The plants of this genus often grow in extensive and rather dense patches on the ground, or on bare rocks, or sometimes on decayed tree-trunks, and though present in most grass or forest areas are more common on the mountains than in the lowlands. Sometimes several similar species are intermixed, and in many places some species continue indefinitely to produce short bushy or discoid sterile plants which under other conditions develop from these discs long slender stems of very different appearance and leafage, which, taken alone, would be considered very distinct species. C. Müller in his old age described (in Hedwigia, p. 38) many species, mostly from either juvenile or not fully developed sterile plants, and further evidence discredits a good many of these, while there still remain other unidentifiable juvenile forms which may be very different when fully developed and whose life-history we do not yet know.

Campylopus is closely related to *Dicranum*—the wider nerve being the main difference—and in Müller's Syn. Musc. and in his subsequent writings it is included as a section of *Dicranum*; consequently most of the species described by Müller were described as *Dicranum* species, and now appear as synonyms consequent upon the separation of these two genera. Rehmann meantime distributed the supporting type specimens of these species, as well as of many other *nomina nuda* of his own, under the name *Campylopus*, and thereby added confusion which can now be eliminated with difficulty, especially where no further specimens have been found.

Almost all the characters vary exceedingly in *Campylopus*, the leaves on juvenile forms often differ from those that are mature on the same species, the hair-point may be absent or variously developed, the lamellae on the nerve may be almost absent on the leaf-base but highly developed half-way up the same leaf, the alar cells may be hyaline or coloured, plane or ventricose, few or numerous, all on one plant, the seta which is always cygneous when young is often (or always) straight when mature, the capsule which is cylindrical and sulcate when dry often becomes turgid and smooth when moist, the teeth which are reflexed when dry become inflexed or erect when moist or disappear with age, and the calyptra which in most species is fringed sometimes becomes bald with age. The genus is consequently one in which intimate acquaintance with all stages of development, and with many specimens, is necessary, and it is easy to understand that many immature conditions came to be described and distributed under redundant names, and that many such conditions fit no descriptions. Fertile and fully developed specimens are necessary if available. In addition to this, though the genus itself is a most natural one, the subdivision of the group has not been happy. Müller (Hedw., xxxviii, 77) introduced a new section, which has since been used as a genus by some, under the name *Microcampylopus*, on these characters, "plantae nanae, calyptra basi integra," including two species from Java and two from South Africa, one of the latter of which, however, must lapse. There thus remains only one South African species with unfringed calyptra, and my investigations have not yet shown any more, though sometimes the fringe is rubbed off and in many species still admitted the fruit is unknown. Brotherus, Pfl., x, 183, defines *Microcampylopus* differently without improving matters. Limprecht found that leaf-sections of each species separated the European forms into three groups by their structure ; here, meantime, I consider that the adoption of that classification would do more to

Thysanomitrium transvaaliense.

hinder than to aid the study of local bryology, and so I have not used it, but depend on what can be seen without the use of the microtome, but it may be explained that of these sections *Pulinocraspis* has stereid cells on both sides of the nerve and *Eucampylopus* has them on the under side only.

The leaves are often so rigid as to be brittle, and in several species broken-off parts of leaves produce new plants, while in some species easily detached buds are produced, which also can produce new plants after detachment.

Brotherus (Pfl., x, 188) introduces a genus *Thysanomitrium*, cut out of *Campylopus*, and having the peristome teeth bifid to the base, the base of the capsule rough, and the leaves without border. In this he includes :—

(*a*) Leaves without hair or hair-point	. . .	*T. delagoae* (C. M.) Br.
(*βa*) Leaves with hair-point ; nerve lamellate	. .	*T. olivaceo-nigricans* (C. M.) Br.
(*ββ*) Leaves with hair-point ; nerve not lamellate .		*T. amplirete* (C. M.) Br., and *T. inandae* (Rehm.) Br.

For these see here under *C. lepidophyllus*, *C. stenopelma*, and *C. purpurascens* ; all were described from sterile specimens, where the above generic characters do not count.

But Mr. Dixon writes (29/12/1924) that he has had from Herzog an obvious *Thysanomitrium* from the Transvaal, which is being described as *T. *transvaaliense* Herz. and Dixon, agreeing almost exactly with the Indo-Malayan and Pacific *T. umbellatum* Walker-Arn. (=*T. Blumei* Doz. and Molk.), but with wider, exceedingly concave leaves, all abruptly ending in a short, straight, toothed hair-point. This differs from anything I have.

Brotherus (Pfl., x, 185) wrongly places *C. leucochlorus* (C. M.) Par. as South African ; it should be East African.

Synopsis of Campylopus.

1. Leaves with hyaline hair-points	2.
Leaves with long green setaceous points	7.
Leaves without regularly hyaline or long green points	10.
2. Hyaline points erect	3.
Hyaline points often reflexed	6.
3. Point long, dentate	4.
Point shorter, nearly entire	5.
4. Point very long, channelled, dentate	237. *C. Symonsii.*
Point long, denticulate, leaf involute	238. *C. griseolus.*
Leaf wider ; point long, strong, scabrid all round . . .	239. *C. trichodes.*
5. Leaves distant, inflexed	240. *C. Edwardsii.*
Stem leaves not pointed ; comal leaves pointed . .	241. *C. pseudojulaceus.*
6. Leaves lanceolate ; point long, often reflexed . . .	242. *C. introflexus.*
Leaf wide, point short	243. *C. lepidophyllus.*
7. Plants less than 1 inch long, leaves very setaceous . .	244. *C. leptotrichaceus.*
Plants usually more than an inch long	8.

237. *Campylopus Symonsii* Sim (new species).*

Growing in lax hairy tufts fully exposed on bare mountain rock, dark green or nearly black below, the hairs white and prominent. Stems 2 inches long, simple or slightly branched, straight or curved, penicillate when dry, but with leaves expanded when moist. Leaves 4–5 mm. long (including half that length of hair-point), imbricate, erect or often arching back but not suddenly bent back ; the lower half ovate-acuminate, channelled, the two sides meeting ; the upper half a strongly channelled hyaline hair-point, strongly toothed by projecting cells along each margin, and also to some extent by the projecting upper end of long cells along both back and front. Nerve wide, lamellate, ending where the hair-point begins ; alar group small, enclosed, ventricose, coloured, composed of about twenty lax, globose-quadrate cells, above which are a few rectangular cells, margined by longer, narrower cells, but the chlorophyllose elliptical cells extend to near the leaf-base ; they are large and crooked near the nerve, small and regular near the margin. The whole plant is yellow except the hair-point. Other parts have not been seen. At the base of the stem separate rosulate comal groups are occasionally to be found having shorter leaves, cucullate at the apex without hyaline hair-points (fig. S), and similar leaves clothe the lower part of young stems which are setose higher up ; the first leaves that have short hair-points having a distinct bend on the nerve from the back to the front of the leaf. The almost black tuft covered by the very long, toothed, and channelled hyaline hair-points make this species unlike all others ; it is most like *C. echinatus,* in which, however, the point is yellow and the cells are different.

Natal. Top of Giant's Castle, 8000 feet, 1912, R. E. Symons (Sim 9843 and 9835) ; Mont aux Sources, Gibb.

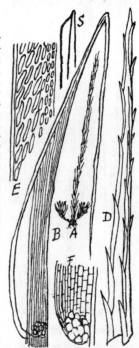

Campylopus Symonsii.
S, hooded point of comal and lower leaves.

238. *Campylopus griseolus* (C. M.), Par., Suppl. Ind., 1900, p. 92 ; Pfl., x, 187.

        = *Dicranum griseolum* C. M., Hedw., 1899, p. 80.

Laxly tufted ; stems 1–2 inches high, stout, the leaves spreading when moist, erect when dry, evenly distributed along the stem except that at intervals they are more numerous as subcomal groups. Stems pointed when dry, and set with very many, long, straight, hyaline bristles giving the plant a grey appearance as lower leaves also have these. Leaves 3–4 mm. long, lanceolate ; nerve half the width of the leaf, ridged on the back ; alar cells few and inconspicuous, hyaline ; lamina of lower half of leaf hyaline and laxly reticulated by thin-walled irregular cells, upper half chlorophyllose, elliptic, and small ; hair-point 1 mm. or more long, flat or channelled below, laxly denticulate along the margin. Floral parts and fruit not seen.

---

* *Campylopus Symonsii* Sim (sp. nov.).

    Cumulis laxis capillosis proveniens in rupibus montanis, capilli albi conspicuique. Caules 2″ longi, plerumque simplices, penicillati si sicci, si humidi, explanati. Folia 4–5 mm. longa, imbricata, erecta, vel saepe retrorsum arcuata sed non abrupte reflexa ; dimidia pars inferior ovato-acuminata, canaliculata, lateribus in unum convenientibus, dimidia pars superior arista hyalina alte canaliculata, fortiter dentata cellis prominentibus secundum marginem utrumque, aliquantum autem et prominente termino superiore cellarum longarum secundum et frontem et tergum. Nervus latus, lamellatus, desinens qua pilum oritur, globus alaris parvus, inclusus, ventricosus, coloratus, ex cellis fere 20 laxis globoso-quadratis constans, supra quas cellae paucae rectagonae, cellis longioribus angustioribus marginatae, lamino-cellae autem chlorophyllosae ellipticae pene ad folii basin pertinent, prope nervum magnae et aduncae, prope marginem parvae et regularer. Cetera non vidi. In Monte Drakensbergensi, Natali, R. E. Symons.

Müller described this from Dr. Wilms' specimens from Lydenburg, Transvaal ; I have it also from the top of Giant's Castle, 8000 feet (Symons), and top of the Sentinel, Mont aux Sources, 10,000 feet (H. Botha Reid), Natal, where it forms one of the conspicuous hyaline-pointed plants characteristic of high altitude and latitude (Sim 9833, 9834).

239. *Campylopus trichodes*, Lorentz, Moosst., p. 159 (1864) ; Dixon, Trans. R. Soc. S. Afr., viii, 3, 183 ; Dixon, S.A. Journ. Sc., xviii, 304 ; Pfl., x, 187.

In loose masses 1–2 inches deep on rock or soil ; stems simple or branched, erect, stout, pointed when dry ; leaves spreading when moist, appressed when dry, 4–5 mm. long, 80μ wide, widest at or near the base, acuminate from there to a long hyaline, toothed hair-point which is scabrid all round and not flattened. Nerve about one-fourth the wideness of the leaf, with about twenty-five pronounced lamellae on the lower surface ; the inner surface covered with oblong cells six times as long as wide. Alar group very variable, sometimes distinctly concave and coloured, sometimes indistinct and flat ; the alar cells shortly oblong, lax, thin-walled ; above them is an area of smaller connate, oblong, empty cells, then the balance of the lamina set with small rhomboid chlorophyllose cells which extend down along the nerve to near the leaf-base.

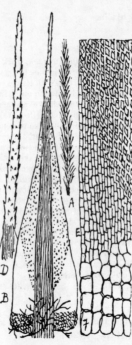

Exceedingly variable as to colour, size of leaf, length of hair-point, and appearance of alar group, but other characters hold good. Sometimes the laminae of the leaf ceases suddenly, giving a rounded or truncate apex from which the hair-point rises, or there may be even auricles there of protruding cells. Free cells in the lamina are sometimes twisted, especially those near the nerve. Not seen fertile.

*C. *latifolius*,* Rehm. 448 (Houtbosh, Transvaal), is a form with wide membranaceous lamina and short hyaline point.

Widely distributed from Table Mountain to Rhodesia, on exposed rocks at altitudes 2500 feet and over ; frequent at higher levels, including 8000 to 10,000 feet on the Drakensberg.

Cape, W. Table Mountain, Sim 3534, 8454, Pillans 3534 ; Cape Town, Rehmann 56, Sim 9553 ; Paarl, Sim 9608, 9622, 9623.

Cape, E. Grahamstown, Miss Farquhar (Sim 9882).

Natal. Zwartkop, 5000 feet (Sim 9878) ; Hilton Road (Sim 9884) ; Little Berg, Cathkin, Miss Owen 13 ; Oliver's Hoek, 6000 feet, Miss Edwards (Sim 9880, 9441) ; Upper Tugela, Dr. Bews ; Mont aux Sources, 8000 feet (Sim 9879) ; Giant's Castle, 8000–10,000 feet, R. E. Symons (Sim 9877, 9883, 9691) ; Scheeper's Nek, under wattles (Sim 9881).

O.F.S. Kadziberg, Rehmann 56β ; Leribe, Basutoland, Mdme. Dieterlin 374a.

Campylopus griseolus.   *Campylopus trichodes.*

Transvaal. Witpoortje, Dr. Moss ; Houtbosh, Rehm. 448, as *C .latifolius*.
Rhodesia. Matopos, 5000 feet, Sim 8550 ; Zimbabwe, 3000 feet, Sim 8829, 8749.

Var. *perlamellosus* Dixon, S.A. Journ. Sc., xviii, 304 ; lamellae more pronounced than usual.

In describing it Dixon says, " A section of the nerve near the base scarcely shows them, but in mid-leaf their height is sometimes actually equal to the thickness of the whole of the rest of the nerve section."
Rhodes' Grave, Rhodesia (Sim 8858) ; Bed of Tugela River, Natal, 6000 feet, Dr. Bews (Sim 8374). Both are robust but sterile.

240. *Campylopus Edwardsii* Sim (new species).*

Laxly caespitose in wet clay ; stems 1–2½ inches long, straight, erect, simple or sparingly branched, the stem-leaves very lax, each spreading outward and then the point turned in toward the stem making an evident chain-like

---

* *Campylopus Edwardsii* Sim (sp. nov.).
Laxe caespitosus ; caules 1–2½″ longi, recti, erecti, plerumque simplices, caulis folia laxissima, patentia singula extrorsum, deinde apex caulem versus inflexus speciem cateniformem efficiens. Folia 3–4 mm. longa, acuminata a basi lata nec non distincte auriculata ad apicem tenuem acutum prope integrum hyalinum. Nervus pro hoc genere

arrangement. Leaves 3–4 mm. long, acuminate from a wide and distinctly auricled base to a slender, acute, nearly entire, hyaline point. Nerve very narrow for this genus, of about fifteen ridges ; alar group very pronounced, ventricose, of about 15×15 small, roundly quadrate cells filling the whole base ; above them are a few lines of rectangular oblong cells, and the rest of the lamina is composed of very numerous small elliptical cells. When dry the incurved leaf remains more or less expanded and undulated, and the hair-points at the apex straight. When moist the margins of the lamina stand erect. Other parts not seen. A remarkable mud-species, the only one of this nature in this genus in South Africa.

Transvaal. Johannesburg, 6000 feet, 1917, Miss Edwards (Sim 9836).

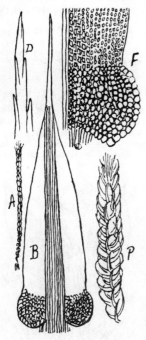

Campylopus Edwardsii.

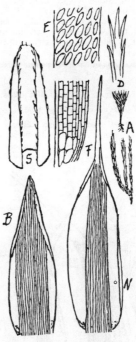

Campylopus pseudojulaceus.
S, apex of hood.

241. *Campylopus pseudojulaceus* Sim (new species).*

=*Campylopus *julaceus*, Rehm. 58, altered to *C. *pseudojulaceus* by Paris, in view of *C. julaceus* (Hpe.),
Jaeg., Pfl., x, 188.

Gregarious, yellow ; sterile stems ½ inch long, slender, julaceous, penicillate, with leaves 1–2 mm. long, closely appressed, boat-shaped, cucullate, upper half involute and with the blunt inbent apex making a hood which is denticulate along the incurved margins and apex, and the turgid cells also make the back rough. Nerve wide, lamellate ; alar cells few, hyaline, and inconspicuous ; supra-alar cells short and square, much longer and narrower toward the

angustissimus, sulcorum circiter 15 ; globus alaris distinctissimus, ventricosus, cellarum circiter 15×15 parvarum rotundo-quadratarum totam basin obtinentium ; cellae supra-alares paucae, oblongae ; lamino-cellae plurimae, parvae, ellipticae, cetera non vidi.

Apud Joannesburgium, Miss Edwards (Sim 9836).

* *Campylopus pseudojulaceus* Sim (sp. nov.).

Gregarius, flavus ; caules steriles ½″ longi, graciles, iulacei, foliis 1–2 mm. longis, arte adpressis, cymbiformibus, cucullatis, parte dimidia superiore involuta, apice obtuso inflexo cucullum efficiente denticulatum secundum margines incurvatos apicemque ; cellae praeterea turgidae dorsum asperum reddunt. Nervus latus, lamellatus ; cellae alares paucae, hyalinae atque inconspicuae, cellae supra-alares breves ac quadratae, multo longiores angustioresque marginem versus, omnes hyalinae ; lamino-cellae superiores rhomboideae, exiguae, multis ordinibus positae ; caules comales ½″ longi, foliis parvis similibus secundum caulem, coma autem foliorum differentium laminam latiorem membranaceam gerentium, nervum latum nec non aristam longam hyalinam leviter denticulatam cum margine inflexo denticulato. Partes florales non vidi.

Apud Kadzibergium, O.F.S., Rehm. 58.

margin, all hyaline ; upper lamina cells rhomboid, small, in many rows ; comal stems ½ inch long, with similar small leaves along the stem, but with a comal group of different leaves having wider membranaceous lamina, wide nerve, and long, hyaline, slightly denticulate hair-point having inflexed denticulate margin. No floral parts seen. Kadziberg, O.F.S., Rehm. 58 (Wager's list gives Katberg, probably in error for Kadziberg).

Not previously described ; Brotherus includes it and other three South African names in subgenus *Palinocraspis* (*i.e.* capsule symmetrical, nerve with small green cells on each side), § *Rigidi* (stem red, tomentose, leaf without hair, when dry collected or sparingly spreading, seldom secund, stiff, usually not shining ; basal cells rectangular ; many marginal rows, small, almost quadrate ; seta short, twisted). Two of these, viz. *C. tabularis* Thér. and *C. mollis* Rehm., are unpublished and not distributed, and the other is *C. Bryhnii*, Broth. (=*Dicranodontium laxitextum* Br. and Bry., changed in view of the Javanese *C. laxitextus* Lac.), and I am not aware that either of the four has been seen fertile, so the twisted seta cannot place them so far.

C. Bryhnii Broth., as described by Bryhn (1911, p. 7) from sterile specimens collected at Ntingwe, Zululand, in 1907, by Titlestad, is shining, with patent, subsecund, lanceolate-subulate leaves about 5 mm. long, the nerve and subula minutely serrated by mamillate cells, and the leaf-base and other characters as in *C. pseudojulaceus* so far as known. More specimens of each are required, and fertile specimens are desired.

242. *Campylopus introflexus* (Hedw.), Mitt., Mus. Aus. Amer., p. 81 (non Brid.) ; Pfl., x, 187 ; Dixon, 3rd ed., p. 100.
=*Dicranum introflexum* Hedw., Sp. M., p. 147, t. 29 (1801) ; C. M., Syn. Musc., i, 405.

Densely caespitose ; tufts 2–5 cm. high, yellow, often grizzled by the abundant white, spreading hair-points. Stems rather slender, simple, or branched below, leafless at the base, club-shaped ; the leaves clasping, with white and denticulate or toothed hair-point often suddenly reflexed when dry, more erect when moist, hyaline, and very variable in length and in development of teeth ; leaf widely lanceolate, closely clasping the stem except near the apex where

Campylopus introflexus.
S, section of upper part of leaf.

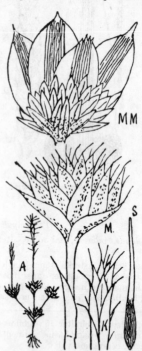

Campylopus introflexus.
Comal tuft (S, archegonium).

the margin is incurved in the upper portion ; nerve about one-third the wideness of the leaf, the back of long, narrow ridges upward except near the leaf-apex where these ridges are covered by small chlorophyllose cells similar to those adjoining. Alar cells few (up to twelve), elliptical, not tumid, and hardly larger than those above, the supra-alar cells oblong, pellucid, and somewhat margined by longer, narrower, hyaline cells ; the upper lamina cells rhomboid, free, chlorophyllose, minute, and abundant. Female in separate tufts ; stems slender, clavate at first, and the leaves crowded in a terminal coma which ultimately is disciform, the outer leaves like the stem-leaves, the inner shorter, much wider, and not or only shortly pointed ; within these are numerous small hairless or shortly haired leaves with a few long archegonia intermixed, these leaves ultimately forming part of many buds with red stems, of which only one or two innovations grow out to an inch long or less, clavate or discoid at the apex, the hairs less recurved there.

C. introflexus extends to Europe and Britain, but the reflexed hair-point does not seem to be noted as a character there ; here it is *the* character, and in some dry old specimens every hair stands at a right angle to its leaf, though in many younger and smaller and possibly fresher specimens the angle is much less and the hair is sometimes even erect when moist. Müller (Syn. Musc., p. 405) includes *D. introflexum* Hedw. and *D. lepidophyllum* C. M. in his group having the hair-point reflexed.

C. Müller describes the fruit from foreign specimens (Syn. Musc., i, 405) thus : " Perichaetal leaves subvaginate, apex sinuate or somewhat piliferous, very laxly areolated below ; capsule on an arched yellow seta, globose-oval, the base strumose, olive-coloured, glabrous ; lid obliquely conical, brown ; peristome teeth narrow, the segments narrow, hyaline, rugulose ; calyptra fimbriated with white, slender, nodulose filaments. Male plant small, simple ; the apex floriferous."

C. *clavatus, Rehm. 451, from Macmac, Transvaal (MacLea), agrees entirely with this. The "name *C. clavatus* Rehm. is antedated by *C. clavatus* (R. Br.) H. f. and W. and altered to *C. pseudoclavatus* by Paris, but both names remain *nomina nuda* " (Dixon, Kew Bull., No. 6, 1923, p. 451), so *C. *clavatus* Rehm. and *C. *pseudoclavatus* (Rehm.), Paris, now sink.

Mitten places *C. *capensis* W. P. Sch. as a synonym here (*fide* Paris' Index, 2nd ed., p. 313), but Paris places it under *C. lepidophyllus* Jaeg. (which see herein).

D. leucobasis C. M., Hedw., 1899, p. 78 (Rehm. 71 from Montagu Pass), belongs here in part, as also its var. *bartramiaceum,* Rehm. 55 (as *Campylopus *subbartramiaceus* C. M.), from Cape Town.

Dixon (Trans. R. Soc. S. Afr., 1920, p. 184) admits *C. leucobasis* (C. M.), Par., but adds, "The plant is to all intents and purposes, however, a form of *C. introflexus,* with the back of the nerve smooth and not lamellate, and I doubt if it be really different from *C. pudicus* Horns."

Abundant, sterile, from Cape Town to Transvaal, usually on nearly bare rock ; when the hair-points are widely reflexed they give it a distinctive appearance ; this bend is at one point only—at the base of the hyaline part—and is not a continuous spiral twist.

Cape, W. Table Mountain, Sim 9252, 9155, 9303, 9293, 9307, 8474, 9870, Rehm. 55 ; Devil's Peak, Sim 9242 ; Stellenbosch, Sim 9575 ; Tokai, Sim 9508 ; Newlands, Sim 9432.

Natal. Zwartkop, 5000 feet, Sim 9873 ; Mont aux Sources, 7000 feet, Sim 9874, 9876 ; Giant's Castle, 8000 feet, R. E. Symons (Sim 9540) ; Tugela Gorge, 6000 feet, Dr Bews.

O.F.S. The Kloof, Bethlehem, Sim 9541.

Transvaal. Johannesburg, Sim 9871, 9875 ; Macmac, Rehm. 451 (MacLea).

243. *Campylopus lepidophyllus* (C. M.), Jaeg. Ad., i, 139.

=*Dicranum lepidophyllum* C. M., Syn. Musc., i, 407.

=*D. *pudicum,* of Drege's list (not Hoch. ?).

A small plant ½–⅔ inch high, gregarious or more or less tufted, growing on soil ; stems almost leafless below, with a rosulate coma round which the leaves rather incurve, then end in a short denticulate hair-point which is more or less reflexed, especially when dry. These and stem-leaves oval-lanceolate, 3 mm. long, 1 mm. wide ; nerve one-third that widens, of long narrow cells, not or hardly ridged on the back upward ; alar cells few, pellucid and indistinct ; supra-alar cells numerous, shortly oblong, pellucid ; cells of upper lamina minute and chlorophyllose, those on and nearest the nerve rhomboid, those nearer the margin smaller and elliptic ; hair-point variable in length and usually not highly developed, erect when moist, somewhat reflexed when dry. Inside the coma are many involucral buds, having leaves about as long as the others but narrow, convolute and sheathing, and either green to the apex or with short hair-point. Capsules small, two to four from each coma, the ½-inch pale seta much twisted when dry, and the capsule arched-cylindrical, with or without struma, but somewhat irregular at the mouth, and with the teeth reflexed and the body of the capsule deeply 16-sulcate as well as many-striate longitudinally when dry, but immediately it is moistened the seta becomes regularly cygneous, the capsule becomes shortly pyriform or nearly oval, many-striate but not sulcate, with regular mouth and erect or incurved red teeth from deep inside the mouth, the teeth short and wide, many-barred, divided for the upper one-third into two short, few-barred, hyaline segments. Lid conical, straight ; calyptra fimbriated with short white segments.

This species closely resembles *C. introflexus* in its clavate coma, but the coma is less succulent, the leaves and hair-points are less developed, the point less reflexed, and the involucral leaves are different.

The difference between the dry and moist condition of the seta and capsule is very marked (as shown in the illustration).

Müller described it from Ecklon's specimens from Palmiet River and Caledon ; Rehm. 66 is from Cape Flats ; Zimbabwe, S. Rhodesia, 3000 feet (Sim 8785), on which Dixon remarks, "The guide-cells of the nerve are very small compared with the cells on the ventral face " (S.A. Journ. Sc., xviii, 304).

Paris places *C. *capensis* (W. P. Sch. in Breut. Musc. Cap.) as a synonym of this, but also quotes Mitten, who places it under *C. introflexus* ; and Brotherus makes a new *C. *capensis* Broth. (Pfl., x, 188) of *Dicranodontium perfalcatum* C. M. in Musc. Rehm.

D. amplirete, C. M., in Hedw., 1899, p. 81 (*Thysanomitrium amplirete* (C. M.), Broth.), from Montagu Pass, Rehm. 60 ; as *Campylopus ampliretis* C. M.

D. weisiopsis, C. M., in Hedw., 1899, p. 79 (*C. weisiopsis* (C. M.), Ther. ; Broth., Pfl., x, 184), from Table Mountain, Rehm. 61, and *D. olivaceo-nigricans,* C. M., in Hedw., 1899, p. 81 (*Thysanomitrium olivaceo-nigricans*

(C. M.), Broth.), from Lydenburg (Dr. Wilms), are probably all immature conditions of this species, judged by the descriptions.

Cape, W.   Cape Flats, Rehm. 66.

Natal.   Ngoya Forest, Zululand, 1000 feet, 1915 (Sim 9888).

Transvaal.   Johannesburg, Sim 9885 and 9887 ; Jessievale, Ermelo Dist., Sim 9889, 9886.

S. Rhodesia.   Zimbabwe, 3000 feet, July 1920, T. R. Sim.

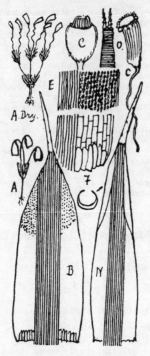

Campylopus lepidophyllus.

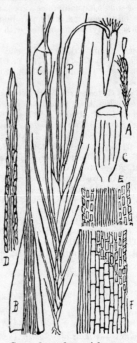

Campylopus leptotrichaceus.

244. *Campylopus leptotrichaceus* (C. M.), **Par.**, Suppl. Ind., 1900, p. 93 ; Pfl., x, 185.

=*Dicranum leptotrichaceum* C. M., Hedw., 1899, p. 84.

=*Campylopus *flavescens*, Rehm. 445.

Yellowish green, in dense tufts on decayed wood or humus, often abundantly fertile.   Stems 5–15 mm. high, slender, simple or branched, the lower leaves short, the upper and perichaetal leaves 2–3 mm. long, sheathing up the seta, and all with erect or slightly spreading or sometimes homomallous setaceous leaf-tips.   Leaves lax, with short membranaceous sheath and long convolute setaceous green tip ; the nerve one-third width of leaf-base, excurrent through the tip, nearly smooth on the back ; alar cells not present but lower cells oblong, pellucid, and fairly large ; upper cells oblong, smaller, chlorophyllose, and the marginal line still smaller and rhomboid, but distinct as a marginal line.   Apex acute, the upper cells linear and protruding somewhat at the back as well as at the margin of the convolute subula.   Seta stout, at first cygneous, afterwards erect and straight, 1 cm. long ; capsule ovoid, thick-necked ; ring red, prominent ; lid conical, beaked ; old capsule (which is present along with young calyptra-bearing seta) widely cylindrical, striated, and the teeth gone.   Calyptra long-fringed at the base.   Much like *Ditrichum flexifolium*, but the wide nerve places it in *Campylopus*.   Abundant in eastern forests.

Cape, W.   Knysna, July 1915, Miss Duthie (Sim 9852, 9858).

Natal.   Fuller's Farm, Nels Rust (Sim 9846) ; Mount Gilboa (Professor Reid), Ngoya Forest (Sim 9848) ; York (Sim 9849) ; Lynedoch, Dr. Bews ; Port Shepstone (Sim 7450) ; Giant's Castle, 7000 feet, R. E. Symons ; Tugela Gorge, 7000 feet (Sim 9854) ; Bulwer (Sim 9855, 9856) ; Lidgetton (Sim 9859).

Transvaal.   Witpoortje, Dr. Moss, Johannesburg Zoo, 5000 feet (Sim 9857) ; Macmac, Rehm. 445 (as *C. flavescens* Rehm.).

245. *Campylopus chlorotrichus* (C. M.), Rehm. 53 and 449 ; Broth., Pfl., x, 188.

=*Dicranum chlorotrichum* C. M., Hedw., 1899, p. 87.

=*Dicranodontium chlorotrichum* (C. M.), Par., Ind., 2nd ed.

Tufts large, loose, 1–3 inches deep, green above, rusty below, and often tomentose ; stems simple or forked, usually curved inward at the apex ; leaves numerous, 5 mm. long, falcate-uncinate, especially those at the apex,

70μ wide at the base, acuminate from there, channelled by the margins being erect, or toward the upper half somewhat convolute, the upper half forming a long green subula when dry; the upper portion occupied entirely by the percurrent nerve. Subula somewhat toothed by exserted clear mamillate cells; below that for some distance the lamina is slightly toothed and occupied by small oblong-trapezoid cells which increase in size downward and are then irregularly oblong. In the lower half there is usually a slight margin of long, narrow, hyaline cells, sometimes rather extensive, sometimes absent. Nerve about one-third the wideness of the leaf, or less; alar cells a distinct group of much larger, tumid, quadrate cells, coloured deep yellow; adjoining cells smaller and not coloured. Capsule on $\frac{1}{2}$-inch seta, ultimately lateral by innovation; seta erect, straight; capsule oval, usually high-backed and slightly irregular; lid conical; calyptra dimidiate, lobed at the base and freely ciliated. Teeth forked half-way, with about thirty crossbars and many longitudinal striae. Perichaetal leaves like the others.

The most common moss on old stumps or on stones in eastern forests, often fertile; seldom found exposed to sunshine except in forest climate, and most abundant where rainfall is 30–70 inches per annum.

Cape, W. Table Mountain, Miss Michell 12; Tyson 2330, Dr. Bews, Pillans 3361, Sim 8602, 8605, 8606, 8493; Chaplin Point (Sim); Montagu Pass, Rehm. 53 (type); Knysna, Rehm. 53β; Blanco, Rehm.

Cape, E. Perie, 4000 feet, Sim 7247, 7169; Tyson 12,758; Grahamstown, Miss Farquhar 76 and 58β.

Natal: Cathkin, Miss Owen 10; Zwartkop (Sim).

Transvaal. Kaapsche Hoek, Wager; Pilgrims Rest, Miss Barry.

Basutoland. Morija, Dr. Stoneman (Sim 8450).

Campylopus capensis Broth.=*Dicranum* (*Dicranodontium*) *perfalcatum* C. M., Hedw., 1899, p. 87, has no distinguishing character except that the upper part of the nerve is somewhat toothed on the back. This is common on young, sterile, vigorous plants, and is not distinctive.

Müller either discarded or overlooked this name when he named a South American species *D. perfalcatum* C. M., Hedw., 1900, p. 250 (see further under *C. lepidophyllus* herein). *C. chlorotrichus* form *orthophylla*, Rehm. 449, from Inanda, Natal, does not differ.

C. Müller described *Dicranum chlorotrichum* and his first *D. perfalcatum* under section *Dicranodontium*, but these species, according to Brotherus, now come under *Campylopus* and not under the genus *Dicranodontium* now in use, and Brotherus has similarly transferred *Dicranodontium laxitextum* Broth. and Bryhn. into *Campylopus* as *C. Bryhnii* Broth.

246. *Campylopus inchangae* (C. M.), Paris; Dixon, Trans. R. Soc. of S. Afr., viii, 3, 185 (1920); Brotherus calls it *C. inchangae* (C. M.), Ther.; Pfl., x, 186 (mis-spelled *Dicr. inerangae* C. M., Hedw., 1899, p. 83; *C. imerangae*, Par., Ind., 2nd ed.; *C. inczangae*, Rehm. 42; *C. incrangae*, Par., Ind., 1st ed.; concerning all these, see Dixon, as above).

=*C. *Macleanus*, Rehm. 447 (from Pilgrims Rest); Pfl., x, 186.

=*C. altovirescens* (C. M.), Par., Suppl., 1900, p. 88, and 2nd ed., p. 297; Pfl., x, 186.

=*Dicranum altovirescens* C. M., Hedw., 1899, p. 152 (from Krantzkop).

Loosely tufted, in very large cushions 2–3 inches deep, with a somewhat stoloniferous habit, old prostrate and dominated stems producing crown innovations which become vigorous plants; stems leafless and tomentose below, abundantly leaved

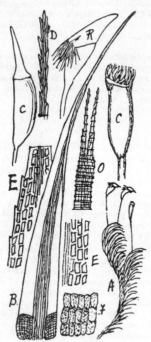

Campylopus chlorotrichus.

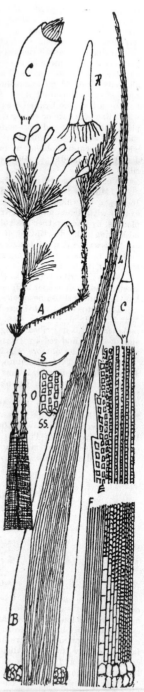

Campylopus inchangae.

S, section of leaf; SS, cells on surface of nerve ridges.

along the stem upwards on the sterile branches, on which the inner leaves are clasping and make the stem pointed, and sometimes this point is falcate-homomallous; on the fertile stems, which are rather smaller, the leaves are more crowded into a lax coma, which is not pointed. Leaves green throughout, erect or somewhat spreading, 5–6 mm. long, straight or arching, often homomallous, acuminate from the base, firm, somewhat channelled, the lamellate nerve half the leaf-wideness at the base, or more or less, extending to the setaceous apex, but margined half-way down or more by a narrow and toothed lamina, and the keel also is often toothed; alar cells roundish, very variable, sometimes few, small and pale, in other cases dilated into a pronounced auricle and orange-brown; supra-alar cells small, oblong, and pellucid, extending for only a short distance along the nerve; cells of the lamina numerous, rounded or rhomboid-elliptical, chlorophyllose, extending down to the alar group, but sometimes bordered toward the base by a few long hyaline cells. Nerve lamellae shallow, sometimes almost flat, in each case each covered outside with a line of square or oblong cells, and there is another such line of cells in the valley between. Leaves of the fertile plant often shorter and more suddenly acuminate, the inner perichaetal leaves widely sheathing and pellucid, suddenly subulate, with green toothed subula. Seta numerous up the stem, two to four per branch, pale and somewhat twisted when dry; capsule oblong, arched, not sulcate nor strumose, suberect, mouth rather irregular; teeth brown, striate, deeply placed inside the mouth, cut one-third into slender, pale, or red, jointed segments. Calyptra fringed with white cilia.

A common species in forestal areas on stumps or stones or even on the ground.

C. Müller's description (Hedw., 1899, p. 83) does not correspond with Rehmann's specimen No. 42 (sterile), from Inchanga, on which it is founded. On this Dixon says (Trans. R. Soc. S. Afr., viii, 3, 185), " Sim's plant agrees exactly with Rehmann, M. Austr. Afr., p. 42. Wager's plants are a little smaller, but agree structurally. The fruiting plant is, as often, rather less robust than the sterile ones. C. Müller's description (Hedw., xxxviii, 83) is decidedly misleading; the description of the stem-leaves as *minuta* is absurd, while that of the lamina as *summitate denti-culato-abruptam* is quite inaccurate, as the subula (in Rehmann's plant) is sharply denticulate for some way down, both on the back and the margins. The perichaetia are aggregate, the capsules rather large, the calyptra deeply fringed at the base."

Wager distributed this as *C. inandae*, which is different. *Campylopus* *variegatus* Dixon and Wager (Kaapsche Hoek, Transvaal, Wager 209) is less lamellate on the nerve-back than usual and the leaf-apex is not much toothed, but the plant is the same species.

Brotherus (Pfl., x, 187) credits to Transvaal as well as Madagascar *C. rigens* Ren. and Card., in Bull. Soc. Roy. Bot. Belg., 1896, i, 300; Ren. Prod., p. 94, a sterile plant, of which the Transvaal record probably belongs to *C. inchangae*; it is described as rigid, stems 1–3 cm. long, leaves 3·5 mm. long, widely lanceolate, suddenly acuminate, almost entire, and red-auriculate, with lamellate nerve half as wide as the leaf, having three strata of large empty cells, with smaller obscure cells intermixed. Lamina cells rectangular, incrassate.

Cape, W. Table Mountain, Sim 9201; near Chaplin Point, Sim 9867; mountains between Muizenberg and St. James; Cape Peninsula, Pillans 3982; Keurbooms River, Jan. 1917, Burtt-Davy 17,037.

Cape, E. Springs, Uitenhage, 9042; Eveleyn Valley, Sim 7069, 7076; Egossa, E. Pondoland, Sim 7077, 7387.

Natal. Inanda, Dr. J. M. Wood; Maritzburg, Sim 9864; Buccleuch, Sim 9860, 9866, 9544; Ihluku, Sim 9862; Scheeper's Nek, Sim 8188; Mont aux Sources, Sim 9832; Giant's Castle, 7000 feet, Sim 8249.

Transvaal. The Downs, Pietersburg, Rev. H. A. Junod 4010; Kaapsche Hoek, Wager 209 (as *C. variegatus*); Woodbush, Wager 94; Pilgrims Rest, Rehm. 447, as *C. Macleanus* Rehm.

Portuguese E. Africa. Magude, Rev. H. A. Junod 334.

S. Rhodesia. Salisbury, Eyles 3442.

247. *Campylopus stenopelma* (C. M.), in Rehm. 52; Broth., Pfl., x, 188.

> = *Dicranum stenopelma* C. M., Hedw., 1899, p. 84.
> = *Campylopus longescens* C. M., var. *firmus*, Rehm. 41.
> = *Campylopus* *transvaaliensis*, Rehm. 450; Pfl., x, 187.
> = *Dicranum delagoae* C. M., Hedw., 1899, p. 86.
> = *Thysanomitrium delagoae* (C. M.), Pfl., x, 188.
> = *Campylopus delagoae* (C. M.), Par., Suppl. Ind., 1900, p. 91.
> = *Dicranum tenax* C. M., Hedw., 1899, p. 83.
> = *Campylopus tenax* C. M., in Rehm. 54; Pfl., x, 188.

Stems 1–4 inches long, loosely tufted, bronze-yellow; the leaves erect when dry, somewhat spreading when moist, often somewhat falcate-secund and the point penicillate or uncinate. Leaves imbricated, 4–5 mm. long, sheathing below, with long, setaceous, acute, denticulate point, usually green but occasionally shortly hyaline; denticulations extending nearly half-way down, and produced by one or more lines of elliptical cells along the nerve of which some of the outer protrude. Nerve half the width of the leaf-base, lamellate half-way up, not much involute; alar cells fairly large, few, forming a distinct coloured group but not always present; adjacent cells oblong-hyaline; upper cells abundant and obliquely elliptical. It fruits freely, the perichaetal leaves similar to others but more clasping, the capsule oblong, nearly regular; lid conical, beaked, straight; and the calyptra fringed.

This species closely approaches *C. chlorotrichum* which is strongly falcate and green, also *C. inchangae* which is larger and green, and has smaller lamina cells.

Cape, W. Table Mountain, Sim 8518; Stellenbosch, Miss Duthie 47; Blanco, Rehm. 54 (as *C. tenax*); Knysna, Rehm. 41 (as *C. longescens* var. *firmus*).

Cape, E. Perie Forest, Sim 7067, 7068.

Natal. Little Berg, Cathkin, Miss Owen 22 ; Mont aux Sources, Miss Edwards (Sim 9544) and Sim 9545.

Transvaal. Macmac, Rehm. 450 (as *C. transvaalensis*).

248. *Campylopus echinatus* Sim (new species).*

=*C. *echinatus* Rehm., in all its varieties (not published).

Fairly strong, loosely tufted, yellow ; stems ½–2 inches high, simple or innovated from a comal tuft, slender ; leaves 3–4 mm. long, widely oval-acuminate from a wide base, and with very long toothed point; when dry convolute above, when moist fully expanded. Nerve one-fourth the wideness of the leaf, or less, extending to the apex, the lamina membranaceous and very laxly set with irregularly quadrate, hexagonal, pellucid supra-alar cells ; alar cells few and inconspicuous, sometimes coloured, often hyaline ; upper cells small and rhomboid, in many lines ; hair-point nearly half the length of the leaf, flat below, ridged with the nerve to the apex, pale liquid yellow, strongly toothed toward the apex, usually erect, but sometimes somewhat reflexed as in *C. introflexus.* Comal leaves rather shorter and wider than others. Inflorescence and fruit not seen. From what Dixon says (Bull. Torrey Bot. Club., xliii, p. 65) concerning *C. introflexus* Mitt. having forms with yellow hair-point, it appears as if he does not separate this from it.

Rehmann's localities are all from Knysna to Cape Town, and I do not find that his varieties can stand, though some tufts are more vigorous than others. These are—

C. echinatus, Rehm. 67, Cape Town.
    „   „   var. *brevipilis*, Rehm. 68, Cape Town.
    „   „   var. *turgescens*, Rehm. 69, Camps Bay.
    „   „   var. *vallis - gratiae*, Rehm. 70, Knysna.
    „   „   var. *leucobasis*, Rehm. 71, Montagu Pass.
    „   „   var. *umbrosus*, Rehm. 72, Knysna

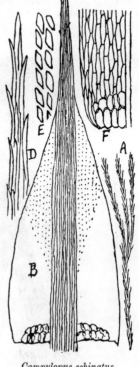

Campylopus stenopelma.

Campylopus echinatus.

Müller (Hedw., 1899, p. 79) places *C. vallis-gratiae* Hampe (in Musc. Cap. Breutelianus), which is *C. echinatus* var. *vallis-gratiae*, as synonym of *C. leucobasis* C. M. (Hedw., 1899, p. 78) var. *longescens*, though the former has pellucid-yellow hair-point and the latter hyaline hair-point, and gives localities Swellendam to Somerset East, but he seems to have been rather mixed about *C. leucobasis*.

I have *C. echinatus* also from Table Mountain, Dr. Bews, Sim 8604 ; Perie Forest, Sim 8581β ; and Murchison Flats, Natal, Sim 9837.

---

* *Campylopus echinatus* Sim (sp. nov.)=*C. echinatus* Rehm. ined.

Fortis, laxus, flavus ; caules ½–2″ alti, graciles ; folia 3–4 mm. longa, late ovato-acuminata a basi lata, apice dentato longissimo; supra convoluta, si sicca ; si humida, explanata. Nervus ad ¼ latitudinis folii minusve, ad apicem pertinens, lamina membranacea itemque cellis irregulariter quadrato-hexagonis pellucidis supra-alaribus laxissime ornata ; cellae alares paucae et inconspicuae, interdum coloratae, saepe hyalinae ; cellae superiores parvae ac rhomboideae in multis lineis. Arista fere folii dimidium longa, infra plana, nervo dorsata ad apicem, pallide flavescens, apicem versus fortiter dentata, plerumque erecta, interdum autem reflexa. Folia comalia paulo breviora ac latiora quam cetera. Reliqua non vidi.

In Monte Tabulari, Peria, Natalia et alibi.

249. *Campylopus Bewsii* Sim (new species).*

Common on eastern mountains in large tufts bordering moist bare rock, itself usually more or less moist ; stems 1–3 inches long, slender, parallel, penicillate when dry, with no clavate or disciform leaf-groups. Leaves usually appressed, imbricate, the lower dark, the upper yellowish green, 3–4 mm. long, acuminate from a wide and auricled base, channelled in front, the sides erect or involute when dry. Nerve wide, lamellate, extending to the apex, where there is usually no hyaline point. Apex subacute and coloured on most leaves, the upper leaves often very shortly hyaline hair-pointed, that point only slightly denticulate ; alar group very pronounced as a small, very swollen, coloured auricle, consisting of about twelve lax, irregular cells. Supra-alar cells hyaline, oblong ; then hexagonal, bordered by long narrow cells ; lamina cells small, yellowish, obliquely elliptical. Other parts not seen. The most common species on moist rocks along the Natal and E. Griqualand Drakensberg from 4000 feet upward. Much more slender than *C. trichodes.*

Natal. Zwartkop, Sim 9838 ; Mont aux Sources, Dr. Bews (Sim 9891) ; Giant's Castle, R. E. Symons.

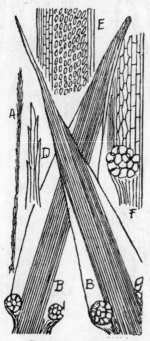

Campylopus Bewsii.

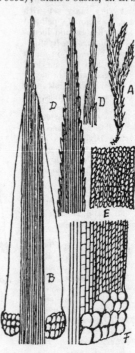

Campylopus purpurascens.

250. *Campylopus purpurascens* Lorentz, Moosst, 1864, p. 118 ; Pfl., x, 187.

=*C. inandae*, Rehm. 43.
=*Dicranum inandae* C. M. (misprinted *imandae*), in Hedw., 1899, p. 95 (sterile).
=*C. *hygrometricus*, Rehm. 59.
=*Dicranum purpureo-aureum* C. M., Hedw., 1899, p. 82.
=*Campylopus purpureo-aureus* (C. M.), Par., Suppl. Ind., 1900, p. 96 ; Pfl., x, 187.
=*Thysanomitrium inandae* (C. M.), Broth., Pfl., x, 189.

Laxly cushioned on exposed rocks or stumps, erect, firm, rigid ; stems simple or somewhat branched, closely and equally leaved from the base, pointed, with erect leaves when dry, usually bronze-yellow and shining ; leaves

---

* *Campylopus Bewsii* Sim (sp. nov.).

Caespitosus ; caules 1–3″ longi, graciles, paralleli, penicillati si sicci, comis nullis clavatis vel disciformibus. Folia adpressa, imbricata, 3–4 mm. longa, acuminata a basi lata et auriculata, a fronte canaliculata, latera erecta involutave si sicca ; nervus latus, lamellatus, ad apicem pertinens, ubi nulla fere cuspis hyalina adest. Apex subacutus idemque coloratus in plerisque foliis, folia superiora saepe in aristam brevissimam hyalinam desinentia, quae arista tantum leviter denticulata est. Globus alaris distinctissimus ut auricula parva tumidissima colorata, ex cellis fere duodecim laxis irregularibus constans. Cellae supra-alares hyalinae, oblongae ; lamino-cellae parvae, subflavae, oblique ellipticae.

Passim in Monte Drakensbergensi, Natali et Transkeia.

3-4 mm. long, erect, stiff, clasping, imbricate, spreading when moist, acuminate from a rather wide base, the upper third or fourth subulate, green, more or less toothed, straight; the nerve quite one-third the wideness of the leaf-base, extending to the apex, thick, ridged on the back; alar group variable from indistinct and pale to swollen and brown; supra-alar cells shortly oblong, pellucid, mostly near the nerve; lamina cells small, rhomboid-elliptical, tumid, chlorophyllose, nearly uniform except that the marginal lines are rather smaller. Seldom fertile; the involucral leaves not different; seta ½ inch long, pale; capsule oval-oblong, somewhat ridged when dry, inclined and high-backed, hardly strumose; teeth fairly long, coloured; lid conical, straight; calyptra not seen. Variable in size and in length and development of point.

Müller's description of *D. purpureo-aureum* mentions a more or less long, strong, serrulate hair-point, but Rehm. 59, which he cites, does not have the hair-point hyaline, though it is sometimes fairly long, green, and toothed, and I have not seen the hair-point hyaline on this species.

Dixon uses the name *C. purpureo-aureus* (C. M.), Par., in Trans. R. Soc. S. Afr., viii, 3, 184, for Natal specimens, Wager 232; Giant's Castle, 8000 feet, Sim 8690; Zwartkop, 5000 feet, Sim 8700, and adds, "both the latter are elongate slender forms, but structurally they present no difference."

He also keeps *C. inandae* (Rehm.), Par., separate (S.A. Journ. Sc., xviii, 304) on my Rhodesian specimens, and remarks: "The anatomical structure is distinct, the nerve in section being thin, with the guide-cells very near the front, the ventral layer extremely thin, of very small stereid or substereid cells. The auricles are large and conspicuous, the supra-alar cells all shortly rectangular, not or little narrowed at the margin."

Müller's types are *C. inandae*, Rehm. 43; *C. purpureo-aureus*, collected by Dr. Wilms; and *C. hygrometricus* (a juvenile form), Rehm. 59.

Cape, W. Table Mountain (as *C. purpurascens* Lotz.); Cape Flats, Tyson 12,756.

Cape, E. Eveleyn Valley, Perie, Sim 7076. 7165, 7284; Dohne Hill, 1893, Sim 7197.

Natal. Inanda, Rehm. 43 (as *C. inandae*); York, J. M. Sim; Giant's Castle, R. E. Symons, 1915 (Sim 8690).

Transvaal. The Downs, Pietersburg, Junod 4001; Johannesburg, Sim 9891; Duivel's Krackler, Lydenburg, Dr. Wilms.

O.F.S. Liebenbergsvley, Rehm. 59 (as *C. hygrometricus*).

S. Rhodesia. Zimbabwe and Matopos, Sim 8474, 8739, 8740, 9890, 8827β, 8874, 8784, 8827α, 8739, etc.

251. *Campylopus chlorophyllosus* (C. M.), Jaeg. Ad., i, 132; Pfl., x, 187.

=*Dicranum chlorophyllosum* C. M., Syn. Musc., i, 395.

Laxly tufted, shining, green above, dark below; stems ½-2 inches long, erect, simple or branched, slender and pointed when dry, leaves more spreading when moist, the lower leaves about 2 mm. long, the upper leaves about 3 mm. long, sometimes in a comal group, but comal groups also occur lower—sometimes one to three to a stem. Leaves closely imbricated in the comal groups, laxly imbricated elsewhere, lanceolate, narrowed to the base, the lower half somewhat channelled, the upper half with the margin involute, often somewhat twisted. Nerve very wide, extending to the apex and ridged or lamellate on the back; alar cells inconspicuous, usually thin-walled, hyaline and somewhat swollen and placed along the nerve, the upper hyaline, but sometimes a few are larger and coloured; adjoining cells rectangular-oblong or subhexagonal; the marginal cells long and hyaline as a border; the upper cells obliquely elliptical, chlorophyllose, sometimes twisted. Apex blunt, chlorophyllose except the marginal cells, or on some leaves reduced to a slightly denticulate, short, acute, smooth, hyaline point. The subula is channelled and fully occupied by nerve. It is not known fertile.

Müller's types are from the Cape (Gueinzius, 1842); Table Mountain, Rehm. 46; and Houtbosh, Transvaal, Rehm. 446 and 446β.

I also have it from Table Mountain, common, and from Zimbabwe, S. Rhodesia, Sim 8820; Zwartkop, Natal, 5000 feet (Sim 9868); and Dohne Hill, Cape Province (Sim 9869).

Dicranum basalticolum C. M., Hedw., 1899, p. 82=*Campylopus basalticolus* (C. M.), Par. (named *C. *subchloro-phyllosum* C. M., in Macowan's herbarium), apparently belongs here, though Brotherus places it (Pfl., x, 187) in § *Palinocraspis brevipili*.

Var. *tristis*, Rehm. 47 (Devil's Peak), differs only in being rather more margined near the apex by hardly protruding hyaline cells.

Forma *pilifera* Dixon (in letter), a small form from Slongoli, Table Mountain (Sim), corresponds exactly in structure, and most of the leaves also in apex-form, but some leaves have a short, slender, hyaline point (fig. S).

Rehm. 46β from Table Mountain is stronger (2 inches long) and more branched, but the structure is the same. I have not seen var. *rivularis*, Rehm. 48, from Table Mountain, or var. *compactus*, Rehm. 49, fr̩n Montagu Pass.

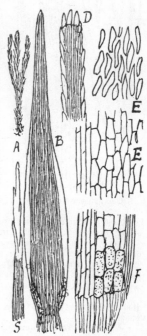

Campylopus chlorophyllosus.
S, hyaline point of some leaves.

12

252. *Campylopus longescens* (C. M.), Broth., Pfl., p. 186.

=*Dicranum clavatum* R. Br., in Schw., Suppl., III, ii, 1, t. 255 ; C. M., Syn. Musc., i, 412.
=*Dicranum longescens* C. M., Hedw., 1899, p. 86.
=*Campylopus clavatus* (R. Br.), Hk. f. and W. ; Fl. N.Z., 69 p.p.
=*Dicranum pudicum* Hsch., in Sieb. M. Nov. Holl., 24 p.p. (*fide* Mitten).

Tufts 1–1½ inch deep, shining, yellow at the surface, dark below ; stems slender at the apex, more robust below ; leaves lax, closely appressed when dry, spreading when moist, imbricate toward the apex, lax below, 2–3 mm. long, acuminate from the wide clasping base, drawn out to a slender, slightly denticulate, straight, acute, green apex, the margins upturned but not involute except in the subula which is convolute ; nerve half the wideness of the leaf-base, not lamellate on the back, extending to the apex ; alar cells lax, small, hyaline, placed along the nerve ; neighbouring cells rectangular, oblong, smaller toward the margin ; in the upper lamina cells oblong or elliptical, separate. Apical cells long with protruding points all round. Capsule not seen.

Müller (*re* Rehm. 40) states he had only seen one old capsule, which he describes, "perichaetal leaves like others, capsule on a short, terminal, slender, yellow seta, minute, oval, somewhat sulcate"; but in Syn. Musc., i, 413, he describes from old capsules on foreign specimens, "perichaetal leaves oblong-acuminate, cuspidate ; capsule on a very arcuate seta, oval, regular, slightly sulcate."

Knysna (as *C. longescens* (C. M.), Rehm.), Rehm. 40 ; also Australasia. (C. Müller cites Rehm. 41 for *D. longescens*, which should be Rehm. 40, No. 41 being different. Both are from Knysna.)

Sim 9284, from Platteklip Ravine, Table Mountain, has green spreading leaves like *C. inchangae* and usually slightly hyaline entire apex, but may belong here.

253. *Campylopus atroluteus* (C. M.), Par., 2nd ed., p. 288 ; Rehm. 63 ; Pfl., x, 187.

=*Dicranum atroluteum* C. M., Hedw., 1899, p. 80.
=*Campylopus *brevis*, Rehm. 44 ; Par., Ind., i, 241.
=*Campylopus bartramiaceus* (C. M.), Rehm. 37 ; Ther., Pfl., x, 185.
=*Dicranum leucobasis*, var. *bartramiaceus* C. M., Hedw., 1899, p. 86.
=*Dicranum nano-tenax* C. M., Hedw., 1899, p. 82.
=*Campylopus nano-tenax* (C. M.), Jaeg. Ad., ii, 761 ; Pfl., x, 186.
=*Campylopus *natalensis*, Rehm. 57.

Usually laxly tufted or gregarious, bronze-yellow ; stems ½–1½ inch long ; leaves spreading except those at the apex which are erect and penicillate, or sometimes slightly falcate-uncinate ; leaves imbricate, 2 mm. long or less, lanceolate, narrowed to the base ; margin somewhat involute above. Apex acute or subacute, slightly denticulate, erect and hyaline at the very apex. Alar group alongside the nerve, hyaline and inconspicuous, neighbouring cells oblong, upper cells rhomboid ; nerve two-thirds the wideness of the leaf-base, extending almost to the apex.

Very near *C. chlorophyllosus* from which the yellow colour, smaller size, and more acute apex are the notable differences.

Dixon (Trans. R. Soc. S. Afr., viii, 3, 183) maintains *C. nano-tenax*, refers to its variability, and mentions that the capsule, which he has had from Knysna, may occasionally be very turgid, glossy, and almost entirely without striae as in the Indian *C. Goughii*, Mitt.

Dicranum aureo-viride (W. P. Sch., in Musc. Cap. Breutel), C. M., Hedw., 1899, p. 85, appears from description to differ only in having a few large, flat, *deciduous* alar cells *on one side of the nerve*, apparently an unnatural condition.

Cape, W. Table Mountain, Rehm. 63, Sim 8419, 9842 ; Cape Town, Rehm. 37 ; Stinkwater, Rehm. 44 ; Stellenbosch, Sim 9588 (*C. nano-tenax*, *fide* Dixon).

Natal. Tugela Gorge, Mont aux Sources, 5500 feet, Sim 8381 ; Giant's Castle, 8000 feet, R. E Symons (Sim 7839).

Transvaal. Melville, Johannesburg, 5000 feet, Sim 9892.

S. Rhodesia. Zimbabwe Acropolis, 3250 feet, Sim 8820 (*C. nano-tenax*, *fide* Dixon), mostly regrowth by detached buds.

Campylopus longescens.

Campylopus atroluteus.

254. *Campylopus catarractilis* (C. M.), Par., Ind. Suppl., 1890, p. 90; Pfl., x, 187.

=*Dicranum catarractilis* C. M., Hedw., 1899, p. 79 (written *catharractitis* by Rehmann, evidently in error).
=*Dicranum serridorsum* C. M., Hedw., 1899, p. 84.
=*Campylopus serridorsus* (C. M.), Par., Pfl., x, 185.
=*Campylopus *semidorsum*, Rehm. 45 (in error).

Exceedingly variable in aspect, sometimes ½-1 inch long, dark, and with rather short subcomal leaves with coloured alar groups, but from the disciform apices of these under very wet or dripping conditions come very slender, simple, straight, julaceous stems 4–5 inches long, with or without comal groups, bright green in colour and without alar groups, or with these hyaline and inconspicuous, quite fit to make a separate species if not known in growth or in connection with the other   Müller apparently saw only the dark condition.   Leaves of the comal groups 2–3 mm. long, closely imbricated, with long-oblong base and acuminate convolute subula ending in a flat, blunt apex having a few protruding mamillate cells at the end and front and back and even down the subula; the nerve wide and fluted on the back, especially upward, and the alar groups few-celled but coloured and ventricose; neighbouring cells smaller, rectangular, and hyaline, while outside these are many lines of linear, narrow, hyaline cells, which extend half up the leaf; higher cells hexagonal-elliptical, and those in the subula minute, rhomboid, chlorophyllose, usually with a margin of hyaline, narrow, oblong cells. Innovations one to three from a comal group, long and very slender; leaves few, small toward the base, larger upward, closely appressed, but similar in form to the comal leaves except that alar cells are absent, though the lower cells near the nerve are rather large, lax, tumid, and pellucid.   The long, narrow leaf-base is characteristic, and tomentum is frequent.   Not seen fertile.   Near *C. chlorophyllosus.*

Frequent round or near the waterfalls of Table Mountain, not known elsewhere.   Rehm. 64; Sim 9416, 9118, 9154, 9156, 9176; Rehm. 45 (as *C. semidorsum*).

255. *Campylopus pulvinatus* (C. M.), Rehm. 62; Pfl., x, 187.

=*Dicranum pulvinatum* C. M., Hedw., 1899, p. 80.

Compactly tufted on soil, green above, dark below; stems slender, 1 cm. long, simple or forked; lower leaves small, hardly imbricate; upper leaves up to 2 mm. long, erect, straight, subulate, the base sheathing somewhat, the convolute subula occupied by nerve except a marginal line of hyaline cells; nerve wide, ridged on the back; apex entire, suddenly acute; the sides inflexed almost to the apex.   Leaf-base pellucid but the swollen alar cells evident, the supra-alar cells oblong, rather smaller toward the margin, and a few long hyaline cells border the alar cells.

Campylopus catarractilis.

Campylopus pulvinatus.

Stinkwater, Cape (Rehm. 62, sterile), is still the only known specimen; in it the alar cells are evident and the subula convolute, but further evidence and calyptra may connect this species with *C. nanus.*

256. *Campylopus bicolor* (Hsch.), Handbook of the N.Z. Flora, 1867, p. 115; Dixon, Trans. R. Soc. S. Afr., viii, 3, 184; Pfl., x, 187.

=*Dicranum bicolor* Horns., in Musc. Sieber., No. 9; C. M., Syn. Musc., i, 392; Flora of N.Z., ii, 69 (1855).

Laxly caespitose on soil, yellowish, shining; stems 5–10 mm. high, stout, rigid, aloe-like; leaves erect when dry, suberect when moist, 2 mm. long, boat-shaped, the lower half oblong and nearly flat, the upper half having the raised margins involute, the apex thick, blunt, and cucullate, with a few protruding cells which make it rough; the nerve half the width of the stem, ridged on the back, extending to the apex; alar cells rather small, rounded, pellucid, about three lines each side of the nerve-base; above these rhomboid and then obliquely elliptical chlorophyllose cells occupy the lamina, but the lower half has a wide margin of pellucid long cells extending to below the alar cells.   Not seen fertile in South Africa, but Müller describes it from Australia as: " Perichaetal leaves membranaceous, amplexicaul, suddenly acuminate, acute, the back rough; thin-nerved; capsule on a short, very arched seta, slightly immersed, oval, rigid, verrucose above; lid with curved beak; calyptra short, fimbriated at the base, with short, thick, yellow, acute cilia. . . . Capsules several together."

Platteklip Rock, Cape Town, 500 feet, Jan. 1919, T. R. Sim 9388 ; Cape of Good Hope, 1912, coll. S. W. Hall, Nos. 4, 5, 9 (*fide* Dixon).

Dixon says, " Like the Australian species also, the Cape plant, while showing obtuse muticous leaves nearly throughout the stem, has a distinct short hyaline point on the floral leaves, and occasionally the comal leaves of sterile branches are acute or shortly hyaline-tipped," and he thinks it probable that *C. pseudobicolor* C. M., of Madagascar, may be conspecific.

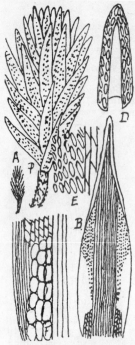

Campylopus bicolor.

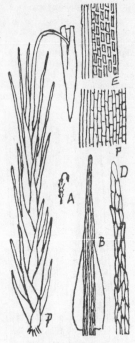

Campylopus nanus.

257. *Campylopus nanus* C. M., in Bot. Zeit., 1847, 804 p.p.

=*Dicranum nanum* C. M., Syn. Musc., i, 383, and in Bot. Zeit., 1859, p. 189.

=*Microcampylopus nanus* (C. M.), C. M. ; Broth., x, 183.

=*Dicranum flexuosum* var. *calyptra cucullata basi integra* Horns. (Linn., xv, 124) (though there is no relationship between *D. nanum* and *D. flexuosum*, Hedw., further than that of genus).

Gregarious, green ; stems 3–6 mm. long, simple or slightly branched, erect, often few-leaved at the base, or leaves distributed along the stem but not numerous ; leaves suberect, somewhat sheathing, lanceolate-subulate ; apex acute, entire, or with a few protruding cells ; nerve one-third the wideness of the leaf-base, ridged on the back, occupying all the subula but not evident at the apex. Alar cells not differentiated, lamina membranaceous, its cells oblong and pellucid below, smaller and chlorophyllose upward. Perichaetal leaves more sheathing, and with longer filiform point ; seta solitary, short, pale, at first cygneous ; calyptra dimidiate, not fringed. Capsule not seen, but described by Müller as " oblong, smooth, almost regular ; ring red ; lid obliquely subulate-conical ; teeth short, narrow, bifid, the segments hyaline. . . . Male plant smaller ; antheridia large, clavate, mixed with a few slender, yellow paraphyses."

Rondebosch, Rehm. 34 ; Stellenbosch, Sim 9589.

Campylopus *subpusillus Par., Suppl. Ind., 1900, p. 98=*Dicranum (Microcampylopus) pusillus* (W. P. Sch., in Breutel Musc. Cap.) C. M., Hedw., 1899, p. 78, from Montagu and Houteniqua Pass, from description does not seem to differ except in being stronger ; the character on which he relies, *i.e.* " long erect seta," counts for nothing in this genus where a cygneous seta when young becomes an erect straight one with age (*C. pusillus* W. P. Sch. is a Mexican species, so Paris changed the South African name to *C.* *subpusillus, which, however, sinks).

Campylopus flexuosus (Hedw.), Brid., i, 469 ; Dixon, Handb., p. 99 and pl. 16E=*Dicranum flexuosum* Hedw., Sp. Musc., pp. 38, 146 ; C. M., Syn. Musc., i, 400 (mentioned above), is a robust European species. It is mentioned under *Thysanomitrion* by Harvey, in Genera of S.A. Plants, 1st ed., 1838, as recorded for South Africa and as found in all countries, and it is again included in Drege's Zwei-Pfl. Doc., 1843, as collected by him, but these records were in the early days before the South African species of *Campylopus* began to be separated from those of Europe, and might be any species of *Campylopus*, and has not been in use for a South African species since then.

258. *Campylopus (Microcampylopus) perpusillus* Mitt., in Journ. Linn. Soc., xxii, 146, 301, and pl. 15, figs. 9–12 ; Pfl., x, 187.

=*C. angustinervis* Dixon, S.A. Journ. Sc., xviii, 303, corrected in S.A. Journ. Sc., 1924.

A small plant, 5–8 mm. high, caespitose but not crowded ; stems bare below, densely leaved above, and aloe-like with spreading leaves when moist, but when dry more or less clavate with closely appressed leaves. Leaves erect or spreading, oblong-lanceolate, cucullate, 2 mm. long, the lamina in the upper half or often for the whole length incurved, though the margin is frequently erect or patent outward ; when dry the whole leaf convolute and tubular. Nerve narrow for the genus, with twelve to twenty lines of cells at the base, extending to the apex, and in the inner leaves, excurrent as a short, hyaline, slightly denticulate point, the rounded green point in other leaves being mucronate, with a few strongly projecting cells along the margin near the apex. My Rhodesian specimens all have the hair-point well developed on most leaves ; my Natal specimens from Sweetwaters are so also, but others (probably younger) have almost no hair-point and are more cucullate and more toothed. Alar group occasionally well developed and yellow, in most leaves it is almost absent, but the lower cells are usually larger than others, and pellucid. Supra-alar cells obliquely elliptical, placed along the nerve, smaller and oblong outward, with usually a few long, narrow, hyaline, marginal cells toward the base. In the upper portion of the leaf the cells are thick-walled and beautifully elongate-hexagonal, or when tumid, elliptical and chlorophyllose. Only once found fertile in South Africa, viz. at Town Bush, Maritzburg.

Natal. Winter's Kloof Station, Sim 9897; Sweetwaters, Sim 10,183; Town Bush, Sim 7633.

Transvaal and Rhodesia on soil; Matopos, Zimbabwe, etc., as cited by Dixon, who says, "The nerve shows, as far as I have examined it, only three layers of cells, all subequal, but varying in relative size to some considerable extent either in different leaves or at different positions in the leaf. The guide-cells and those of the ventral layer are sometimes subequal, at other times the latter are much smaller, and the row of guide-cells is then very much nearer the front of the nerve. The dorsal cells are substereid and obscure, forming a rather thicker layer, but I have not found actual stereid cells, and am inclined to place the plant in the subgenus *Pseudocampylopus*," which genus Brotherus describes (Pfl., i, 3, 331) as "Nerve without stereid cells, but the upper surface of loose, empty, and thin-walled cells, the other surface of similar chlorophyllose and fairly thickened cells." Mitten had fertile specimens when describing Bishop Hannington's Central African plant, and thus describes the parts: "Perichaetal leaves scarcely different from others. Capsule minute, pendulous on an arched seta; lid conical-acuminate; calyptra short, fimbriate at the base; peristome teeth slender, incomplete. . . . The fruit quite invisible till sought for." The fruits found in Maritzburg were old and damaged, but were erect, small, ovate, narrow-mouthed, on seta 5 mm. high; the peristomes had been broken off.

C. dicranelloides Ren. and Card. (Renauld's Prod. de la Flor. Bryol. de Madag., etc., p. 99) is apparently the same plant, though he mentions that the nerve there occupies one-third of the leaf-base.

There remain these three (probably unpublished) mentioned in Pfl., vol. x, concerning which I have no further information than that given there, viz.:—

C. *contiguus* Ther. (§ *Trichophylla*, with hair-point, leaves entire, nerve smooth).
C. *Rehmanni* Ther. (§ *Trichophylla*, with hair-point, leaves entire, nerve smooth).
C. *tabularis* Ther. (§ *Palinocraspis rigidi*).

Thériot apparently included what he intended to publish some time.
Rehmann's *nomina nuda* are stated elsewhere herein.

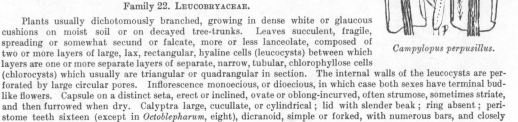

Campylopus perpusillus.

Family 22. Leucobryaceae.

Plants usually dichotomously branched, growing in dense white or glaucous cushions on moist soil or on decayed tree-trunks. Leaves succulent, fragile, spreading or somewhat secund or falcate, more or less lanceolate, composed of two or more layers of large, lax, rectangular, hyaline cells (leucocysts) between which layers are one or more separate layers of separate, narrow, tubular, chlorophyllose cells (chlorocysts) which usually are triangular or quadrangular in section. The internal walls of the leucocysts are perforated by large circular pores. Inflorescence monoecious, or dioecious, in which case both sexes have terminal bud-like flowers. Capsule on a distinct seta, erect or inclined, ovate or oblong-incurved, often strumose, sometimes striate, and then furrowed when dry. Calyptra large, cucullate, or cylindrical; lid with slender beak; ring absent; peristome teeth sixteen (except in *Octoblepharum*, eight), dicranoid, simple or forked, with numerous bars, and closely striate on the outer, and papillose on the inner surface. Spores small, globose.

The white colour of the plant is produced by the presence of the unusual number of leucocysts surrounding a very small number of narrow tubular chlorocysts, and the leucocysts are also connected with the highly hygroscopic properties of the group.

In some respects they resemble Sphagnum, but their relationship is quite with Dicranaceae, although they have peculiarities of their own which warrant their being placed in a separate family.

Synopsis.

Peristome teeth regularly sixteen; chlorocysts 4-angled in section.
    82. Leucobryum. Capsule unequal, often strumose; calyptra cucullate.
    83. Schistomitrium. Capsule regular; calyptra campanulate-cylindrical.
Peristome teeth usually eight; chlorocysts 3-angled in section.
    84. Octoblepharum.

Genus 82. Leucobryum Hampe.

Habit of the family. Plants growing in tufts or cushions several inches deep, the stems dichotomously branched. Leaves without distinct nerve, or the whole leaf nerve only. Leaves convolute toward the leaf-apex. Chlorophyllose ducts of leaves (chlorocysts) 4-angled in section. Calyptra dimidiate, cucullate; capsule irregular, often strumose, furrowed when dry. Peristome teeth sixteen, purple, each unequally forked to about the middle. No species has as yet been seen fertile in South Africa.

Synopsis.

259. *Leucobryum perfalcatum* Sim (new species).*

=*Leucobryum* *Rehmanni* C. M., in Rehm. 75, but not as described in Hedw., 1899, p. 58.

Densely tufted, 3–5 cm. high, light green ; stems much-branched, robust ; leaves clasping at the base, spreading elsewhere, very falcate-secund, often distichous below, imbricate, widely lanceolate, entire, very concave, with a contracted semi-lunar attachment, the margins much incurved toward the shortly pointed apex. Nerve occupying the whole leaf except a hyaline margin of several lines of long narrow cells in the upper half and a wide similar formation in the lower half. Leucocysts two layers deep in the upper half of the leaf with the chlorocysts between, but two to three layers deep in the middle and lower portion of the leaf, the chlorocysts then under the upper hyaline layer ; intercellular pores evident. Leucocysts 50μ long, 20μ wide, bordered by narrow hyaline cells, those in the lower portion of the leaf larger, except those at the very base which are much smaller. Chlorocysts 5μ wide. Only known sterile.

The specimens of *L. Rehmanni* C. M. distributed by Rehmann as his No. 75 belong here, but Müller's description (Hedw., 1899, p. 58) is unrecognisable, hence the new name given above.

Cape, W. Esternek, Knysna, Rehm. 75.
Cape, E. Umtentu River, Burtt-Davy 15,345 ; East Pondoland, 1899, Sim 9895.
Natal. Ngoya Forest, Sim 9894 ; Maritzburg Town Bush, Sim 9893.
Transvaal. Woodbush, Wager (as *L. Gueinzii*).

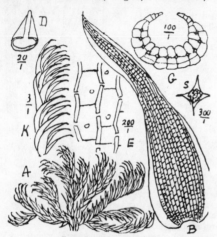

Leucobryum perfalcatum.
S, section of chlorocyst.

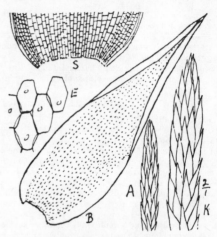

Leucobryum madagassum.
S, base of leaf.

260. *Leucobryum madagassum* Besch., Fl. Br. Reunion, p. 49 (337), not C. M. ; Cardot, Recherches sur les Leucobryacees, p. 26 ; and Mousses de Madag., p. 142 ; Renauld, Flor. Bryolog. Madag., p. 107 ; Dixon, Trans. R. Soc. S. Afr., viii, 3, 186 ; Broth., Pfl., x, 224.

=*L. selaginoides* C. M., in Wright, Mus. Mad. Journ. Bot., 1888.
=*Schistomitrium* *africanum*, Rehm. 456.

Tufts large and dense, one to several inches deep, compact ; stems erect, julaceous, pointed ; leaves ovate-lanceolate, entire, rigid, closely imbricated and inflexed, the upper half with involute margins and consequently rounded smooth back. Margins in the lower half widely hyaline-bordered, the outer cells long and narrow, the inner very short and numerous. Leucocysts two to three times as long as these small cells, three to four layers deep, especially in the upper half of the leaf, apex blunt or mucronate. Fruit not seen.

Dixon says (*loc. cit.*), "It differs from *L. Gueinzii* C. M. notably in the chlorocysts in the nerve section being distinctly hypercentric in the upper part of the leaf. The foliation is rigid and somewhat regularly seriate, whence the name *L. selaginelloides* applied by C. Müller to the species" (Journ. Bot., 1884, p. 264). The name *selaginelloides* should be *L. selaginoides*, C. M. (see Renauld, Fl. Br. Mad., p. 108).

---

* *Leucobryum perfalcatum* Sim (sp. nov.).
Dense caespitosum ; folia connexa ad basin, alibi patula, admodum falcato-secunda, saepe disticha infra. Leucocystae 2-stratosae in dimidio folii superiore, 2–3-stratosae in parte media et inferiore.

Transvaal. Macmac, 8000 feet, MacLea ; Kaapsche Hoek, Wager 337.
Rhodesia. Salisbury, 4500 feet, Eyles 3439.
(*L. madagassum* C. M. is = *L. Rutenbergii* (C. M.) Besch., a Madagascar species.)

261. *Leucobryum Gueinzii* C. M., in Hedw., 1899, p. 58 ; and in Rehm. 74 (not Rehm. 75 as stated by Paris, which is
L. perfalcatum), Broth., Pfl., x, 224.

= *L. *Macleanum*, Rehm. 455 ; Broth., Pfl., x, 224.

Tufts large, ½–2 inches deep, compact, the stems usually parallel, erect, somewhat branched ; leaves numerous,
erect, imbricate, ovate-lanceolate from a narrow attachment, acute, entire, straight, 3–4 mm. long, 1 mm. wide,
mucronate, apparently acuminate because very convolute except
near the base, but wide almost to the apex ; the long chlorocysts
diamond-shaped in section and placed below the upper layer of
empty cells. Section of leaf almost flat near the base, convolute
above, with about three layers of leucocyst cells in the lower part
and two layers upward.

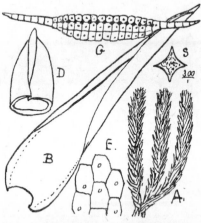

C. Müller (Hedw., 1899, p. 58), who includes this in his section
Juniperella, remarks on a similarity it has to the European *Leucobryum glaucum* Hampe, but I find nothing closer than the general
resemblance of the family.

Cape, W. Montagu Pass, Rehm. 74.
Natal. Buffelsbush, Lidgetton, Sim 9896.
Transvaal. Macmac, MacLea (Rehm. 455) ; Houtbosh, Rehm.
454β ; Sabie, Dr. Moss ; The Downs, Pietersburg, Rev. H. A. Junod
4013β ; Pilgrims Rest, MacLea (Rehm. 454, where it is spelled
Guinzii in error).

Leucobryum candidum Brid. (= *L. brachyphyllum* Hampe) appears
in Shaw's list as collected barren by Shaw and Macowan, at Somerset
East, in 1871. C. Müller's description of *L. brachyphyllum* Hampe,
in Syn. Musc., i, 76 (an Australian species), does not agree with any
of our known species, and Shaw's identifications are too unreliable
to be taken alone. It may have been any other species.

Leucobryum Gueinzii.
G, section of leaf near leaf-base ;
S, section of chlorocyst.

Genus 83. SCHISTOMITRIUM Dz. and Mk.

Habit as in *Leucobryum*. Leaves of two layers of hyaline cells
(leucocysts) and between them one layer of scattered chlorocysts
of quadrangular section as in *Leucobryum*. Inflorescence terminal, dioecious. Capsule regular ; calyptra campanulate-
cylindrical. Peristome teeth sixteen, entire, papillose.

C. Müller (Syn. Musc., p. 80), dealing with the one Javan species then known, states that the calyptra recalls
Encalypta, but that the relationship is here.

262. *Schistomitrium acutifolium* Mitt., Journ. Linn. Soc., xxii., 146, 302 ; Renauld's Flore Bryol. de Madag., p. 109.

= *Leucobryum acutifolium* (Mitt.) Card. ; Broth., Pfl., x, 224.

In Mitten's description he cites " Natal, Mrs. Saunders," as well as Usagara Mountain and Madagascar ; Renauld
gives a locality in Madagascar without comment, and though Brotherus (Pfl., i, 3, 344), in 1901, says that according to
Cardot it is a *Leucobryum*, still in Paris' Index, 2nd ed. (1904), the three countries above mentioned are repeated for
it as *Schistomitrium*. In Pfl., x, 224, Brotherus keeps it in *Leucobryum*. I am not aware that this species has ever
been found fertile to show what is its genus, and so far as South Africa is concerned I know of no other record, and
doubt Mitten's Natal record which accompanies an insufficient description of a sterile plant which would quite answer
Leucobryum Gueinzii C. M. Mitten describes his species as caespitose, small, and many-leaved ; leaves erecto-patent
or secund, concave from an oval base, subulate channelled with the incurved margins. Apex hooded, smooth, entire,
with a small narrow mucro ; perichaetal leaves longer, somewhat subulate. Stem 1 cm. high, together with the
leaves scarcely 3 mm. wide. Upper leaves glaucous green, lower whitish, 2 mm. long.

Genus 84. OCTOBLEPHARUM Hedw.

Habit very much as in *Leucobryum*, but the leaves, which are formed of several layers of leucocysts, have the
separate chlorocysts between these, triangular in section and placed near mid-leaf. Leaves lorate, not involute, usually
rounded above and below, and often reflexed. Inflorescence monoecious, the male near the female flowers. Capsule
small, oval, on a short seta ; lid with a long beak rising erect or subobliquely from a nearly flat base. Peristome
teeth eight, short, erect, distant, widely lanceolate, each with a mid-line or sometimes split. Male flower without
paraphyses.

263. *Octoblepharum albidum* Hedw., Musc. Frond., iii, 15 ; C. M., Syn. Musc., i, 86 ; Broth., Pfl., x, 226.

Usually 1 cm. or less in height, occasionally up to 1 inch, erect, simple or slightly branched, scattered or con-
gregate or caespitose ; lower leaves often somewhat distichous, upper leaves spirally surrounding the stem and leaving
an open star-shaped crown ; leaves from a wide base with membranaceous basal margin, soon lorate to near the quickly

acuminate then mucronate apex, succulent, rounded on the upper and lower surfaces with the chlorocyst layer about mid-leaf and sometimes occasional chlorocysts scattered below. Membranaceous margin near the base only, of large, lax, hexagonal, hyaline cells, with a border of long, narrow, hyaline cells. Leaves spreading, the upper half usually reflexed, which character and the absence of convolution easily distinguish sterile specimens from *Leucobryum.* Perichaetal leaves not different. Capsule shortly oval, erect on ¼-inch straight seta; teeth short, yellow, set inside the mouth and inflexed when dry; lid conical, rostrate from a nearly flat base; calyptra conical, entire, not fringed. Usually fertile.

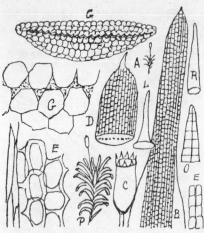

Octoblepharum albidum.

Frequent in Asia, Africa, and America, usually on trees or decayed wood.

Cape, W. Knysna, *ex* Rehm. 280.

Natal. Great Noodsberg, J. Wood (Rehm. 453β).

Transvaal. Macmac, MacLea, Rehm. 453.

Portuguese E. Africa. Mozakwen Forest, Lourenzo Marques, Junod; Bijon River, Mageuja da Costa, 1908, T. R. Sim; Inhareme River, Sim 7386.

S. Rhodesia. Makoni Forest, Eyles 829; Zimbabwe, Sim 8832; Victoria Falls, Sim 8509, 8520, and Brunnthaler, i, 25.

Wager includes in his check-list *O. africanum* (Broth.), Card., a Madagascar species, concerning which Brotherus says (Pfl., i, 3, 349) its distinctive character is chlorocysts 4-sided, only toward the leaf-apex 3-sided. Renauld, earlier, gives what may be the same species as *Arthrocormus africanus* Broth., Bot. Cent., 1888, as Mascarene, but that genus has the chlorocysts irregular, 3–7-sided, at least three layers upward, and no such plant has been recorded from South Africa. Wager probably included this in error for *Schistomitrium africanum* Rehm. (*Leucobryum madagassum* Besch.), which he omits.

Octoblepharum asperum Mitt. is mentioned in Paris' Index, 2nd ed., from Japan, Himalayas, and Prom. B. Spei, as synonym for *Leucophanes.* Under *Leucophanes* he puts *L. asperum* as synonym for *Exodictyon*, and under *Exodictyon* he maintains *E. asperum* (Mitt.), Card., from Samoa only. It seems to me the Cape record is an error, as I know nothing else to support it.

Family 23. FISSIDENTACEAE.

Plants either small and simple or dimorphic, or larger and more or less branched; the largest, which are several inches long floating from a fixed base in streams, are much branched; all are inhabitants of usually moist or wet places, or grow in water. Leaves distichous and vertically placed, *i.e.* the nerve has no true lateral lamina like most other mosses, but lamina is produced like a dorsal keel below the nerve and like a frontal keel on the upper side of the nerve, these two being often or usually unequal in size and together forming a vertical leaf, but the upper keel is for some distance split to the nerve, the split laminae being usually folded together, clasping the stem at the base, and often including in the sheath the base of the next higher leaf (*i.e.* leaves equitant), while at the upper end of the sheath the attachment is often not marginal, but is fairly regular as to form and position in one species. (Other views regarding the morphology of the leaf are discussed by Salmon, see further on.) Nerve and margin are often either pellucid or ferruginous; in one section the nerve is absent, and classification is partly by the nature of the leaf-border; the cells of the lamina are hexagonal or rounded, frequently papillose, often minute, and also often obscured by bright green chlorophyll. Fruit is terminal or cladocarpous, capsule has long or short seta and is erect and regular or obliquely incurved and unequal; lid large with a slender beak. Calyptra usually small and cucullate, occasionally mitriform or entire. Peristome of sixteen teeth, forked or truncate; archegonia and antheridia few, small, with few paraphyses.

A most natural group, the distichous arrangement of the foliage being distinctive, and produced by a wedge-shaped apical cell instead of the usual 3-sided apical cell of other mosses. The construction of the leaf has been the subject of much controversy, alike in regard to its vertical position, its dorsal wing (which was formerly held to be an expansion of the nerve), and its sheathing portion, one-half of which often does not correspond with the other at the junction, and gives the appearance of a superimposed lobe. In Annals of Botany, March 1899, p. 103, E. S. Salmon discusses the morphology of the leaf, and uses South African species as his types, illustrating leaf-sections. To distinguish the parts, which have no equivalent elsewhere, I propose to use these terms, viz. :—

Sheath-lobe, for the apparently attached portion of the sheath.

Ventral lamina for the whole of the leaf on the upper side of the nerve except the sheath-lobe.

Dorsal lamina for the whole of the lamina on the lower side of the nerve.

The family is represented in every part of the world.

Synopsis of the Fissidentaceae.

Peristome of sixteen subulate teeth; calyptra cucullate or mitriform; stems simple or little branched, terrestrial or amphibious, with central strand in stem. Capsule having stomata; terminal, or terminal on short lateral or basal branches . . . . 85. FISSIDENS.

Peristome of sixteen short truncate teeth; calyptra very small, conical, undivided. Stems much-branched, aquatic, without central strand. Capsules without stomata, very small on short axillary setae . 86. CONOMITRIUM.

Genus 85. Fissidens Hedw., Fund., ii, 94.

Plants simple or sparingly branched, or dimorphous, varying from minute up to two or more inches in length; gregarious and usually crowded in masses, especially when near or under water. Leaves distichous, lanceolate or oval to ligulate, vertically placed, often equitant, giving the stem a frond-like, flat appearance when moist, the leaves usually incurved to the sheathlobe-side when dry and sometimes also crisped, especially when young. Cells hexagonal or rounded, dense or lax, usually small, pellucid or chlorophyllose, often turgid or papillose. Seta terminal, or terminal on short lateral or basal branches, or sometimes lateral through innovation. Capsule erect and regular, or cernuous and unequal, usually oval or shortly cylindrical; sometimes apophysate; calyptra cucullate or rarely mitriform; lid with a long rostrate or subulate beak; peristome of sixteen subulate teeth, each forked into two slender segments having numerous horizontal bars and usually numerous papillae in subspiral or knotted arrangement; the teeth widely spread when dry, but folding horizontally inward when wet, with upturned segments. Ring narrow, spores small; a central strand is usually present in the stem and stomata on the capsule.

A very natural and pretty genus, generally distributed, usually as a pioneer on soil, and showing considerable variation within itself, but no wide divergence from type.

There has been much transference of species between *Fissidens* and *Conomitrium* and between the sections of each (the latter being often included in the former), and the sections cannot be said to be satisfactorily settled yet. Many of the characters are unstable, or have been badly noted; thus the capsule may be ovoid or cylindrical, and a neck may be present or absent in accordance with moisture supply; the capsule and seta may be erect or arched or cernuous; the seta may be straight from a horizontal site, and geniculate from a vertical site; the border is sometimes irregularly intermittent or absent; cells are fairly large or minute on the same plant; obscurely tumid and densely but minutely papillose cells are very easily mistaken; innovation may make a terminal seta appear lateral; and the dimorphic condition of some species was formerly overlooked. The minute dimorphic species, and species partly bordered, are more abundant as pioneers than is usually thought, and either they are subject to local or conditional variation, or the number of these species is considerable.

Synopsis of Fissidens.

1. Leaves nerveless or almost so	2.
1. Leaves nerved distinctly	3.
2. Leaf without border; nerve very weak and short (§ *Weberiopsis*)	264. *F. Wageri.*
2. Leaf strongly bordered; nerve absent (§ *Polypodiopsis*)	265. *F. enervis.*
3. Plants minute, heteroecious (§ *Heterocaulon, i.e.* fertile and sterile branches different)	4.
3. Plants not heteroecious (except *F. cuspidatus* and *F. amblyophyllus*)	8.
4. Leaf without any border	5.
4. Leaf partly bordered	6.
4. Leaves lanceolate, all bordered all round; cells dense, chlorophyllose (§ *Pycnothallia*)	266. *F. pycnophyllus.*
5. Seta long; areolation lax	267. *F. splachnifolius.*
5. Seta short; capsule apophysate; cells minute but pellucid	268. *F. longulus.*
6. Cells pellucid; sheath bordered by a slight margin of short narrow cells	269. *F. calochlorus.*
6. Cells dense; leaves more or less bordered	7.
7. Leaves ovate-acuminulate; capsule cernuous	270. *F. laxifolius.*
7. Leaves lanceolate; capsule straight; only female sheath bordered	271. *F. parvilimbatus.*
8. Leaf of one layer of cells	9.
8. Nerve several layers of cells deep, lamina often so also; leaf not bordered (§ *Pachyfissidens*, see under *F. fasciculatus*)	30.
9. Leaves not bordered (§ *Crenularia*)	11.
9. Leaves more or less bordered	10.
10. Leaf bordered only in the sheath (§ *Semilimbidium*)	14.
10. Leaf bordered all round (§ *Bryoidium*)	21.
11. Leaves lanceolate; capsule cernuous	272. *F. Haakoni.*
11. Leaves subulate, acute; cells tumid	273. *F. pseudoserratus.*
11. Leaves oblong, roundly pointed	12.
12. Cells tumid; seta terminal	274. *F. rotundatus.*
12. Cells many-papillose	13.
13. Seta terminal	275. *F. erosulus.*
13. Seta soon lateral by innovation	276. *F. Borgeni.*
13. Cells pellucid; leaves many, ligulate, acute, crenulate, crisped when dry (§ *Crispidium*)	277. *F. zuluensis.*
14. Leaves shining, border interruptedly thickened	278. *F. nitens.*
14. Leaves somewhat shining, small; sheath bordered	279. *F. linearicaulis.*
14. Leaves more or less opaque	15.
15. Leaves much crisped when dry, falcate when moist	280. *F. megalotis.*
15. Leaves not much crisped when dry nor very falcate when moist	16.
16. Cells papillose	17.
16. Cells smooth	19.
17. Capsule inclined	281. *F. submarginatus.*
17. Capsule erect	18.
18. Seta very short; plant ½ inch long	282. *F. brevisetus.*

18. Seta 3 mm. long; plant minute; leaf rounded, apiculate . . . . . 283. *F. urceolatus.*
18. Seta 3 mm. long; leaf shortly acute . . . . . . 284. *F. papillifolius.*
19. Leaves rounded, apiculate . . . . 285. *F. hyalobasis.*
19. Leaves crowded, lanceolate-acuminate, straight; capsule horizontal . . 286. *F. stellenboschianus.*
19. Leaves few, lax, oblong-acuminate . . . . 20.
20. Leaves straight . . . . 287. *F. perpaucifolius.*
20. Leaves arched backward, synoecious . . . . . . 288. *F. microandrogynus.*

(Brotherus (Pfl., x, 149) includes also in § *Semilimbidium*: *F. eshowensis* Broth. and Bryhn (Videnkops Forhandl., 1911, p. 4), and *F. linearicaulis* Broth. and Bryhn (Videnkops Forhandl., 1911, pp. 4, 9).)

21. Heteroecious; cells lax, clear, smooth . . . . . . 289. *F. cuspidatus.*
21. Not heteroecious . . . . 22.
22. Synoecious . . . . 290. *F. androgynus.*
22. Dioecious . . . . 23.
23. Leaf-border regularly present all round . . . . . . 24.
23. Leaf-border intermittent except on the sheath . . . . . 28.
24. Seta and capsule very short; border narrow, diaphanous; plant very small . 291. *F. subremotifolius.*
24. Seta and capsule normal . . . . 25.
25. Sheath very open . . . . 292. *F. rufescens.*
25. Sheath not open . . . . 26.
26. Leaf from a narrow base . . . . 293. *F. latifolius.*
26. Leaf from a wide base . . . . 27.
27. Leaves lanceolate, acute . . . . 294. *F. marginatus.*
27. Leaves lanceolate; apex aristate . . . . 295. *F. aristatus.*
28. Leaves oblong-lanceolate . . . . 29.
28. Leaves widely elliptic-oblong, acuminate; sheath joined at the margin; border irregular except in the sheath, 298. *F. dubiosus.*
29. Sheath joined at the nerve . . . . 296. *F. remotifolius.*
29. Sheath joined at the margin; seta rather short . . . . . 297. *F. malacobryoides.*

(Brotherus (Pfl., x, 146) maintains in *Bryoidium*: *F. curvatus* Horns.; *F. Gueinzii* C. M.; *F. Rutenbergii* Par.; *F. gracilis* (Hampe), Par.; *F. Macowanianus* C. M.; *F. Breutelii* Sch.; and *F. ischyrobryoides* C. M., all disposed of herein.)

30. Fruit lateral, basal, or terminal; leaves subulate-acute . . . . 299. *F. glaucescens.*
30. Fruit terminal . . . . 31.
31. Leaves linear . . . . 32.
31. Leaves wider than linear . . . . 33.
32. Leaves falcate, dense, linear-acute . . . . . . 300. *F. fasciculatus.*
32. Leaves flabellate, erect, dense, subacute . . . . . . 301. *F. plumosus.*
33. Leaves subulate-acute . . . . (See *F. glaucescens*).
33. Leaves lorate; apex rounded; cells tumid . . . . . 302. *F. amblyophyllus.*
33. Leaves lorate, acute; cells tumid-papillose . . . . . 303. *F. corrugatulus.*

Species known to me by name only, or insufficiently known.

F. brevifolius Hk. and Wils., Fl. N.Z., ii, 61, t. 83, and Handbook of the Flora of N.Z., p. 608, is included in Paris' Index as from South Africa. I know of nothing to support that. Mitten, in Musc. Austr. Amer., mentions that the South African plant is not the same as the New Zealand one.

F. humilis Mitt. is mentioned in Shaw's list, without exact locality, or reference to literature.

F. *Rutenbergii Par., Suppl. Ind. 1900, p. 164, and 2nd ed., p. 225 = *F. pauperrimus* C. M., in Hedw., 1899, p. 54, which was changed from the latter to the former name by Paris in view of the earlier *F. pauperrimus* C. M., in Abhandl. Brem., vii, 213, from Australia, is founded on a sterile scrap of a minute plant without locality, having many minute, obtuse, narrowly bordered leaves, and minute, obscure, round cells, and the border narrow or in places absent. This seems too indefinite to maintain. Brotherus places *F. pauperrimus* in *Heterocaulon* and *F. Rutenbergii* in *Bryoidium*.

Delete *F. Thunbergii* Brid., Bryol. Univ., ii, 699; C. M., Syn., i, 72.
= *F. asplenioides* Hedw. (Horns., Linn., 1841, p. 157).
= *F. asplenioides* var. β *capensis* Brid., Sp. Musc.
= *Hypnum asplenioides* Thunb., Prod. Fl. Cap., ii, 175.

From below Table Mountain, Cape (Thunberg). According to the information to hand this might be any terminal-fruited *Fissidens*, and the type is not in the Thunberg Herbarium, so it must sink. Though deleted in C. M., Syn. Musc., i, 72, it is included in Paris' Index, 2nd ed., p. 230, as *F. Thunbergii* Brid., Mant. M., p. 191, and Bryol. Univ., ii, 699; C. M., i, 72; Lindb., in Oefv., 1864, p. 607. He adds, male flower and fruit unknown. I know no reason for its restoration.

264. *Fissidens Wageri* Dixon, Trans. R. Soc. S. Afr., vol. iv, 1914, pl. 1, A ; Broth., Pfl., x, 145.

Caespitose, dark green ; stems erect, 2–3 mm. long, simple, or with basal innovations ; leaves few (eight to ten), flabellate, very large for the size of the plant, 3–4 mm. long, 0·5–0·75 mm. wide, equitant, widely lorate with shortly acuminate-acute or mucronate, slightly oblique apex ; nerve almost absent or reaching just above the sheath-lobe ; sheath-lobe joined near the margin, extending about half the leaf-length ; border completely absent ; cells very long and thick-walled, 80–100µ long by about 20µ wide, rhomboidal-hexagonal, very narrow when dry ; margin entire, leaf not crisped when dry. Seta terminal, 3–4 mm. long, erect, with inclined, regular, ovate-cylindrical, small capsule ; lid short, conical-rostrate ; peristome teeth not papillate.

Rare ; Natal, Umkomaas, Wager ; Nottingham Road, Sim 9899 p.p. ; Richmond, Sim 9898.

Cape, E., Kentani, June 1906, 1000 feet, Miss Pegler 1358, and Sim 8134.

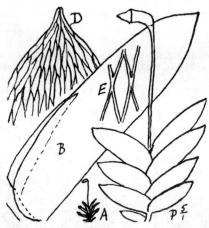

Fissidens Wageri.

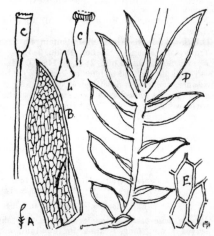

Fissidens enervis.

265. *Fissidens enervis* Sim (new species).*

Plants minute, scattered among other small soil-mosses ; stems weak, up to 2 mm. long ; leaves scattered, yellowish, pellucid, about 4-jugate, 1–1·25 mm. long, 0·3 mm. wide, erecto-patent, lanceolate-acute from a wide base, the point usually slightly deflexed, the upper leaves the larger ; sheath sometimes evident, sometimes hardly present or hardly visible ; nerve never present ; border distinct and clear all round ; cells large, parenchymatous, very open, average 80µ long, 30µ wide, with thick walls and little or no chlorophyll ; seta terminal, 2 mm. long, slender, brown ; capsule very small, 100µ long, dark red, shortly cylindrical, erect, composed of large, tumid, square cells. Capsule wide-mouthed when dry ; peristome red, trabeculate, the segments strong and nodulose ; lid wider than the capsule, as long, and conical. The whole plant is minute, openly cellular, and not crisped when dry ; very distinct in nerveless scattered leaves, much smaller than in *F. Wageri*. Frequent in Natal on damp ditch banks or moist bare soil, but scattered and pale and almost invisible. Mr. Dixon writes : "Your 9899 is well distinct ; the only other African species at all near it is *F. metzgeria* C. M., and that has a much weaker border and more elongate, narrower, chlorophyllose cells."

Natal. Town Bush Valley, Maritzburg, 1917, 3000 feet, Sim 9899 ; Nottingham Road, Dr. V. d. Bijl (Sim 8648) ; Burger Street, Maritzburg, among *F. papillifolius*, Sim 9900, etc.

266. *Fissidens pycnophyllus* C. M., Hedw., 1899, p. 57 ; Rehm. 293 ; Dixon, Trans. R. Soc. S. Afr., viii, 3, 187 ; Broth., Pfl., x, 147.

Minute, monoecious, dimorphous, caespitose ; fertile branch basal with four to six leaves rather larger than those of the sterile branch, which is 2–3 mm. long and has its leaves 5–12-jugate, 0·4 mm. long, 80µ wide, and also some reduced basal ones. Leaves narrowly lanceolate, narrowed gradually to the acute apex from the sheathing or, where imbricate, equitant base, the sheath-lobe extending to beyond the middle and joining at the margin ; the dorsal lamina

* *Fissidens enervis* Sim (sp. nov.) (§ *Polypodiopsis*).

Plantae minutae, sparsae ; caules debiles, ad 2 mm. longi ; folia sparsa, subflava, pellucida, circiter 4-iugata, 1–1·25 mm. longa, 0·3 mm. lata, erecto-patentia, lanceolata, acuta a basi lata, apex plerumque leviter deflexus, folia superiora grandiora. Vagina interdum apparet, interdum vix adest vel vix cernitur ; nervus nunquam adest ; margo distinctus et continuo manifestus ; cellae grandes parenchymatosae, apertissimae, pleraeque 80×30 mm., parietibus densis et chlorophyll aut parvo aut nullo. Seta terminalis, 2 mm. longa, gracilis, fusca. Capsula minima, 100 mm. longa, atro-rubra, breviter cylindrica, erecta, grandi-cellata. Peristomium rubrum, trabeculatum, segmenta fortia ac nodulosa. Operculum latius quam capsula, tam longum et conicum.

Creber in Natalia Media.

gradually reduced at the base or not reaching the base ; the nerve central except at the base ; the dorsal and ventral lamina and sheath-lobe each of three to six rows of minute, hexagonal-rectangular, densely chlorophyllose cells 5μ diameter, and each bordered by a hyaline or yellowish border two to three cells deep of very long empty cells. Margins entire ; nerve strong, striate, hardly reaching the apex, but lost in the long terminal cells where border and nerve combine. Seta 3–4 mm. long, longer than the sterile branch, green ; capsule shortly ovoid, inclined, and high-backed or nearly horizontal. Peristome bright red, the teeth split, segments very slender and very papillose ; lid shortly and obliquely beaked ; male buds (as seen by me) basal with two very short leaves. Müller describes it as dioecious, probably in error, or it may be so sometimes.

F. *Macleanus* Shaw, Cape M. Mag., 1878, p. 314, from Graaff Reinet, as shown in Rehm. 583, is exactly the same. Cape, W. Cape Town, Rehm. 293.

Cape, E. Perie, Sim 8652 ; Port Elizabeth, Wager, distributed by him in error as *F. bifrons*, Schp. ; Graaff Reinet (as above), Rehm. 583.

Natal. Nottingham Road, Dr. P. v. d. Bijl (Sim 8647) ; Burger St., Maritzburg, Sim 9901 ; Mtunzini, Zululand, Miss Edwards.

O.F.S. Wilde Als Kloof, Bloemfontein, Professor Potts.

Transvaal. Macmac, MacLea (Rehm. 584).

F. *curvatus* Hochs., Linn., 1841, p. 148 ; C. M., Syn., i, 59, and ii, 532 ; Broth., Pfl., x, 146, is so closely related to this that I felt inclined to refer *F. pycnophyllus* to that as a synonym, but the description does not agree exactly ; for instance, the seta is described as curved and the capsule erect, whereas *F. pycnophyllus* has the seta straight and the capsule curved. But without further proof I do not think both can stand as species.

F. *pycnophyllus* has by Müller and Brotherus been placed in the section *Pycnothallia*, but its cells, though turgid, are not papillose in the same sense as those of such tropical African species as *F. glaucissimus* Web. and Duby, or *F. subglaucissimus* Broth.

267. *Fissidens splachnifolius* Horns., Linn., 1841, p. 145 ; C. M., Syn. Musc., i, 46 ; Broth., Pfl., x, 148.

=*F. *stolonifer* Rehm. 585 and 294.

Minute, heteroecious ; the fertile stem very short, 3–5-leaved ; the leaves 1 mm. long, ovate-acute, equitant, not bordered ; the sheath-lobe widely open, extending more than half-way ; the dorsal lamina five to ten cells wide upward, narrowed or disappearing downward, and when dry often apparently all absent ; the areolation lax ; cells quadrate-hexagonal ; marginal cells hardly different, seldom the lower marginal cells longer ; nerve yellowish, translucent,

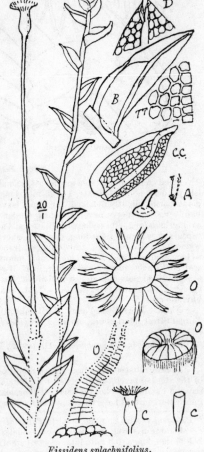

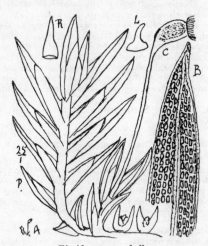

Fissidens pycnophyllus.

Fissidens splachnifolius.

with large oblong-linear cells, and extending to the apex. Sterile stem rising as innovation from the base of the fertile stem, very lax and slender, 5–8 mm. long ; naked or very small-leaved at the base ; all the leaves much smaller than those of the fertile stem, widely separated, spreading, openly sheathing ; the ventral lamina and sheath-lobe equal or almost so, divergent ; the nerve strong, translucent, and either percurrent or ceasing below the apex ; dorsal lamina narrow, its cells smaller. Seta slender, erect, yellowish, 2–5 mm. long ; capsule minute, at first shortly cylindrical and tapering to the base, when mature constricted below the wide mouth ; the teeth spreading when dry, incurved when moist. Additional fertile basal stems without capsules are sometimes present.

Hochsteter describes the nerve on the leaves of fertile stems as disappearing above mid-leaf ; I find it percurrent.

F. bifrons, Sch. MS. ; C. M., Bot. Zeit., 1859, p. 198 ; Broth., Pfl., x, 148, as illustrated in *Pflanzenfamilien* (after Salmon), shows exactly the same leafage of sterile stem as occurs in this, and evidently belongs here ; and Salmon, Annals of Botany, March 1899, sinks *F. stolonifer*, Rehm. 294 and 585, in *F. bifrons*. Wager's specimens distributed as *F. bifrons* are quite different, having linear leaves and arcuate setae.

F. pygmaeus Horns. (Linn., 1841, p. 147) ; C. M., i, 63, and ii, 531 ; Broth., Pfl., x, 148, does not differ except that the sterile stem is short and has only a few large leaves similar to the fertile. It is probable that the development of the longer, slender, small-leaved, sterile stem depends on exposure, and may be induced or drawn through want of light.

Cape, W. Table Mountain, 1827 (Ecklon) ; Cape Town, Rehm. 586 ; Devil's Peak, Rehm. 295 ; Cape Town Gardens, common, Rehm. 585 and 294 (as *F. stolonifer*) ; Papajaisberg, Stellenbosch, Aug., Miss Duthie No. 47.

268. *Fissidens longulus* C. M., Hedw., 1899, p. 56 (not 54 as stated in Par., Ind.) ; Broth., Pfl., x, 148.

A minute, dioecious, dimorphous species without border but with pale percurrent nerve and minute pellucid cells. Leaves of fertile plant about 4-jugate, narrowly lanceolate-acuminate, the inner bent inward or falcate ; leaves of sterile branch much smaller, numerous, distant, very small, equally linear-lanceolate ; capsule on a small yellow slender seta, erect, minute, narrowly oblong, often arched-apophysate ; lid conical, obliquely rostellate, peristome very short, narrow, antennaceous.

Boschberg, Somerset East, Cape, E. (Not seen.)

269. *Fissidens calochlorus* Dixon, S.A. Journ. Sc., xviii, 306 ; Broth., Pfl., x, 149.

A minute, bright green, rhizantoecious species growing in dense little tufts on limestone rocks at the river's-edge at low stream as a first pioneer, just below the Victoria Falls, and probably subject to prolonged annual inundation when the river is in flood, but not found many yards away from the river. Stems usually simple, sometimes 3–4-branched at the base, 2 mm. long, without different central strand ; the fertile stems with about three pairs of rather distant leaves, the sterile stems flexuose, 2–3 mm. long and with leaves 8–10-jugate, distant. Leaves of fertile stem 1 mm. long, lanceolate, acute, without border except a slight border along the sheath ; sheath-lobe as wide as the ventral lamina, almost truncate, extending half-way up, connected at the margin ; dorsal lamina fairly wide, rounded in at the base ; margin almost entire, a few apical cells and basal cells having the upper end protruding as a point ; border absent except a slight border of short narrow cells along the sheath ; nerve strong, flexuose above the junction, ending below the apex, composed of long, narrow, pellucid cells ; cells of lamina hexagonal, $8 \times 10\mu$, pellucid, slightly chlorophyllose, the cell-walls narrow, pellucid. Leaves of sterile stems wider, from an ovate base, widely lanceolate or oblong-acuminate, otherwise similar. Seta 2 mm. long ; capsule erect, small ; lid conical-beaked ; calyptra conical, dimidiate. Male flowers bud-like.

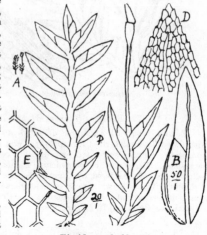

Fissidens calochlorus.

On lime rocks near water's-edge below the Victoria Falls, fertile, Sim 8891, 8882. Included in section *Aloina* by Dixon, and described as without border ; my specimens vary slightly from that. I cannot separate specimens collected at Lo Magundi, 3500 feet, Eyles 3165, though they are not quite similar.

The South American *F. calochlorus* Broth., in Ule Bryoth. Brasil., No. 206, does not appear to have been published, so Dixon's name is valid.

270. *Fissidens laxifolius* Horns., in Linn., 1849, p. 149 ; C. M., Syn. Musc., i, 63 ; Broth., Pfl., x, 149.

=*Conomitrium laxifolium* C. M., Syn. Musc., ii, 527.

Hornschuch's description indicates a minute, deep green, heteroecious plant like *F. splachnifolius*, with dense areolation, lax, widely ovate leaves without border and split to near the apex ; nerve diaphanous, disappearing above the middle ; cernuous capsule on erect seta, and minute dimidiate calyptra. Perichaetal leaves lanceolate, acuminate, bordered, and with excurrent nerve. Sterile branch flabellate ; leaves "like the perichaetal leaves." Peristome teeth cut almost to the base, trabeculate. Said to be not rare among rocks under oaks round Table Mountain (Ecklon), but not collected again since. (Not seen.)

Brotherus places it in § *Semilimbidium*.

Var. *breviseta* Horns. (as above) ; C. M., Syn., i, 64. Seta very short. Below Table Mountain.

271. *Fissidens parvilimbatus* Sim (new species).*

Very minute, 1–3 mm. high, light green, caespitose, heteroecious, the fertile plant about 6-leaved; inner leaves up to 0·75 mm. long, larger and wider than those on the sterile branch, and with the sheath strongly bordered, which is not the case on the sterile branch, the leaves of which, though more numerous, correspond in other respects, the margin being crenulate all round on all leaves except the sheaths on the fertile branch. Leaves about ten on the sterile branch, distant below, crowded above, but occasionally after leaves being close and equitant another equal growth is made with distant lower leaves. Leaves lanceolate, acute, spreading, the dorsal lamina narrowed gradually to the insertion; the sheath-lobe extending hardly to mid-leaf and attached by a point about mid-lamina; nerve translucent, extending almost to the apex; cells dense, round, 7μ wide when chlorophyllose, hexagonal 8μ long when empty. Seta terminal; capsule oval, erect, small (0·3 mm. long); when dry or old wide below, then constricted at the middle. Leaves longer and narrower than *F. submarginatus*, not papillose, and with border only on leaves of fertile branch.

Natal. Greenkopies, York, Jan. 1918, J. M. Sim (Sim 9915); Umhwati, New Hanover, 3000 feet, Aug. 1920, Sim 9905; Albert Falls, Umgeni River, 3000 feet, 1917, Sim 9912.

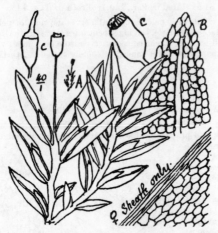

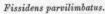

Fissidens parvilimbatus.

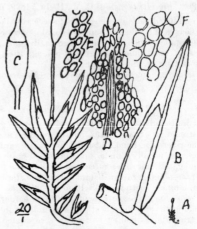

Fissidens pseudoserratus.

272. *Fissidens Haakoni* Br. and Bry., Vidensk. Forh., 1911, iv, 10; Broth., Pfl., x, 151.

Slender, 2–6 mm. long, simple, densely leafy; leaves 5–12-jugate, equitant, erecto-patent, lower leaves very small with dorsal lamina starting far up, upper and comal leaves obliquely lanceolate, shortly pointed, 1 mm. long, 0·3 mm. wide; margins serrulated by prominent cells; sheath to mid-leaf or higher, dorsal lamina rising, rounded from nerve-base. Nerve narrow, not reaching the leaf-apex. Cells minute, roundly quadrate, very chlorophyllose, hardly pellucid. Sheath bordered by elongate-hexagonal hyaline cells, one to two rows above, two to three rows below. Per. leaves like others. Seta 5 mm. long, red; capsule red, cernuous, cylindric-ovate, 0·8 mm. long, when dry constricted somewhat below the mouth. Teeth of peristome purple, spirally thickened.

Eshowe, Zululand, H. Bryhn.

273. *Fissidens pseudoserratus* (C. M.), Jaeg., Enum. Fiss., 1869, p. 24=*F. serratus* C. M., Syn., i, 65 p.p.; Shaw's List; Broth., Pfl., x, 151.

=*Conomitrium serratum* C. M., in Bot. Zeit., 1859, 197 p.p.; C. M., Syn. Musc., ii, 327 p.p.
=*Conomitrium pseudoserratum* C. M., in Bot. Zeit., 1859, p. 197 (included at first in *F. serratus* but afterwards that Asiatic species was separated out under that name and this made *F. pseudoserratus*).

Minute, gregarious, pale green; stems 2 mm. long, often red; leaves spreading, widely lanceolate-acuminate, not bordered, the lower scattered, the upper equitant, straight, 0·5–1 mm. long; dorsal and ventral lamina about equal in the upper half, dorsal gradually reduced to the base; sheath-lobe serrulate, extending half the leaf-length or

---

* *Fissidens parvilimbatus* Sim (sp. nov.) (§ *Heterocaulon*).

Minutissimus, 1–3 mm. altus, caespitosus, heteroicus, planta fertilis circiter 6-foliata, haecce folia ad 0·75 mm. longa, longiora ac latiora quam quae in ramo sterili proveniunt, vagina vero fortiter marginata, quod non fit in ramo sterili, ubi exsistunt folia circiter 10, lanceolato-acuta, patula, lamina dorsalis infra contracta; vagino-lobuli vix ad dimidiam longitudinem pertinentes atque ad mediam fere laminam adiuncti. Margo continuo crenulatus praeter vaginas in ramo fertili. Cellae 7–8 mm. latae. Seta terminalis; capsula ovata, erecta, parva, ad mediam constricta si sicca.

In media Natalia.

more, joined sharply near the middle of ventral lamina ; nerve very strong, straight, striate, lost among loose cells just below the apex. Cells lax, 8μ diameter, globose, and so convex on each surface but not papillose ; those toward the apex more lax and the marginal ones more exserted, making a serrate margin there and a subcrenate margin lower. Monoecious ; male flower basal or axillary ; seta terminal, 2 mm. long ; capsule oval or oval-cylindrical, grey, lid beaked on a convex base.

Müller says, "Monoecious ; archegonia about three ; male flowers axillary, bud-like, numerous ; perigonial leaves subciliato-serrate."

Cape, E. Krakakamma on decayed wood (*fide* Hampe) ; Alexandria Forest, Jas. Sim ; Hogsback, Tyumie, C.P., 1917, D. Henderson 338 (*fide* Dixon, Trans. R. Soc. S. Afr., viii, 3, 187).

Natal. Knoll, Hilton Road, Apr. 1917, Sim 1922.

274. *Fissidens rotundatus* Dixon, S.A. Journ. Sc., xviii, 307 and 332.

Stems scattered on tree-bark, 2–3 mm. long, erect, many-leaved, without different central strand ; leaves 50μ long, bright green, imbricate, distichous when moist, falcate-incurved and crisped above the sheath when dry, oblong, spreading, rounded and obtuse or slightly apiculate-crenulate throughout without border ; sheath-lobe joined above the middle, near the nerve ; dorsal lamina nearly as wide as the other, rounded in below ; nerve pellucid, ending well below the leaf-apex, very flexuous above the junction when dry ; cells minute, numerous, turgid, chlorophyllose, opaque, 5μ wide, all turgid-papillose on both sides, the marginal cells protruding and making the margin crenulate when moist, spiculate when dry. Fertile stems fewer-leaved than the sterile, and inner leaves slightly margined below. Seta 1 mm. long, terminal, erect ; capsule elliptical, small, erect, thin-walled, with large square cells, wide-mouthed when dry ; peristome teeth split more than half-way, segments long and very slender, spreading and incurved ; teeth widely trabeculate within.

Portuguese E. Africa. Sherindgen, Magude, Junod 329a, sterile (Herb. Sim).

S. Rhodesia. Tatagura Valley, Mazoe, 4300 feet, Eyles 653, fertile (Herb. Sim).

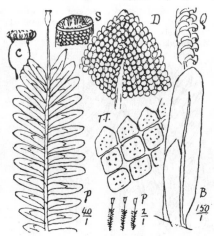
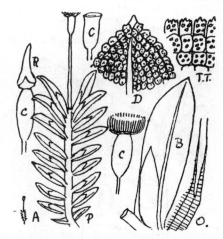

Fissidens rotundatus.
S, mouth of capsule, with incurved teeth.

Fissidens erosulus.

275. *Fissidens erosulus* (C. M.), Par., Ind. ; Dixon, S.A. Journ. Sc., xviii, 307 and 332 ; Broth., Pfl., x, 151.

=*Conomitrium erosulum* C. M., in Linn., 1875, p. 367.

Very minute, gregarious, deep green pioneer on soil ; stems erect, simple, or branched at the base, 2 mm. long, 10–12-leaved ; leaves separate, elliptic-oblong, shortly acute or almost blunt ; the dorsal lamina, which tapers gradually to the base, is usually narrower than the other, and the point consequently is somewhat oblique ; sheath-lobe about half the length of the leaf, less wide than the ventral lamina, and connected by an acute point well inside the margin. Margin without border, and minutely crenulate throughout, the few upper cells pointed, but the other marginal cells, which are slightly larger than those of the lamina, are chlorophyllose and transversely quadrate, with several minute papillae on the exserted apex, while all lamina cells are minute, obscure, and opaque with several very minute papillae on each side. Nerve strong, yellowish, ending below the apex. Seta terminal, 1–1·5 mm. long ; capsule very small, elliptical ; lid much wider, beaked from a convex base ; calyptra dimidiate, conical, hardly covering the lid ; peristome very large, spreading widely then upward ; teeth cleft more than half-way ; segments very slender, lower portion strongly barred inside, almost pectinate. Perichaetal (inner) leaves largest.

Transvaal. Pretoria, Apr. 1918, G. H. Gray.

O.F.S. Kadziberg, above the Caledon River (as *F. pseudorufescens* R., in Hb. Bolus).

Portuguese E. Africa.   Magude, Sim 8995, on lime mud.

Natal.   Stream at Inchanga, March 1915, Sim 8158.

S. Rhodesia.   Zimbabwe, 3000 feet, Sim 8883 ; Victoria Falls, 3000 feet, Sim 8887 ; Matopos, 5000 feet, Sim 8846.

It occurs also in Uganda and Niam-niam.

Dixon compares it with the Cameroon *F. sarcophyllus* C. M.

276. *Fissidens Borgeni* Hampe, in Bot. Zeit., 1870, p. 36 ; Broth., Pfl., x, 151 ; var. *obtusifolius* Dixon, Trans. R. Soc. S. Afr., viii, 3, 187 ; Broth., Pfl., x, 151.

A minute, many-leaved, gregarious plant, 3–7 mm. high, from simple to much-branched, the branches growing as innovations immediately below the terminal inflorescence. Leaves very numerous, not bordered, scattered below, close and equitant above, 0·4–0·5 mm. long, oval, with nearly cordate base, and deflexed, rounded, or apiculate apex ; sheath-lobe extending half-way or more, joined near the margin. Ventral lobe wide, rounded below, dorsal lobe hardly so wide, rounded quickly at the insertion. Nerve strong, deflexed from the lobe-junction, not reaching the apex, translucent, clear, in young plants pale, red in old leaves ; cells very minute, numerous, tumid, chlorophyllose, and opaque, each ending in several papillae. Leaves recurved to one side when dry. Inflorescence terminal, seta 2–3 mm. long, erect, at first terminal, soon lateral by growth of leafy innovation. Capsule inclined or erect, oval-cylindrical, very small ; teeth red, trabeculate, deeply cut ; segments long, spirally thickened, not papillose.

Dixon, in describing this variety, says, " In the type the leaves are acute, variously pointed, with the nerve mostly percurrent." The type, which is unknown to me, also belongs to Natal.

Natal, Van Reenen, Wager 166.

I cannot see from description wherein the sterile *F. subobtusatus* C. M. (Hedw., 1899, p. 56) from Lake Chrissie, Transvaal, differs from this. Müller places it in § *Aloma* ; Brotherus (Pfl., x, 151) places it in § *Crenularia*. Dixon (Bull. Torrey Bot. Club, xliii, 65 (1916)) records it fertile from Pretoria, and later he wrote that *F. *subelimbatus* Wag. and Dixon might be referred to it.

Brotherus includes after § *Crenularia* a § *Crispidium*, described as very slender, more or less elongated, weak plants, with leaves stiff, inrolled, or crumpled when dry, lingulate, without border ; the margin crenulate ; cells of the lamina small, roundish, hardly transparent ; seta terminal ; peristome limb closely knotted.

Several African species are mentioned, but only one South African, viz.

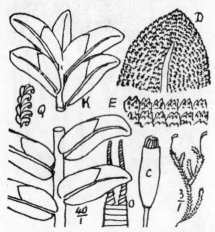

Fissidens Borgeni var. *obtusifolius.*

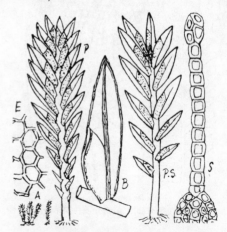

Fissidens nitens (after Salmon).

P, sterile stem ; P.S, fertile stem ; S, section of lamina.

277. *Fissidens zuluensis* Broth. and Bryhn, Vidensk. Forh., 1911, iv, 11 ; Broth., Pfl., x, 151.

Dioecious ?, deep green, opaque ; stems 10 mm. long, mostly simple, the base nude, the apex recurved ; leaves 15–20-jugate, crowded, erecto-patent, slightly accrescent, 2 mm. long, 0·4 mm. wide, ligulate, acute, the margins crenulated by projecting cells, not bordered. Sheath two-thirds as long as the leaf, dorsal lamina rising rounded from the base of the nerve. Nerve thin, yellowish, disappearing below the apex. Cells minute, rounded, chlorophyllose, scarcely pellucid. Entumeni, Zululand, April 1908, H. Bryhn. Near *F. Arbogasti* Ren. and Card., Ren. Prod. Madag., p. 111, from Madagascar. Not known to me.

278. *Fissidens nitens* (Rehm. 289), Salmon, Ann. Bot., March 1899, p. 121, figs. 69–74 on pl. vii, vol. xiii.

Unknown to me, but described as " Dioecious ?, caespitose, tufts olive-green, shining ; stem usually simple, 4–6 mm. long, rigid, curved-prostrate, leafy to near the base. Leaves 7–12-jugate, erecto-patent, crowded upward, lingulate-lanceolate, rather acute, 1 mm. long, the lower smaller and more remote, with margin entire or minutely and irregularly crenulate, thick-bordered ; border 2-stratose of thickened parenchymatous cells ; nerve thick, yellowish red, subflexuous, disappearing below the apex ; sheath-lobes unequal, lobe produced to the middle of the leaf, quickly narrowed at the apex ; dorsal lamina rounded to the base of the nerve ; cells small, about 10μ diameter, round, 4–6-angled, pellucid, smooth. Fertile plant similar to the sterile, but leaves 4–6-jugate, more distant and slightly larger ; perichaetal leaves similar to stem-leaves, paraphyses absent. Other features not seen.

"Inanda, Natal (Rehm. 289)."

He further states that the shining appearance is apparently due to the presence of prominent oil-guttulae in the leaf-cells. These at first look very like papillae or warts, but the cells are quite smooth ; and in regard to the border, he says "the margin is subentire, and the vaginant lamina, as well as the apical, show the sporadic incrassation, the limb of the vaginant laminae being sometimes composed of more or less prosenchymatous cells."

He also describes there from Usambara a var. *neglectus* of this species which is not shining and differs in other respects.

Wager lists *F. nitens* from Knysna.

279. **Fissidens linearicaulis** Br. and Bry. (Videns. Forh., 1911, iv, 9).

Minute, dioecious, gregarious, slightly shining ; stem 2–3 mm. high ; leaves about 10-jugate, homomallous, crowded, erecto-patent, oblong-linear lanceolate, acute, 0·6–0·8 mm. long, 0·20–0·25 mm. wide ; margins minutely serrulated by protruding cells, except that the sheath, extending one-third of the leaf, is narrowly bordered by hyaline cells. Dorsal lamina starts roundly from the base of the leaf-apex. Cells minute, roundly quadrate, very chlorophyllose, densely papillose. Seta terminal, 2 mm. high, pale ; capsule small, 0·8 mm. long, irregular, obliquely oblong, cernuous, pale, when dry constricted below the purple mouth. Plants linear through many small equal leaves.

Ntingwe, Zululand, on red clay, Nov. 1907, Titlestad. Not known to me.

280. **Fissidens megalotis** (J. P. Sch. MSS.) ; C. M., in Bot. Zeit., 1858, p. 154 (*fide* Paris, 2nd ed., p. 214) ; Broth., Pfl., x, 149.

Belongs to § *Semilimbidium*, and Dixon writes : "This is moderate sized, densely tufted ; leaves remarkably crisped when dry, strongly falcate when moist, and with the vaginate lamina open (not conduplicate). Sterile. Dorsal lamina narrowly decurrent. *F. submarginatus* Br. is somewhat intermediate between *F. minutulus* Rehm. and *F. megalotis* Schp. ; leaves not falcate when moist, but not so complanate nor so regularly plumose as in *F. minutulus*, longer, less opaque, with nerve forming an apiculus ; border well marked on all leaves."

I have not seen *F. *minutulus*, Rehm. 296 (from Natal), but the name is antedated by *F. minutulus* Sull. (Cuban), *fide* Kew Bull., No. 6, 1923, p. 209, though Paris sinks the latter name as synonymous with *F. monandrus* Mitt., which may or may not be the case.

Dixon writes : "*F. minutulus* Rehm. appears to me to be a good species, not unlike *F. opacifolius* Mitt., but differing in entire leaves and in having the leaves of the sterile branches almost entirely unbordered, even on the vaginate lamina ; the cells in *F. opacifolius* too are highly papillose."

Re *F. opacifolius* Mitt., in M. hb. Hook., 1860, p. 54, Paris' Index, 2nd ed., p. 218, cites "Niger, Natal." I have no further information, but Mr. Dixon writes : "Mitten gives 'Natal, Gueinzius,' but I find no specimen at Kew, and I strongly suspect an error on Mitten's part."

281. **Fissidens submarginatus** Bruch, in Flora, 1846, p. 133 ; C. M., Syn. Musc., i, 56, and ii, 530 ; Dixon, S.A. Journ. Sc., xviii, 306 ; Broth., Pfl., x, 149.

Gregarious, monoecious ; stems 3–5 mm. long, brown, with basal innovations ; leaves 4–10-jugate, yellowish green, spreading, narrowly elliptical, shortly pointed and mucronate, the lower small, the upper 1 mm. long, 0·25 mm. wide, the sheath-lobe two-thirds leaf-length ; sheath open, strongly bordered, the rest of the leaf not margined. Nerve strong, brown, straight, median in upper half of leaf, ending in the mucro. Dorsal lamina tapering to the leaf-base or slightly decurrent ; cells circular, minute, chlorophyllose, obscure, tumid, the marginal cells pointing outward making the margin crenulate, other cells sufficiently tumid to make the leaf papillose on both surfaces all over. Sheath-border of several rows of very long, narrow, hyaline cells. Vaginula large, red ; seta 3–5 mm. long, pale red ; capsule inclined, small, dark, with red peristome. Male flowers sessile. Leaves incurved when dry, not crisped. Near *F. papillifolius* but larger, with longer sheath-lobe, and tumid not papillose cells.

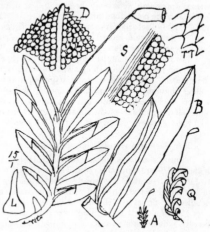

Fissidens submarginatus.

Concerning *F. Breutelii* (W. P. Sch. MSS., in Breutel, Musc. Cap.) C. M., in Bot. Zeit., 1859, p. 198 ; C. M., in Hedw., 1899, p. 154 ; Broth., Pfl., x, 146, Dixon writes : "Very similar to *F. submarginatus* Br. but rather more robust, with broader leaves, and all the leaf narrowly bordered. Concerning *F. submarginatus*, see further under *F. megalotis*."

Natal. York, J. M. Sim ; Mtunzini, Zululand, Miss Edwards ; Ngoya, Hilton Road, and Zwart Kop, Sim 9934, 9935, 9936. Müller described from a Natal specimen collected by Krauss.

O.F.S. Eagles Nest, Bloemfontein, Professor Potts.

S. Rhodesia. Umtali, on an ant-hill in granite country, Eyles 1742 ; Khami Ruins, T. R. Sim.

13

282. *Fissidens brevisetus* Sim (new species).*

Gregarious, erect, light green ; stems 1 cm. high, slender, simple, with scattered leaves about 6-jugate, the upper the longer, all widely lanceolate, 1–2 mm. long, straight or arched inward ; sheath very open, attached sharply in mid-lamina, its lobes clasping almost round the stem and very widely bordered by long hyaline cells ; dorsal and ventral lamina about equal in upper half of the leaf ; dorsal wide at the middle and sharply rounded in just above the leaf-insertion. Nerve strong, lost just below the leaf-apex ; apex subacute ; cells all conically tumid, so papillose on the lamina as well as all round the leaf except the sheath, chlorophyllose (except sometimes a few near the base), very small, 6µ diameter, the margin serrulate or serrate. Seta terminal, very short, 0·5–1 mm. long ; capsule very short, wide-mouthed, erect, 0·3 mm. long ; lid conical, longer than the capsule, erect.

New Hanover, Natal, Jan. 1918, 2500 feet, Sim 9906.

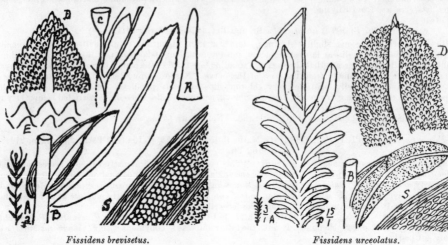

Fissidens brevisetus.   *Fissidens urceolatus.*

283. *Fissidens* (*Semilimbidium*) *urceolatus*, Wager and Dixon (sp. nov.).

" *F. eschowensi* Broth. and Bryhn peraffinis, sed *minor* ; caulis fertilis brevissimus, folia paucijuga, plerumque *obtusa, apiculata* ; seta *multo brevior*, 2–3 mm. tantum longa. Cellulae peropacae, dense papillosae. Theca matura *brevis*, sub ore constricta, *urceolata*.

" *Hab.*—Pretoria, Transvaal, 1914–15, H. A. Wager (264).

" A very small, indeed minute, species, certainly closest to *F. eschowensis*, but smaller, with leaves not acute but obtuse and apiculate, and with shorter seta." (As sent by Mr Dixon.)

The following is additional, from co-type specimens sent by Mr. Wager (T.R.S.).

Plant minute, 3 mm. long, with terminal seta 2–3 mm. long, and small oval urceolate capsule, narrowed below the mouth. Leaves six to seven pairs, equitant, complanate, the lower small, the upper larger, up to 1 mm. long, oblong-lingulate, obtuse-apiculate, mostly arched backward except the upper two, which are erect. Nerve very wide and pellucid ; cells minute, opaque, darkly chlorophyllose, each tumidly papillose on each side all over the leaf, and along all margins except the sheath where about four lines of long, narrow, hyaline cells form the border.

Near *F. papillifolius*, but with wider, clearer nerve, different papillae, and more apiculate leaf.

284. *Fissidens papillifolius*, Dixon, Trans. R. Soc. S. Afr., viii, 3, 187, and pl. xi, p. 3 ; Broth., Pfl., x, 149.

Minute, simple, erect, gregarious, autoecious, deep green plants, growing on moist clay banks. Stems all alike, without different central strand, 2–3 mm. long, 8–10-leaved, the lower leaves small, upper larger, from elliptic-oblong to lanceolate-acute ; the sheath-lobe extending about half-way, joined bluntly near the margin. Dorsal lamina nearly equal to the other but rounded in at the base. Nerve narrow, pellucid, extending to the apex. Cells small, quadrate, 5–6µ wide, well-defined but chlorophyllose and opaque, with one pointed and high papilla on each surface over the whole leaf ; marginal cells with protruding triangular points making the margin crenulate along all the margin of dorsal lamina and along upper margin from apex to junction of sheath-lobe, thence both margins are strongly bordered by long, narrow, hyaline cells sharply separated from the others, with usually one or two such long cells

---

* *Fissidens brevisetus* Sim (sp. nov.) (§ *Semilimbidium*).

Gregaria, erecta ; caules 1 cm. alti, graciles, simplices ; folia circiter 6-iugata, superiora grandiora, late lanceolata, 1–2 mm. longa, recta vel introrsum flexa ; vagina apertissima, in media lamina abrupte adiuncta, lobis prope circum caulem complectentebus, nec non latissime marginata cellis longis hyalinis. Nervus firmus, pene percurrens, apex subacutus ; cellae conice tumidae, unde papillosae in lamina marginibusque praeter vagino-margines. Seta terminalis, brevissima, 0·5 ad 1 mm. longa ; capsula brevissima, erecta, ore lato, 0·3 mm. longa ; operculum conicum, longum, erectum. Apud New Hanover, Natal, Sim 9906.

extending into the lamina among the quadrate cells. Seta terminal, erect, 2–3 mm. long, somewhat sheathed by upper leaves ; capsule erect, oval, small, thin-walled, with large turgid cells ; lid rostrate from an arched base ; calyptra long, conical, entire. Male flowers axillary. When dry the leaves are strongly incurved.

F. *subelimbatus* Wager and Dixon, from Kaapsche Hoek, Transvaal, is very near this (see under *F. Borgeni* Hpe.), and *F. eschowensis* Broth. and Bryhn (Videns. Forh., 1911, iv, 8) seems from description to differ only in the dorsal lamina tapering to the base of the nerve. It is from Eshowe, Zululand (H. Bryhn), and should be seen again. Brunnthaler records it from Usambara, East Africa (pp. 1–24).

Natal. Walls, Umgeni Nook, Albert Falls, 2000 feet, 1917, Sim 8709 ; Burger St., Maritzburg, Sim 8651 ; New Germany, Pinetown, Dr. P. v. d. Bijl (Sim 9902, with more lanceolate leaves than usual) ; College Road and Alexandra Park, Maritzburg, Sim 9923 and 9926 ; Umhwati, New Hanover, 3000 feet, Sim 9925 ; Knoll, Hilton Road, Sim 9911.

Transvaal. Johannesburg, 5000 feet, Jan. 1919, Sim 9924.

Fissidens papillifolius.

H, cells of sheath-lobe.

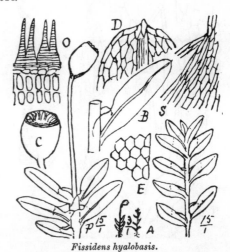

Fissidens hyalobasis.

S, sheath-border of leaf of fertile stem.

285. Fissidens (Semilimbidium) hyalobasis, Dixon (sp. nov.).

" Humilis ; gregarius ; caules perbreves, foliis paucijugis ; *laetevirens*. Folia sicca *parum mutata*, complanata, *stricta*, e basi vaginante *perbrevi*, superne *multo constricta*, inde latiora, oblongo-lanceolata, apice subacuto vel sub-obtuso, raro obtuso ; costa *valida*, percurrens. Cellulae superiores majusculae, laeves, basilares *valde elongatae*, *rectangulares, hyalinae*.

" Seta *perbrevis*, circa 3 mm. alta ; theca *perminuta*, inclinata, matura macrostoma.

" *Hab.*—Moorddrift, Transvaal, H. A. Wager (406).

" Quite distinct in the bright green, minute plants, with leaves little altered when dry, the form of the leaf con-stricted to a narrow waist above the short vaginant lamina, and especially in the lax, hyaline, basal cells, so that the border though well developed is inconspicuous through the similarity of its cells to those of the lamina. The ventral margin of the upper lamina is usually also somewhat bordered. The leaves of the sterile stems are unbordered." (As sent by Mr. Dixon.)

The following is additional, from co-type specimens sent by Mr. Dixon (T.R.S.).

Fertile stems 1–2 mm. high, about 4–5-leaved, with prominent thick vaginula set with several withered archegonia, and one erect seta 3 mm. high, bearing a slightly inclined, very small, globose, urn-shaped, light brown capsule ; the leaves 1 mm. long, very much constricted at the top of the short, open, sheathing base, thence oblong-lanceolate, with rounded, usually mucronate apex, wide percurrent nerve, roundly hexagonal upper cells 15μ wide, rather irregular cells 20–25×10–15μ in the sheath, bordered in the sheath by three to five rows of longer, narrower, hyaline cells ; occa-sionally border cells also occur above the junction, and sometimes on the lower part of the dorsal lamina ; capsule somewhat microstomous, composed of thick-walled cells $35 \times 20\mu$ in straight lines ; a narrow mouth-band of horizontal cells, and the teeth usually arched inward down into the capsule mouth, closely trabeculate and somewhat papillose. Sterile plants rather longer, with more numerous and rather shorter leaves, less constricted above the sheath, and without border.

286. Fissidens (Semilimbidium) stellenboschianus, Dixon (sp. nov.).

" *Pallide virens*, subrobustus, caules usque ad 2 mm., seta ad 1 cm. alta. Folia *conferta*, siccitate parum mutata, madida nec patentia nec valde complanata, *latiuscula, acuta*, costa *valida, concolor, carinata* ; cellulae sat *magnae, pellucidae*, irregulariter hexangulares, 7–14μ latae, parietibus *tenuibus*. Lamina vaginans *peranguste* limbata, lamina dorsalis ad folii basii haud vel vix attingens. Seta pallida, *longa*, flexuose ; theca *horizontalis, gibbosa* ; operculum conico-rostellatum. Caules steriles, microphylli, plumosi saepe iuveniuntur.

" *Hab.*—Stellenbosch, Cape Province, H. A. Wager (612, type) ; East London, Cape Province 1918, H. A. Wager (777) ; a starved form.

" A quite distinct species with lax cells, unlike those usual in *Semilimbidium*, a stout nerve, not pellucid as usual in this group, but concolorous long seta and inclined curved capsule. The colour, wide leaves, and rather open, not closely conduplicate, vaginant lamina also give it a marked habit." (As sent by Mr. Dixon.)

The following is additional, from co-type specimens sent by Mr. Wager (T.R.S.).

Fertile plant 2 mm. high, with erect pale seta 1 cm. high, and small, horizontal, oval, microstomous capsule. Leaves few, crowded, basal, equitant, 2 mm. long, widely lanceolate, acuminate from the sheath junction ; sheath open, short, distinctly bordered, the lobe united at the margin, dorsal lamina hardly reaching the point of attachment ; nerve strong, concolorous, almost percurrent ; cells smooth, 12–20μ long, irregularly hexagonal, pale green ; leaf-margin entire, seta long, straight, with inclined pale capsule, having very slender, crimson trabeculate, and papillose teeth ; lid short, pointed. Sterile stems not seen, but described above as " with small leaves, often plumose."

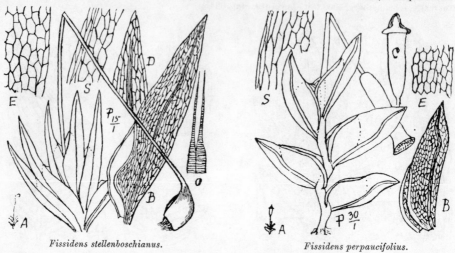

Fissidens stellenboschianus.   *Fissidens perpaucifolius.*

287. *Fissidens (Semilimbidium) perpaucifolius* Dixon (sp. nov.).

" Stirps *pernanus*, caules fertiles foliis 3–4-jugis solum, laete virides. Folia eis species praecedentis subsimilia, sed minora, *minus acuta*, cellulis *minoribus*, 6–8μ latis, magis regularibus ; seta *circa 3 mm. alta*, pallida, basi geniculata, *crasseuscula* ; theca pro plantae magnitudine *magna*, suberecta, *anguste oblonga*, 1–1·25 mm. longa, operculo conico-rostellato, rubro. Autoicus. Flos ♂ basilaris.

" *Hab.*—Stellenbosch, Cape Province, H. A. Wager (647).

" In the leaf structure and form this is somewhat a miniature of *F. stellenboschianus*, but the cells are much smaller and the fruiting characters quite different. The capsule is noticeably narrow, and markedly large in relation to the size of the plant." (As sent by Mr. Dixon.)

The following is additional, from co-type specimens sent by Mr. Wager (T.R.S.).

Plant 1–2 mm. high, autoecious ; seta and capsule 3 mm. high ; leaves few (four to five), about 1 mm. long, three times as long as wide, widely oblong, acuminate, spreading, with open and bordered sheath, uniting half-way up at the margin ; apex straight, shortly acute from a wide lamina ; leaf not bordered ; dorsal lamina narrow, ending at or above the attachment ; nerve strong, concolorous, flexuous, percurrent ; cells smooth, green, regularly hexagonal, 6–8μ wide ; border cells of sheath about four rows, long and narrow ; seta erect, pale ; capsule erect or almost so, long, oblong-cylindrical, wide-mouthed when dry, pale ; lid shortly conical from a wider, over-lapping base. Calyptra not seen ; teeth bright red, slender, usually incurved. Sterile stems not seen but may be different.

288. *Fissidens microandrogynus*, Dixon, S.A. Journ. Sc., xviii, 306 ; Broth., Pfl., x, 149.

Scarcely 0·5 cm. high ; inflorescence synoecious ; leaves 5–7-jugate, distant, not bordered except on the sheath where it is narrowly bordered. Cells chlorophyllose, clear, smooth. Seta 3 mm. long ; capsule minute, short, erect or inclined, urceolate when dry ; lid conical, acute. This looks like *F. androgynus* on a small scale and destitute of most of its border.

Dixon adds : " The very narrow border is frequently to be found only on the upper or the perichaetal leaves, the lower leaves being entirely unbordered." (Not seen, T.R.S.)

Bulawayo, S. Rhodesia, Wager 895.

289. *Fissidens cuspidatus* C. M., in Linn., 1843, p. 558 ; C. M., Syn. Musc., i, 48, and ii, 530 ; Dixon, S.A. Journ. Sc., xviii, 305 ; and Trans. R. Soc. S. Afr., viii, 3, 186 ; Broth., Pfl., x, 146.

Minute, gregarious, heteroecious, light green mosses ; fertile stem 1 mm. long, about 6-leaved ; the leaves 0·6 mm. long, lanceolate from the very open sheaths, pellucid, usually twisted, and with border all round. Sterile stem erect, rather longer ; leaves 10–12-jugate, 0·6 mm. long, lanceolate, erecto-patent, complanate, very lax and pellucid ;

sheath-lobe joined at the margin ; in all the leaves the nerve strong, joining the border at the apex ; border strong all round ; younger cells rounded, subchlorophyllose ; older cells hexagonal, 6μ long, usually empty. Seta terminal, 3 mm. long, very slender, with erect cup-shaped capsule and shortly conical straight lid. Calyptra not seen.

A very small plant, not unlike *F. pycnophyllus*, but larger in all its parts, the leaves laxly areolate and pellucid, and the capsule erect.

F. cuspidatus is included by Müller in the group having terminal setae. On this Dixon says (Trans. R. Soc. S. Afr., viii, 3, 186), " This species is heteroecious. The fertile flower may be apical, or terminal on a very short basal branch. The male flowers are principally lateral, either on a fertile stem or on a separate stem (which, however, might develop ultimately a terminal female flower) ; they are also occasionally basal. They are rather large and numerous. The leaf-cells are unusually elongate and rather large, with thin walls."

Cape, W. Tokai, Sim 9519 ; and Dixon cites George, (Wager 548).

Cape, E. Gwacwaba River, King William's Town, 1897, Sim 7112 ; Eveleyn Valley, Perie, 4500 feet, 1893, Sim 7094.

Natal. Nels Rust, Sim 9908 ; Rosetta Farm, on tree-bark, Sim 9914 ; Greenfields, Mooi River, Sim 9919 ; Blinkwater, Sim 9920 ; Burger St., Maritzburg, Sim 8901 p.p. ; Edendale, Sim 8111 ; Vryheid, July 1915, Sim 9910.

Mr. Dixon writes that *F. gracilis* (Hpe.), Jaeg. (Enum. Fiss., p. 12 (1869) = *Conomitrium gracile*, Hampe MSS. ; C. M., in Bot. Zeit., 1859, p. 197 ; and mentioned by C. M., in Hedw., 1899, p. 154), is a small form of *F. cuspidatus* ; and that the African plant referred to the South American *F. Lindigii* Hpe. by Salmon, and by Paris and by Brotherus, is also *F. cuspidatus*. It is also *F. *Eckloni* W. P. Sch. MSS., in Herb. Kew (see Salmon, Annals of Botany, March 1899, p. 126).

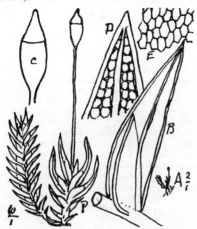

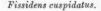

Fissidens cuspidatus.

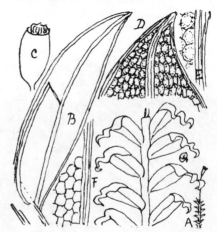

Fissidens androgynus.

290. *Fissidens androgynus* Bruch, Flora, 1846, p. 34 ; C. M., Syn. Musc., i, 57, and ii, 532 ; Dixon, S.A. Journ. Sc., xviii, 305 ; Broth., Pfl., x, 146.

Gregarious or tufted on soil, under very humid conditions, synoecious, dark green ; stems ¼–½ inch long, erect, simple, but often with many basal branches ; stems few- to many-leaved, usually red, with bright red vaginula, pale red seta and capsule, and bright red peristome. Lower leaves and leaves on few-leaved stems distant ; upper leaves larger and approximate or often imbricate, spreading, with somewhat falcate apex ; much crisped when dry ; lanceo-ate, acuminate to an acute, usually deflexed apex. Sheath-lobe extending half-way and joined well inside the margin ; ventral and dorsal lamina about equal above the junction ; below that the dorsal lamina is gradually narrowed but joins the stem below the nerve. Nerve strong, striated, rusty red, curved, and often flexuose in upper half, and ending in the apex. Border strong all round the leaf, pale red, of three to four lines of long translucent cells ; cells of lamina minute (about 6μ), quadrate-hexagonal, translucent in the tumid centre, minutely papillose. Seta terminal, 5–7 mm. long, erect ; capsule oval, erect, and regular, or, through drought, twisting of seta often inclined, and when old cylindrical and wide-mouthed. Peristome teeth usually erect so far as united, with horizontal segments ; teeth strongly pectinate on inner surface and highly papillose on outer surface ; segments blunt, more solid than usual, red and opaque, strongly papillose. Müller states there are sometimes two to three capsules from one flower, which condition I have not seen.

The margin is very strong, and it is a common occurrence at Victoria Falls for the lamina to be washed out, leaving the nerve and border as a complete pale red skeleton on the older leaves all up the stem.

Frequent in eastern forest country ; less known from the west.

Cape, W. Stellenbosch, Miss Duthie ; and Burtt-Davy (Sim 8721).

Cape, E. Uitenhage, Sim 9029, 9018, 9909, 9006 ; Umtentu River, East Pondoland, Burtt-Davy 15,347 ; King William's Town, Sim 285 ; Kentani, Miss Pegler 1356 p.p.

Natal. Maritzburg, Sim 7399, 9940, 9646 ; Goodoo Pass, Dr. Bews ; Rosetta Farm ; Hilton Road ; York ; Acutts Stream, Rosetta ; Bush below Cathkin Peak ; Zwart Kop Valley ; Edendale Falls ; Epworth High School ;

Scheeper's Nek ; Edgehill, Mooi River ; St. Ives ; Merrivale, Sim 9937, 9938, 9939, 9941, 9943, 9944, 8119, 9945, 9946, 9948, 9949, 9952.

Transvaal. Zoo Stream, Johannesburg, Sim 9947 ; Melville, Sim 9942 ; Johannesburg, J. Burtt-Davy ; Macmac, MacLea ; Pretoria, Wager ; Lemona Wood, Spelonken, Junod 15.

O.F.S. Rensberg Kop, Wager ; Eagles Nest, Bloemfontein, Professor Potts.

S. Rhodesia. Victoria Falls, Sim 8881, 8903, 8926, 8929, 8946 ; Zimbabwe, Sim 9950 ; Salisbury, Eyles 1574.

291. *Fissidens subremotifolius* C. M., Hedw., 1899, p. 54 ; Broth., Pfl., x, 146.

Described as dioecious, very small ; leaves about 8-jugate, small, oblique, shortly and acutely acuminate or upper leaves oblique or falcate ; base narrow ; border narrow all round, pale or diaphanous. Nerve thick, ferruginous, ending in a small mucro ; cells round ; seta very short ; capsule short, erect, oblong, constricted when dry ; lid rostellate. Peristome teeth short, narrow, antennaceous. Lydenburg, Transvaal, Dr. Wilms. (Not seen.)

Dixon (S.A. Journ. Sc., xviii, 305) compares *F. latifolius* with it as probably its nearest relation, but he had not seen this.

292. *Fissidens rufescens* Horns., in Linn., 1841, p. 153 ; C. M., Syn. Musc., i, 54, and ii, 532 ;
Dixon, Trans. R. Soc. S. Afr., viii, 3, 186 ; Broth., Pfl., x, 146.

= *F. Macowanianus* C. M., Hedw., 1899, p. 53 (from description) ; Broth., Pfl., x, 146.

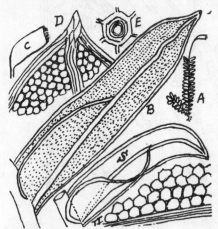

Fissidens rufescens.

Gregarious or tufted, dark green ; dioecious ; innovations light green ; stems $\frac{1}{2}$–$\frac{3}{4}$ inch long, the younger erect and simple but branches frequently occur lower ; stems few- or many-leaved ; leaves not crisped when dry, seldom in contact, erecto-patent, 1–1·5 mm. long, oblong, oblique, shortly drawn to a point which is reflexed in the lower leaves ; sheath-lobe extending half-way, joined at the nerve, always gaping open, especially in the older and in the lower leaves, which latter are shorter and consist mostly of sheath, with less dorsal lamina than the upper leaves. Ventral lamina wider near the base than above the junction ; dorsal lamina nearly as wide above but tapering off to the base ; nerve strong, rusty red, flexuose upward, and ending in the mucro. Border of three to four long cells, wide, rusty red, extending all round the leaf, but sometimes absent at the apex, or reduced to one line of long cells or even intramarginal near the base of the dorsal lamina. Cells numerous, roundly hexagonal, small (7μ diameter), hardly papillose, opaque but translucent at the centre, and with clear walls. Leaf-points inflexed when dry. Seta terminal, pale ; capsule inclined, small, oval, or when dry oblong or substrumose ; lid straight, bluntly conical ; calyptra long-conical ; peristome teeth strongly trabeculate inside ; segments pale red, strongly knotted. The open sheath is a remarkable feature of this moss ; the nerve and border of old leaves are usually rusty brown, but not always so. The leaves also, which are light green when young, often have a rusty tint when old. I have seen this with sheath-border intramarginal and a strongly serrate margin three to four cells wide outside it in some leaves. Very near *F. marginatus* Schp., but the leaves are less acuminate.

Macowan's specimen from Boschberg (Rehm. 587), which is fertile in all stages, has the leaves short and the sheath-lobe in all the leaves, including even the perichaetal leaves, connected above the middle, and the sheath full open. But in MacLea's specimen (also fertile), Rehm. 588, from near Graaff Reinet, which Rehmann distributed as var. *orthopyxis* Rehm., this is the case with most of the leaves, only the upper being as here illustrated, and I cannot separate the variety, as is also the case with Rehm. 291*e* from Rondebosch, which he himself (under 588) connects with that specimen.

Hornschuch (Linn., 1841, p. 154) remarks on its likeness to the European *F. bryoides* Hedw., and at p. 157 he describes his *F. bryoides* Hedw. var. *capensis* from Table Mountain (Ecklon), which description agrees with *F. rufescens*. Harvey, in Genera of South African Plants, 1st ed., introduces *F. bryoides* as " common in all parts of the world," and in Shaw's list it is included from Rondebosch (Zeyher), and Wager lists it as common, but later investigations exclude it from South Africa. Though much like *F. rufescens*, it is autoecious. *F. Gueinzii* C. M., in Linn., 1871–73, p. 168, is described as hermaphrodite and very near *F. bryoides*, but I doubt its being different from this, but if so, then its place is in *F. androgynus* Br.

F. pseudo-rufescens, Rehm. 589, from Kadziberg, O.F.S. (on which label he states that as Rehm. 291 the same form was confused with *F. rufescens* Horns.), is represented in the Bolus herbarium (the only specimen I have seen of either number) by a much smaller species without border, *i.e. F. erosulus* C. M. Rehmann is not likely to have made this mistake, so I presume the specimen in Bolus herb. does not correspond with others distributed. But Dixon says (Kew Bull., No. 6, 1923, No. 228), with reference to Rehmann's note above mentioned, " Whatever this may be worth, the name is antedated by *F. pseudo-rufescens* C. M., and is altered to *F. austro-africanus* by Paris," so both names sink until the latter is proved to exist.

F. rufescens is common in the western province, but always sterile ; eastward it is occasionally fertile. Salmon mentions this or an allied plant in Ann. Bot., March 1899, p. 121.

Cape, W. Common everywhere on Table Mountain and round Cape Town (Sim) ; Kenilworth, Mrs. Duncan ; Stellenbosch, Sim 9574, 9586 ; Garside No. 9 ; and Burtt-Davy (all sterile) ; Knysna, Burtt-Davy 17,014 (sterile).

Cape, E. Graaff Reinet, MacLea (Rehm. 588) ; Boschberg, Macowan (Rehm. 587).

Natal (Gueinzius).

293. *Fissidens latifolius* Dixon, S.A. Journ. Sc., xviii, 305 ; Broth., Pfl., x, 146.

Gregarious or tufted, on stones or soil or bark, dark green ; stems 0·5–1 cm. high ; young and fertile stems usually erect but often rising from a terminal or side innovation from a prostrate old stem. Leaves ten to twenty, seldom in contact, 0·6 mm. long, 0·2 mm. wide, distichous and plane when moist, incurved and crisped when dry ; oblong-elliptical, shortly drawn to a point ; the ventral lamina arched with the base wide ; the sheath not very open ; the sheath-lobe extending half-way or more, and united at a point well inside the margin ; the dorsal lamina also arched, but reduced gradually to the base. Nerve strong, pale, straight to the junction, then somewhat deflexed and flexuose, ending in the apex, which on account of the nerve deflection is somewhat oblique. The border consists of one row of long hyaline cells all round, occasionally replaced at the apex by shorter green cells. Cells numerous, in lines, smooth, chlorophyllose but translucent, rounded, about 7μ diameter. Seta 6–8 mm. long, red, straight ; capsule inclined, slightly gibbous, oblong, pale ; peristome teeth red. (See remarks under *F. subremotifolius*.)

S. Rhodesia. Zimbabwe, 3000 feet, Sim 8766, 8807, 8753, 8761, 8768 ; Khami Ruins, 5000 feet, Sim 8841 ; Matopos, 5000 feet, Sim 8856 ; Salisbury, 4900 feet (Eyles 1574).

F. Meynharti C. M., in Hedw., 1899, p. 54 (not 1894, p. 154, as in Paris, Index, 2nd ed.) ; Broth., Pfl., x, 146 from Zambesi at Barome (Rev. Meynhart), possibly belongs here, though that can only be proved by comparison.

Müller describes it as very small, having minute, crisped, oblong-acuminate, falcate leaves very narrowly bordered ; nerve excurrent in a short mucro ; cells very minute, green, obscure, round ; and in other respects as described above.

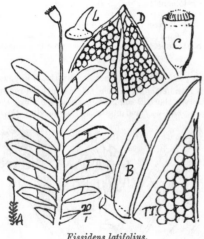

Fissidens latifolius.

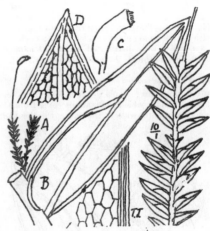

Fissidens marginatus.

294 *Fissidens marginatus* (Schp. MS.), C. M. in Bot. Zeit., 1858 ; Broth., Pfl., x, 146.

=*F. ischyro-bryoides* C. M., Hedw., 1899, p. 55 ; Broth., Pfl., x, 146.

A gregarious, dark green, dioecious soil-moss ; stems 5 mm. long, or by terminal innovation two to three times that length, with terminal red seta ⅜ inch long, and inclined, arcuate, small, pale capsule. Basal branches frequent. Leaves 1 mm. long, strongly bordered all round, lanceolate, straight or slightly falcate, seldom in contact except the upper leaves, which are often imbricate and equitant. Apex acuminate from the junction, mucronate ; sheath-lobe joined mid-way up, at or near the nerve, often rather open (as in *F. rufescens*) ; border pale, strong, thick ; nerve strong, pale, striate, ending in the mucro ; cells 8μ diameter, hexagonal, often empty, when chlorophyllose tumid, circular, and opaque. Perichaetal leaves rather longer and narrower than others. Seta long, capsule arcuate, teeth short, red, trabeculate, the segments blunt and strongly papillose. Vaginula bright red, large ; lid beaked from a convex base. Leaves turned to one side when dry ; lower leaves smaller and consist mostly of sheath.

Cape, W. Newlands Ravine and Window Gorge, Kirstenbosch, Sim 9453 and 9495 ; Stellenbosch, Sim 9584 and 9580 ; Devil's Peak, Rehm. 290 (as *F. ischyro-bryoides*, C. M.).

295. *Fissidens aristatus* Sim (new species).*

Caespitose, light green ; stems 2–5 mm. high, often on remains of previous plants ; leaves 3–15-jugate ; lower leaves separate, upper equitant, all lanceolate-acuminate from a rounded base, 1·5–2 mm. long, greenish pellucid ;

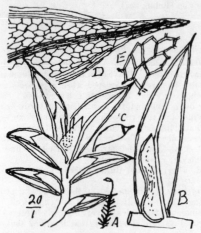

sheath-lobe extending nearly half-way, connected by a point near the nerve, very open ; dorsal and ventral lamina about equal above mid-leaf, the former ending at the insertion. Nerve grey or brown, very strong, striate, straight to the lobe-junction, then either straight or arched or bent backward, extending as a distinctly excurrent point. Border very strong all round the leaf and joining the nerve in the point, one to four cells wide ; these cells long, narrow, and hyaline ; cells of lamina widely and irregularly 5–6-sided, about 30μ long, 20μ wide, greenish, pellucid, thick-walled. Female inflorescence of about four archegonia almost invariably terminal, but occasionally basal with a few large leaves. Capsule inclined or horizontal on a long pale seta ; lid short, beaked.

Very similar to *F. androgynus*, but the leaves more regularly straight, equal and sharply aristate, with very open sheath and larger cells.

Natal. West St., Maritzburg ; Alexandra Park, Maritzburg ; Zwart Kop Stream ; Knoll, Hilton Road, Sim 9903, 9907, 9909, 1916.

Fissidens aristatus.

296. *Fissidens remotifolius* C. M., Syn. Musc., i, 60, and ii, 530 ; Broth., Pfl., x, 146.

Fairly large ; leaves 10–20-jugate, lax, undulate, yellow-bordered along the sheath and dorsal lamina ; apex unequal, not bordered, denticulate ; sheath-lobe half as long as the leaf, open, attached to the nerve ; dorsal lamina decurrent ; nerve thick, excurrent, mucronate ; cells minute but pellucid and smooth. Seta terminal, erect ; capsule horizontal, oblong-cylindrical. Leaves of old plant straight, of young plant crisped. Cape, Zeyher, No. 41. (Not seen.) Rehmann's specimens 590 and 591 from Transvaal, which were distributed under this name, have the sheath-lobe attached at the margin and consequently do not belong here (but to *F. malaco-bryoides* C. M.).

297. *Fissidens malaco-bryoides*, C. M., Hedw., 1899, p. 55 (not p. 25 as in Paris' Index) ; Broth., Pfl., x, 146.

Tufted, dioecious ; stems ½–1 inch long or more, erect, with terminal inflorescence below which an innovation often starts, besides which innovations come from the base. Leaves short (2 mm. long), broadly oblong, roundly acute and mucronate, wide at the base ; the sheath-lobe extending half-way and united at the margin ; the dorsal

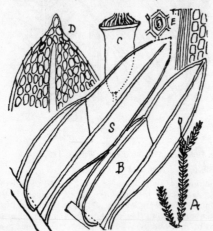

Fissidens malaco-bryoides.

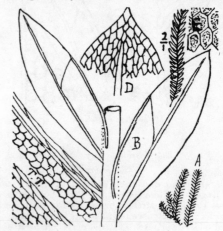

Fissidens dubiosus.

---

* *Fissidens aristatus* Sim (sp. nov.) (§ *Bryoidium*).

Dioicus, caespitosus ; caules 2–5 mm. alti ; folia 3–15-jugata ; folia superiora equitantia, lanceolato-acuminata a basi rotundata, 1·5 ad 2 mm. longa, recta, vagino-lobulus prope ad medium pertinens, prope nervum connexus, apertissimus ; lamina dorsalis et ventralis fere pares supra folium medium, illa qua inseritur desinens. Nervus firmissimus, fuscus, striatus, in apicem excurrentem desinens. Margo firmissimus continuo circum folium, ex 1–4 cellis longis angustis hyalinis ; lamino-cellae 30×20 mm. Infl. feminea plerumque terminalis, non numquam basalis. Capsula inclinata vel horizontalis in seta longa, pallida ; operculum breve, rostratum.

lobe rather narrow, decurrent, border strong, of several lines of long cells all round except that at the apex this border is sometimes replaced by a single row of shorter cells which render that portion subdenticulate. Nerve strong, yellowish, percurrent or almost so ; cells large, roundly hexagonal, chlorophyllose, smooth, those toward the leaf-base separate and square or oblong, often empty ; perichaetal leaves rather longer and narrower than others. Capsule small, erect, on a rather short, yellowish seta ; lid shortly rostellate.

Müller described this from Boschberg, Somerset East (Macowan), but I find that Rehm. 590 and 591 from Pilgrims Rest, Transvaal, distributed as *F. remotifolius*, belong here also.

298. *Fissidens dubiosus* Dixon, S.A. Journ. Sc., xviii, 305 ; Broth., Pfl., x, 146.

Semi-aquatic or aquatic, gregarious ; stems ½–1 inch long, about 3 mm. wide, usually simple, plumose, and pectinate, occasionally branched from below ; leaves all of equal length (about 1·5 mm.), erecto-patent, hardly in contact, elliptic, usually acute ; the sheath-lobe extending half-way, brown on old leaves, with diagonal end connected at the margin ; dorsal lamina equal to the other in the upper half but tapering gradually to the base ; nerve strong, ending below the apex, light green and translucent on younger leaves, tinged with red in old leaves ; border of several lines of long, narrow, hyaline cells on both sides of sheath as far as the junction, beyond that and on the dorsal lamina usually no border or one of shortly elongate, oblique, chlorophyllose cells ; cells of lamina large, quadrate or irregularly 4–6-sided, 12μ wide, densely chlorophyllose, with thin clear walls. Leaves not much different when dry. Fruit unknown. Dixon says it " might be considered to belong to *Semi-limbidium*, but the occasional full development of the border together with the character of the areolation leave little doubt that its true place is in *Bryoidium*."

Mixed with *Conomitrium* in a pool in Palm Grove, Transvaal ; Pretoria Kopjes, 3000 feet, Apr. 1921, Sim 9954. S. Rhodesia, Victoria Falls, 2500 feet, Sim 8819 ; Khami Ruins, 5000 feet, Sim 9953.

299. *Fissidens glaucescens* Horns., Linn., 1841, p. 154 ; C. M., Syn. Musc., i, 51, and ii, 529 ; Broth., Pfl., i, 3, 359, and x, 151 ; Rehm. 283 and *b, c* ; Dixon, Kew Bull., No. 6, 1923, 208, p.p.

=*F. *involvens* W. P. Sch. hb. (*fide* Salmon, Ann. Bot., March 1899, p. 125).
=*F. lanceolatus* Brid., in Flora, 1846, p. 134 ; C. M., Syn., i, 73, name only ; Rehm. 592.
=*F. Rehmanni* C. M., Hedw., 1899, p. 56 ; Rehm. 282 and *b, c, d*.
=*F. cymatophyllus* C. M., Hedw., 1899, p. 57, and in Musc. Macowanianus, No. 46.
=*F. mucronatus* Schimp, MS. ; C. M., in Bot. Zeit., 1858, p. 151, *fide* C. M., in Hedw., 1899, p. 58.
=*F. glaucescens* var. **minor*, Rehm. 284.
=*F. glaucescens* var. **crispulus*, Rehm. 287.
=*F. glaucescens* var. **remotifolius*, Rehm. 288.

In dense tufts on wet rocks and swampy ground, glaucous above, brown below ; stems ½–2 inches long, simple or slightly branched, or with several basal branches, erect, often red when old, many-leaved, pectinate ; roots usually or always red ; leaves rigid, lanceolate, widest up to the junction, rather narrower and acuminate from there to the acute, equal, mucronulate apex ; upper leaves usually equitant ; sheath-lobe two-thirds length of leaf, joined inside the margin at an acute point ; dorsal lamina about half the width of the other, slowly or suddenly rounded to or at the base or decurrent ; nerve strong, pellucid brown, composed of several layers, flexuose above the sheath-junction and extending to the mucro. Margin nowhere bordered ; marginal cells like others but with exserted points, especially near the leaf-apex ; cells small, 7μ diameter, roundly quadrate, chlorophyllose, tumid, with clear thin walls, lower cells larger, square or oblong, up to 30μ long. Female inflorescence either axillary, lateral, usually below mid-stem, or terminal on small basal branches, or terminal ; perichaetal leaves four to six, longer and narrower than the stem-leaves, more vaginate and cellular than other leaves ; vaginula cylindric, red ; seta ½ inch long, straight ; capsule small, oblong-cylindric, erect or inclined, somewhat curved and irregular, brown ; mouth red ; peristome teeth well developed, each split about half-way, the lower half freely trabeculate, the segments very slender, red, trabeculate, very papillose, somewhat striate ; lid very large, convex and beaked, red. Leaves when dry more or less crisped and curved back to one side or hooked. The most common *Fissidens* all over South Africa, present in almost every stream, not very variable in its foliage, but still it seems to have troubled Müller and Rehmann and even Brotherus, as their synonyms show, but it is variable in its fructification. It fruits very rarely. In addition to my 1919 gatherings near Cape Town, I have only twice found it fertile within thirty years, *i.e.* in 1893 at Perie and in 1899 at Ugie, and in both these gatherings terminal as well as basal setae are present. But specimens collected by Miss Owen at Cathkin Peak, Natal, and by Rev. H. A. Junod at The Downs, Pietersburg (Junod 4016 and 4017), are abundantly fertile, with terminal setae only, and still have no other difference except that the marginal cells are more hyaline than usual (2nd fig., p. 202). Hornschuch, on one fertile and many sterile specimens, described it in 1841 as having axillary lateral fruit, but I consider the position variable.

Var. *crispulus*, Rehm. 287, from Inchanga, Natal, does not differ except that the cells are less tumid, but under Rehm. 592 Rehmann connects it with *F. lanceolatus*, Hpe.

Var. *crispus* C. M., in Rehm. 286 (as shown in Albany Museum), from Devil's Peak Cataract, is an immature state of the strongly bordered *F. rufescens*, an unusual mixture from Rehmann.

Var. *minor*, Rehm. 284, from Rondebosch, is not even particularly small, stems being ¾ inch long, but mostly buried in mud so that points appear as short plants.

F. *Rehmanni C. M., Rehm. 282 and *b, c, d* ; Broth., Pfl., x, 151, does not differ, and in Rehm. 592 he distributed it as *F. lanceolatus* Hpe.=*F. cymatophyllus* C. M., MSS., in Musc. Macowanianus, No. 46.

F. lanceolatus Hpe., in Bot. Zeit., 1870, p. 36 ; Broth., Pfl., x, 151, is from Madagascar, and *fide* Brotherus, South Africa.

F. lanceolatus Bruch., in Regens., Bot. Zeit., 1846, p. 134, from Devil's Peak (Krauss), was not known to Müller, and is not known to me.

Müller (Hedw., 1899, p. 58) also connects with *F. Rehmanni F.* (*Orthothallia*) *mucronatus* Schimp., in Musc. Breutel., and C. M., in Bot. Zeit., 1858, p. 154; Broth., Pfl., x, 151, from Claremont (Cape, W.) and Lydenburg (Transvaal).

F. taxifolius Hedw. (Sp. Musc., p. 155, t. 39), a synoecious species with lateral fruit and unbordered leaves, was formerly reputedly South African (Shaw's list, from Graaff Reinet, etc.) as well as European and American, but is not now so recognised; probably *F. glaucescens* was mistaken for it.

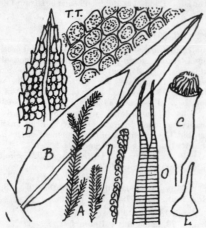

Fissidens glaucescens.

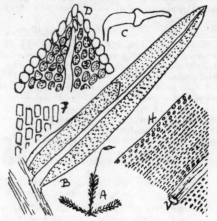

Fissidens glaucescens (form).

300. *Fissidens fasciculatus* Horns., Linn., 1841, p. 155; C. M., Syn. Musc., i, 67 (as *Eufissidens*), and ii, 528 (as *Pachyfissidens*); Broth., Pfl., x, 153.

=*F. linealis* Br. and Sch., Bry. Eur., 1, Mon., p. 12, t. 6; C. M., Syn., i, 46, and ii, 528.

A tufted, dark green, or glaucous green, falcate moss, growing on damp ground; stems 1–2 inches high, erect, robust, rigid, simple or branched below, often red, closely leaved; the leaves distichous, sheathing, imbricate, mostly very secund-falcate, especially the longer perichaetal leaves; leaves linear-lanceolate, 3–5 mm. long; the sheath-lobe extending one-third, joined near the nerve; beyond the junction the leaf is linear-acuminate with the very wide nerve in the middle, apex narrow but bluntly rounded and entire; below the junction the dorsal lamina sometimes tapers out, but it usually extends narrowly to the leaf-base, and both happen on the same plant. Nerve striate, very strong and wide, composed of several layers of cells; the leaf-lamina is of one layer toward the margin, but of more layers and of longer cells near the nerve. Margin entire, not bordered. Seta terminal, pale, ¾ inch long, somewhat curved; perichaetal leaves much longer and more falcate and more acuminate than others, showing in occasional tufts up the stem where terminal setae once have been; their sheaths wide, short, membranaceous, several layers deep near the nerve, green; the portion of leaf beyond the sheath longer and more linear than in other leaves; lower leaves short, and sheathed to the apex; capsule inclined, small, somewhat irregular at base and mouth, with bent back, varying much with condition and moisture from oval-oblong when young and moist to cylindrical and wide-mouthed when old, with peristome either hid inside or horizontal with segments erect, or when dry spreading with segments incurved. Peristome red, with two to three segments, or sometimes perforated and with two segments, barred and closely striate inside; the segments slender and papillose. Hornschuch describes the inflorescence as dioecious and terminal, the male bud-like, globulose, pointed; perigonial leaves four, sheathing, with linear subula; antheridia thirty, large, oblong-cylindric, turgid, without paraphyses.

There is much confusion in the group to which this species belongs. Hornschuch (in Linn., 1841, pp. 151, 153, and 155) described *F. plumosus*, *F. glaucescens*, and *F. fasciculatus* with no indication that either belonged to *Pachyfissidens*; Bruch and Schimper (in Bry. Eur., p. 12, t. 6) described *F. linealis* C. M. (in Syn. Musc., i, 46), placed it under *Pachyfissidens*, but in vol. ii, p. 528, he placed all four in *Pachyfissidens* (having the nerve often composed of many layers of cells).

Brotherus (Pfl., i, 3, 359, and x, 153 and 151) further reverses the order by placing *F. fasciculatus* in *Pachyfissidens*, and *F. glaucescens*, *F. plumosus*, and several others from South Africa in *Eufissidens*, § XI *Amblyothallia*, together with "*F. linearis* Hsch.," which latter name is evidently a mistake and should be *F. linealis* Bry. Eur. The previous descriptions do not warrant transposing *F. linealis* with *F. fasciculatus*, and that seems to have been done in error. But no specimens that I have seen support *F. fasciculatus* C. M. and *F. linealis* Bry. Eur. as distinct. In Müller's Syn. *F. linealis* is said to be monoecious, while *F. fasciculatus* is given as dioecious, but I have not seen male flowers, and unless Ecklon and Zeyher's original specimen supports a separate species, I consider *F. linealis* should be sunk in this species.

Between *F. fasciculatus* and *F. plumosus* there is no difficulty; the former has the leaves shortly falcate-secund when dry, the nerve striate and very wide and the margin entire; the latter has the leaves closely flabellate and erect, or only incurved when dry, and the marginal cells protruding, often acutely mucronulate.

So far as the section is concerned, all the five species I have here included in *Pachyfissidens* have the nerve of several layers, and I find no use in South Africa for § *Amblyothallia*, which name so far as leaf-form is concerned only fits one South African species put into it by Brotherus.

Salmon, Annals of Botany, March 1899, p. 103, uses *F. fasciculatus* largely in dealing with the morphology of *Fissidens* leaf.

Abundant on the Cape Peninsula, Rehm. 281 ; Dr. Bews ; Sim 8514, 8461, 8599, 9382, 9469, 9455, 9265, 9107, 9109 ; Bolus herb. 14,824 p.p., etc. ; also Montagu Pass, Rehm. 281*b*.

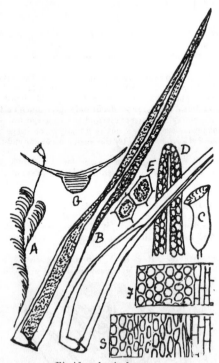

Fissidens fasciculatus.

S, cells on sheath, margin to nerve.

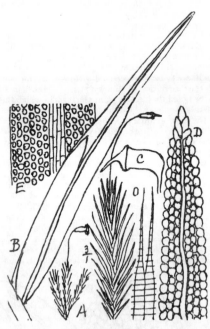

Fissidens plumosus.

301. *Fissidens plumosus* Horns., in Linn., 1841, p. 151 ; C. M., Syn. Musc., i, 68, and ii, 528, Rehm. 280, 280β ; Dixon, in Trans. R. Soc. S. Afr., viii, 3, 188 ; Broth., Pfl., x, 151.

=*F. plumosus* var. *serrulatus Wager.

Gregarious or tufted, dioecious, glaucous soil-mosses ; stems erect or suberect, rigid, ½–1 inch high, simple or more usually branched at the base, with terminal fructification ; leaves in flabellate arrangement, homomallous or incurved when dry, when moist very erect, parallel, numerous, distichous, translucent, 4–5 mm. long, those lower spreading more and the basal leaves shorter. Leaves lanceolate up to the junction, beyond that linear-acuminate, subacute, with central striate nerve to the apex ; the dorsal lamina usually gradually narrowed to the base, and the sheath-lobe joined acutely below the middle and near the margin. Apex subacute, the lamina below it usually wider in this species than in *F. fasciculatus*, nerve almost reaching the apex, margin serrulate through exserted, rounded, or pointed cells, rather larger than those within them, which are rounded and clear, 7μ diameter; margin not present anywhere. Seta red, somewhat curved ; capsule inclined or horizontal, pale, oblong, with widened throat; lid obliquely conical from convex base, red ; calyptra cucullate. Inner perichaetal leaves with wide sheath and very narrow subula, rather shorter than other leaves. Peristome teeth lanceolate-subulate, red, cleft below the middle, the bands strong and almost pectinate, the segments also trabeculate. See note under *fasciculatus* concerning its subgenus.

Cape, W. Skeleton Ravine, Bolus Herb. 14,827 ; Devil's Peak, Rehm. 281 ; Constantia Nek, Sim 9353 ; Montagu Pass, Rehm. 280β ; Knysna, Rehm. 280 ; Miss Duthie 15 ; Stellenbosch, Miss Duthie.

Cape, E. Paradise Kloof, Grahamstown, Miss E. T. C. Farquhar 102 and Miss Hilner 2171.

Transvaal. Kaapsche Hoek, Wager (as var. *serrulatus*).

S. Rhodesia. Inyanga, in Wet Forest, 6000 feet, J. S. Henkel.

302. *Fissidens amblyophyllus* C. M., Hedw., 1899, p. 57 ; Broth., Pfl., x, 151.

=*F. glaucescens* var. *natalensis*, Rehm. 285, 285*b*, 593, 593*b*.

Gregarious, seldom tufted, green above, dark below ; stems 1–2 inches long, with one to two innovations above and with fruit terminal on a short basal branch ; leaves scattered below, equitant above, 3–4 mm. long, 0·5 mm. wide, erecto-patent, complanate, the stems usually with a narrow pectinate outline, sometimes with a wide, closely leaved outline ; leaves lorate, widest in the lower half, rounded at the apex, not bordered ; the dorsal lamina about equally wide to near the base, then rounded in ; the sheath-lobe two-thirds the leaf-length, attached near the middle of the ventral lamina, with a rounded or bluntly pointed end. Nerve yellowish, very strong and thick, arched and flexuose above the junction, ending below the apex ; cells minute, dense, tumid, chlorophyllose, with translucent walls, hexagonal but apparently circular ; the marginal cells making the margin crenulate all round, but are more pointed and make it denticulate at the apex. Fertile branch short, basal, few-leaved ; seta terminal on it, $\frac{1}{2}$ inch long ; capsule inclined, small, red.

Dixon (Kew Bull., No. 6, 1923, under No. 282) reduced *F. amblyophyllus* to a synonym of *F. glaucescens*, but these two differ considerably in the leaf-apex and do not seem to grade into one another, as he has since seen. Rehmann went wrong in placing it as a variety of *F. glaucescens*, and Müller evidently saw this and described it as *F. amblyophyllus*, but his description is weak in several respects.

Dixon, in 1920, in Trans. R. Soc. S. Afr., viii, 3, 187, records *Amblyophyllus* from Kaapsche Hoek, Transvaal, and Knysna Forest, Cape Province (Wager 387 and 10) ; near Hogsback, C.P., 4000–6000 feet, Tyumie (Rev. Jas. Henderson 215 and 198), and says, " The latter is so much larger a plant than the other gatherings that it would seem likely to be *F. procerior* Broth. and Bryhn ; but the apex and dorsal lamina (here decurrent) agree much better with the description of Müller's plant. I think in all probability they may be forms of one and the same species. It may be noted that in this plant the form in which the dorsal lamina terminates below is not as in many or most species a reliable character. On the same fertile stem I have seen all stages between an abruptly rounded, auricle-like ending, and the most gradual narrowing or decurrence."

F. procerior Broth. and Bryhn (Videnskaps Forhandl., 1911–14, p. 9) seems to differ mostly—or only—in vigour, and is described as having stems 25 mm. high and with the leaves 4 mm. wide. It is from Zululand (Bryhn) and Victoria Falls (Brunnthaler).

Common in forest country of Natal and Transvaal.

Cape, W. The Wilderness, George, Miss A. Taylor ; Knysna, Wager (as above).

Cape, E. Hogsback, Dr. P. v. d. Bijl ; and Rev. Jas. Henderson (as above).

Natal. Impolweni Bridge ; Cathkin Peak, 7000 feet ; Mont aux Sources ; Greenfield, Mooi River ; Umwhati, New Hanover ; Ngoya Forest, Zululand (Sim 9927, 9928, 9929, 9930, 9931, 9933) ; Van Reenen's Pass and Inanda (Rehm. 285 and 285*b*, as *F. glaucescens* var. *natalensis* Rehm.).

Transvaal. Rustenburg, Playford ; The Downs, Pietersburg, 4000 feet, Rev. H. A. Junod 4011 ; Reitfontein (Wager) ; Kaapsche Hoek (Wager, as above) ; Pilgrims Rest, MacLea ; Macmac (MacLea) and Houtbosh (Rehm.), Rehm. 593 and 593*b*, as *F. glaucescens* var. *natalensis*.

S. Rhodesia. Inyanga, 6000 feet, Aug. 1920, J. S. Henkel.

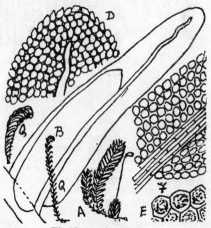

Fissidens amblyophyllus.

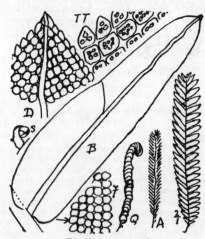

Fissidens corrugatulus.

S, dry leaf.

303. *Fissidens corrugatulus* Dixon, **S.A. Journ.** Sc., xviii, 307 ; Broth., Pfl., x, 151.

Laxly tufted, dark green ; stems usually simple, 1 inch long, erect, many-leaved, pectinate ; the leaves in contact or somewhat imbricate, erecto-patent, 2 mm. long, oblong, very wide, not narrowed to the base, often rather wider at the sheath than above that, shortly rounded to the apiculate point ; the sheath-lobe extending more than halfway up, joined at the margin, with obliquely truncate end ; the dorsal lamina with widely rounded, or auriculate base. Nerve strong, flexuose upward, lost in the apex, rusty brown, and showing as an exserted keel on the lower surface. Margin not bordered anywhere, the marginal cells pointed outward making the margin crenulate, but in other respects the marginal cells correspond with the others, which are quadrate, about 10μ wide, all highly papillose, especially when dry, chlorophyllose in circular form but pellucid, the lines of tumid cells running transversely as well as longitudinally. When dry the stem is curved and leaves curved inward or hooked, all to one surface. Dixon gave the name from and remarks on the leaves being transversely waved ; he placed it in § *Serridium*, but Brotherus has transferred it to *Amblyothallia*.

S. Rhodesia. Victoria Falls, 3000 feet, Sim 8885, 8904, 8924, 8928 ; Matopos, in stream, 5000 feet, Sim 8878.

Genus 86. CONOMITRIUM Mont.

Plants slender, flaccid, several inches long, often much branched and usually aquatic, growing on rocks in running streams. Stems without a central strand. Leaves distant, linear-lanceolate, distichous, vertically placed, equitant, not bordered, and nerve not excurrent. Cells hexagonal. Inflorescence monoecious. Capsules minute, often numerous, erect and regular, without stomata, on very short, straight setae from axillary perichaetia along the stem; lid conical; calyptra very small, conical, undivided, but fringed at the base. Ring absent; peristome of sixteen short lanceolate teeth.

Conomitrium Mont. (in Ann. Sc. Nat., 1837, p. 250), distinguished by its small, conical, entire calyptra, as used by C. Müller, included as one section the older genus *Octodiceras* Brid. (Bryol. Univ., ii, 675 (1827)) to which our plant belongs, together with other small terrestrial species from his sections *Reticularia* and *Sclarodium*, which are now included in *Fissidens*. Indeed *Fissidens*, according to some botanists, includes the whole group, but Paris maintains *Conomitrium*, while Dixon and Jameson use *Octodiceras* for our plant, though in later papers Mr. Dixon uses that as a section name within *Fissidens*.

304. *Conomitrium julianum* (Savi), Mont., in Ann. Sc. Nat., 1837, p. 246, t. 4; C. M., Syn., ii, 524; W. P. Sch., Syn., 2nd ed., p. 122.

=Fissidens julianus W. P. Sch., in Flora, 1838, pl. 1, p. 271; C. M., Syn., i, 44; Dixon, Trans. R. Soc. S. Afr., viii, 3, 188; and S.A. Journ. Sc., xviii, 308; Broth., Pfl., x, 153.
=Fontinalis juliana Savi, Bot. Etr., iii, 107.
=Octodiceras julianum Brid., ii, 678; Dixon and Jameson, 3rd ed., p. 135.
=Skitophyllum fontanum, La Pyl., Journ. Bot. Desv., 1813, v, 52, t. 34.
=Fissidens Dillenii C. M., Syn., i, 45 (=Octodiceras Brid.; Conomitrium Mont.; C. M., Syn., p. 524, and Dregé's list).
=Fissidens Berteri C. M., Syn., i, 45 (=Conomitrium, Mont.; Shaw's list, and C. M., Syn., ii, 524).
=Conomitrium capense C. M., Syn., ii, 524.
=Fissidens capensis (C. M.), Broth., Pfl., x, 154.
=Fissidens Muelleri Hampe.
=Conomitrium nigrescens Rehm., in Rev. Bryol., 1878, p. 69.
=Octodiceras nigrescens. Rehm. 278.
=Fissidens nigrescens (Rehm.), Par., Ind. Suppl., 1900, p. 162, and 2nd ed., p. 216; Wager's list.
=Octodiceras capensis C. M., Rehm. 581 and 581b.
=Octodiceras capensis C. M. var. nigrescens, Rehm. 278 and 278b, and various other foreign synonyms.

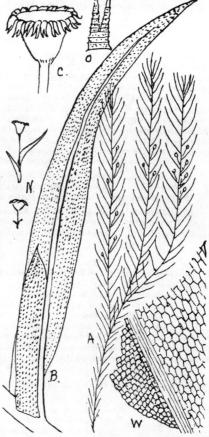

Conomitrium julianum.

Usually floating in running water or semi-submerged; monoecious; stems 2–6 inches long, branched, often much branched; leaves numerous, distichous, distant, 3–6 mm. long, straight, inflexed or reflexed, lanceolate, acuminate, acute, without border; the nerve ceasing below the apex; the sheath-lobe about one-third the length of the leaf, united far from the margin. Cells fairly lax, hexagonal, those of the dorsal lamina smaller and more dense. Perichaetia axillary, one per leaf. Seta very short (*i.e.* 1 mm. long), straight, usually on a minute, 1–2-bracted, axillary, floral branch; capsules very small, shortly obconic, with wide mouth; teeth short, usually 2-fid half-way; lid conical, straight; calyptra conical, entire. Müller says, "Male flower terminal on a very short axillary branch, rarely subsessile."

In Trans. R. Soc. S. Afr. Dixon clears up the synonymy of this widely distributed species, which previously had local names in Europe, North America, South America, Australia, and South Africa. It is often so encrusted in lime or other substances as to be hardly recognisable, and it is sometimes covered with diatoms.

Widely distributed in streams but seldom common; not recorded from the Western Cape Province.

Cape, E. Perie Forest, 1892, Sim 7118; Gwacwaba River, King William's Town, Sim 7096; Kentani, 1000 feet, Miss Alice Pegler 1353; Nadouwsberg, MacLea (Rehm. 581b); Dregé 9379a and b.

Natal. Thornville Stream, Dr. Bews; Umsundusi, Sim 7466; Van Reenen's Pass, Rehm. 562b; Cathkin, 6000 feet, T. R. Sim; Durban, 1887, J. M. Wood; Albert Falls, Sim 7405; Sweetwaters, June 1916, T. R. Sim.

O.R.C. Kadziberg, at Caledon River, Rehm. 278.

Transvaal. Macmac, MacLea (Rehm. 581, 582).

S. Rhodesia. Palm Grove, Victoria Falls, Sim 8517; Mazoe, 4000 feet, F. Eyles (Sim 8147); Umtali, 3800 feet, Eyles 1663; Sinoia Cave, Lomagundi, 4000 feet, Eyles 3160.

Family 24. GRIMMIACEAE.

Rock-growing plants, usually found in dense cushions except the few terrestrial species which are loosely caespitose. Stem short to long, simple to much-branched, rooting at the base, acrocarpous or cladocarpous, or with numerous innovations below the inflorescence, and with numerous, often crowded leaves placed all round the stem. Leaves usually lanceolate, ovate-lanceolate or subulate, always concave, usually acute, and often with hair-points which are a continuation of the lamina rather than of the nerve. Lamina often several cells deep, the cells toward the apex being small, dense, hexagonal, thick-walled, and abundantly chlorophyllose; those toward the base longer, sometimes linear, often hyaline, with in most cases (except *Ptychomitrium*) thick sinuose cell-walls. Inflorescence bud-like; capsule oval or cylindric on a short seta, sometimes immersed, and either erect and regular or slightly unequal on an arcuate seta. Calyptra mitriform, or sometimes cucullate or campanulate-lobed, smooth or furrowed or papillose. Peristome of sixteen coloured, rather wide teeth, entire or more or less split, sometimes to the base, without vertical striae; the outer plates usually thickest and often transversely barred. No basal membrane is present.

Schimper included in this family the Orthotrichaceae, where the peristome is normally double, though in some cases simple or absent by reduction.

Synopsis.

Calyptra not plicate, covering only the top of the capsule; leaf-cell walls sinuate.
    Peristome teeth entire, or perforated, or partly split .   .   .   .   .   87. GRIMMIA.
    Peristome teeth cut to the base into filiform segments .   .   88. RHACOMITRIUM.
Calyptra campanulate-lobed, covering most of the capsule; leaf-cells not papillate and walls not sinuose,
                                  89. PTYCHOMITRIUM.

Genus 87. GRIMMIA Ehrh.

Small, dichotomously branched plants, forming close or lax round cushions on nearly bare rocks. Leaves from ovate-lanceolate to subulate, having entire or sometimes thickened margins, one of which is often recurved along its central or lower portion; leaf usually nerved to near the apex, but the nerve not extending into the long hyaline hair-point which is more or less produced in many species, and which may be flat or round, toothed or smooth. Leaves often much crisped or twisted when dry. Lower cells vary from short to oblong, with more or less sinuose cell-walls; upper cells always short and dense, often opaque, and sometimes 2-stratose. Capsule oval, shortly exserted or sometimes included, smooth or striated. Seta straight or curved; calyptra covering only the upper part of the capsule, not plicate, variously and inconstantly cut at the base. Peristome teeth sixteen, entire, perforated or more or less forked or perforated (rarely absent), purple or bright red, spreading, suberect, or incurved when dry. Most of our species inhabit the mountains and uplands.

The genus divides into several sections, of which we have to do with the following :—

    A. *Schistidium.* Capsule immersed on a short seta; leaves comparatively few.
    B. *Eugrimmia.* Capsule exserted on a straight or arcuate seta, smooth; leaves crowded.
    C. *Rhabdogrimmia.* Seta cygneous when young; capsule ribbed.

The difference between *Grimmia* and *Rhacomitrium* rests mostly on the length of the lower leaf-cells and on the depth to which the peristome teeth are split, both rather unsatisfactory characters, so it is not surprising that some species described as *Grimmia* by Müller are now included in *Rhacomitrium*.

Synopsis of Grimmia.

1. Capsule immersed; leaves not crowded, usually not hair-pointed .   .   .   305. *G. apocarpa.*
1. Capsule exserted; leaves crowded .   .   .   .   .   .   2.
2. Leaves ovate; nerve almost obsolete .   .   .   .   306. *G. senilis.*
2. Leaves shortly ovate-acuminate; seta straight; hair-point rough .   .   307. *G. campestris.*
2. Leaves lanceolate, crowded .   .   .   .   .   3.
3. Leaf-margin revolute or thickened; seta cygneous when young .   .   308. *G. pulvinata.*
3. Margin not revolute; leaf dense, 2-stratose above; hair-point smooth .   .   4.
4. Leaves seldom hair-pointed, usually rounded .   .   309. *G. drakensbergensis.*
4. Leaves strongly hair-pointed, the comal leaves very long .   .   310. *G. commutata.*

305. *Grimmia apocarpa* (Linn.), Hedw., Descr., i, 104, t. 39; C. M., Syn. Musc., i, 776; Dixon, 3rd ed., p. 138, t. 17*h*; Broth., Pfl., x, 311.

        =*Bryum apocarpum* Linn.
        =*Schistidium apocarpum*, Bryol. Eur., iii, Mon., p. 7, t. 233.
        =*Grimmia boschbergiana* C. M., Hedw., 1899, p. 119; Broth., Pfl., x, 311.
        =*Grimmia oranica* C. M., Hedw., 1899, p. 119; Broth., Pfl., x, 311.
        =*Grimmia depilis* C. M., Syn., i, 778; Broth., Pfl., x, 311.
        =*Grimmia maritima* Turn. var. *australis Hpe., in Hb. Kanz.

Exceedingly variable from small dense tufts of compact plants ½ inch high to loose cushions 1–2 inches deep; stems branched or forked, often repeatedly so, with terminal capsule, below which one or two innovations start, leaving the capsule terminal or in the fork, where it remains dry for several years, thus the capsules along the stem then appear to be lateral. Lower leaves absent or small; stem-leaves ovate-lanceolate, imbricate, keeled in the upper half, erect and appressed when dry, spreading or erecto-patent when moist ; the basal margin convex or recurved on one or both sides ; nerve strong, ending just below the apex ; apex bluntly pointed, somewhat cucullate, entire or slightly toothed, or occasionally somewhat hyaline-pointed ; cells numerous, in straight lines, quadrate, 10μ diameter, chlorophyllose, the basal cells toward the nerve oblong and about 15μ long, often empty. Perichaetal leaves larger, thinner, and more spreading, forming a clavate apex in which the capsule is immersed and almost hid ; the chlorophyllose cells in these are larger than in stem-leaves, irregular in form, about 25μ long ; the basal cells oblong-cylindrical, regular, empty, 80μ long, and the lower margin is usually not reflexed but the upper margin often is. Capsule inversed bell-shaped, 1·5 mm. long, 1 mm. wide, on a very short, erect seta ; ring evident, red ; teeth sixteen, rising inside the mouth, reflexed when dry, erect when moist, small, conical-lanceolate, with or without long, slender, paler points, undivided or sometimes somewhat perforated, subspirally few-trabeculate ; lid short, red, convex, with short suboblique beak, and with the columella attached ; the cells of the lid small and quadrate, in many rows ; calyptra

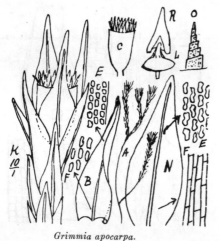

Grimmia apocarpa.

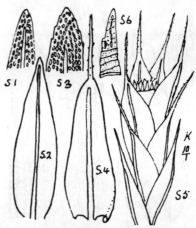

Grimmia apocarpa (forms).

S1, common form ; S2–3, narrow leaf and toothed apex ; S4, hair-pointed form ; S5, same, $\frac{10}{1}$; S6, perforated peristome tooth.

conical, dimidiate, small, hardly covering the lid, its cells lax and pellucid. Inflorescence monoecious, the male flower at first terminal.

This describes the South African plant ; in Europe and elsewhere various forms have been named, especially in reference to habit and leaf-point, which latter is more often hair-pointed there than here. Here the hairless and the hair-pointed conditions are usually more or less separate, but I have found all the conditions shown in the 2nd figure on one plant, also entire and perforated peristome teeth, and long and short teeth, and occasionally 2–3-fid teeth, so the forms separated out by Müller cannot stand as distinct species or even as varieties ; they are conditional forms subject to variation.

Var. *rivularis* (*G. rivularis* Brid.) is a large subaquatic form with very large capsule, with which var. *stricta*, Rehm. 132, and *G. caffra* C. M. (Hedw., 1899, p. 118), Rehm. 130, Broth., Pfl., x, 311, both from Kadziberg, O.F.S., correspond.

Widely distributed over the world, but I have no records from the Western Cape Province nor from Rhodesia ; all my records are from 4000 to 10,000 feet altitude, so it is quite a mountain species.

Cape, E. Oudeberg, MacLea (Rehm. 494*b*).

Natal. Goodoo and Tugela Gorge, Dr. Bews (Sim 8383, 8397) ; Van Reenen's Pass, Rehm. 494*c* ; Giant's Castle, 8000 feet, R. E. Symons ; Little Berg, Cathkin, Miss Owen 12, 24, 29 ; Mont aux Sources, Sim 8996, and at 10,000 feet altitude, Miss Edwards.

O.F.S. Morija, Basutoland, May 1916, Dr. Stoneman ; Bethlehem, Sim 9955 and Rehm. 131 ; Witteberg, above Kadziberg, Rehm. 130 (as *G. caffra*) and 132 (as var. *stricta*).

Transvaal. Mahaliesberg, Rehm. 494.

306. **Grimmia senilis** (Shaw, ined., Cape Monthly Mag., Dec. 1878, p. 380) Sim (new species).*

Densely caespitose on igneous karroo-dykes, the tufts usually large and closely packed with fine sand or dust ; stems from simple to much-branched, often closely tomentose below, leafless below with clavate many-leaved tops,

---

* *Grimmia senilis* (Shaw, ined.) Sim (sp. nov.).

Dense caespitosa ; caules foliis carentes infra, pluri-foliati supra, cani folii-apicibus hyalinis ; folia ovata, admodum concava, seta longa hyalina fortiter denticulata ; margines erecti ; nervus prope absens.

very hoary with hyaline leaf-tips ; leaves ovate, very concave, quickly rounded to the long, hyaline, strongly denticulate bristle ; the margins erect, one sometimes thickened upward. No nerve present, but the mid-cells in the lower half of leaf are long and narrow, and then gradually shorten and become densely packed, chlorophyllose and quadrate outward and upward, 7μ diameter, only slightly sinuose, the marginal ones more or less regularly transverse, all chlorophyllose. I find it abundant but sterile in the Karroo ; Shaw describes as, " Seta erect, short ; calyptra covering the capsule entirely when young, when ripe only half."

Similar to *G. pulvinata* in general appearance, but more hoary, the leaves wider and more concave, the bristles drawn to a stem-point when dry, and the nerve is very different.

Cape, E. Graaff Reinet, Bolus ; MacLea ; Colesberg, Shaw ; Cookhouse, Middelburg, Naauwpoort, Beaufort West, etc. (Sim 9956, etc.).

O.F.S. Springfontein, Sim.

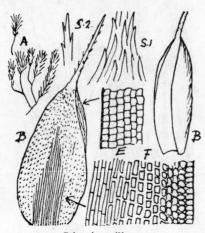

Grimmia senilis.

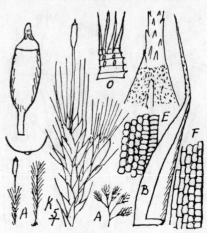

Grimmia campestris.

307. *Grimmia campestris* Burch., in Hook., Musc. Exot., 1818–1820, t. 129 ; Broth., Pfl., x, 307.

=*Grimmia leucophaea* Grev., Trans. of the Werner Soc., iv, t. 6 (1822) ; C. M., Syn., i, 749 ; Dixon, 3rd ed., p. 156, and Tab. 20c.

=*Dryptodon leucophaeus* Brid., Bryol. Univ., i, 773.

=*Dryptodon campestris* Hook., in Brid., Bryol. Univ., i, 774.

Rather a rough, very dark, dioecious moss, growing in very open tufts on rocks, the stems $\frac{1}{2}$–$\frac{3}{4}$ inch long, simple or branched above, rigid, clad in old leaf-base, except the top where there is a clavate coma of leaves, each ending in a long, flat, diaphanous, denticulate hair-point, the inner hairs twice as long as the leaf, but the lower leaves short and appressed, without pilum. Leaves numerous, crowded, imbricate, appressed when dry, erecto-patent when moist, triangular-ovate-acuminate, 1–2 mm. long, concave, the inner longest ; nerve often rather indistinct, but present from the base up to the end of the lamina ; point made denticulate by many cell-points on the surface as well as on the margin (not serrulate as stated in Musc. Exot.). Leaf-margins erect, not recurved nor thickened ; cells densely chlorophyllose, crowded, roundly quadrate, the outer cells often transverse ; at the leaf-base the outer cells are quadrate and gradually longer and more empty inward but not sinuose-incrassate. Seta rather over-topping the hair-points of the sheathing perichaetal leaves ; capsule erect, 1·25 mm. long, oblong when dry, oblong-ovoid when moist, not furrowed when old ; lid convex and shortly rostellate ; calyptra conical, lobed ; ring present ; peristome teeth irregularly divided, often split into three to four slender segments to near the base, the lower portion red and trabeculate, the segments clear.

In Musc. Exot. the illustration shows entire teeth, but the text describes them as often split half-way, which is usual.

Nearly cosmopolitan in suitable rock-localities ; not recorded from Transvaal or Rhodesia.

Cape, W. Cape Town, Rehm. 135 ; Kloof Nek and Table Mountain slopes as lithophytic pioneer, Dr. Bews (Sim 8420, 8455) ; Chaplin Point, Sim 9369 ; Paarl 9603, 9618, 9619, 9655 ; Roggeveld between Jakhalsfontein and Kuylenberg, Burchell ; Matjesfontein, Brunnthaler.

Cape, E. Graaff Reinet, MacLea (Rehm. 492).

Natal. Top of Giant's Castle, R. E. Symons 1915.

O.F.S. Bloemfontein, Rehm. 135b.

308. *Grimmia pulvinata* (Linn.), Sm. Eng. Bot., t. 1728 ; C. M., Syn., i, 783 ; Dixon, 3rd ed., p. 145, and Tab. 18*e* ;
Broth., Pfl., x, 310.

=*Bryum pulvinatum* Linn., Sp. Pl., p. 1121.
=*G. pulvinata* var. *africana* Hook. f. and Wils.
=*Grimmia leptotricha* C. M., in Hedw., 1899, p. 120 (from Boschberg, Macowan) ; Broth., Pfl., x, 310.

A small autoecious moss, hoary with hyaline hair-points, forming dense little round cushions ½–1 inch high and wide, on stones, and usually fertile, with all stages of maturity present at once. Stems ½–⅔ inch high, branched, many-leaved ; the leaves imbricate, erecto-patent, somewhat concave, lanceolate or the upper leaves lanceolate-oblong, each ending in a long, slender, slightly denticulate, hyaline hair. Nerve distinct, often ferruginous, extending to the base but with long cells adjacent there, ending below the hair-point ; margin reflexed on one or both sides near the base, and reflexed or thickened below the hair-point ; cells in lines, roundly quadrate, somewhat tumid, 8μ diameter, translucent, with irregularly sinuose, incrassate walls, the lower cells longer, narrower, and usually empty, those nearest the nerve largest, 30×8μ. Seta terminal, at first cygneous, with hanging, ovoid, smooth capsule ; afterwards the capsule becomes erect on a crooked seta, oblong, 8-striate when dry ; lid concave and very shortly rostellate, often bright red ; calyptra conical, tight on the lid, lobed, each sinus ending a furrow. Ring wide, white ; teeth red, fairly long, often 2–3-cleft, or perforated for that. Although the leaves are usually all piliferous I have specimens from Leribe, Basutoland, in which the lower leaves are rounded at the apex and only the upper leaves are piliferous. Also, in high localities the bristle is often destroyed by snow or wind, but traces of such destruction are evident.

Var. *obtusa* Brid. is rather smaller and has the lid convex-mucronate rather than beaked. This is the common form in the Cape Peninsula and is *Fissidens pulvinatus* var. *b africanus* Hedw., Sp. M., p. 159, t. 40. *G. orbicularis* Bruch (=*G. africana* W. Arn.) is a very closely allied European and North African form, having the margin not thickened, the calyptra cucullate, and the lid blunt.

G. *MacLeana, Rehm. 493, is a sterile specimen from Graaff Reinet which may belong here and has no distinctive character. *Grimmia Eckloni* Spreng., Syst. Veget., iv, ii, 321 ; C. M., Syn., i, 787 (not 78 as printed in Paris' Index), was described from Ecklon's specimens from Lion's Mountain and from near Zwellendam. Mr. Dixon writes that he has examined an original specimen of Ecklon's from C. Müller's herb., at Kew, and finds it to be *G. pulvinata* var. *obtusa.* Rehmann's 133 and 491 belong there also, and, no doubt, Brunnthaler's record from Caledon.

G. pulvinata is almost cosmopolitan, and in South Africa, though seldom common, it occurs from the coast to the mountains and Karroo, in Cape, W. and E., Natal, Basutoland, and O.F.S. I have no records from Transvaal or Rhodesia.

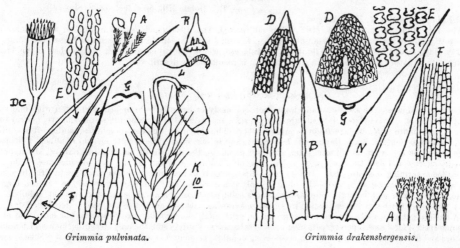

Grimmia pulvinata.   *Grimmia drakensbergensis.*

309. *Grimmia drakensbergensis* Sim (new species).*

Forming dense wide cushions ½ inch deep, on rocks on top of the Drakensberg, the tips yellowish, the foliage dark below, the leaves more or less drawn to a stem-point, and either rounded, or shortly pointed, or with a fairly long diaphanous, smooth hair-point, all these appearing on one plant, but usually the central leaves are hair-pointed or show signs whence the hair has been broken. Leaves narrowly lanceolate, 1·5 mm. long, concave, the margins erect or hardly recurved, bistratose above ; the nerve very strong, acting as a keel, and disappearing just below the end of the lamina. Cells lax, mostly translucent or lightly chlorophyllose, 8μ diameter, quadrate, with very crooked outline,

---

* *Grimmia drakensbergensis* Sim (sp. nov.).

In toris densis, supra subflava, infra nigra ; folia lanceolata, apice rotundato, breviter acuminata vel pilata, omnia in una planta. Margo 2-stratosus supra, erectus vel planus ; cellae basales interiores longae claraeque ; ceterae quadratae claraeque.

lower inner cells longer and clearer, more or less sinuose, except those near the nerve at the base which are compact, clear, and straight-edged.

Dixon compares it thus with *G. pulvinata* : " This has the margin bistratose above, plane throughout, and the basal cells are narrower ; in fact, it is much more like *G. orbicularis* (of Europe) except in the bistratose margin ; the habit, however, is different, the plant being smaller and much more compact. The colour, too, is (if constant) distinct. It is unfortunate that there is no fruit, but it is certainly, I think, a new species having the bistratose margin of *G. pulvinata*, the basal areolation of *G. orbicularis*, and the habit different from both."

Natal. Top of Giant's Castle, R. E. Symons (Sim 9962) ; Mont aux Sources, Sim 9963, and Miss Edwards.

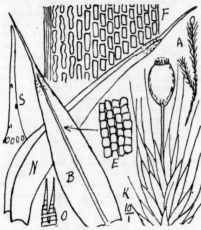

Grimmia commutata.

310. *Grimmia commutata* Hüb., Musc. Germ., p. 185 ; Broth., Pfl., i, 3, 449, and x, 308, and fig. 256 ; Dixon, 3rd ed., p. 155, Tab. 19*i*.

A fairly robust, dioecious moss, growing in dark green cushions 1–1½ inch high ; the stems usually branched, sometimes fastigiate near the top, bare below, leafy above ; the leaves erecto-patent when dry, patent when moist, very concave and consequently apparently narrower than they really are, lanceolate-acuminate, 2–3 mm. long ; the perichaetal leaves longer and narrower than the others, all hyaline-tipped, but the comal and perichaetal leaves have much longer hairs than those below ; the hair hardly denticulate ; nerve distinct but narrow, extending more faintly to the tip of the lamina ; margins erect ; cells quadrate, crowded, obscure, 2-stratose in the upper half of the leaf ; the lower cells empty, those near the nerve much longer than wide, sinuose, gradually shorter and more oblong outward, oblong and smaller near the margin. Seta terminal, short, straight, erect, with erect, ovoid, smooth, brown capsule and short peristome ; lid rostrate, oblique ; calyptra cucullate ; peristome teeth split half-way, barred and often perforated.

Widely distributed, mostly in the Northern Hemisphere. Here it is rare.

Cape, E. Dohne Hill, 5000 feet, 1898, Sim 7218.

O.F.S. Rydal Mount, 1910, Wager.

The very similar *G. ovata* Schw. (Dixon, 3rd ed., p. 154, Tab. 19*h*) which occurs in Europe and at least as far south as Uganda is autoecious ; it has not yet been recorded from South Africa. *Grimmia *assurgens* Shaw, Cape Monthly Mag., Dec. 1878, belongs no doubt to Tortulaceae, where the papillose leaves of described shape occur, as nothing of that nature is known here in *Grimmia*. It is insufficiently described.

Genus 88. RHACOMITRIUM Brid., Mant., p. 78.

Loosely caespitose, the stems usually long, dichotomously or irregularly branched, and often with many small lateral branchlets. Leaves lanceolate to subulate-pilose, equally distributed along the stem, concave, nerved to near the apex, with more or less recurved entire margin where chlorophyllose, but the apex of the lamina often continued as a more or less dentate, long, diaphanous hair-point, and in one species the serrate margin hyaline. Cells usually in one layer, the cell-walls all sinuose, often strongly papillose, the lower cells all long and narrow, the upper cells gradually shorter and near the apex more or less quadrate, the marginal cells at the base in some cases pellucid and thin-walled or forming small auricles. Inflorescence dioecious ; fruit terminal or cladocarpous ; seta straight ; capsule erect and regular, oblong or cylindrical, smooth ; calyptra covering only the upper part of the capsule, large, not plicate, mitriform or with lacerate base, and having a solid papillose or scabrid beak. Lid straight, subulate ; ring wide ; peristome teeth sixteen, connected in a basal membrane ; usually forked to the base ; erect when dry.

A natural genus, usually rock-grown or grown on gritty soil, closely related to *Grimmia*, but less dense in habit and some species form loose open carpets over considerable areas of gritty soil, the long straggling stems with numerous branches and often apparently lateral fruit, resembling in habit the pleurocarpous mosses though not otherways related to them.

Grimmia and *Rhacomitrium* connect the South African mountain moss floras with the subarctic and subantarctic floras in all other lands.

Synopsis of Rhacomitrium.

Leaf rounded at the apex	.	.	.	.	.	311. *R. aciculare.*
Leaf bluntly pointed ; alar pockets pronounced		.	.	.	.	312. *R. nigroviride.*
Leaf acutely lanceolate	.	.	.	.	.	313. *R. crispulum.*
Leaf ovate below, acuminate and hyaline-pointed	.		.	.	314.	*R. drakensbergense.*
Leaf lanceolate, long hyaline-pointed, serrate and papillose	.	.	.	.	315.	*R. hypnoides.*

311. *Rhacomitrium aciculare* (Linn.), Brid., Mant., p. 80 ; Bryol. Univ., i, 219 ; Schimp., Syn., 2nd ed., p. 174 ; Dixon, 3rd ed., p. 161, Tab. 20 ; Broth., Pfl., x, 312.

   =*Bryum aciculare* Linn., Sp. Pl., p. 1118.
   =*Grimmia acicularis* C. M., Syn., p. 801.
   =*Rhacomitrium pseudoaciculare* (C. M.), Par., Suppl. Ind. (1900), and 2nd ed., p. 157 ; Broth., Pfl., x, 312.
   =*Grimmia pseudoacicularis* C. M., in Hedw., 1899, p. 120.

A strong-growing, roughly matted moss, often subaquatic ; stems 1-2 inches high, dichotomous, bare below ; leaves oblong, 2 mm. long, rounded at the broad, entire or almost entire apex ; nerve wide below, channelled, indistinct above, ending well below the apex ; margin recurved on one or both sides, thickened above. Cells very dense and quadrate-sinuose toward the leaf-apex ; longer downward, and at the base hyaline and cylindrical-sinuose, with a few oblong, shorter cells at the base of the marginal line. Seta ½ inch long, terminal or by innovation lateral ; capsule shortly cylindric, smooth ; lid subulate, long ; calyptra smooth ; peristome teeth irregularly 2-3-fid.

This is only recorded from South Africa once, viz. by Dregé (fertile), and was described by Müller as *Grimmia pseudo-acicularis*, but he discloses no points of difference except that the leaves are not secund while in the European plant they *occasionally* are. *R. aciculare* is widely distributed in the Northern Hemisphere, and to some extent is known south of that, so there is no reason to doubt the record. I have not seen the specimen, and above description and figure are from European plants. It is possible but unlikely that Dregé's specimen was *Hyophila atrovirens* Broth.

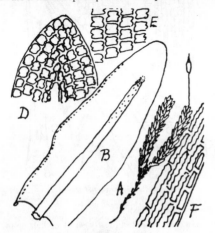

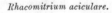

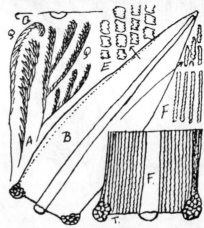

Rhacomitrium aciculare.   *Rhacomitrium nigroviride.*

312. *Rhacomitrium nigroviride* (C. M.), Par., 2nd ed., p. 155 ; Rehm. 139 ; Broth., Pfl., x, 313.
   =*Grimmia nigroviridis* C. M., Hedw., 1899, p. 121.

Laxly tufted or intermingled among other swamp mosses, dark green, the dominated portions black ; stems 1-2 inches high, ultimately leafless below, simple or variously branched, the points curved downward when dry ; leaves imbricate, usually secund, spreading when moist, appressed when dry, 2 mm. long, broadly ovate-oblong and narrowed gradually to the obtuse point ; nerve wide, strong, extending almost to the apex ; margin reflexed on one or both sides ; cells densely chlorophyllose, quadrate-sinuose at the apex, longer and more crenulate at mid-leaf, while toward the leaf-base the long, crenulate, hyaline lines show with no evident cross-divisions ; alar pockets very pronounced, with small round cells.

Capsule not seen, but Müller says : Dioecious ; perichaetal leaves similar ; capsule on a very short, erect seta, large, cylindrical, small-mouthed ; lid straight, rostrate.

Var. *robusticula* C. M. Larger and greener, leaves more mucronate and areolation longer. Devil's Peak, Rehm. 139c.

Cape, W. Table Mountain, Rehm. 139 ; mountain above Worcester, Rehm. 139β ; The Wilderness, George, April 1924, Miss Taylor.

313. *Rhacomitrium crispulum* Hook. f. and Wils., Fl. Tasm., p. 181, t. 173.
   =*Dryptodon crispulus* Hook. f. and W., Fl. Antarct., i, 57 ; Lond. Journ. Bot., 1844, p. 544.
   =*Grimmia crispula* C. M., Syn. Musc., i, 804.
   =*Rhacomitrium symphyodontum* (C. M.), Par., Suppl. Ind. (1900), and 2nd ed., p. 159.
   =*Rhacomitrium symphyodon*, Jaeg. Ad., i, 375 ; Mitt., Fl. Tasm., ii, 181, t. 173.
   =*Grimmia symphyodonta* C. M., Syn. Musc., i, 809.
   =*Rhacomitrium nigritum* (C. M.), Jaeg. Ad., i, 368.
   =*Grimmia nigrita* C. M., Syn. Musc., i, 801.
   =*Rhacomitrium austropatens* (C. M.), Broth. ; Dixon, Trans. R. Soc. S. Afr., viii, 3, 197 ; Broth., Pfl., x, 313.
   =*Grimmia austropatens* C. M., Hedw., 1899, p. 121.

Tufts lax, spreading, dark green, 1-3 inches deep ; stems eventually bare below, dichotomous, but also with many irregularly placed small branches ; leaves numerous, imbricate, spreading when moist, erecto-patent and often with reflexed tips when dry, narrowly oblong-acuminate, somewhat plicate ; margin revolute in the lower half ; apex acute, hyaline, but not piliferous ; nerve narrow, extending to the acute apex, not lamellate on the back ; cells quadrate and dense above, cylindrical crenulate and hyaline below, with a few oblong marginal cells at the base. Fruit not seen by me, but Dixon, in reference to Henderson's No. 240, says, " One or two old and imperfect capsules are present and a single immature one ; these show a very short, straight seta, only a few millimetres long, and a small elliptic capsule."

Widely distributed in Australasia, South America, and subantarctic lands.

Cape, W. Table Mountain, Rehm. 137 ; Blinkwater Ravine above Camps Bay, Dr. Bews (Sim 8633) ; Slongoli and Newlands Ravine, Sim 9193, 9427, 9433 ; Paarl Rock, Sim 9631.

Cape, E. In Trans. R. Soc. S. Afr., viii, 3, 197, Dixon cites Gaika's Kop, Tyumie, 6000 feet, 1916, Rev. Jas. Henderson 240 (fertile, as above).

I have not seen specimen or description of *R. capense* Ltz., Moost., p. 163, which Brotherus (Pfl., i, 3, 454, and x, 313) groups with *R. nigroviride* and *R. austropatens*, as having pointed but not hair-pointed leaves, and Paris calls it corticolous, but I have no doubt it belongs here.

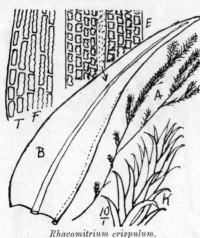

Rhacomitrium crispulum.

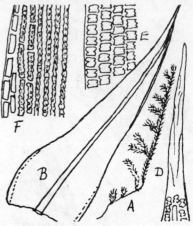

Rhacomitrium drakensbergense.

314. *Rhacomitrium drakensbergense* Sim (new species).*

Lax open cushions 1–3 inches high, subpinnately many-branched, the branches mostly short and irregular as to position ; leaves numerous, imbricate, entire, somewhat concave, ovate below, thence acuminate and ending usually in a smooth, hyaline, entire hair-point ; nerve narrow, channelled, extending to near the apex but not into the pilum ; margins reflexed toward the base ; cells densely quadrate, crenulate and nodulose toward the leaf-apex, longer downward to cylindrical, crenulate, papillose, and hyaline, with a few wider and shorter, oblong, marginal cells at the base, where the lamina is corrugated. Fruit not seen.

Known only from the top of the Giant's Castle, Natal. Drakensberg, 8000 feet altitude, collected in 1915 by R. E. Symons (Sim 8697). It was at first thought to belong to *R. crispulum* (see Dixon, Trans. R. Soc. S. Afr., viii, 3, 197, where the name *R. symphyodontum* is used), but fuller acquaintance with that species, as found in Cape Province, convinces me that they are distinct, the leaf-form, papillosity, hair-point, and branching being all different. On the many specimens of *R. crispulum* collected in the western province not one leaf is pilose ; on this species almost every leaf, from down the stem as well as from its summit, carries a hair-point, sometimes one-third the length of the leaf.

315. *Rhacomitrium hypnoides* (Linn.), Lindb. id Oefv. at K. Vet.-Akad. Förh., 1866, p. 552 ; Broth., Pfl., x, 314.

=*Bryum hypnoides* Linn., Sp. Pl. 1119, p.p.
=*Grimmia hypnoides* Lindb., M. Scand., p. 29.
=*Rhacomitrium lanuginosum* (Ehrh. ; Hedw.), Brid., Mant. M., p. 79, and Bryol. Univ., i, 215 ; Dixon, 3rd ed., p. 166, Tab. 21E.
=*Trichostomum lanuginosum* Hedw., Descr., iii, 3, t. 2.
=*Grimmia lanuginosa* C. M., in Syn., i, 806.
=*Rhacomitrium incanum* C. M., in Verh. d. K.K. zool. bot. Ges. in Wien, 1869, p. 224.

Our form is var. *pruinosum* (=*R. lanuginosum* var. *pruinosum* H. f. and W., Fl. N.Z. and Handbook, p. 427 = *R. pruinosum* C. M., in Verh. d. K.K. zool. bot. Ges. in Wien, 1869, p. 224).

---

* *Rhacomitrium drakensbergense* Sim (sp nov.).

Laxum habitu ; caules 1–3″ alti, subpinnatim pluri-ramosi, rami breves et irregulares ; folia imbricata, integra, concava, infra ovata, deinde acuminata, plerumque in aristam levem hyalinam integram desinentia ; nervus angustus, sub ipsum apicem desinens, margines basin versus reflexi ; cellae quadratae, crenulatae ac nodulosae folii apicem versus, deorsum longiores, cellae inferiores cylindricae, crenulato-papillosae, hyalinae ; cellae basales marginales paucae latiores. Fructum non vidi. Apud Giant's Castle, Natal (Sim 8697).

Growing in large, scattered, hoary mats in abundance on gritty soil on top of Table Mountain, Cape ; mats 1–3 inches deep, from lax to crowded ; stems 1–4 inches long, irregularly many-branched ; branches subpinnate or at the apex fastigiate, mostly short ; leaves crowded, spreading when moist, appressed when dry, lanceolate-acuminate, undulated at the base, 3–4 mm. long to end of hair, the lower part convex, the upper half flat and hyaline, usually crisped, and when dry much crisped ; nerve thin, broad, channelled, chlorophyllose to the diaphanous hair-point, often 2-ridged at mid-leaf on the back ; margin entire in the lower half of the leaf, strongly and irregularly serrate from there to the base of the hyaline-point ; the serratures and adjoining lamina as well as the hair-point densely papillose on both surfaces, with long, conical papillae ; the lower part of the leaf corrugated at every line of crenulated cells and papillose on the ridge on each surface ; cells obscure wherever chlorophyllose, but papillose on the ridges, all crenulated ; in the lower part of the leaf the long, hyaline, crenulate lines are corrugated and appear continuous without cross cell-divisions, and are papillose or nodulose on the ridges also ; and there is sometimes a short marginal line of oblong, wider, smooth cells. Capsule not found in South Africa, but said to be elsewhere oval or elliptic-oblong, on a short seta ; peristome teeth split to the base into long slender segments.

Widely distributed in northern and southern lands, into the colder regions ; in South Africa it is recorded only from Table Mountain, Cape, where it is abundant but sterile.

Rhacomitrium *Dregeanum Schpr. and *R.* *tomentosum Brid. (Dregé, V, *b*, p. 66) are both names appearing in Dregé, Zwei Pfl. Doc., 1843, as South African, which I cannot trace further.

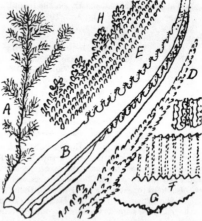

Rhacomitrium hypnoides.

Genus 89. PTYCHOMITRIUM (Bruch), Fürnr., in Flora, 1829 ; Ergänz, ii, 19.

In dense cushions on stones ; stems from short up to 2 inches long, branched below ; leaves numerous, evenly distributed, more or less lanceolate, acute, obtuse or cucullate, not hyaline-pointed nor papillose, entire (in our species), channelled, sometimes plicate toward the base, strongly nerved, the nerve rigid and protruding as a backbone in the strongly circinate incurvature when dry, pliable when moist ; cell-walls not or hardly sinuate ; lower cells long, narrow, and pellucid ; those toward the margin usually shorter, with a wider marginal line ; those in the upper part of the leaf small, opaque, roundly quadrate, in regular rows. Capsule erect on a long or short, straight seta, symmetrical ; calyptra widely campanulate, enclosing the whole or most of the capsule, more or less plicate, deeply lobed at the base. Peristome of sixteen teeth, usually split to near the base or more or less deeply into two papillose, red, hardly trabeculate segments, but sometimes nearly entire ; peristome rising inside the mouth and the hyaline ring. Monoecious.

This genus, distinguished in Grimmiaceae by its plicate and deeply lobed, bell-shaped calyptra, includes *Brachystelium* Rchb., and is more or less synonymous with *Glyphomitrium* Brid. in accordance with the views held by different authors. The characters separating these generic groups are unsatisfactory, and I prefer to use *Ptychomitrium* for all our species, which are, on the whole, closely related, and are also very difficult in sterile specimens to distinguish from *Trichostomum* and other Tortulaceae. The deeply lobed, campanulate calyptra covering most of the capsule is distinctive, and here the leaf-cells are smooth and the nerve ceasing at or below the apex, while in these others they are often papillose and the nerve percurrent or excurrent. Brotherus (Pfl., xi, 7) makes a family Ptychomitriaceae to hold this group, which he places next Erpodiaceae.

Synopsis of Ptychomitrium.

1. Leaves distinctly and roundly cucullate	316.	*P. cucullatifolium.*
1. Leaves not or hardly cucullate		2.
2. Basal margin widely recurved ; nerve filling the subula . .	317.	*P. depressum.*
2. Basal margin not recurved ; lamina cellulose . . .		3.
3. Margin unistratose		4.
3. Margin all or partly bistratose		6.
4. Upper cells minute, opaque	318.	*P. crispatum.*
4. Upper cells larger, distinct		5.
5. Leaf tapering gradually from near the base upward . .	319.	*P. crassinervium.*
5. Leaf suddenly narrowed above the wide base . . .	320.	*P. eurybasis.*
6. Seta short ; capsule oval	321.	*P. subcrispatum.*
6. Seta ⅓–½ inch long ; capsule cylindrical . . .	322.	*P. marginatum.*

Brachystelium convolutifolium Shaw (Cape Monthly Magazine, Dec. 1878) is insufficiently described and might belong anywhere in this genus, and as no specimens are available it must be dropped.

Dixon says (S.A. Journ. Sc., xviii, 316) *P. mucronatum* Schimp., C. M., in Hedw., xxxviii, 122, does not belong to this genus. Original specimens in herb. Schimper at Kew show it to be a *Trichostomum* or allied genus.

P. Krausei W. P. Sch., in Breutel, M. Cap. (=*Glyphomitrium Krausei* Par., Ind., 1895, p. 514) is an unpublished name. I have not seen the type.

316. *Ptychomitrium cucullatifolium* (C. M.), Jaeg. Ad., i, 382 ; Broth., Pfl., xi, 9.

=*Glyphomitrium cucullatifolium* (C. M.), Broth. ; Dixon, Trans. R. Soc. S. Afr., viii, 3, 196.
=*Brachysteleum crispatum* var. *brachycarpum* Horns., 1841, p. 126.
=*Brachysteleum cucullatifolium* C. M., Syn. Musc., i, 769.

Plant usually 1–2 cm. long, occasionally much stronger, branched by innovations from below the setae ; leaves numerous, 1–1·5 mm. long, erect or erecto-patent, short, lanceolate, with rounded cucullate apex, often mucronate, but hooded behind the mucro ; nerve strong, bent in at the hood, and extending to the mucro ; margin plane, entire, unistratose ; cells small, dense, quadrate ; lower cells oblong-hexagonal, six times as long as wide, hyaline. Perichaetal leaves not different or slightly longer. Seta short, erect, 2–3 mm. long ; capsule oval, crowned by a large hyaline annulus, and with short peristome rising inside the mouth ; teeth deeply split, very papillose. A larger form occurs on the Natal Drakensberg, up to 2 inches long, strong and vigorous, with the leaves rounded without mucro at the wide cucullate apex ; the cells are small and dense, and there is no border or thickening.

Widely distributed, mostly on the mountains.
Cape, W.   Camps Bay, Jan. 1919, Sim 9327.
Cape, E.   Wittebergen, 7500 feet, Dregé ; Graaff Reinet, MacLea (Rehm. 496) ; Wittebergen, Herschel, 1917, Hepburn (Sim 8708).
O.F.S.   Wittebergen, Rehm. 142*b* ; Wilde Als Kloof, Bloemfontein, Professor Potts.
Natal.   Little Berg, Cathkin, 6000 feet, Miss Owen ; Van Reenen, Wager ; Koodoo Pass, 8000 feet, Mont aux Sources, 7000 feet, and Tugela Valley, 7000 feet, Sim 9965, 9964, and 9966.
S. Rhodesia.   Matopos, 5000 feet, March 1902, Eyles 1051.

317. *Ptychomitrium depressum* (C. M.), Par., Suppl. Ind., 1900 ; Broth., Pfl., xi, 9.

=*Brachysteleum depressum* C. M., Hedw., 1899, p. 153.
=*Glyphomitrium depressum* (C. M.), Dixon, Trans. R. Soc. S. Afr., viii, 3, 196.

Although a Natal plant this has not been seen by me, but Müller's description indicates a distinct species about 1 inch high, in cushions, the leaf narrow, having long, narrow, *vaginate base*, with lanceolate-acuminate, firmly subulate, acute, entire, green lamina ; the margin below *very revolute* ; nerve thick, *percurrent and occupying the whole subula* ; upper cells minute, round, lower cells lax and narrow. Capsule erect, ovoid, on a very short, slender, red seta.

Jammerlappen, Natal, J. Dittrich, 1898.
Dixon, who records it from Hogsback, Tyumie, C.P. (D. Henderson 324, c.fr.), says, "The fruiting characters are in the main the same as those of *G. cucullatifolium* ; the calyptra is, however, much more longly beaked, as is also the capsule lid. The reflexed margin is also characteristic, an unusual feature in the genus." Had Dixon not seen it I would have had a doubt, from the marked characters, if it belonged to this genus.

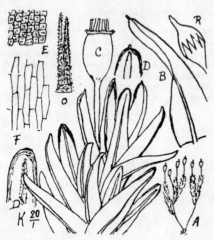 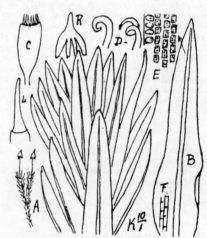

Ptychomitrium cucullatifolium.   *Ptychomitrium crispatum.*

318. *Ptychomitrium crispatum* (Hk. and Gr.), W. P. Sch., Syn., 2nd ed., p. 291 ; Dixon, S.A. Journ. of Sc., xviii, 315 ; Broth., Pfl., xi, 9.

=*Orthotrichum crispatum* Hk. and Gr., in Edinb. Journ. Sc. (1824), i, 115.
=*Grimmia crispata* Spreng., Syst. Veget., iv, 1, 154.
=*Glyphomitrium crispatum* Brid., Mant. M., p. 30 ; Dixon, Trans. R. Soc. S. Afr., viii, 3, 195 ; Bryhn, p. 12.
=*Encalypta crispata* Schwaegr., Suppl., i, 1, 60.

=*Brachysteleum crispatum* Horns., in Linn., 1841, p. 126 ; C. M., Syn. Musc., i, 768 ; Horns., in Endl. and Mart. Fl. Braz., i, 20.
=*Brachypodium crispatum* Brid., Bryol. Univ., i, 147 and 717.
=*Notarisia capensis* Hampe, in Linn., 1837, p. 379.
=*Ptychomitrium nigricans* Br. and Sch., Bryol. Eur., ii, and iii, 5.

Monoecious, densely caespitose, usually on stones, the cushions compact and rounded, $\frac{1}{2}$–2 inches wide ; stems more or less forked, $\frac{1}{3}$–1 inch long, usually fertile, and with old capsules at the forks or down the stem as well as terminal young capsules. Leaves numerous all along and all round the stem, erecto-patent or spreading, lanceolate from a wider or oval base, concave but not cucullate, entire, strongly nerved to near the apex, the nerve smooth on the back and more translucent than the lamina ; cells small, quadrate, and densely chlorophyllose in the upper portion, much larger and rectangular at the base, gradually larger between, and hyaline for one-third of the leaf-length ; the basal cells nearest the nerve often cylindrical, shorter outward, with a basal marginal line of wider cells. Leaves when dry very circinate with their inflexed margins ; when half-dry bent outward and then inward ; involucral leaves longer and narrower but with base wider. Seta 3–5 mm. long, erect ; capsule erect, oval, or shortly cylindrical, with long rostrate beak and short papillose peristome, the lanceolate teeth usually forked to near the base. Calyptra smooth campanulate, nearly always present, covering about half the capsule, many-furrowed and deeply many-lobed at the base, the lobes often separate to the beak. Peristome and lid red. Male flower axillary, bud-like, with three to four short, ovate-acuminate leaves and about ten antheridia. Usually the leaves are lanceolate and flat toward the apex, but occasionally they are wider and somewhat concave or subcucullate, with shortly acuminate, acute apex, in which condition they seem from description to be *B. obtusatum* C. M. (Hedw., 1899, p. 122), which Brotherus (Pfl., xi, 9) places with *P. cucullatifolium* (C. M.) Jaeg. in a cucullate group sufficiently distinct to merit separation.

The above description represents the more restricted species known as *P. crispatum*, but all the following species are so closely related to it that with *P. crispatum* as an aggregate name they could be included as subspecies or varieties in a slightly wider definition.

Frequent in every province, extending from the coast to the mountains, usually on doleritic or granite rock.

Cape, W. Abundant all over the Cape Peninsula and half up Table Mountain (Sim, many numbers), Rehm. 191. Tokai ; Kluitjes Kraal, Paarl, Paarl Rock, Stellenbosch, Sim 9503, 9973, 9604, 9974, 9578 ; Eerste River, Garside ; Camps Bay, Professor Bews ; Knysna, Miss Duthie.

Cape, E. Hankey, Jas. Sim ; Pirie, 1889, Sim 120 ; Grahamstown, Miss Britten ; Victoria East and Port Elizabeth, Miss Farquhar ; Kimberley, Miss Wilman.

Natal. Zwart Kop, 5000 feet ; Blinkwater ; Falls near Hilton College ; Albert Falls, 2500 feet ; Mont aux Sources, 8000 feet (Sim 9967, 9969, 9968, 9970, 9971) ; Imbezane, 50 feet, Eyles 1420 ; Little Berg, Cathkin, Miss Owen ; Eshowe, Bryhn.

Transvaal. Johannesburg, 5000 feet, Miss Edwards ; Lemona Wood, Spelonken, Junod.

O.F.S. Rydal Mount, Wager.

S. Rhodesia. Khami Ruins, 5000 feet, Sim 8865 and 9972 ; Great Zimbabwe Temple Ruins, Sim 8754, 8798, 8813 ; Matopos, F. Eyles 1098.

Brachypodium *flexum* Nees occurs in Drege's list next to *B. crispatum*. I cannot trace it.

319. *Ptychomitrium crassinervium*, Jaeg. Ad., i, 384 ; Dixon, Trans. R. Soc. S. Afr., viii, 3, 196 (as *Glyphomitrium*) ; Broth., Pfl., xi, 9.

=*Brachystelium crassinervium* C. M., in Hedw., 1899, p. 121.

A vigorous plant, growing in dense cushions on wet stones, usually in streams. Stems $\frac{3}{4}$–1$\frac{1}{2}$ inch long, simple or with one to two branches, with terminal and ultimately lateral setae, many-leaved ; leaves spreading or erecto-patent when moist, circinate when dry ; the strong nerve very marked as a backbone ; leaves 2–3 mm. long, lanceolate-acuminate, not specially widened toward the base, entire, acute, somewhat concave ; nerve strong, wide below, yellowish ; areolation lax ; cells in the upper half large, roundly irregular, 10–12μ wide, smooth, chlorophyllose but separate ; in the lower half cells are lax and hyaline, large, laxly and irregularly long-oblong, the walls thin but deep, making the cell like a rectangular box, the outer line of cells alone not showing that ; intermediate cells shortly oblong, separate, and pellucid ; margin not thickened. Perichaetal leaves up to 4 mm. long ; capsule on $\frac{1}{3}$-inch seta, erect, cylindric, tapering to the base, furrowed when dry ; lid with linear beak from a convex red base ; calyptra deeply cut, covering half the capsule or more at first ; peristome teeth rigidly erect, deeply 2-fid, not trabeculate, but with many longitudinal striae of papillae, which are abundant and scattered in the blunt red segments.

Cape, W. Kluitjes Kraal, Wolseley, 2500 feet, T. R. Sim.

Cape, E. Buffalo River and Gwacwaba River, King William's Town, Sim 7150 and 7111 ; Kokstad, Tyson.

S. Rhodesia. Zimbabwe, 3000 feet, Sim.

320. *Ptychomitrium eurybasis* Dixon, S.A. Journ. Sc., xviii, 315 ; Broth., Pfl., xi, 9.

Pulvinate ; stems $\frac{3}{4}$–1 inch long, not much branched ; leaves numerous, very circinate and rigid when dry, erecto-patent and pliable when moist ; base shortly very wide, concave, with erect or often undulate margins, and with the inner cells oblong or linear, thick-walled, and pellucid ; the upper half of the leaf quickly reduced to ligulate-acuminate, concave to the apex, with ascending entire margins, not thickened ; point acute, not cucullate ; nerve very wide, 80μ at the base, and extending to the apex. Cells of lamina large, 10–12μ diameter, very chlorophyllose, tumid but not papillose, unistratose ; seta 3–4 mm. long ; capsule small, nearly ovate, with erect or incurved peristome when moist,

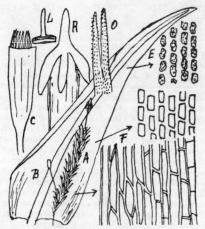

Ptychomitrium crassinervium.

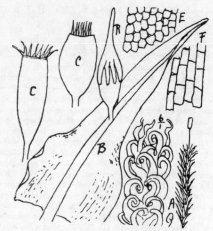

Ptychomitrium eurybasis.

elliptical with spreading peristome when dry ; lid shortly beaked ; calyptra longer than the capsule, covering the upper half of it, rostrate above and furrowed and lobed below ; the lobes incurved.

S. Rhodesia. Matopos, Sim 8851 ; Zimbabwe, 3000 feet, Sim 8808 ; on granite rocks, Macheke, 5000 feet, Eyles 1994.

321. *Ptychomitrium subcrispatum* Theriot and P. de la Varde, Rev. Gen. de Botanique, xxx, 65 ;
Dixon, Trans. R. Soc. S. Afr., viii, 3, 196 ; Broth., Pfl., xi, 9.

A vigorous plant, 1–1½ inch high, growing in cushions on moist stones ; stems considerably branched, with old capsules remaining for years down the stem, giving the appearance of lateral fructification ; leaves rather lax, 2 mm. long, lanceolate, concave, strongly keeled by the nerve ; apex acute ; cells lax and fairly large, roundly quadrate, almost pellucid except at the margin where two to four lines of cells width is 2-stratose for almost the whole length of the leaf, giving a dark green border. Basal cells all hyaline, the inner long-oblong to cylindrical, gradually shorter outward, the outer line wider and clearer. Seta 2–3 mm. long, frequently two from the same perichaetium ; capsule shortly oval ; peristome teeth densely papillose ; lid rostrate ; calyptra lobed at the base.

Natal. Weenen County Drakensberg, 1898, J. M. Wood 7123*b* ; Van Reenen, Goodoo Pass, and Ladysmith, Wager 714 and 681 (*fide* Dixon).

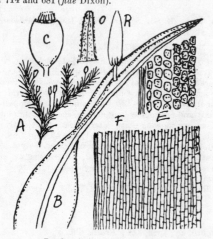

Ptychomitrium subcrispatum.

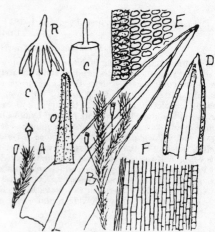

Ptychomitrium marginatum.

322. *Ptychomitrium marginatum* Wag. and Dixon, in Trans. R. Soc. S. Afr., viii, 3, 196, pl. xii, fig. 23 ;
Broth., Pfl., xi, 9.

Vigorous, in loose, dark green cushions on moist stones ; stems 1–2 inches high, simple or branched, many-leaved ; leaves erecto-patent, 2 mm. long, lanceolate, somewhat concave, consequently subcucullate, shortly acuminate, the margins entire and several-stratose, forming a dark border, the apical portion of leaf also sometimes 2-stratose. Nerve strong through the whole leaf ; cells large, pellucid, smooth, transversely elliptical-quadrate ; basal cells long-oblong, pellucid, with a wider quadrate marginal line. Seta ⅓–½ inch long, mostly lateral ; capsule narrowly cylindrical, narrowed to the base, somewhat furrowed when dry. Peristome teeth strong, red, densely papillose ; lid an erect beak from a wide flat base ; calyptra deeply many-lobed.

P. suberispatum comes very near this and may eventually be merged after further specimens have been seen.

Cape, W. Riversdale, 1200 feet, Dr. Muir 3718 ; Bak River, 1925 feet, K. H. Barnard.

Cape, E. Springs and Hell's Gate, Uitenhage, Sim 9036, 9044, 9005 ; Hogsback, Tyumie, 4000–6000 feet, Rev. Jas. Henderson (*fide* Dixon).

Natal. Cathkin, 6000 feet ; Zwart Kop, 4000 feet ; Karkloof ; Hilton Road ; St. Ives ; Mont aux Sources ; Polela, Sim 9975, 9407, 9976, 9977, 9978, 9979, 9980 ; Muden, T. R. Sim.

Transvaal. Kaapsche Hoek, 6000 feet, Wager 298 (*fide* Dixon).

S. Rhodesia. Makoni, Forest Hill Kop, 5000 feet, F. Eyles 838.

Family 25. TORTULACEAE.

Mostly small erect plants, gregarious in wide patches or more or less tufted, usually growing on soil, frequently on rocks, and occasionally on tree-bark. Leaves all round the stem, often rosulate, varying from linear or lanceolate to broadly spathulate, usually strongly nerved, and the nerve frequently excurrent as a mucro or as a long slender hair-point, even from a rounded leaf-apex. Cells usually more or less quadrate, the lower cells pellucid, rectangular, and thin-walled, those in the lamina small, obscure, and usually minutely papillose. Seta straight, capsule erect or almost erect, terminal, oval to cylindrical, symmetrical. Calyptra cucullate or conical, narrow, smooth ; ring sometimes present ; peristome absent in the lower forms, rudimentary or inconstant in some, usually present, and of sixteen slender, papillose, straight or spirally twisted teeth, sometimes connate and tubular toward the base : in others having a short basal membrane, and often split to the base into thirty-two more or less equal segments or teeth. Spores large, often granulose. Perichaetal leaves often sheathing.

This family corresponds in part to the Pottiaceae of Schimper's Synopsis, which, however, excludes *Weisia* and some other genera here included. It corresponds with Brotherus' Pottiaceae, in Nat. Pflanz., p. 380, except that Cinclidoteae and Encalypteae are there included, of which the former is not South African.

A large family, having a common habit, but including wide variations of character, especially in regard to the peristome, which in some species is inconstant or caducous, and thus connects the peristomate with the eperistomate forms.

Sections of the Tortulaceae.

1. POTTIEAE. Leaves rather wide, ovate to spathulate-lanceolate ; cells seldom minute (for genera, see below).
2. TRICHOSTOMEAE. Leaves narrow (lanceolate or narrower) ; lamina leaf-cells minute (for genera, see further on.)

§ 1. POTTIEAE. Annual or perennial, usually gregarious, often rosulate ; leaves wide, ovate to spathulate-lanceolate: upper areolation rather lax and pellucid, or smaller and more obscure, often more or less papillose, seldom minute. Peristome absent or present ; when present consisting of sixteen teeth, entire or forked or cleft to the base (then apparently thirty-two), straight or spirally twisted and often united at the base, sometimes connate and tubular below, or up to half-way up.

Genera of the Pottieae.

Capsule cleistocarpous (*i.e.* bursting irregularly).

90. SPHAERANGIUM. Capsule globose, hardly pointed, included ; calyptra conical, with torn base.

91. PHASCUM. Capsule ovate or ovate-oblong, mucronate, included or exserted ; calyptra cucullate.

Capsule either gymnostomous or peristomate.

Peristome teeth when present straight, rather imperfect.

92. POTTIA. Leaf-margins more or less flat or reflexed ; areolation very dense.

Leaf-margins ascending or involute, especially when dry ; areolation rather lax.

93. PTERYGONEURUM. Nerve lamellate.

Nerve not lamellate.

94. HYOPHILA. Capsule thick-walled, without peristome.

95. WEISIOPSIS. Capsule thin-walled, peristome of sixteen teeth.

Peristome teeth usually twisted and of thirty-two equal linear segments.

96. TORTULA. Peristome teeth more or less connate below, sometimes half-way.

97. ALOINA. Peristome teeth free on a very short basal membrane ; nerve of the leaf bearing granulose threads on its upper surface.

Genus 90. SPHAERANGIUM W. P. Sch., Syn., 1st ed., p. 12.

Plants minute, bud-like, gregarious on moist soil, forming protenema as well as rhizoids. Lower leaves very small, the few inner leaves very much larger, convolute-imbricate, inflated, ovate-oblong, nerved, minutely papillose, and sometimes denticulate ; cells large, sparingly chlorophyllose, quadrate. Monoecious, the male flower on a basal branch. Capsule included and rolled in the upper leaves, cleistocarpous, spherical, hardly pointed. Calyptra erect, conical, with torn base. Spores subglobose, rather large, faintly granulated.

323. *Sphaerangium muticum* C. M.

Müller records (Syn. Musc., i, 22) under *Acaulon muticum* C. M. that he has received that species along with *Astomum (Pleuridium) Pappeanum* (which was from Swellendam). *A. muticum* is well described and illustrated in Dixon's Handbook, 3rd ed., p. 174 and Tab. 22C, as a minute acaulescent plant having only a few rosulate leaves of which the inner two are larger, convolute, not keeled, enclosing and covering the capsule; with distinct nerve excurrent as a short mucro; margin plane; cells smooth, the basal cells larger and hyaline; capsule spherical or with a minute obtuse apiculus; calyptra small, torn. Male flowers on a basal branch or basal radicles.

Brotherus, however (Pfl., x, 284), separates several species from *A. muticum* on account of their having the perichaetal leaves flat, with entire margin, and includes among these *A. capense* C. M. from South Africa. This is *Sphaerangium capense* (C. M.), Jaeg. Ad., i, 184 = *Acaulon capense* C. M., Bot. Zeit., 1856, p. 415, and is evidently included in Müller's record of *A. muticum* mentioned above. I have seen no specimen, nor does any appear to have been collected or distributed since, except that Dr. Shaw in his list mentions the above and then describes as follows in English what he calls a new species, viz. "*Acaulon *sphaericum* Shaw. Monoicous; leaves lanceolate, subulate, serrated toward the apex, base of the leaf papillose and very loosely cellular. Capsule spherical with a slightly conical summit; calyptra dimidiate. Discovered by MacLea near Graaff-Reinet, 1872; fruiting in July."

In the absence of any specimens *S. muticum* C. M. may be held to cover what is known meantime.

Genus 91. Phascum (Linn. p.p.) Schreb.

Small, gregarious, perennial, earth-grown plants, not forming protenema; stem simple or branched; leaves ovate-lanceolate, acute, entire, with strong excurrent nerve. Areolation laxly rectangular and hyaline at the base, densely quadrate and green upward, sometimes minutely papillose. Inflorescence monoecious, male flowers axillary or basal, bud-like or naked. Calyptra cucullate; capsule roundly ovate or ovate-oblong, with a distinct point, included or shortly exserted, having a short seta, cleistocarpous, and often red or brown at the apex. Columella persistent, spores small. Slightly larger plants than the last genus, smaller than the next, the small red capsules being often evident though hardly exserted.

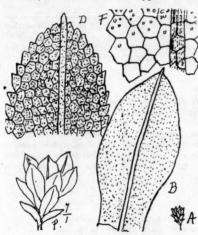

Phascum leptophyllum.

324. *Phascum leptophyllum* C. M., in Flora, 1885, No. 1; Dixon, S.A. Journ. Sc., xviii, 315; Broth., Pfl., x, 284 (§ *Leptophascum*).

Plants gregarious, 2–5 mm. high, few-leaved, the lower leaves scattered and small, the upper rosulate and about 1 mm. long, nearly ½ mm. wide, spathulate, narrow below, wider above mid-leaf, and rounded or tapering at the subacute mucronate apex, green, with numerous chlorophyll granules, tumid; the marginal cells pointed outward toward the leaf-apex, making the leaf denticulate, and crenulate where they are rounder lower down, the rounded more or less pellucid and thin-walled lamina cells about 30µ diameter, the basal ones more quadrate-hexagonal and rather larger, but chlorophyllose almost to the very base; old plants, however, are entirely pellucid. Nerve narrow, fairly strong, green or rusty, covered with lamina cells on the upper surface, and percurrent or slightly excurrent as a somewhat reflexed mucro; leaf nearly flat, convex when dry, with plane or suberect margin, sometimes slightly reflexed toward the base. Other parts not seen.

S. Rhodesia, Zimbabwe, 3000 feet, Sim 8711; Matopos, 5000 feet, Sim 8857, both sterile. Dixon gives the general distribution as Cape Province, Transvaal, and Rhodesia.

Phascum cuspidatum Schreb. (de Phasco., p. 8, t. 1); C. M., Syn., i, 25; Dixon's Handbook, 3rd ed., p. 176, and Tab. 22E, is stated by Müller to grow as an annual on clay soil over the whole temperate world, but I have no record of its having been found in South Africa. It is a minute autoecious moss, with the capsule immersed or almost so, the leaf oval-lanceolate, concave, entire, cuspidate or piliferous, and the leaf-margin often reflexed.

325. *Phascum splachnoides* Horns. (Hor. Phys. Berol., p. 57, t. 12); C. M., Syn., i, 27; Schwaegr., Suppl., iii, t. 203 = *Physedium splachnoides* Brid., p. 151; C. M., in Bot. Zeit., 1847, p. 102, is not known to me. Müller's description is: " Dioecious, caespitose, low, subsimple; leaves oval, wide; nerve thick, shortly cuspidate, subobtuse, crowded, very concave, sometimes glaucous; upper cells large, subquadrate, lower hexagonal-rectangular; capsule oval, regular, erect, brown, shortly apiculate, obtuse. Calyptra dimidiate, yellowish, apex brown."

Localities given are: Kimberley (Bergius); Cape (Ecklon). Fruits, July.

Brotherus places it as *Pottia splachnoides* (Horns.), Broth., Pfl., x, 290, in *Pottia*, § *Pottiella*, which has seta as long as or a little longer than the capsule. Capsule without throat, with small shortly pointed or obliquely beaked permanent lid.

Phascum peraristatum C. M., in Flora, 1888, No. 1, is known to me by name only, but is included by Brotherus (Pfl., x, 284) in § *Euphascum*, having entire, pointed leaves and cucullate calyptra on an included, almost sessile capsule.

Genus 92. POTTIA Ehrh.

Small, erect, annual or biennial, mostly unbranched plants with rosulate foliage ; leaves softly succulent, wide, oblong-obovate, oblong or spathulate-acuminate, papillose or smooth, nerved usually to or beyond the apex. Areolation dense, quadrate and small ; that toward the base lax, rectangular, and pellucid. Calyptra cucullate ; capsule oval or shortly cylindric, erect, exserted on a short seta ; lid with or without a beak, rough or smooth, sometimes attached to the columella and not readily shed, or taking the columella with it. Capsule usually without peristome (*gymnostomous*), or with a more or less imperfect peristome of sixteen more or less imperfect forked teeth, or sometimes cleistocarpous. Spores granulated.

Synopsis.

326. *P. afrophaea.* Leaves ligulate-acuminate.
327. *P. Macowaniana.* Leaves ovate.

326. *Pottia afrophaea* C. M., Hedw., pp. 38, 97.

=*Trichostomum afrophaeum*, Rehm. 120 and 473.
=*Hyophila afrophaea* (Rehm.), Warnst. ; Broth., Pfl., x, 270.

Caespitose, minute, dark green ; stems up to ½ inch long ; lower leaves lax and small, upper leaves imbricate, with a comal tuft of ascending ligulate-acuminate leaves 1 mm. long, clasping at the base, concave above, with short, dense, subacute, mucronate apex ; the nerve fairly strong, often red, ending in the mucro ; margin flat or suberect, entire ; cells minute, quadrate, 6μ wide, densely chlorophyllose ; the marginal line often semipellucid or less chlorophyllose ; the basal cells elliptic-hexagonal, hyaline , seta terminal, 3–6 mm. long ; capsule small, oval, erect, yellowish, gymnostomous ; lid shortly and subobliquely rostrate.

O.F.S. Bethlehem, Rehm. 120 (as *Trichostomum afrophaeum* Rehm.).
Transvaal. MacLea, Rehm. 473 (as *Trichostomum afrophaeum* Rehm.).

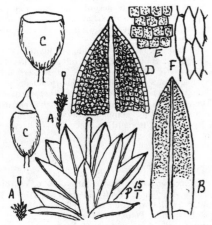

Pottia afrophaea.

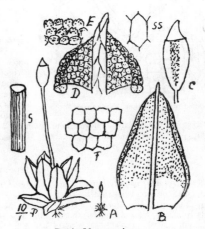

Pottia Macowaniana.
S, part of seta ; SS, cells of capsule.

327. *Pottia Macowaniana* C. M., Hedw., 1899, p. 98.

A minute, gregarious, monoecious, deep green or brownish, almost acaulescent moss, usually simple or with a small basal male branch ; leaves rosulate, eight to twelve, erecto-patent, the leaves 1 mm. long or less, broadly ovate, wide above the base and rounded at the apex, except that the nerve is excurrent as a strong, red mucro, or short bristle, the rest of the nerve pale or light red ; margin very evidently revolute from near the base almost to the mucro ; cells quadrate, 4μ diameter, minutely but evidently 2- or many-papillose and chlorophyllose, lower cells rather longer and mostly hyaline ; perichaetal leaves not differentiated. Seta pale, 2–4 mm. long ; capsule erect, elliptical, gymnostomous, with hexagonal cells 10μ long ; lid shortly conical, slightly oblique. Calyptra short, cylindrical, cucullate.

Cape, W. Aughrabies Falls, Aug. 1925, K. H. Barnard.
Cape, E. Klyn Fisch River, 1878, Macowan ; Kowie, 1912, Wager.
Brotherus (Pfl., x, 291) includes *P. *verrucosa*, Rehm. 460, from the Eastern Cape, in § *Eupottia*, having exserted eperistomate capsule with deciduous lid, long-keeled spreading leaves with reflexed margin, papillose cells, shortly conical, erect lid, and papillose spores. I have no further information concerning it, and there is no specimen in South African herbaria.

Paris cites Rehm. 460 for *P. Macowaniana* C. M. in error for 461.

P. eustoma Ehrh., stated in Shaw's list to have been collected near Graaff Reinet by MacLea, is probably *P. Macowaniana*, which differs from *P. eustoma* Ehrh. (*P. truncatula* Lindb. of Europe) in the leaf-form, the revolute margin, papillose cells, longer capsule, etc.

Genus 93. PTERYGONEURUM Jur. Laubm., 1882, p. 95.

Minute, autoecious mosses, with very concave, ovate-acute leaves, having the excurrent nerve widened and often lamellate on the upper half inside ; cells translucent, tumid, papillose below when dry. Capsule erect, oval, regular, on short erect seta. Peristome absent (in our species) ; spores large ; lid shortly beaked.

328. *Pterygoneurum Macleanum* (Rehm.), Warnst., in Hedw., 1916, pp. 58, 69 ; Broth., Pfl., x, 292.

=*Pottia* *Macleana Rehm., p. 461 (ined.).

A very small, caespitose, autoecious, dark green moss, growing in dry country ; stems 2–6 mm. long, rosulate, or when dry and the leaves infolded almost globose, but sometimes producing successive whorls ; leaves imbricate,

Pterygoneurum Macleanum.

subtubular, cucullate, suberect, 1 mm. long, exceedingly concave, the margins entire and inflexed but not involute ; the nerve narrow below and wide above, excurrent as a mucro, and bearing two wide lamellae on the upper surface of the wide portion on many leaves but not on every leaf. Cells transversely oblong, 20μ long, nearly pellucid, smooth or very minutely papillose ; basal cells laxly and longitudinally oblong, 25μ long, hyaline. Perichaetal leaves not different ; seta 2–4 mm. long, yellowish, with short, gymnostomous, oval capsule, very rugose when dry, narrow-mouthed, and with short rostellate lid, having its cells in straight rows and no trace of peristome. Very closely allied to the European *Pterygoneurum cavifolium* (Ehrh.) Jur. (=*Pottia cavifolia* Ehrh.=*Tortula pusilla* Mitt.), but Warnstorf, in Hedwigia, lviii, 69, maintains it as a good species, having leaf-cells 15–25μ diameter, while those of *Pterygoneurum cavifolium* are, according to Limpricht, about 12μ, and the leaves as figured by Warnstorf are without the hair-point characteristic of *P. cavifolium*. As it is known from only one gathering further observation is necessary.

These and some allied more or less lamelliferous plants have been placed in many genera and are unsatisfactory in each. The lax cell-formation differs from *Pottia* and *Gymnostomum* ; from *Tortula* this is excluded by being eperistomate though some closely allied European species have rudimentary peristome, and even in *Hyophila* the small size is unusual. The widened nerve recalls *Aloina*, which is peristomate and does not have lamellae but has filaments instead.

Cape, W. Klein Karroo, Riversdale, Dr. Muir 3717B ; Bak River, K. H. Barnard.
Cape, E. Open ground, Graaff Reinet, MacLea, Rehm. 461.

Genus 94. HYOPHILA Brid., Bryol. Univ., i, 760.

Slender plants, usually tufted ; stems often forked, leafy ; leaves spreading, linear, lanceolate or spathulate ; margin entire or somewhat toothed toward the apex, plane, ascending or involute ; firmly involute when dry. Nerve strong, ceasing in or near the apex or in some species excurrent. Areolation rather lax ; cells roundly quadrate, slightly papillose, seldom smooth, those at the base rectangular and hyaline. Perichaetal leaves often smaller. Seta long, slender, erect. Capsule erect, narrowly cylindric or sometimes oval. Ring dehiscent. Peristome imperfect or absent, seldom present. Lid with a slender beak, calyptra cucullate. Dioecious or autoecious.

Hyophila is now divided into two genera, viz. :
(A) *Hyophila* without peristome ; (B) *Weisiopsis*, peristomate.

Synopsis of Hyophila (§ A).

1. Nerve excurrent as a mucro	2.
1. Nerve not excurrent as a mucro	3.
2. Margin plane (when moist)	329. *H. Zeyheri.*
2. Margin strongly involute. Plant ½ inch high or more ; capsule cylindrical	330. *H. cyathiformis.*
3. Leaves oblong-spathulate, obtuse	4.
3. Leaves lingulate	5.
4. Leaves often dentate toward apex	331. *H. atrovirens.*
4. Margin entire	332. *H. baginsensis.*
5. Some marginal cells protruding	333. *H. erosa.*
5. Margin entire	334. *H. basutensis.*

329. *Hyophila Zeyheri* (Hpe.), Jaeg. Ad. i, 201 ; Dixon, in Trans. R. Soc. S. Afr., viii, 3, 193.

=*Hymenostomum Zeyheri* (Hampe, of Wager's list) ; Brunnthaler, i, 25.
=*Gymnostomum euchlorum* Zeyh., M. Cap., No. 480.
=*Pottia Zeyheri* (Hpe., in litt.) C. M., Syn. Musc., i, 561.
=*Weisia capensis* Spreng., in sched.
=*Trichostomum *oocarpum*, Rehm. 117 (Kadziberg, O.F.S.).
=*Hymenostomum opacum* Wager, Trans. R. Soc. S. Afr., iv, 1 (1914), p. 3, pl. 1, *e* ; Broth., Pfl., x, 254.
=*Weisia (Hymenostomum) brachycarpa*, of Wager's distribution, but not of C. M.
=*Hymenostomum (Weisia) *hyophiloides* Wag. and Br., M.S. ; and Wager's list.
=*Trichostomum *eperistomatum* Dixon M.S., in sched.

Caespitose, forming large tufts or carpets inches to yards across, on compact soil, turf banks, etc., fully exposed to light ; bright green when moist and expanded, dark green below ; leaves conduplicate, involute and circinate when dry, flat when moist, but taking a long time to unroll. Plants standing apart, usually simple and 3–6 mm. high, occasionally forked or tufted and up to ½ inch or even an inch high ; innovations when present starting immediately below the base of the seta. Leaves rather sparse, 2–3 mm. long, lingulate, tapering slightly to both ends but widest one-third up, more or less keeled (especially when dry), and curved quickly to the mucronate point. Old leaves flat, rigid, usually appressed and imbricate ; fresh leaves spreading, flat or channelled, and crowded. Nerve very strong, forming a keel, yellowish, smooth on the back, excurrent as a subulate, strong, hyaline or yellowish point or mucro, and often incurved in the upper lamina, making it somewhat hooded. Margin flat, entire, often somewhat undulated, never revolute, but often more or less involute especially when dry. Cells minute and dense, quadrate, 10µ wide, those at the base of the leaf long or oblong, lax, rectangular, pellucid, and up to 50µ long and 15µ wide. Seta terminal, 8–12 mm. long, much twisted when dry, erect, straight, yellowish ; capsule erect, oval-oblong, yellowish, sulcate when dry, gymnostomous, when dry microstomous ; lid rostrate, suboblique, from a wide base ; calyptra cucullate, beaked, often fixed horizontally on the beak of the lid. Peristome absent ; spores minute. Perichaetal leaves more lanceolate than the others. It varies considerably, and occasionally has the keel of the nerve papillose (Sim 8725).

Regarding the generic position of this species, Dixon says (p. 193), " The vegetative characters are those of *Hyophila*, while the short oval capsule which may have the mouth closed on de-operculation with a membrane suggests *Hymenostomum*."

Sim 10,021 grows on wet banks of the Mooi River ; usually it is on dry banks. The most common moss on sparse grass-lands in Eastern Cape, Natal, Transvaal, extending to Portuguese East Africa and South Rhodesia, Sim 9044, 7348, 8574, 198, 7334, 10,021, etc.

Var. *lanceolata* (=*Trichostomum oocarpum* Rehm. var. *lanceolatum*). Leaves more lanceolate.
Pilgrims Rest, MacLea (Rehm. 474) ; Müller's Farm, MacLea, Rehm. 474*b*.

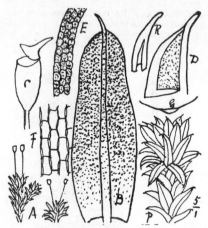

Hyophila Zeyheri.

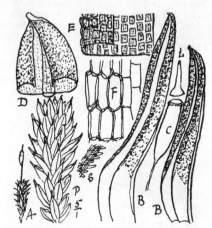

Hyophila cyathiformis.
S, mamillae on mid-keel.

330. *Hyophila cyathiformis* (Dixon) Sim.

=*Trichostomum cyathiforme*, Dixon, S.A. Journ. Sc., xviii, 310.

Caespitose, dark green ; stems 6–12 mm. high ; leaves numerous, rather lax below, closely imbricate above, involute and circinate when dry ; the short, wide, hyaline base sheathing, thence the leaf erecto-patent, apparently lanceolate but always so involute and boat-shaped along the whole lamina that the true form is not seen ; these involute margins meeting in the rather incurved nerve-apex so that the leaf is cucullate and slightly mucronate. Nerve strong, 120µ wide at the base, 60–80µ wide near the apex, percurrent, sometimes densely papillose but sometimes mostly smooth or often with a few large white mamillae (S) about the mid-keel ; cells separate, quadrate, 8μ wide, very densely chlorophyllose, smooth or minutely papillose toward the leaf-apex, but papillose on upper surface about mid-leaf or lower ; changing quickly where the sheath begins to very laxly hexagonal, thin-walled cells 60×20μ, those nearest the nerve rather wider and rectangular, and the outer cells narrower, and ascending up the margin some distance in several rows. Capsule cylindrical, gymnostomous, 1·5–2 mm. long, when young wide-mouthed and constricted below the mouth. Lid with long subulate beak.

S. Rhodesia. Victoria Falls, 3000 feet, Sim 8934 ; Salisbury Commonage, Feb. 1917, 4900 feet, F. Eyles, 684 p.p., fertile (along with *Fabronia*, also fertile) ; Matopos, 5000 feet, Madme. J. Borle 40.

Cape, E. Queenstown, Miss E. A. Graham.

Natal. Muden, 2000 feet, Oct. 1919, Sim 10,094, 10,093.

Transvaal. Melville, Johannesburg, 5000 feet, Sim 10,044 ; Pretoria, G. H. Gray.

331. *Hyophila atrovirens* (C. M.), Jaeg. ; Broth., Pfl., i, 3, 403, and x, 270 ;
Dixon, Trans. R. Soc. S. Afr., viii, 3, 193

=*Trichostomum atrovirens* C. M., Hedw., 1899, p. 100 ; Rehm. 119 and 475 (not 495 as given in Paris' Index).
=*Trichostomum *ruparium*, Rehm. 121.
=*Trichostomum *dentatum*, Rehm. 468 and 468b (*fide* Dixon).

Caespitose, dioecious, usually growing on stones in or close to stream-water ; stems ¼–1 inch in length, usually interruptedly rosulate, or more or less evenly leaf-clad, with terminal seta and hardly changed perichaetal leaves. Leaves 2–2·5 mm. long, spathulate, brown, dense, very obtuse or rounded at the apex, sometimes subacute, narrowed toward the base, clasping but not sheathing, spreading or erecto-patent, concave in the lower half, nearly flat in the upper half, or when dry with slightly inflexed sides and circinate bending. Nerve strong, usually red, ending in the apex but not mucronate, covered with lamina cells, smooth on the back ; margin plane, not involute though inflexed when dry, entire toward the base, often more or less dentate toward the apex ; cells minute, 8μ wide, quadrate, smooth, dense, chlorophyllose, separate ; basal cells few, hexagonal, 20×10μ, hyaline. Seta ½–¾ inch long, terminal, pale ; capsule 1·5 mm. long, cylindrical, brown, with no peristome, distinct white ring and erect rostellate lid. Lower leaves usually the shorter, wider and subacute. This species varies in vigour, etc., and in dentation very considerably ; sometimes the margin is almost entire ; indeed Müller described the margin as entire, though usually it is distinctly toothed.

Present in nearly every stream in eastern South Africa, from Port Elizabeth to Portuguese East Africa (Sim 8991, 8992, 8994) and Southern Rhodesia (Sim 10,071), seldom fertile in water, but frequently so on adjoining flat rocks.

Hyophila per-robusta Broth. (Denkschr. der Math. Nat. K.K. Akad., 1913, p. 25 ; and Pfl., x, 270) is recorded sterile as from Victoria Falls, on tree trunks, coll. Brunnthaler (Dixon, S.A. Journ. Sc., xviii, 312). Brotherus describes the leaf as spathulate-oblong, "subrotundata-obtusa mutica vel brevissima mucronata, circa 3 mm. longa et 0·9 mm. lata." Stem is only 1 cm. long and it seems to me to be *H. atrovirens*.

I have a bark-grown *Hyophila* from Victoria Falls Knife-edge (Sim 8924), which is rather narrower-leaved than is usual in *H. atrovirens*, and with the margin almost entire, but I cannot separate that specifically from *H. atrovirens*.

Pottia per-robusta C. M., Hedw., 1898, p. 233, is, so far as I know, a bark-grown, eperistomate, North-American *Pottia*, and different from this.

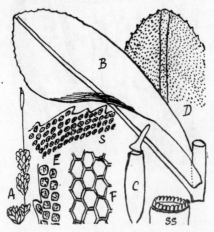

Hyophila atrovirens.

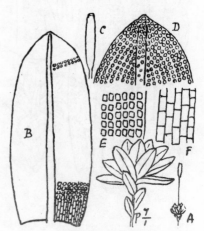

Hyophila baginsensis.

332. *Hyophila baginsensis* C. M., Musci. Schweinfurthiani, Linnaea, 1875, p. 399 (sterile) ;
Dixon, S.A. Journ. Sc., xviii, 312 ; Broth., Pfl., x, 270.

Laxly caespitose, grey-green ; stems 4–5 mm. long, nearly leafless below ; leaves rosulate above, oblong-spathulate, 1–1·5 mm. long, rounded to the almost blunt apex, flat or slightly concave ; the margin plane and entire ; cells quadrate, smooth (tumid or subpapillose when dry), separate, chlorophyllose, 10μ diameter, alike almost to the leaf-base where a small area has hyaline cells $20 \times 10\mu$, with a border line of shorter square cells. Nerve strong, ending just below the apex, the upper surface of the nerve covered with lamina cells. Perichaetal leaves not differentiated. Seta ½ inch long ; capsule dark, cylindrical, narrowed to the mouth. No peristome was seen but only old capsules were available. Leaves involute and more tumid and opaque when dry.

S. Rhodesia. Zimbabwe, 3000 feet, Sim 8831 (fertile) ; Rhodes Grave, Matopas. 5000 feet, Sim 8868 ; Victoria Falls, 3000 feet, Sim 8880 and 8883.

Apparently from Dixon's remarks this is near *H. Potierii* Besch. (Fl. Bryol. Reunion, p. 341 ; Renauld, Fl. Bryol. Madag., p. 121), belonging to East Africa, Madagascar, etc., though not yet recorded from South Africa, which is described as having narrower, strongly mucronate, subcrenulate leaves and narrowly cylindrical, gymnostomous, annulate capsule.

333. *Hyophila erosa* Sim (new species).*
=*Trichostomum *erosum*, Rehm. 472.

Loosely caespitose : stems 10–25 mm. high, robust, branched by innovations below the seta which thereby appears lateral a year afterwards. Leaves 1·5 mm. long, lingulate, slightly widened at the base, rounded at the apex, somewhat concave upward ; margin plane or slightly flattened outward on the inflexed basal sides ; nerve very strong, disappearing below the apex, smooth on the keel, covered with lamina cells in front ; cells small, quadrate, minutely papillose on the upper surface, tumid on the lower surface, very dense, the marginal cells irregular, especially in the upper half of the leaf, about every second or third cell projecting and giving an erose appearance. Leaves keeled and twisted when dry. Seta 1 cm. long, pale ; capsule 1·5 mm. long, brown, oval-oblong, with a short conical lid. Ring present. Peristome apparently absent or quite rudimentary, but may have been broken off, as all the capsules seen were old. (See *Oreoweisia erosa* (Hpe.), Par., which is apparently different.)

Natal. Müller's Farm, Drakensberg, MacLea (Rehm. 472), c.fr. ; Greytown, 2000 feet, Sim 10,058.

O.F.S. Leribe, Basutoland, Mme Dieterlin, c.fr.

Transvaal. Johannesburg, 5000 feet, Sim 10,059.

Much like *H. atrovirens* in general appearance, but the leaves are narrower, more acute, crenulate instead of dentate ; cells more tumid, and capsule different.

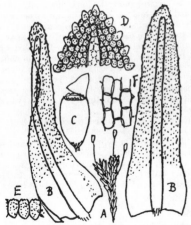

Hyophila erosa.

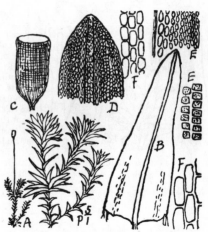

Hyophila basutensis.

334. *Hyophila basutensis.* Rehm. 458, Sim.†
=*Barbula *basutensis*, Rehm. 100.

Closely caespitose, and known specimens are abundantly fertile ; stems ½–¾ inch high, branched, the branches starting as strong red innovations, with small scattered leaves in the lower part, and imbricate, almost rosulate leaves

---

* *Hyophila erosa* Sim (sp. nov.)=*Trichostomum erosum* Rehm. (ined.).

Laxe caespitosa ; caules 10–25 mm. alti, ad apicem innovantes ; folia 1·5 mm. longa, lingulata, ad basin leviter dilatata, ad apicem rotundata. Margo planus supra, latera basalia versus inflexus, margine extrorsum obtuso. Nervus firmus, evanescens, levis ; cellae quadratae, minute papillosae in superficie superiore, in inferiore tumidae, densissimae, cellae marginales irregulares in parte dimidia folii superiore, nonnullae extrorsum prominentes. Seta 1 cm. longa, pallida ; capsula 1·5 mm. longa, fusca, ovato-oblonga ; operculum breve, conicum. Peristomium non vidi.

In montibus Natalibus, Transvaalensibus, Basutensibus.

† *Hyophila basutensis* Sim (sp. nov.)=*Barbula basutensis* Rehm. (ined.).

Arte caespitosa ; caules ½–¾″ alti, ramosi, innovantes ; folia lingulato-acuminata vel latiora infra, ubi latera erecta paucaeque cellae hyalinae, supra prope plana ; apex obtusus rotundatusve, nervus firmus, percurrens, lamino-cellis tumidis tectus ; cellae 10 mm. latae, quadrato-ellipticae, exteriores transversae ; quae prope nervum sunt, grandiores. Margo integer, inflexus si siccus, non involutus. Seta ½–¾″ longa, rubra ; capsula levis, 1 mm. longa, ovato-oblonga. Peristomium non vidi. Apud Rhenosterberg (Cape, E.) et Witpoortje (Transvaal).

at the apex, and these rosulate whorls recur.   Leaves apparently lingulate-acuminate but actually much wider below than above because the sides there turn up sharply, making the leaf deeply concave below, where it is hyaline, less concave or nearly plane above, and highly chlorophyllose.   Apex blunt, or on some terminal leaves rounded ; nerve strong, extending to the apex, covered upward by lamina cells. Cells quadrate, elliptical, 10μ wide, chlorophyllose, tumid, smooth, separate, the outer transverse and those near the nerve larger ; basal cells few, hyaline, unequal, up to $25 \times 10\mu$.   Margin inflexed when dry, not involute.   Perichaetal leaves not different ; seta $\frac{1}{2}$–$\frac{3}{4}$ inch long, red ; capsule small, 1 mm. long, oval-oblong, rather wide-mouthed, formed of many rows of oblong cells.   No peristome has been seen but only old capsules have been available.   The nerve is strongly connected by fibres down the stem, and the leaf-margins are so also somewhat.

Cape, E. Rhenosterberg, Middleburg, Cape, May 1873, MacLea (Rehm. 458).

Transvaal. Witpoortje, 5000 feet, Dr. Moss.

I have not seen *Barbula basutensis*, Rehm. 100, from the Caledon River, O.F.S., which in Rehmann 458 he makes the same as this, but his 459 distributed as *H. basutensis* Rehm. var. *tenella* is different, and is *Gymnostomum afrum*.

Paris in error spells the name *bassoutensis*.

Genus 95. Weisiopsis Broth.

Dixon, in reference to *W. plicata*, says (S.A. Journ. Sc., xviii, 313), "This little species . . . is placed by Brotherus in the new genus *Weisiopsis*, distinguished from *Hyophila* by the thin-walled, more or less plicate, peristomate capsule, and comprising five known species, the remaining four being confined to east Asia."

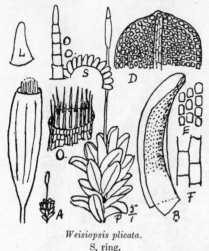

Weisiopsis plicata.
S, ring.

335. *Weisiopsis plicata* (Mitt.), Broth., in Oefv. af Finsk. Vet. Soc. Foerh., p. 62, No. 9 (1920) ; Dixon, S.A. Journ. Sc., xviii, 313.

=*Hyophila plicata*, Mitt., Linn. Soc. Journ., 1886, No. 146, p. 305, and pl. xv, pp. 13–16 ; Broth., Pfl., x, 270.

Monoecious, caespitose ; stems 2–4 mm. long, erect, mostly simple ; lower leaves smaller and more scattered than those of the rosulate coma ; leaves erecto-patent, the inner erect, all spathulate-concave, with rounded apex and entire erect margins ; nerve strong, percurrent, smooth ; cells roundly quadrate, 10μ diameter, smooth but chlorophyllose, toward the base rather larger, those at the very base up to $35 \times 10\mu$, laxly rectangular, hyaline.   Perichaetal leaves hardly different ; seta 8–12 mm. long, pale ; capsule oval-cylindrical, tapering below, slightly 8-ridged ; lid conical, erect ; ring dehiscing, white.   Peristome of sixteen slender, short, erect, smooth, red, subulate teeth, inserted below the mouth ; male flower terminal. In this species I find the capsule so indistinctly plicate or even ridged that that character is of little value.

On stone, mouth of cave, mostly in shade, Salisbury, S. Rhodesia, altitude 5200 feet, June 1920, Eyles 2282 (Herb. Sim).

Rehm. 103 is "*Barbula plicata* Rehm., n. sp., Basutoland, at the Caledon River," an unpublished name, and not represented in South African herbaria.   It evidently has nothing to do with the present species.   But C. Müller rather mixed matters in quoting in Hedw., 1899, p. 70, Rehm. 103, for *Bryum porphyreothrix*, which was a clerical error and should likely have been Rehm. 243.

Genus 96. Tortula Hedw.

Either small, gregarious, earth-grown plants, or larger and loosely tufted, and sometimes found on decayed tree-stumps, mud-walls, or moss-clad stones ; stems simple or dichotomously branched, and with leaves either evenly distributed along the stem, or getting larger upward, or rosulate.   Leaves oblong or obovate-oblong, usually rounded or shortly acuminate at the apex, and with a strong excurrent nerve sometimes produced into a bristle or hair-point. Lower cells rather large, quadrate and pellucid ; the upper opaque and chlorophyllose.   Capsule erect on a long seta, oblong or cylindric, usually peristomate, with thirty-two filiform equal papillose segments which are more or less united, sometimes half-way, into a tubular membrane, and are usually spirally twisted, though occasionally straight. Perichaetal leaves often sheathing the seta.

C. Müller in his Syn. Musc. used the name *Barbula* instead, because *Tortula* had been sunk on account of a previous phanerogamic genus of that name, but as that phanerogamic genus has since become obsolete *Tortula* again holds priority here, while part of the original genus has been separated as *Barbula*, and other groups have also been separated into or as new genera, consequently the characters of *Tortula* are now modified and the synonymy is considerable.

Synopsis of Tortula.

1. Synoecious	.	.	.	.	.	.	.	.	336. *T. Mülleri.*
1. Dioecious or autoecious	.	.	.	.	.	.	.	.	2.
2. Leaf piliferous	.	.	.	.	.	.	.	.	3.
2. Leaf mucronate, acute or obtuse	.	.	.	.	.	.	.	.	11.
3. Leaf-margins incurved ; keel papillose	.	.	.	.	.	.	.	.	4.

336. *Tortula Mülleri* (Brid.), Wils., Bryol. Brit., p. 134 and t. 44.

=*Barbula Mülleri* Br. in F. Müller, Musc. Sard., Flora, 1829 ; W. P. Sch., Syn., 2nd ed., p. 232.
=*Tortula princeps*, De Not. Specim. d. Tortula, ital. et epil. ; Dixon, p. 204 and Tab. 26E ; Broth., Pfl., x, 302.
=*Barbula princeps* C. M., Syn., i, 656.
=*Syntrichia princeps* Mitt., Journ. Linn. Soc., i, Suppl., p. 39 (1859).

A strong, vigorous, synoecious moss, growing in loose mats, usually on bark ; stems branched, the branches short and crowded ; stems ¼–¾ inch long, light green above, somewhat red below ; leaves erecto-patent or spreading when moist, tightly concave, twisted over the stem-top, and somewhat channelled when dry, rosulate at each season's apex, 2 mm. long, oblong-elliptical, rounded, emarginate or shortly acuminate at the apex where the strong red nerve is excurrent as a short, hyaline, straight, almost entire hair-point, somewhat narrowed at the base ; margin smooth, revolute in all or part of the lower half, particularly about mid-leaf ; leaf somewhat keeled-channelled elsewhere ; nerve strong, 60μ wide at the base. Cells distinct but rather obscure, rounded, 15–18μ wide, papillose on the back, the marginal lines more transverse ; basal cells hyaline-oblong, thin-walled, $50 \times 30\mu$, the outer four to five lines are of much shorter cells (20μ long and wide), sometimes chlorophyllose. Capsule cylindrical, slightly curved, 3 mm. long, dark brown, on a long seta. Ring of two rows of cells ; peristome tube long.

An almost cosmopolitan species, little known here ; probably many specimens passed as *T. erubescens* (=*T. brachyaechme*) belong here ; the synoecious inflorescence separates them.

Shaw's list includes *B. mollis* (Br. and Sch.) C. M. ; evidently instead of *T. Mülleri.*

Cape, W. Kloof Road, Cape Town, Jan. 1919, Sim 9554 ; Stellenbosch, Jan. 1919, Sim 9572 ; Cape Town, Wager 350 (*fide* Dixon).

Cape, E. Graaff Reinet, MacLea, *fide* Shaw as *B. mollis* B. and S. ; Toise River, on Cycads, Uitenhage, Sim 10,100.

Natal. Rosetta, Sim 10,083.

337. *Tortula reticularia* (C. M.), Broth., Pfl., i, 3, 434, and x, 300.

=*Barbula reticularia* C. M., Hedw., pp. 38, 101, and Rehm. 106.

Laxly caespitose ; stems up to half-inch high, usually simple, with erecto-patent, very concave leaves, almost tubular at mid-leaf, but expanded and nearly rounded at the apex, and more or less narrowed toward the base ; margin erect or inflexed, not closely involute, crenulated by rounded papillose cells ; nerve strong, brown, lost in the long cells below the hair-point, set with conical papillae on the back of the upper two-thirds, often bearing gemmae on the upper surface near the apex. Hair-point smooth, more or less brown, the point usually pellucid, the awn-cells extending down to meet those of the nerve but differing from them. Cells large, roundly hexagonal, 20–25μ wide, nearly pellucid, but with usually two high conical papillae per cell ; in the lower third of the leaf gradually more hyaline and less chlorophyllose, through square and oblong to large lax, elliptical, basal cells 60μ long, the marginal ones rather smaller and sometimes chlorophyllose. Inflorescence and fruit not seen.

Rehmann's Cape Town specimen (Rehm. 106) sterile, seems distinct, but I have doubts as to whether this will not eventually be found to be a condition or form of *T. papillosa*, having wider and more spathulate leaves with erect margin. Müller's description does not describe Rehmann's plant well, and seems more like *T. erubescens.* Both require further field-knowledge in Cape Town. No other specimen known.

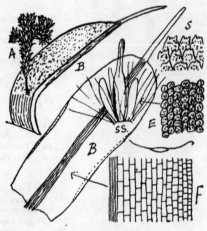

Tortula Mülleri.

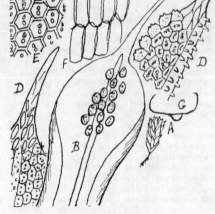

Tortula reticularia.

338. *Tortula papillosa* (Wils. MSS.), Spr. in Hook., Lond. Journ., iv, 193 ; Dixon, 3rd ed., p. 205, Tab. 26C.

=*Barbula papillosa* C. M., Syn., i, 598, and ii, 630 ; W. P. Sch., in Besch. Fl. Bryol. Antill. Fr., p. 23.

=*Tortula *hystricosa* Dixon MSS., in letter.

Caespitose on bark ; stems 5–10 mm. high, seldom branched ; inner leaves larger and more piliferous than lower leaves ; all suberect, or when dry erect ; ovate from a rather wide base, concave with involute margin ; apex rounded or subacute ; nerve thick, ferruginous, with long clear papillae in the upper half, the nerve ending just beyond the leaf-apex or extending as a short, smooth, ferruginous hair-point ; cells 20μ wide, hexagonal, pellucid, greenish, with one papilla on the back of each ; the upper marginal cells irregular ; nearly the lower third of the leaf hyaline, of which the upper cells are oblong and papillose, and the lower cells 60μ long, hexagonal-oblong, with thin deep walls. Small gemmae are produced on the upper surface of the nerve, but are not always present. This species is known from Europe, North and South America, Australia, and now from Africa, but fruit has only been found in Australasia ; showing a short capsule on a short reddish-brown seta.

Tokai, Cape, W., on oaks, 1000 feet altitude, Jan. 1919, Sim 9499.

Tortula papillosa.

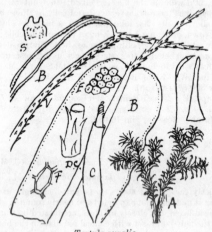

Tortula ruralis.

S, one cell with two papillae.

339. *Tortula ruralis* (Linn.), Ehrh., Pl. Crypt., No. 184 ; Bietr., viii, 100 ; Dixon, 3rd ed., p. 202, Tab. 26A; Broth., Pfl., x, 301, and fig. 230.

=*Bryum rurale* Linn., Sp. Pl., 1st ed., ii, 1116.

=*Barbula ruralis* Hedw., Fund., pp. 11, 92 ; C. M., Syn. Musc., i, 639.

Loosely caespitose, dioecious, green above, brown below ; stems 3–6 cm. high, usually more or less branched, the lower parts rooting when in contact with humus. Leaves numerous on the young growth, 3–4 mm. long, squarrose or recurved when moist, erect when dry, oblong-elliptical, concave-keeled, hardly narrowed to the base, each side rounded at the apex, the hair-point rising from a sinus ; the nerve very strong, 60μ wide at the base, brownish red up to the end of the lamina, excurrent as a strong, *spinulose*, hyaline hair-point 2–3 mm. long, and also spinulose with hyaline spines along the greater part of the keel ; margin slightly recurved the whole length of the leaf, the margin showing the rounded papillose outline of each cell ; cells rather large, 17μ wide, rounded hexagonal, green and obscure, strongly papillose with two papillae in each cell ; cells at base of the leaf large, hyaline, oblong-hexagonal, $60 \times 40\mu$. Perichaetal leaves not differentiated but long-piliferous. Seta 2–3 cm. long, straight, yellowish ; capsule cylindrical, 3 mm. long, slightly curved, with a long conical lid ; peristome teeth long, united below.

When dry the leaves are erect or appressed but hardly twisted, and usually show two longitudinal furrows. Male flowers terminal, bud-like, with outer leaves like small stem-leaves, and inner smaller and without pilum. Antheridia and paraphyses numerous. A very distinct but variable species.

The following forms are too close for separation :—

1. *T. afroruralis* (C. M.), Broth., Pfl., i, 3, 435, and x, 302 = *Barbula afroruralis* C. M., Hedw., 1899, p. 101 (§ *Syntrichia albipilae*), Rehm. 114 (Stinkwater, Cape, W.), in which the leaf is emarginate-auriculate at the apex, the back of the nerve is smooth but the hair-point closely denticulate.

2. *T. leucostega* (C. M.), Broth., Pfl., i, 3, 435, and x, 302 = *Barbula leucostega* C. M., in Syn. Musc., i, 641 ; C. M., Hedw., 1899, p. 102 (§ *Syntrichia albipilae*), has the same leaf-outline and the nerve scaberulous at the apex ; peristome with a long *white* tube. Zwellendam, 1829 (Mundt, Ecklon) ; Genadendal (Breutel).

3. *T. trachyneura* Dixon, Trans. R. Soc. S. Afr., viii, 3, 195, and Pl., xi, 7 ; Broth., Pfl., x, 302, has back of the nerve scabrid in the upper part, with a few large papillae. Dixon says, " This may possibly belong to *T. ruralis*, but the roughness at the back of the nerve is different from anything I have seen in that species. There it may be quite smooth or markedly but finely muriculate ; here it is coarsely and sparsely papillose or almost hispidulous. None of the other South African species show this character of nerve." It could hardly be more hispidulous than Rehm. 477, but I find the presence of sharp papillae in all stages.

4. *T. montana* (Nees), Lindb., M. Sc., p. 20 ; Par., Ind., xxv, 48 = *T. intermedia* Berk., Handb. Br. M., p. 251 ; Dixon, 2nd ed., p. 204, is credited by Paris to Cape, as well as to Europe, Asia, North Africa, and North America. I find nothing to support the record. In Europe it is considered very close to *T. ruralis*, but is distinguished by the leaves being erecto-patent instead of squarrose, the cells smaller and more indistinct, the margin almost plane, and the arista less toothed. It may be a good species in Europe but is not known to me here.

5. *T. erythroneura* (Sch.), Broth., Pfl., i, 3, 435, and x, 302 = *Barbula erythroneura* (W. P. Sch., in Breutel Musc. Cap , as *Syntrichia*) C. M., in Hedw., 1899, p. 102, is described as dioecious, but my specimens answering it in every other respect are synoecious, and so belong to *T. Mülleri*. Dixon, Trans. R. Soc. S. Afr., viii, 3, 198, in recording a specimen of it collected at the Cape in 1912 by S. W. Hall, says, " This agrees well with Schimper's specimens (Coll. Wilms) at Kew. But is it anything more than a *ruralis* form ? " I think not ; but these synoecious specimens deserve further investigation as they indicate the proximity of *T. ruralis* with *T. Mülleri*, even though they differ in the roughness of the hair.

T. ruralis is known from Europe, Asia, N. and S. Africa, N. and Mid America, and Australasia.

Cape, W. Table Mountain, Sim 9185, 9216 ; Devil's Peak, Sim 10,040 ; Tokai, Sim 9504 ; Zwellendam, 1829 (Mundt ; Ecklon, 1828), and Genadendal (Breutel), as *T. leucostega* ; Stinkwater (Rehm. 114) as *T. afroruralis*.

Cape, E. Cave Mountain, Graaff Reinet, MacLea (Rehm. 477) ; Tyumie, 1917, D. Henderson 330 (as *T. trachyneura*) ; Port Elizabeth, Sim 9062.

340. *Tortula erubescens* (C. M.), Broth., Pfl., i, 3, 434, and x, 301 ; Dixon, Smithsonian Misc. Coll., pp. 69, 214 ; Dixon, Trans. R. Soc. S. Afr., viii, 3, 195, and S.A. Journ. Sc., xviii, 315 (and as *T. brachyaechme*).

= *Syrrhopodon *obscurus* Rehm. (*fide* C. M., in Hedw., pp. 38, 104) ; Kew Bull., 6 (1823), No. 126.
= *Barbula erubescens* C. M., in Nuov. Giorn. bot. Ital., iv, 14 (1872).
= *Barbula exesa* C. M., Hedw., 1899, p. 103 ; Dixon, Trans. R. Soc. S. Afr., viii, 3, 195.
= *Tortula exesa* (C. M.), Broth., Pfl., i, 3, 434, and x, 301.
= *Barbula Macowaniana* C. M., Hedw., 1899, p. 103.
= *Tortula Macowaniana* (C. M.), Broth., Pfl., i, 3, 434, and x, 301.
= *Barbula oranica* C. M., Hedw., 1899, p. 102 ; Dixon, Trans. R. Soc. S. Afr., viii, 3, 195.
= *Tortula brachyaechme*, Broth., Pfl., x, 300 ; Brunnthaler, i, 26.
= *Barbula meruensis* C. M., Flora, pp. 73, 480 ; Broth., Pfl., x, 301.
= *Barbula Hildebrandti* C. M., Linn., pp. 40, 204 ; Broth., Pfl., x, 301.
= *Barbula Lepikiae* C. M., Flora, pp. 73, 480 ; Broth., Pfl., x, 301.
= *Barbula brachyaichme* C. M., Hedw., p. 102 ; Rehm. 107.
= *Barbula brachyacme* (C. M.), Broth., Pfl., i, 3, 434 ; Par., Ind., 2nd ed.
= *Barbula brachyaechme* (C. M.), Dixon, Trans. R. Soc. S. Afr., viii, 3, 195.

Dioecious, dull green, loosely tufted, usually on bark. Stems ½–1 inch high, robust, simple or slightly branched ; leaves rigid, spreading or erecto-patent, appressed and somewhat twisted when dry, 2 mm. long, spathulate, twice as long as broad or more, usually much narrowed in the lower half, concave-keeled, the apex almost rounded or emarginate ; the nerve very strong, reddish, usually smooth on the back, and excurrent as a short hyaline or pale or sometimes bright red, smooth point ; the margin plane or sometimes somewhat recurved toward the apex, entire ; cells large, lax, quadrate, 17μ long, opaque, minutely many-papillose ; those at the base much larger ($50 \times 20\mu$), regular, oblong-hexagonal, hyaline, but suddenly meeting the chlorophyllose quadrate cells. Seta short ; capsule small, widely cylindrical, with long conical lid. Lower leaves shorter than those above ; leaves variable in form and size, even on the same plant, often bearing gemmae on the surface of the upper leaves. Perichaetal leaves not differentiated.

The forms included in the synonymy given above differ rather widely in their extremes, but range into inseparable similarity, even on the same plant, and sometimes approach closely to *T. brevimucronata* on the one hand and *T. eubryum* on the other, but these appear to be distinct, the former chiefly in the hyaline cells toward the basal margin, the latter in the tapering undulated leaf.

Widely distributed from Abyssinia southward, common throughout moist and even subxerophytic South Africa, from the coast to the mountains, and in Orange Free State, Portuguese East Africa, and Rhodesia.

The leaves are often damaged, and seem to be very fragile or probably easily damaged under xerophytic conditions, the midribs frequently standing alone along a whole stem.

341. *Tortula eubryum* (C. M.), Dixon, Bull. Torrey Bot. Club., xliii, 66.

　　　=*Barbula eubryum* C. M., Flora, pp. 62, 379 (1879).

A very vigorous, caespitose moss, usually on soil, and often under xerophytic conditions (as at Bulawayo, Khami, etc.); the upper parts deep green, the lower leaves

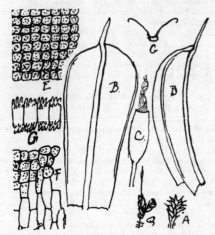

Tortula erubescens.

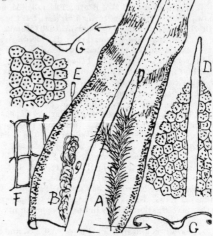

Tortula eubryum.

rusty brown. Stems ¾–1½ inch high, usually simple, the leaves spreading when moist, but crisped and circinate, and twisted round the stem to a point when dry, the red nerve then very prominent as a backbone. Leaves 2–4 mm. long, the upper larger than the lower, all crowded and imbricate, clasping but not sheathing, widely oblong-acuminate, usually much wider near the base than near the apex; the margin widely revolute in the lower half, usually not revolute above but crenate through densely papillose, rounded, marginal cells, which correspond quite with those of the lamina, all densely papillose and more or less chlorophyllose; this condition extends about two-thirds down, below which the cells are hyaline and quadrate, gradually larger, till near the base the cells are oblong-hexagonal, deep but thinly walled, $70 \times 30\mu$, except those of the revolute margin which are papillose and similar to the lamina cells. Nerve very strong, rusty brown on old leaves, deep green on the younger leaves, smooth on the back, but showing as a strong ridge on dry leaves, and extending as an exserted short hair or often as a long mucro, red or green in accordance with age. Gemmae frequently present on the leaf-surface, so between this species and *T. erubescens*, Brotherus, group separated by production of gemmae breaks down. Indeed this species often has leaves which would pass as belonging to *T. erubescens*, but the most of its leaves are long and tapering upward and have the revolute margin in the basal half, whereas in *T. erubescens* it is toward the apex. Capsule shortly ovoid or ovoid-cylindrical on ½–¾ inch pale seta, the mouth deep red and slightly contracted.

For reasons stated (in above paper) Dixon prefers Müller's section *Rhystobarbula* (having undulate leaf) for this over his previous section *Bulbibarbula*.

Common all over Rhodesia and northward; scarce further south.

S. Rhodesia. Zimbabwe, Sim 8736, 8700; Matopos, Sim 8459, 8469, etc.

Transvaal. Pretoria Kopjes, Apr. 1921, Sim 10,088.

342. *Tortula pilifera* Hook., Musc. Exöt., Pl. 12.

=*Barbula pilifera* (Hk.), Brid., Bryol. Univ., i, 572; C. M., Syn. Musc., i, 641; Dixon, Trans. R. Soc. S. Afr., viii, 3.
　　　194 (and as *B. torquescens*); Broth., Pfl., x, 280.
=*Barbula crinita* Schultz, in Nov. Act. Acad. Leop., xii, 1, 226, t. 34.
=*Barbula* *flavipila* W. P. Sch., in Breutel M. Cap.

=*Barbula torquescens* W. P. Sch., in Bot. Zeit., 1853, p. 163 ; C. M., in Hedw., 1899, p. 105 ; Broth., Pfl., x, 280 ; Dixon, Journ. of Bot., Aug. 1824, p. 234.

=*Barbula *Mauchii*, Rehm. 109.

=*Barbula *bokkefeldiana*, Rehm. 108 ; Kew Bull., No. 6, 1923, No. 108, p. 200.

Dioecious, loosely caespitose, dull green or grey-green, hoary when dry ; stems 1–3 inches long, robust, simple or somewhat branched, either near the base or higher, crisp or twisted when dry. Leaves spreading, 2–3 mm. long, concave-carinate, oblong-lingulate ; the lower portion wider ; the apex rounded ; the nerve very strong, excurrent as a hyaline smooth arista half as long as the leaf ; margin widely revolute the whole length of the lamina ; the surface minutely papillose. Basal cells large, lax, oblong, and hyaline ; cells of the lamina small, rounded, and very dense toward the leaf-apex. Perichaetal leaves about six, 5 mm. long, widely lanceolate, tubular with a long arista, hyaline, with very lax oblong areolation and yellowish faint nerve, each sheathing those within, all erect and closely surrounding the seta. Seta 2–3 cm. long, pale ; capsule cylindrical ; lid conical ; peristome red, the teeth separate to the base, much twisted, very papillose.

In the sterile juvenile state it is laxer, more torquescent, more pellucid, and has acuminate tips without hair-point, but the nerve is percurrent and the margin strongly revolute the whole length, even to the base.

B. pilifera var. *aquatica* C. M. (*aquatilis* per Par., Ind.), from cataract above Rondebosch (Rehm. 111), differs only in being longer and more branched, and the leaves rather more long and lax.

B. pilifera var. *gracilis* C. M., Syn., i, 642, is a slender form not distinguishable here.

B. pilifera var. *senilis*, Rehm. 113, Cape Town (var. *sessilis* C. M., in Rehm. 113, per Par., Ind., 2nd ed.), hardly differs.

T. pilifera was formerly confused with the European *T. laevipila* Schw., which though similar is autoecious, has large basal cells, and has the peristome teeth connate in a tube for one-third of their length.

Tortula pilifera.
S, papillose cell.

B. torquescens W. P. Sch. (in Bot. Zeit., 1853, p. 163 ; C. M., in Hedw., 1899, p. 105), according to Dixon (Journ. of Bot., Aug. 1924, p. 234), is synonymous. I have not seen the type, but Rehm. 109, distributed as *B. Mauchii* Rehm. (from mountains above Worcester, fertile) and said by Müller in Hedw., 1899, p. 105, to be the same, is a dense little form with nothing but size to separate it. In Kew Bull., No. 6 (1923), under Rehm. 105, Dixon states that the Kew specimen named *B. torquescens* Schp. in Rehmann's Exsiccatae is a totally different thing from *B. torquescens* Schp.—possibly *B. unguiculata* Broth. MSS., or near it. Other specimens may be the same ; there is no specimen of that number in South Africa. But there seems to have been another plant named *Trichostomum torquescens* W. P. Sch. MSS. ; C. M., in Bot. Zeit., 1859, p. 229, also credited to South Africa by Paris, and which probably is *Tortella torquescens* (W. P. Sch.), Broth., Pfl., i, 3, 397, and x, 263, and confusion may have come thereby.

Trichostomum systylium C. M., Syn., i, 589, is recorded by Shaw as collected by MacLea near Graaff Reinet. No specimen is available, and as it is an alpine species in Europe, and more or less resembling *Tortula pilifera* but with shorter teeth, the probability is that that record belongs there, and that the species is not South African.

T. pilifera is the most common moss on soil under or adjacent to bushes all over the Western and Eastern scrubveld ; present but less common on open lands or in forest, or in Natal, Transvaal, Rhodesia, and Portuguese East Africa.

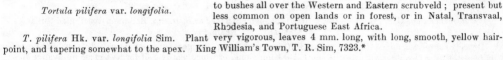

Tortula pilifera var. *longifolia.*

T. pilifera Hk. var. *longifolia* Sim. Plant very vigorous, leaves 4 mm. long, with long, smooth, yellow hair-point, and tapering somewhat to the apex. King William's Town, T. R. Sim, 7323.*

---

* *Tortula pilifera* Hook. var. *longifolia* Sim (var. nov.).

Planta admodum vigens ; folia 4 mm. longa, arista longa, levi, flava, apicem versus paulo decrescentia. Apud King William's Town, Sim 7323.

343. *Tortula muralis* (Linn.), Hedw. Fund., pp. 11, 92 ; Dixon, 3rd ed., p. 198, Tab. 25 B ; Broth., Pfl., x, 297.

=*Bryum murale* Linn., Sp. Pl., 1st ed., pp. 11, 1117.

=*Barbula muralis* Timm., Fl. Megap. Prod., p. 240 ; C. M., Syn., i, 625 ; W. P. Sch., 2nd ed., p. 205, and many others.

=*Barbula chrysoblasta* C. M., in Hedw., 1899, p. 104 (*fide* Broth., Pfl., i, 3, 431, and x, 297).

Autoecious, gregarious or tufted, dull green or bright green and hoary ; stems usually unbranched and 3–4 mm. high, occasionally erect and branched and 8–10 mm. high. Leaves rosulate, 2–3 mm. long, patent or erecto-patent, when dry curled inward with the hair-points surrounding the seta.

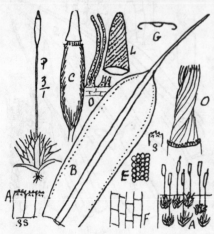

Tortula muralis.

S, papillose surface of cell ;

SS, papillose cells of ring.

Leaves oblong, shortly acuminate, usually widest below the apex, to which the margin tapers quickly, rather narrowed to the base ; surface flat except that the margins are shortly recurved the whole length of the leaf ; nerve very strong, yellowish brown, excurrent in a long, usually hyaline, smooth arista, in some forms half the length of the leaf, in others one-fourth to one-eighth the length of the leaf. Though usually hyaline, some leaves on the same plant may have it ferruginous. Basal cells oblong-rectangular and hyaline, 50μ long, gradually shorter upward, then small, dense, quadrate, 15μ wide, and minutely papillose elsewhere. Seta 10–15 mm. long, straight, purple ; capsule 3 mm. long, cylindrical, brown, with red conical lid, in which the areolation is twisted to the right. Peristome also twisted to the right, of sixteen teeth, each split to the low basal membrane into two slender hyaline segments, or frequently four or more segments are connate. Male inflorescence has about six long red archegonia without paraphyses. Spores minute. Per. leaves not differentiated. Abundant everywhere round Cape Town, Stellenbosch, Paarl, and Port Elizabeth on lime-pointing on stone or brick walls ; probably introduced as almost absent in the country districts. Eagles Nest, Bloemfontein, Professor Potts ; Eshowe, Zululand, *fide* Bryhn, p. 12, which I doubt.

Var. *brevipedunculata*, Rehm. 88, from Montagu Pass, is ½ inch or more high, with ½-inch seta, and the hyaline hair-point somewhat decurrent into the leaf-apex, but agrees in other respects.

Barbula (Eubarbula pungentes) afro-inermis C. M. (Hedw., 1899, p. 105) has the leaves widely ligulate and the nerve shortly apiculate, sometimes long and strong, and seems to be only a small short-haired form of this.

344. *Tortula longipedunculata* (C. M.), Broth., Pfl., i, 3, 432.

=*Barbula longipedunculata* C. M., Syn., i, 630.

Unknown to me. Brotherus (Pfl., i, 3, 432) places it in Section III *Zygotrichia*, described as " low plants ; leaves more or less spathulate, awned," and it has tube of the peristome low and leaves bordered by long narrow cells ; and C. Müller, in Syn., i, 630, puts it in the group with *margin erect, leaves margined*, and compares it with the European and North African *T. marginata* Spruce (Dixon, p. 199, and t. 27) =*Barbula marginata* B. and S., of C. M., Syn. M., i, 629, which is very similar to *T. muralis* Hedw., but distinguished by the pale red seta, the bordered leaves, and the short mucro. His description of *B. longipedunculata* is " dioecious, low, simple, caespitose, rigid ; leaves twisted when dry ; when moist erecto-subappressed ; base oblong, concave, and not ventricose ; cells lax and yellow ; lamina shortly and bluntly lanceolate, margin flexuose, more or less involute, with a narrow margin of thick brown diaphanous cells. Leaf-cells green, minute, opaque, hexagonal ; leaves entire ; nerve thick, reddish, slightly papillose, often ending in a mucro. Perichaetal leaves wider. Capsule cylindrical on a long flexuous, erect, red seta ; lid conical, acute, arcuate ; ring simple ; peristome without a membrane, much twisted." Zwellendam, Pappe, 1838.

Under *B. afro-inermis* C. M. (Hedw., 1899, p. 105) Müller describes the leaves of this as oblong-acuminate, mucronate.

345. *Tortula deserta* (C. M.), Broth., Pfl., i, 3, 430, and x, 297.

=*Barbula deserta* C. M., Hedw., 1899, p. 108. Though published as *B. deserta*, Rehmann distributed this as B. *desecta C. M., Rehm. 96.

Densely caespitose, mixed in sand ; stems 3–10 mm. high, erect, straight, not much branched ; stem and branches bare below, with a clavate coma ; leaves short, 1–1½ mm. long, of which nearly half forms a clasping hyaline sheath narrower than the ovate lamina, the whole leaf forming a short, somewhat concave spatula with a very strong red nerve, which maintains its wideness to the apex though covered with chlorophyllose lamina upward. Margin inflexed at the bend, spreading or erect elsewhere, entire, the marginal cells mostly transverse. Cells densely chlorophyllose, obscure, round, 15μ diameter, minutely papillose ; sheath cells lax, oblong, hyaline, the inner wider and longer than the outer.

I only know this from Rehm. 96 from bare ground, Cape Town (sterile), and it is possible that age may develop it into something else. In the condition shown in that specimen it is, however, quite distinct. The leaf-lamina form resembles somewhat that of *Gymnostomum afrum*, which, however, does not have the long sheath nor the strong percurrent nerve.

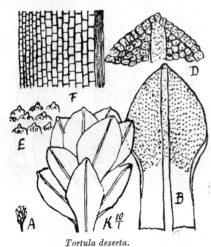

Tortula deserta. *Tortula irregularis.*

346. *Tortula irregularis* Sim (new species).*

Loosely caespitose on bark and stones; stem ½–1 inch long, single or branched, erect; leaves numerous, light green, imbricate, 2–3 mm. long, 1 mm. wide; widely spathulate, rounded at the apex; nerve very strong, rufescent, smooth to the short hair-point which is strongly spinose; margin irregular and deeply crenate in the upper half where it is plane and also minutely crenated by tumid papillose cells; the lower half plane or often slightly revolute; cells minute, 8μ diameter, tumid, mostly 2-papillate, chlorophyllose, these often descending along the margin to the leaf-base; basal cells rectangular, thin-walled, 50 × 15μ, occupying the inner lower third of the leaf. Other parts not seen.

Resembles *T. brevimucronata* in leaf-size and leaf-form and in having small outer cells toward the leaf-base, but differs from all in the exceedingly irregular margin and in the short spinose point.

Natal. Bark and rocks above Edendale Falls, 2500 feet, Aug. 1918, Sim 10,064.

347. *Tortula brevimucronata* (C. M.), Broth., Pfl., i, 3, 434, and x, 301.

= *Barbula brevimucronata* C. M., Hedw., 1899, p. 104.
= *Tortula brevitubulosa* Broth., in Brunnthaler, 1913, i, 26; Broth., Pfl., x, 301.

Loosely tufted, bright green or tinged red. Stems single or slightly branched, robust, 6–15 mm. high, without gemmae; leaves spathulate or oblong-acuminate from rather a narrow base, 3 mm. long, erecto-patent or nearly horizontal when moist, flat, entire, somewhat undulate; when dry incurved and somewhat torquescent; nerve strong, reddish, extending to the apex, and forming a slight hyaline mucro. Cells hexagonal, rather large, 16μ, but with globose surface appearing quadrate, freely but minutely papillose; the marginal cells in a close line and papillose on the margin; the basal cells large, 60 × 30μ, oblong-rectangular, smooth and hyaline, those toward the margin smaller, but quadrate and hyaline. Dioecious. Perichaetal leaves not differentiated. Seta 12–14 mm. long, red; capsule 2·5 mm. long, cylindrical, pale; peristome rather short, the teeth free almost to the base.

Cape, E. Perie, 1888, Sim 269.
Natal. Karkloof, Sim 10,086; Mooi River, Sim 10,087; Donnybrook, Feb. 1920, Sim 9981; Oliver's Hoek, Miss Edwards (Sim 10,056); on Cycads, Falls near Hilton College, Sim 10,057; Mont aux Sources, Sim 10,084. 10,085; Van Reenen, Brunnthaler.
O.F.S. Rydal Mount, Dec. 1910, H. A. Wager.
Transvaal. Spitzkop, near Lydenburg, Feb. 1888, Dr. Wilms (sterile), *fide* C. M.

---

* *Tortula irregularis* Sim (sp. nov.).
Laxe caespitosa; caules ½–1″ longi, erecti; folia crebra, imbricata, 2–3 mm. longa, 1 mm. lata, late spatulata; apex rotundatus cuspidatusque; nervus firmus, ruber, levis, excurrens ut arista fortiter spinosa; margo planus, irregulariter alteque crenatus in parte dimidia superiore nec non minute crenulatus cellis tumidis, pars dimidia inferior saepe leviter revoluta; cellae minutae, tumidae, pleraeque 2-papillatae, hac marginem usque ad folii basin descendentes; cellae basales interiores 50 × 15 mm., tertiam folii partem obtinentes.
Apud cataractam ad Edendale, Natal, Sim 10,064.

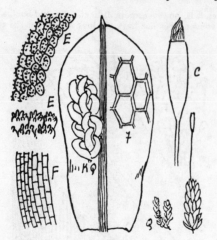

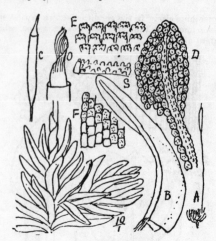

Tortula brevimucronata.   *Tortula Rehmanni.*
                                             S, back of nerve.

348. *Tortula Rehmanni* (C. M.) Sim.

=*Barbula Rehmanni* C. M., Hedw., 1899, p. 106, and Rehm. 101 ; Broth., Pfl., x, 280.

Dioecious, almost acaulescent, but caespitose and branching freely at the crown, the central stem fructiferous, the side innovations coming on for future use. Leaves numerous, imbricate, clasping but not sheathing, 1–1·5 mm. long, erecto-patent when moist, somewhat circinate when dry, lingulate, rather wider below, channelled toward the base, almost flat at the apex when moist, somewhat acuminate, and either obtuse or mucronate on the same plant, the margins neither involute nor revolute, crenulated by rounded, densely papillose cells ; the strong nerve pale in young leaves, deep red in older leaves, set along the back with clear conical papillae, and ending in or just below the apex ; the cells about 12μ diameter, very densely chlorophyllose and minutely papillose, all the cells alike to the very base where a few of the outer cells are usually rather larger, quadrate or oblong, and hyaline. Perichaetal leaves convolute round the seta, partly membranaceous, short, wide, rounded, oblong, incrassate, with longer hyaline cells, though some of these leaves have papillose dense lamina, narrow nerve, and short mucro. Seta an inch long, pale ; capsule 2 mm. long, cylindrical, red, with twisted peristome on rather a high tubular base. Lid subulate ; ring caducous. Evidently a xerophyte.

Touwsrivier, Cape, W., Rehm. 101 ; S. Rhodesia, Eyles 1720.

Müller's description in Hedwigia does not well describe Rehm. 101, on which it was founded, especially since he omits all reference to the densely papillose nature of the leaf and of the nerve, and describes the lamina as linear-lanceolate (which is only the case when it is dry) and the cells as diaphanous and the perichaetal leaf-cells as indistinct. He seems to have had the same plant, but either it is variable, or he missed points.

349. *Tortula atrovirens* (Sm.), Lindb., *de Tortula*, p. 236 ; Dixon, 3rd ed., p. 195, Tab. 24H ; Dixon, Kew Bull., No. 6 (1923), under Rehm. 89 ; Broth., Pfl., x, 296.

=*Grimmia atrovirens* Sm., Eng. Bot., xxviii, 2015.
=*Barbula atrovirens*, W. P. Sch., Syn., 2nd ed., p. 194.
=*Barbula circinalis*, C. M., Linnaea, 1884, p. 703.
=*Tortula recurvata* Hook., Musc. Exot., t. 130 (Broth., Pfl., x, 296).
=*Trichostomum recurvatum* (Hk.), C. M., in Hedw., 1899, p. 101.
=*Barbula recurvata* Brid., Bryol. Univ., i, 530 and 825 ; C. M., Syn., i, 627.
=*Didymodon capensis* Spreng., Syst. Veg., iv, 2, 323.
=*Trichostomum convolutum* Brid., Sp. Musc., i, 232 ; C. M., Syn., i, 590, and ii, 629 ; Shaw's list.
=*Weisia Breutelii* W. P. Sch., in hb. Hampe (see C. M., in Hedw., 1899, p. 101).
=*Barbula parvula* Spreng., Syst. Veg., 16th ed., iv, 1, 179.
=*Desmatodon nervosus* B. and S., etc.

Crowded on the surface of the ground, monoecious (*fide* C. M.) ; plants 2–4 mm. high, simple or with several almost sessile crowns ; leaves numerous, rosulate, stiff, erecto-patent, 1·5–2 mm. long, widely lanceolate or lanceolate-spathulate, widest upward, subacute, concave-keeled ; the nerve very strong, brown, excurrent as a short hyaline mucro ; on old leaves the margin much reflexed or recurved except at the base, on young leaves usually not reflexed but light green ; cells small, hexagonal but on account of chlorophyll appearing globose, minutely 2-papillose, the outer usually transverse ; basal cells hyaline, lax, oblong-hexagonal. Seta 1 cm. long, yellowish ; capsule erect, brown, oval-oblong to cylindrical, small (1·5–2 mm. long) ; lid shortly conical ; peristome twisted, of thirty-two rather short segments united by a narrow membrane at the base. Hooker's otherwise excellent description and figure describe the seta as long and capsule cylindrical, which I do not find. Male inflorescence bud-like, on basal branches, perigonial leaves laxly areolated at the base, densely areolated above; antheridia few, large; paraphyses filiform.

Not unlike *T. muralis* in habit, size, and leaf-form, but without arista, and with different capsule and peristome.

Barbula recurvata var. *aristatula* C. M. (Rehm. 90, Cape Town, and Rehm. 481, Müller's Farm, Quathalamba) do not differ and are not more aristate.

B. atrovirens W. P. Sch. var. *edentula* W. P. Sch., Syn., 2nd ed., p. 195, is described as " Plant smaller ; lid straight, conical ; peristome rudimentary. Cape Town. Many specimens."

B. atrovirens is known from Europe, Asia, Africa, S. America, and Australasia.

Cape, W. Matjesfontein, and Caledon (Brunnthaler, i, 26); Cape, not rare (C. Müller's Syn.) ; Cape Town (Rehm. 90, as above, and Sim 9108) ; Camps Bay, Sim 9320 ; Walls, Cape Town, Sim 9548 ; Blinkwater, Sim 9263.

Cape, E. Mountain above Graaff Reinet, MacLea, Rehm. 480; Longhope, Cookhouse, Sim 10,081 ; Port Elizabeth, Sister Theola.

Natal. Müller's Farm, Rehm. 481, as above ; Blinkwater, Sim 10,082.

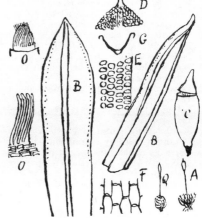

Tortula atrovirens.

Genus 97. ALOINA, C. M.

Small erect plants with short stems and small, usually obtuse, concave or cucullate leaves, sheathing at the base, and with the nerve often ceasing below the apex, occasionally excurrent, broad, and producing on its upper surface abundant granulose threads. Seta short ; capsule erect or inclined, cylindrical ; lid rostrate or rostellate ; calyptra cucullate ; ring broad ; peristome of thirty-two filiform segments on a very short, broad, basal membrane, and having about one spiral twist. The threads on the upper surface of the nerve are an unusual feature among mosses, and are not on every leaf even on this.

350. *Aloina rigida* (Hedw. p.p. ; Schultz), Kindb. Laubm. Schwed. u. Norw., 136 p.p.

=*Barbula rigida* Schultz (Rec. Barb., t. 32) ; Hedw. Descr., i, 65 p.p.
=*Tortula rigida* Schrad., Dixon, 3rd ed., p. 192, Tab. 24D.
=*Tortula stellata* Lindb. ; Braithwaite, Br. M. Fl.
=*Aloina stellata* (Schr.), Lindb. ; Broth., Pfl., x, 294.

Barbula rigida, var. β *pilifera* Horns., in Linn., xv (1841), 125, is mentioned in C. M., Syn. Musc., i, 596, as having been sent from South Africa by Dregé, as well as from many other localities elsewhere, and as it occurs in Europe, Asia, and North America, as well as in Algeria, there seems no reason to doubt the record, though I am not aware of its having been collected since, and it is not mentioned as South African by either Paris or Brotherus. In Dixon's Handbook it is included in *Tortula*. It is described as small, dioecious, with oblong, obtuse, entire, mucronate, concave leaves (or the upper leaves piliferous), subtubular or with involute margin ; the nerve very wide above and set with granulose filaments on its inner surface ; capsule small, cylindrical, on $\frac{1}{4}-\frac{1}{2}$-inch seta ; the lid subulate, half the length of the capsule ; calyptra covering half the capsule ; ring broad, dehiscent ; peristome fairly long, spiral ; male plants minute. It has much in common with *Pterygoneurum Macleanum* (see fig. on page 220), but is dioecious, has filaments instead of lamellae, and has a distinct peristome. The concave or subtubular leaf of *Tortula reticularia* and of *T. papillosa* are somewhat similar, but the cells and nerve in these are papillose. *A. rigida* is sometimes obtuse or hardly mucronate ; in other forms it is hoary with piliferous leaf-tips, and it is evident from Hornschuch's name that that is the form Dregé had collected. But Hornschuch's reference is not convincing that he had *B. rigida*. Dregé's locality is Silverfontein, Namaqualand, 2000 feet altitude, Aug. 1830.

§ 2. TRICHOSTOMEAE. Caespitose or gregarious perennials, often more or less rosulate. Leaves usually lanceolate and tapering, more or less involute, flat or revolute, strongly nerved to near or beyond the apex, with margin usually entire ; areolation rather lax and hyaline below, and minute, opaque, and papillose upward. Peristome normally with sixteen straight or twisted teeth on a very short basal membrane, the teeth usually split to the base and appearing as thirty-two slender papillose segments, but sometimes only partly divided, or entire, or absent ; when cut to the base the segments are either equal or the alternate ones smaller.

Genera of the Trichostomeae.

1. Leaves 3-seriate ; capsule peristomate	106. TRIQUETRELLA.
1. Leaves many-seriate	2.
2. Leaves serrate or dentate ; capsule peristomate . . .	103. LEPTODONTIUM.
2. Leaves minutely serrulate by prominent acute hyaline cells ; capsule peristomate .	104. OREOWEISIA.

2. Leaves entire or almost so . . . . 3.
3. Peristome present.
4. Margin of the leaf distinctly recurved, especially at the base . . . . . 5.
5. Peristome teeth usually short, erect or almost so . . . . . . 102. DIDYMODON.
5. Peristome teeth long and much twisted spirally . . . . . . 98. BARBULA.
4. Margin of leaf flat or slightly incurved . . . . 6.
6. Peristome teeth much twisted, long, and slender . . . . . . 99. TORTELLA.
6. Peristome teeth erect and straight, often imperfect . . . . . . 100. TRICHOSTOMUM.
6. Peristome teeth perforated . . . . 101. EUCLADIUM.
4. Margin of leaf usually incurved or involute ; outer layer of peristome thicker than the inner one, with conspicuous
   transverse ridges . . . . 107. WEISIA.
3. Peristome teeth minute or often absent ; ring dehiscing . . . . . 105. GYROWEISIA.
3. Peristome absent . . . . 7.
7. Lid more or less persistently attached or cleistocarpous . . . . . 108. TETRAPTERUM.
7. Lid separating . . . . 8.
8. Mouth of the capsule closed by a membrane . . . . . 109. HYMENOSTOMUM.
8. Mouth not closed by a membrane . . . . 9.
9. Nerve not reaching the apex of the leaf ; lid falling, with columella attached . . 110. HYMENOSTYLIUM.
9. Leaves nerved to or beyond the apex, the margin not involute ; columella permanent, lid remaining attached ;
   ring permanent . . . . 111. GYMNOSTOMUM.
(Leaf margin involute, see 94, HYOPHILA.)

Genus 98. BARBULA Hedw.

Rather slender, dull green or brownish-green, loosely tufted plants ½–2 inches high, not much branched, having rather small leaves loosely placed along the stems ; these are usually spreading when moist and either twisted and curled or closely appressed when dry. Leaves more or less lanceolate from a wide or clasping base, usually narrowed gradually to a point, channelled above, having a strong but seldom excurrent nerve and usually entire margin, which is distinctly recurved, at least at the base ; cells small, usually subquadrate or rectangular, toward the apex very small and thick-walled. Capsules small, oblong or cylindric, erect, on a fairly long seta ; peristome from a very short basal membrane, of thirty-two long, filiform segments, usually more or less twisted to the left, sometimes as much as two complete turns. Other characters as in *Tortula*, from which the narrow leaves and the general habit are the best distinctive characters.

As explained under *Tortula*, C. Müller described all his species of *Tortula* under *Barbula*, hence synonyms are numerous.

Synopsis of Barbula.

1. Seta 2 inches long, leaves linear-lanceolate, perichaetal leaves sheathing-convolute . . . 351. *B. flexuosa.*
1. Seta less than an inch long . . . . 2.
2. Nerve excurrent in a long cuspidate point . . . . . . 352. *B. torquatifolia.*
2. Nerve percurrent or shortly mucronate . . . . 3.
3. Nerve with tumid cells or papillae on the back . . . . . . 4.
3. Nerve smooth on the keel, margin much revolute . . . . . . 6.
3. Cells and nerve smooth, margin erect . . . . . 353. *B. afrofontana.*
4. Plant ½ inch high or less . . . . 5.
4. Plant 1–2 inches high . . . . 354. *B. elongata.*
5. Peristome teeth short, erect . . . . . . 355. *B. indica.*
5. Peristome teeth long and twisted . . . . . . 356. *B. Stuhlmannii.*
6. Cells minutely papillose, leaves narrow above, and nerve not widened upward . . 357. *B. xanthocarpa.*
6. Cells not papillose . . . . 7.
7. Leaves wide above, nerve wide near the apex . . . . . 358. *B. trichostomacea.*
7. Leaves narrow or linear near apex . . . . . . 359. *B. salisburiensis.*
7. Leaves ovate-lanceolate, shortly acuminate . . . . . 360. *B. Hornschuchiana.*

Barbula (*Senophylla pungentia*) *anoectangiacea* C. M., Hedw., 1899, p. 105, from Boschberg, Somerset East, is stated by Brotherus (Pfl., i, 3, 409) to be *Ceratodon*, sterile, with strong excurrent nerve, and the description answers that.

Barbula (*Senophylla obtuso-acuminata*) *filicaulis* C. M., Hedw., 1899, p. 107 =*Grimmia?* *suspecta*, Par., Ind., ii, 289, is said by Brotherus (Pfl., i, 3, 411), after examination of specimens, to be a sterile *Schistidium* in which case it would seem to be *Grimmia apocarpa*. Müller's incomplete description answers that so far as it goes. It is from Rhenoster River, Eastern Cape Province.

Barbula lurida Harv. is included in Shaw's list, " On oak trees near the Cape, by Harvey, and collected by Macowan near Somerset East, in Boschberg." *B. lurida* Horns. is a South American moss, and no *B. lurida* Harv. is known.

Known to me by name only.

B. eucalyx W. P. Sch., in Breutel M. Cap. ; Jaeg. Ad., i, 305.
B. Laureriana Ltz., Moosst, p. 161 ; Pfl., i, 3, 411, and x, 280, Prom. B. Spei. Rehm. 95 was distributed as *B. Laureri* Ltz. (from Cape Town), which authorship is apparently wrong, *B. Laureri* Kindb. being a northern

hemisphere moss (= *Desmatodon*), and presumably Rehmann's label was intended for *B. Laureriana* Ltz., but there is no description of that in South Africa, nor any specimen of Rehm. 95.

B. *Wageri Broth. (Pretoria) ; Wager's list, 277.

351. *Barbula flexuosa* (Hk.), Schultz, Rec., p. 16, t. 32 ; Brid., Bryol. Univ., i, 573 ; C. M., Syn., i, 609 ;
Shaw's List ; Broth., Pfl., i, 3, 411, and x, 280.

= *Tortula flexuosa* Hook., Musc. Exot., t. 125.
= *Barbula Hookeri* Steud., Nom. crypt., p. 72.

Not known to me, but the description and illustration in Musc. Exot. indicate a distinct plant, not otherwise included herein, and evidently a *Barbula*. Hooker's description is :—

" Stem elongate, branched ; leaves linear-lanceolate, keeled, the nerve excurrent. Perichaetal leaves long-sheathing ; capsule cylindrical ; lid long, subulate. Sent from the Cape by D. Menzies in 1791.

" Stem scarcely an inch long, erect, thick, branched near the apex ; leaves dark green, much imbricate, erecto-patent, rigid, linear-lanceolate, keeled ; margin subundulate ; nerve strong, very dark, excurrent a little beyond the apex. Perichaetium elongate, sheathing the base of the seta and twisted round, dark yellowish, attenuate, the nerve obscure ; seta 2 inches long, flexuous, pale red. Capsule cylindrical, very dark. Calyptra subulate, split to the base. Lid subulate ; capsule subequal, long ; peristome long, rusty ; teeth or cilii spirally twisted.

" This plant in the curious structure of the perichaetium will rank with *T. convoluta, revoluta,* and *pilifera* of this work (Tab. XII), differing from all in its flexuous fruit stalks and their great length. The operculum and the peristome, too, are remarkably long, as well as the perichaetal leaves."

Hooker's very excellent figure bears out his description. Müller adds to this that the margin is erect and entire, and peristome teeth cut to the base.

Barbula flexuosa (after Hooker).

352. *Barbula torquatifolia* Geheeb, in Bull. hb. Boissier, iv (1896), 409 ; Broth., Pfl., x, 280, is mentioned in S.A. Journ. Sc., xviii, 315, by Mr. Dixon, as found sterile on the Matopos by Wager (898). He says, " A distinct species, hitherto only known from S.-W. Africa, the leaves strongly twisted when dry, with broadly recurved margins, rather large cells, and stout nerve excurrent in a long cuspidate point. *B. acutata* C. M. is described as having the nerve only shortly mucronate and the leaves lanceolate-acuminate ; here they are oblong-lanceolate and very little narrowed above." The earlier record is from Amboland, Oshando (Schinz), " Deutsch Sudwest Afrika," von Hans Schinz, Zurich, p. 1 (1911).

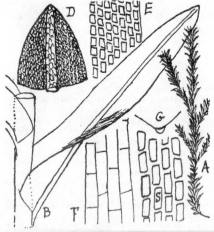

Barbula afrofontana.

Barbula acutata C. M., in Hedw., 1899, p. 109, is insufficiently and badly described, and Mr. Dixon, who has examined an original from Müller's herbarium, informs me that the leaves are " not lanceolate-acuminate but linear, not indeed acuminate at all." It is near *B. Hornschuchiana* and is founded on a specimen from Spitzkop, near Lydenburg, Transvaal, Dr. Wilms, 1887.

353. *Barbula (Hydrogonium) afrofontana* (C. M.), Broth.,
Pfl., x, 280.

= *Didymodon afrofontanus* (C. M.), Broth., Pfl., i, 3, 407.
= *Trichostomum afrofontanum* C. M., Hedw., 1899, p. 99 (not *T. fontanum* C. M., as cited in Par., Ind., 2nd ed., p. 65).
= *Didymodon fontanus*, Rehm. 82.

A soft, flaccid water-moss, growing in pale green tufts 2–3 inches deep ; stems branched by innovations which at first are short-leaved ; leaves lax, erecto-patent, clasping at the base, oblong-lingulate above, drawn to a blunt apex, the sides ascending, with erect entire margins, making the leaf keeled-concave ; nerve strong, often rusty, hardly reaching the apex. Apex sometimes subcucullate ; cells smooth, 8–10μ diameter, quadrate-elliptical, almost pellucid ; lower basal cells rectangular, thin-walled, 120 × 20μ, upper basal cells oblong-separate, pellucid, gradually smaller. Fruit not seen. A very lax, pellucid moss.

Rehmann's name *D. fontanus* is antedated by *Trichostomum fontanum* C. M. (Linn., 1876, p. 295) from Somaliland, so even the name *afrofontanus* is not a happy choice, both being African.

Natal. Van Reenen's Pass, Rehm. 82 (as *D. fontanus* Rehm.), which specimen is somewhat lime-encrusted ; Giant's Castle, 8000 feet, March 1915, R. E. Symons (Sim 8246).

354. *Barbula elongata* Dixon, S.A. Journ. Sc., xviii, 314 ; Broth., Pfl., x, 279.

Laxly caespitose, completely filling paths and open spaces where present ; stems strong, 1–2 inches long, erect, rigid, brown ; leaves lax, equally scattered, spreading, rigid, naturally always moist but when dry bent inward, 2 mm. long, clasping but not sheathing at the rather wide base, thence lanceolate, somewhat cucullate, shortly acute or mucronate, channelled ; the nerve very strong and stiff, somewhat muricate on the back in the upper portion by large rounded cells, percurrent or often ceasing in the hood, narrowly plicate near the margin ; the margin widely revolute for the whole length of the leaf, or sometimes not revolute ; cells quadrate, 8 mm. wide, separate, chlorophyllose ; cells of the leaf-base larger, the inner oblong $20 \times 12\mu$, hardly pellucid ; fruit unknown.

Present everywhere and almost the only moss at and round Danger Point, Victoria Falls, for 100 yards distance, subject to almost perpetual spray and hardly ever dry. This spot is on the top of the rocks enclosing the gorge below the falls ; it is the only portion not tree-clad of the famous " rain forest," the first point on which the spray falls, and is almost destitute of soil, Sim 9495, 9497, 9498. The back of the nerve shows its relationship to *B. indica*, but it is much more robust and quite a hydrophyte, though seldom in standing water.

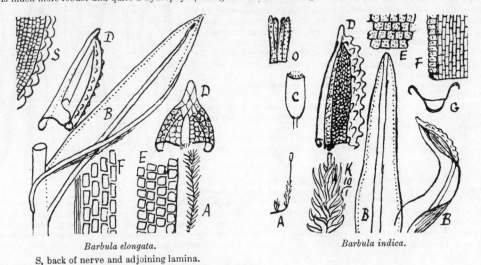

Barbula elongata.
S, back of nerve and adjoining lamina.

Barbula indica.

355. *Barbula indica* Brid., Bryol. Univ., i, 544, 828, p.p. ; Dixon, S.A. Journ. Sc., xviii, 313 ; Broth.,Pfl., x, 279.

= *Tortula indica* Hook., Musc. Exot., ii, 7, t. 135, p.p.
= *Trichostomum indicum* Schw., Suppl., iii, 142, t. 36.
= *Trichostomum orientale* Willd. hb. ; C. M., Syn., i, 568.
= *Barbula natalensis* C. M., Hedw., 1899, p. 106, and in Rehm. 104.

A slender, little, light green moss 5–10 mm. high, simple or branched from near the base ; the lower leaves rather sparse, 1 mm. long, sheathing at the rather wide base, then apparently narrowed (through being more concave) into a bluntly lanceolate, channelled lamina with exserted mucro ending the very strong wide nerve ; the margins revolute for the whole length of the leaf, forming a dark border, but the lamina all densely chlorophyllose except the sheath, which is hyaline, though its reflexed border is often more or less chlorophyllose and has the cells transverse, the oblong sheath cells being longitudinal. Lamina cells small and quadrate, exceedingly obscure, but finely papillose. Nerve wide and deep, yellowish, very verruculose on the back, with very large pimples, especially on the upper half ; the mucro smooth. Perichaetal leaves much longer and narrower than stem-leaves, more channelled and more pointed, often narrowly ligulate and somewhat cucullate. Capsule 1 mm. long on a brown seta 4–5 mm. long, elliptic-cylindrical, with very short, stout, straight, conical, striate-papillose teeth. Some forms have small green gemmae. *B. Kiaerii* Broth. of Madagascar is very near this. S.-E. Asia and islands, N. Australasia, Mascarenes, and South Africa.

Natal. Guldats, Elandskop, 5000 feet, Sim 10,054 ; Fox Hill, Sim 8725 ; Durban, Rehm. 104 (as *B. natalensis* Rehm.) ; Muden, on lime, Sim 10,089 ; Mont aux Sources, Sim 10,090.

S. Rhodesia. " Covering vertical sides of narrow trench cut in soft travestine, 70–90 per cent. carbonate of lime," Mazoe, 4300 feet, June 1917, F. Eyles 711, p.p. ; Bulawayo, Sadler 982 (*fide* Dixon) ; Zambesi, near Victoria Falls, 3000 feet, F. Eyles 1309.

Barbula (Senophylla pungentia) afro-unguiculata C. M., Hedw., 1899, p. 105, is only a juvenile condition of this, from bark of trees on Zambesi islands, Chiambe, P.E.A. (Rev. Menyhärt, 1890).

356. *Barbula Stuhlmannii* Broth., Pfl., i, 3, 410, and x, 279.

= *Anoectangium Stuhlmannii* Broth., in Engl. Bot. Jahrb., 1897, p. 177.
= *Tortula *pretoriensis* Sim, in letters.

Laxly caespitose in extensive mats; stems ½-¾ inch long, erect, usually simple; leaves regularly placed, imbricate, erecto-patent, somewhat concave upward, lingulate, 1·5 mm. long, with rounded and usually mucronate apex; nerve very strong all its length, ending in the mucro, covered on both surfaces with tumid lamina cells, consequently strongly muriculate along the ridge from base to apex; margin entire or lightly crenulated by tumid cells; upper part plane; below the middle to near the base often slightly revolute; cells 10μ wide, rounded, tumid, with a small quadrate central portion chlorophyllose or lightly papillose; basal cells rectangular, hyaline, few, 20×12μ; the chlorophyllose cells sometimes extending to the base toward the margin. Perichaetal leaves not differentiated. Seta yellowish, ½ inch long; capsule erect, narrowly pyriform; the peristome as long as the capsule, twisted, of sixteen teeth split almost to the base into long, very slender, strongly papillose segments, the two segments of a tooth adhering together as if united in places; lid long, conical, with twisted striations. Other parts not seen.

Transvaal. Abundant on the Kopjes, near Pretoria, Sim 10,061 (10,006?). Differs from *T. Rehmanni* and *T. atrovirens* in its caulescent habit and from *B. indica* in its very long twisted peristome, nearly plane leaves, absence of differentiated perichaetal leaves, tumid rather than papillose cells, etc.

This apparently comes exceedingly near *B. indica*, and Dixon (S.A. Journ. Sc., xviii, 314) records it from the packet mentioned above (Eyles 711), from Mazoe, S. Rhodesia, and says concerning that: " The long peristome shows that this is distinct from *B. natalensis* and that group; it agrees very well with *B. Stuhlmannii* from Zanzibar, the only difference that I can detect being that the cells are very slightly more distinct than in the specimens I have seen of that species. The leaves are much narrower than in *B. natalensis*, the apex more acute, slightly cucullate, and the cells less opaque."

My specimen shows only the short erect peristome of *B. indica* (*B. natalensis*), but since the long twisted Barbuloid peristome occurs intermixed, along with the slight vegetative differences mentioned, it seems to me that these indicate rather one species variable as to its peristome being perfect or not, rather than two species, just as happens in *Weisia viridula*; but if so, then a difficulty crops up as to which genus it should be placed in; in my opinion the more perfect form is the better, even though it be less common, since the other is depauperate, so I include both in *Barbula* and leave further investigation to decide generic and specific rank.

Weisia (*Eu-weisia*) *vallis-gratiae* (Hpe., in Musc. Cap. Breutel, 1858); C. M., Hedw., 1899, p. 111; Broth., Pfl., x, 255; Brunnthaler, i, 25, from Genadendal (Cape, W. Prov.), probably belongs here; it is described as very small, with revolute leaf-margin, and very short, narrow peristome teeth. See under *W. crispata* C. M.

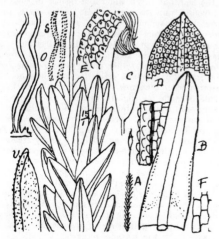

Barbula Stuhlmannii.

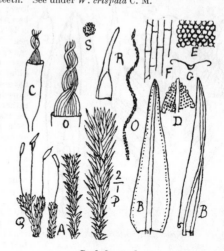

Barbula xanthocarpa.
· D, apex of young leaf of Rehm. 482.

357. *Barbula xanthocarpa* C. M., in Linn., 1843, and Syn., i, 619; Par., Ind., 2nd ed., v, 33; Broth., Pfl., i, 3, 410, and x, 279.

= *Tortella xanthocarpa* (C. M.), Broth., Pfl., i, 3, 397, and x, 263; Dixon, Trans. R. Soc. S. Afr., viii, 3, 191.
= *Trichostomum xanthocarpum* (Schimp. MSS.), C. M., in Hedw., 1899, p. 100.
= *Trichostomum leiodontium* C. M., in Hedw., 1899, p. 100.

Loosely caespitose, autoecious, dull green above, dark below; stems 1–3 cm. long, simple or branched; leaves 1·5–2 mm. long, lanceolate, tapering from near the wide base to the pointed apex, erecto-patent, rather rigid, seldom recurved, concave-keeled; the margin usually revolute the whole length of the leaf; the nerve very strong, yellowish brown, smooth on the keel, excurrent as a strong hyaline mucro. Cells quadrate to oblong at the base of the leaf, elsewhere minute, dense and obscure, and minutely papillose. Leaves incurved when dry but hardly contorted; perichaetal leaves not specialised except occasionally one shorter, rounded at the apex, and sheathing, with margin flat and cells longish, 8μ diameter; archegonia about seven, very long. Seta terminal, 1·5–2 cm. long, erect, yellowish; capsule erect or inclined, shortly cylindrical, oval or oblong, straight, 2·5 mm. long, pale yellow, lid half as long, conical; calyptra cucullate, conical, with a long beak. Peristome of twenty to thirty-two long, slender segments, separate to the narrow connecting membrane at the base, and several times twisted. Spores minute, papillose.

Wager 119 was wrongly distributed as *B. vinealis* Brid., a European species having a longer, narrower, recurved leaf, and which is not known to occur in South Africa.

B. xanthocarpa is near *B. rigidula* Mitt. and *B. spadicea* Mitt. in leaf-form and aspect, but differs in having long peristome teeth; from *B. fallax* Hedw., *B. cylindrica* Schp., and *B. vinealis* Brid. it differs in the leaves, though spreading, being rigid or rather incurved at the apex, not recurved.

B. Macleana, Rehm. 482 (Rhenosterberg, Middelburg, E. Cape Prov., 6500 feet altitude, MacLea, 1873), shows the youngest leaves to have rather wide apex, with projecting marginal cells giving a toothed appearance (fig. D), but this soon disappears; otherwise the plant is identical.

B. perlinearis C. M. (Hedw., 1899, p. 107), founded on a sterile specimen from Spitzkop, Lydenburg, Transvaal (Dr. Wilms), agrees except that the leaf is described as "linear-acuminate" with erect margin. Brotherus (Pfl., i, 3, 408, and x, 278) places it in § *Asteriscium*, along with *B. trichostomacea.*

Barbula (*Senophylla obtuso-acuminata*) *per-torquata* C. M. (Hedw., 1899, p. 108), from Cape Town, in which the scattered leaves when dry are very torquescent and "make a circle round their own axes," probably belongs here, but is insufficiently described. The leaves are described as linear-oblong acuminate, the margin everywhere revolute, the thick ferruginous nerve disappears at the apex, and the seta is long, red, straight, with cylindrical capsule, and it grows partly covered with sand (as if that were a character).

Cape, W. Cape Town, Rehm. 94; Paarl Rock, Sim 9634; Robertson, Sim 9595, 9596; Kluitjes Kraal, Sim 9597; Sir Lowry's Pass Mountains, Miss Stephens.

Transvaal. Heidelberg, July 1912, Wager 119.

Brotherus' inclusion of *B. xanthocarpa* C. M. and also *Tortella xanthocarpa* (C. M.) seem to refer to the same plant.

358. Barbula trichostomacea C. M., Hedw., 1899, p. 108; Broth., Pfl., x, 278.

A small, caespitose, dioecious, brownish plant, having stems 4–6 mm. high, usually simple and erect; leaves rather lax, oblong-acuminate, hardly narrowed at the base, channelled above, the apex made mucronate by the

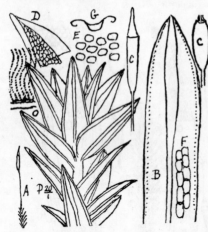

Barbula trichostomacea.

slightly excurrent yellowish-brown nerve, which is wide in the lower half of the leaf, but much wider and deeper in the upper half; margin revolute from the apex to near the leaf-base; upper cells 17μ diameter, diaphanous, quadrate, and separate; those nearest the nerve oblong and rather larger; below mid-leaf the cells are quadrate in contact, gradually becoming rather longer and laxer at the base where they are 20μ long. Perichaetal leaves not different. Seta $\frac{1}{2}$ inch long or rather more, pale; capsule small, narrowly cylindrical; lid subulate, erect, half the length of the capsule. Peristome with a narrow basal membrane and having rather long, pale, straight or slightly twisted, trichostomaceous teeth separate and split to the membrane, and Müller adds a note: "On account of peristome this is better in *Trichostomum*, but on account of the spirally cellulose lid it belongs to *Barbula*." Peristome teeth very papillose. Lid long, conical, spirally areolated; calyptra very long, narrow, dimidiate. Like a young and slender form of *B. xanthocarpa* in appearance, but leaves larger, more red, wider upward, and laxly reticulated without papillae; nerve wide near the apex, and the peristome teeth are different.

B. trichostomacea C. M. var. **chlorophyllosa*, Rehm. 99, though cited by Müller as what he founded *B. trivialis* C. M. on (Hedw., 1899, p. 106), does not differ either in the leaves being more sheathing or in having the margin regularly erect; it sometimes is so, but many of the leaves have the margins recurved, and Rehm. 98, distributed by Rehmann under the same name, also agrees entirely with *B. trichostomacea*, though strangely Müller cites 99 but not 98 for his *B. trivialis*, which Brotherus maintains (Pfl., x, 279).

Rehm. 484, distributed as *Barbula* sp. (old walls round Cape Town), is the same, and Kew Bull. (6, 1923, No. 484) states that its Kew label is written up by Brotherus as "*subxanthocarpa* Broth., n. sp."

Cape, W. Rondebosch, Rehm. 97; hills above Cape Town, Rehm. 483; Kloofnek, 500 feet, Sim 9564 and 9567; Constantia Nek, 300 feet, Sim 9356; Tokai, Sim 9515; Cape Town, Rehm. 98; Montagu, Miss Stephens.

Natal. Mooi River, Sim 10,055.

O.F.S. Kadziberg, Rehm. 99.

359. Barbula salisburiensis Dixon, S.A. Journ. Sc., xviii, 314; Broth., Pfl., x, 279.

Caespitose; stems erect, 2–8 mm. long; leaves rather sparse below, crowded and imbricate above; the inner leaves rather longer and narrower than those below; base clasping but not sheathing; from the rather wide base narrowed gradually to the apex, the upper part appearing almost linear through being deeply channelled ; nerve very strong, yellowish, extending to the mucro, with smooth back ; margin strongly revolute from the base to the mucro ; cells oblong, separate and lax from the apex down, even on the margin, 17μ long, little altered toward the base except that the inner cells are rather larger, confluent, and thick-walled, and a few large lax cells sometimes occur at the very base. Seta ½ inch long, pale red ; capsule erect, elliptical to narrowly cylindrical, 1·5 mm. long, its walls formed of large, irregular, thin-walled cells. Peristome teeth nearly as long as the capsule, slender, papillose, and several times twisted. Perichaetal leaves not differentiated.

· Near *B. xanthocarpa*, but with the leaf narrower, the margin all strongly revolute, the cells all pellucid, the lower cells hardly different from others, and the cells of the capsule-wall large and thin-walled.

S. Rhodesia. Salisbury, East Commonage, on schist formation, 4900 feet altitude, Eyles 596.

Barbula salisburiensis.

D and F, back view ; S, cells of capsule wall.

Barbula Hornschuchiana.

360. *Barbula Hornschuchiana* Schultz, Rec., in Nov. Act. Acad. Leop., xi, 217, t. 33, f. 25 ; C. M., Syn., i, 608 ; Broth., Pfl., x, 278 ; Dixon, 3rd ed., p. 217, Tab. 27.

=*Tortula Hornschuchiana* De Not., Syll., p. 179.

Loosely caespitose on or near rock, firm but low, yellowish green ; stems 5–10 mm. high, usually simple ; leaves erecto-patent, sparse and small below, crowded and larger above, ovate-lanceolate, shortly acuminate, keeled-concave, with strongly revolute margin the whole length, pellucid, yellowish ; the nerve very strong, deeper yellow, extending beyond the leaf apex as a shortly exserted, blunt mucro ; margin entire : cells lax, separate, pellucid, smooth, irregularly quadrate, 6–9μ diameter ; inner basal cells longer, up to 20×8μ ; outer basal cells smaller, 6–10μ long, quadrate, thick-walled. Leaves incurved and twisted when dry.

Fruit not seen in South Africa ; recorded in Europe as : Capsule small, narrowly elliptic or subcylindric ; the seta red below, pale above ; lid long-beaked ; peristome teeth from a very narrow membrane, rather long, purple, much twisted. Perichaetal bracts larger than the stem-leaves, longer, sheathing, with plane margins, longly acuminate, with narrower nerve. Dioecious.

Cape, W. Stellenbosch, 1000 feet, Jan. 1919, Sim 9571 ; Kloofnek, Cape Town, Sim 9561.

Genus 99. TORTELLA C. M.

Plants varying up to large size, loosely tufted and dichotomously branched ; leaves long, linear-lanceolate, clasping at the base, rigid, flexuose, keeled, with the margin flat or slightly incurved, but neither strongly involute nor ever revolute ; undulate, when dry crisped and twisted, the areolation lax and hyaline toward the base, giving a white colour, and minute, opaque, and papillose upward. Seta long, somewhat twisted to the right when dry ; capsule oblong-cylindrical, straight or curved, somewhat thin-walled ; the basal membrane of the peristome usually none ; peristome highly coloured, much twisted to the left, in some cases falling off in one piece ; its teeth long, slender, and papillose. Spores small and smooth.

This is § *Tortella* of *Barbula* in C. Müller's Syn. Musc., and § *Tortuosae* of *Barbula* in Schimper's Synopsis ; it was made a genus by Limpricht (Laubm. Deutchl., i, 599 (1888)), and is maintained by Brotherus (Nat. Pfl., i, 3, 396) and by Paris. The group is included by Dixon and Jameson under *Trichostomum* where the spiral peristome does not agree with the generic name, nor with the original definition of that genus.

Dixon (S.A. Journ. Sc., xviii, 312), dealing with *T. obtusifolia*, says, "Some African species of the same affinity have been placed under *Barbula* (*Streblotrichum*) on account of the long tubular perichaetium ; but the plane-margined leaf and dense obscure upper cells appear to me to indicate *Tortella*; otherwise I do not see on what grounds the two genera can be kept apart."

Synopsis of Tortella.

361. *T. opaca.* Leaf ligulate, margin involute.
362. *T. obtusifolia.* Leaf ligulate, margin plane.
363. *T. eutrichostoma.* Leaf oval-lanceolate, margin plane.
364. *T. caespitosa.* Leaf lanceolate-subulate, margin plane, chlorophyllose.
365. *T. Petrieana.* Leaf lanceolate-subulate, lower margin hyaline.

361. *Tortella opaca* Dixon, S.A. Journ. Sc., xviii, 311.

=*Tortella opacifolia* Dixon, in Sched., but it got published as *T. opaca.*

A small, dark green, caespitose moss 4–5 mm. high; leaves rather lax and small at the stem-base, imbricate and longer in the comal head, erecto-patent when moist, circinate when dry, 2–3 mm. long, ligulate but narrowed upward, the base hyaline with oblong, firm-walled cells except the inner which are longer and narrower, up to 40μ long, the lamina gradually tapering toward the apex, with minute, opaque, rounded cells 4–5μ diameter, the lower third with the margin more or less flat, but in the upper two-thirds the margin is widely involute and dense, often separating from the leaf as if thickened; the leaf in the upper part is channelled and somewhat hooded at the hyaline mucronate apex, the nerve very strong below, 70μ wide at the base, and extending into the mucro, pale in the upper part, brown below and in the old leaves.

Dixon mentions having sometimes seen terminal rosettes of elliptical brood-leaves, having minute elliptical gemmae; these I have not seen.

Victoria Falls, 3000 feet, Sim 8884, 8890.

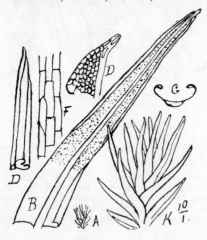

Tortella opaca.

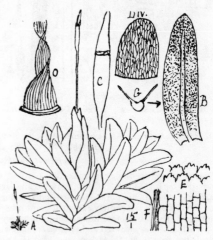

Tortella obtusifolia.

362. *Tortella obtusifolia* Dixon, S.A. Journ. Sc., xviii, 311.

Very small for this group, only 4–6 mm. high, caespitose, dark green, and with rather conspicuous light-coloured sheathing perichaetia terminal on the stem, below which innovations rise on each side. Leaves small, 0·5–1 mm. long, spreading or somewhat reflexed when moist, crisped when dry, lingulate, somewhat narrowed to the base and to the shortly apiculate apex, channelled above, with erect margins; the nerve strong, extending to near the apex, with smooth keel; the cells minute and very dense, papillose above but the margin entire; at the very base the cells are longer, oblong, and pellucid, but the dense cells extend almost to the base. Perichaetal leaves rather longer than others, convolute, the inner wide, with rounded apex and large cells, without nerve, the outer similar to the above toward the base except that the nerve is present, but the upper part of each is narrower and dense, similar to the stem-leaves. Seta ½ inch long, slender, yellow; capsule small, shortly cylindrical, with wide ring and conical lid, and conical dimidiate white calyptra. Peristome comparatively long, the teeth twisted.

Claybank, Umtali, S. Rhodesia, 4200 feet altitude, June 1919, F. Eyles 1741.

363. *Tortella eutrichostoma* (C. M.), Broth., Pfl., i, 3, 397, and x, 263.

=*Barbula eutrichostoma* C. M. (Hedw., 1899, p. 110), and in Rehm. 91.

Loosely tufted; stems 6–12 mm. high, deep green, mostly simple. Leaves from a wide sheathing-base lanceolate or oblong-acuminate, 1·5–2 mm. long, spreading or ascending, flat above, concave in the lower half, the margin not recurved ; nerve strong, yellow, extending into a very short mucro ; cells small, 15μ, hexagonal, densely chlorophyllose, globular, and papillose on the surface ; basal cells much larger, up to 50$\times$20μ, oblong, hyaline ; leaves appressed, somewhat twisted or circinate when dry ; perichaetal leaves longer, 3 mm. long, rather narrow, sheathing at the base only. Seta 10–15 mm. long, red ; capsule 2–3 mm. long, cylindrical, slightly curved, pale ; lid half as long ; calyptra very long ; peristome teeth very long and slender, twisted, connected by a very narrow basal membrane within the mouth. Perichaetal leaves resemble those of *T. caespitosa*, but the stem-leaves are longer and narrower, and the lax basal cells are different.

Cape, W. Hills above Blanco, Rehm. 91 ; Stellenbosch, S. Garsyde.

Cape, E. Knysna, Miss Duthie ; Alexandria Forest, James Sim.

Natal. Umfolosi, Aug. 1911, H. A. Wager.

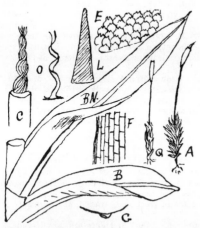

Tortella eutrichostoma.

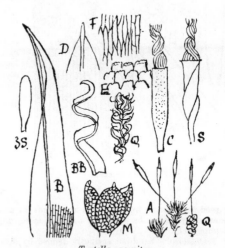

Tortella caespitosa.

S, twisted dry capsule.

364. *Tortella caespitosa* (Schw.), Limpr., Laubm. Deut., i, 600 ; Dixon, S.A. Journ. Sc., xviii, 311 ; Kew Bull., No. 6, 1923, No. 479.

=*Barbula caespitosa* Schw., Suppl., i, 1, 120 ; Schimp., Syn., 1st ed., p. 177 ; 2nd ed., p. 216 ; Broth., Pfl., x, 263.

=*Barbula afrocaespitosa* C. M. (Hedw., 1899, p. 109).

=*Tortella afrocaespitosa* (C. M.), Broth., Pfl., i, 3, 1397, and x, 263.

=*Barbula caespitosa* Schw. var. *angustifolia*, as per Rehm. 479.

Gregarious or caespitose, autoecious, dark green ; stems 4–10 mm. high, simple or divided into several almost sessile crowns ; leaves linear-lanceolate or lanceolate-subulate, keeled, the upper portion conduplicate ; margin crenulated by cells, somewhat undulate, sometimes revolute ; nerve strong, yellow, excurrent in a short hyaline point ; cells 15μ wide, dense, minutely papillose ; those at the base much larger, oblong, hyaline, the outer there small and rounded. Leaves when dry much contorted ; perichaetal leaves longer and narrower ; perigonia axillary, almost sessile, with three very short, oval, laxly cellular, mucronate leaves, without nerve, surrounding several clavate antheridia. Seta 10–15 mm. long, red or pale ; capsule cylindrical, brown ; peristome two to three times twisted, consisting of sixteen long, slender, purple teeth, separate to the base ; lid subulate, half as long as the capsule ; calyptra pale, dimidiate, as long as the capsule. Spores very small, greenish.

I cannot separate *B. syrrhopodontioides*, Rehm. 476, which is a strong, bark-grown, tufted form from Houtbush, Transvaal, with brown foliage in the dry specimens.

B. natalensi-caespitosa C. M., in Linn., 1899, p. 110 ; Pfl., x, 263, from description agrees, except that the antheridia are said to be separate, axillary, without perigonial leaves. Natal (Gueinzius).

T. eutrichostoma C. M. is very near—perhaps too near—to *T. caespitosa*.

Cape, W. Rocks, Camps Bay, *ex* Rehm. 355 (Sim 10,050) ; Blinkwater Ravine, Professor Bews (Sim 8596).

Cape, E. E. Cape Prov., MacLea (Rehm. 479) ; Port Elizabeth, July 1913, Wager ; Knysna, J. Burtt-Davy 17,006 ; Keurbooms River, J. Burtt-Davy 17,036 ; Perie, 4000 feet, Sim 7271 ; Sterkspruit, Herschel, Hepburn.

Natal. Gueinzius, as *B. natali-caespitosa* C. M., Zwart Kop, Sim 7406 ; Sweetwaters Stream, Sim 10,049 ; Hilton Road, Sim 10,051 ; Ladysmith, Sim 10,068.

Transvaal. Lechlaba, Houtbush, Rehm. 476, as *B. syrrhopodontioides* Rehm.

S. Rhodesia. Mazoe, on Fatagura River, 4300 feet, Eyles 712.

Portuguese E. Africa. On white sand in Mozakwen Forest, Rev. H. A. Junod (Sim 10,052).

16

365. *Tortella Petrieana* Sim (new species).*

Laxly caespitose on bark and soil, dark green, autoecious; stems ½ inch high, erect, mostly simple, with axillary male buds high up. Leaves numerous, imbricate, sheathing, 2–2½ mm. long, oblong-acuminate, rather wider at the

long hyaline sheath, which occupies one-third of the leaf and extends up the sides further as a hyaline border; leaves nearly plane or somewhat channelled; when dry strongly circinate, showing the smooth shining keel; nerve strong to the mucro, ferruginous, smooth on the keel; margin plane, crenulated by tumid papillose cells; frequently undulated; cells 8μ diameter, tumid, densely chlorophyllose and minutely papillose; sheath cells 100×20μ, hyaline, with a small alar group of rounded cells; the hyaline sheath cells changing almost suddenly into quadrate chlorophyllose cells except at the margin where hyaline cells ascend often half-way. Seta ¾ inch long, red when mature; capsule cylindrical, 2 mm. long, with twisted peristome half that length, the teeth cut to the base into free, densely papillose segments. Lid conical; calyptra as long as the capsule, dimidiate.

Glynn Falls near Hilton College, 3500 feet, May 1921, Sim 10,063. Near *T. caespitosa*, but easily distinguished by the hyaline cells ascending far up the margin. Named after Professor Petrie, who has assisted very largely in the production of this work. A sterile plant might be mistaken for *Trichostomum brachydontium*.

Tortella Petrieana.

S, part of a peristome segment.

Genus 100. TRICHOSTOMUM Hedw.

Plants tall and strong; leaves long and narrow, curled or inflexed when dry, with flat or slightly incurved margin, not recurved nor strongly involute; the lower cells lax and pellucid, the others minute, obscure, and papillose. Capsule oblong or cylindric, the peristome having slender, deeply forked teeth, erect and straight, the segments more or less imperfect.

Trichostomum has been variously defined by different authors, and Dixon and Jameson now include *Tortella* in it, and change its definition so as to include the twisted peristome.

Tortella is here kept separate, consequently *Trichostomum* contains the allied species which have straight and often somewhat imperfect peristome teeth, but as explained under *Barbula Stuhlmannii* this is an unreliable character. A further difference is quoted from Dixon under *Tortella*.

Synopsis of Trichostomum.

1. Leaves widely lanceolate; nerve excurrent; lamina nearly flat   .   .   .   366. *T. brachydontium.*
1. Leaves long, acuminate from the base   .   .   .   .   2.
1. Leaves linear-ligulate; nerve percurrent   .   .   .   .   3.
2. Leaves 2–3 mm. long, nearly flat; nerve smooth; seta long, red   .   .   .   367. *T. rufisetum.*
2. Leaves 4 mm. long; nerve-keel papillose toward the apex   .   .   .   .   368. *T. tortuloides.*
3. Leaves narrowly linear, nearly flat   .   .   .   .   .   .   369. *T. Rehmanni.*
3. Leaves linear-lanceolate or ligulate; margin notched or irregularly dentate; areolation not very obscure,
370. *T. cylindricum.*
3. Leaves widely linear, crenulate at margins and nerve   .   .   .   .   371. *T. rhodesiae.*
3. Leaves sheathing below, boat-shaped; hyaline cells extending up the margin   .   .   372. *T. inclinatum.*

366. *Trichostomum brachydontium* Bruch, in Flora, 1829, ii, 393; Broth., Pfl., x, 261.

=*Trichostomum mutabile* (Bruch MSS.), Dixon, 2nd ed., p. 238.
=*Trichostomum *Bainsii*, Rehm. 469, 469β.
=*Mollia brachydontia* Lindb., M. Scand., p. 21.

Caespitose, vigorous, deep green; stems ½–1 inch long, erect, usually simple, or extending by innovations below the capsules, these innovations having small leaves at the base and normal leaves above. Leaves laxly arranged, circinate and keeled when dry, erecto-patent when moist, 2 mm. long, numerous, imbricate, clasping and hyaline at

---

* *Tortella Petrieana* Sim (sp. nov.).

Laxe caespitosa, autoica; caules ½″ alti, erecti, simplices, cum gemmis masculinis axillaribus alte positis. Folia plurima, imbricata, vaginata, 2–2½ mm. longa, oblongo-acuminata, paulo latiora ad vaginam longam hyalinam quae tertiam folii partem obtinet et longius secundum latera sursum pertinet marginem hyalinum efficiens. Nervus firmus ad mucronem, ferrugineus, levis in carina; margo planus, crenulatus lamino-cellis tumidis papillosis, vagino-cellae 100×20 mm., hyalinae, cum globo parvo cellarum alarium rotundatarum. Seta ¾″ longa, rubra; capsula cylindrica, 2 mm. longa, peristomio torto, dentes fissi ad basin in segmenta libera dense papillosa. Operculum conicum; calyptra conica, dimidiata.

Apud cataractam ad Hilton College, Natal, Sim 10,063.

the base, lanceolate and keeled above, somewhat twisted, concave ; the strong, smooth, yellowish nerve extending to an excurrent acute point, or in some forms percurrent ; cells very densely chlorophyllose, tumid, obscure, and minutely papillose, 6–8μ diameter, the cells of the sheath laxly oblong, pellucid, those nearest the nerve up to $40 \times 15\mu$, the outer ones smaller but oblong and pellucid ; margin straight or undulate, plane or somewhat incurved in accordance with dryness, entire except for one or two protruding cells at the base of the point which consists of large hyaline cells ; involute when dry. Capsule oval-oblong or narrowly elliptical on a 10–12 mm. yellow seta ; ring elastic, white ; peristome yellow, erect, short, of sixteen slender papillose segments, or more or less imperfect. Lid long, rostrate. Calyptra long, cucullate.

Occurs in Europe, Asia, N. Africa, Mascarenes, and Australasia, and varies there into several forms, including one with more rounded apex.

Cape, W. Cape Town, Sim 9546, 9540 ; Camps Bay, Sim 9325, 9322 ; Platteklip, Sim 9280 ; Elgin, Sim 9594 ; Peak, Sim 10,022 ; Stellenbosch, Sim 8719.

Cape, E. Sharks River, Port Elizabeth, Sim 9053 ; Boschberg, Somerset East, Jas. Sim ; Hankey, Jas. Sim ; Hogsback, Tyumie, 4000–6000 feet, Rev. Henderson 194 (*fide* Dixon) ; Hell's Gate, Uitenhage, Jan. 1922, Sim 9012, 9007 ; Wilderness, George, Feb. 1924, 1700 feet, Miss Stephens ; Gwacwaba River, King William's Town, 2500 feet, Sim 7114.

Natal. Zwart Kop Location, Sim 10,037 ; Giant's Castle, 8000 feet, Sim 10,038 ; Mtunzini, Zululand, Miss Edwards ; Adamshurst, Merrivale, Sim 8698 ; Inchanga, Rehm. 469β (as *T. Bainsii* Rehm.) ; Muden, 2000 feet, Sim 10,031 ; Mont aux Sources, Sim 10,029 ; Camperdown, 2000 feet, G. Webster 926 (*fide* Dixon) ; Cathkin, Sim 10,030 ; Overwood, Polela, Sim 10,023 ; Ngoya, Zululand, 1915, Sim 10,024 ; Benvie, York, 4000 feet, May 1920, Sim 10025 ; Rosetta, Sim 10,032, 10,028 ; Umkomaas, Sim 10,033.

Transvaal. Pilgrims Rest, Rehm. 469 (as *T. Bainsii* Rehm.).

O.F.S. Kadziberg, Rehm. 118 (as *T. Bainsii* Rehm.).

S. Rhodesia. Matopos, Dec. 1920, Mme J. Borle.

Trichostomum mahaliesmontanum, Rehm. 470, from Aapies River, Pretoria, Transvaal (MacLea), is too broken in the specimens available to me to decide upon, but appears to belong here.

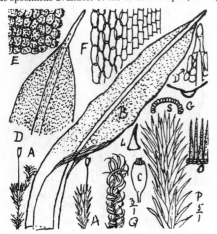

Trichostomum brachydontium.
S, ring.

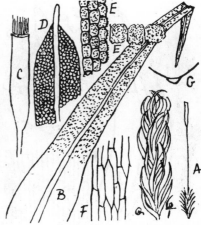

Trichostomum rufisetum.

367. *Trichostomum rufisetum*, C. M. (Hedw., 1899, p. 99, and Rehm. 115, not 4500 as stated by Müller).

=*Tortella rufiseta* (C. M.), Broth., Pfl., i, 3, 397, and x, 263 ; Dixon, Trans. R. Soc. S. Afr., viii, 3, 191, 192.

Laxly caespitose, dioecious, bright green ; stems 3–6 mm. high, simple or much branched at the base, with crowded, stout, almost acaulescent branches ; leaves 1–2 mm. long, nearly flat or slightly concave, spreading when moist, circinate when dry, lanceolate from a rather wider, sheathing, hyaline, laxly long-celled base, the upper portion not cucullate, and tapering gradually its whole length to the acuminate apex, with strongly excurrent nerve ; margins neither involute nor revolute, but rather incurved at the apex ; nerve strong, smooth on the back, pale ; cells of lamina minute, slightly papillose on both sides and margin ; perichaetal leaves not different except that they are rather longer. Capsule red, on slender red seta 3–3·5 cm. long, cylindrical, hardly regular ; lid conical-beaked, red ; peristome teeth irregular, pale, straight or slightly twisted to the left, papillose, some 50μ long, others present but short as if rudimentary ; basal tube very narrow, red. This straight peristome brings it into *Trichostomum*. The long red seta is distinctive.

Cape, W. Tokai, Sim 9521 ; on shady hills above Blanco, Knysna, Rehm. 115 ; George, 1916, and Knysna, 1916, H. A. Wager, 505 and 528.

Cape, E. Elands River Reserve, Uitenhage, James Sim.

Natal. Mont aux Sources, 7000 feet, July 1921, Sim 10,070.

368. *Trichostomum tortuloides* Sull. and Lesq., in Proc. of the Amer. Acad., 1859, p. 277.

=*Tortella tortuloides* (S. and L.), Broth. ; Broth., Pfl., x, 263 ; Dixon, Trans. R. Soc. S. Afr., viii, 3, 192.

Caespitose on soil or rock, autoecious, glaucous green ; stems ½–1½ inch long, with comal tufts of long leaves separated by portions having shorter, mostly broken leaves. Leaves tubular when dry, but when moist plane or slightly concave, 4 mm. long, sheathing below, spreading widely with the points ascending, widest at the hyaline sheath, thence lanceolate, or acuminate from the base, densely chlorophyllose ; margin neither revolute nor involute, crenated by rounded cells, the lamina cells rounded, 8–10µ wide, densely chlorophyllose, minutely papillate but mostly with two larger papillae, the sheath cells hyaline and oblong-hexagonal ; nerve strong, green, covered with lamina cells above, usually smooth on the back except near the apex where there are scattered, conical, hyaline cells. The apex varies much, sometimes shortly rounded, sometimes long and slender. Sheath equally areolate with hyaline, rounded, oblong cells about 50µ long. Fruit not seen.

A much stronger plant than *T. rufisetum*, its nearest relation, differing also in colour and in the back of the nerve being papillose upward, and in the form and size of the basal cells.

This has been placed in *Tortella*, but in the absence of capsules it is better to keep it in *Trichostomum* where it seems among its relatives, until the peristome is seen.

Trichostomum squarrosum Schw. (Suppl., ii, 1, 78, t. 113) ; C. M., Syn., i, 578, of Shaw's list, is not *Barbula squarrosa* Brid. (= *Pleurochaete squarrosa* Lindb., Dixon, 2nd ed., p. 245) related hereto, but is *Leptodontium.*

Natal. Giant's Castle, 8000 feet, 1915, R. E. Symons (Sim 8682).

Transvaal. Barberton, 3000 feet, Jan. 1917, W. Hendry (Sim 8684).

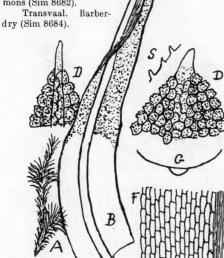

Trichostomum tortuloides.
S, papillae on back of nerve.

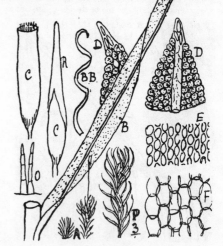

Trichostomum Rehmanni.

369. *Trichostomum Rehmanni* Sim (new species).*

=*T. *sulphureum*, Rehm. 471 (*nomen nudum* and antedated by the American *T sulfureum* C. M.).

Loosely caespitose, pale green ; plants 3–10 mm. long, erect, slender ; leaves linear, narrow, up to 3 mm. long, somewhat twisted, acute, clasping and sheathing slightly, nearly flat ; nerve smooth, percurrent, losing itself in a hyaline. acute, somewhat inflexed point ; margin plane ; cells very minute and dense, 10µ wide, quadrate ; those toward the base larger and elliptical, chlorophyllose ; those at the very base (few) laxly hexagonal, hyaline, 40×20µ. When dry the leaves are still spreading but much twisted. Seta 10–12 mm. long, pale ; capsule 1·5–2 mm. long, oblong or shortly cylindrical, with a long cucullate calyptra and short, slender, 3-celled teeth, in pairs.

---

* *Trichostomum Rehmanni* Sim (sp. nov.)=*T. sulphureum* Rehm. (ined.).

Laxe caespitosum ; caules 3–10 mm. longi, erecti, graciles ; folia linearia, angusta, ad 3 mm. longa, torta, acuta, complectentia, prope plana ; nervus levis, percurrens, apex cuspidatus ; margo planus ; cellae minutissimae et opacae, basin versus maiores, cellae basales grandiores et hyalinae ; seta 10–12 mm. longa, pallida ; capsula 1·5 ad 2 mm. longa, oblonga vel breviter cylindrica ; dentes breves, graciles, 3-cellatae, in paribus positi ; calyptra longa, cucullata.

In montibus Natalibus et Transvaalensibus.

Natal. Cathkin, 5000 feet, 1919, Sim 10,034 ; Hilton Road, Apr. 1917, Sim 10,039 ; Mont aux Sources, Sim 10,080.

Transvaal. Pilgrims Rest, MacLea (Rehm. 471).

Dixon (Trans. R. Soc. S. Afr., viii, 3, 192) places Rehmann's specimen as a small state of *T. tortuloides* ; I have compared the two carefully and find sufficient differences to keep them apart. *T. Rehmanni* is smaller, more laxly leaved, the leaves shorter and linear, the nerve smooth and not excurrent, the cells though dense and tumid not papillose, the basal cells few, but especially the teeth of the peristome very short, smooth, and straight, about 10µ long, so distinctly not *Tortella*. *T. rufisetum* is more or less intermediate, but that also has comparatively short straight teeth.

Dixon (as above) also explains : " Paris' reference of this to *Leptodontium*, as *L. transvaaliense* Par., is due to some confusion merely ; *Trich. sulphureum* Rehm. is antedated by *T. sulphureum* C. M. (*Leptodontium sulfureum* Mitt.), and this, no doubt, led to the confusion on the part of Paris ; Rehmann's plant could under no circumstances be considered a *Leptodontium*," so to avoid further confusion I have preferred a new name rather than that given by Paris.

370. *Trichostomum cylindricum* Bruch, C. M., Syn., i, 386, and ii, 627 ; Broth., Pfl., x, 260.

=*Didymodon cylindricus* B. and S., Sch. Syn.
=*Weisia cylindrica* Bruch, in Bry. Germ., ii, 2, 58, t. 29.
=*Weisia tenuirostris*, Hook. and Tayl., Musc. Brit., 2nd ed., p. 83 ; Suppl., t. 3.
=*Trichostomum tenuirostre* Lindb., *de Tort.*, p. 225 ; Dixon, 3rd ed., p. 239.
=*Trichostomum leptotortellum* (C. M.), Broth., Pfl., i, 3, 394, and x, 260.
=*Barbula leptotortella* C. M., Hedw., 1899, p. 110.

Stems ½–1 inch long, laxly caespitose, erect, simple or forked ; leaves laxly imbricate, spreading and flexuose, the lower small, the upper gradually longer and more acuminate, linear-lanceolate, keeled, very openly areolated ; the base not widened but shortly sheathing ; the apex mucronate ; nerve strong but not wide, pale, striated, not covered by lamina cells, ending just below the mucro ; margin plane, somewhat waved, especially near the base, sometimes entire, sometimes irregularly dentate or crenate, always crenulated by the marginal cells which protrude and are 2-papillate ; marginal cells usually transverse and often larger than others ; lamina cells 8µ diameter, quadrate, separate, tumid, chlorophyllosely many-papillose, and thus irregular in form and with much clear space between the green bundles ; toward the base the cells are longer and gradually join the longer, oblong, thin-walled, hyaline cells of the sheathing base. Seta ½ inch long, yellow, erect, with a yellowish cylindrical capsule 1·5 mm. long, having a red dehiscent ring, a small-celled red band round the neck only, and short peristome of sixteen separate, linear, pale, 6-jointed, slender, smooth, straight segments. Leaves twisted and crisped or circinate when dry. The leaf-margin is very irregular, sometimes quite entire, at other times irregularly dentate, on the same specimen.

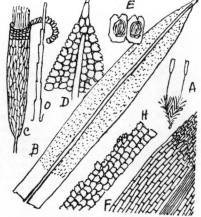

Trichostomum cylindricum.

T. circinatum Broth., Pfl., x, 260 (=*Symblepharis circinata* Besch., Fl. Bry. Reunion, etc., p. 20), known only from the Mascarenes and Central Africa, is recorded by Bryhn (p. 12) as found at Ekombe, Zululand, by Titlestad, but it belongs to the same group as *T cylindricum*, which, no doubt, was mistaken for it.

Müller described his *T. leptotortella* from Boschberg.

Cape, W. Stellenbosch, Miss Duthie (Sim 10,041).

Cape, E. Somerset East, collected by Professor Macowan in 1878 ; Hogsback, Tyumie, 4600 feet, 1916, Rev. Jas. Henderson 202, and again 1917, D. Henderson 328 p.p. ; Humewood, Port Elizabeth, Sim 9063 ; Boschberg, Somerset East, James Sim.

Natal. Little Berg, Cathkin, 6000 feet, July 1922, Miss Owen 17 (Sim 9669) ; Wilsons,' Donnybrook, Feb. 1920, Sim 9989 ; Rosetta Farm, 4000 feet, Sim 9999, 10,043 ; Misty Home, Hemu-hemu, 5000 feet, Jan. 1920, Sim 10,035 ; Giant's Castle, Sim 10,042 ; Mont aux Sources, Sim 10,065 ; Polela, Sim 10,067 ; Inanda, 1000 feet, 1875, J. M. Wood.

Trichostomum rigidulum (Sm.)=*Didymodon rigidulus* Hedw. is recorded by Shaw as collected at Graaff Reinet by MacLea and at Katberg by Shaw. No specimens are available, and there is no reason to think that the identification is correct. It belongs to Europe, Asia, and N. America, and its general appearance may be gathered from the fact that a gemmae-bearing form was formerly named *Ceratodon purpureus*, forma *gemmipara* Jaap. For description and illustration, see Dixon, 2nd ed., p. 214, Tab. XXIXG, as *Barbula rigidula* Mitt. It comes near *T. cylindricum*, but has shorter firmer leaves.

371. *Trichostomum rhodesiae* Broth., in Denkschr. der Math. Naturwissensch K.K. Akad., Bd. 88, p. 735 ; Dixon, S.A. Journ. Sc., xviii, 311 ; Broth., Pfl., x, 260.

Fairly vigorous in loose mats, dark green above, brown below ; stems ½–¾ inch long, mostly simple, erect ; leaves lax below and about 1 mm. long, dense and imbricate above, 2–2·5 mm. long, exceedingly fragile so that almost every leaf is broken off at half-length ; linear from a hardly wider, clasping, pellucid base, and with short acute mucro ; conduplicate in the upper half, the sides flat with erect margin ; nerve fairly strong, ending in the mucro, covered in the upper half of the leaf with tumid lamina cells; cells tumid, fairly large, 12μ wide, quadrate, almost pellucid but minutely papillose, the margin being rendered crenulate by the outer cells; basal cells long and narrow, 70×20μ, thin-walled, hyaline. Other parts not seen.

Named *Gymnostomum *fragillifolium* in letter by Mr. Dixon, a very apt specific name considering the fragility of the leaves without evident cause, but afterwards found to have been previously named as above. Brunnthaler (i, 25) says cells smooth.

S. Rhodesia. Victoria Falls, on trunks of trees, Brunnthaler (i, 25); Victoria Falls, on bark, 3000 feet, Sim 8901, both sterile.

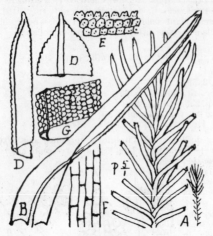

Trichostomum rhodesiae.

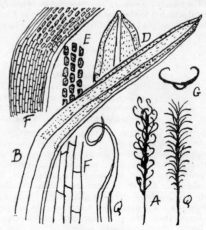

Trichostomum inclinatum.

372. *Trichostomum inclinatum*, Dixon, 3rd ed., p. 242 and Tab. 30A.

=*Tortella inclinata* (Hedw. f.), Limpr.
=*Tortella inclinata* var. *leptotheca* (Brid.), Par., Ind., 2nd ed., v, 30.
=*Barbula inclinata* var. *leptotheca*, Brid., Bryol. Univ., i, 833.
=*Tortula inclinata* β, Hk. and Gr., in Brew., Edin. Journ. Sc., i, 298.

Plant vigorous, 1–2 inches high, mostly simple; leaves lax, spreading or reflexed, bent back above the sheath, ligulate, hardly narrowed upward, carinate-conduplicate, the margin much involute above, producing a short, somewhat cucullate apex in which the nerve is percurrent or shortly mucronate; nerve fairly strong, shining yellowish, the back many-papillose or mamillate below and covered with tumid lamina cells toward the apex; cells irregularly quadrate, 8μ wide, chlorophyllose and papillose, with pellucid interspaces; suddenly changed above the sheath into the long rectangular cells, which ascend the leaf-margin about half-way as a border and occupy the whole of the sheath. Margin entire and plane or slightly irregular below, entire but much involute above, the cells hardly reaching the margin. Fruit not seen in South Africa, but from elsewhere described as having "seta reddish, capsule oval-oblong, more or less curved; peristome teeth long, spirally twisted, fugacious." Leaves very involute when dry, the then spreading, strong, tubular subula making about one twist. The hyaline cells ascending the margin and the involute leaf-margin are the characteristics of this species.

Cape, W. Steenbrass Valley, Sir Lowry's Pass Mountains. Oct. 1917, Miss Stephens, Sim 10,036.

Natal. Ngoya Forest, Zululand, 1000 feet, 1914, T. R. Sim 9987; Umhwati, New Hanover, 3000 feet, Aug. 1920, T. A. Sim 10,011; Blinkwater, Sim 10,095.

There is a previous very old record in the above-mentioned old works.

Var. *brevifolia* Sim (new variety). Leaves 1–1½ mm. long, involute only when dry, plane when moist; sheath wide, leaf-margin bordered beyond the middle. Table Mountain, Dr. Bews (Sim 10,047). Perhaps a good species, but not much of it seen.

Genus 101. Eucladium Br. Eur.

Loosely tufted, glaucous mosses, branching more or less by innovations and having the leaves lax in the lower part and crowded in the upper part of each growth; the stem tender, without central strand; leaves erecto-patent, more or less lanceolate from a rather wider, somewhat sheathing, pellucid base having large hyaline cells; the margin toothed in the upper part of the sheath only; nerve very wide. Capsule (not seen so far in South Africa) erect, regular, oblong; ring permanent; peristome rising from deep inside the mouth, with rather wide and often perforated or somewhat split papillose teeth. This description does not include everything which has formerly been placed in *Eucladium* and is now placed elsewhere.

373. *Eucladium verticillatum* (Linn.), Br. Eur., i, 40 ; Schimp., Syn., 2nd ed., p. 45 ; Par., Ind., 2nd ed., p. 160 ;
Broth., Pfl., x, 258 and fig. 212.

=*Weisia verticillata* Brid., Sp. Musc., i, 121 ; Dixon and Jameson, 3rd ed., p. 235, Tab. 29G.
=*Bryum verticillatum* Linn., Sp. Pl., ii, 1120 (1753).
=*Mollia verticillata* Lindb., M. Scan., p. 21 ; Braithw., Br. M. Fl.

A lax, glaucous, somewhat branched moss ½-2 inches high, having erecto-patent or recurved, spreading, oblong-lanceolate or linear-lanceolate, subacute, fragile leaves 2 mm. long, with pellucid, somewhat sheathing base, and the margin entire except at the upper part of the sheath, where pro-truding cells produce sharp teeth. Lamina cells lax, quadrate, 8–10μ long, chlorophyllose, obscure, hardly papillose ; nerve very wide, percurrent, the upper half covered by lamina cells. Sheath cells 40–50μ long, hyaline. The teeth on the sheath-shoulder remind one of *Dicranella subsubulata*, which, however, differs in all other respects. Leaves slightly keeled, when dry more so and incurved, but not otherwise different. Widely distributed (usually on lime) over Europe, Asia, N. America, and Africa.

Natal. Giant's Castle, 8000 feet, R. E. Symons (Sim 10,097).
Transvaal. Johannesburg, Sim 10,060.

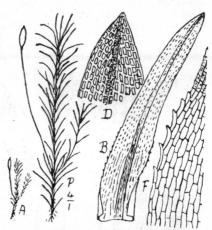

Eucladium verticillatum.
A and P, capsule from European specimens.

Genus 102. DIDYMODON Hedw.

Corresponding with *Barbula* in habit and foliage, the leaves having the same lanceolate-tapering, concave-carinate form from a wide base, the nerve not excurrent, the margin distinctly recurved, areolation rather lax and hyaline toward the leaf-base, and minute, obscure, and densely papillose upward, the papillae being 2-pointed, or smooth and chlorophyllose, but usually obscure. Capsule on a long seta, cylindrical, regular or somewhat arched. Calyptra cucullate, long ; lid rostrate ; ring distinct. Perianth segments thirty-two, slender, short or long, sometimes irregular, distantly barred, not obviously united by a basal membrane, tender and fugacious, erect or slightly inclined to the right.

This is included by Dixon and Jameson in *Barbula*, with which the group agrees except in regard to the twisting of the peristome. Brotherus (Pfl., i, 3, 404) describes the genus as peristomate, and then includes several groups without peristome as well as others with peristome. I have here removed the eperistomate species to *Gymnostomum*, which renders both more intelligible.

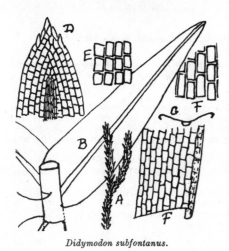

Didymodon subfontanus.

Synopsis of Didymodon.

Leaf-margins slightly revolute ; cells smooth 374. *D. subfontanus.*
Leaf-margins strongly revolute ; cells smooth 375. *D. knysnae.*
Leaf-margins not revolute ; cells papillose ;
    plant ½ inch high or less   .   . 376. *D. afrorubellus.*

374. *Didymodon subfontanus* Dixon (new species).*

Densely tufted, semi-aquatic, on a damp bank below falls ; stem 2–3 cm. high, crowded, straight, simple or slightly branched ; leaves lax, incurved and complicate when dry, erecto-patent when moist, 2 mm. long, lanceolate from a rather narrower, clasping base, somewhat concave below, pellucid, smooth ; the nerve very strong at the base, extending almost into the apex, smooth ; margin expanded but narrowly revolute almost the whole length, margin entire except at the very apex where a few cells protrude more or less ; apex subacute ; upper cells with oblong lumen 12–15μ long, lower cells also with oblong lumen 35–40μ long, all empty, pellucid, and smooth. Other parts not seen. Witpoortje Kloof, Transvaal, Mrs. Moss 10,322.

---

* *Didymodon subfontanus* Dixon (sp. nov.).
Caespitosus, semi-aquaticus ; caules 2–3 cmm. alti, plerumque simplices, folia laxa, erecto-patentia, lanceolata, basi rotundata amplectento ; margo anguste revolutus ; apex leviter denticulatus ; cellae omnes leves atque oblongae ; superiores 12–15 mm., inferiores 35–40 mm. longae. Cetera non vidi.

375. *Didymodon knysnae* Sim (new species).*

=D. *knysnae Rehm., Exsicc. M.A.A., No. 83.

A vigorous plant, only known sterile; stems 1–1½ inch long, erect, simple; leaves small at the stem-base, up to 2 mm. long above, laxly imbricate, lanceolate, keeled-concave in the upper half, concave and somewhat furrowed toward the base, clasping at the base; acute and mucronate at the apex; nerve very strong, deep and wide (120μ) at base, gradually narrower but deep to apex, smooth, yellowish red, excurrent in the reflexed mucro; margin strongly revolute the whole length of the leaf and remaining so, however much it is soaked. Cells smooth, minute, so densely chlorophyllose when green that they cannot be distinguished, but in the lower old leaves all cells are empty, 6–8μ diameter with very thick opaque walls; lower cells quadrate, hyaline, hardly larger except a few at the very base which are laxly long-hexagonal, up to 70×20μ. Leaves incurved and circinate when dry. Other parts not seen.

Knysna, Rehm. 83; Port Elizabeth, Sim 9061; Sharks River, Port Elizabeth, Sim 9056.

Though not seen fertile the areolation and strongly revolute margin indicate *Didymodon*. The leaf has general resemblance to *Hyophila Zeyheri*, but the revolute margin present on every leaf distinguishes it at once.

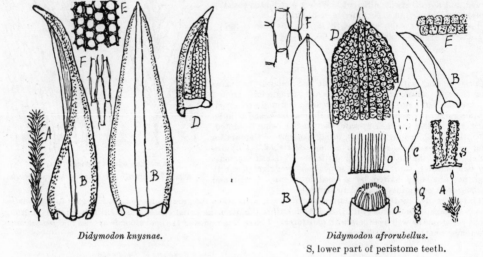

Didymodon knysnae.

Didymodon afrorubellus.
S, lower part of peristome teeth.

376. *Didymodon afrorubellus* Dixon, Trans. R. Soc. S. Afr., viii, 3, 194, and pl. xi, fig. 6 (subgenus *Erythrophyllum* Limpr.); Broth., Pfl., x, 273.

Densely pulvinate, dark green, synoecious; stems crowded, 3–8 mm. high or more, simple or branched, the old fertile stems having leaves twice as long and wide as those of the innovations. Leaves erecto-patent, rigid, channelled above, green or pale red, widely lingulate from a wider clasping base, the leaves on old stems 1–2 mm. long; nerve extending to the mucro, slightly papillose on the back toward the apex; margin plane or slightly revolute, crenulate by protruding cells or sinuate; cells rather large, quadrate, strongly chlorophyllose and papillose, the marginal cells showing their rounded outline and papillae; basal cells large, hyaline, oblong-rectangular, the outer smaller. Leaves much incurved and more or less twisted when dry. Seta yellowish, 6–9 mm. long; capsule small, ovate, with shortly conical beak. Peristome not twisted, of sixteen rather short teeth, each cut to the base into two filaments, strongly papillose, with short basal membrane.

Named from some resemblance to the European and Asiatic *D. rubellus*, Br. Eur., but I find scarcely any trace of the red colour characteristic of that species. *Gymnostomum lingulatum*, Rehm. 437 (sterile), having the nerve very papillose may eventually be found to belong here.

Natal. Van Reenen's Pass, H. A. Wager 419 (79, *fide* Dixon); Scheeper's Nek, 4000 feet, Oct. 1919, Sim 10,092; Benvie, York, 4000 feet, Sim 9668; Rosetta Farm, 4000 feet, Sim 9994, 10,008.

Dixon adds, "Resembles *D. rubellus* in miniature, but with wider, shorter, widely pointed, obtuse leaves, which are not denticulate at summit, but at the most obscurely sinuate or notched."

---

* *Didymodon knysnae* Sim (sp. nov.).

Caules 1–1½″ longi, erecti, simplices; folia superiora 2 mm. longa, lanceolata, basin versus concava, supra carinata, acuta et mucronata; nervus firmissimus ad basin, levis, excurrens in mucrone reflexo, ferrugineus; margo per totam longitudinem fortiter et constanter revolutus; cellae minutae, leves; cellae basales quadratae, hyalinae; folia incurvata circinataque si sicca. Cetera non vidi.

Apud Knysna, Sharks River, Port Elizabeth.

Genus 103. Leptodontium Hampe.

Loosely tufted plants 1–2 inches high, with spreading or squarrose-recurved leaves, which are ovate-oblong to oblong-lanceolate, concave or concave-carinate, mostly flexuose, with roughly serrulate or spinulose-dentate margins, and usually crisped when dry. Nerve reaching to near or beyond the apex, in some forms producing gemmae on the excurrent point and elsewhere. Lower cells shortly rectangular, pellucid, those toward the apex rounded-quadrate, papillose and obscure, and a hyaline margin goes round the leaf. Perichaetium often sheathing. Capsule cylindrical, erect ; peristome of sixteen irregular teeth, each having several smooth, straight, erect, filiform segments, not regularly equal, and occasionally connected in pairs, or threes or fours.

377. *Leptodontium squarrosum* (Hook.), Par., Ind. ; Dixon, Trans. R. Soc. S. Afr., viii, 3, 192 ; Broth., Pfl., x, 268.
=*Didymodon squarrosus* Hook., Musc. Exot., No. 150.
=*Trichostomum squarrosum* Schw., Suppl., ii, 1, p. 78 ; Shaw's list ; C. M., Syn., i, 578.
=*Trichostomum epunctatum* C. M., Syn., i, 579.
=*Leptodontium epunctatum* (C. M.), Par., 1st ed., p. 731 (Broth., Pfl., x, 268).
=*Leptodontium armatum*, Rehm. 466 and 466β, and its var. *brevifolium*, Rehm. 467 (not 466 as cited by Paris);
　　Broth., Pfl., x, 268.
=*Leptodontium squarrosum* var. *paludosum* (R. and C.), Card. (=*L. armatum*, Rehm. 466, *fide* Dixon).
=*Zygodon Simii* Dixon, Trans. R. Soc. S. Afr., viii, 3, 198.
=*Holomitrium Maclenani* Dixon, Smithson. Misc. Coll., 72, 3, 2, and pl. 1, fig. 2.
=*Leptodontium Maclennani* (Dixon), Broth., Pfl., x, 268.
=*Holomitrium acutum* C. H. Wright, Lond. Journ. Bot., 1892, p. 263.
=*Leptodontium acutum* (C. H. Wr.), Broth., Pfl., x, 268.
=*Leptodontium radicosum* Mitt., Journ. Linn. Soc., xxii, 146, 301 =*Didymodon radicosum* Mitt., Journ. Linn. Soc., vii, 149.

Loosely tufted, 1–3 inches high, simple or more usually branched, pale green, often tomentose below, dioecious ; leaves usually lax, deeply sheathing, hyaline there so as to make the stem sheathed in white, ovate-lanceolate, acuminate, squarrose, undulate or twisted, carinate-conduplicate ; apex acute ; margin erect above, entire and revolute toward the base, and showing there strongly the mamillate, many-pointed papillae of the upper surface ; in the upper half of the leaf it is strongly but irregularly toothed or sometimes biserrate, the teeth usually containing one to four cells ; nerve not very strong, straight, ceasing just below the apex, often covered in upper half by papillose lamina cells on both surfaces, and smooth or slightly papillose lower ; cells toward the leaf-apex oval, separate, strongly papillose ; sheath cells longer, thin-walled, hyaline, smooth, or often the marginal cells chlorophyllose to the base. Perichaetal leaves convolute, long, sheathing very high, but of the same structure as others, *i.e.* strongly serrate, nerved to the apex, and with papillose oval cells in all above the sheath. Seta 1–1½ inch high, pale ; capsule cylindrical, 3–5 mm. long, with long subulate lid. Peristome teeth sixteen, each irregularly cut into several slender, unequal, pale segments, trabeculate only above and somewhat papillose. Ring of several rows. When dry the leaves are much crisped and undulated.

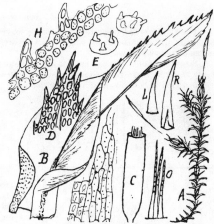

Leptodontium squarrosum.

In June 1917 I visited the top of the Zwart Kop Mountain, Natal, 5000 feet altitude, and found the bare doleritic flow which surrounds the trigonometrical beacon (½ acre or thereby) had every joint in the rock full of this, strong and vigorous, 2 inches high, and much branched but all sterile. Two years later I again visited and found no trace of this or any moss in these cracks, where only some bulblets and some plants of *Selaginella rupestris* were alive ; evidently the rock-surface had dried out completely during the interval, and every moss had not only died but disappeared. As *L. squarrosum* has not at any time been found elsewhere within twenty miles of that rock, one wonders where all the spores had come from which made it abundant on an unusual nidus, for other gatherings have been off tree-stumps. Unfortunately I have not been able to visit the locality again.

Var. *paludosum* (Ren. and Card.) Card.=*L. armatum*, Rehm. 466, is a form having the leaves rather longer and more doubly serrate, but it is not worthy even of varietal rank. The species is exceedingly variable as to length of leaf and amount of serration, but not so as to form varieties.

Hooker's excellent figure, though from a foreign specimen, shows the convolute perichaetal leaves. The species is now found to be widely distributed in Asia, Africa, and Mascarenes.

Natal. Fuller's Bush, Maritzburg, 4000 feet, July 1917, Sim 8705 ; Benvie, York, 4000 feet, May 1920, Sim 9667 ; Zwart Kop Mountain, 5000 feet, on rock, 8698.

Transvaal. Rosehaugh, 4000 feet, Dec. 1914, Sim 8540 ; Woodbush, Wager 121, and T. Jenkins 20/6/1910; Macmac, MacLea, Rehm. 466 (as *L. armatum* Rehm.) ; Mount Lechlaba, above Houtbosh, Rehm. 466β (as *L. armatum* Rehm.) and Rehm. 467 (as *L. armatum* Rehm. var. *brevifolium* Rehm.).

S. Rhodesia.   Umtali, on rock in shade of mountain bush, 5000 feet, Eyles 1720; Dzomba, Zambesia, 5000–7000 feet, Sir John Kirk ;   Eastern Zambesia (as *Holomitrium acutum* Wright).

Leptodontium transvaaliense Par. has already been dealt with under *Trichostomum Rehmanni* Sim.

The Central African genus *Leptodontiopsis* Broth. is closely allied, but has the leaves almost entire, the perichaetal leaves hardly different, and the peristome absent.

Genus 104. OREOWEISIA (W. P. Sch.) De Not.

Its dozen species are scattered over Europe, Asia, and S. America, and in Bot. Zeit., 1858, p. 163, C. Müller described from South Africa *Weisia erosa* (Hpe. MSS.), which in Par., Ind., 1896, was transferred to *Oreoweisia*. No specimen is available in South Africa, nor is the description of the species available here, but the description of the genus as known elsewhere is : Plants soft, densely cushioned, fairly large, up to one or more inches high, and dichotomously branched ; leaves soft, long, linear, sheathing below, keeled above, the upper part distinctly chlorophyllose and highly papillose, with the areolation obscure, but the margin minutely and densely serrulated by prominent, acute, hyaline cells ; the lower part thin and hyaline, with elongated, horizontal, thin-walled cells. Perichaetium not sheathing. Capsule on a slender suberect seta, ovate to oblong, slightly incurved, not striate. Ring absent. Peristome teeth shortly subulate from a lanceolate base, transversely barred, otherways smooth, reddish yellow ; spores large, warted. Brotherus (Pfl., i, 3, 316, and x, 198) placed the genus in Dicranaceae where its papillose surface is an unusual feature, and he maintains this species.

378. *Oreoweisia erosa* C. M.

Mr. Dixon, to whom I referred the matter, writes (26/8/24) : " I also examined the type of *Weisia erosa* Hpe. It is a true *Oreoweisia*, the single capsule very imperfect, but clearly it is very close to our *O. serrulata* but—for one thing at least—with a much stouter nerve. Cells markedly papillose at the back, dense and obscure, with a single marginal row clearer and paler ; nerve quite one-fourth to one-third width of leaf at base, broad to apex ; leaves lingulate, broadly pointed, and incurved at point, denticulate near apex."

Genus 105. GYROWEISIA, W. P. Sch. ; Broth., Pfl., x, 256.

Plants very small, caespitose, dioecious ; leaves linear, obtuse, spreading, not crisped when dry. Capsule elongate-elliptical ; lid short-beaked ; ring very wide, dehiscent ; peristome absent or composed of minute narrow teeth ; spores large, slightly warted.

Since the South African species are without peristome, the wider leaf, rounded leaf-apex, and laxer cells most easily separate them from *Gymnostomum*. *G. tenuis* (Schrad.) W. P. Sch., the one species formerly recorded from South Africa, is here included in *Gymnostomum*.

Synopsis.

379. *G. amplexicaulis.*   Leaf-base clasping ; margins all revolute.
380. *G. latifolia.*   Leaf-base hardly clasping ; margins flat except the lower half in some.

379. *Gyroweisia amplexicaulis* Sim (new species).*

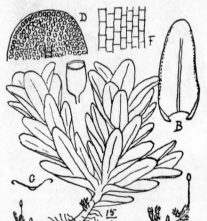

Gyroweisia amplexicaulis.

A minute moss, 3–4 mm. high, branched from near the base, growing in dense tufts ; leaves ascending, 0·7 mm. long, widely clasping at the base, loosely concave upward, rounded at the apex ; the nerve disappearing below the apex ; margin narrowly revolute on every leaf for nearly the whole length and often round the apex, entire ; cells wide apart, translucent, irregular as to form, smooth ; basal cells rectangular ; perichaetal leaves similar. Seta 5–10 mm. high ; capsule 1·5 mm. long, oval-oblong ; peristome absent.

Wager distributed this as *Didymodon afer* C. M. var. *minor* (an undescribed variety), but besides being a much smaller plant than *D. afer* (=*Hymenostylium ceratodonteum*), its leaves are more clasping, every one rounded at the apex, and the areolation is more translucent.

From *G. latifolia* the more numerous leaves, more clasping leaf-base, and every leaf-margin revolute are the best distinctions.

Natal. Van Reenen, July 1911, H. A. Wager 414.

* *Gyroweisia amplexicaulis* Sim (sp. nov.).

Minuta ; 3–4 mm. alta, a prope basin ramosa, caespitosa. Folia plurima, ascendentia ; 0·7 mm. longa ; late complectentia ad basin, late linearia, ad apicem rotundata, nervus non ad apicem pertinens ; margo revolutus per totam longitudinem in quoque folio, plerumque etiam circum apicem ; cellae pellucidae, leves ; folia perich. similia. Seta 5–10 mm. alta ; capsula 1·5 mm. longa, ovato-oblonga ; peristomium abest.

Apud Van Reenen, Natal. Wager 414.

380. *Gyroweisia latifolia* Dixon, S.A. Journ. Sc., xviii, 309.

A dwarf caespitose moss, 3–5 mm. high, usually branched by innovations which leave the perichaetium terminal and start leafless but then have scattered leaves all along the stem, or lax comal tufts where inflorescence is coming; leaves very small, 0·3–0·4 mm. long, the lower shortly elliptical, the upper oblong-lingulate, not narrowed upward, but with rounded, entire, erect apex in which the fairly strong nerve, which is wide below and tapers upward, disappears without reaching the margin. Leaves suberect, somewhat incurved and cucullate and more separate when dry; somewhat concave below, but with the margin shortly and closely reflexed in the lower half on some leaves. Cells rounded, the upper in contact, about 5μ diameter, dense and tumid with clear cell-walls, separate below, and gradually replaced at the base by pellucid square cells, up to 8μ diameter, but with no very large basal cells. The dry condition shows the back of the nerve to be papillose and the lamina cells' mamillate, though when moist they only appear tumid and the nerve cannot easily be seen. Perichaetal leaves few, rather larger, but not otherwise different. Seta about ½ inch high, red, twisted when dry; capsule 1 mm. long, elliptical, brown, with leathery walls of large hexagonal cells. My specimens are all rather old and show no trace of peristome; on which Dixon assumes that they are gymnostomous. Lid and calyptra not seen.

Dry area, Victoria Falls, S. Rhodesia, Sim 8931.

Gyroweisia latifolia.

Genus 106. TRIQUETRELLA C. M., in Österrbot. Zeitzshr., 1897, p. 420.

Slender, thread-like, dioecious plants, growing in compact, rigid, fragile, shining tufts, or sometimes scattered and straggling; stems bluntly 3-edged, forked or variously branched; leaves 3-ranked, loosely placed, slightly reflexed when moist, closely appressed when dry; keeled-concave, ovate or ovate-lanceolate, with the margin entire and in the lower half reflexed; nerve disappearing below the apex, rounded above, papillose on both sides. Cells roundish, chlorophyllose, on both sides papillose with one to two papillae each; those toward the base shortly rectangular and yellowish. Perichaetal leaves sheathing half-way; seta erect, twisted, yellow; capsule erect, cylindrical, smooth; ring breaking off in fragments; peristome inserted below the mouth; teeth sixteen, short, irregularly linear, transparent, without markings, somewhat unequal and sometimes split in places as open holes. Lid beaked, from a cone-shaped base.

Brotherus (Marloth's Flora of South Africa) characterises this genus as having both layers of the peristome similar, without transverse ridges, and the leaves 3-seriate; the latter character distinguishing this from everything else related to it. As it has not been found fertile in South Africa the peristome character is of little value meantime. That *Triquetrella* has troubled systematists is evident from the synonymy which includes *Leskea* (Tayl.); *Didymodon* (Hk. and W.); *Zygodon* (C. M.); *Anomadon* (Sch.); and *Leptodon* (Mitt.), etc.

Brotherus (Nat. Pfl., i, 3, 398) informs us that most of the species are only known sterile, and that all except the Australian species are only known as fragmentary specimens. That need not be the case in regard to *T. tristicha*, which is abundant in places in South Africa.

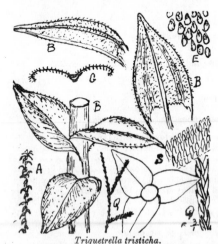

Triquetrella tristicha.
S, papillose cells on reflexed bend.

381. *Triquetrella tristicha*, C. M., in Öst. Bot. Zeitschr., 1897, n. 11, 12; Brunnthaler, 1913, p. 26; Broth., Pfl., x, 265.
= *Anomodon tristichus* W. P. Sch., in Breutel M. Cap.; Jaeg. Ad., ii, 306.
= *Anomodon Macleanus*, Rehm. 638 (*fide* Dixon).
= *Zygodon tristichus* C. M., Bot. Zeit., 1855 (p. 764, *fide* Mitt.; p. 704, *fide* Par.).
= *Triquetrella strictissima* (Rehm.), C. M., in Öst. Bot. Zeit., 1897; Broth., Pfl., x, 266.
= *Zygodon strictissimus*, Rehm. 144.

Mitten (Journ. Linn. Soc., xx, 146, 301) says, "Seems identical with *Z. Priessianus* Hampe (Linn., 1860, p. 683) = *Didymodon papillatus* Hk. f. and W. = *Leskea rubricaulis* Taylor."

Plants straggling and loosely cushioned as pioneers on rock or shingle, or on dry soil; stems ½–2 inches long, usually branched; leaves usually short, open, and very lax below, longer, closer, imbricate, and more conduplicate toward the apex of the stem, arranged in three lines, the older leaves spreading, squarrose, or reflexed when moist, the younger erecto-patent and conduplicate, all closely appressed when dry, forming three distinct lines, but the leaf-apices often twisted to the left; the leaf ovate-acuminate to ovate-lanceolate, far decurrent at the margins of the wide base, channelled along the centre or widely conduplicate, the margin openly revolute in the lower half, plane elsewhere, entire; areolation very lax, pellucid; the cells in straight lines, 8μ wide, irregularly quadrate, mamillate on both surfaces, but these mamillae are often difficult to see except at the revolute part (fig. S) where only the mamillae are visible; nerve a very strong keel, covered on both surfaces by lamina cells, but often hardly papillate on the keel, ceasing just below the apex; stem usually red, often muriculate between the decurrent leaf-ridges. The plants vary exceedingly in size, and the leaves in density and in leaf-form, amount of conduplication and papillosity, but the lower margin is always bent backward or revolute, and the mamillae are visible there if not elsewhere.

Rehmann's 145 and *b* and *c*, named *Zygodon tristichus* C. M., are fairly compact, short-leaved specimens from full exposure, and his 144, named *Z. strictissimus* Rehm., a longer, more lax, long-leaved and conduplicate condition, but I find these conditions occur together or even on the same plant, and that under shade, or if kept long in a box travelling to me, much more etiolated and almost epapillose conditions occur, showing that the slight differences are due to exposure, not to species.

Dixon (Trans. R. Soc. S. Afr., viii, 3, 192) mentions another condition which he had found mixed among *Thuidium promontorii* (a most unusual mixture), from Hogsback, C.P., in which evidently etiolation had taken place, and which then agreed with the South American *T. filicaulis*, Dus., and also with the Australian *T. scabra*, C. M., but in absence of fruit it could not be placed. Although the South African plant has not been found fertile its vegetative variation under conditions of exposure as against even temporary shade are sufficient to account for all here. Naturally it occupies shingle or soil fully exposed in hot dry country subject to occasional rains; really a pioneer on flushes.

If Mitten be correct in his surmise mentioned above, then the papillosity shown in Brotherus' illustration (Pfl., i, 3, 399, fig. 255, and x, fig. 217) of *T. papillata* (Hk. f. and W.) and *T. scabra* C. M. is much more pronounced than anything seen here, where the papillae sometimes have to be searched for, especially in the dry condition where they usually show most.

Shaw included *Didymodon papillatus* Hk. f. and W. (*Triquetrella* Broth.) in his list as collected by Zeyher at Uitenhage; I have collected round Uitenhage and found plenty of *T. tristicha*, but nothing else. Shaw does not appear to have known it, as he mentions *Zygodon tristichus* also—with doubt.

Cape, W. Abundant on and round Table Mountain and in every locality visited in the south-west, Sim 9505, 9333, 9182, 9160, etc., Rehm. 145, 145β, 144.

Cape, E. Abundant in the scrub-veld, rare in grass or forest-veld. Uitenhage, Sim 9013, 9665; Fort Murray, King William's Town, Sim 7318; Alicedale, Professor Potts; Port Elizabeth, Mrs. T. V. Paterson; Grahamstown, Miss Britten; Oudtshoorn, 1824, Miss Taylor, etc.

Genus 107. WEISIA Hedw.

Usually small, terrestrial on rock plants, sometimes forming cushions, sometimes closely gregarious over considerable areas on banks or among grass, the stems though short usually dichotomously branched through innovations below the inflorescence. Leaves lanceolate, linear-lanceolate or subulate, more or less concave, twisted when dry, dark green, with margin usually involute or incurved, and with distinct nerve to near or beyond the apex; the lower cells fairly large, quadrangular, and hyaline; the upper cells minutely quadrate, densely chlorophyllose, and minutely papillose. Inflorescence usually autoecious; capsule pedicellate, small, erect, oval or oblong, symmetrical, normally peristomate, but the peristome is often imperfect and sometimes practically absent; the peristome, when present, of sixteen short, erect, lanceolate or truncate, simple or more or less forked teeth. Outer layer of peristome thicker than the inner one, with conspicuous transverse ridges. Spores large, opaque, minutely papillose.

Synopsis.

382. *W. crispata.* Leaves widely lanceolate, involute.
383. *W. viridula.* Leaves narrowly lanceolate or the upper part linear, very involute.

382. *Weisia crispata*, C. M., Syn., i, 622; Dixon, 3rd ed., p. 230 and Tab. 28N; Broth., Pfl., x, 255.

$=Hymenostomum crispatum$, Bry. Germ., i, 204.
$=Gymnostomum oranicum$, Rehm. 19.
$=Weisia (Hymenostomum) latiuscula$ C. M., Hedw., 1899, p. 111.
$=Weisia oranica$ C. M., Hedw., 1899, p. 112.
$=Hymenostomum latiusculum$ C. M., Par., Ind., p. 189 (1900); Broth., Pfl., x, 254.

Caespitose, dark green; stems 2–4 mm. long, simple or branched low, bare below and with rather sparse imbricate leaves above; leaves when dry cirrhate and very narrow, and twisted like Koodoo horns; when moist spreading or erecto-patent, lanceolate from a wider sheathing base; margin in upper half widely involute and somewhat cucullate at the apex; nerve strong, 70μ wide at the base, smooth, tinged red, acting as a keel along the leaf, curled inward at the apex and shortly mucronate; cells roundly quadrate, 8μ wide, densely chlorophyllose and minutely papillose; sheath-cells oblong-rectangular, hyaline; intermediate cells many, of various sizes of oblong-quadrate cells; perichaetal leaves rather longer and narrower than others; capsule oval-cylindrical on a 10-mm. pale seta, red, slightly constricted at the mouth, and furnished with very irregular and imperfect, short, erect teeth (or with short, erect, slender teeth, *fide* Dixon), but no ring; lid subulate from a wide base.

Much like *Hymenostomum humicolum*, but minutely papillose and shortly peristomate.

C. Müller describes his *Weisia (Hymenostomum) oranica* as of Rehmann, whose name was *Gymnostomum oranicum*; Brotherus (Pfl., x, 255) calls it *W. oranica* Rehm.; and as Dixon points out (Journ. Bot., Apr. 1919, p. 75), a short peristome is present.   Dixon says, " The sixteen teeth are very minute, very little exserted above the capsule-mouth, and sometimes not at all, very narrow and pale, but they are regular, articulate, linear, smooth, and hyaline.   It is therefore a true *Weisia*, not a *Hymenostomum*.   The dioecious inflorescence appears to be the principal character by which it can be separated from *W. viridula*."   Other specimens in my herbarium have the teeth wide, short, irregular, and imperfect as illustrated herein, which quite corresponds with Dixon's illustration of *W. crispata* (Handbook, 2nd ed., p. 230 and Tab. 62F).   It occurs also in Europe, Asia, and North America.

Cape, W.   Slongoli, Table Mountain, 2000 feet, Sim 9180.

Natal.   Mont aux Sources, 7000 feet, Sim 10,048.

O.F.S.   Bloemfontein, Rehm. 19 (as *G. oranicum* Rehm.); Leribe Mountain, Basutoland, Mme Dieterlen 789. (Is this *W. Dieterlenii* Ther., Pfl., x, 255, said to be paroecious while the others are autoecious ?)

Transvaal, Mittel Komati, near Lydenburg, Dr. Wilms, 1887 (as *W. latiuscula* C. M.).

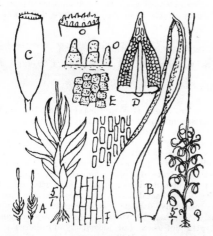

Weisia crispata.

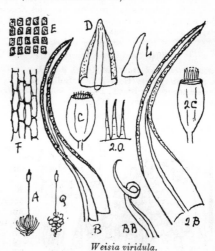

Weisia viridula.

2B, 2C, 2O, var. *longifolia.*

383. *Weisia viridula* (Linn.), Hedw. Fund., ii, 90 ; C. M., Syn. Musc., i, 651 ; Dixon, 3rd ed., p. 230 and Tab. 29A ; Broth., Pfl., x, 255, and vars. *sclerodonta* Hpe. ; *stereodon* C. M., Rehm. 20 ; and *tortilis* Spr. ; all recorded from South Africa.

=*Bryum viridulum* Linn., Sp. Pl., ii, 1119.

=*Mollia viridula* Lindb. ; Braithw., Br. M. Fl.

Laxly caespitose, monoecious, light green ; plant 3–5 mm. high and wide, simple or branched ; leaves much contorted when dry, narrowly lanceolate or linear, from a wide sheathing base, curving outward, then upward ; the margin regularly involute and very dense in the upper two-thirds, thereby making the leaf appear more linear than it actually is.   Margin entire ; nerve excurrent in a short mucro.   Cells minute, dense, opaque, papillose ; those of the sheathing-base oblong and pellucid.   Seta terminal, 6–12 mm. long, yellow ; capsule oval, rather narrowed at the mouth, sulcate when dry ; lid shortly conical, rostrate, somewhat oblique ; peristome present, teeth variable, linear-lanceolate (or sometimes almost absent), rising from inside the mouth.   Male flower terminal on innovations, perigonial leaves small, mostly pellucid, nerve short.

Var. *longifolia* Wag. and Br. (Trans. R. Soc. S. Afr., iv, 1, 6).   A rather stronger plant, with longer leaves, linear-lanceolate toward the apex, and running gradually into the nerve-point instead of being mucronate.   Seta up to 15 mm. long.   Capsule with a hyaline band at the mouth, and peristome of sixteen long, linear, entire or cleft, red teeth (fig. 355, 2B, 2C, and 2O).

Table Mountain, Cape, Sept. 1910, H. A. Wager 122.

Rehmann's No. 20, var. *sclerodonta* Hpe., is put as var. *stereodon* C. M. in Paris' Index.   Concerning *Weisia* (*Hymenostomum*) *brachycarpa* C. M., Hedw., 1899, p. 112 (Broth., Pfl., x, 254), there is nothing in the description which does not fit *W. viridula*, except that no peristome is mentioned, and that Müller included it in his section *Hymenostomum* ; but *W. oranica*, described next preceding this, is in the same position, and is now found to be peristomate. Moreover, in *W. viridula* the peristome is sometimes very rudimentary and soon absent, and I consider that *W. brachycarpa* cannot stand.   Wager wrongly distributed *Hyophila Zeyheri* under this name.

W. linguaelata Shaw, Cape Mon. Mag., Dec. 1878, agrees with *W. viridula* Hedw.

W. viridula is recorded from Europe, Asia, N. and S. America, and Africa.

Cape, W.   Rondebosch, Rehm. 20 (as var. *sclerodonta* Hpe.) ; Cape Town, Rehm. 21 (as var. *tortilis* Spr.) ; Stellenbosch, J. Burtt-Davy ; Peak, Sim 10,076.

Cape, E.   Uitenhage (Zeyher, *fide* Shaw) ; Graaff Reinet (MacLea, as *W. linguaelata* Shaw) ; Perie, Sim 7340.

Natal :—Natal, Wager 435 ; Mount Prospect, 6000 feet, Sim 10,074 ; Maritzburg, Sim 10,075 ; Imbezane, Eyles

1412; Mont aux Sources, Sim 10,077; Little Berg, Cathkin, Sim 10,078, and Miss Owen 27 and 44; Muden, Sim 10,091.

　　Transvaal. Lydenburg, Rehm. 438; Johannesburg, Miss Edwards.
　　O.F.S. The Kloof, Bloemfontein, Sim 10,073.
　　S. Rhodesia. Matopos, Eyles 1049.

Genus 108. TETRAPTERUM Hampe.

　　Small terrestrial plants, annual or perennial, simple or branched. Lower leaves small, distant; upper leaves larger, in a tuft, more or less lanceolate, inflexed when dry, nerved to or beyond the apex; the lower cells hyaline, others densely chlorophyllose, opaque, minute, and papillose. Inflorescence monoecious. Capsule included or shortly exserted, with a short seta, erect, regular, and cleistocarpous. Spores globose, granulate. A small genus, all Australasian except the one South African species, which is illustrated by Brotherus in Marloth's Flora of South Africa (fig. 6G) as *Astomum tetragonum*. Its capsule is roundly quadrangular, hence the name.

384. *Tetrapterum capense* (Hpe. MSS.), Jaeg. M. Cleist., p. 26; Broth., Pfl., x, 253.

=*Phascum tetragonum* Harv., in Hook., Bot. Misc., i, 124, t. 31; Schw., Suppl., iv, 303; C. M., Syn., i, 29, and ii, 521.
=*Systegium tetragonum* (Harv.), Par., Ind., iii, 354.
=*Astomum tetragonum* Broth., Pfl., i, 3, 385, and in Marloth's Cape Flora, pl. 6G; Wager's list 207.
=*Astomum capense* Harv., in Wager's list 206.

　　Caespitose, dioecious; stems 4–5 mm. high, branched or much branched; the lower leaves small and reflexed, upper leaves subrosulate, up to 1 mm. long, imbricate, erecto-patent or somewhat reflexed, widely lanceolate or oblong-acuminate, rather wider in the basal hyaline portion; the apex acute or mucronate, the very wide, usually brown nerve ending in or just below the apex; cells quadrate, very obscure, and set with small papillae; in the lower quarter of the leaf the cells are hyaline and laxly quadrate, 4μ long; border erect when moist; the leaves erect or inflexed with channelled, somewhat convolute lamina when dry. Perichaetal leaves not different; the seta 3–4 mm. long, cleistocarpous, capsule just emergent, grey, oblong-elliptical, loosely 4-angled, or when young or dry irregularly 8-ridged. Calyptra not seen, but shown by Brotherus to be conical and dimidiate, while C. Müller says subulate with the base many-lobed.

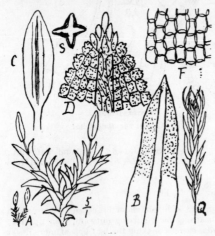

Tetrapterum capense.
S, section of capsule.

　　Described by Harvey long ago, and characteristic specimens were distributed as Rehm. 76 (Cape Town), under the name of *Tetrapterum capense* Harv.; in Shaw's list the name is misprinted *Phascum artragonum* C. M., and he cites as localities Tiger Bay (Harvey), Tigersberg (Pappe), near Zwellendam (Ecklon), and describes an Australian variety (var. *b, cylindricum*) which does not appear to me to belong here, and Brotherus separates it as another species.

Genus 109. HYMENOSTOMUM R. Br.

　　Plants small, the comal leaves much larger than the others; crisped and twisted when dry; opaque, densely but minutely papillose; cells very small and very chlorophyllose, roundly hexagonal; those toward the base pellucid and rectangular. Inflorescence monoecious. Lid subpermanent, long-beaked; capsule often oblique, without peristome; with a narrow mouth, or with the mouth more or less closed after the lid drops by a membrane connected with the columella. Ring present. Spores large, globose, minutely warted.

　　Hymenostomum is maintained by Paris and by Brotherus in Pfl., x, 253, and appears in Brotherus' list (Marloth's Flora of South Africa), but is included in *Weisia* by Dixon and Jameson, 2nd ed.

Synopsis.

385. *H. humicolum.* Leaves mucronate, not papillose.
386. *H. eurybasis.* Leaves obtuse, minutely papillose.

Known to me by name only.

H. cucullatum (C. M.), Par., Ind., p. 595 (1895); Broth., Pfl., x, 254=*Weisia* C. M., in Bot. Zeit., 1858, p. 163 =*Gymnostomum*, Jaeg. Ad., i, 278.

385. *Hymenostomum humicolum* (C. M.), Paris; Dixon, S.A. Journ. Sc., xviii, 309; Broth., Pfl., x, 254.

=*Weisia humicola* C. M., Hedw., 1899, p. 112.

　　Densely caespitose, dioecious, dark green, acaulescent or nearly so; plant only 2–3 mm. high, simple, or branched at the base; leaves rosulate, 1–1·5 mm. long, erecto-patent, the outer short and wide, the inner twice as long, with a rather wide, clasping, pellucid base, and narrower boat-shaped upper half in which the margins are strongly involute, making a somewhat hooded apex. Nerve strong below, bending upward in the hood, and extending as an acute hyaline mucro. Cells 10μ wide, roundly quadrate, lightly chlorophyllose, not papillose; basal cells rectangular, 60×15μ with shorter cells forming the outer lines. Perichaetal leaves longer and narrower than others. Seta pale, 5 mm. high; capsule small, oval, gymnostomous; ring wide; lid conical; calyptra long, white, cylindrical. Occasional specimens have longer, wider-mouthed capsules (fig. S).

The leaf resembles in form that of *Hyophila cyathiforme* but is smaller, and the plant habit is quite different. Dixon says, " The rather large, clear cells separate this species markedly from most of its congeners."

Cape, E. Described by Müller from Macowan's specimen from shaded moist soil in woods, Boschberg, Somerset East, 1876.

S. Rhodesia. Rhodes Grave, Matopos, 5000 feet, Sim 8863 and Wager 892, 915, 890; Zimbabwe, Sim 8813; Rua River, near Salisbury, 5000 feet, on stem of Vellozia, Eyles 1345 and 3330; Lo Magundi, Darwindale, 4500 feet, Eyles 697.

O.F.S. Eagles Nest, Bloemfontein, Professor Potts.

S. Rhodesia. Odzani River, Umtali, Teague 167; Lo Magundi, Eyles 697; Matopos, Eyles 1051, 1099, all fertile. (These four were at first considered by Mr. Dixon to be *Hymenostomum socatranum* (Mitt.), Broth., Pfl., i, 386, which he now connects with *H. tortile* C. M., almost an eperistomate form of *Weisia crispata* C. M., but he decided later that *H. humicolum* has the wider leaves and very markedly larger and more distinct cells found in these specimens.)

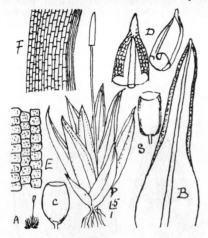

Hymenostomum humicolum.
S, longer capsule occasionally seen.

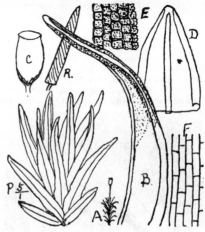

Hymenostomum eurybasis.

386. *Hymenostomum eurybasis* Dixon, S.A. Journ. Sc., xviii, 333.

Caespitose, deep green; stems 3–7 mm. high, mostly simple, with small leaves below and larger leaves above; leaves lanceolate-acuminate, up to 2 mm. long, with rather wide, sheathing, pellucid base for one-third of its length, thence the leaf is narrower, with widely incurved margin and dense, minute, chlorophyllose cells, 7–8μ wide, which show minutely papillose on the leaf-margin and surface, the chlorophyllose cells extending further down near the nerve than at the margin. Apex usually blunt and slightly cucullate, except on central or perichaetal leaves, which may be apiculate. Sheath cells rectangular, hyaline, thin-walled, the lower 80–100×20μ. Nerve rather narrow, hardly percurrent. Seta 4 mm. long, slender, yellowish. Capsule oval-oblong, peristome absent but veil present at first. Calyptra 1 mm. long, white, conical, with areolation twisted to the left. Leaves circinate when dry.

Dixon adds, " The apex of the leaf is generally quite obtuse and subcucullate, with the margin strongly incurved; and the leaves not shining at all and not strongly crisped when dry seem marked features, though these characters are to some extent shared by *H. socotrana* Mitt." Near *W. viridula*, but larger and gymnostomous, with shorter seta.

S. Rhodesia. Mchelele Valley, Matopos, 4700 feet, Eyles, 940, 941.

P.E.A. Hellett's Concession, Magude, 500 feet, Sim 8989.

Genus 110. HYMENOSTYLIUM Brid. (Bryol. Univ., ii, 81 (1827)).

This genus stands in Par., Ind., 2nd ed., with twenty species from all parts of the world, and appears in Brotherus' list (Marloth's Flora of South Africa) as containing the only narrow-leaved and small-celled gymnostomous species in his Pottiaceae, having the mouth of the capsule open when the lid falls off. In Nat. Pfl. he adds to this that the lid falls with the columella attached. C. Müller's description of his section *Hymenostylium* of *Pottia* is: " Leaves narrow, lanceolate, keeled, with erect or revolute margins; capsule gymnostomous," and in his description of *P (H.) ceratodontea* C. M., from Kat River, he states that it resembles *Ceratodon purpureus* but is Pottioid, has conduplicate-keeled, oblong-lanceolate leaves, with minute round cells upward and rather larger quadrate-hexagonal cells toward the base. almost smooth, with a thick yellowish nerve not reaching the apex.   Perichaetal leaves longer ;   capsule erect, elliptical on a long flexuose purple seta ; without ring ; the lid obliquely subulate, long, attached to the columella.

Synopsis.

387. *H. crassinervium.*   Leaves ligulate. margin plane.
388. *H. ceratodontium.*   Leaves lanceolate-acuminate, margin often revolute.

387. *Hymenostylium crassinervium* Broth. and Dixon, in Smithson. Misc. Coll., lxix, No. 2, p. 12, and pl. 1, fig. 2 ; Dixon, in S.A. Journ. Sc., xviii, 310 ; Broth., Pfl., x, 257.*

Caespitose, vigorous ; stems 1–2 inches high, red, often tomentose, and usually branched by small-leaved innovations from near the base which eventually become normal ; without central strand ; foliage light green above, brown below ; leaves scattered below, imbricated above, 2 mm. long, squarrose, spreading, incurved, and somewhat circinate-crisped when dry, ligulate, shortly acuminate, acute, concave-keeled, with short clasping base.   Nerve strong,.smooth, 15μ wide at the base, hardly reaching the apex, covered on the upper surface by lamina cells ; margin plane, slightly irregular in the upper half through some marginal cells rather projecting and others somewhat sunk ; cells small, 8μ diameter, roundly quadrate, chlorophyllose when young, with two to three small white papillae each ; pellucid and empty on old leaves ; lower cells shortly quadrangular and chlorophyllose, pellucid ; basal cells few, laxly hexagonal. 40×10μ.   Capsule on a half-inch pale seta, erect, elliptical, the mouth contracted and slowly 4-ridged, the ridges red. Lid long, conical-rostrate.
Formerly known from Mount Kenia, now from S. Rhodesia and southward.
Cape, W.   Montagu, July 1918, Miss Stephens.
Cape, E.   Umtentu River, E. Pondoland, July 1915, J. Burtt-Davy 15,351 ; Oudtshoorn, 1924, Miss A. Taylor (Sim 10,046).
Natal.   Rosetta Farm, 4000 feet, Aug. 1918, Sim 9998 ; Mont aux Sources, Sim 10,079, 10,045.
Transvaal.   Melville, Johannesburg, 5000 feet, July 1920, Sim 10,026 ; Pilgrims Rest, MacLea.
S. Rhodesia.   Victoria Falls, 3000 feet, Sim 8899, 8902.

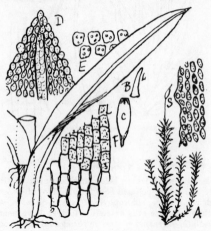

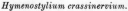

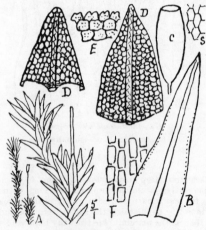

Hymenostylium crassinervium.   *Hymenostylium ceratodonteum.*
S, cells of capsule wall.

388. *Hymenostylium ceratodonteum* (C. M.), Broth., Pfl., i, 3, 389 : Broth., Pfl., x, 257.
=*Pottia ceratodontea* C. M., Syn., i, 564 ; Shaw's list.
=*Gymnostomum ceratodonteum* Par., Ind., p. 543 (1895).
=*Didymodon afer* (C. M.), Broth., Pfl., i, 3, 406, and x, 273.
=*Trichostomum afrum* C. M., in Hedw., 1899, p. 98.
=*Pottia afra*, Par., Suppl. Ind. (1900), p. 283.

Caespitose, light green, dioecious ; stems ¼–¾ inch long, erect, simple or branched by subterminal innovations ; leaves erecto-patent, laxly imbricated, oblong-lanceolate from a rather. wider base, keeled and laxly conduplicate ; apex usually acute, with mucro, but often rounded with nerve hardly reaching the apex ; nerve fairly strong, yellowish,

---

* *Hymenostylium crassinervium* Br. et Dixon (addendum).
Capsula in seta pallida, ½″ longa, erecta, elliptica, os contractum et breviter 4-iugum ; dorsa rubra.   Operculum longum, conicum, rostratum.

smooth on the back where not covered by lamina cells. Margin almost entire or slightly crenulate, revolute in some leaves, especially toward the lower half of leaf, plane in others; cells 7μ wide, quadrate, slightly papillose, with usually two papillae on each, lightly chlorophyllose, and on old leaves all pellucid; basal cells oblong, separate, up to $20 \times 8\mu$. Perichaetal leaves rather longer; capsule on a half-inch purple seta, elliptical, brown, gymnostomous, with walls of lax thin-walled cells; ring absent; lid subulate. See description of *Hymenostylium.*

Dixon suggests in letter that it should be *Didymodon ceratodonteus* (C. M.).

Cape, W. Stellenbosch, Jan. 1919, Sim 10,000, and Burtt-Davy (Sim 10,001).

Cape, E. Kat River, Phillipstown (Ecklon, *fide* C. M., *fide* Dixon); E. Province, MacLea, as *Hyophila basutensis* R. var. *tenella*, Rehm. 459.

Natal. Slangspruit, Maritzburg, 2500 feet, Sim 10,005; Umsundusi, above Mason's Mill, Sim 10,027.

S. Rhodesia. Victoria Falls, 3000 feet, July 1920, Sim 8952, p.p.

H. rigescens Broth. (Pfl., i, 3, 389) = *Gymnostomum rigescens* B. and S., in Musc. Abyss. Schimp., p. 447 = *Weisia rigescens* C. M., Syn., i, 659, is only known as South African on Shaw's record as collected by MacLea near Graaff Reinet, and the identification is extremely doubtful. C. Müller's description of it from Abyssinia would allow it to be *H. ceratodonteum*; it is evidently not *Weisia* as here understood, since the margin is revolute. Müller describes the inner perichaetal leaves as suddenly subulate from a wider sheathing, large-celled base. *Weisia rigescens* Broth. is different and is from Japan.

Genus 111. GYMNOSTOMUM Hedw.

Small slender mosses with erect stems dichotomously forked or having innovations clustered from below the inflorescence, these innovations rooting at the base. Leaves small at the base and increasing in size upward in each year's growth, lanceolate or linear-lanceolate, acute or blunt, channelled in the middle, with entire margins, not involute but sometimes revolute; nerve prominent on the back and extending to the leaf-apex. Cells minute, quadrate, those at the base hexagonal-elongate or rectangular. Inflorescence monoecious or dioecious; male flower terminal, bud-like. Perichaetium more or less specialised, in some species sheathing. Capsule small, on rather a short erect seta, ovate, subglobose or elliptical, symmetrical, rarely subincurved. Peristome absent, ring permanent; lid long or subulate-rostrate, remaining attached to the columella after dehiscence; calyptra cucullate, split to the beak; spores large or small, smooth.

Gymnostomum has been variously defined by different authors; it has been dropped and its species placed in other genera by many. I have transferred hereto some eperistomate species formerly included in *Didymodon.*

Brotherus, who includes some of the species in *Didymodon*, rather confuses matters since in the synopsis in Pfl., x, 248, he uses the separating characters " Per. absent " and " Per. usually present, more or less developed," and at p. 272 describes *Didymodon* as peristomate and describes the peristome; then at pp. 272–273 includes groups without peristome.

Synopsis.

1. Leaf-margins not or hardly revolute . . . . . . 2.
1. Leaf-margins revolute . . . . 4.
2. Innovating to 1 inch height in dense cushions; leaves ligulate . . 389. *G. Bewsii.*
2. Innovating to 2 inches height; leaves lingulate, densely papillose . . 394. *G. lingulatum.*
2. Minute, 2–4 mm. high . . . . 3.
3. Perichaetal leaves almost like others . . . . . 390. *G. gracile.*
3. Perichaetal leaves long and linear . . . . . 391. *G. calcareum.*
4. Leaves ovate-lanceolate . . . . . . 392. *G. dimorphum.*
4. Leaves lanceolate-lingulate, plane above, revolute below . . . 393. *G. Pottsii.*

389. *Gymnostomum Bewsii* Dixon, Trans. R. Soc. S. Afr., viii, 3, 190, and Pl. xi, 4.

= **Eucladium africanum* Wag. and Br. of Wager's list.

Densely caespitose, autoecious, dark green; tufts very compact, up to a foot across, 1–1½ inch deep, level on the surface, growing on soil, usually above wet rock. Stems erect, slender, growing upward by innovations, each growth 3–6 mm. long, with few and short leaves at the base and numerous comal leaves upward, the seta terminal on this branch while one or more innovations again proceed from among the leaves of the rosette. Up to six growths have been seen on one stem, each bearing a persistent capsule, and the stem branched more or less through innovations, singly, or in pairs, or more. Leaves numerous, keeled, clasping at the base, 1–1·7 mm. long, erecto-patent, bluntly oblong-lanceolate or ligulate, the apex rather suddenly brought to a point, margin entire, not or seldom revolute, but often arched backward and toward the apex, the margin often 2-stratose; nerve very strong, wide at the base, extending into the point, often muriculate on the back where covered with lamina cells; cells small, quadrate, chlorophyllose, 5–8μ wide, dense, opaque, minutely papillose; the basal cells hyaline, rectangular, very irregular as to size, up to $40 \times 20\mu$ wide; seta 4–5 mm. long, erect, brown; capsule erect, yellowish, oval, not striate; its cells very lax with about five lines of small, transverse, coloured cells surrounding the mouth; peristome absent; hymenium absent; lid short with a long beak; spores 10–12μ punctulate. Male flowers near the female flowers. The youngest capsules rise somewhat above the youngest innovations, the older capsules are persistent but overgrown. Very similar to *Anoectangium Wilmsianum*, but the mouth of the capsule is distinct. Closely allied to the European lime-loving *W. calcarea* C. M., but a larger plant with longer, more lanceolate, more acute leaves which twist when dry, and with obliquely rostrate lid. In Sim 8713 all the leaves are over 1·5 mm. long.

Dixon says (Trans. R. Soc. S. Afr., viii, 3, 190) : " Its generic position is not quite certain. The bistratose cells are almost unique among these small gymnostomous Trichostomoid mosses; the autoecious inflorescence is rare if not unknown in *Gymnostomum*. On the other hand, the entire absence of hymenium precludes *Hymenostomum*, of which also it has not the habit, and it seems most at home here ; it is not, indeed, unlike some forms of *G. calcareum* in appearance and fruit. . . . The entire absence of peristome appears to me, on the whole, to preclude *Eucladium* as well as *Weisia*."

Common above 7000 feet altitude, absent below that.

Cape, E. Drakensberg, above Ongeluks Nek, Griqualand East, 1915. Bro. Mayol (Sim 8251); Wittebergen at Herschel, 6000 feet, Hepburn 5.

Natal. Tugela Gorge and Goodoo Pass, Mont aux Sources, Professor Bews (Sim 8660, 8713) ; Upper Bushman's River and Giant's Castle, R. E. Symons (Sim 8658, 8661); Mont aux Sources, Miss Edwards ; Cathkin Peak, 7000 feet, Jan. 1918, Sim 9995, and Miss Owens, No. 6 ; Van Reenen Pass, Wager 7 ; also Sim 9988, 9990, 9991, 10,007 ; Olivers Hoek Valley, Miss Edwards.

O.F.S. Rydal Mount, Wager 423, as **Eucladium africanum* W. and B., and as *Gymnostomum calcareum* Broth.

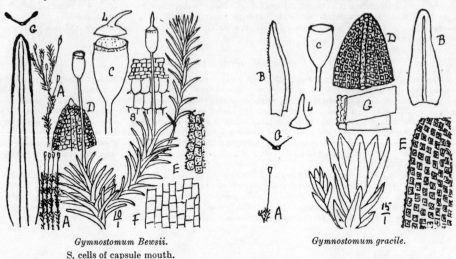

Gymnostomum Bewsii.
S, cells of capsule mouth.

Gymnostomum gracile.

390. *Gymnostomum gracile* Dixon, Trans. R. Soc. S. Afr., viii, 3, 191.

=*Weisia gracilis* Wager, in Trans. R. Soc. S. Afr., iv, 4, and pl. iiA.

Caespitose, small, pale green, growing on soil ; stems erect, 3–4 mm. high, simple or sparingly branched ; the leaves small, 1 mm. long, spreading, clasping at the base, keeled above, bluntly lanceolate-ligulate, rounded at the apex, sometimes hooded and upper leaves sometimes acute ; texture rather open ; margin entire, slightly papillose upward ; nerve strong, reaching almost to the apex, muricate in upper half so far as covered with tumid cells ; cells small, 8–10 mm. long, more or less rectangular, distant, tumid, smooth, opaque, hardly different at base of leaf. Seta terminal, twisted when dry, ½ inch long, erect, yellow ; capsule erect, ovate, brown, gymnostomous ; peristome absent, lid obliquely rostrate ; leaves erect or crisped when dry. Perichaetal leaves more obtuse, and with more lax areolation. Wager describes the stem as unbranched, but I find innovations from the base on his specimens.

Very similar to *Gyroweisia*, from which the opaque cells continuous to the base differ.

Differs from *G. Bewsii* in habit, the nerve more scabrid, seta longer, and leaves wider and more obtuse.

Natal. Blinkwater, Sim 10,003 ; Mont aux Sources, 6000 feet, Sim 10,053 ; Umkomaas, Wager ; Muden, 2000 feet, Oct. 1919, Sim 10,091.

P.E.A. Magude, on lime, 500 feet, March 1921, Sim 8990.

391. *Gymnostomum calcareum* Bry. Germ., i, 153, t. 10, f. 15 ; W. P. Sch., Syn., 2nd ed., p. 40 ;
Par., Ind., 2nd ed., p. 293 ; Broth., Pfl., x, 256.

=*Weisia calcarea* C. M., Syn., i, 659 ; Dixon, 3rd ed., p. 232, Tab. 29D.
=*Mollia calcarea* Lindb., in Braithw., Br. M. Fl., i, 239.

(Very near *G. tenue*, Schrad. (*Weisia tenuis* C. M.), which, however, has longer, laxer, basal cells and constantly obtuse leaves.)

Dioecious ; in neat, dense, close patches, usually on lime or cement walls, brownish green in colour ; stems 2–4 mm. long, much-branched, the branches small and slender as compared with the perichaetal stems, which have longer, more pointed, and quite different leaves. Branch leaves lax, 1 mm. long, ligulate, arching backward, keeled but nearly flat, somewhat narrowed upward, obtuse, entire, except that the minute dense papillae are abundant along the margin and on the cells and nerve where that is covered with lamina cells. Nerve strong, hardly reaching the apex. Cells quadrate, 6–8μ wide, very densely chlorophyllose and minutely many-papillose, except near the leaf-base where they are gradually larger up to 20μ long, and hyaline. Perichaetal leaves 2 mm. long, sheathing and nerved at the base, linear-pointed and nerveless above, erecto-patent when moist, erect when dry, of various sizes, and forming whole branches distinct from the others. Seta 4 mm. long, yellowish; capsule brown, gymnostomous, without hymenium, 1 mm. long, elliptic-cylindrical, somewhat striate when dry, tapering below, rather constricted at the mouth by a band of about five rows of transverse small cells; cells elsewhere very large, lax, and pale. Lid conical.

Cape, E. Recorded (as *Weisia tenuis* Schr.) by Shaw as found by MacLea near Graaff Reinet; Sharks River, Port Elizabeth, Jan. 1922, Sim 9057.

Natal. Mont aux Sources, 7000 feet, July 1921, Sim 9992; Tugela Gorge, 7000 feet, Sim 9993; Koodoo Pass, 7000 feet, Sim 10,010.

O.F.S. The Kloof, Bethlehem, 4000 feet, Jan. 1920, Sim 9997.

Found also in all the continents.

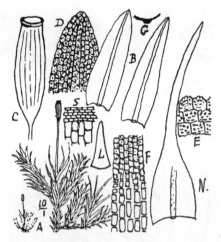

Gymnostomum calcareum.

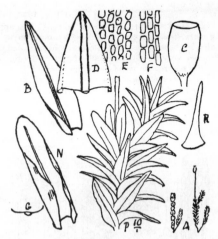

Gymnostomum dimorphum.

392. *Gymnostomum dimorphum* (C. M.) Sim.

=*Barbula dimorpha* C. M., Hedw., 1899, p. 106, and in Rehm. 102.
=*Didymodon dimorphus* (C. M.), Broth., i, 3, 407, and x, 274.

Densely caespitose, dark green, growing on clay; stems 8–12 mm. long, erect, rather slender, sparingly branched by innovations below the capsule, also by basal innovations. Stem-leaves shortly lanceolate or ovate-lanceolate or almost ovate-acuminate, spreading, hardly recurved, bluntly pointed, 1 mm. long, somewhat concave-keeled, clasping below, slightly papillose; margin plane or slightly reflexed, usually undulate; nerve very strong and wide, extending to the apex. Cells minute, quadrate, often diaphanous; basal cells somewhat larger, hyaline, oblong-rectangular. Seta 1 cm. long, purple, terminal but apparently lateral through growth of an innovation; perichaetium of several or many leaves, these leaves rather wider than others, clasping, rounded at the apex, the nerve not extending quite to the apex. Capsule oval-oblong, brown; peristome absent; lid rostrate; calyptra conical; spores minute, slightly papillose. Leaves laxly incurved when dry, not circinate nor much inrolled. Basal innovations have leaves scattered, smaller, and more acute, except the comal portion.

Brotherus (Pfl., x, 274) places this in a group in *Didymodon* which has "fruit unknown," but my specimens show it to be eperistomate.

Müller's description is partly misleading as he speaks of "leaves linear-acuminate, acute, or obtuse," but Rehmann's specimens identify the plant. The name dimorphum is a very appropriate one, since the rounded perichaetal leaves are quite different from the others, and sometimes are rather numerous, but the leaves do not answer Müller's description of linear-acuminate.

Hymenostylium ceratodonteum (C. M.) Br. is very similar.

Cape, W. Camp's Bay Road, Newlands Ravine, Tokai, Cape Town pavements, and Kloof Road, Sim 9311, 9437, 9513, 9545, 9556; Claremont and Kenilworth, Sept. 1922, Mrs. Duncan (Sim 10,012, 10,013); Stellenbosch, 1919, J. Burtt-Davy (Sim 10,014, 10,020); Bishops Court, Pillans 3541, Cape Town, Rehm. 102.

Cape, E. Sterkspruit, Herschel, July 1913, Hepburn 7; Perie, Sim 7332.

Natal. Edendale, May 1919, Sim 10,015; Cathkin, 5000 feet, 1919, T. R. Sim 10,098.

Transvaal. Pretoria, Nov. 1910, H. A. Wager 125; Potehefstroom, 5000 feet, Sept. 1920, Gibbs.

O.F.S. Eagles Nest, Bloemfontein, Professor Potts (Sim 10,017, 10,018).

393. *Gymnostomum Pottsii* Sim.* (New combination.)

=*Didymodon Pottsii* Dixon, Trans. R. Soc. S. Afr., viii, 3, 193 ; Broth., Pfl., x, 273.

Densely caespitose, dioecious, dark green ; stems 4–10 mm. high, mostly simple, slender ; leaves lax but equal below, pale green and closely imbricate toward the apex, 1 mm. long, erecto-patent, keeled, lanceolate-lingulate from a rather wider sub-basal portion ; apex rounded and subacute ; nerve wide all its length, ending just below the apex, covered on both sides by tumid lamina cells, hence apparently papillose or scabrid for the greater part, smooth below ; margin usually flat except near the base, but sometimes the whole length narrowly reflexed, crenulated by protruding cells. Cells irregularly roundly quadrate, 8μ diameter, lightly chlorophyllose, very tumid and slightly papillose, the marginal cells protruding ; basal cells up to $12 \times 8\mu$, oblong, pellucid ; old leaves pellucid throughout. Leaves suberect when dry, not twisted. Perichaetal leaves larger and less chlorophyllose, the inner shorter and widely rounded at the apex. Seta 1 cm. long, slender, red ; capsule small, 1·5 mm. long, elliptical, brown, shining, somewhat sulcate when young, composed of large lax cells except a band of small cells round the mouth ; lid half as long as the capsule, conically beaked, with cells in straight lines. Peristome absent.

O.F.S. Eagles Nest, Bloemfontein, Dec. 1916, Professor Potts (Sim 8663).

Dixon, in describing this species, adds : " This appears from the foliage to belong to *Didymodon* (subgenus *Didymodon sensu stricto*, Limpr.), and being gymnostomous is allied to *D. afer* C. M., but differs widely from that species, *inter alia*, in the leaf-apex, which while little narrowed is usually acute or subacute, rarely rounded. In many respects it resembles very closely the smaller forms of the European *D. tophaceus* (Brid.), which, however, is peristomate and has not the scabrous nerve of the present species."

Didymodon tophaceus (Brid.), referred to above, is credited in Par., Ind., 2nd ed., p. 77, to South Africa and elsewhere, but I find Brotherus does not include South Africa as a habitat, and I have nothing to support the South African record. For description and illustration, see Dixon, 3rd ed., p. 209 and Tab. 26 I (as *Barbula tophacea* Mitt.), who adds : " Easily known by its colour (olive-green or brown), the obtuse lingulate leaves with comparatively short nerve, and the very distinct, rounded, upper cells, and when in fruit by the dark, rather wide-mouthed capsule and the short slender peristome."

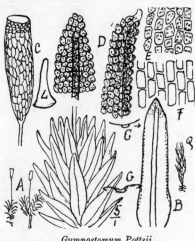

Gymnostomum Pottsii.

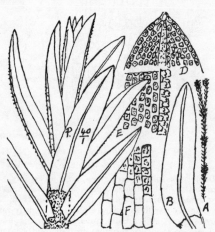

Gymnostomum lingulatum.

394. *Gymnostomum lingulatum* Sim (new species).†

=*G. *lingulatum*, Rehm. 437.

Densely matted in close tufts 1½–2 inches deep, having several level layers of comal growth, apparently annual. Plants slender, slightly branched ; the stem succulent, red ; the leaves crowded and imbricate in comal tufts, sparse between these, lingulate, somewhat concave, 1½ mm. long, minutely but openly areolated ; the apex shortly drawn to a rounded or almost mucronate point ; the base amplexicaul, slightly contracted where it joins ; the nerve fairly

---

* *Gymnostomum Pottsii* Sim.

Dense caespitosum ; caules 4–10 mm. alti, plerumque graciles ; folia imbricata supra, 1 mm. longa, lanceolato-lingulata a latiore parte sub-basali ; margo planus supra, ad basin revolutus.

† *Gymnostomum lingulatum* Sim (sp. nov.)=*G. lingulatum* Rehm. (ined.).

Dense caespitosum artis cumulis ad 2″ altis ; plantae graciles, caulis succidus, ruber ; folia imbricata cumulis comalibus, lingulata, 1½ mm. longa ; cellae minutae, dense papillosae, cum paucis grandibus cellis hyalinis ad folii basin concurrentes ; nervus non ad mucronem pertinens, pallidus, dense papillosus in carina si siccus ; margines integri, plani ; cetera non vidi.

Apud Lechlaba, Transvaal, Rehmann 437.

strong, ending just below the apex, pale, covered on both surfaces with elongated tissue cells ; densely papillose on the keel when dry or newly moistened, less papillose when saturated ; margin entire, plane ; cells minute, chlorophyllose and papillose, separate, extending almost to the leaf-base, where a few large hyaline cells occur, the lowest line often red. Nerve and leaf-margins attached to the stem by long fibres. Other features not seen. Known only from Rehm. 437 from Lechlaba, Houtbosch, Transvaal, collected apparently from a swamp or spring. This resembles and may eventually prove to be *Didymodon afrorubellus* under special conditions of growth, but without fruit this is difficult to decide.

Family 26. SYRRHOPODONTACEAE.

Erect caespitose mosses of vigorous habit and brown colour, from bark or damp rocks, having simple or branched leafy stems ½–2 inches long and inflorescence at first terminal, ultimately lateral ; the leaves imbricate, with long. white, hyaline sheaths, sometimes widened upward, thence the lamina extends to a rather abrupt point ; lamina concave or conduplicate-keeled, much twisted when dry, often bordered, and having minute, tumid, usually papillose cells ; the sheath rigid, erect, pellucid, white, with lax, rectangular, smooth cells. The strong nerve percurrent, sometimes papillose on the keel, sometimes gemmae-bearing at the apex. Capsule on a short straight seta ; erect ; capsule short, with calyptra clasping the capsule neck ; obliquely beaked lid, and the peristome usually present from deep inside the mouth, consisting of sixteen short, widely lanceolate, unsplit teeth, having six to eight transverse bars but no longitudinal division, and usually inflexed, often inside the mouth.

Widely distributed in the tropics and Southern Hemisphere, but absent from Europe, North Asia, and North America, represented in tropical Africa by many species.

Cavers places Leucophanaceae, Syrrhopodontaceae, and Calymperaceae in a new group of the Haplolepideae, viz. Monocranoideae, characterised by having "Teeth undivided, outer layer with papillae, the two layers of about equal thickness, no basal membrane, preperistome often present."

Syrrhopodontaceae and Encalyptaceae as here used constituted Calymperaceae in my check-list, as also in C. Müller's Synopsis Muscorum.

Synopsis of the Genera.

112. CALYMPERES. Calyptra conical, fluted.
113. SYRRHOPODON. Calyptra dimidiate, not fluted.

Genus 112. CALYMPERES Sw.

Erect plants having the general appearance of *Barbula*, but with the hyaline sheath-cells ascending the nerve by steps, and not reaching the margin where a border of small chlorophyllose cells extends to the base ; and also provided with *teniola*, *i.e.* secondary nerve-like rows extending in among the lamina cells. Gemmae sometimes present in an apical cluster on the nerve. Fruit not seen as yet in South Africa, but described as erect, regular, cylindrical, narrowed to the base ; peristome *absent* ; lid conical, beaked ; calyptra conical, furrowed, covering the capsule.

Known from Central and North Africa, East Asia, and Australasia ; from South Africa it is only known by the one species here described.

395. *Calymperes rhodesiae* Dixon, S.A. Journ. Sc., xviii, 303.

Gregarious, on bark ; plant 4–5 mm. high, almost acaulescent ; the lower leaves short and oval with rounded apex ; the nerve ceasing below the apex ; upper leaves sheathing, oblong-lanceolate, 2–3 mm. long, expanded or with slightly involute margin when moist, but circinate and convolute when dry. Sheath varying in length to one-third of the leaf, the basal hyaline cells ascending the nerve by short steps, the apex varying from rounded to acute, often with exserted nerve, on the top of which a cluster of gemmae is usually developed though not always present. Nerve fairly strong, papillose along the whole keel ; lamina cells small, 5μ diameter, rounded, chlorophyllose, papillose on the back, the upper marginal two rows bistratose and denticulate, sometimes detaching, the lower margin usually involute and consisting of several rows of small chlorophyllose cells to the leaf-base ; between these and the nerve the laxly quadrate, hyaline cells occupy the sheath in about six rows, but a well-marked continuation like a nerve (teniola), consisting of two to four rows of long narrow cells, extends half-way up the leaf near its margin in many of the longer leaves. Fruit not seen.

S. Rhodesia. Victoria Falls, 3000 feet, Sim 8879.

Calymperes rhodesiae.

Genus 113. SYRRHOPODON Schw.

Erect, caespitose, usually dioecious plants, ½–2 inches high, the stem apparently succulent through being clad in long, rigid, hyaline, white sheaths, each ending in a twisted or undulated chlorophyllose lamina, usually more or less bordered by long, narrow, smooth, hyaline cells, the lamina cells being small, opaque, and papillose. Teniola not present : gemmae sometimes present on the nerve. Perichaetal leaf-sheaths forming a white cylinder from which the crisped lamina spreads ; the sheaths being more appressed when dry and suberect when moist. Capsule short, oval or subglobose, on a short terminal or ultimately lateral seta ; peristome of sixteen short unsplit teeth, folding into the capsule-mouth when moist ; lid rostrate ; calyptra long, smooth, covering the capsule, and clasping its neck, but split thence upward half-way or more. A large genus in the tropics and Southern Hemisphere, absent in northern countries, frequent in Central Africa and African islands.

Brotherus (Pfl., x, 234) restores *Hypodontium* as having sheathing per. leaves and cylindrical capsule, but our species are almost alike in that.

Synopsis.

1. Nerve-keel papillose	2.
1. Nerve-keel not papillose	3.
2. Leaf bordered all round	396. *S. pomiformis.*
2. Leaf bordered little above the sheath	397. *S. Dregei.*
3. Dry leaf-margins undulated	398. *S. obliquirostris.*
3. Dry leaf-laminae much contorted	399. *S. uncinifolius.*

S. recurvifolius Hsch., in Linn., xv, 118 ; C. M.. Syn., i, 546, is placed by C. Müller (Syn., ii, 641) in *Orthotrichum.*

S. obscurus, Rehm. 126, from Bethlehem, is *Tortula erubescens* C. M.

S. tortuosus Horns., Linn., 1841, p. 117, is *Amphidium cyathicarpum* (*fide* Paris).

396. *Syrrhopodon pomiformis* (Hk.), C. M., Syn., i, 531; Shaw's list.

= *Weisia pomiformis* Hook., Musc. Exot., p. 131.
= *Grimmia pomiformis* Brid., Bryol. Univ., i, 187.
= *Hypodontium pomiforme* C. M., in Hedw., 1899, p. 97 ; Broth., Pfl., x, 234.
= *Syrrhopodon *pyriformis* C. M., in Rehm. 127 and 488 (not 448 as cited by Paris).

Vigorous dioecious plants growing in dense, rigid, flat, light green to dark brown cushions, 1–2 inches deep, usually on rock, sometimes on bark ; fruit rare ; stems 1–2 inches high, simple or forked or fastigiate near the top, erect, rigid when dry ; leaves very numerous, closely imbricate, spreading or squarrose, 3 mm. long, lingulate-acute with rather wider hyaline sheath and usually gradually narrowed to the acute apex, somewhat concave ; nerve very strong at the base, gradually smaller upward, strongly papillose or mamillose on the keel, percurrent ; margin entire, flat or erect when moist, inrolled when dry, with a hyaline border of long narrow cells almost the whole length ; lamina cells minute, 3–5μ diameter, opaque, descending to the sheath where they suddenly meet the long, narrow, hyaline sheath-cells, of which the border is a narrower-celled continuation. Dry leaves bent inward with waved and involute margins, and showing papillose keel mostly. Perichaetal leaves forming a sheathed cylinder standing high above the tuft, then the laminae spreading, very papillose. Seta 1 cm. long, yellowish ; capsule oval or subglobose, with constricted dark mouth-band, from behind which, low down, rise the sixteen short, unsplit, conical, smooth, red teeth, of about six trabeculae, which are erect when dry, but incurved and included in the mouth when moist. Lid obliquely rostrate ; calyptra clasping the base of the capsule, but thence split to near its apex.

C. Müller (Hedw., 1899, p. 97) describes from Boschberg, Somerset East, a variety *Macowanianum* C. M., as "beautifully pulvinate, ferruginous, dense ; leaves small" ; and Rehm. 124, named var. *truncicola* Rehm., from Montagu Pass (sterile), is compact and with short erect leaves, but these do not materially differ. The former seems to have been named in Macowan's Schedule No. 9, *Syrrhopodon Macowanianus* C. M.

Cape, W. The Saddle, Table Mountain ; Devil's Peak ; Platteklip Ravine ; Blinkwater Ravine ; Disa Ravine, Sim 9404, 9244, 9290, 9153, 9268, 9167, and 9135; top of Table Mountain, Dr. Bews (Sim 8664), and Pillans 3535 ; St. John's Peak, Pillans 4050 ; Table Mountain, Rehm. 123 ; Montagu Pass, Rehm. 123β, 124 ; Esternek, Knysna, Rehm. 127 ; Krombeks, Zwellendam, Burchell ; Zeyher, 493 ; Ecklon.

Transvaal. Mount Lechlaba, above Houtbosch, Rehm. 488.

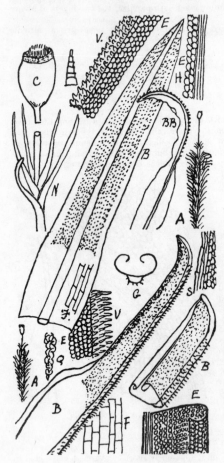

Syrrhopodon pomiformis (above).
Syrrhopodon Dregei (below).—S, denticulate hyaline margin.

397. *Syrrhopodon Dregei* Horns., in Linn., xv (1841), 116 ; C. M., Syn. Musc., i, 530 ; Shaw's list.

=*Hypodontium Dregei* C. M., in Hedw., 1899, p. 97 ; Broth., Pfl., x, 234.
=*Syrrhopodon perichaetialis* Brid., in Regents. Bot. Zeit., 1846, p. 152.

Densely caespitose, rigid, dioecious, dark green, on bark or stones ; stems ½–1 inch high, simple or forked or fastigiate, rigid when dry ; leaves numerous, spreading when moist, inflexed when dry ; sheath fairly long, appressed, white, widely oblong, reduced quickly at its top to the strongly and permanently involute or almost convolute lamina which, when moist, is straight to near the apex and then turned inward, making a cucullate apex ; when dry it is, much crisped ; nerve very strong, extending to the apex as a stiff backbone, densely mamillate from the base to the bend near the apex where it becomes smooth ; the mamillae long and white ; margin involute from the sheath to the apex, entire ; lamina cells 5µ diameter, round, turgid, papillose (though that is difficult to see except at the curves), extending down into the upper part of the sheath and rather further down the nerve, but replaced by narrow, hyaline, slightly denticulate cells on the margin to the top of the sheath, not up the leaf. Sheath cells hyaline, oblong, the outer narrower.

Perichaetal leaves many, long-sheathing and subulate, standing erect and distinct ; seta 1 cm. long, pale ; capsule oval, red, with constricted mouth ; teeth sixteen, lanceolate, trabeculate, yellowish, folding into the capsule when moist ; lid obliquely rostrate ; male flowers axillary.

Cape, W. Slongoli Stair, 1500 feet, Sim 9206 ; Camps Bay, Rehm. 125 ; Paarl Rock, Jan. 1919, T. R. Sim 9630.
Cape, E. On Encephalartos, on summit of the Windvogelberg, Shiloh, Cathcart, Rev. Bauer (Rehm. 487).
Transvaal. Mount Lechlaba, Houtbosch, Rehm. 487*b*.

Wager's 169 (to Natal Museum) distributed under this name is a *Trichostomum*.

Hornschuch's description contains a confusing error ; at first he describes the seta as short and the per. leaves extending beyond the capsule ; afterwards this latter statement is repeated, together with " Seta erect, 2 inches long . . . capsule subimmersed." I find,the seta considerably longer than any of the sheaths, and the lamina then spreads outward or becomes crisped. Hornschuch also refers to the nerve as vanishing below the apex ; I find it percurrent. He compares the species with *S. clavatus* Schwaeg. (C. M., Syn., i, 532), an Australian species which Shaw records in Hb. Hook. as from the Cape, but this seems to have been an error of identification.

398. *Syrrhopodon obliquirostris* C. M., Syn., i, 543 ; Shaw's list ; Broth., Pfl., x, 231.

=*Syrrhopodon erectifolius* C. M., in Hedw., 1899. p. 96 ; Rehm. 128 ; Broth., Pfl., x, 231.

Densely caespitose, dioecious, dark green ; stems ½ inch high, mostly simple or fastigiate near the top, apparently succulent through its clavate form from long white sheaths crowned by small crisped green lamina ; leaves numerous, closely imbricated, erecto-patent, the sheath always expanded and rigid, but the lamina only fully expanded after long saturation ; when dry the smooth-keeled nerve is circinate and the leaf-margins much crisped and undulated. Sheath erect, nearly as long as the lamina, narrow at the base, much wider above ; the lamina thence linear-lanceolate, often incurved, finally erect, keeled and channelled ; the nerve and margins yellowish ; nerve smooth, percurrent ; the margin undulate or erect, bordered everywhere, entire except at the apex ; apex acute, with top of keel, nerve-point, and margins dentate there ; cells 5µ, hexagonal, with central chlorophyll, or often opaque, strongly papillose (though that is difficult to see), rectangular hyaline cells 20µ long occupying the whole sheath except its margin, where a narrow border of long, narrow, hyaline cells (fig. F on p. 264) occurs, which extends as a border from the leaf-base to near the leaf-apex, the chlorophyllose cells descending the nerve a little, and descending as a narrow belt inside the marginal border half down the sheath. Perichaetal sheaths erect as a long white cylinder, with crisped, serrulate, nerved, green lamina. Seta 6–10 mm. high, red, with erect, shortly cylindrical, brown, regular, somewhat striate capsule, slightly constricted at the mouth ; teeth short, red (probably damaged in specimens examined) ; lid and calyptra not seen. The leaf-margin apparently is stronger than the lamina, as it often remains in position when the lamina is washed out.

Müller seems to have had doubt if this were distinct from the South American *S. Gaudichaudi*, Mont., which I have not seen.

Cape, W. Montagu Pass, Rehm. 128 ; Grootvadersbosch, Swellendam, Ecklon.
Natal. Blinkwater, 5000 feet, J. M. Sim (Sim 8728).
O.F.S. Witteberg, above Kadziberg, Rehm.

399. *Syrrhopodon uncinifolius* C. M., in Linn., 1899, p. 96 ; Rehm. 129 ; Broth., Pfl., x, 231.

Laxly caespitose, usually on bark, dioecious ; stems ½–1 inch long, erect, simple or fastigiate, widely clad in erect, white, rigid, expanded sheaths, topped when dry with small twisted corkscrew-form lamina, which when moist is ligulate but usually still somewhat screw-form. Leaf sheaths 1 mm. long, narrow below, wider above, quickly reduced to the narrow twisted or undulated lamina ; apex shortly acuminate and toothed, or often that part is replaced by a gemmae-disc ; nerve yellowish, smooth except at the apex ; margin bordered from leaf-base to apex by long, narrow, yellowish cells ; the overlapping ends of these cells make it sometimes minutely denticulate ; lamina cells small, opaque, chlorophyllose and somewhat papillose, occupying the leaf down to the sheath, and thence as a narrow line inside the margin often to near the base ; hyaline cells laxly rectangular, not changed when dry. Perichaetal cylinder long and white, with narrow, twisted, spreading lamina. Seta 4–6 mm. long, the young ones terminal but older ones remain and become lateral through innovation. Capsule oval-elliptical, dark, small ; teeth short, slender ; lid rostrate ; calyptra clasping the base of the capsule, but thence split high. Gemmae elliptical, of about four cells when mature, but not always present.

Distinguished from *S. obliquirostris* principally by its twisted instead of undulated lamina, but further proof is necessary as to whether that character is constant and sufficient.

Concerning *S. cirrifolius* W. P. Sch. in Breutel, Musc. Cap., I have no further information than the name only, but that is an unpublished name, and indicates that it belongs here.

Cape, W. Window Gorge Waterfall, Kirstenbosch, Sim 9466 ; Montagu Pass, Rehm. 129.

Cape, E. Perie Forest, 4000 feet, 1888, Sim ; Deepwall, Knysna, 1500 feet, Phillips.

Natal. Inanda, J. M. Wood (Rehm. 486β).

Transvaal. Lydenburg, MacLea (Rehm. 486 and *ex* Rehm. 434 and 453).

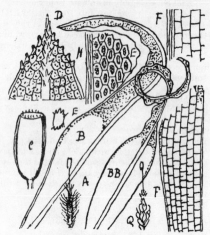

Syrrhopodon obliquirostris.

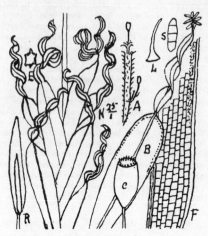

Syrrhopodon uncinifolius.

Family 27. ENCALYPTACEAE.

Erect leafy plants, varying from short to 1 inch in height, gregarious or caespitose, and usually growing on soil or bare turfy banks. Leaves broadly lanceolate, oblong-acute, or spathulate, very large, strongly nerved to or beyond the leaf-apex ; the areolation at the base is rectangular and hyaline, with several rows of marginal cells longer and narrower, and often more coloured ; the rest of the leaf has smaller, roundly quadrate cells, more or less papillose. Capsule erect on a long straight seta, cylindrical ; the calyptra campanulate-cylindrical, usually fringed at the base, enclosing the whole capsule, and remaining long adherent, removing the long, conical, straight-beaked lid with it when it falls ; the peristome very variable, but in the South African species of sixteen lanceolate, short, entire teeth sometimes more or less rudimentary or fragile and fugacious, hence absent ; when present they become permanently inflexed when moistened, so not easily visible.

This family stands alone as an intermediary group between the Aplolepideae and the Diplolepideae, some of its members having single or no peristome, others (though not South African species) having more or less developed double peristome, consequently it has by many been given a place to itself as Heterolepideae, though so far as South Africa is concerned it is conveniently retained in Aplolepideae.

Cavers states, " Along with great uniformity in all other characters there is great diversity in the structure of the peristome, which may be constructed either on the Haplolepidean or the Diplolepidean plan, and in the latter case the superposed exostome and endostome are sometimes coherent, owing to incomplete resorption of the horizontal walls of the peristome-forming cells. According to Philibert, *Encalypta* represents the synthetic ancestral type from which the Haplolepideae and the Diplolepideae have arisen."

Dixon and Jameson also refer to the near resemblance of the peristome in some species to those of *Polytrichum*, thus making this family a meeting-place for these three groups.

Genus 114. ENCALYPTA Schreb.

The only genus ; characters those of the family as described above. In the South African species the calyptra is smooth or almost so, but in another European section it is ribbed and striated, longitudinally or spirally.

400. *Encalypta ciliata* (Hedw.), Hoffm., Deutschl. Fl., ii, 27 ; C. M., Syn., i, 517 ; W. P. Sch., Syn. Musc., 2nd ed, ii, 343 ; Dixon, 3rd ed., p. 252 and Tab. 31B ; Broth., Pfl., x, 242.

=*Leersia ciliata* Hedw., Descr., i, 49.
=*Leersia laciniata* Hedw., Fund. ii, 103, t. 4 ; Braithw., Moss. Fl.

Autoecious, gregarious, up to an inch high, on turfy soil at high altitudes ; stems mostly simple ; seta at first terminal, afterwards lateral through innovation. Leaves very large, 4–5 mm. long, over 1 mm. wide, imbricate, somewhat crisped when dry and then showing the keel prominently; leaves oblong-elliptical, acuminate, narrowed somewhat to the base, and at the apex very variable, either rounded with evanescent nerve or more frequently on the same plant sharply cuspidate by an excurrent nerve. Nerve smooth, yellowish, strong; margin undulate, sometimes slightly revolute near the middle, crenulated by tumid, many-papillose cells; cells roundly quadrate, 15µ diameter, tumid and closely papillose, very opaque; basal cells up to 80µ long, hyaline, rectangular, the inner wide, the outer few lines longer and narrower. Seta 10 mm. long, pale ; capsule cylindrical, smooth, greenish brown, with darker mouth-band. Teeth sixteen, very short, lanceolate, unsplit, of about six trabeculae, erect at first but permanently inflexed when moistened. Lid long, rostrate; calyptra white, conical, scabrid at the apex, covering the capsule, much too wide for it at the base, with about eight free unequal segments there, which are either closely inflexed or widely arched outward in accordance with condition. These segments become detached with age, which has led to Rehm. 489 being distributed as *E. ciliata* and Rehm. 490 as *E. vulgaris* (which has no fringe); both are from Rhenosterberg and are identical, so *E. vulgaris* (Hedw.) Hoffm. drops out of the South African record in Paris' Index.

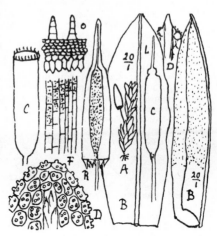

Encalypta ciliata.
B, leaf, $\frac{20}{1}$.

Cape, E. Rhenosterberg, 6800 feet, May 1873, MacLea (Rehm. 489 and 490); Wittebergen, above Herschel, 8000–9000 feet, on southern slopes of a high mountain, June 1917, Ian Hepburn (Majuba Nek).

O.F.S. Rydal Mount, Dec. 1909, H. A. Wager.

It occurs also on mountains in Europe, Asia, N. Africa, N. America, and Australasia.

Tribe II. DIPLOLEPIDEAE.

Peristome (absent in a few cases) consisting normally of two series of teeth, though in some only one series is present. In the outer peristome each tooth is usually transversely barred, and consists of an outer membrane having a longitudinal division down its centre, and an inner membrane without such division. The inner peristome when present is usually of more delicate tissue, variously divided or plaited, its segments usually alternating with the teeth of the outer peristome. A very large tribe, represented everywhere.

Dixon and Jameson say (2nd ed., p. 254), " Philibert has shown that the single peristome of the Aplolepideae is the homologue not of the outer, but of the inner peristome of the Diplolepideae. It is, therefore, misleading to speak of the outer layer of teeth in the latter as the peristome and the inner as the endostome, a term occasionally used ; it would, indeed, be more in accordance with the actual facts to term the latter the peristome and the outer layer the exostome."

Synopsis of Subtribes.

Subtribe A. Diplolepideae Acrocarpae.

The fruit normally terminal on the stem or branches (except in a few cases), though by innovation subsequent to fertilisation it sometimes appears to be lateral. Stems usually erect, seldom pinnate.

Subtribe B. Diplolepideae Pleurocarpae.

Fruit produced from a *lateral* bud on the side of the stem or branches, not from the apex. Stems usually prostrate, and often more or less pinnate.

Subtribe A. Diplolepideae Acrocarpae.

Characters as above.

Synopsis of Families.

28. ORTHOTRICHACEAE. Capsule erect, terminal or lateral, symmetrical, sometimes hardly exserted, often striate ; leaf-cells small, dense, rounded-hexagonal, seldom pellucid, often papillose.

29. EUSTICHIACEAE. Capsule erect, basal, striate ; outer peristome absent ; inner of sixteen teeth united at the base ; leaves distichous, horizontally attached ; cells quadrate, papillose on both sides.

(SPLACHNACEAE. Capsule erect, symmetrical, not striate, on a long seta, with distinct apophysis. Peristome teeth usually in pairs. Leaf-cells very large, hexagonal, smooth, pellucid.)

30. GIGASPERMACEAE. Plant rhizomatous ; capsule sessile or nearly so, globose, eperistomate.

31. EPHEMERACEAE. Protenema abundant ; capsule sessile, globose, cleistocarpous.

32. FUNARIACEAE. Capsule erect or cernuous, often pyriform, terminal, usually with distinct neck but no apophyses. Peristome sometimes absent, usually single or double and well developed. Leaf-cells parenchymatous, smooth.

33. MEESIACEAE. Capsule curved or inclined, clavate or oval-oblong. Peristome double; outer of sixteen teeth; inner of sixteen processes alternate to these teeth, and with basal membrane usually attached to the outer teeth.

34. AULACOMNIACEAE. Capsule cylindrical, regular, ribbed, furrowed when dry. In *Leptotheca* cells smooth, nerve protruding.

35. BARTRAMIACEAE. Capsule usually globose, ultimately striate, erect or cernuous; peristome, when present and developed, of sixteen outer teeth and sixteen inner cleft processes; leaf-cells small, usually short, narrow, and papillose, the papillae often at one or each end of cell.

36. BRYACEAE. Capsule pendulous or inclined, pyriform oval, cylindrical or clavate, not striated, and usually with a neck; peristome, when present and developed, of sixteen outer teeth, and the inner of sixteen often perforated processes, more or less connate below, and alternating with the outer teeth; leaf-cells of very different form in different genera, but smooth and usually long-hexagonal. Intermediate cilia present or absent from inner peristome.

37. MNIACEAE. Smooth; capsule without a neck; sterile stems prostrate; leaf-cells very large, roundly hexagonal.

38. RHIZOGONIACEAE. Cells small, fertile branch basal or nearly so.

Family 28. ORTHOTRICHACEAE.

Plants often erect, dichotomously branched, and cushioned on the bark of trees or on rocks; in South Africa frequently provided with long, prostrate, vegetative growth, pinnately branched and sending up short, erect, floriferous branches, forming wide flat mats. Leaves abundant, mostly equal, spreading when moist, often imbricate or crisped when dry, more or less lanceolate, often from a wide base, very hygroscopic, with the upper areolation roundly hexagonal, small, dense, thick-walled, sometimes 2-layered, and often papillose; and the lower more rectangular or elongated, thin-walled, and more or less hyaline; the nerve strong, reaching to or near the apex, seldom beyond, and the margin often revolute. Vaginula oblong, tube membranaceous, often hairy. Seta erect, short; capsule immersed or exserted, erect, regular, with or without longitudinal striae, usually eight. Calyptra cucullate and smooth, or more frequently campanulate and plaited; often hairy, with jointed hairs. Lid usually short and beaked. Ring none; peristome double or single or fugacious or absent, often reflexed when dry; the thirty-two short, usually papillose teeth of the outer peristome frequently united in pairs or fours so as to seem sixteen or eight, with a few transverse bars; and the inner peristome of eight to sixteen free, equal or alternately shorter cilia. Part of this group closely resembles *Weisia*, the balance has a well-marked general appearance of its own.

Brotherus (Pfl., xi, 11) places Ptychomitriaceae and Orthotrichaceae between Erpodiaceae and Rhacopilaceae.

Synopsis of Orthotrichaceae.

1. Plants erect, slender, cushioned; inflorescence truly lateral. Peristome none . 115. ANOECTANGIUM.
1. Inflorescence terminal, though the seta is often apparently lateral through subsequent innovation . . 2.
2. Peristome none; seta very short . . . . 116. AMPHIDIUM.
2. Peristome usually present, single or double . . . . 3.
3. Calyptra cucullate, not furrowed and not hairy . . . . 4.
3. Calyptra campanulate, sometimes furrowed or hairy . . . . 5.
4. Erect; leaves narrow or lanceolate-acuminate; perianth normally double . . 117. ZYGODON.
4. Erect; leaves oblong; peristome single . . . . . . 118. RHACHITHECIUM.
   DRUMMONDIA.)
4. (Prostrate in flat masses . . . . 6.
5. Calyptra furrowed . . . . 6.
5. Calyptra not furrowed . . . . 9.
6. Stems erect or mostly erect . . . . 7.
6. Stems prostrate with erect fertile branches . . . . 8.
7. Leaves crisped, leaf-margin hyaline; capsule well exserted; stomata exposed; calyptra covered with erect yellow hairs . . . . 119. ULOTA.
7. Leaves not crisped, leaf-margin not hyaline; capsule mostly immersed; stomata usually sunk; calyptra naked or sparsely hairy . . . . 120. ORTHOTRICHUM.
   DASYMITRIUM.)
8. (Capsule pear-shaped . . . . 121. MACROMITRIUM.
8. Capsule oval-cylindrical . . . . COLEOCHAETIUM.)
9. (Capsule pear-shaped; lid conical, blunt . . . .
9. Capsule oval-cylindrical; lid subulate . . . . 122. SCHLOTHEIMIA.

Genus 115. ANOECTANGIUM (Hedw.), Br. Eur.

Stems slender, 2–3 inches high, dichotomously branched, erect, densely matted in large, soft, smooth, light green cushions on damp rocks. Leaves small, lanceolate, keeled, densely papillose, spreading, when dry incurved or twisted at the apex; the margin almost entire, but crenulated by papillae below. Cells very small, obscure, quadrate, those at the base rectangular and more translucent. Dioecious. Perichaetium *distinctly lateral*, thus differing from all its congeners. Male plant more slender; capsule erect, small, ovate, smooth, without peristome. Lid nearly horizontal, with a long, slender, subulate beak. Ring very narrow; calyptra smooth, cucullate.

401. *Anoectangium Wilmsianum* (C. M.), Par., Suppl. Ind., p. 13 (1900) ; Broth , Pfl., x, 246.

=*Zygodon Wilmsianus* C. M., Hedw., 1899, p. 113.
=*Anoectangium *mamabolense*, Rehm. 435.

Dioecious, densely caespitose, green above, brown below ; stems ⅓–¾ inch long, very slender, branched by innovations either basal, lateral, or terminal, these innovations usually few-leaved below, then sparsely leaved, but more regularly and closely leaved above like the main stem, or often crowded in a coma. Leaves laxly imbricate, erecto-patent, 0·5–1 mm. long, mostly short but with a few longer ones mixed in, clasping at the base, ligulate, acuminate, arched backward and deeply channelled above ; the keel very strong and placed outside the lamina, smooth except where covered with lamina cells near the shortly rounded apex, percurrent in a slight mucro ; the margin folded outward in the upper half, almost entire, but minutely crenulated by the numerous small papillae ; cells 15μ wide, obscure, densely and minutely papillose, these cells extending almost to the leaf-base, where a few hyaline oblong cells replace them, but are not larger. Perichaetium truly axillary, composed of about six sheathing leaves having separate oblong cells and little or no nerve. Seta 4–5 mm. long, erect, pale ; capsule small, ovoid, gymnostomous, brown, with short neck and rostrate lid. Calyptra not seen. Male flower short leaved, on basal or lateral innovation. Leaves appressed and twisted when dry.

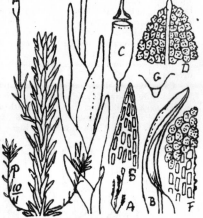

Anoectangium Wilmsianum.

Very much like *Gymnostomum Bewsii*, and easily mistaken for it, but smaller in all parts.

C. Müller's description errs in describing the perichaetal leaves as similar to others, but Wager added to this by describing (Trans. R. Soc. S. Afr., iv, 1, 5, and Pl. IID) as *Anoectangium assimile* Broth. and Wag. the same plant, as is shown in his specimen 127 in the Natal Museum. " Distinguished by its smaller stature, shorter and less-pointed leaves, short capsule without neck, and perichaetal bracts not distinct," in not one of which particulars does his specimen differ from the usual form of *A. Wilmsianum*. Brotherus still includes *A. assimile* Br. and W., and places it in a group not fitting the description.

A. *mamabolense, Rehm. 435, is sterile and much branched, but shows no differentiation and is the oldest name but was not published. It appears to have the nerve papillose, but the nerve seldom shows itself, and it is the convex lamina which is papillose as usual.

A. Wilmsianum is a mountain species entirely.

Natal. Mont aux Sources, 8000 feet, Gibb ; Goodoo Stream, Upper Tugela, 6500 feet, Sept. 1915, Professor Bews (Sim 8380) ; Giant's Castle, 8000 feet, 1915, R. E. Symons (Sim 8659) ; Natal, Wager 433 ; Van Reenen, Wager 127 (as *A. assimile* B. and W.).

O.F.S. Basutoland, top of the Giant's Castle, 8000 feet, R. E. Symons (Sim 8662).

Transvaal. Lydenburg, Dr. Wilms, Aug. 1887 (*fide* C. M.) ; Lechlaba Mountain, above Mamabola, Houtbosch, Rehm. 435 (as *A. mamabolense* Rehm.).

Concerning the mid-African *A. pusillum* Mitt. (Journ. Linn. Soc., xxii, 146, 305)=*A. scabrum* Broth., see Journ. of Bot., Apr. 1919, p. 75.

Genus 116. AMPHIDIUM (Nees in Sturm, Deutschl. Fl., ii, 17, 1819) Schimp., emend. in Bry. Eur. Consp., 1855 (=*Amphoridium* W. P. Sch., Syn., 1st ed., p. 247).

Rather slender, erect, dichotomously branched mosses, growing in dense cushions several inches deep on moist rocks, with leaves all along the stem, often intermixed with rhizoid tomentum. Leaves more or less lanceolate, acute, keeled, minutely papillose, twisted and crisped when dry ; margin subentire, and the nerve hardly reaching the apex. Cells much as in *Anoectangium* but rather larger. Inflorescence terminal, soon apparently lateral through innovation ; capsule oval and when dry urn-shaped, on a very short seta, almost included, deeply grooved, and without peristome. Calyptra cucullate, smooth ; lid obliquely beaked.

Included by Dixon and Jameson (2nd ed.) in *Zygodon*, but kept separate by Paris and by Brotherus.

402. *Amphidium cyathicarpum* (Mont.), Broth., Pfl., 1, i–iii. 1, 460 ; x, 193 ;
Dixon, Smithson. Misc. Coll., lxix, 2 (1918).

=*Zygodon cyathicarpus* Mont., in Ann. Sc. Nat., 1845, p. 106, and Syll., p. 37 ; C. M., Syn. Musc., i, 682 ; Shaw's list.
=*Didymodon cyathicarpus* Mitt., in Journ. Linn. Soc., 1859, p. 70.
=*Gymnostomum linearifolium* Tayl., in Lond. Journ. Bot., 1846, p. 42.
=*Oncophorus cyathicarpus* Mitt., in Voy. H.M.S. " Challenger," Bot., iii, 78.
=*Syrrhopodon tortuosus* Horns., in Linn., xv (1841), 117.
=*Amphoridium cyathicarpum* (Mont.), Jaeg. Ad., i, 386 ; Par., Ind., 2nd ed., i, 31.
=*Zygodon *subcyathicarpus*, Rehm. 146 ; Par., Ind., 1378 (1898), and 2nd ed., p. 32.
=*Zygodon kilimandscharicus* C. M., Flora, lxxiii (1890), 482.
=*Zygodon *Macleanus*, Rehm. 501.

A vigorous, green, caespitose, autoecious moss, growing usually on lime soil in mountain localities, tomentose below ; stems erect, ½ inch high, slender, much branched by fasciculate innovations, with minute terminal capsule on a very short seta. Leaves mostly lax, but with fastigiate crowded clusters ; sheathing somewhat, thence erecto-patent even when dry but then twisted and crisped ; 2 mm. long, channelled, narrowly linear, with acuminate-acute apex ; sheath rather wide and hyaline toward the base ; nerve strong, percurrent or almost so, smooth ; margin plane when moist or slightly revolute in the lower half, mostly entire, but with occasional irregular teeth or steps inward all along, especially near the apex ; cells quadrate, minute, 6–10μ long, chlorophyllose and minutely papillose, dense toward the apex, sparse lower, and gradually merging into the larger, oblong, hyaline, basal cells, 20μ long, in the lower part of the sheath ; marginal lamina cells rather larger and transverse. Seta 1–2 mm. long, erect ; capsule 1 mm. long or less, erect, almost immersed, the perichaetal leaves extending considerably beyond it ; oval when moist, when dry narrowed from a wide mouth (like an inverted bell) and 8–16-ribbed, gymnostomous, brown ; lid rostrate from a convex base ; calyptra dimidiate, smooth ; perichaetal leaves longer, more sheathing, more toothed, and the lower part more laxly areolated than other leaves. Male flowers bud-like at base of fertile stem ; its leaves short, ovate-subulate, pellucid below.

Brotherus now includes *Amphidium* in his subfamily Rhabdoweisioideae of Dicranaceae along with Rhabdo-weisia (Pfl., x, 192).

Cape, W. At the waterfall on the eastern side of Devil's Peak, Ecklon (*fide* Hornschuch) ; cataract at Devil's Peak, Rehm. 146 (as *Z. subcyathicarpus* Rehm.).

Transvaal. Pilgrims Rest, Rehm. 501 (as *Zygodon Macleanus* Rehm.).

Found in North, Central, and South Africa, New Zealand, Australasia, and South America, but rare in South Africa. My drawing is from a Mount Elgon specimen.

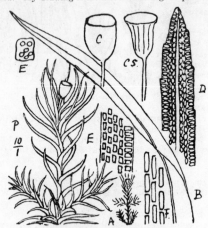

Amphidium cyathicarpum.
CS, capsule, when dry.

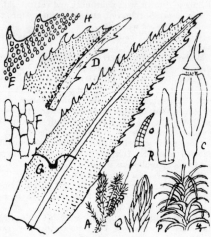

Zygodon runcinatus.

Genus 117. ZYGODON Hook and Tayl.

Strong or slender erect mosses, dichotomously or fastigiately branched by innovation, growing in dense or lax cushions on moist rocks, the stems often interlaced by red rhizoid tomentum below. Leaves numerous, more or less lanceolate or spathulate-acute, spreading when moist, incurved or twisted when dry, keeled to or near to the apex, and the margin plane, entire, crenulate or serrate ; cells very small and dense, rounded, papillose, those at the base roundly quadrate, thick-walled. Capsule 8-furrowed on rather a long seta, well exserted, oval pyriform with distinct neck ; narrow when dry ; peristome double, single, fugacious or absent ; lid obliquely beaked ; calyptra cucullate, smooth. Often confused with *Weisia*.

Synopsis of Zygodon.

1. Leaves serrated . . . . 403. *Z. runcinatus.*
1. Leaves not serrated . . . . 2.
2. Vigorous plants ; leaves 2 mm. long . . . . 3.
2. Small or slender ; leaves 1 mm. long or less . . . . 4.
3. Leaves lanceolate-acuminate, margins undulate . . . 404. *Z. trichomitrius.*
3. Leaves linear-acuminate, margins plane . . . . . 405. *Z. africanus.*
4. Leaves oblong-acuminate, channelled . . . . . 406. *Z. leptobolax.*
4. Leaves ligulate-acute, margins arched back . . . . 407. *Z. Dixoni.*
4. Leaves narrowly ligulate-acuminate, plane . . . . 408. *Z. transvaaliensis.*

403. *Zygodon runcinatus* C. M., in Hedw. (1899), p. 114 ; Broth., Pfl., 1, iii, 1, 463 ; Rehm. 150, 150β ; Broth., Pfl., xi, 15.

Vigorous, laxly caespitose, dioecious, brownish green, when old or dry more brown ; stems erect, freely branched, ¼–1 inch high, laxly leaved below ; leaves crowded above, appressed when dry, patent-squarrose or reflexed when moist, channelled but with the margins spreading, more channelled-complicate toward the apex, 1–2 mm. long, lanceolate, slightly narrowed and clasping at the base, acuminate towards the apex, roughly and irregularly serrate in the upper half and less so in the lower half, these teeth containing several cells. Nerve strong as a keel, smooth to half-way up or more, but thence to the apex covered with lamina cells, sometimes in two lamellae, and with occasional strong spinose teeth like the marginal ones, usually present or numerous near the apex. Cells round, 8μ diameter, green, obscure, but separate and in rows; basal cells 20×10μ, hyaline, roundish oblong. Perichaetium at first terminal, its leaves not different. Seta ½ inch long, pale; capsule cylindrical, 8-ribbed, tapering below in a rather long neck; peristome very short, white; lid long, rostrate; calyptra dimidiate, white, smooth. Concerning the peristome, Brotherus places this species in the section, "peristome double or rudimentary, cilia eight." That may be its position, as cilia are often fugacious, and my specimens had capsules too young or rather old; I find the sixteen smooth, short, white, trabeculate, outer teeth in pairs, but not the cilia. Specimens differ much in size, and the leaf serration resembles that of *Leptodontium*, but the colour and appearance is different, the perichaetal sheath is absent, and the capsule is different.

Cape, W. On Paarl Rock, Sim 9636; Disa Ravine, Sim 9140 and 9150; Blinkwater Ravine, Platteklip Ravine, and The Saddle, Table Mountain, Sim 9266, 9289, and 9398; Table Mountain, Rehm. 150; Devil's Peak, Rehm. 150β.

Rehmann's numbers quoted by Müller in Hedw., 1899, p. 114, are wrong; they may be his collecting numbers, but are not his distribution numbers in Exsicc. Musc. Afr. Austr.

404. *Zygodon trichomitrius* Hk. and Wils., Lond. Journ. of Bot., 1846, p. 143, t. 4*b*; C. M., Syn., i, 677, and ii, 637, and Hedw., 1899, p. 115; Broth., Pfl., xi, 14.

=*Z. Rehmanni* C. M., in Rehm. 149 and 498 (reduced to synonym by C. Müller in Hedw., 1899, p. 115).

Dioecious, laxly caespitose in large tufts on bark, very light green, grey below, and with much-branched red tomentum at the base; stems 1–2 inches long, sparsely dichotomously branched; leaves yellowish green, varying from erecto-patent to recurved, channelled, undulated, 2 mm. long, openly areolated, widely lanceolate-acuminate, with excurrent pale nerve, which is strong from the base and smooth on the keel; margin suberect or spreading, or subrecurved toward the base, entire but minutely crenulated by papillose marginal cells. Cells separate, elliptical, 20μ long, lightly chlorophyllose and papillose, all more or less transverse, the marginal rather larger, the basal cells to 80μ long, narrowly hexagonal; the outer make the margin denticulate where they join; the chlorophyllose cells descending rather further down the margin than elsewhere. Dry leaves are considerably crisped and twisted, and more erect and channelled than when moist. Other parts not seen, but described by Müller (Syn., i, 677) as: "Perichaetal leaves wider, erect; vaginula pilose, theca erect on very short seta, elliptic-oblong, gymnostomous, the mouth narrowed, plicate when dry, dark grey; lid conical-subulate, suboblique, equal half the length of the capsule. Calyptra hairy, with yellow hairs." It is included among those with peristome absent, gemmae not produced, and capsule sulcate, and Brotherus describes it as dioecious.

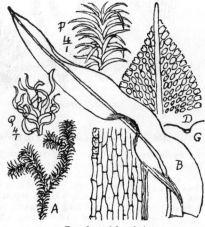

Zygodon trichomitrius.

Mitten (Journ. Linn. Soc., xxii, 146, 305), in describing the Kilimanjaro *Z. erosum* Mitt., includes *Z. runcinatus* in error as a synonym of *Z. trichomitrius*; the two are quite distinct, though there may be a doubt as to whether this *Z. erosum* Mitt. is distinct from *Z. trichomitrius*. Mitten describes it as having the leaves narrower and the basal cells quadrate-elongate; I have not seen the type, but a sterile specimen I have from Mount Elgon, Uganda, 13,000 feet altitude (Dummer 3412β), agrees in every respect with *Z. trichomitrius* except that the leaves are slightly wider instead of narrower. The crenation is identical.

Cape, W. Mount Outeniqua, on bark, Rehm. 149 and 498 (both as *Z. Rehmanni* C. M.).

Cape, E. Zuurberg-Griqualand East, 3500 feet, 1883, W. Tyson 1575; Toise River, 3000 feet, 1919, Dr. Brownlee.

Natal. Mistyhome, Hemuhemu, on bark, 5000 feet, Jan. 1920, Sim 9666.

Zygodon cernuus C. M. (in Hedw., 1899, p. 114) is described from Professor Macowan's Boschberg specimen (1878) as being very similar but smaller, and has the leaves smooth, the capsule *semilunari cernua*, and the calyptra smooth. There is no type specimen in South Africa and no such plant has been seen, and the description of the capsule suggests either that it is wrongly placed or that it is some unnatural freak—probably due to fungoid action, at least it is not a normal *Zygodon* capsule.

405. *Zygodon africanus* Sim (new species).*

=*Z.* *affinis*, Rehm. 497 (mentioned in Broth., Pfl., xi, 15).

A vigorous plant, 1–2 inches high, laxly branched, and so far seen only sterile. Stems red, rather laxly leaved below and red-tomentose toward the base ; leaves when dry laxly twisted or torquescent, not appressed ; when moist erecto-patent, flexuous, channelled, linear-acuminate, acute ; the fairly strong, smooth, yellowish nerve excurrent as a short mucro ; margin erect, entire, but crenulate-papillose. Cells 12μ long, elliptic-hexagonal, minutely but abundantly papillose, the walls pellucid ; basal cells hyaline, long-hexagonal toward the nerve, up to 60μ long, the outer shorter and oblong. Floral parts and fruit not seen.

This resembles *Z. trichomitrius* in size and general appearance and in the torquescent dry condition, but the leaves are longer and narrower, more linear in form, less undulate but more twisted.

Transvaal. Macmac, MacLea (Rehm. 497), as *Z. affinis* Rehm.

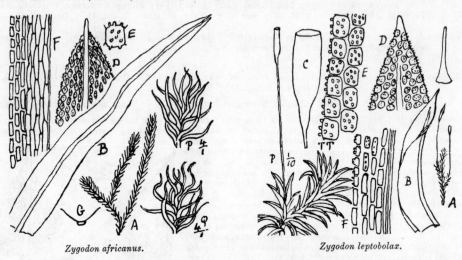

Zygodon africanus.   *Zygodon leptobolax.*

406. *Zygodon leptobolax* C. M., in Hedw., 1899, p. 113, and Rehm. 147 and 499 ; Broth., Pfl., xi, 14.

Synoecious, caespitose on bark, yellowish green ; stems ½ inch long, slender, mostly simple, sometimes tomentose below, with eventually lateral fruit ; leaves 1 mm. long, erecto-patent or some patent or reflexed, arched backward, oblong-acuminate, channelled, the apex acute or mucronate, the nerve strong, smooth, yellowish, and not reaching the apex ; margin erect, entire, except that it is densely papillose ; cells 12–15μ wide, many-papillose, larger toward the base ; basal cells roundly oblong, separate, hyaline, the upper slightly papillose. Perichaetal leaves hardly different ; seta pale, ½ inch long or longer ; capsule erect, brown, 2 mm. long, 8-furrowed, with distinct neck when dry, less evident when moist ; lid narrowly subulate ; calyptra smooth. Concerning the peristome Brotherus places it as probably belonging to the group having outer peristome absent, cilia sixteen, his specimen being too young ; my specimens have capsules too old to decide that on, they have neither outer nor inner segments remaining.

Cape, W. On shaded stones above Rondebosch, Rehm. 147 and 499.

Harvey's Genera of South African Plants, 1st ed. (1838), includes *Zygodon* as having sixteen teeth and eight or sixteen cilia, on which he says : " *Z. conoideum*, a small moss with slender, straggling, pale green branches, growing on trees, is found near Wynberg, but rarely." In those early days fine distinctions were not drawn, and it is likely that *Z. leptobolax* C. M., which in a general way resembles *Z. conoideus* (Dicks.) Hk. and Tayl., was the plant intended.

C. Müller also described (Hedw., 1899, p. 115) *Z. perreflexus* C. M. from Claremont, Oct. 1876 (said to be Rehm. 297, which is wrong, but that name was distributed with Rehm. 148, sterile, from Rondebosch). He admits its close proximity to *Z. leptobolax* on account of the minute recurved leaves, but describes it as having closely imbricate leaves making terete branches when dry, and some other slight differences, all of which appear to me to be insufficient. There is no type specimen in South Africa, and Brotherus places this and *Z. cernuus* C. M. in the section producing gemmae and having peristome absent (though unknown in these two). Müller does not mention gemmae, and I consider the name should sink.

---

* *Zygodon africanus* Sim (sp. nov.).

Vigens, 1–2″ altus ; folia lineo-acuminata acuta, torquescentia ubi sicca, 2–3 mm. longa, flexuosa, **sed non** undulata ; margines plani ; fertilis parum compertus.

407. *Zygodon Dixoni* Sim (new species).*

Caespitose, small, deep green, synoecious (or polyoecious) ; stems erect, mostly simple, 2–3 mm. long ; leaves ten to twenty, imbricate, hardly 1 mm. long, the lower ovate-acuminate, the upper ligulate, nearly equal in width all along and shortly drawn to an acute point ; surface channelled, the margin arched backward below ; nerve very strong and deep, ending just below the pointed apex ; margin entire, the cells hardly reaching the margin, cells roundly oblong, $8 \times 4\mu$, in straight lines each way, nearly pellucid, with one or two chlorophyll granules (hardly papillae) in each ; lower cells oblong-rectangular, pellucid, thick-walled, $20 \times 10\mu$; on old leaves all cells hyaline ; perichaetal leaves hardly different, rather longer ; vaginula bright red, very pronounced, not hairy ; seta 4 mm. long, twisted when dry ; capsule 0·75 mm. long, pyriform, the neck quite pronounced ; mouth small, 8-furrowed when dry, the furrows showing as crenations in section (fig. S) ; lid shortly and obliquely rostrate from hemispherical base too wide for the mouth; calyptra longer, conical, cucullate ; peristome of sixteen white teeth, of two segments each, four segments being usually connate up to the top, making eight bundles, erect when dry, incurved when moist ; the teeth trabeculate, and with horizontal lines of small papillae ; cilia eight, slender, shorter than the teeth.

A very distinct species, named after Mr. H. N. Dixon, who has been an enthusiastic worker in England upon African and other mosses.

Natal. Below Cathkin Peak, 7000 feet, Jan. 1918, Sim 10,004.

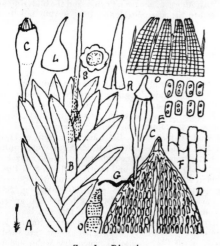

Zygodon Dixoni.
S, section of a capsule.

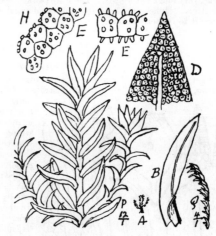

Zygodon transvaaliensis.

408. *Zygodon transvaaliensis* Sim (new species).†

=Z. *transvaaliensis*, Rehm. 500.

A minute plant, 2–4 mm. high, densely caespitose on bark, the main stem erect and laxly leaved, the leaves 0·5–0·75 mm. long, pellucid, clasping at the base, thence erecto-patent, lingulate-acuminate or oblong-lanceolate, nearly flat across, with plane, entire, papillose margins ; the apex widely acute, subrecurved, densely areolated, the nerve smooth on the back, very pellucid, and ending some distance below the apex ; cells roundly 5–6-sided, opaque, many-papillose, the papillae high and clear ; lower cells similar but rather larger, all chlorophyllose and papillose to the leaf-base , innovations from near the base are small-leaved or bract-leaved toward the base, all leaves smaller than those of the main stem and mostly more reflexed but otherwise similar. When dry the leaves are 2–3-stichous, secund, hardly crisped, and the point of the stem bends over to the same side as the leaves. Other features not seen.

Paris, in error (presumably misled by the name), makes this a synonym of *Rhachithecium transvaaliense* (C. M.) Broth., though there is no close resemblance, and that is taken over in Kew Bull., No. 6, 1923, No. 500.

On trees, Lechlaba, Houtbosch, Transvaal, Rehm. 500.

---

* *Zygodon Dixoni* Sim (sp. nov.).
Caespitosus, parvus ; folia imbricata, vix 1 mm. longa ; inferiora ovato-acuminata, superiora ligulata, ad acutum apicem breviter contracta ; margines retrorsum arcuati infra, cellae vix papillosae, oblongae ; cellae inferiores rectangulares, pellucidae ; capsula erecta, pyriformis, 8-sulcata ubi sicca ; operculum breviter rostratum ; peristom. dentes albi, 16, in 8 fasciculis, erecti ubi sicci.

† *Zygodon transvaaliensis* Sim (sp. nov.).
Minutus, tenuis, caespitosus in cortice ; folia 0·5–0·75 mm. longa, connexa, anguste lingulato-acuminata, plana ; cellae omnes pluri-papillosae. Homomallus ubi siccus.

Genus 118. Rhachithecium Broth. (in Acta Soc. Sc. Fenn., xix, No. 5, p. 20; and Pfl., 1, iii, 1, 1198).

=*Zygodon*, sps., Thwait and Mitt., Trans. Linn. Soc., 1872, p. 303.
=*Decadon* (C. M.), Broth., Pfl., 1, iii, 1, 1198.
=*Hypnodon* C. M. (in Hedw., xxxvi, 113 (1897)).

Little, light green, shining, autoecious mosses, more or less matted, on bark of living trees, and resembling the fertile branches of *Fabronia* in size and appearance. Stem without central strand, very short, erect, closely leaved, simple or freely branched at the base, and sometimes set with smooth rhizoids toward the base. Leaves spreading, keeled-concave, oval or oblong from a spathulate base, blunt, mucronate, with flat margin without border; when dry closely appressed; nerve fairly strong and wide, smooth, ending just below the leaf-apex. Cells very lax and thin-walled, roundly hexagonal, smooth, chlorophyllose, the marginal row often transverse, smaller and quadrate; the basal cells rectangular, hyaline, and thin-walled. Per. leaves sheathing, lanceolate-pointed, entire, with weak nerve and long hyaline cells except toward the leaf-apex, where they are shorter. Vaginula distinct and naked. Seta 2 mm. long, smooth, much twisted when dry. Capsule erect, widely oval, with small mouth, eight ribs, very thin-skinned, at first pendent, afterwards erect. Ring wide, caducous; peristome single, inserted far below the mouth; teeth sixteen, in pairs, lanceolate, when dry meeting in a cone, when moist curved inward, smooth, thin, closely articulated, yellowish brown; spores large. Lid flatly conical, shortly pointed; calyptra cucullate, reaching the middle of the capsule, very rough at the apex. There are two species from South America, one from Ceylon, and the under-mentioned from Africa. Müller considered that this peculiar genus belonged to Erpodiaceae (Hedw., 1899, p. 126), but Brotherus restores it to Orthotrichaceae.

409. *Rhachithecium transvaaliense* (C. M.), Broth., Pfl., 1, iii, 1, 1199, and xi, 17.

=*Hypnodon transvaaliensis* C. M., in Hedw., 1899, p. 126.

A very small, peculiar, autoecious, caespitose bark-epiphyte, formerly placed in Erpodiaceae and compared with *Fabronia*, but erect in habit, without prostrate vegetative growth; the capsule is deeply grooved when dry, and the

peristome and areolation are different. Stems very short, usually freely branched at the crown; leaves numerous, imbricate, spreading, channelled, 1 mm. long, oblong, with rounded mucronate apex. Nerve strong, smooth, ending just below the mucro; margin erect, entire, smooth; cells 20μ wide, irregularly 5–6-sided, laxly chlorophyllose, the marginal cells more transverse, all tumid when dry, but not papillose; the basal cells hyaline and shortly oblong, $80 \times 40\mu$, thin-walled: perichaetium terminal, erect, of several white, sheathing leaves having long, narrow, hyaline cells ($80 \times 20\mu$, fig. SS) and usually short nerve, but the apex is often subulate and green, with usual lamina cells. Seta 2 mm. longer than the per. leaves, much twisted when dry, at first cygneous, afterwards erect (figs. S and S). Capsule ultimately erect, widely oval when moist, crowned with distinct white ring till that dehisces; when dry wider below and deeply 8-grooved; peristome of sixteen smooth, yellowish-brown, closely trabeculate teeth united in eight pairs, starting deep within the throat, meeting above the capsule when dry, infolding when moist; lid concave and pointed; calyptra dimidiate; spores large, oval, dark.

Transvaal. On old trees near Utombi, between Cook River and Sand River, Aug. 1884, Dr. Wilms (*fide* C. M.).

Rhodesia. Mazoe, Ironmask, on shaded side of tree trunks, 5000 feet altitude, June 1915, Eyles 616*β*; Salisbury, 4900 feet, Eyles 1573. It has been found also in the Cameroons and the Niger basin.

Rhachithecium transvaaliense.
S and S, twisted and cygneous dry or young seta; SS, cells of perichaetal leaves.

R. demissum (C. M.) Broth. appears in error in Wager's list; it is South American.

Genus 119. Ulota (Mohr. MSS.), Brid., Mant. M., p. 112.

Small mosses, usually of more or less erect habit, growing in dense tufts, or sometimes scattered, on the bark of living trees, or occasionally on moist rocks. Leaves more or less lanceolate from a concave oval base, spreading when moist, crisped when dry; upper cells small, roundish, opaque and somewhat papillose, the central basal cells narrow, linear, and opaque, the marginal basal cells wider and elliptic-rectangular, and usually hyaline, then forming a hyaline border extending more or less up the leaf. Calyptra yellowish, conical-campanulate, deeply lobed at the base, plaited, usually covered with numerous long, yellow, erect, jointed hairs. Capsule erect, oval, with or without a tapering neck, 8-striate, on a fairly long seta; peristome simple or double, the outer of sixteen more or less bifid teeth united in pairs, inner of sixteen narrow processes, alternating with the pairs of teeth, or absent. Capsules persistent for years. Stomata on the capsule superficial. The curled dry leaves, when moistened, at first suddenly become reflexed-squarrose, then gradually erecto-patent.

This genus is not satisfactorily separated from *Orthotrichum*, and some authors include it in that, which may account for Brotherus not having shown *Ulota* in his list in Marloth's Flora. C. Müller (Syn. Musc.) places it as section of *Orthotrichum*, using "leaves crispate when dry" as its character. See *U. Eckloni*.

410. *Ulota Eckloni* (Hsch.), Broth., Pfl., 1, iii, 1, 472, and xi, 25 ; Par., Ind., 2nd ed., p. 99.
=*Orthotrichum Eckloni* Horns., in Linn., xv (1841), 129 ; C. M., Syn., i, 715.

This species, which is very fully described by Hornschuch (condensed by C. M. in Syn.), appears to have been collected only by Ecklon, in " rock fissures on top of Table Mountain " about a hundred years ago, and does not agree with anything I have seen, but until the type is re-examined or the plant found again it is advisable to retain it. Hornschuch describes it as having erect ½-inch stems, fastigiately branched ; leaves linear-lanceolate from oval-concave base ; contorted-crisped when dry ; margin plane ; nerve vanishing below the apex ; basal marginal cells quadrate, inner cells linear, lamina cells minute and round.   Perichaetal leaves sheathing half up the leaf, longer than the capsule ; capsule small, apophysate, 8-striate ; outer peristome of eight pairs of white, split teeth, reflexed when dry ; cilia eight, shorter ; lid convex, apiculate ; calyptra laciniate at the base, very hairy, with brown point.

Orthotrichum afrofastigiatum is the only species which in any way meets this ; that also has exposed stomata, but not hyaline leaf-margin, both required in *Ulota.*

Genus 120. Orthotrichum Hedw., Descr., ii, 96.

Usually cushioned and suberect (in South African species always so), growing on trees or occasionally on rocks. Leaves widely lanceolate, bluntly lanceolate, or more or less oblong, from a not expanded base, concave-connivent, sometimes hooded, strongly nerved at the base ; the nerve weaker to or near to the apex ; the point subacute, acute or hair-pointed, the point part of the lamina rather than of the nerve ; leaves very hygroscopic, spreading when moist ; when dry straight, imbricated, and rarely curled or crisped ; margin often revolute, usually entire, or denticulate near the apex only ; cells small, roundish, chlorophyllose, usually minutely papillose ; basal cells rectangular, thin-walled, and hyaline.   Capsule immersed or almost so, rarely on a longer seta, usually 8–16-striated, rarely smooth, often constricted below the mouth, elliptic with or without a tapering neck, on which or on the capsule stomata are more or less abundant, either sunk or superficial.   Calyptra campanulate, lobed at the base, deeply ridged, naked or sparsely hairy (always hairy when young in South Africa) ; lid conical or convex with short erect beak.   Ring absent or very narrow ; peristome simple or double ; outer peristome of sixteen separate or eight pairs of forked, broadly lanceolate teeth, with fine striae or papillae or both ; inner when present of eight or sixteen slender jointed processes, often unequal.   The dry outer peristome may be erect, or arching outward or closely reflexed against the capsule ; these characters, together with the position of the leaf-margin and the position of the stomata on the capsule, define the different groups into which the many species are arranged, but I find (in *O. afrofastigiatum*) that the spreading peristome of a mature capsule may be closely reflexed against the capsule in a specimen having shrunk capsule several years old, so that character is unreliable.

On all the South African species jointed hairs are rather numerous on the calyptra when young ; with age they are often fewer.

Synopsis of Orthotrichum.

1. Leaves twisted and crisped ; stomata superficial . . .	411. *O. afrofastigiatum.*
1. Leaves not crisped ; stomata sunk.	2.
2. Leaves pilose	3.
2. Leaves acute, subacute, or subobtuse	4.
3. Leaves long-pilose	412. *O. piliferum.*
3. Leaves shortly pilose	413. *O. glaucum.*
4. Leaves long, acute, denticulate	414. *O. pseudotenellum.*
4. Leaves long, acute, entire	415. *O. Macleai.*
4. Leaves short, subacute, widely lingulate, 1·5 mm. long .	416. *O. transvaaliense.*
4. Leaves rather short, subobtuse, 2 mm. long ; margin irregular . .	417. *O. subexsertum.*

Shaw's list includes *O. affine* Schw. (of Europe, etc.) from Cape Mountain, Graaff Reinet.   MacLea (Hornschuch, Linn., xv (1841), 131) includes *O. pumilum* Schw. (of Europe) as collected at Cape Town in 1826 by Ecklon, and at Paarl in 1828 by Dregé ; and *O. leiocarpum*, Br. Eur. (of Europe, Asia, N. Africa, and N. America), is doubtfully included by Paris as from the Cape, but there is no reason to think that either of these identifications were correct, and only reference to the original types can prove that they are.   All are related to *O. afrofastigiatum*, but the localities given suggest other species.

See under *Ulota* concerning *O. Eckloni* Horns., which also may be *O. afrofastigiatum.*

411. *Orthotrichum afrofastigiatum* C. M., in Hedw., 1899, p. 113 ; Broth., in Pfl., 1, iii, 1, 470, and xi, 17.

On bark, green, autoecious ; stems 3–10 mm. high, often tomentose, laxly leaved below, fastigiately branched and many-leaved above ; leaves 1·5–2 mm. long, openly areolated, widely oblong-acuminate, acute, laxly imbricate, often contorted near the apex, spreading and nearly flat across when moist (except the more erect per. leaves), twisted or torquescent and somewhat crisped when dry, yellowish green ; the margin entire or sometimes with irregular steps, revolute from the shoulder downward, expanded toward the apex ; nerve narrow, somewhat ferruginous, smooth in the lower half, covered with lamina cells in the upper half, weak upward, disappearing below the apex ; cells circular or roundly hexagonal, tumid, 12μ wide, in contact or nearly so, minutely 2-papillose, not affecting the margin, extending down as far as the leaf is revolute and sometimes to the base, also extending somewhat down the nerve ; the basal cells hyaline, up to 60μ long, and often extending up the lamina some distance between the marginal and costal chlorophyllose cells.   Inner perichaetal leaves shorter and rounded at the apex, without point.   Capsule half-immersed or almost immersed, elliptical, without neck, and suddenly contracted to the very short seta, brown, 8-striate, or when dry 8-sulcate near the mouth ; mouth-band red and made of small round cells ; stomata exposed below mid-capsule (fig. S). Peristome white rather long, of eight triangular-conical bundles, with few cross-lines but numerous papillae ; collected in a dome when moist, suberect and star-like when dry, completely reflexed on old shrunken capsules ; cilia eight, slender, nearly as long ; lid shortly conical ; calyptra hairy ; spores elliptical, 17μ long, papillose. Antheridia in a cluster in the axil of an ordinary leaf, without paraphyses. Almost like *Zygodon Dixoni* in its white star-like peristome, but differing in all other respects, especially in the length of the seta.

Cape, E. Müller described it from Boschberg, Somerset East (Macowan, 1878), which specimen I have not seen.

Natal. Rosetta Farm, on wattles, 4000 feet, Aug. 1918, Sim 10,103 ; Arcadia, Scheeper's Nek, Oct. 1919, Sim 10,104 ; Chard, Rietvlei, 4500 feet, 1920, Sim 10,109.

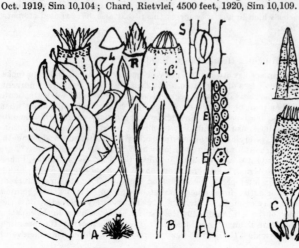

Orthotrichum afrofastigiatum.
S, stomata and surrounding cells.

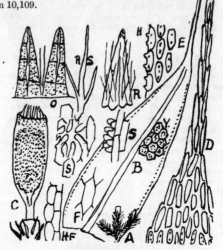

Orthotrichum piliferum.
S and S, stomata, front and side view ;
RS, hair of calyptra.

412. *Orthotrichum piliferum* Sim (new species).*

=*O.* *piliferum Schp. MSS., in herb. ; Rehm. 516 ; Shaw's Catalogue (not fully published).

On tree-bark, monoecious, yellowish green ; stems 3–6 mm. high, branched, many-leaved, but with occasional basal innovations which have the leaves very laxly placed. Leaves yellowish green when young, more pellucid when old, nearly erect and not much altered when dry ; when moist spreading, nearly flat across, oblong-lanceolate so far as reflexed, larger-celled and acuminate beyond that into a long denticulate, flat, or twisted hair-point, the upper hair-points often half the length of the leaf, but all leaves distinctly piliferous, the point sometimes twisted corkscrew-fashion when dry. Leaf-base apparently much narrowed, but this is aided by plication ; the cells there hyaline, hexagonal, $60 \times 10\mu$, except the outer which are $30 \times 10\mu$, the outer row roundly oblong, those within rectangular and drawn downward to a raised point ; lamina cells varying from quadrate to hexagonal or elliptical, mostly $15 \times 10\mu$, on older leaves closely packed and translucent, on younger leaves beautifully yellow, mostly 2-papillate, the papillae small ; chlorophyll grains more numerous. Apical cells longer, up to $25 \times 10\mu$, hyaline, the point of the cell often protruding, especially on the upper surface. Margin not bordered, but often with transverse cells ; revolute from the shoulder to near the base, and irregularly denticulate where not revolute. Nerve not reaching the apex, smooth on the lower half, covered with lamina-cells upward. Perichaetal leaves not different, except somewhat longer-haired. Capsule subimmersed on a pronounced vaginula and very short seta, the vaginula set with perished archegonia intermingled with a few jointed paraphyses ; capsule brown, widely cylindrical, hardly contracted when dry, longitudinally striated by brown ridges but not deeply sulcate, and only slightly narrowed below the mouth ; the stomata placed about mid-capsule and hid by six to ten projecting mamillae ; mouth-band red, of small round cells, the other capsule cells large, hexagonal, and thin-skinned. Peristome erect when moist, spreading when dry, few-barred, but densely papillose, often split or perforated nearly to the base ; cilia sixteen, in pairs, irregular or partly rudimentary ; lid shortly rostrate ; calyptra long, white, red-pointed, shortly 8-lobed at the base, and often 8-plicate, set with a few nearly basal, wide, many-celled, and often several-branched hairs nearly as long as the calyptra.

An apparently similar piliferous species is *O. mollissimum* C. M. (Linnaea, 1875, p. 407) from the Red Sea, but that has no peristome.

---

* *Orthotrichum piliferum* Sim (sp. nov.).

Folia oblongo-lanceolata, prope apicem acuminata, pila longa denticulata saepe ad dimidiam folii longitudinem, cellae laminares 2-papillatae ; cellae apicales longiores ac prominentes. Capsula sub-immersa ; stomata depressa.

Cape, E.   Rehmann's 516, collected in the Eastern Province by MacLea (Shaw says near Graaff Reinet); and at Genadendal (Breutel); King William's Town, 1893, Sim 7053; on mulberry tree, De Jager's Drift, Longhope, Cookhouse, April 1921, Sim 10,099; on oaks in Mr. Dolley's garden, Uitenhage, Jan. 1922, Sim 9001. Shaw's specimen is in Kew Herbarium.

413. *Orthotrichum glaucum* Spreng., Syst. Veg., iv, ii, 323; C. M., Syn., i, 695; Broth., Pfl., 1, iii, 1, 468, and xi, 22.

Monoecious; stems 3–5 mm. high, simple or branched, many-leaved; the leaves 2 mm. long, oblong-lanceolate, acute, shortly hair-pointed, imbricated so that they are much longer than they appear to be; somewhat sheathing but the entire margin revolute its whole length, not bordered; point diaphanous, of long loose cells not connected with the nerve; lamina cells roundly quadrate, densely packed, tumid, somewhat granular-chlorophyllose but not opaque nor papillose, 17μ wide, covering the nerve also for most of the leaf; basal cells hyaline, rather few, those nearest the nerve narrowly oblong or longer; the outer few lines, which are revolute, are quadrate and shorter. Nerve weak throughout, covered by lamina cells except at the base, not papillose, disappearing below the apex. Perichaetal leaves similar but shorter and wider, ovate-acuminate. Vaginula large, smooth; seta 0·5–0·75 mm. long; capsule 1 mm. long, oval-oblong, brown, half-exserted; stomata sunk, mouth-band small-celled; peristome teeth erect or irregular when dry, meeting in a dome when moist, the outer peristome of sixteen unsplit teeth in lanceolate pairs, trabeculate and closely papillose in horizontal lines; cilia sixteen, in pairs, each pair united below, with free tips, papillose, the pairs alternate with the outer pairs, and opposite the capsule ridges; lid convex-mucronate; calyptra short, hairy, with jointed hairs. Dry capsule narrowed below the mouth and sharply 8-ridged. Male inflorescence terminal on short basal branches; outer perigonial leaves ovate-acuminate, piliferous; inner wider, without nerve or hair.

C. Müller remarks on its resemblance to *O. diaphanum* Schrad., but it differs in having the hair-point entire or almost entire (*i.e.* less denticulate), in having the calyptra distinctly hairy, and in the leaves being hardly papillose.

Cape, W.   Paarl, on oaks, Sim 9614; Cape Town, on oaks, Sim 9543, 9544; Elgin, Sim 9591 p.p.; Table Mountain, Ecklon; Stellenbosch, Dr. Shaw.

Cape E.   Kokstad, Burtt-Davy; on acacias, Kat River at Fort Beaufort, July 1829, Mundt.

Natal.   Sweetwaters, 2500 feet, Sim 10,106; Cathkin, on bark, 6000 feet, Sim 10,102; Scheeper's Nek, Sim 10,105.

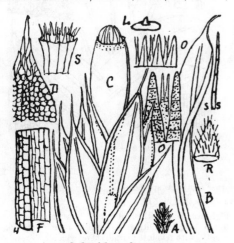

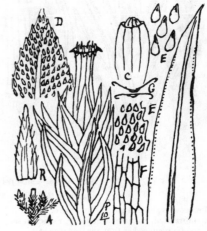

Orthotrichum glaucum.

F, back-view at leaf-base; S, dry peristome;
SS, jointed hair of calyptra.

Orthotrichum pseudotenellum.

414. *Orthotrichum pseudotenellum* (Hpe. MSS.), C. M. in Bot. Zeit., 1859, p. 230; Broth., Pfl., 1, iii, 1, 468, and xi, 22; Rehm. 515.

Autoecious, in tufts ½ inch high, on bark; stems simple or dichotomously branched, yellowish green above, rufescent below, closely leaved from the base upward; leaves lanceolate-acuminate, 2–3 mm. long, erecto-patent when dry, and nearly flat across, or sometimes concave-channelled, spreading widely when moist; apex acute, hyaline, strongly denticulate but hardly hair-pointed. Areolation very lax and pellucid throughout, the upper cells elliptical but drawn to an upraised point or papillae at the upper end of each; these stand in straight lines each way, drawing away from the nerve which disappears below the apex, is smooth on the upper surface and covered with these cells to near the base on the under surface, and many cells project as denticulations on the margin throughout the upper half of the leaf, particularly near the acute apex, making the margin and surface scabrid; basal cells laxly rectangular or irregularly hexagonal, 40×10μ, the outer shorter but hyaline and not revolute; margin revolute from near the base to near the apex; vaginula smooth; seta very short; capsule long, elliptic-cylindric; ridge-marked when moist, shrunk and strongly 8-ridged when dry; stomata sunk; lid shortly pointed; calyptra straw-coloured, long, shortly hairy when young. Outer peristome teeth in eight triangular bundles, spreading when dry, met in a cone when moist, papillose on both surfaces ; cilia erect, slender.

Cape, W. Cape Town, Jan. 1919, Sim 10,118.

Cape, E. Eastern Cape Province, MacLea (Rehm. 515).

415. *Orthotrichum Macleai* Sim (new species).*

=*O.* *Macleai*, Rehm. 514.

=*O.* *Macleanum*, Shaw's Catalogue, Cape Monthly Mag., 1879, p. 378 (not fully published).

Monoecious, plant vigorous, 1–2 inches high, branched, bare below, abundantly leaved above. Leaves suberect when dry, 2 mm. long, widely lanceolate-acuminate, more or less conduplicate in the upper half, with strongly revolute entire margin the whole length ; apex acute, entire, the nerve covered with lamina cells almost the whole distance and vanishing well below the apex, often more or less red or ferruginous. Cells lax, 12μ long, elliptical or irregular, mostly 2-papillate ; basal cells thin, rectangular, $60 \times 15\mu$, the outer (revolute) oblong $20 \times 10\mu$, pellucid, but the oblong, sub-basal cells papillose. Perichaetal leaves not different. Capsule oblong-ovate, with thick neck and red mouth-band, 8-ridged when dry ; outer peristome teeth in eight bundles, met in a cone when moist, spreading when dry, transversely many-striate ; cilia not seen ; lid shortly pointed ; calyptra long, red-pointed, many-haired.

Cape, E. On trap rocks near Graaff Reinet, MacLea (Rehm. 514).

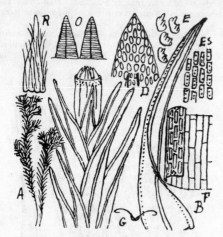

Orthotrichum Macleai.
ES, sub-basal cells.

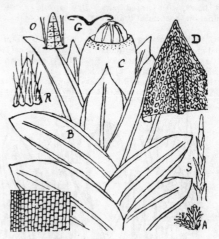

Orthotrichum transvaaliense.

416. *Orthotrichum transvaaliense* Sim (new species).†

=*Orthotrichum* *transvaaliense*, Rehm. 517.

Monoecious ; stems caespitose or gregarious, 3–7 mm. high, simple or with similar male basal branches ; leaves closely imbricated, suberect when dry, rather numerous, 1·5 mm. long, clasping at the hyaline base, ovate-lingulate but with sides incurved so that the leaf is boat-shaped at the middle, with revolute margin there, which reflex, however, does not extend to the base nor to the apex. Apex shortly acute or acutely submucronate, or subacute, entire, laxly areolated with lamina cells ; margin entire, strongly revolute in the middle portion ; nerve fairly strong, disappearing some distance below the apex, covered with tumid lamina cells ; cells separate, roundly irregular with translucent spaces between, minutely papillose, these cells descending to near the base where hyaline quadrate-hexagonal cells 20μ long appear ; inner basal cells 25μ long, the outer few lines about 15μ long. Perichaetal leaves rather longer than others, as long as the capsule ; seta very short, at first terminal, then pseudo-lateral. Capsule oval, thin-walled, with a red band of small cells at the mouth, no preperistome, and sunk stomata ; lid shortly beaked, red ; calyptra white, conical, 8-furrowed, and at the base shortly 8-lobed, set with numerous, many-celled, denticulate, yellowish, upright hairs (fig. S) starting mostly from near the base and about as long as the calyptra. Male inflorescence on

---

* *Orthotrichum Macleai* Sim (sp. nov.).

Vigens, 1–2″ altum. Folia 2 mm. longa, late lanceolato-acuminata a lata basi ; margo revolutus, integer ; cellae parvae, 2-papillatae ; oblongae sub-basales cellae papillosae. Capsula sub-immersa ; calyptra capillosa.

† *Orthotrichum transvaaliense* Sim (sp. nov.).

Pumilum, monoecum ; folia 1·5 mm. longa, ad basin connexa, ovato-lingulata, breviter acuta, canaliculata ; margo revolutus ad medium folium ; cellae minute papillosae praeter parvas quadratas basales. Capsula ovata, sub-immersa ; stomata depressa ; calyptra capillosa.

a basal branch, with usual outer leaves and shorter, thinner, inner bracts. Dry capsule deeply sulcate and shrunken to cylindrical or less, with teeth spreading and star-like and nearly horizontal ; cilia suberect.

On Faurea trees, Lechlaba, Houtbosch, Transvaal, Rehm. 517.

417. *Orthotrichum subexsertum* (W. P. Sch. MSS.), C. M., in Bot. Zeit., 1858, p. 154 ;
Broth., Pfl., 1, iii, 1, 470, and xi, 21.

Autoecious ; plant 4–8 mm. long, simple or branched from the base ; many-leaved ; the leaves appressed when dry, spreading much more when moist, very much concave boat-shaped, with rather incurved apex ; 2 mm. long, oblong from a rather wider lower part and narrower base ; the obtuse or subobtuse apex quickly reached either by steps or by rounded apical margin ; the steps irregular or the margin either entire or subentire ; margin of the leaf (except at the apex) strongly revolute to the base ; nerve not reaching the apex, covered by tumid lamina-cells in the upper half ; areolation very lax ; cells separate, roundly elliptical, 15μ long, nearly pellucid, very tumid, and with several chloroplasts in each, but not papillose though appearing so when dry ; cells near the base longer and more irregular in form ; the basal cells hyaline, 35μ long, narrowly hexagonal ; the chlorophyllose cells descending the

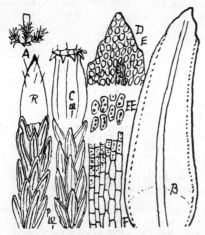

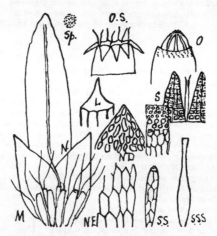

Orthotrichum subexsertum   *Orthotrichum subexsertum* (male flowers, etc.).

O, peristome, moist ; OS, peristome, dry ; S, part of capsule-mouth with two teeth and two ciliae ; Sp, spore ; SS, antheridium ; SSS, archegonium ; EE, cells toward leaf-base ; NE, cells of perigonial leaves.

margin further than inside the leaf, and the revolute basal portion with oblong cells shorter than the basal cells. When dry the two revolute margins and the nerve show as three distinct ridges for each leaf. Perichaetium at first terminal ; per. leaves hardly different and covering only the base of the capsule which is long for the plant (2 mm.), oval-cylindrical when moist ; shrunken, wide-mouthed, and deeply 8-ridged when old ; cells of capsule large and thin-walled except several rows at the mouth which are small and round and of deeper colour. Capsule white when young, brown when old. Peristome rising deep inside the mouth, consisting of sixteen outer teeth united in pairs into eight conical bundles, which are trabeculate, densely papillose outside, collected in a dome when moist, spreading or reflexed when dry, the inner of sixteen slender, erect ciliae collected in pairs, alternate with the pairs of outer teeth. Lid shortly pointed ; calyptra conical, covering almost the whole capsule, the point red, the balance white, at first freely set with erect jointed hairs, but with age many or all of these disappear. Male inflorescence terminal on basal branches, the outer leaves almost like cauline leaves but more laxly areolated, the inner bracts much shorter, ovate-acute, mostly made of basal cells, or with a few at the apex elliptical and chlorophyllose. Antheridia about ten, turgid, set among numerous long, slender, red-jointed paraphyses.

Cape, W. On old oaks, Main St., Stellenbosch, Sim 9569 ; on oaks, Experiment Farm, Stellenbosch, Sim 9568 ; oaks and pines, Disa Ravine, Table Mountain, Sim 9220 ; Bishop's Court, Sim 9461 ; Cape Town, Rehm. 170 ; Elgin, Sim 9591 p.p. ; Tokai, on oaks, Sim 9498 ; Peak Forest Station, Sim 10,107 ; Hels Hoogte, Stellenbosch, S. Garside ; Oranjezicht and Skeleton Ravine, Brunnthaler.

(Genus DRUMMONDIA Hook.

Growing in flat masses on tree trunks, like *Macromitrium*, the branches prostrate, rooting, much-branched ; the stem and branches producing many erect, little, tufted branchlets on which the inflorescence is produced terminally. Stem-leaves densely imbricated, ovate-lanceolate, concave, the margin erect ; keel not reaching the apex ; cells everywhere rounded. Capsule subglobose, erect, *smooth* ; lid obliquely beaked ; ring none ; calyptra widely inflated, cucullate, large, covering the young capsule ; peristome single, with sixteen very short, truncate, entire, smooth teeth, each showing a longitudinal line, and closely barred.

Of this genus, which is known to be represented only in North America and West Asia, Harvey mentions (Genera of S. Afr. Plants, 1st ed., 1838) that *D. clavellata*, a North American and Japanese species, is said by Sprengel to be a native of the Cape. No further record is known to me, and as the general appearance of the plant is exceedingly like *Macromitrium*, it is probably a case of mistaken identity, but it is mentioned here to draw attention to it so that it may be confirmed or otherwise.)

(Genus DASYMITRIUM Lindb. [in Oefv. af. K. Sv. Vet. Akad. Foerh., 1861, xxi, 421-423] is included in Par., Ind., as having four species from China and Japan, one from Taiti, and one from the Transvaal, viz. *D. Macleai* Rehm. (from Macmac, MacLea), but the latter is distinctly *Macromitrium Mannii* and has no generic distinction, so *Dasymitrium* goes out from the South African list.)

(Genus COLEOCHAETIUM Besch., Fl. Bryol. Reunion, etc., p. 66 (as section of *Macromitrium*), 1880–1881 ; Ren. and Car. in Bull. Soc. Roy. Bot. Belg., 1894, ii, 120, as a genus.

In Nat. Pfl., 1, iii, 1, 474 (and also xi, 26), Brotherus places *Leiomitrium* Mitt. as a synonym of *Coleochaetium*, but in Marloth's Flora of South Africa he uses the generic name *Leiomitrium*, which has priority. But in so far as the two sterile South African species are concerned, I find no valid reason for separation and have included them in *Macromitrium*, so *Coleochaetium* is not required as a South African generic name.)

Genus 121. MACROMITRIUM Brid.

Prostrate and living on bark or on stones, and having extensive vegetative, often filiform, but sometimes wider stems, more or less pinnately branched, from which shorter, erect, or suberect branches rise, bearing the inflorescence, the whole mat often dense and intricate and covering many square inches. Leaves often small, sometimes long, more or less lanceolate or acuminate, with concave basal wings ; usually keeled or channelled, crisped or inflexed or cirrhate when dry, spreading when moist, flexuose, somewhat papillose, with minute, roundish, obscure cells, those at the base more or less hexagonal and more pellucid. Capsule more or less oval or ultimately cylindrical, striate, much striate when dry, sometimes constricted at the mouth ; on a seta 1–2 cm. long. Peristome single, double, or absent ; outer peristome with sixteen lanceolate teeth, separate or in pairs ; the inner a thin truncate unreliable membrane, or also having slender processes. Ring absent or narrow ; calyptra campanulate, plicate, more or less deeply cut into segments in accordance with age.

A large genus (probably much overestimated), belonging to the tropics and the Southern Hemisphere, abundant throughout South Africa and north to Abyssinia, Ceylon, Mexico, and Georgia, U.S.A., but absent elsewhere in the Northern Hemisphere.

The section *Macrocoma* was long ago subjected to an undue amount of hair-splitting, considering its variability, and that has caused trouble ever since. Its species form a distinct group, in which, in South Africa, variation with degree of development is a constant feature. These species are common and are present in every bush and on nearly every soft moist stone in South Africa ; some plants are vigorous in every respect, while others are dwarfed by surrounding circumstances ; specimens can consequently be obtained from one tuft to answer several descriptions without having permanent characters, and I find it necessary to sink several and change others. Without seeing the original type specimens this may seem unfair, but after examining very many specimens I have not yet found one with the calyptra shorter than the capsule, though I find that the same specimen which has cylindrical, entire calyptra when young often has it split on one side later and multifid when mature, the hair covering being equal in each case or reduced by exposure or age. Vigour gives no character, the calyptra gives no character, cells vary with vigour and age and induration ; the capsule changes much in form and in plication with age ; leaves vary in form on the same stem and between *M. tenue* and *M. lycopodioides* the main line of separation is monoecious and dioecious, and even that has been questioned. Hornschuch was lost among "intermediates" ; C. Müller and Dr. Rehmann divided unnecessarily ; even Dixon admits difficulty (S.A. Journ. Sc.), and the three species given below easily cover all that occur.

Synopsis of Macromitrium.

§ *Macrocoma.* Stems linear-julaceous when dry ; peristome inner only or with weak outer teeth.
  418. *M. lycopodioides.* Dioecious ; per. inner only ; leaves 1 mm. long or less, usually acute.
  419. *M. pulchellum.* Dioecious ; per. double ; leaves 1·5 mm. or thereby.
  420. *M. tenue.* Monoecious ; per. an inner membrane only ; leaves 1 mm. long or less, usually obtuse.
§§ *Eumacromitrium.* Leaves very rugulose. Dry leaves straight, secund, crisped, incurved or cirrhate, not julaceous. Capsule microstomous. Peristome double or with the outer strong.
  421. *M. Mannii.* Leaves long, tapering ; dry leaves crisped-cirrhate.
  422. *M. secundum.* Leaves shorter, tapering from a wide base, secund.
  423. *M. rugifolium.* Leaves lax, tapering from a very wide decurrent base, erose-denticulate.
  424. *M. serpens.* Leaves lingulate-acute, incurved when dry, papillose at the base ; outer peristome well developed ; per. leaves erect.
  425. *M. macropelma.* Leaves linear-ligulate, acute, not papillose at the base.

418. *Macromitrium lycopodioides* (Burch.), Schw., Suppl., II, ii, 2, 141, t. 193 ; C. M., Syn., i, 721.

=*Orthotrichum lycopodioides* Burch., Cat. Geogr. Pl. Afr. aust. extra-trop., No. 5141–47.

=*Macrocoma elegans* Hook., in Sched.

=*Maschalocarpus Eckloni* Spreng., Syst. Veg., iv, ii, 321.

=*Macromitrium confusum* Mitt., in Journ. Linn. Soc., 1886, p. 305 ; Dixon, S.A. Journ. Sc., xviii, 317 ; Broth., Pfl., xi, 30.

=*Macromitrium dawsoniomitrium* C. M., Rehm. 160.

=*Macromitrium caespitans* C. M., Rehm. 159, 159C (which C. M. makes *M. tenue* Brid., in Hedw., 1899, p. 116).

=*Macromitrium tenue* var. *brachypus* C. M., Rehm. 162 (*fide* Mitten).

Dioecious ; cushions large, widely spread by very slender, thread-like, much-branched, prostrate stems 2–3 inches long, the branches thereon 3–10 mm. long, subpinnately arranged, all the older branches branched, and the oldest abundantly fertile ; when dry the leaves on these branchlets are straight, closely julaceous, in spiral rows, making a narrowly cylindrical, pointed branch ; when moist the leaves are erecto-patent, very numerous, lingulate-acuminate and expanded from a wider conduplicate base ; the margin entire, expanded or slightly recurved ; apex acute ; nerve strong, ending some distance below the apex, sometimes red, and occasionally producing numerous rhizoids ; areolation lax, cells in straight rows, rounded, 6μ diameter, the basal hardly different but rather larger and tumidly 1-papillose on the upper surface ; those nearest the nerve rhomboid-elliptical ; under-surface and upper cells smooth. Perichaetal leaves numerous, erect, more subulate, enclosing the red vaginula which is subtended by numerous jointed hairs or paraphyses much longer than itself. Seta about 1 cm. long ; capsule elliptical, tapering somewhat below, small-mouthed and somewhat wrinkled there ; cylindrical and more wrinkled with age ; peristome single, inner only, and that reduced to a granular membrane irregularly crenated on the upper margin, or absent ; always gone when old (Schweingrichen and also Müller describe it as without peristome). Calyptra enclosing the whole capsule, at first entire, afterwards many-fid, set with long, simple, narrow hairs. Spores granular, 25μ wide, deep green. Schw. describes inflorescence as monoecious and dioecious ; C. M. as dioecious only, which is what I find.

M. abyssinicum C. M. (from Abyssinia) is similar but has leaves more papillose, rough, and denticulate at the base. Brotherus separates *M. dawsonimitrium* as autoecious, which I doubt.

Abundant everywhere ; among the many localities are :—

Cape, W. Blinkwater Ravine, Sim 8592 ; Camps Bay and Table Mountain, Rehm. 159, 160.

Cape, E. Valley of Desolation, Graaff Reinet, MacLea (Rehm. 513β ; Despatch, Brunnthaler ; Knysna, J. Burtt-Davy 1704, and Rehm. 160 ; Cave Mountain, Bolus 65 ; Grahamstown, Hepburn.

Natal. Bulwer, J. W. Haygarth (Sim 8053) ; Van Reenen, Sim ; Nottingham Road, Dr. v. d. Bijl.

Transvaal. Macmac, MacLea (Rehm. 513) ; Spelonken, Rev. H. A. Junod 17.

S. Rhodesia. Zimbabwe, 3000 feet, July 1920, Sim 8822, 8777, 8802.

Macromitrium lycopodioides.

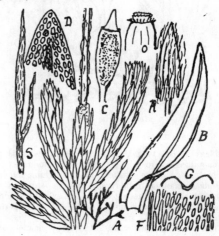

Macromitrium pulchellum.

419. *Macromitrium pulchellum* Brid., i, 313 and 737 ; C. M., Syn., i, 722 ; Broth., Pfl., xi, 30.

=*Schlotheimia pulchella* Hsch., in Hor. phys. Berol., p. 61, t. 12 ; Schw., Suppl., III, i, 2, t. 249.

=*Orthotrichum Hornschuchii* Hk. and Gr., in Brew. Ed. Journ., i, 129, t. 6.

Dioecious, matted, light green, mostly erect and much-branched dichotomously, but with prostrate pinnate extension-growths ; leaves 1·5 mm. long, lanceolate-acuminate, subacute, laxly areolated, the margin arched backward in the lower half ; one leaf-base usually ventricose ; cells roundly quadrate, nearly pellucid, turgid or slightly papillose, the lower cells more elliptical but not much larger, except those next the nerve which are longer. Perichaetium usually terminal at a fork, with erect, rather wider leaves enclosing the long red vaginula subtended by paraphyses, and a thin transparent membranaceous covering usually extends along the seta.   Seta ½ inch long ; capsule small, ovoid, slightly wrinkled with age, with short conical lid, and with distinct, erect, ring-like, crenulate membrane as inner peristome, together with short rudimentary membranaceous teeth of outer peristome which spread and are evident when dry, but most difficult to find when moist.   Calyptra covering the whole capsule, its base made up of numerous, long, flat, scale-like hairs (frequently branched) which with age separate, when the calyptra base becomes multipartite.   No specimen has been seen with the short calyptra covering only half the capsule, which Müller makes a feature.

Cape, W.   Blinkwater Ravine and Table Mountain, Sim 9253, 9269 ; Devil's Peak, Rehm. 164.

420. *Macromitrium tenue* (Hk. and Gr.), Brid., i, 740 ; C. M., Syn., i, 720, and Rehm.

=*Orthotrichum tenue* Hk. and Gr., in Brew. Ed. Journ. Bot., i, 120, t. 5.
=*Macromitrium Dregei* Hsch., in Linn., xv (1841), 131 ; C. M, Syn., i, 721 ; Broth., Pfl., xi, 30
=*Macromitrium microphyllum* Brid., i, 737 ; C. M., Syn., i, 721 ; Broth., Pfl., xi, 30.
=*Maschalocarpus declinatus* Spreng., Syst. Veg., iv, i, 158.
=*Orthotrichum microphyllum* Hk. and Gr., in Brew. Ed. Journ. Bot., i, 121, t. 6.

Monoecious ; cushions large on stones or bark, with usually pinnate or multi-pinnate prostrate slender stems extending outward where there is space, and erect or suberect, simple, dichotomous or pinnate fertile branches on which the perichaetia are usually lateral and abundant.   Leaves 0·5–1 mm. long, ovate-lanceolate or lingulate from a wider base, suberect when moist, julaceous when dry ; the apex usually obtuse but varying from that to acute on the same stem ; the base often ventricose on one side ; margin entire, often flattened but hardly reflexed : nerve strong, coloured

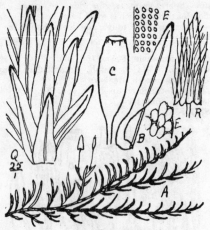
Macromitrium tenue.

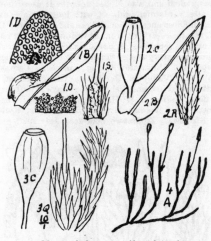
Macromitrium tenue (four forms).
See *forms* below.

in old leaves ; cells turgid, ovoid, crowded or scattered, pellucid on young leaves, incrassate and coloured in old leaves ; perichaetal leaves erect, longer and more acuminate than the others ; vaginula red, subtended by paraphyses ; seta 1–2 cm. long ; capsule elliptical, 1·5–2 mm. long, microstomous, slightly wrinkled, with no outer peristome and only a crenulated granular membrane as inner peristome (which often disappears with age) ; lid convex-rostrate ; calyptra hairy, enclosing the whole capsule, at first cylindrical and fitting closely at the base, later split or multifid, and spreading at the base.   Male flowers small, bud-like, below the perichaetia ; perigonial leaves few, ovate-acuminate, thin ; margin crenulate.

The following forms occur, but are not permanent enough to be varieties :—

1. Form *obtusum* Sim.   Leaves mostly obtuse (figs. 1B, D, O, S).
2. Form *acutum* Sim.   Leaves mostly acute and rather larger (figs. 2B, C, R)
3. Form *microphyllum* Sim (=*M. microphyllum* Brid.).   Branches very slender ; leaves 0·5 mm. long or less ; per. leaves 1 mm. long, erect, acute ; capsule pyriform, 1 mm. long (figs. 3C and Q).
4. Form *dichotomum* Sim.   Branches longer, less-branched, usually dichotomous only (fig. 4A).

From Cape Town to Zambesia, abundant wherever there is bush or scrub, and often on rocks, stone walls, etc., from the coast to the mountains and to the Karroo kopjes ; said to be found also in Australasia and in the East African islands.

M. *elegans* Horns. (ined.) probably belongs here.   See Herb. Kew.

421. *Macromitrium Mannii* Jaeg., Ad., i, 421 ; Dixon, S.A. Journ. Sc., xviii, 317 ; Broth., Pfl., xi, 43.
=*M. Menziesii* Mitt., Journ. Linn. Soc. Bot., vii, 152 (1863).
=*M. undatifolium* C. M., in Flora, 1886, p. 278 ; Broth., Pfl., xi, 43.
=*M. rugifolium* C. M., in Dusen Musc. Cameroons, p. 263 ; Broth., Pfl., i, 3, 493, and xi, 43 ; Broth., in Engl. Bot. Jahrb., 1897, p. 241 (see Dixon, S.A. Journ. Sc., xviii, 317, and see herein under *M. rugifolium*).
=*Dasymitrium *Macleai*, Rehm. 508 and 508β ; Par., Ind., 2nd ed., p. 377.
=*Macromitrium *Macleai*, Par., Ind. (1896), p. 780.
=*Macromitrium *insculptum* Mitt. (ined.), in Hb. Kew.

Dioecious ; main stems prostrate, small-leaved, several inches long, the leaves usually on its upper side and the under-surface is often set with red rhizoid tomentum ; from these stems there rise on the upper side numerous, erect, usually simple, fertile branches 1–1½ inch long, the whole making a dense mat an inch deep and often a foot across. Leaves numerous, imbricate, translucently yellow-green except the youngest, which are nearly white, 2–3 mm. long, acuminate from a fairly wide base ; straight, spreading and waved somewhat, or incurved when moist ; spreading and much crisped and waved when dry, often twisted like a corkscrew, the perichaetal leaves alone remaining straight when dry ; lamina cells roundly hexagonal, 10μ diameter, close, chlorophyllose, and with a few low papillae near the apex, but much longer papillae at and below mid-leaf ; toward the base separate, longer, oblong, papillate cells appear, and at the base long, narrow, hyaline, 2-stratose cells with single mamillae are common though not always mamillate. Apex acute, the margin and surface there denticulate with projecting elongated cells, but the leaf is not bordered ; on some leaves this denticulate condition extends to the base, on others it is present at apex and base but absent at mid-leaf, where the margin is revolute on old leaves and showing turgid cells. Nerve strong, red, percurrent, raised on the under-surface, smooth except where covered with lamina cells, usually smooth toward the base, but there are often present two secondary nerves or wrinkles which are coloured and abundantly mamillate. Undulations are common over the leaf. Perichaetium always found lateral, even when young ; per. leaves shorter, wider, and more subulate than others, but standing erect when dry, usually with basal cells only or mostly, a few others being in the subula.

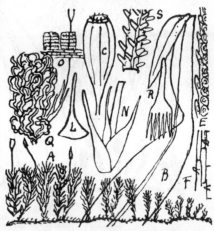

Macromitrium Mannii.
S, wrinkles, with mamillae.

Vaginula red, smooth, subtended by numerous red, withered archegonia and paraphyses. Seta red or pale, smooth, twisted ; capsule microstomous, pear-shaped, erect, 8-ridged when dry. Outer peristome of short, truncate, barred, grey teeth ; inner peristome a fairly high membrane, without cilia. Lid convex, red, with long subulate beak. Calyptra striated or wrinkled, without hairs, and not rough, bellshaped-beaked, covering half the capsule, the dilated lower portion which is half its length many-laciniated when mature. Old capsules permanent for years, then more oblong-cylindrical and deeply ridged.

Common on forest trees in Natal, forming rough vigorous tufts ; occurs also in Central Africa, West Africa, St. Thome, Madagascar, etc.

Natal. Zwart Kop, 5000 feet, Sim 8579 ; Blinkwater, Sim 8729 ; Town Bush, Pietermaritzburg, Sim 10,113, 10,117 ; Bulwer, J. W. Haygarth, No. 7 (as *M. insculptum* Mitt.) ; " Natal, 4000 feet, Miss Armstrong " (as *M. insculptum* Mitt., in Hb. Kew, in Mitten's hand-writing).

Transvaal. Macmac, MacLea (Rehm. 508), and Lechlaba, Houtbosch, Rehm. 508b (both as *Dasymitrium Macleai* Rehm.).

S. Rhodesia. Inyanga, 6000 feet, J. S. Henkel.

422. *Macromitrium secundum* C. M., in Bot. Zeit., 1856, p. 420 (sterile), and in Rehm. 168, 168β, 168c (all sterile).
=*Coleochaetium secundum* Broth., i, 3, 475, and xi, 26 ; Par., Ind., 2nd ed.
=*Leiomitrium*, Cardot, Mousses de Madag., p. 230 ; Kew Bull., No. 6 (1923), under No. 151.

Plant prostrate on bark in large cushions ; main stems prostrate, 1–3 inches long, tomentose on the under surface, clothed with small leaves on the upper surface, with numerous erect, simple or branched branches ¼–1 inch long rising irregularly. Leaves shrunk and longitudinally furrowed but not crisped when dry, numerous, imbricate, often some or all twisted to one side, giving a secund arrangement, in which case if in contact with or near the ground numerous red rhizoids are produced from the stem, and also from the lower half of the leaf-nerve. Leaves on prostrate stems always small but very variable in size and form, from triangular to lanceolate-acuminate, those on the younger portions almost pellucid, on older portions greener or rufescent, and often with red nerve. Erect stems many-leaved, leaves produced all round the stem, more or less secund, often rhizoidiferous, 1½–2 mm. long, tapering from the base to the acute apex, yellowish at the branch-point (straw-coloured when young), green or brown when old, and then usually with red nerves and often red rhizoids. Upper cells rather lax, quadrate-elliptical, tumidly papillose, those on the nerve more tumid ; lower cells more densely packed, irregularly quadrate, mamillate, as best shown on the basal margin ; hyaline cells absent ; nerve very strong, and usually coloured at the base, smooth in the lower half, covered with tumid lamina cells upward, the nerve vanishing a little below the apex. Apex acute, margin often or usually denticulate apart from the cells, or papillose when dry, sometimes apparently entire. Inflorescence and fruit have not yet been found.

Cape, W.　Stinkwater, Rehm. 168 ; Constantia Nek, Jan. 1919, Sim 9364 ; Montagu Pass, Apr. 1924, Miss Taylor.
Cape, E.　Gwacwaba River, King William's Town, 1895, Sim 7103.

C. appendiculatum Ren. and Card., in Bull. Soc. Roy. Bot. Belg., 1894, ii, 120 ; Ren., Fl. Bryol. Madag., 1897, p. 144 ; Ren., Musc. Mad. Exsicc., No. 218 (Madagascar), seems to be closely allied, but is described as having the leaves distinctly in five lines and the basal cells large, with tooth-like cilia.

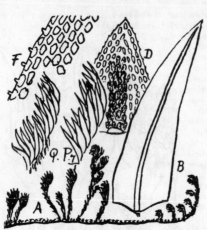

Macromitrium secundum.　　　　　　　　*Macromitrium rugifolium.*

423. *Macromitrium rugifolium* C. M., in Hedw., 1899, p. 115.

=*Coleochaetium rugifolium*, Broth., Pfl., i, 3, 475, and xi, 26.
=*Leiomitrium* (see Kew Bull., No. 6, 1923, under No. 151).
=*Macromitrium* **schlotheimiaeforme*, Par., Ind., Suppl., p. 241.

Closely allied to *M. secundum*, but with leaves comparatively wider (*i.e.* ovate-acuminate from a narrow, fibrously decurrent base), more sparsely distributed, more mamillate at the margin and upper surface near the base, and more erose in its denticulation upward. Leaves are 2×0.75 mm. long and wide, somewhat secund, the nerve strong, flexuous, and ferruginous, ending below the apex in a wide, densely tumid-papillose cushion. The upper leaves are acute, the lower round at the apex (fig. D,S) ; margin mostly expanded, occasionally partly reflexed toward the base, where the mamillae show as in fig. S, and the margin is there similarly mamillate ; toward the apex protruding cells make it irregularly erose. Müller makes a feature of its wrinkled leaf condition, but in that I find it does not surpass other species here included in § *Eumacromitrium*.

Known only from Rehm. 151 from Devil's Peak (sterile). Confusion may arise in regard to the name. A different plant from Brazil, *Orthotrichum rugifolium* Hk., Musc. Exot., p. 128, now *Schlotheimia rugifolia* (Hk.) Brid., Bry. Univ., i, 322 p.p. =C. M., Syn., i, 760 (1849), was published by C. Müller, in Bot. Zeit., 1845, p. 543, as *Macromitrium rugifolium* C. M. The present plant was published by C. M. as *M. rugifolium* C. M., in Hedw., 1899, p. 115, after having been distributed as *Zygodon rugifolius* C. M. in Rehm. 151, and presumably on account of that double entry Paris changed it to *M.* **schlotheimiaeforme* Par., Suppl. Ind., p. 241 (1900), but in his second edition transferred this into *Coleochaetium*, and introduced another *Macromitrium rugifolium* C. M. in Dusen M. Camer., No. 263 ; Broth., in Engl. Jahrb., 1897, p. 241, which Dixon now places under *M. Mannii* (S.A. Journ. Sc., xviii, 317).

The present position of these three plants named *M. rugifolium* by C. M. is :

1. *Schlotheimia rugifolia* (Hk.) Brid., from Brazil.
2. *Macromitrium Mannii* Jaeg.=*M. rugifolium* C. M., from Cameroons.
3. *Macromitrium rugifolium* C. M., from South Africa, which has been *Coleochaetium* and *Leiomitrium*, but being sterile may be either, or may grow out of either into something fertile.

424. *Macromitrium serpens* (Hk. and Gr.), Brid., Bry. Univ., p. 736 ; C. M., Syn., i, 739 ; Broth., Pfl., i, 3, 485, and xi, 35.

=*Orthotrichum serpens* Hk. and Gr., in Brew. Ed. Journ. Bot., p. 119, t. 5.
=*Leiomitrium capense* Broth., in Brunnthaler, 1913, i, 26 (sterile).
=*Coleochaetium capense* (Broth.), Broth., Pfl., xi, 26.
=*Dasymitrium* *Rehmanni C. M., in Rehm. 157.
=*Macromitrium* *Rehmanni (C. M.), Par., Ind. ; and probably
=*M.* *serpentinum Mitt. MSS. ; Shaw's Catalogue, which specimen is in Kew Herbarium.

Dioecious ; main stems prostrate, 1–3 inches long, dark brown, with very many short, stout, yellowish branches closely packed, ovoid when dry, densely bristling with spreading leaves when moist ; and a few older, erect, branched, fertile branches ½–1 inch high, the whole forming a dense cushion on bark. On the short branchlets leaves are very numerous, 1·5 mm. long, spreading when moist, but bent inward when dry, the per. leaves then standing erect as a separate group and of different form. Leaves lingulate-acuminate acute, yellowish, deeply conduplicate at the base and often channelled to the apex ; the margin entire or sometimes somewhat in steps toward the apex, expanded, or slightly revolute about mid-leaf. Nerve strong, smooth, ferruginous, ending just below the apex. Cells minute, 6–8μ diameter, irregularly rounded, somewhat turgid, irregularly and minutely papillose, the basal cells pale yellowish, pellucid, somewhat wider where they meet, up to 50μ long × 4μ wide, forming long lines for one-fifth of the leaf, gradually shorter and wider upward. Perichaetal leaves lanceolate-subulate, 2 mm. long, the basal cells as in cauline leaves but the upper mid-leaf and subula cells long elliptical (fig. NE), quite distinct from those of the cauline leaves, the nerve percurrent. Perichaetium usually found to be lateral. Seta ½ inch long or more, pale or red, capsule small, ovoid-pyriform, smooth except that it is furrowed at the contracted mouth ; peristome single, erect when dry, met in a dome when moist, of sixteen separate, columnar, blunt, grey or nearly white teeth, closely but finely transversely striate and papillose ; inner peristome not seen. Vaginula red, without hairs, but subtended by paraphyses. Lid convex-rostrate ; calyptra straw-coloured, glabrous, split at the base into many segments, these at first incurved, afterwards spreading.

Müller describes the calyptra as set with few simple hairs ; beyond an occasional hair I find them usually glabrous and smooth, but multifid for one-third of their length, the balance being subulate.

Cape, W. Knysna, Rehm. 157 (as *Dasymitrium Rehmanni* C. M., see C. M. in Hedw., 1899, p. 116) ; George, Burchell ; Montagu Pass, Apr. 1924, Miss Taylor.

Cape, E. Oliphant's Hoek, Uitenhage, Ecklon ; Alexandria Forest, Longbushkop, Jas. Sim ; Perie Forest, 4000 feet, 1892, Sim 7141.

Natal. Polela, 5000 feet, J. W. Haygarth (Sim 7810).

P.E.A. Magude, March 1919, Junod 328.

I cannot see wherein *Dasymitrium borbonicum* Besch. (Fl. Bryol. Reunion) differs from this, except perhaps in finer and more lax habit, a character here of no importance, but in any case the above name holds long priority.

I have not seen *M. elegans* Duby, Choix de Crypt ex. (1867), p. 6, which apparently is very near this if not identical. Mr. Dixon informs me that from description and figure it differs in the long and narrow, narrowly lingulate leaves and glabrous calyptra. It was collected by Pappe at the Cape—*teste* Duby. Broth., Pfl., xi, 35, still maintains it.

Macromitrium serpens.

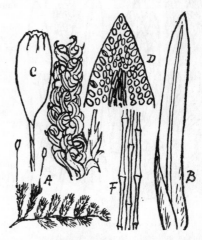

Macromitrium macropelma.

425. *Macromitrium macropelma* C. M., in Bot. Zeit., 1856, p. 420 ; Broth., Pfl., i, 3, 481, and xi, 34 ; Rehm. 167.

Dioecious ; habit very similar to *M. serpens*, with prostrate stems having many erect branchlets, of which some of the stronger are fertile. Leaves very numerous, 2 mm. long, linear-lanceolate from a rather wider clasping base ; suberect when moist but cirrhate when dry, acuminate to the acute apex ; laxly areolated ; margin entire, expanded, conduplicate at the base ; nerve strong, smooth, ending shortly below the apex ; cells pellucid, scattered, rounded or elliptical ; about mid-leaf oblong, hyaline, and toward the base long, narrow, smooth, and hyaline. Perichaetal leaves erect even when dry, subulate, with nerve throughout and mostly elongated smooth cells. Seta ½ inch long, much twisted when dry ; capsule oval-elliptical, tapering below, narrow-mouthed, and furrowed at the mouth only. Outer peristome well developed, inner not seen.

Nearly related to *M. serpens*, but the leaf longer, narrower, and smooth. Known only from Rehm. 167. Knysna Woods.

M. nitidum Hk. f. and Wils., a related species from Brazil and Australia, is recorded in Shaw's Catalogue as collected by him at Katberg, May 1869, presumably a case of mistaken identity.

Genus 122. SCHLOTHEIMIA Hedw.

Vegetative growth prostrate and rooting, several inches long, usually on bark or wood-humus, from which growth erect short branches grow in more or less clustered fashion, bearing the inflorescence terminally and having the leaves twisted round them in spiral fashion (torquate). Leaves ligulate-acuminate from a somewhat oblique ventricose base, concave, the keel strong to near the apex; the cells minute, roundish, and turgid, arranged in many regular lines, and much alike to the base, except the lowest which are rather larger and more rectangular, but all chlorophyllose. Capsule erect, cylindrical on a straight seta, furrowed when dry; lid cup-shaped below, subulate above, erect; calyptra campanulate, shining, straw-coloured, not plaited, the lobes at the base incurved at first, afterwards spreading; outer peristome teeth sixteen, more or less in pairs, well developed; inner a basal membrane with more or less rudimentary processes. Ring absent. Near *Macromitrium* but distinguished by more torquate leaves, better outer per. teeth, and the calyptra not furrowed.

Synopsis of Schlotheimia.

Perichaetal leaves not exserted.
   426. *S. Grevilleana.*
Perichaetal leaves exserted and standing erect when dry.
   427. *S. rufo-aeruginosa.* Leaves short, oblong-mucronate.
   428. *S. percuspidata.* Leaves longer; per. leaves long; per. teeth very long.
   429. *S. ferruginea.* Leaves longer; per. leaves long; per. teeth medium length.

426. *Schlotheimia Grevilleana* Mitt., M. Ind. Or., p. 52; Broth., Pfl., xi, 47.

=*Orthotrichum squarrosum* Griff. Not., p. 143.

Dioecious; a vigorous dark green moss having long prostrate main stems on bark, from which rise numerous erect, simple stems ¼–2 inches long, 1·5–2 mm. diameter, on which the capsules are borne; leaves lingulate-acuminate

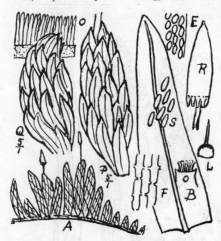

Schlotheimia Grevilleana.

S, cells of mid-leaf; O, inside view of part of outer and inner peristome.

acute, 2 mm. long, twisted round the stem to a point when dry, erect and almost appressed when moist, rugulose, the lower much indurated, the upper laxly areolated; the perichaetal leaves not longer or different from others except that they are thinner, the nerve stops below the apex, and the membranaceous margin is crenulate. Nerve narrow, deeply channelled below, covered with lamina cells on both surfaces upward, and ending in the mucro. Margin expanded or arched back; cells 15μ long, tumid; those of mid-leaf longer, those toward the base oblong, hyaline, 2-stratose, with pores between. Seta terminal, 5–7 mm. long, straight, pale; capsule erect, elliptic-cylindric; lid erect, rostrate from a cup-shaped base. Calyptra straw-coloured, smooth, shining, without hairs, 6–8-fid at the base, the segments at first clasping, later spreading. Outer peristome of sixteen linear, brown teeth, free and somewhat reflexed when dry and apparently connate in pairs when moist, trabeculate and papillose; inner of a white granular membrane; no cilia seen. The prostrate extension stems are light green, 2–3 inches long, abundantly branched, with twisted leaves on the upper side and abundant red rhizoids on the under side. In old cushions these extension stems are lost and the upright stems stand parallel in dense clumps, each stem twisted to a point.

S. Macleai, Rehm. 503, is only a vigorous form.

S. ventrosa C. M., Syn., i, 756, may belong here, but the description is evidently unreliable since four segments in the calyptra can only be a condition, not a full description of that part. Bryhn records it from Entumeni, Zululand, and Shaw calls it *S. ventricosa* C. M.

S. subventrosa Br. and Bry. (Vidensk. Forh., 1911, iv, 13) is described as near *S. ventrosa* C. M., but has wider ovate-ligulate, much more rugulose leaves and the nerve ending below the leaf-apex, which is the usual condition in *S. ventrosa*, so I see no use for it. Brotherus maintains both with very rough calyptra in Pfl., xi, 47.

Cape, W. Newlands Avenue, 2000 feet, Sim 9421; Platteklip Ravine, Sim 9288.

Cape, E. Oliphant's Hoek, Uitenhage, Ecklon (*S. ventrosa* C. M.).

Natal. Giant's Castle, 7000 feet, 1915, R. E. Symons (Sim 10,111); Van Reenen, Jan. 1911, Wager; Inanda, Dr. J. M. Wood.

Transvaal. Pilgrims Rest and Macmac, MacLea (Rehm. 502*b* and 502*c*).

It occurs also in India and S.-E. Asia.

427. *Schlotheimia rufo-aeruginosa* C. M., Musc. Schweinf. Linn., 1875, p. 410; Broth., Pfl., xi, 48.

=*S. rufoglauca* C. M., in Hedw., 1899, p. 118, and Rehm. 153; Broth., Pfl., xi, 48.

Dioecious, prostrate, forming a dense compact brown cushion 3–10 mm. deep, on bark, with very numerous, crowded, erect, short branchlets 3–10 mm. long, ovoid or cylindrical when dry, bristling with short spreading leaves when moist, and occasional longer branches tomentose with red rhizoids below and with scattered leaves above. Leaves on the branchlets very numerous, pale at first, brown or yellowish brown later, 1 mm. long, lingulate, shortly pointed, mucronate, expanded and pellucid when young, and excessively incrassate when old. Margin entire. often revolute about mid-leaf; nerve strong, percurrent, red with age, smooth; cells 8–10μ diameter, separate, the upper roundish, then rhomboid to near the base, the inner cells near the base 20×8μ, the outer cells there like the upper lamina cells, and all smooth; with age all cells become in contact and exceedingly indurated with deep yellow walls and translucent lumen, the upper cells then tumid and the inner basal cells then elliptical or vermicular. Leaves spirally arranged, erecto-patent when moist; when dry twisted spirally round the stem, not crisped and not closely appressed; perichaetium terminal, the per. leaves twice as long as others, rather wider and less pointed, somewhat sheathing and convolute, nerved nearly to the apex, laxly areolated. Vaginula stout, red, without hairs, but at the base are a few or many withered archegonia and some jointed paraphyses. Seta 3–4 mm. long, half wrapped in the perichaetal leaves, pale, very thick. Capsule cylindrical, tapering below, wide-mouthed, smooth, brown, and indurated to the mouth where there is a small-celled red band, but with white extension of similar small cells; the peristome rising from inside the mouth, consisting of sixteen pale, conical, banded and minutely papillose, separate teeth, split or translucent down the middle; inner per. teeth slender, shorter, with low membrane; when dry the outer teeth all bend over into the mouth and the cilia stand erect; when moist both are erect but the inner processes are difficult to find, and with age they disappear and the teeth assume a permanently inbent position, with the appearance of having numerous bullate indurated segments (fig. OS). Lid subulate; calyptra long, straw-coloured, smooth, without hairs, entire to near the base, but with the base multifid, the segments at first inbent, afterwards spreading. Varying much in appearance between the young pellucid expanded condition, with double peristome, and the old, exceedingly indurated, brown condition with revolute margin and no processes, but all are to be found on one plant.

Cape, W. Newlands Avenue, Sim 9429.

Cape, E. Oudtshoorn, Miss Taylor; Portland, Knysna, Rehm. 153 (as *S. rufoglauca*).

Natal. Giant's Castle, R. E. Symons (Sim 10,112); Mont aux Sources, Sim 10,115; Benvie, York, Sim 10,118; Ngoya Forest, Zululand, Sim 10,116; Burdon's Bush, Karkloof, J. M. Sim.

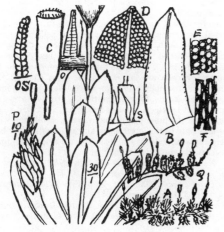

Schlotheimia rufo-aeruginosa.
OS, indurated peristome tooth.

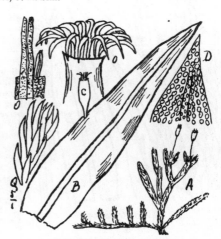

Schlotheimia percuspidata.

428. *Schlotheimia percuspidata* C. M., in Hedw., 1899, p. 117, and Rehm. 155 and 501; Broth., Pfl., xi, 48.

    =*S. *cuspidata* C. M., in Rehm. 155, 504, 504β, 505.

    =*S. *cuspidata* C. M. var. *brevipedunculata*, Rehm. 156, and var. *tenuis*, Rehm. 505.

    =*S. rufopallens* C. M., in Hedw., 1899, p. 117; Broth., Pfl., xi, 48.

    =*S. *mollis*, Rehm. 506, and forma *procera*, Rehm. 507.

    =*S. exrugulosa* C. M., in Hedw., xxxviii, 118 (sterile).

    =*S. *Woodii* Brid. (see below).

Dioecious, rusty to very dark brown, vigorous, tufted, mostly erect, with some prostrate extension stems. Erect stems ½–1 inch long, repeatedly branched; leaves rather sparse, more or less twisted when dry, rugulose, almost straight and suberect when moist, lanceolate from a narrow base; nerve ferruginous, hardly extending into the pointed or sometimes aristate apex; margin usually expanded, sometimes suberect; lamina undulated; cells roundly quadrate, smooth, 10μ wide, the lower cells hardly larger. Perichaetal leaves longer, more pointed, somewhat twisted round the seta. Seta short, red, terminal, usually occupying a fork; capsule oval-cylindrical, when old cylindrical and sulcate; lid rostrate on a cup-shaped base. Calyptra glabrous, straw-coloured, the apex rough, the base widely laciniate. Perianth teeth sixteen, longer and narrower than in other species, reflexed singly when dry, erect in pairs when moist; inner membrane high, fugacious, with occasional short slender processes, either singly, or doubly, or absent.

Rehmann's specimens distributed as *S. cuspidata* C. M. all show rather shorter leaves, wider above than here illustrated, and just as in *S. rufo-aeruginosa*, but with a longer point into which the nerve does not reach, but have the narrower capsule-mouth and long teeth of this species, and in other respects agree. My type for this is a specimen collected below the Natal Drakensberg by Dr. J. M. Wood (Wood 7118), which is labelled *S. Woodii* Brid. (Auct. V. F. Brotherus), but I am unable to trace that name, and it seems more like C. M.'s plant than do Rehmann's specimens, referred to in his description of *S. percuspidata.*

Müller's description of *S. rufo-pallens* differs in no respect except in being very torquescent, with narrower leaves, though he compares it with *S. rufo-aeruginosa* on account of habit.

Cape, W. Saddle, Table Mountain, Pillans 3358; Montagu Pass, Rehm. 155, 156, 505; Table Mountain, Rehm., sterile (as *S. exrugulosa* C. M.).

Cape, E. Knysna, Rehm. 504; Blanco, Rehm. 154 (as *S. rufo-pallens*, C. M.).

Natal. Weenen County, 3000–4000 feet, March 1898, Dr. J. M. Wood; New Germany, Pinetown, April 1917, Dr. v. d. Bijl 22; Little Berg, Cathkin, Miss Owen, No. 7.

Transvaal. Lechlaba, Houtbosch (sterile), Rehm. 506 and 507 (as *S. mollis*).

Rhodesia. Inyanga, July 1918, Dr. E. A. Nobbs (Eyles 1358).

429. *Schlotheimia ferruginea* (Hk. and Gr.), Brid., Bry. Univ., i, 743; C. M., Syn., i, 755.

=*Orthotrichum ferrugineum* Hook. and Grev., in Brew. Ed. Journ. Bot., i, 118, t. 5.
=*Macromitrium ferrugineum* C. M., in Bot. Zeit. (1845), p. 544.

Dioecious; habit mostly erect, the decumbent extension branches being lost under the others, the erect branches branched or repeatedly branched, brown above, rusty tomentose or darker below; leaves very much spirally twisted, ending each branch in a point whether fertile or not, the perichaetium terminal or axillary, but there also the leaves are spirally twisted. Leaves 1·5–2 mm. long, brown, shining, smooth or rugulose, erect when moist, numerous, imbricate, oblong-lingulate, widest below the middle, somewhat rounded or acuminate to the acute point; nerve strong, rusty with age, ending in the mucro; margin expanded or somewhat reflexed; upper cells obliquely elliptical, 10μ long, smooth, lower longer; basal cells oblong or longer, hyaline, thin-walled; perichaetal leaves longer, thinner, often twice as long, from a wider base tapering throughout to an acuminate point; the cells are larger and more lax than in the stem-leaves, and the nerve ends below the apex. Seta pale, ½ inch long or more; capsule 2 mm. long, cylindrical, 8-striate, narrowed below the mouth; teeth sixteen, more or less in pairs, reflexed when dry, erect when moist, the inner peristome a narrow membrane with short processes alternate to the outer pairs, and standing erect when dry. Lid long, subulate; calyptra conical, straw-coloured, shining, smooth, with eventually spreading segments.

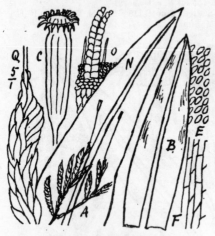

Schlotheimia ferruginea.

Cape, W. Newlands Ravine, 2000 feet, Sim 9433; "Cape, on trees, not rare, Burchell, Ecklon, Drege, etc." (C. M.).

Cape, E. Perie Forest, 4000 feet, 1893, Sim 7235; Umtentu River, E. Pondoland, J. Burtt-Davy 15,352.

Natal. Town Bush, Maritzburg, Sim 10,110; Burdens Bush, Karkloof, Jan. 1917, J. M. Sim; Giant's Castle, 6000 feet, Nov. 1915, R. E. Symons (Sim 8534).

S. Brownei Schw., Suppl., II, ii, 1, 52, t. 167; Brid., i, 799 = *M. Brownei* C. M., in Bot. Zeit., 1845, p. 544, is the Australian representative of this type and very near this species. Shaw uses the name *S. Brownei* Schw. for Zeyher's Cape specimens, now placed here.

Family 29. EUSTICHIACEAE.

Small, slender, dioecious plants growing on moist rocks or on the ground in shining, bright green tufts; stems erect, more or less branched upwards, closely leaved, the leaves distichous-imbricate or equitant, horizontally inserted, very concave, blunt, shortly mucronate, with strong nerve extending into the mucro; margin erect and wide, finely papillose-serrate; cells quadrate or roundish quadrate, on both sides papillose, those at the base thicker and shortly rectangular. Inflorescence lateral, perichaetal leaves longer, subulate, subserrate; seta erect, slender; capsule erect, regular, oval, 8-striate, when dry furrowed; ring absent. Outer peristome absent; inner rising from below the mouth, of sixteen lanceolate-subulate teeth, united at the base, occasionally perforated, striated, hardly papillose. Lid obliquely subulate from a conical base; calyptra cucullate. Only seen sterile in South Africa.

Genus 123. EUSTICHIA Mitt.

=*Diplostichium* Mont., Fl. Chil., vii, 67.

Characters of the family. A small genus, native almost entirely of the Southern Hemisphere, of which two species belong to the South African mountains, and have not been seen fertile there.

C. Müller (Syn. Musc., p. 40) placed *Eustichia* in Distichiaceae, but remarked upon its wide divergence from

Distichium; Brotherus placed it in Orthotrichaceae (Nat. Pfl., i, 3, 456), but later made Eustichiaceae a separate order (Nat. Pfl., i, 3, 1198) with *Eustichia* as the one genus, which arrangement he maintains in Pfl., x, 421.

430. *Eustichia longirostris* (Brid.), C. M., Syn., i, 42, and ii, 523.

=*Diplostichium longirostre* (Brid.), Mont., in Ann. Sc. Nat., iv, 116, and Syll., p. 27.
=*Phyllogonium longirostris* Brid., Bry. Univ., ii, 674.
=*Pterigynandrum longirostrum* Brid., ii, 195.

Plants 5–15 mm. long, slender, scattered or caespitose on wet stones, erect, simple or irregularly much-branched; leaves distichous, often closely equitant but sometimes more or less distant, 0·4–0·6 mm. long, boat-shaped, closely conduplicate, somewhat cucullate, with a wide translucent keel extending as a short, often incurved point, the flattened aspect being oblong-acuminate; cells separate, quadrate, 10–12 × 7μ, in straight lines each direction, usually with four to six papillae on each, the cells of the keel larger, the keel and the leaf-margin both crenate through the presence of papillae. The leaves being distichous and closely conduplicate, with a lighter coloured keel, make the plant resemble *Fissidens*. Sexual parts not seen. The leaves are numerous and usually longer and more spreading in the lower than in the upper half of the stem, and the points all directed upward and of equal length give a continuous outline to the whole.

Natal. Bushman's River, 7000 feet, Nov. 1906, Sim 8535; Giant's Castle, R. E. Symons (Sim 10,125); Mont aux Sources, 7000 feet, Sim 10,119, 10,122, 10,123; National Reserve, 7000 feet, Professor Bews (Sim 8334); Tugela Gorge, 6000 feet, Sim 10,120, 10,124.

Occurs also in Tristan d'Acunha, and this or a similar species in Madagascar.

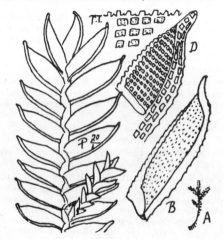

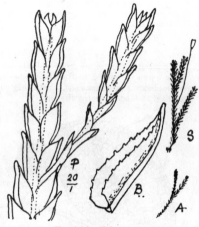

Eustichia longirostris.

Eustichia africana.
S, fructification of *Eustichia* from a South American plant.

431. *Eustichia africana* (C. M.), Broth., Pfl., x, 421.

=*Diplostichium africanum* C. M., in Hedw., 1899, p. 53.
=*Fissidens* *eustichium, Rehm. 279, 485, see Dixon, in Kew Bull., 6 (1923), No. 279.

Similar to *E. longirostris* but smaller in all parts; leaves minute, erect, and appressed; mucro straight or bent outward (not inward), and the cells are smaller (6–9μ).

Natal. Sweetwaters Waterfall, 3000 feet, 1917, Dr. Bews (Sim 8636); Wittebergen, above Kadziberg, Rehm. 485, as *Eustichia longirostris* Brid., and Rehm. 279, as *Fissidens eustichium* Rehm.

(Family SPLACHNACEAE.

Erect, gregarious, or tufted mosses, usually growing on animal excrement, with leafy, simple, or forked stems, long setae, and erect regular capsules having (in *Splachnum*) distinct apophyses, and the peristome absent or single, with the teeth usually in pairs. Leaves large, wide, somewhat succulent, with large, very open, hexagonal-elliptical, smooth cells. Calyptra entire and conical, or cucullate. A very distinct group.

Genus SPLACHNUM Linn. Gregarious or loosely caespitose, with more or less obovate-acute leaves, with large thin-walled cells, and very flaccid when dry. Capsule oval or globose, with a wide, globose, pear-shaped or umbrella-shaped apophysis on a long succulent seta. Peristome teeth sixteen, in pairs, usually reflexed when dry, three layers thick, the mid-layer larger. Male inflorescence discoid; discs stalked, terminal. The climatic conditions in South Africa are usually such that excreta disappears before any moss can germinate upon it, hence probably the scarcity of this group.

Splachnum longicollum Schw., Suppl., II, ii, 2, 85, t. 178, is described and illustrated there from a specimen said to have been collected by Menzies on Table Mountain long ago, and C. M., in Syn., i, 148, placed it among *Splachnaceae suspectae*, and suggested that it might be an *Entosthodon*, but as Paris' Index still retains the record, it is mentioned here and *Splachnum* described. In Bull. de l'herbier Boissier, 1899, Cardot refers to *S. longicollum* Dicks. (and cites the above as synonym) to clear up the North American record. It appears that Menzies' Table Mountain specimen can still be seen among Hedwig's and Schwaegrichen's types which are housed in the Herbarium Boissier, Geneva, and that Cardot saw the one which is figured by Schw. on t. 178, and which has the label " *S. longicollum* D. ex Cap. b. sp. ; ex herbar. Dicksoni, Turner," and he says slight examination separates it from the others completely, but he was interested only in the American plant which he states is not *S. longicollum* though a *Splachnum*. Menzies' specimen or its illustration requires re-examination, but even if proved correct a further doubt is cast on it because Dickson's localities were not always correct, *e.g.* he cited this same species from Scotland where its presence is denied. *S. longicollum* Schw. must sink as a South African moss until its presence is established. Müller's brief description (Syn., i, 148) is " Caulis simplex ; folia spathulato-oblonga acuta ruptinervia denticulata ; theca subglobosa, apophysi attenuata. Peristomi dentes 16, aequidistantes erecti." Paris gives *S. longicollum* (Dicks.) as different and as being *Tayloria tenuis*, W. P. Sch., a European and N. American moss.)

Family 30. GIGASPERMACEAE.

Very similar to Funariaceae in the large, lax, smooth cell-formation, but small rhizomatous plants ; stem without central strand ; capsule erect, regular, globose or nearly so ; peristome absent ; spores very large ; calyptra very small, caducous. Till lately included in Funariaceae.

Leaf with nerve.
 Leaves ovate-spathulate ; capsule sessile . . . . . . 124. OEDIPODIELLA.
 Leaves orbicular, some aristate ; capsule exserted . . . . 125. CHAMAEBRYUM.
Leaf without nerve . . . . 126. GIGASPERMUM.

Genus 124. OEDIPODIELLA Dixon.

Short-stemmed, rhizomatous ground plants, with short, erect, rosulate branches having obovate, succulent, entire leaves ; wide nerve extending to the apex, large roundly hexagonal areolation, more quadrate in the marginal row, and sometimes bearing large gemmae. Seta very short ; capsule hardly emergent, small, subglobose, shortly pointed ; lid permanent (?) beaked ; spores 40–50μ.

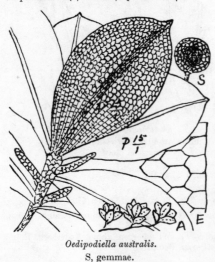

Oedipodiella australis.
S, gemmae.

432. *Oedipodiella australis* Dixon, in Journ. of Bot., lx, 105 (1922) ; Broth., Pfl., x, 315.

=*Oedipodium australe* Wag. and Dixon, Trans. R. Soc. S. Afr., iv, 1, 3, and pl. 1F ; Dixon, Bull. Torrey Bot. Club, xliii, 67.

A succulent, little, rosulate, large-leaved, light green soil-moss, sometimes crowded, sometimes scattered, spreading by succulent rhizomes axillary from the lower leaves, and also by gemmae from a terminal disc, the rhizomes sending up rosules at 5–10 mm. distance, and each rosule containing in its centre several to many somewhat flattened, erect, very large, circular gemmae shortly stalked from one side. Leaves spreading, the inner largest, surrounding the disc, few, spathulate from a narrow base or sometimes ovate-spathulate, 4–5 mm. long, rounded to a strong yellowish mucro ; nerve wide, extending to the mucro, but hardly evident upward, smooth, its cells same size as lamina cells, and chlorophyllose ; margin entire, undulated, not ciliated at the base ; areolation very lax and thin ; cells very large, 50×40μ. Smooth, irregularly hexagonal, smallest and closest toward the apex, those at the base quadrate or oblong and rather larger, all chlorophyllose or pellucidly green and thin-walled ; leafless axillary rhizomes unlike those in any other moss, and the gemmae remarkably large, up to 300μ diameter, dark green, with a transparent marginal wing and thickened centre. The leaves shrink somewhat when dry and are then more compact and erect. I have not seen the capsules, but they have been found and described as above. Resembles the European *Oedipodium Griffithianum* (Dicks.) Schw. in growth. Recorded by Wager from Perie Forest, C.P., Umkomaas (Natal), Waterval, and Pretoria (Transvaal). Abundant in Alexandra Park, Maritzburg, Natal, April 1917, and in Nov. 1924, Sim 10,127.

Genus 125. CHAMAEBRYUM Thér. and Dixon, Journ. of Bot., 1922, p. 106.

Minute plants with creeping, subterranean, succulent rhizomes, from which small, erect, globose branches ascend, on which the leaves are closely imbricated, very concave, spathulate-orbicular, with weak nerve ceasing below the apex on sterile branches, and extending as a twisted (where dry) hair on fertile branches. Seta short, capsule erect, mouth-band distinct, ring and peristome absent ; lid large, flat, mamillate. Leaf-cells very lax.

433. *Chamaebryum pottioides* Thér. and Dixon, Journ. Bot. (1922), p. 106 ; Broth., Pfl., x, 315.

A minute creeping plant having the generic characteristics ; branches globose ; leaves closely imbricated, the fertile stems having turgid, onion-shaped vaginula of spongy texture, with the upper or perichaetal aristate leaves densely crowded round it, the arista twisted when dry, flat when moist ; the sterile branches with leaves not aristate but obtuse or apiculate with weak nerve ending below the apex ; margins flat, entire ; basal cells very lax, $20 \times 10\mu$, widely rectangular, hyaline, one marginal line shorter, quadrate ; upper gradually smaller, hexagonal, roundly quadrate, 10μ wide, marginal rather smaller, all walls thin, pellucid. Capsule globose or urn-shaped, erect, 1 mm. long, its cells very lax, with five to six rows of smaller transverse cells at the mouth. Spores $30–35\mu$. Male flower not seen.

Cape Town, 1917, Wager 633 (fertile), 654, 655 (sterile).

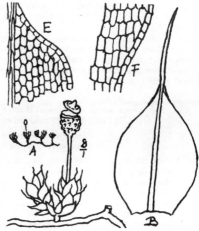

Chaetobryum pottioides (after Dixon).

Genus 126. GIGASPERMUM Lindb., in Öfvers. af. K. Vet.-Akad. Förh., 1864, p. 599.

Small earth-grown plants with rhizomatous, prostrate, nearly leafless stems, from which erect, fertile, club-shaped rosulate, leafy stems are produced, bearing the almost sessile capsule erect in the centre of the rosule ; leaves ovate-acute or pointed, concave, imbricating ; the lower small, the upper larger and longer-pointed ; cells laxly hexagonal or rhomboid ; nerve absent ; seta very short ; capsule globose, urn-shaped when open, lax-celled, without neck ; peristome absent ; spores very large, granular ; lid flatly conical ; calyptra campanulate, entire, smooth, very small, evanescent.

There are three Australian and one South African species.

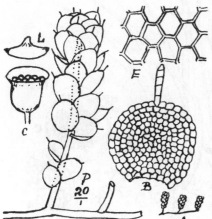

Gigaspermum repens.

434. *Gigaspermum repens* (Hook.), Lindb., as above ; Broth., Pfl., x, 316.

=*Gigaspermum Breutelii* (C. M.), Par., Ind., p. 511 (1895).
=*Physcomitrium Breutelii* C. M., in Bot. Zeit., 1855, p. 749.
=*Leptangium Breutelii*, Jaeg. Ad., ii, 71.
=*Anoectangium repens* Hook., M. exot., ii, 8, t. 106 (1820).
=*Hedwigia repens*, Hook. f. and W., Fl. of N.Z., ii, 92, and Handbook of the Fl. of N.Z., p. 424 ; and Fl. Tasmania, ii, 178.
=*Leptangium repens*, Jaeg. Ad., ii, 71.
=*Physcomitrium repens* C. M., Syn., ii, 544.
=*Schistidium repens* Brid., Bryol. Univ., i, 120.

A very minute and remarkable moss having branched, underground, leafless rhizomes, from which erect, club-shaped, leafy branches rise to a height of 2–3 mm. ; leaves few and small at the base, but numerous, imbricated, and larger upward, nearly circular, concave, entire, nerveless, spreading widely when moist, erecto-patent when dry, with a long hair-point of single cells on end, and with lax hexagonal areolation. Capsule terminal, almost sessile, mostly included among the more lanceolate perichaetal leaves, at first globose, urn-shaped when open. Lid nearly flat, shortly pointed. Calyptra small, campanulate.

Wilde Als Kloof, Bloemfontein, O.F.S., Professor Potts (Sim 8645), among *Entosthodon* and *Fimbriaria*.

Family 31. EPHEMERACEAE Broth., Pfl., x, 317.

Minute soil-mosses with abundant protenema ; areolation very lax ; outer leaves very small, inner longer ; male flowers very small, bud-like, on protenema. Capsule almost sessile, globose or ovoid, apiculate, cleistocarpous : spores very large. Formerly included in Funariaceae as its lowest type.

Genus 127. EPHEMERUM Hampe.

Minute, almost invisible, or small annual or perennial plants consisting mostly of a much-branched green protenema, from which the stemless rosette of little leaves rises like an open bud, surrounding the almost sessile capsule. Leaves few, lanceolate or ovate-lanceolate, the outer small and the inner more acute and larger ; the cells large and lax, rhomboidal-hexagonal, smooth ; nerve present in the upper half of the leaf only, or absent. Capsule almost sessile globose or rounded, shortly mucronate, cleistocarpous ; spores fairly large ; calyptra conical-campanulate, torn at the base. These plants are gregarious on moist soil, and the protenema is often more evident than the plants. *Ephemerum* and *Sphaerangium* are very much alike, each the lowest grade in its own line of relationship.

19

Synopsis.

Calyptra campanulate.

435. *E. serratum.* Nerveless ; leaves shorter than the capsule.
436. *E. sessile.* Nerve present ; leaves longer than the capsule.
437. *E. diversifolium.* Outer leaves nerved, inner nerveless ; margins entire.

Calyptra large, cucullate.

438. *E. Rehmanni.* Leaves subulate, nerved.

Dr. Shaw described in English, in his Catalogue, a new species *E. piliferum* which I cannot place, viz. " Dioecious ; leaves ovate, becoming suddenly attenuated at the apex into a very long bristly hair-point, nerveless. Capsule perfectly enclosed in the concave perichaetal leaves. Calyptra cucullate. Collected by MacLea in the Oudeberg, near Graaff Reinet, April 1872."

435. *Ephemerum serratum* (Schreb.) Hampe, in Flora, 1837, p. 285 ; C. M, Syn., i, 31, and ii, 521 ; W. P. Sch., Syn., 2nd ed., p. 3 ; Dixon, 3rd ed., p. 293 and Tab. 35H ; Broth., Pfl., i, 3, 513, and x, 318.

=*Phascum serratum* Schreb., " de Phasco," p. 9, t. 2.
=*Ephemerum capense* C. M., in Flora, 1888, p. 12 ; Broth., Pfl., x, 318.

A very minute dioecious plant, having abundant repeatedly dichotomously branched protenema growing as a dense green carpet and very evident on the surface of the soil when the plant itself cannot be seen except when the capsules are mature and red. Plant erect, consisting of about four cellular nerveless leaves 0·3 mm. long, in the midst of which the capsule is almost sessile, larger than the leaves, and being red is more easily seen than they are. Leaves elliptical, few-celled, cells large, longer than wide, the marginal cells protruding, making the leaf serrate. Capsule globose, 400μ diameter, mucronate, terminal, cleistocarpous, splitting irregularly ; calyptra shortly campanulate, acuminate, somewhat lobed at the base ; spores large, 35μ diameter, dark brown, minutely papillose. I cannot separate this from the European and North American *E. serratum*, though Müller did so as *E. capense*.

King William's Town, Cape Province, May 1893, Sim 7094.

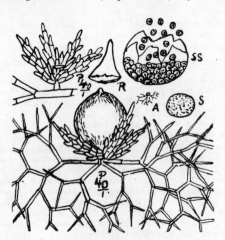

Ephemerum serratum.
S, spore ; SS, burst capsule.

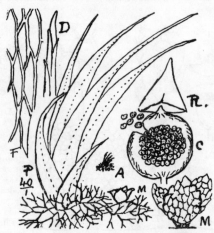

Ephemerum sessile.

436. *Ephemerum sessile* Br. Eur., 1, Mon., p. 7, t. 2 ; C. M., Syn., i, 31, and ii, 522 ; Dixon, 3rd ed., p. 295, Tab. 36C.

=*Phascum stenophyllum* Voit., in Sturm. Deutsch. Fl., ii, 14.
=*Ephemerum stenophyllum* (Voit.), W. P. Sch., Syn., 1st ed., p. 5, and 2nd ed., p. 6.

A small, erect, perennial, autoecious plant, with much more leafage and less protenema than *E. serratum*. Protenema dichotomously very dense and short, producing large-leaved fertile stems and small-leaved male crowns on the same mass ; on the fertile crown the leaves are six to ten, slightly falcate, lanceolate subulate, laxly cellular, the nerve wide but not very evident, extending weakly to the apex ; margin entire except that some cells protrude slightly at the long narrow apex ; cells very large, long-hexagonal, incrassate when old, capsule terminal, central, brown, much shorter than the leaves, almost sessile, globose, apiculate, cleistocarpous, opening irregularly, and with shortly conical-campanulate calyptra, somewhat torn at the base. Spores 25μ diameter, dark. Male inflorescence in small disciform 3–4-bracted flowers on the protenema, the bracts small, ovate, serrate, nerveless, and the antheridia short and turgid. The fertile plant dies off after producing spores, and several innovations (also fertile in due time) rise from below it.

Cape, E. Gwacwaba River, King William's Town, June 1893, Sim 7097.

437. *Ephemerum diversifolium* Mitt., in Harv. Thes. Cap., p. 100, fig. A.

Mitten describes this from Cape as " Monoecious, with the habit of *E. sessile*, outer leaves spreading, lanceolate, gradually narrowed, entire. with percurrent nerve ; the inner longer, more erect, wide above, entire, without nerve ; capsule sessile, ovate, rostellate ; male flowers bud-like, collected at the base of the stem. In a field near Zwart Kop River, Uitenhage (Zeyher. herb. Mitt.). The distinctly nerved outer and nerveless inner leaves of this species, as well as their entire margins, distinguish this from all its allies ; it produces an abundance of stolons which are much branched at their extremities and almost bury the plants when in a fertile state in a stream of confervoid filaments."

His illustration shows cucullate calyptra and that the inner leaves are wider than in *E. sessile*, especially in the upper half.

438. *Ephemerum Rehmanni* (C. M.), Broth., Pfl., x, 319.

=*Ephemerella Rehmanni* C. M., in Flora, 1888, p. 12.

Brotherus (Pfl., i, 3, 513) maintained a genus *Ephemerella*, distinguished by cucullate calyptra, in which this species was included along with some others, but in Pfl., x, 317, that genus disappears, and this species is included in *Ephemerum* and described as having the leaves gradually extending as a long subula from a narrowly elliptical base ; toothed upward ; nerve faint, filling the whole subula ; the calyptra cucullate, and protenema present.

The difference between a lobed or fringed calyptra and a cucullate calyptra is very slight among these minute mosses ; this species has, I believe, been only once collected, and it may prove to be *E. sessile* if re-examined.

Family 32. FUNARIACEAE.

Mostly short-stemmed, gregarious plants, growing on soil, with more or less rosulate, wide, thin leaves, and very lax areolation, terminal inflorescence, usually straight seta, and globose, oval, or pyriform capsule, either erect and regular, or cernuous and oblique, frequently having a tapering neck. Capsule cleistocarpous, gymnostomous, or peristomate. Outer peristome, when present, of sixteen teeth, usually twisted to the left, and sometimes united at their tips ; inner peristome, when present, of sixteen separate processes, opposite to the outer teeth, and without connecting membrane. Calyptra inflated, lobed or cucullate, smooth, beak often oblique.

Synopsis of Funariaceae.

1. Capsule cleistocarpous		128. PHYSCOMITRELLOPSIS.
1. Capsule gymnostomous		2.
1. Capsule peristomate (normally) or peristome weak or absent		4.
2. Capsule almost sessile		3.
2. Capsule on a well-developed seta		131. PHYSCOMITRIUM.
3. Calyptra shorter than the lid		129. MICROPOMA.
3. Calyptra covering the capsule		130. GONIOMITRIUM.
4. Peristome rudimentary or single, or absent		132. ENTOSTHODON.
4. Peristome double and well developed		133. FUNARIA.

Genus 128. PHYSCOMITRELLOPSIS Broth. and Wag., in Journ. of Bot., lx, 107 (1922) ; Broth., Pfl., x, 321.

Plants small, green, gregarious, autoecious, without protenema ; leaves lanceolate-subulate ; capsule nearly globose, cleistocarpous ; calyptra swollen at the base enclosing the capsule ; lid not separating. Only one species so far.

439. *Physcomitrellopsis africana* Wag. and Broth., in Journ. of Bot., lx, 107 ; Broth., Pfl., x, 321.

Stems 5 mm. high ; upper leaves spreading, lax, lanceolate from a long spathulate base ; apex shortly subulate, usually toothed ; nerve extending to or beyond the leaf-apex ; cells very lax, the upper elongated 6-sided. Seta very short ; capsule slightly pointed ; calyptra large, shortly pointed, containing the capsule and swollen at the base. Spores about 30μ.

Natal, Wager.

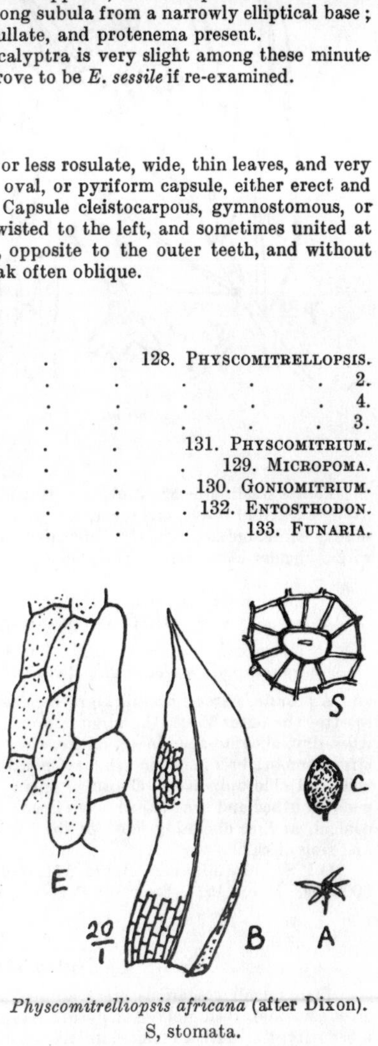

Genus 129. MICROPOMA Lindb., in Not. Sallsk. Faun. et Fl. Fenn. Förh., xi, 56.

Very small, erect mosses (there are only two species in the genus so far, both African), gregarious on soil ; stems simple, with few lanceolate-acuminate leaves, and terminal almost sessile, hemispherical capsule, with flat mucronate lid, no peristome, short calyptra torn at the base, and narrow ring.

Physcomitrelliopsis africana (after Dixon).
S, stomata.

440. *Micropoma niloticum* (Del.) Lindb. (as above) ; Broth., Pfl., x, 322.

=*Gymnostomum niloticum* Del., in Descr. de l'Egypte, Hist. Nat., ii, 289 ; C. M., Syn., i, 165.
=*Physcomitrium niloticum* C. M., in Bot. Zeit., 1858, p. 154 ; Par., Ind., p. 937 (1896).

A small gregarious moss 2–4 mm. high, with simple few-leaved stems and terminal hemispherical capsule, all very much shrivelled when dry. Leaves lanceolate-acuminate from a narrow base, the outer 2 mm. long, the inner shorter and narrower, concave with raised undulate margin, very laxly areolated, with strong nerve extending to the apex. Lower cells rectangular, upper irregularly elliptical, the inner largest, the outer forming a line with slightly projecting tips, all thin-walled ; seta short, slender ; capsule shortly cup-shaped, thin-celled, wider than long, with flat mucronate lid, and short calyptra ; spores small, numerous.

Originally found in Lower Egypt long ago, now at Antioka, near Magude, Portuguese East Africa, Rev. H. A. Junod 323 ; and at Lomogundi, S. Rhodesia, 3500 feet, July 1921, Eyles 3166. The other and closely related species is *M. bukobense* Broth. from Central Africa.

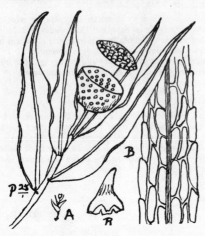

Micropoma niloticum.

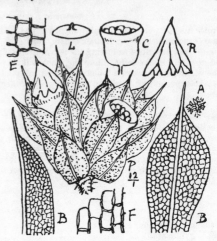

Goniomitrium africanum.

Genus 130. GONIOMITRIUM Wils.

Plants small, almost stemless, with wide, rosulate, obovate-lanceolate, laxly celled chlorophyllose leaves, having strong excurrent nerve, and erect, almost sessile regular capsule without peristome, an almost flat apiculate lid, and a shortly campanulate, regular, 8-furrowed calyptra, nearly enclosing the capsule, shortly many-lobed at the base, and with a slender little beak. The genus contains only three species, two of which are Australian.

441. *Goniomitrium africanum* (C. M.), Broth., Pfl., i, 3, 521, and x, 324.

=*Rehmanniella africana* C. M., Bot. Centralb., 1881, p. 347 ; Par., Ind., p. 1067 (1897), and Rehm. 518.
=*Sphaerangium africanum*, Rehm. 171.

A minute, paroecious, dark green moss, almost stemless, but forked at the crown ; leaves rather numerous in a rosette, the outer short, the inner 1 mm. long, suberect or erecto-patent when moist but bent in over the capsule when dry, obovate-acuminate, entire, smooth, keeled, concave ; the nerve very strong and yellowish extending as a strong smooth bristle beyond the leaf-point ; areolation very lax, the cells quadrate, the basal cells rather longer than wide, all chlorophyllose. Capsule central and terminal, almost sessile, thin-walled, laxly celled, globose or oval, wide-mouthed and urn-shaped when open ; spores very large, 75μ wide, smooth ; lid flat, apiculate ; calyptra shortly conical, at first enclosing most of the capsule, 8-furrowed, eventually irregularly 8-lobed at the base. Peristome and ring absent.

O.F.S. Bloemfontein, Rehm. 171, and as *Rehmanniella africana* C. M. ; Rehm. 518 ; Eagles Nest, Bloemfontein, 5000 feet, March 1917, Professor Potts (Sim 10,128, 10,129).

Genus 131. PHYSCOMITRIUM (Brid.) Br. and Sch.

Plants small, gregarious, erect, autoecious, with more or less rosulate, wide, large-celled leaves, sometimes bordered ; erect seta, and erect, globose or pyriform regular capsule, with pointed or beaked lid and small, fugacious, symmetrical, lobed calyptra, neither oblique, inflated, nor cucullate, and not reaching below the middle of the capsule when mature. The cells of the lid are in straight lines from the centre to the circumference, and no peristome is present. Seta about 1 cm. long or shorter. Frequent on clay banks.

Known to me by name only.

P. *Harveyi* Mitt. (collected at Uitenhage by Harvey, *fide* Shaw's Catalogue).

Entosthodon rivularis Geh., in Bull. de l'herb. Boissier, 1896, p. 410, from Great Namaqualand =*P. rivale* (Geh.), Broth., Pfl., x, 324 (apparently near *P. spathulatum*).

442. *Physcomitrium spathulatum* (Horns.), C. M., in Linn., xviii (1884), 695 ; C. M., Syn., i, 118 ; Broth., Pfl., x, 324.

=*Gymnostomum spathulatum* Horns., in Linn., 1841, p. 115.

=*Physcomitrium brachypodium* C. M., in Hedw., 1899, p. 59 ; Broth., Pfl., x, 324.

=*Physcomitrium* *Macleai, Rehm. 519.

=*Physcomitrium* *poculiforme Mitt. MS. (see Dixon, S.A. Journ. Sc., xviii, 334).

An erect, little, autoecious, annual plant, found singly or in gregarious condition on moist alluvial soil or on stream banks. Stems 2–3 mm. high, usually simple, with somewhat distantly rosulate leaves, the lower distant and small, the upper imbricated and 1·5–2 mm. long ; widely spathulate-acuminate, very laxly areolated ; the nerve frequently red, extending almost to the acute apex ; cells irregularly hexagonal, thin-walled, nearly pellucid, thinly chlorophyllose, the lower rectangular, sometimes larger, the marginal line all round of long narrow cells, of which the upper end is usually exserted in the upper part of the leaf, and often for its whole length, this line forming an indistinct border, and the cell-points making the margin more or less denticulate. Young plants sometimes nerveless. Seta terminal, red, 3–5 mm. long, with erect gymnostomous capsule, at first green, ultimately red, almost globose, with distinct thick neck until open, when the capsule becomes urn-shaped, wide-mouthed, and the neck often shrinks somewhat, or it may become oblong (fig. CS). Lid convex with short conical beak ; calyptra white, campanulate-acuminate, at first enclosing the capsule, afterwards only half of it, and with torn base ; capsule-mouth of three to eight lines of transverse oblong cells ; below them are small round cells (fig. SS). The lid varies much ; specimens arrested by drought during growth often have the lid flat or sunk, with or without apiculus (fig. S), and seem to me to be *P. leptolimbatum* C. M. (Hedw., 1899, p. 59), though Brotherus suggests that that may be an *Entosthodon*.

Cape, W. Knysna, 500 feet, Miss Duthie 43 and 11.

Cape, E. Sunday's River, near Graaff Reinet, Rehm. 519 (as *P. Macleai* Rehm.) ; Philippstown Pappe, Ecklon ; and between the Kei and Bashee Rivers, Dregé, 1832 (*fide* C. M.) ; Great Fish River, June 1877, Professor Macowan (as *P. brachypodium*) ; Springs, Uitenhage, Sim 9049.

Natal. Blinkwater and York, Sim 10,134, 10,135 ; Maritzburg, 8539 ; Edgehill, Mooi River, Sim 10,133 ; Hilton College, Sim 10,136 ; Hilton Road, Sim 10,130, 10,136.

Transvaal. Kuilen, near Lydenburg, Feb. 1888, Dr. Wilms (as *P. leptolimbatum* C. M.) ; Melville, Johannesburg, Sim 10,131.

S. Rhodesia. Salisbury, Eyles 1446, 1590, and 1746 ; Victoria Falls, 3000 feet, Sim 8941.

P.E.A. Shirindgen, near the Limpopo, 1920, Rev. H. A. Junod 332.

Var. *brevicollum* C. M., Syn., i, 118. Leaves very wide, much denticulate, nerve often red, capsule short and wide, almost immersed, with short neck and very short seta =*P.* *Macleai R. var. *humilis*, Rehm. 520, from Graaff Reinet =*P.* *sessile, Shaw's Catalogue, from Graaff Reinet, MacLea, and from Colesberg, Dr. Shaw.

Physcomitrium spathulatum.

S, capsule of starved plant ; SS, part of the capsule-mouth ; CS, old capsule as sometimes found.

Genus 132. ENTOSTHODON Schwaegr.

Habit of plant, foliage, and calyptra as in *Funaria*, but capsule pear-shaped or club-shaped, usually erect or inclined, symmetrical or almost so, and having the peristome usually absent or rudimentary, or only the outer present. All South African species are autoecious. Often united with *Funaria*, but I follow Paris in keeping it separate, though Dixon (2nd ed.) points out that any distinction based on the peristome separates species allied by the form of the capsule and *vice versa*, these two characters not regularly going together. For a local flora they are better separate. Rehmann spells the name *Entostodon* throughout.

Shaw's list has *Entosthodon* *Menziesii Mitt. (from " Menzies at the Cape "), concerning which name I know nothing

Synopsis of Entosthodon.

1. Peristome present	3.
1. Peristome absent ; capsule ultimately inclined, symmetrical	2.
2. Leaves ovate-acuminate, shortly pilose	443. *E. Rottleri.*
2. Leaves lanceolate, long-pilose	444. *E. urceolatus.*
3. Capsule erect, symmetrical	4.
3. Capsule somewhat irregular, inclined ; outer peristome fairly strong, inner absent or weak . . .	7.
4. Peristome very weak or brittle, outer only	5

4. Outer peristome fairly strong, inner absent . . . . 6.
5. Neck short ; leaf flat, spreading . . . . 445. *E. micropyxis.*
5. Neck short ; leaf erect, very concave . . . . 446. *E. carifolius.*
5. Neck long . . . . 447. *E. Dixoni.*
6. Leaves ovate-acuminate, not bordered ; nerve often short . . . . 448. *E. Bergianus.*
6. Leaves ovate-obtuse or subacute, bordered ; nerve almost percurrent . . 449. *E. marginatus.*
7. Leaves entire . . . . 450. *E. plagiostomus.*
7. Leaves toothed . . . . 451. *E. spathulatus.*

443. *Entosthodon Rottleri* (Schwaeg.), C. M., i, 121.

= *Gymnostomum Rottleri* Schwaeg., Suppl., I, i, 24, t. 3.
= *Physcomitrium Rottleri* Hpe., in Linn., 1844, p. 696.
= *Rottleria gymnostomoides* Brid., Bryol. Univ., i, 105.
= *Entosthodon longicollis* W. P. Sch., in Breutel M. Cap.
= *Funaria Rottleri* (Schw.), Broth., Pfl., i, 3, 525, and x, 329 ; Dixon, Trans. R. Soc. S. Afr., viii, 3, 199.
= *Entosthodon *chlorophyllosus*, Rehm. 522 and 522b.
= Rehm. 179, distributed as *E. clavatus* Mitt.

A small, erect, rosulate, laxly areolated, autoecious plant with red roots, usually gregarious ; the leaves spathulate, widest near the apex, shortly rounded to a long hair-point ; cells very lax, hexagonal, chlorophyllose, the marginal similar or more quadrate, often hyaline, the upper point of each sometimes projecting, making the leaf denticulate ; basal cells more oblong ; all thin-walled ; nerve red, very strong, excurrent as the hair-point ; seta central, erect, pale red, 3–5 mm. long ; capsule clavate-pyriform with long tapering base, at first erect and included in the then cylindrical white calyptra, soon more or less inclined and slightly arched with a then obliquely placed, inflated, dimidiate, rostrate calyptra with truncate base ; capsule somewhat microstomous, when open more cylindrical and somewhat wide-mouthed ; peristome almost or quite absent ; lid flat. Leaves bending over the coma when dry. Dixon says (Trans. R. Soc. S. Afr., viii, 3, 199), " The leaves vary on the same stem from entire or subentire to sharply denticulate, and the nerve is at times excurrent. *Funaria campylopodioides* (C. M.) is, I suspect, the same thing " ; and later he added in his Herbarium, " Yes, nerve is often much excurrent." I find the nerve not only excurrent but usually extending as a hair-point, although I agree that *E. campylopodioides* C. M. (Broth., Pfl., x, 329) is the hardly pilose condition of the same species. *E. clavatus* Mitt., in Harvey's Thes. Cap., p. 100, may be the same with nerve disappearing below the apex. The Australasian *Funaria apophysata* (Tayl.) Broth. hardly differs except in the capsule being permanently erect.

E. Rottleri occurs in Asia as well as in South Africa.

Cape, W. Bains Kloof, Drakenstein Mountains, Rehm. 522 (as *E. chlorophyllosus* Rehm.) ; Cape Town, Rehm. 179 (distributed as *E. clavatus*, Mitt.) ; Knysna and Camps Bay, Wager 560 and 600.

Cape, E. Mountain Drive, Grahamstown, and Lilyvale Farm, Victoria East, Miss Farquhar 83 and 79.

O.F.S. Wit Als Kloof and Eagles Nest, Bloemfontein, Professor Potts ; Rydal Mount (Wager), Kadziberg (as *E. chlorophyllosus* Rehm.), Rehm. 177.

Transvaal. Spelonken, Rev. H. A. Junod 20 and 15 p.p. ; Mahaliesberg, MacLea (Rehm. 522b).

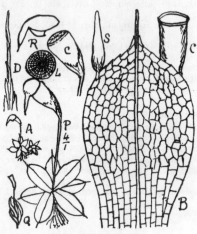

Entosthodon Rottleri.
S, young calyptra.

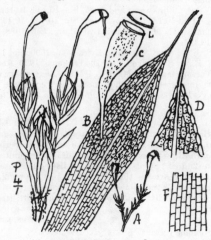

Entosthodon urceolatus.

444. *Entosthodon urceolatus* Mitt., in Harvey's Thes. Cap., pl. 100, fig. C.

Monoecious ; stems ½ inch high, erect, crowded, branched at the base, stout, bare below, then a few small leaves, and then a coma of five to six lanceolate, piliferous, suberect, concave, laxly areolated leaves, 2–3 mm. long, crisped

and somewhat twisted round the seta when dry ; nerve rusty red from leaf-base to point of pilum or sometimes ending below the leaf-apex ; lamina somewhat undulated and with margins usually erect, making the leaf appear narrower than it actually is ; the basal cells rectangular, up to 80μ long ; upper cells irregularly hexagonal, chlorophyllose, with a marginal line of hyaline elliptical cells. Male inflorescence terminal on a similar stem, with shorter leaves and longer hair-points than the others, the hair-point sometimes as long as the lamina and very strong and prominent, bent over the disc. Seta 6–12 mm. long, erect, brown ; capsule almost erect or slightly inclined, regular or almost so, ultimately brown, pyriform with equally long neck ; lid flat, with its cells in straight lines ; mouth hardly contracted, of five to six lines of transverse cells ; peristome absent ; calyptra ultimately inflated and dimidiate, then sitting obliquely. Rhizoids red. The leaf is narrower and the pilum longer than in any other South African species.
E. Rottleri has a much wider and flatter leaf.

Mitten describes it from specimens from near East London, collected by Capt. Rooper.

Cape, E. Sharks River, Port Elizabeth, Jan. 1922, Sim 10,126. Wager distributed the same species under the erroneous name *F. spathulata* Schrb. (which has double peristome) and lists it from Table Mountain and Port Elizabeth.

445. Entosthodon micropyxis C. M., in Hedw., 1899, p. 60 (*Funaria*, Broth., Pfl., x, 328).

=*Funaria* *gymnostoma Dix. (ined.), S.A. Journ. Sc., xviii, 318.
=*Physcomitrium succuleatum* Wager and Wright, in Trans. R. Soc. S. Afr., iv, 1, 3, and pl. 1D ; see *P. succulentum* (W. and W.), Broth., Pfl., x, 324.

A small, erect, annual, laxly areolated, autoecious plant, usually gregarious ; leaves rosulate on a short leafless stem, usually simple, with few large flat leaves (shrivelled when dry), 1·5–2 mm. long, widely spathulate or widely oblong-acuminate, the lower smaller leaves ovate and nerveless ; nerve narrow, yellowish, extending to near or into the apex ; apex shortly acute ; cells irregularly hexagonal, up to 40μ long, thinly chlorophyllose ; the basal cells hardly different ; the marginal cells not different but slightly protruding upward, making the leaf slightly denticulate there, or sometimes not denticulate (as in fig. TT). Seta terminal, 5–7 mm. long, yellowish, erect ; capsule brown, pyriform, tapering to the base but without separate neck ; mouth constricted ; ring absent ; lid almost flat ; calyptra at first erect, ultimately oblique and inflated, cucullate, with truncate base ; peristome very rudimentary, of sixteen undeveloped teeth from inside the mouth. Male inflorescence on a lateral branch.

Exceedingly near the European *E. fascicularis* (Dicks.), C. M., Syn., i, 120, which, however, is nearly always a more robust plant, and the leaves are decidedly narrower and more strongly toothed.

I might have used for the present plant the name *E. limbatus* C. M. (in Bot. Zeit, 1858, p. 155)=*Funaria limbata* Broth., in Pfl., i, 3, 524, and x, 328, as having precedence ; but that species is described as gymnostomous probably in error, for Paris unites with it *E.* *submarginatus W. P. Sch. (in Breutel M. Cap.), which Mr. Dixon lately examined at Kew and found to have the peristome teeth fairly well developed and persistent, as if either a highly developed form of *E. micropyxis*, or a lowly developed form of *E. marginatus*. The capsules seen were urn-shaped, with little neck, almost as in *Physcomitrium spathulatum* (see fig., p. 293).

Cape, E. Knysna, Blanco, Rehm. 175.
Natal. Van Reenen, Wager.
O.F.S. Rydal Mount, Dec. 1910, Wager (as *Physcomitrium succuleatum* W. and W.).
Transvaal. Waterval Onder, Wager.

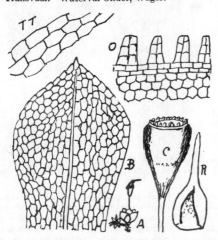

Entosthodon micropyxis.

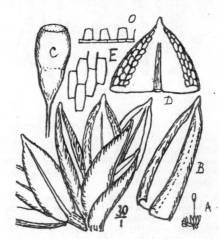

Entosthodon cavifolius.

446. Entosthodon cavifolius, Mitt. ; Harv. Thes. Cap., p. 100.

A minute, gregarious, monoecious plant, vigorous while moist, almost disappearing when dry, almost sessile, but with branched crown. Leaves few, erecto-patent when moist, erect when dry, widely oval, but concave with strongly incurved margins and mucronate apex; margin entire, of rounded cells hardly distinguishable as a border; inner cells more hexagonal; seta 3–4 mm. long; capsule erect, 1 mm. long, subpyriform, including the short tapering neck; peristome teeth present but very rudimentary.

Near Cape Town, Harvey (Hb. Mitten); Kirstenbosch, June 1924, Pillans 4750.

447. *Entosthodon Dixoni* Sim.

=*Funaria longicollis* Dixon, S.A. Journ. Sc., xviii, 318 (which specific name could not be used in *Entosthodon* as it is preoccupied there by the South American *E. longicollis* Mitt.); Broth., Pfl., x, 327, and xi, 530.

Gregarious on soil, erect; stems up to ½ inch high, firm, few-leaved below, and these leaves small, but with a terminal few-leaved coma of leaves 1½–2 mm. long, widely obovate, subacute or acute, very laxly areolated, and inclined to be undulated and shrivelled when dry; leaves not bordered, hyaline, the marginal cells like the others but projecting somewhat, giving a bluntly denticulate margin, with occasional more projecting cells intermixed. Nerve yellowish, extending as a keel to the apex; cells irregularly hexagonal, chlorophyllose; basal cells hardly different, or in more regular lines. Seta terminal, erect, red, about 1 cm. long; capsule erect or almost erect, 2–3 mm. long, brown, pyriform (cylindrical when old), with an equally long neck, at first microstomous with flat lid and white, long-beaked calyptra clasping below the capsule, but later the mouth is wider and the calyptra inflated and split on one or two sides but quickly dehiscent, the cells of its beak in spiral arrangement; capsule mouth of about four lines of transverse cells; peristome of sixteen separate, small, conical, papillose, pale teeth lying horizontally in the mouth of the capsule, exceedingly brittle, and old capsules which have rather wide bell-shaped mouth show no trace of peristome, hence it was described at first as having no peristome. Spores about 20μ diameter. Apparently dioecious; male inflorescence not found on fertile plants and not seen. Fertile disc with about eight archegonia, without paraphyses or small bracts.

S. Rhodesia. Zimbabwe, altitude 3000 feet, Sim 8735, 8796, 8797; Khami Ruins, altitude 5000 feet, Sim 8842.

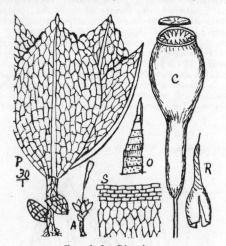

Entosthodon Dixoni.

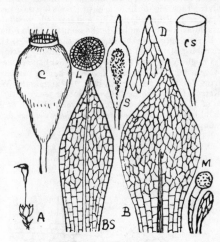

Entosthodon Bergianus.

CS, old capsule; S, young calyptra.

448. *Entosthodon Bergianus*, Bry. Eur., iii, Mon., p. 4 (1841); C. M., Syn., i, 126.

=*Physcomitrium Bergianum* C. M., in Linn., 1844, p. 696.
=*Funaria Bergiana* (Horns.), Broth., x, 328.
=*Weissia Bergiana* Hsch., in Hort. Phys. Berol., p. 59, t. 12, and Linn., 1841, p. 116.
=*Funaria *rhomboidea* Shaw, Cape Monthly Magazine, May 1878, and as shown in Rehm. 523 and 523β as *Entosthodon rhomboideus* Shaw.

Monoecious, erect, gregarious, small; stems with a few rather small, lanceolate-spathulate, red-nerved, and excurrent leaves below the coma, in which the leaves are 1·5 mm. long, widely ovate-acuminate from spathulate base, the nerve extending half-way or less; margin entire; cells smaller than in most other species of *Entosthodon*, and thin-walled but pellucidly and laxly areolated in the upper leaves, more incrassate in the lower. Seta erect, red, 1 cm. long; capsule red, pyriform, microstomous, and peristomate; when empty obconical and eperistomate; lid, nearly flat, with cells in straight lines; calyptra white, at first clasping below the capsule, ultimately inflated, dimidiate, and oblique. Peristome of sixteen slender, narrow, obtuse, entire teeth from inside the mouth. Male inflorescence on small basal branches with ovate-acuminate small leaves, few antheridia, and few clavate paraphyses.

Hornschuch proposed the generic name *Desmostomum* and Fürnr. *Bergia* for this species, but neither was adopted.

E. Schinzii Geh., in Bull. de l'herb. Boissier, 1896, p. 410 (=*Physcomitrium*, Hampe =*Funaria*, Broth., Pfl., x, 328), from wet granite rocks in Great Namaqualand, is closely allied but has the upper part of the leaf toothed.

Funaria Dieterleni Thér.; Broth., Pfl., xi, 530 (from Basutoland), is, so far as I know, unpublished, but comes into the same group as *E. Bergianus.*

Cape, W. Slongoli, Table Mountain, Jan. 1919, Sim 9171; Cape Town, Rehm. 172, and as var. *minor*; Lion's Head (Berg.), Devil's Peak, and on garden walls in Cape Town, Ecklon, Dregé, etc. (*fide* C. M.).

Cape, E. Herschel, Hepburn; Graaff Reinet, MacLea (Rehm. 523 and β, as *E. rhombirdeus*).

Natal. Nottingham Road, Aug. 1917, P. v. d. Bijl; Knoll, Hilton Road, Sim 10,137.

O.F.S. Eagles Nest, Bloemfontein, 5000 feet, Professor Potts.

Transvaal. Johannesburg, Miss Edwards.

449. *Entosthodon marginatus* C. M., in Syn., i, 125, and ii, 548.

=*Funaria marginata* (C. M.), Broth., i, 3, 525, and x, 328.
=*Entosthodon gracilescens* C. M., in Hedw., 1899, p. 59 (*Funaria*, Broth., Pfl., x, 328).
=*Entosthodon ampliretis* C. M., in Hedw., 1899, p. 60 (*Funaria*, Broth., Pfl., x, 329).

Plants erect, usually gregarious, autoecious, few-leaved; leaves mostly rosulate, ovate-acuminate with blunt apex; clasping below, flattened above; suberect, laxly areolated, 1–2 mm. long; shrivelled and undulate-crispate when dry, flat when moist; nerve extending almost to the apex, often pale red; border consisting of one or two (fig. TT) lines of oblique, long, narrow, yellowish cells; inner lamina cells hexagonal, 40μ long; basal cells hardly different but larger. Seta terminal, slender, erect, pale red, 5–10 mm. long; capsule brown, erect, oblong, and rather wide-mouthed when mature, more pyriform and microstomous when young, but with hardly a distinct neck. Outer peristome well developed, the sixteen teeth of four to eight cells each, standing separate and erect; inner peristome absent; ring absent; lid flat; calyptra white, erect, 4-sided, and enclosing the capsule when young; inflated, dimidiate, and somewhat oblique when mature, with beak as long as the capsule; male inflorescence terminal on lateral branch, more flatly rosulate, with numerous antheridia mixed with longer club-shaped paraphyses. Lower part of the stem and the leaves and nerves often red if growing where well exposed. The leaves are obtusely pointed or on the lower leaves more or less rounded at the apex; sometimes a cell stands out as a mucro.

Var. *obtusatus.* Leaves all rounded, hardly acute, with nerve ceasing a little below the apex. Umhwati, New Hanover, Natal, Sim 8585.

Cape, W. Blinkwater Ravine, Table Mountain, 1917, Professor Bews; Zwellendam, Ecklon.

Cape, E. Kentani, rare, 1000 feet, Sept. 1906, Miss Alice Pegler 1360 (Sim 8138).

Natal. Knoll, Zwart Kop, Sim 10,138; Hilton Road, Sim 8244; Little Berg, Cathkin, 6000 feet, Miss Owen 25; above Maritzburg, Rehm. 174 and 521, as *E. ampliretis*; Burdon's Bush, Karkloof, Jan. 1917, J. M. Sim.

Transvaal. Kuilen, Lydenburg, Feb. 1883, Dr. Wilms (as *E. gracilescens*).

S. Rhodesia. Inyanga, 6000 feet, Aug. 1920, J. S. Henkel.

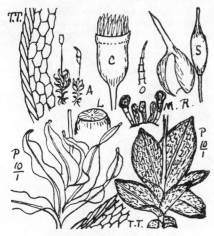

Entosthodon marginatus.
S, young calyptra.

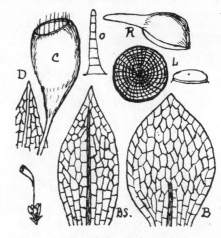

Entosthodon plagiostomus.

450. *Entosthodon plagiostomus* Sim.

=*Funaria plagiostoma* C. M., in Bot. Zeit., 1855, p. 748; Broth., Pfl., i, 3, 528, and x, 330.

Stems strong, usually red, up to ¼ inch long, erect, leafy; leaves more or less crowded at the top, ovate-acute or ovate-acuminate, lower leaves lanceolate, very thin and openly areolated; nerve very variable in the upper leaves from excurrent to reaching hardly half-way, all on the same plant; lower leaves usually nerved to near the apex; margin entire or crenulated by rather narrower cells; cells 80–100μ long, hexagonal; lower cells rectangular; seta pale, nearly erect, ¼ inch long; capsule pyriform, somewhat unsymmetrical, smooth, inclined or nearly erect, with distinct neck, nearly flat lid having somewhat twisted cell-lines, large dimidiate ultimately oblique calyptra, and more or less developed outer peristome, but the teeth very brittle and usually absent. No inner peristome seen, but it is described as being present. Capsule mouth-band strong, red, of about five lines of cells ; no ring seen.

This has the general habit of *Funaria hygrometrica* but with inclined or nearly erect capsules, exceedingly variable leaves, outer peristome of separate short teeth, and inner peristome absent or caducous.

Cape, W. Klein Karroo, Riversdale, on rocks in very dry places, common, Dr. Muir 3717.
Natal. Oliver's Hoek, 6000 feet, Aug. 1922, Miss Edwards.
O.F.S. Leribe Mountains, Basutoland, Mme. Dieterlin 686.

451. *Entosthodon spathulatus* Sim = *Funaria spathulata* W. P. Sch., in Breut. M. Capens.; C. M., in Hedw., 1899, p. 61 ; Broth., Pfl., x, 330, agrees with the last except that the leaves are toothed, and that C. M. describes the inner teeth as membranaceous. I doubt its qualification to stand. Groenkloof, Cape, W., Breutel.

Genus 133. FUNARIA Schreb.

Short-stemmed, erect, gregarious plants, often almost caespitose, with leafy stems often rosulate above, the leaves widely obovate-lanceolate, nerved to near the apex, frequently more or less bordered, and with lax rhomboid cells. Capsule (in our species) pear-shaped with a tapering neck ; more or less curved, with very oblique mouth, wrinkled surface, and arched seta. Lid convex, hardly pointed, the cells arranged in somewhat spiral lines from the centre outward. Calyptra large, ultimately inflated in the lower half with narrow beak ; split at one side and with the rest of the base entire, then oblique and straw-coloured. Peristome double, well developed, the outer teeth twisted obliquely to the left and united at their points, nearly flat over the mouth ; inner peristome of sixteen membranous processes opposite the outer teeth, without basal membrane. Usually autoecious ; male flower discoid on a lateral branch. A large genus, of which our species is almost cosmopolitan. *Entosthodon* is included in *Funaria* by many authors.

Funaria *transvaaliensis* Wager and Broth. is not published and not distributed, but listed by Wager.

452. *Funaria hygrometrica* (Linn.), Sibth., Fl. Oxon., p. 288 ; Brid., Bry Univ., ii, 51 and 738 ; C. M., Syn., i, 107 ; W. P. Sch., Syn., 2nd ed., p. 384 ; Broth., Pfl., i, 3, 528, and x, 331 ; Dixon, 3rd ed., p. 302, Tab. 37β.

= *Mnium hygrometricum* Linn., Sp. Pl., 1st ed., ii, 1110 (1753).
= *F. gracilescens* Schimp. MSS. ; C. M., in Bot. Zeit., 1858, p. 154 ; Broth., Pfl., x, 332.
= *F. calvescens* Schwaegr. ; Broth., Pfl., x, 332 ; Bryhn, p. 14.
= *F. lonchopelma* C. M., in Rehm. 181 ; Broth., Pfl., x, 332.
= *F. hygrometrica* var. *planifolia*, Rehm. 528, and forma *latifolia*, Rehm. 527.

Autoecious ; usually caespitose, often in large patches on the ground or on charcoal formed by forest or wattle fires. Stems, several from the same root, usually simple, ¼–½ inch long ; lower leaves small, upper sometimes scattered, sometimes more rosulate and larger, 2 mm. long, spreading when moist, more or less drawn together under ordinary

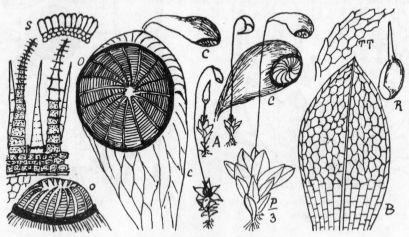

Funaria hygrometrica.

O, part of peristome, with central outer tooth removed to show position of inner tooth.

conditions, and folded over when dry. Leaves oval-oblong, flat or somewhat concave, drawn to a short point ; the nerve strong and extending to but not beyond the apex. Cells quadrate or the upper hexagonal, 60–80μ long, very lax, often sparingly chlorophyllose ; the margin either entire and similar to the lamina or slightly by rather longer and narrower cells having the upper ends exserted, these two conditions being found on the same plants ; basal cells rectangular, up to 100μ long. Upper leaves acuminate, and when dry clasping the seta. Capsule at first erect and enclosed in the then erect white calyptra, but as it progresses the seta becomes cygneous, 1–2 inches long, frequently flexuose or twisted, the capsule then green and more or less pendulous, obliquely pyriform, the mouth small and very oblique, facing downward, and the calyptra more oblique, swollen below, cucullate and obliquely beaked; when fully mature the capsule is furrowed or wrinkled, yellowish red, horizontal or hanging, and with red peristome, and it becomes very wrinkled and wide-mouthed with age. Lid convex; ring wide, breaking off in curled pieces and having one outer line of large, elliptical, hyaline cells; mouth constricted, red; peristome double; outer of sixteen free, lanceolate, red teeth, trabeculate, without mid-line, slightly twisted obliquely to the left, arched over and at first connate in the depressed centre by the appendages of the slender apices · inner peristome teeth opposite the outer, not so long, mem-branaceous, faintly trabeculate; spores small, very numerous.

A very distinct but very variable plant which takes possession of burned soil or charcoal, or sometimes fresh railway cuttings and similar places within a year of their being available, and seta varying from $\frac{1}{2}$–2 inches in length may be found together. The names cited above have no permanent differentiating characters, though sometimes apparently distinct. There are many localities in each region, and the species is almost cosmopolitan.

Family 33. MEESIACEAE.

Swamp plants of upright, somewhat forked, tufted habit, with foliage various, seta long, capsule incurved-pyriform with long tapering neck, oblique mouth, and double peristome, the outer of sixteen short, pyramidal, barred teeth, the inner of sixteen longer, thin, lanceolate-keeled processes alternating with the outer teeth, or both peristomes of equal length, the inner having a basal membrane which is more or less connate with the outer peristome. (This family is probably included in error.)

Synopsis of Meesiaceae.

134. AMBLYODON. Outer peristome short, inner longer; leaf-cells smooth.
135. PALUDELLA. The two peristomes of equal length; leaf-cells papillose.

Genus 134. AMBLYODON P. Beauv.

453. This genus, containing only one species, *A. dealbatus* P. B., which answers the above family and synoptic description, so far as these go, and has stems up to an inch high, with ovate-lanceolate thin leaves, nerve very wide at the base, vanishing below the apex, and large, lax, rectangular, smooth areolation, is mentioned by Harvey (Genera of S. Afr. Plants, 1st ed.), under the name of *Messia dealbata* Sw., as being recorded by Sprengel as South African, and under that name it appears as No. 453 in my check-list. I am not aware of its having been found again since, and am inclined to think that it was recorded in error.

Genus 135. PALUDELLA Ehrh.

Stems erect, slightly branched, several inches long, rhizoid-tomentose below, growing in light green tufts in swamps; leaves in five rows, spreading-squarrose, decurrent, oval-acuminate, recurved, with a faint nerve to near the apex, and the margin serrate toward the apex; cells small, roundly hexagonal, dense and papillose, those at the base narrowly rectangular and hyaline; seta several inches long, straight; capsule oval-oblong, somewhat curved, cernuous or suberect, with short neck; lid flatly conical, mamillate; calyptra cucullate, outer and inner peristomes of equal length, without cilia. Dioecious.

454. This genus has only one species *P. squarrosa* (Linn.) Brid., which species is known from northern Europe and northern North America only, except that C. Müller (Syn. Musc., i, 468) adds, that specimens from South Africa collected by Bergius agree quite with the European plant. The South African locality is not mentioned by Brotherus (Pfl., i, 3, 627), but Paris quotes it from Müller. I am not aware that it has been collected again.

Family 34. AULACOMNIACEAE.

Swamp plants of erect habit, lanceolate leaves scattered round the stem, long, slender, erect seta; capsule erect or almost so, cylindrical, striate, when dry furrowed; peristome double, free, as in *Bryum*, lid conical. Two genera only, and few species.

Genus 136. LEPTOTHECA Schwaegr., Suppl. II, ii, 135; Broth., Pfl., i, 623.

Dioecious, erect, swamp plants, with nearly flat, widely lanceolate, erecto-patent, pointed or aristate leaves laxly arranged all round the stem; margin entire or almost so; nerve protruding, strong; cells lax, roundish-quadrate, tumid, smooth, hardly different below but the lowest rather longer. Capsules not seen on South African specimens, but elsewhere they are found to be suberect or curved, cylindrical, slender, regular, striate, and when dry furrowed or angled, on long slender seta; ring wide; outer peristome with lanceolate yellow teeth, papillose on the back, and with low lamellae; inner peristome teeth hyaline, papillose, with narrow basal membrane, perforated processes, and strong knotted cilia; lid shortly conical.

455. *Leptotheca Gaudichaudii* (Spreng.), Schwaeg., Suppl. II, ii, 135, t. 137 ; Brid., Bry. Univ , i, 838
C. M., Syn., i, 183.

=*Aulacomnium Gaudichaudii* Mitt., in Journ. Linn. Soc., 1859, p. 94.
=*Bryum Gaudichaudii* Spreng., in L. Syst. Veg., iv, 212.
=*Hymenodon ovatus* C. M., Syn., ii, 557.
=*Brachymenium ovatum* Hf. and W. ; C. M., Syn., i, 180.

Leaves lax, erecto-patent and flat when moist, erect or bent inward and somewhat revolute when dry, widely lanceolate, rounded below to a wide clasping base ; shortly pointed, and with the strong nerve protruding along the lower surface and extending to or well beyond the apex ; margin entire or almost so, or denticulate near the apex or in the upper half, flat, except at the base; lamina cells lax, in one layer, the lumen appearing as if separate, tumid, roundish quadrate, smooth. Stem strong, 1–1½ inch long, caespitose, repeatedly innovated from below terminal inflorescence. A few sterile stems were found by Mr. Dixon among *Campylopus* collected by me in Disa Ravine, Table Mountain, Jan. 1919, Sim 9164β and 9162, and I found it again at Devil's Peak, Cape Town, Sim 9246; and Mr. Pillans has recently sent me fine much-denticulate and aristate specimens from Echo Valley, Table Mountain, under the name *Rhizogonium capense* Dixon MSS., Pillans 4900 (1925). It occurs also in Australia, Tasmania, and New Zealand. Capsules have not been seen on South African specimens, and the capsule illustrated herein is from a New Zealand specimen kindly sent me by Mr. Dixon; there it is curved and the plant repeatedly branched. Brotherus describes the capsule as erect, but illustrates it (Pfl., i, 3, 623) as horizontal or subpendulous and straight. When dry it is bluntly 4-angled, and the peristome disappears with age.

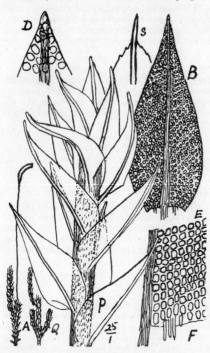

Leptotheca Gaudichaudii.

Family 35. BARTRAMIACEAE.

Plants slender or robust, short or long, simple or branched at the base, or often with whorled innovations at the apex; leafy along the stem and usually growing on wet or moist rocks or soil in dense, light green masses one or more inches deep ; but some species grow on moss-clad decayed tree stumps. Leaves usually linear-lanceolate or ovate-lanceolate, almost always acute, often serrate, with narrow, more or less rectangular, papillose cells, the papillae often at the ends of the cells. Inflorescence terminal ; seta well developed, usually straight, sometimes short and arcuate ; calyptra small, fugacious, cucullate, often striate, and when dry furrowed. Capsule globular and glaucous green when fresh, erect or cernuous, usually striate, and with age becoming brown and furrowed. Peristome usually double, the outer peristome of sixteen equidistant, barred, lanceolate teeth ; the inner shorter, consisting of a membrane having sixteen keeled, split processes, occasionally alternating with imperfect cilia ; peristome in some cases absent or single. Lid convex, apiculate ; calyptra cucullate. Male inflorescence usually discoid in the dioecious species.

Synopsis of Bartramiaceae.

1. Branches irregular, not whorled ; leaves narrow ; male flower bud-like . . . 137. BARTRAMIA.
1. Branches when present whorled below the inflorescence ; leaves usually wide and short ; male flower usually discoid in dioecious species . . . . 2.
2. Leaves not furrowed . . . . 3.
2. Leaves furrowed . . . . 140. BREUTELIA.
3. Peristome absent . . . . 138. BARTRAMIDULA.
3. Peristome present, single or double . . . . 139. PHILONOTIS.

Genus 137. BARTRAMIA Hedw.

Plants usually caespitose, light green, with more or less rigid, dichotomously branched stems, not whorled. Leaves rigid, keeled, mostly narrowly lanceolate from a wider and often sheathing base ; more or less serrate, usually papillose, with small, roundly quadrate or rectangular cells. Capsule globose or almost so, erect, inclined or pendulous, usually smooth when young, deeply striate when dry ; mouth regular or somewhat oblique ; peristome double, single, or absent ; cilia usually absent. Calyptra dimidiate, small. Dioecious, synoecious or autoecious; male flowers usually bud-like. Frequent in moist mountain localities, more rare in drier or coastward localities.

(*Glyphocarpus* (R. Br.) Brid. and *Anacolia* Schimp. have not come into general use with definitions which include South African species, though each (and particularly the former) was at one time more or less in use for such specie of *Bartramia* as have symmetrical, erect, smooth capsules, without peristome, but the definitions have been so varied and unsatisfactory that *Bartramia* here more satisfactorily covers all which may have been included under these.)

Synopsis of Bartramia.

1. Leaves sheathing (*Vaginella*). Dioecious . . . . . . 456. *B. Hampeana.*
1. Leaves not sheathing; peristome single or absent (*Strictidium*) . . . . 2.
2. Outer peristome present, inner absent . . . . . . 457. *B. substricta.*
2. Peristome unknown . . . . 458. *B. squarrifolia.*
2. Without peristome. Synoecious . . . . 3.
3. Capsule subimmersed; subula long . . . . . . 459. *B. Macowaniana.*
3. Leaves very lax, short; papillae mostly in lower angle of cell . . 460. *B. sericea.*
3. Leaves numerous, long; cells almost smooth . . . . . 461. *B. compacta.*

Not known to me: *B. inserta* Sull. and Lesq., in Proc. Amer. Acad., 1859, p. 279; Broth., Pfl., x, 457=*Glyphocarpus*, Jaeg. Ad., i, 524. *Bartramia Breutelii* (W. P. Sch., in Breut. M. Cap.); C. M., in Bot. Zeit., 1858, p. 162, is known to me by name only; the brief description in Shaw's list conveys little except that specimens from Schimper are in the Kew Herbarium.

456. *Bartramia Hampeana* C. M., in Bot. Zeit., 1858, p. 162; Par., Ind., p. 110; Broth., Pfl., i, 637, and x, 453.

= *Bartramia *vaginans*, Rehm. 201 and 534.
= *Glyphocarpus asperrimus* Hampe, in Jaeg., Ad., p. 512.
= *Bartramia asperrima* (Hpe.), C. M., in Hedw., 1899, p. 94; Broth., Pfl., x, 453.
= *Bartramia penicillata* C. M., in Hedw., 1899, p. 94 (sterile); Broth., Pfl., x, 453.
= *Bartramia *somerseti*, Par., Suppl. Ind., p. 36 (1900).
= *Bartramia ramentosa* C. M., in Hedw., 1899, p. 94 (sterile).
= *Bartramia ramentosa* forma *subcompacta* C. M.
= *Bartramia tecta* C. M., Hb.; Hedw., 1899, p. 94.

Loosely tufted or sometimes closely tufted, light green, dioecious; stems equal in one tuft, but varying from ½–2 inches in height; simple or forked, with the seta in the fork or ultimately lateral where there is none. Leaves abundant, 1–3 mm. long, with oblong and often pellucid and white sheath, suddenly contracted into a linear-acuminate denticulate lamina ending in a narrow serrulate subula mostly occupied by nerve. Nerve very wide and strong, scabrid on the back upward; margin expanded with strong hyaline teeth toward the apex, smaller lower as protruding papillae above the sheath; margin entire in the sheath; lamina cells 20μ long, chlorophyllose, almost all shortly papillate on one or both ends, especially those covering the nerve. In the sheath the cells are pellucid and linear, and where the sheath narrows a few small round cells connect those above and below. Perichaetal leaves not different. Seta ½–¾ inch long, red, straight; capsule oval or globose, gymnostomous, microstomous, sulcate when dry; lid shortly conical; calyptra not seen. Terminal leaves often drawn together to a point. Stems often tomentose at the base. Brotherus (Pfl., x, 453) still retains *B. asperrima*, but see Kew Bull., 6 (1923), under No. 198.

Cape, W. Slongoli, Sim 9212; Blockhouse Forest Station, Sim 10,182; Blinkwater Ravine, Dr. Bews (Sim 8598 and 8597) and Sim 9256; Table Mountain, Rehm. 200 and Rehm. 198, as *B. asperrima* Hpe.

Cape, E. Rhenosterberg, MacLea (Rehm. 534, as *B. vaginans* Rehm.); Boschberg, Macowan, 1878, No. 41, as *B. penicillata* C. M., and as *B. ramentosa* C. M., forma *subcompacta*.

Natal. Mont aux Sources, 10,000 feet, July 1921, Sim 10,181; Giant's Castle, 8000 feet, R. E. Symons.

O.F.S. Kadziberg, Rehm. 201β (as *B. vagans* Rehm.); Leribe, Basutoland, Mme. Dieterlin.

Transvaal. Lydenburg, 1887, Dr. Wilms.

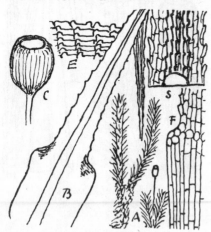

Bartramia Hampeana.
S, back view of part of leaf.

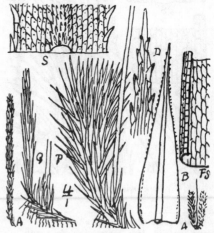

Bartramia substricta.
S and FS, back view.

457. *Bartramia substricta* (Schimp. MSS.), C. M., in Bot. Zeit., 1858, p. 162 ; Dixon, Trans. R. Soc. S. Afr., viii, 3, 205 ;
    Broth., Pfl., x, 457.

    =*Bartramia afrostricta* C. M., in Hedw., 1899, p. 94, and in Rehm. 203, 204, 205 : Broth., Pfl., x, 457.
    =*Bartramia* *marginalis, Rehm. 202 and 533.

Loosely tufted ; stems mostly simple or forked at the basal inflorescence or occasionally above ; erect, ½–2 inches high, brownish yellow, the leaves when dry lightly appressed, when moist erecto-patent, very numerous and closely imbricated forming a densely bristling catkin, the leaves 1·5–3 mm. long, not sheathing, rigid, straight, lanceolate-acuminate, or acuminate direct from above the somewhat narrowed base ; nerve very strong, extending as a short subula, smooth toward the base, usually lightly papillose upward ; margin somewhat revolute toward the base, entire there, with quadrate smooth cells ; from mid-leaf upward strongly dentated by exserted, nearly triangular, hyaline cells pointing upward in a single or double row ; cells at and above mid-leaf quadrate or oblong, 20μ long, chlorophyllose, lightly papillose ; subula cells longer ; basal cells oblong, 50–70×8μ, smooth or almost so, the marginal cells shorter and wider (12×10μ). Fruit not seen, but old setae found are basal, with hardly differentiated leaves ; all specimens known to me are without capsules. Brotherus includes it as dioecious in the group of § *Strictidium* which has capsule erect or slightly bent, regular, furrowed, inner peristome absent ; seta 1–2 cm. high. A very vigorous form of this is common on the higher Drakensberg, usually 1½–2 inches long, very laxly caespitose, closely appressed when dry, the colour being usually bright brownish red though sometimes green ; it is unlike the Table Mountain plant, but not structurally different.

Cape, W.   Camps Bay, Sim 9526 and 9317 ; Cape Town, Rehm. 203, 204, 205 (as *B. afrostricta*, C. M.) ; Matjesfontein, Brunnthaler, 1913, i, 28 (as *B. afrostricta* C. M.).

Natal.   Mont aux Sources, Sim 10,151, 10,158, 10,159, 10,160 ; Giant's Castle, Sim 10,161, and R. E. Symons (Sim 10,161) ; Ladysmith Drakensberg, Rehm. 533 (as *B. marginalis* Rehm.).

O.F.S.   Kadziberg, Rehm. 202 (as *B. marginalis* Rehm.).

458. *Bartramia squarrifolia* Sim (new species).*

=*B. substricta* Schp. var. *squarrifolia* Dixon, in Sched.

Laxly caespitose, glaucous green or very light green, with stems mostly simple, 1–1½ inch high, having numerous long green subulate leaves, not sheathing, suberect or spreading even when dry, and more so when moist. Leaves lanceolate-subulate, chlorophyllose, and lightly papillose, 2–3 mm. long ; the nerve very strong, excurrent as a long serrulate subula, almost smooth ; the margin entire and recurved at the base and to about mid-leaf, above that expanded, and serrulated by long, protruding, sharp, hyaline cells ; lamina cells quadrate and 12μ long in upper part of the leaf, oblong below mid-leaf, and larger, up to 25μ long, toward the base, except on the revolute margin where the cells are nearly square ; all cells very lightly 1-papillose ; capsule erect, small, oval, microstomous, on a red straight seta 5–6 mm. long. Specimens all too damaged to show peristome. Other parts not seen. Evidently semi-aquatic. Differs much from *B. substricta* in general appearance and in its permanently spreading leaves and long subula.

Cape, W.   Disa Gorge and Paarl Mountain, Sim 9166 and 9639 ; Schoone Kloof, east side of Lion's Rump, Pillans 4089 ; Table Mountain, Miss Michell and Professor Bews.

Transvaal.   Witpoortje, Professor Moss.

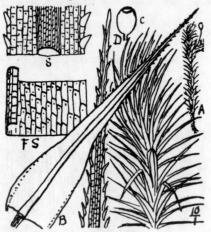

Bartramia squarrifolia.

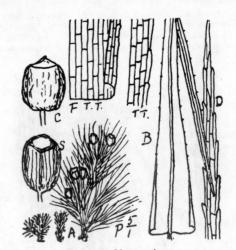

Bartramia Macowaniana.
S, old empty capsule.

---

* *Bartramia squarrifolia* Sim (sp. nov.).
Semi-aquatica, laxe caespitosa ; folia sub-laxa, patula, lanceolato-subulata. Capsula erecta, parva, ovata. Differt a *B. substricta* specie generali nec non foliis constanter patulis subulaque longa.

459. *Bartramia Macowaniana* C. M., in Hedw., 1899, p. 95 ; Broth., Pfl., x, 457.

About 5 mm. high, green ; stems simple, branched or much branched, with erecto-patent leaves 4 mm. long, clasping at the base, then soon reduced to a long, narrow, acuminate, slightly denticulate, nearly pellucid subula, the narrow nerve extending to near the apex of the subula ; margin flat or erect, clasping at the base, entire there, slightly denticulate elsewhere ; basal cells quadrangular, 30μ long, except the outer line which are 10–20μ long ; lamina cells mostly about 20μ long, oblong or quadrate, smooth or with slightly projecting points on the under-surface, but the marginal cells sometimes twice that length. Synoecious ; perichaetal leaves not different. Seta 2–4 mm. long ; capsules *almost immersed*, numerous, thin-walled, oblong or globose, erect, wrinkled when young, sulcate and quadrangular when old and dry ; lid nearly flat ; peristome not seen (believed to be absent), certainly absent in all old capsules.

Cape, E. Boschberg, Somerset East, June 1921, Jas. Sim (Sim 10,149) ; and C. M. described Macowan's No. 24 from there (including capsule), and states, " common, always very fructiferous." Apparently through error, Paris renders that " sterile," which misses the main character.

460. *Bartramia sericea* Horns., Hor. Phys. Berol., p. 63, t. 13 (1820) ; C. M., Syn., i, 502 ; Broth., Pfl., i, 3, 641, and x, 457.

=*Gymnostomum capense* Hook., Musc. Exot., t. 165 (1820).
=*Glyphocarpa capensis* R. Br., in Trans. Linn. Soc., xii, 575 (incompletely described).
=*Glyphocarpus sericeus* (Hsch.), Jaeg., Ad., i, 524 ; Par., Ind., p. 259.

Laxly caespitose, yellowish green ; stems mostly prostrate or suberect, branching irregularly, red, ½–2 inches long ; leaves lax, erecto-patent even when dry, spreading more when moist, 2 mm. long, widely lanceolate-acuminate, very concave at the base, flat above ; nerve strong, extending through and occupying the serrulate subula, slightly papillose on the back ; margin expanded and crenate-denticulate, the lower portion erect with short, quadrate, slightly papillose cells ; upper cells oblong, chlorophyllose, drawn somewhat to each end, and lightly papillose, usually at the lower angle ; basal cells not much larger, also showing slight papillae in the lower angle. Autoecious ; perichaetal leaves rather longer than others, not otherwise different. Seta ½ inch long, red ; capsule erect, red, at first globose, ultimately oval and subquadrate, deeply sulcate when dry, gymnostomous, with small mouth and small conical lid. Antheridia small, paraphyses few, yellow.

Cape, W. Table Mountain and neighbourhood, frequent ; Disa Ravine, Sim 9157, 9165 ; Platteklip Ravine, Sim 9283 ; Newlands Ravine, Sim 9410 ; Slongoli Stair, Sim 9200, 9186 ; Miller's Point, Cape Peninsula, Pillans 4060 ; Stellenbosch, Miss Duthie ; kloof behind the top of French Hoek Pass, May 1917, Miss Stephens.

Cape, E. Boschberg, Somerset East, June 1921, Jas. Sim.

Natal. Top of Zwart Kop, 5000 feet, Sim 10,155.

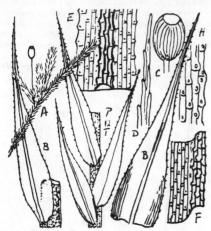

Bartramia sericea.

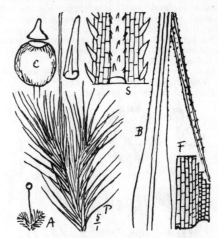

Bartramia compacta.

461. *Bartramia compacta* Horns. (Hor. Phys. Berol., p. 63, t. 53) ; C. M., Syn., i, 501 ; Broth., Pfl., x, 457.

=*B. subasperrima* C. M., in Hedw., 1899, p. 95 ; Broth., Pfl., x, 457.
=*B. Kraussii* Br. and Sch., Fasc., xii, 10, t. 1.
=*B. stricta* Schw., Sp. Musc., p. 104.

Laxly caespitose, synoecious, glaucous green ; stems varying from very short up to 1 inch long, usually forked, erect, red, with seta about ½ inch high in the fork, or ultimately lateral if there be no fork, with very small, erect, gymnostomous microstomous brown capsule about 80μ long, short conical lid, and conical-cylindrical split calyptra. Leaves numerous, rigid, imbricate but not closely imbricate, 3 mm. long or more, erecto-patent or more widely spread both when dry and wet, lanceolate-subulate below with long narrow subula mostly occupied with the strong nerve which is somewhat scabrid on the back and strongly denticulate along the margin, as is also the margin of the leaf except the wider basal portion. Cells oblong, about 20μ long, almost smooth ; marginal sharp hyaline protruding cells longer ; basal cells also oblong, about 30μ long except the marginal there, which are quadrate and 10μ long. Basal lamina not sheathing but somewhat concave, with flattened or slightly reflexed margin. Perichaetal leaves not different. Capsule striate, somewhat furrowed when old ; if immaturely dried more or less quadrate, and it seems to me that *Bartramia quadrata* Hook., Musc. Exot., cxxxii ; Broth., Pfl., x, 457 (*Gymnostomum quadratum* Hook., in synopsis and index thereof), is a vigorous specimen of this species. It is from Postberg, George (rare, Burchell), and in the illustration is shown with terminal seta and erect or inclined quadrate capsule. Müller describes *B. compacta* as having paraphyses and antheridia few and small.

Cape, W. Camps Bay, 1919, Sim 9329, 9331 ; Paarl, Sim 9606 ; Lions Mountain, Bergius, Dr. Krauss, July 1838 ; Cape Town, Rehm. 199 (as *B. subasperrima* C. M., not Rehm. 213 as cited by C. M., in Hedw., 1899, p. 95).

Cape, E. Perie Forest, 4000 feet, 1893, Sim 7148 ; Fort Murray, 1893, Sim 7325.

Genus 138. BARTRAMIDULA Br. Eur.

Plants laxly caespitose, small, slender ; stems rising from a prostrate base, verticillately branched under the female inflorescence ; leaves somewhat spreading, lanceolate, denticulate, light green, subpapillose, with lax-rectangular or hexagonal-rectangular areolation, and nerve not reaching the apex. Inflorescence unisexual or bisexual. Capsule globose-pearshaped, with a short neck, on an arched seta, thin-skinned, not striate, pale, and with lax areolation. Lid small, plano-convex ; peristome absent ; spores large, granular. The species grow on moist soil.

Synopsis.

462. *B. comosa.* Dioecious ; capsule warted.
463. *B. globosa.* Synoecious ; capsule smooth.

462. *Bartramidula comosa* Hampe and C. M., in Bot. Zeit., 1859, p. 221 ; C. M., in Hedw., 1899, p. 90 ;
and see Dixon, S.A. Journ. Sc., xviii, 322 ; Broth., Pfl., x, 460.
=*Glyphocarpa *pilulifera* C. M., in Rehm., 187.

Dioecious : very small and slender ; stems ½ inch long or less, with terminal seta and often with a whorl of about three branches below the inflorescence, glaucous green ; leaves numerous, 0·5–1 mm. long, suberect, lanceolate-acuminate, narrowed at the base, keeled, denticulate above or often to near the base by the marginal cells having the upper end protruding ; all cells rectangular, the upper cells drawn to a papilla at the upper end, those of mid-leaf separate (E), those of the leaf-base connate and empty ; apex acutely pointed, the rather narrow nerve ceasing below the leaf-apex ; perichaetal leaves longer, wider below, subulate and keeled, but denticulate, and the nerve not reaching the apex. Seta terminal, red, 4–6 mm. long, erect, with erect, globose, warted capsule having small gymnostomous mouth, flat lid, and large granular spores. Calyptra not seen. Male inflorescence terminal, disciform, with leaves like the perichaetal leaves, or more spreading ; paraphyses numerous, club-shaped. Leaves spinoso-dentate when dry. A small plant easily recognised when fertile by the warted capsule. Müller's citation of Rehmann's numbers does not agree with Rehmann's labels.

Cape, W. Montagu Pass, Breutel 1856, and Rehm. 188 and 187 (as *Glyphocarpa pilulifera* C. M.) and Rehm. 189 (as *Glyphocarpus comosus* var. *nanus*, C. M.) ; Langebergen Mountains, July 1923, Miss Stephens.

Cape, E. Boschberg, Somerset East, Jan. 1921, Jas. Sim (Sim 10,156).

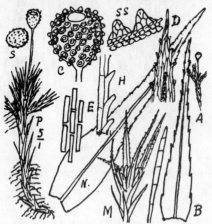

Bartramidula comosa.
S, spore ; SS, warts on surface of capsule.

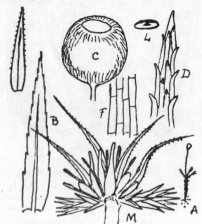

Bartramidula globosa.

463. *Bartramidula globosa* (C. M.), Broth., Pfl., i, 3, 644, and x, 460.

=*Bartramidula* *Breuteliana W. P. Sch., in Breutel. M. Cap.
=*Glyphocarpus Breutelii* Jaeg., Ad., i, 521 (*Glyphocarpa Breutelii* Hampe).
=*Bartramia globosa* C. M., in Hedw., 1899, p. 90.
=*Glyphocarpus* *humilis Rehm., which Paris changed to *G.* *Macleanus on account of the S. American *G. humilis* (Mitt.) Jaeg.

Synoecious; very small and slender, seldom ½ inch long, pale green, with terminal red seta, subtended often by about three innovations. Leaves very numerous, densely imbricated, rigid, 0·5–1 mm. long, usually small and narrow, linear-lanceolate, somewhat keeled; nerve brown, ceasing below the apex; margin denticulate through all upper cells, being papillate at the upper end, and on the margin these project. Cells narrowly oblong, dense; the basal quadrangular, empty, and smooth. Seta 5 mm. long, erect; capsule globose, erect, brown, regular, smooth, very small, with small microstomous mouth and flat lid. Calyptra not seen. Inflorescence synoecious, with longer similar or more spreading leaves, oblong antheridia, and slender club-shaped paraphyses. Seldom fertile.

Var. *tenuicaulis* C. M. Stem subcapillary; leaves narrow, densely imbricate=*B.* *sordida C. M., Hb.=*Glyphocarpa* *Schimperi Hpe., Hb.=*Bartramidula* *capensis Schimp., Hb.

Cape, W. Gnadendal, Breutel, 1856.

Cape, E. Oudeberg, near Graaff Reinet, 3700 feet, May 1872, MacLea (Rehm. 529, as *Glyphocarpus* *pilulifer C. M.); Boschberg, June 1921, Jas. Sim; Kloof Valley, Port Elizabeth, Miss Farquhar 114; Eastern Province, MacLea (as *Glyphocarpus* *humilis, Rehm. 530).

Natal. Nottingham Road, Dr. v. d. Bijl (Sim 10,178); Little Berg, Cathkin, Miss Owen 11.

Transvaal. Lemona Wood, Spelonken, March 1918, Junod 14.

O.F.S. Leribe, Basutoland, Mme. Dieterlin 797.

S. Rhodesia. Matopos, Eyles 1050 and 5030.

Genus 139. PHILONOTIS Brid.

Plants usually erect, often tall, sometimes densely caespitose and rhizoid-tomentose below, with whorled innovations usually subtending the inflorescence. Leaves small, equally distributed, erect or somewhat secund, ovate-lanceolate, short, bluntly toothed, papillose; nerve strong, often excurrent; cells dense and linear-vermicular toward the leaf-apex, gradually shorter and wider downward, at the base loosely rectangular. Capsule globose, cernuous, peristomate, striate, when dry furrowed, with a long seta. Peristome double, or single, or badly developed.

The species inhabit wet localities, on soil or rock; usually dioecious; the male inflorescence usually discoid, and the male plant often more slender, with the leaves more sparsely placed and less acute.

Synopsis.

1. Cells roundly hexagonal with one central papilla each	464. *P. scabrifolia.*
1. Cells smooth or with papillae in the angles	2.
2. Autoecious	3.
2. Dioecious	4.
3. Peristome very rudimentary; leaves nearly smooth . . .	465. *P. laeviuscula.*
3. Peristome double, outer strong, inner weak; leaves papillose .	466. *P. androgyna.*
4. Cells smooth or almost so	5.
4. Cells with papillae in upper angle	7.
5. Leaves imbricate, nerve serrate on the back	467. *P. zuluensis.*
5. Leaves sparse; cells very lax; basal marginal cells quadrate .	468. *P. laxissima.*
5. Some leaves rounded; basal marginal cells oblong . . .	469. *P. Dregeana.*
5. All leaves concave and rounded at apex	470. *P. obtusata.*
5. Male inflorescence has deep red bracts	6.
6. Bract subula slender; peristome double, the outer well made, the inner irregular .	471. *P. imbricatula,*
6. Bract subula leaf-like, spreading horizontally . . .	472. *P. afrocapillaris.*
7. Leaf-margin expanded; peristome double, well developed . .	473. *P. afrofontana.*
7. Leaf-margin revolute	474. *P. africana.*

Known to me by name only: *Philonotis transvaalo-alpina* (C. M.), Broth., Pfl., i, 3, 598; Par., Ind., 2nd ed., p. 386=*Bryum* (*Eu-brya alpina*) *transvaalo-alpinum* C. M., in Hedw., 1899, p. 73; Par., Suppl. Ind., p. 74 (1900), sterile, from between Middelburg and Lydenburg, Transvaal (Dr. Wilms, 1889) is stated by Brotherus (Pfl., i, 3, 598) to be a *Philonotis*, but in Pfl., x, 402, it is omitted from both *Bryum* and *Philonotis*. The description does not indicate a *Philonotis*, but rather *Bryum alpinum* with evanescent nerve.

464. *Philonotis scabrifolia* (Hk. f. and Wils.), Broth., Pfl., i, 3, 649, **and** x, 464.

=*Hypnum scabrifolium* Hk. f. and Wils.; C. M., Syn., ii, 487.
=*Philonotis hymenodon* (C. M.), Jaeg., Ad., i, 540; Dixon, Trans. R. Soc. S. Afr., viii, 3, 205.
=*Bartramia hymenodon* C. M., in Bot. Zeit., 1895, p. 220.
=*Glyphocarpus hymenodon* C. M., in Rehm. 186.

A minute swamp-moss 4–8 mm. long, very slender, with long, ferruginous, branched, tomentose rhizomes, from which stand the erect stems, which are more or less irregularly branched as if pleurocarpous; leaves on main stem

20

up to 1 mm. long, lax ; those on the branches much smaller and closely imbricated, but all widely ovate-acuminate, acute, strongly nerved to the apex, keeled, and closely set with roundly hexagonal chlorophyllose cells 15μ diameter, each having one central mamillate papilla on each surface ; the marginal cells have the papilla protruding, making a crenulate-denticulate margin throughout. Leaves when dry have the point bent inward and margin narrowly incurved. Inflorescence and capsules have not been seen in South Africa, but Brotherus (Pfl., x, 464) uses § *Catenularia* (C. M., in Flora, 1885, p. 411, as section of *Bartramia*) consisting of this one species, in which he includes ten so-called species from South America, Australasia, and South Africa, which he describes as having dioecious inflorescence, discoid male flowers with few spreading bracts. Seta 2 cm. long, strong ; capsule large, inclined, irregular, deeply furrowed ; peristome double (illustrated in Pfl., x, fig. 406, p. 464).

Paris places *Hypnum scabrifolium* under *Philonotis appressa* (Hk. f. and W.), Mitt. (in Journ. Linn. Soc., 1859, p. 81), and maintains *P. hymenodon* (C. M.), Jaeg.

Cape, W. Cataract, Devil's Peak, Rehm. 186 (as *P. hymenodon*) Newlands Ravine, Sim 9414, 9408, 9411.

Cape, E. Gaika's Kop, 6000 feet, Henderson 228 (tall and robust, 5–6 cm. high and of Philonotoid habit, *fide* Dixon).

Natal. Giant's Castle, 8000 feet, 1915, R. E. Symons (Sim 8683 and 10,152).

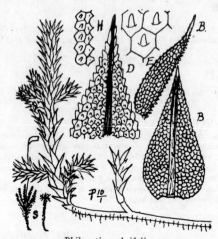

Philonotis scabrifolia.
S, male and female plant from foreign specimens.

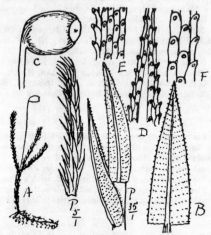

Philonotis laeviuscula.

465. *Philonotis laeviuscula* Dixon, in S.A. Journ. Sc., xviii, 322.

A fairly robust, autoecious, swamp-moss growing in erect, compact, light green tufts 1½–2 inches deep, and with usually a whorl of subfloral branches ; leaves more or less regularly in rows, often but not always falcate-secund ; lanceolate from a wide base, keeled-concave with expanded margin and subulate apex. Margin closely denticulate its whole length ; nerve strong, extending to the apex ; cells all lax and pellucid, 100–120μ long, the basal cells not much longer but much wider, all bearing one short papilla each at the upper end. Perichaetal leaves wider at the base ; male flower bud-like, near the perichaetium, sometimes terminal, or ultimately lateral. Seta 1·5–2 cm. long, pale red, slender ; capsule subglobose, horizontal, thin-skinned, green, usually smooth or almost so, with small mouth and flat lid. Peristome very rudimentary.

Transvaal. Melville, Johannesburg, July 1920, Sim 10,154.

S. Rhodesia. Odzani Valley, Umtali, A. J. Teague 253.

Dixon adds : " A very distinct species, closely allied in the foliation and inflorescence to the widely spread Asiatic species *P. falcata* (Hook.), but differing in the thin-walled capsule, smooth or lightly plicate only when dry at times, and the peristome rudimentary or at least ill-developed."

466. *Philonotis androgyna* (Hpe.), Jaeg., Ad., i, 551 ; Dixon, in Kew. Bull., 6, 1923, No. 532 ; Broth., Pfl., x, 461.

=*Bartramia androgyna* Hampe, Bot. Zeit., 1870, p. 34.

=*Bartramia afro-uncinata* C. M., in Hedw., 1899, p. 91, and in Musc. Macowanianus, No. 10, and in Rehm. 532, and its vars. *gracilescens* C. M. and *breviseta* C. M. (=B. *boschbergiana* C. M., MSS.) ; Broth., Pfl., x, 463.

=*Bartramia (Philonotula) pernana* C. M., in Hedw., 1899, p. 92.

=*Glyphocarpus pernanus* C. M., in Rehm. 191 (not 146 as in Paris' Index).

Autoecious ; hardly caespitose or straggling on moist soil, light green, fertile stem very short, subtended by a small whorl of 4–6 mm. sterile branches, and further short fertile inflorescences starting later below these, almost basal. Branches very short and slender, mostly erect, many-leaved, the leaves suberect or the apical leaves drawn together ; leaves 0·5–1 mm. long, lanceolate, subulate, keeled ; the margin expanded, denticulate ; the upper cells long, crowded,

1-papillose, pellucid ; the lower cells wider, quadrangular, the outer basal cells almost square ; nerve coloured, weak above, and ending in the subula. Perichaetal leaves few, long and thin, with larger lax cells. Vaginula exposed and prominent, red ; seta red, 2 cm. long, erect ; capsule inclined, oblong, slightly bent, many-ridged, especially when dry, thin-skinned, brown, the skin made of roundly hexagonal small cells, with about three rows of smaller transverse cells forming the mouth, inside which is a fairly strong red peristome having subspiral bars, and the inner peristome is thinly membranaceous and of shorter processes. Lid and calyptra not seen. "Male flower very inconspicuous with rather long antheridia" (Dixon, in Journ. Bot., Aug. 1924). Very small for the group, with long seta and thin leaves.

Müller describes *B. afro-uncinata* as dioecious, but Rehm. 532 is autoecious and belongs here. Paris (2nd ed.) in error places the two varieties of *B. afro-uncinata*, viz. var. *breviseta* C. M. and var. *gracilescens* C. M., under *Philonotis afro-fontana*.

*P. *natalensis*, Rehm. 197, from Inchanga, Natal, is unsatisfactorily represented in my herbarium, but seems to belong here.

Cape, W. Claremont, 1924, M. Duncan ; near Belvedere, Knysna, Rehm. 191 (as *G. pernanus* C. M.).

Cape, E. Springs, Uitenhage, Sim 9048 ; Boschberg, June 1921, Jas. Sim ; Boschberg, common, Macowan 10 and Rehm. 532 (as *B. afro-uncinata* C. M.), and Macowan (as var. *gracilescens* C. M.), and with short seta (as *P. boschbergiana* C. M., and as var. *breviseta* C. M.).

Natal. Rosetta Farm, 4000 feet, Sim 10,169, 10,170 ; Mooi River, Sim 10,179 ; Elands Kop, 5000 feet, Sim 10,171 ; Pentrich, Blinkwater, Hilton Road, top of Zwart Kop, and Zwart Kop Valley Stream, Sim 10,173, 10,174, 10,176, 10,177, 10,175 ; Imbezana, Sept. 1918, F. Eyles 1414.

Transvaal. Melville, Johannesburg, July 1920, Sim 10,172.

S. Rhodesia. Victoria Falls, Sim 8944 ; Zimbabwe Ruins, Sim 8765 (young) and 8805. Brotherus cites Usambara also for it.

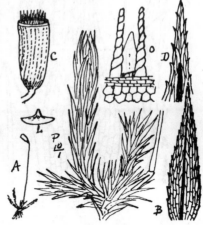

Philonotis androgyna.

O, part of peristome, one outer tooth removed to show inner tooth.

467. *Philonotis zuluensis* Br. and Bry. (Vedensk. Forh., 1911, iv, 16) is described as slender and opaque. Fertile stems up to 10 mm. high, much branched above, closely leaved ; leaves erecto-patent, imbricate when dry, 1 mm. long, 0·3 mm. wide, margin expanded or subrecurved, below simply and above doubly serrated, nerve serrate on the back in three rows, ending below the apex. Cells lax, oblong, smooth, or with low mamilla in upper angle ; sterile stems simple, leaves subsecund. Male flowers bud-like, bracts lanceolate-subulate from a narrow base, 1·5 mm. long, margins acutely serrate, nerve smooth, not percurrent. Per. leaves similar, erect. Seta 2 cm. high, red. Capsule subhorizontal, obliquely oblong-globose, 2 mm. long, 1 mm. wide, plicate when dry. Eshowe, Zululand, H. Bryhn.

468. *Philonotis laxissima* (C. M.), Bry. Jav., i, 154, t. 124 ; Renauld Exsic., p. 175 ;
Ren., Prod. Fl. Br. Madag., p. 173 ; Par., Ind., p. 377.

=*Bartramia laxissima* C. M., Syn., i, 480.
=*Hypnum hastatum* Duby, in Moritzi Verz. d. Zolling., Pfl., v, Java, p. 132.

Dioecious ; tufted, in water or swamp. Stems ¾–1½ inch long, simple or whorled, tomentose below, light green, with the leaves laxly arranged, the whorl branches when present very slender, with tender succulent stems and more lax and smaller leaves, homomallous at the point, the leaves variable, usually appressed when dry and spreading when moist, but sometimes spreading widely even when dry. Leaves widely lanceolate, ½–1 mm. long, pointed or with rounded point on the same plant, with strong nerve to the apex, margin flat or slightly revolute, and with very lax areolation, the basal cells hexagonal-quadrangular, 25–30μ long, except the outer line which toward the base are quadrate and about 15μ diameter ; lamina cells 25–30μ, irregularly hexagonal, the outer making the margin crenulate rather than denticulate ; all cells pellucid and smooth, without papillae, but overlapping upward on the underside which is thus rough. Capsule not seen, but Müller describes "Perichaetal leaves long-acuminate with nerve excurrent, denticulate ; capsule globose, short, horizontal, on an elongated flexuous seta ; lid hemispheric, mamillate ; peristome double, short, the inner rugulose." He describes the margin of the stem-leaves as "duplicato-serratus," which condition I have not seen. Stems often simple, ½ inch long, slender, with disc-like apex in which several young stems start, which when furnished with five to seven leaves break off and start separate life as propagula. Probably only a form of *P. imbricatula* Mitt., which also is too near *P. Dregeana* (C. M.) Jaeg.

Cape, W. Chaplin Point, Sim 9377.

Natal. Albert Falls and Impolweni Bridge, Sim 8704 and 10,148 (with abundant propagula and some leaves acute and others rounded) ; Alexandra Park and Prince Alfred Street, Maritzburg (Sim 10,146 and 10,147).

S. Rhodesia. Floating on surface of Senoia Cave Pool, Lomagundi (Eyles 3159) (with very lax foliage).

Known also from E. Asiatic and E. African islands.

The East African *P. marangensis* Broth. is exceedingly similar.

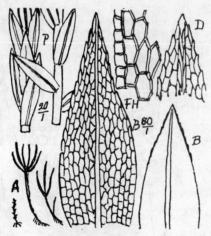

Philonotis laxissima.

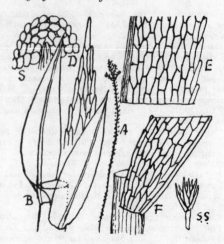

Philonotis Dregeana.

S, rounded apex of some leaves ; SS, propagula.

469. *Philonotis Dregeana* (C. M.), Jaeg., Ad., i, 534 ; Broth., Pfl., x, 463.

=*Bartramia Dregeana*, in Bot. Zeit., 1856, p. 419 ; C. M., Musc. Schw., p. 395 ; C. M., in Hedw., 1899, p. 154.

Stems 1–2½ inches long, slender, succulent, few-leaved ; a moss of waterfall drip or similar surroundings, usually simple, occasionally with terminal whorl of short branches, or with numerous gemmae ; leaves scattered, sometimes slightly secund, 1–1·5 mm. long, concave, pellucid, lanceolate, and usually with long acute apex, but in the same tuft and even on the same stem there are usually obtuse or rounded leaves having shorter nerve, very lax round cells, marginal exserted round cells, and no apical mucro. Nerve pale, extending almost to the apex ; margin expanded, slightly out-folded near the clasping-base, almost entire, but marginal cells slightly exserted all along except at the base where there are a few oblong cells ; cells long-hexagonal, 80μ long, smooth above, but often with the point projecting on the lower side. Short detachable branchlets with six to seven small leaves (propagula) sometimes produced in the crown. Other features not seen. Near *P. laxissima* but larger and more simple stemmed, with larger leaves, rougher on the lower surface and with some rounded leaves mixed. Exceedingly near *P. laxissima* and *P. imbricatula*, which see.

Cape, W. Newlands Ravine, 1000 feet, Sim 9447 ; Devil's Peak, Sim 9235.

Natal. Falls, Gwyn Stream, Hilton College, 3500 feet, May 1921, Sim 10,150 ; Allerthorpe, Dargle Road, Sim 10,180.

470. *Philonotis obtusata* C. M. *apud* Wright, in Journ. Bot., xxvi, 265 (name only) ; Ren. and Card., Musc. Exot. in Bull. Soc. Bot. Belg., xxxiv, 2nd part, pp. 61, 189.

=*P.* *cavifolia* Sim, MSS.

Densely caespitose, growing in drip or moisture ; stems ½ inch high, laxly clad all along and all round with pellucid, very concave leaves 1–1·5 mm. long ; leaves almost boat-shaped, elliptical, not decurrent, strongly nerved to near the widely rounded apex ; cells hexagonal, 70×35μ, smooth, thick-walled ; marginal cells roundly semi-exserted, pellucid, 30×15μ ; basal cells hexagonal, 80×40, margin erect ; stem very strong ; other parts absent.

Almost like a *Bryum* in its areolation, and similar in leaf-form to the occasional leaves found in *P. Dregeana* and *P. laxissima.*

Natal. War Dept. Stream, Maritzburg, 2000 feet, Aug. 1917, Sim 10,215.

Recorded from several Madagascar localities and also from the Azores.

471. *Philonotis imbricatula* Mitt., Musc. Ind. Or., p. 59 ; Broth., Pfl., x, 462.

A widely distributed and variable dioecious species, usually small, but easily recognised by its whorled innovations subtending bright red male inflorescence, in which the six to ten large red bracts are ovate-concave, with large, lax, hexagonal cells 40×20μ, and suddenly narrowed to a membranaceous, acuminate, slightly denticulate subula having oblong smooth cells 20×10μ, the whole capitulum forming a prominent feature, even without the use of a lens. The bracts are almost 1 mm. long, and have a wide but rather weak nerve, which extends even more weakly the whole length of the subula. Stems thick, often red, laxly leaved ; stem-leaves lanceolate or wider or narrower, acute, erect or loosely appressed, often less than half the length of the bracts ; sometimes up to 1 mm. long in strong forms, somewhat keeled, pellucid, laxly reticulated, the cells roundly oblong, about 20μ long, smooth or only slightly papillose ; the marginal cells narrower, with out-bent points, making the margin slightly denticulate, and the nerve extending to the apex, and in strong forms the lower part of the margin is expanded or slightly revolute, then showing papillae on all cells. Fruit rare ; perichaetal leaves rather long and narrow ; capsule subglobose, brown, on ½-inch red seta, inclined, microstomous, eventually wrinkled. Perianth double, the outer well made, the inner irregular. Antheridia numerous, club-shaped, 500μ long, mixed with slender paraphyses about 6-celled. The smaller forms are about ½ inch high and hardly caespitose, the larger forms are up to 1½ inches high, much stronger in all parts, and densely caespitose as a dark green swamp-moss. Many forms or conditions, mostly male or sterile, have elsewhere been named as species, and the Transvaal *Bartramia delagoae* C. M. (in Hedw., 1899, p. 92, sterile) probably belongs here,

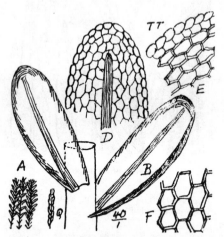

Philonotis cavifolia.

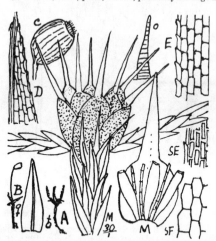

Philonotis imbricatula.

though it has little to identify it by, and no type specimen is available in South Africa ; it is described from Omtombi, between Delagoa Bay and Lydenburg (Dr. Wilms, 1884), but Dixon records it (in Trans. R. Soc. S. Afr., viii, 3, 205) on a fertile specimen from George, C.P. (Wager), as having recurved margin and double peristome, " the inner reddish below, processes imperfect, shorter than the outer teeth, smooth or nearly so." He also connects *P. simplex* (C. M.) Jaeg. here ; see under *P. africana. P. mauritiana* Aongst. ; *P. zuluensis* Broth. and Bryhn ; *P. delagoae* (C. M.) Par., and *P. imbricatula*, Mitt., are all placed together by Brotherus (Pfl., x, 462) in his group § *Philonotula* β in which the dioecious inflorescence, the male usually orange-red, is the connecting character.

P. mauritiana Aongst., in Oefv. af. K. Vet. Akad. Förh., 1873, p. 140, from tropical Africa and its eastern islands, is recorded by Dixon (Trans. R. Soc. S. Afr., viii, 3, 205) from Hogsback, Eastern Cape Province, 4000–6000 feet altitude, Rev. Henderson 193 and D. Henderson 373. He had earlier identified it with *P. luteoviridis* Besch., but found that Cardot identifies that with *P. mauritiana*, but later writes his suspicion that that is *P. Dregeana*, and that *P. imbricatula* is probably identical (including *P. laxissima*) " owing, I have little doubt, to the vagaries of one widespread and very variable species, usually sterile." If so, *P. Dregeana* is the earliest name.

Bartramia tenella C. M., of Shaw's list, from near Graaff Reinet, is probably *B. tenella* Wils., which belongs here.

Cape, W. Intake, Woodhead Tunnel, Table Mountain, Sim 9126.

Cape, E. Dohne Hill (Kakazella), 4500 feet, 1898, Sim 7188.

Natal. Mooi River, Sim 10,140.

S. Rhodesia. Zimbabwe, Sim 8810, 8830, 8788 ; Victoria Falls, 8527, 8539, 8496 ; Khami, Sim 8466, 8788 Matopos, Sim 8447 and Mme Borle 39 and 1389.

472. *Philonotis* (*Euphilonotis*) *afro-capillaris* (Dixon, in letter), Sim, new species.*

Laxly caespitose, straggling or suberect, weak, tender, dark green, with dark red male inflorescence ; stems usually simple, sometimes 3–4-branched in a whorl below the inflorescence ; leaves scattered, 2–3 mm. long, lanceolate, with long, narrow, denticulate subula ; nerve narrow, percurrent in the subula, smooth ; margin expanded, crenulate below, more denticulate above ; cells hexagonal, 50–70μ long, the lower wide and empty, the upper narrower and chlorophyllose, sometimes with low papillae at the upper end ; dioecious ; male capitulum dark red, of four to five concave bracts, extending outward in horizontal lamina similar to the leaves or more membranaceous. Female plant not seen. The Wilderness, George, Cape, East, Apr. 1924, Miss A. Taylor (Sim 10,153). Mr. Dixon writes, " I know nothing like it, except some forms of our European *P. capillaris*."

---

* *Philonotis* (*Euphilonotis*) *afro-capillaris* (Dixon in ep.) Sim (sp. nov.).

Dioeca, sparsa vel sub-erecta, debilis, inflor. atro-rubra mascula. Folia lanceolata, subula longa angusta denticulata ; allae non numquam papillis humilibus ad extremum superius. Bracteae masculae 4–5, concavae, supra patentes ut lamina horizontalis frondiformis. Feminea non visa.

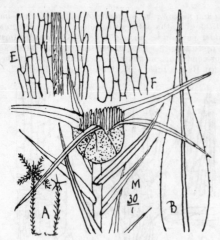

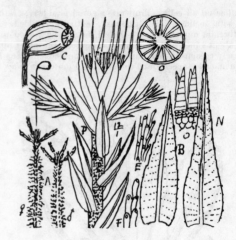

Philonotis afro-capillaris.

Philonotis afro-fontana.

473. *Philonotis afro-fontana* (C. M.), Par., Suppl. Ind., p. 264 (1900) ; Broth., Pfl., x, 465 ;
    Par., Ind., 2nd ed., p. 368, where two varieties are included in error.

    =*Bartramia afro-fontana* C. M., in Hedw., 1899, p. 94.
    =*Philonotis *oraniae*, Rehm. 192=*P. *oranica*, Rehm. 535 and 536, forma *atrovirens*.
    =*Philonotis *Woodii* Rehm. 537.

    Plants dioecious ; densely matted in large, loose, light green tufts growing in or near water, mostly in mountain streams ; stems red, strong, ½–3 inches long, abundantly tomentose below among old leaves ; sterile stems simple ; inflorescent stems whorled below the inflorescence. Leaves lanceolate, nearly pellucid, erecto-patent ; closer when dry ; strongly keeled but openly concave, not sulcate ; the keel extending into the short subula and slightly papillose upward ; the margin everywhere erect and denticulate ; the basal cells roundly oblong, $20 \times 10\mu$; the upper cells $25 \times 5\mu$, each papillose in its upper angle on both surfaces ; perichaetal leaves hardly wider but longer and more acuminate-subulate, denticulate and keeled (fig. N). Archegonia numerous, red, as long as the leaf-lamina, with clavate paraphyses ; archegonial leaves with ovate-acuminate base and subulate-denticulate nerved apex, sometimes ferruginous or red. Fruit rare ; capsule inclined, globose-oval, thick-walled, striate when young, sulcate when old ; the peristome at first folded over the mouth, ultimately erect ; inner and outer teeth well developed ; lid shortly conical ; calyptra dimidiate, evanescent. Male plants rare, discoid, with wide, shortly ovate, nerved, green bracts ; antheridia numerous in a disc, turgidly elliptical, with longer, slender paraphyses.

    A very common and exceedingly variable moss, always in beautiful cushions full of water, and with pellucid leaves with erect margins. The leaves are sometimes in five to eight regular rows, sometimes more numerous and not in lines, and often subfalcate or even falcate, especially on the strongest specimens, but all these conditions occur together, or even on one stem. Harvey (Genera of S. Afr. Plants, 1st ed.) and other early writers used the names *Bartramia fontana* and *Philonotis fontana* Brid., from which widely distributed species its erect leaf-margin and the form of the male bracts separate it.

    Philonotis *molmonica, Rehm. 196 (Molmonspruit, O.F.S.), contains this and other things mixed, and is too scrappy to determine.

    Cape, W. Blinkwater Ravine, Dr. Bews, 1917 (Sim 8615) ; Cape Peninsula, above St. James', Pillans 3590.

    Cape, E. Cataract at top of Boschberg, Macowan ; East Pondoland, 1000 feet, July 1899, Sim 10,142, and in 1915, J. Burtt-Davy 15,298 ; Hell's Gate, Uitenhage, Jan. 1922, M. Duncan ; Waterfurrows, Kokstad, 1919, J. Burtt-Davy.

    Natal. Giant's Castle, 5000 feet, R. E. Symons (Sim 8667 and 10,141) ; Rosetta Farm, 1918, Sim 10,165 ; Greenfields, Mooi River, Sim 10,166 ; Sweetwaters, Sim 10,168 ; Mont aux Sources, 7000 feet, Sim 10,167 ; Inanda, Rehm. 531, as *P. Woodii* Rehm.

    O.F.S. Rydal Mount, Wager 424 ; Caledon, above Kadziberg, Rehm. 192, as *P. oraniae* Rehm.

    Transvaal. Sand River, Aug. 1884, Dr. Wilms ; Lechlaba, above Houtbosch, Rehm. 536 (as *P. *oranica*, forma *atrovirens*) ; Magaliesberg, MacLea (Rehm. 535, as *P. oranica*) ; Mbabane, Swaziland, April 1917, Miss Edwards.

    Not recorded from S. Rhodesia.

474. *Philonotis africana* (C. M.), Par., Ind., and C. M. in Rehm. 193 ; Broth., Pfl., x, 461.

    =*Bartramia africana* C. M., in Hedw. 1899, p. 94.

    Stems ½–¾ inch long, simple or whorled, with thick and often red stem, imbricate, ovate-acuminate, somewhat falcate, nearly pellucid, yellow leaves 1–1·5 mm. long, in which the strong nerve protrudes as a keel on the lower surface and extends to beyond the apex as a denticulate subula; margin strongly revolute from the base to near the apex, showing the upper surface as very rough with papillae and the lower surface also papillate; in each case the papilla is at the upper end of the cell. Upper cells $50 \times 7\mu$, basal cells more lax but shorter, $20 \times 12\mu$, all more or less papillose. This species, which is only known from Rehm. 193 (sterile) and Müller's description from that, has been unfortunate in that Müller's description is very incomplete and not good, while Brotherus (Pfl., i, 3, 645, and x, 461) puts it in § *Leiocarpus*, into which, on the known vegetative characters, I cannot see that it fits. If, however, it is correctly placed there, then it should be dioecious with bud-like male flowers, erect, regular, globose or shortly ovoid capsule without swollen throat, smooth or lightly striate when dry, and with the inner peristome absent. All these characters are unknown. It is, however, distinct from other species.

Natal. Inanda, Rehm. 193; Giant's Castle, 7000 feet, R. E. Symons.

C. Müller, in Musci Schweinfurthiani (Linnaea, 1875, p. 396), mentions a *Bartramia pellucida* Hpe. from Natal as being allied to *B. baginsensis* C. M.; I have no further information concerning the former, but the latter is included by Brotherus (Pfl., x, 461) in the same group as *P. africana* C. M., as also is *P. simplex* (C. M.), Jaeg., Ad., ii, 700 (= *Bartramia simplex* C. M., in Linn., 1875, p. 398), which Paris records from N'Gama only. Dixon writes, "*P. simplex* C. M. has nothing in description or in Schweinfurth's specimens to distinguish it from *P. Dregeana*."

Philonotis africana.
S, upper part of leaf, back view; SS, part of same showing revolute margin.

Genus 140. BREUTELIA W. P. Sch.

Plants robust, more or less suberect or procumbent, rhizoid-tomentose at the base; irregularly branched, the branches scattered; often also with whorled innovations from below the inflorescence. Leaves long, more or less widely lanceolate from a sheathing base, deeply furrowed, rough, shining, the nerve weak and not excurrent, and the cells narrow. Inflorescence dioecious, the male discoid. Capsule hanging from a short arched seta, or inclined and subpendulous from a long seta, globose or ovate-oblong, striated, when dry furrowed; the lid small, convex, mucronate. Peristome double, very seldom absent; the inner shorter, usually faintly papillose. Cilia rudimentary or absent.

Synopsis of Breutelia.

1. Leaves plicate at the base	2.
1. Leaves plicate for the whole length	3.
2. Opaquely green; basal marginal quadrate cells many	475. *B. afroscoparia.*
2. Shining yellowish; quadrate alar cells few	476. *B. aristaria.*
3. Leaves spreading, narrowly lanceolate, aristate; alar cells rectangular . .	477. *B. angustifolia.*
3. Leaves suberect, acuminate from wide base; alar cells quadrate in five to eight rows	478. *B. gnaphalea.*
3. Leaves squarrose, long, base widened above, thence acuminate; alar cells few, roundly elliptical,	479. *B. subgnaphalea.*
3. Very large; leaves ovate-lanceolate below, lanceolate above; alar cells few . .	672. *B. tabularis*

475. *Breutelia afroscoparia* C. M., in Hedw., 1899, p. 91; Dixon, in Trans. R. Soc. S. Afr., viii, 3, 206 p.p.; Brotherus, Pfl., x, 470.

= *Bartramia *laetevirens*, Rehm. 206.
= *Breutelia *Breutelii* (Sch. MSS.), Broth., Pfl., x, 470.

Tufts fairly strong, dense, opaque; stems 1–1½ inch long, green or yellowish green, thick, terete through appressed leaves, simple or irregularly branched, red, tomentose below; leaves erect even when moist, plicate at the base, ovate-acuminate, 2 mm. long, including a long, slender, denticulate-pointed, hyaline arista; nerve narrow, coloured, excurrent in the arista; margin revolute from base to top of lamina, lightly denticulate; upper cells narrowly elliptical, lower linear with low papillae at lower end; inner basal cells rectangular-linear, marginal ten to twelve rows quadrate and short, extending one-fifth up the margin; perichaetal leaves not different; seta 1 cm. long, red; capsule erect or almost so, globose or shortly oval, sulcate when old.

Müller's description is misleading and not supported by the specimens cited.

Cape, W. Mountains above Worcester (sterile), Rehm. 266 (as *Bartramia laetevirens* Rehm.); Newlands Ravine; Devil's Peak; Window Gorge, and lower plateau, Table Mountain, Sim 9422, 9237, 9497, 9292; Lions Head, Pillans 4086.

Cape, E. Boschberg, south-east side, Macowan 1877 and 1883.

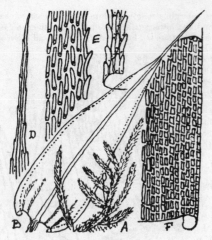

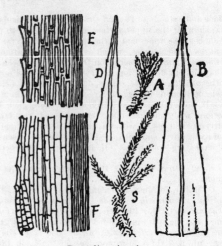

Breutelia afro-scoparia.   *Breutelia aristaria.*

S, var. *plumosa.*

476. *Breutelia aristaria* (C. M.), Broth., Pfl., i, 3, 657, and x, 471.

=*Bartramia aristaria* C. M., in Hedw., 1899, p. 93.
=*Glyphocarpus aristarius* C. M., in Rehm. 184, 184*b*, 184*c*.
=*Philonotis subcordata*, Rehm. 194.

Weak for this genus but stronger than *Bartramia* ; stems ½–2 inches long, tomentose below, irregularly or fasciculately branched, shortly cuspidate, red ; the leaves erecto-patent, lax, shining yellow, lanceolate-acuminate, acute, 1·5–2 mm. long, plicate at the base ; nerve varying from narrow to strong, pale red ; margin expanded, denticulate, the basal margin slightly revolute ; upper cells long and narrow, papillose in the lower angle, basal cells shortly rectangular, with a small alar group of small quadrate cells. Fruit not seen. C. M. guardedly describes the fruit of *Bartramia Spielhausii* C. M. (see var. *plumosa* hereunder) as erect, shortly oblong, on a very short, straight, red seta.
Cape, W. Montagu Pass, Rehm. 184 ; Table Mountain, Rehm. 184*b* ; Worcester, Rehm. 184*c*.
Natal. Inanda, Rehm. 194, as *Philonotis subcordata* Rehm. (*fide* C. M., in Hedw., 1899, p. 93).

Var. *plumosa.* Leaves long-aristate, patent ; nerve usually strong.
=*Glyphocarpus aristarius* var. *plumosus*,* Rehm. 185.
=*Bartramia Spielhausii* C. M., in Hedw., 1899, p. 91 (as *Breutelia*, in Par., Ind., and in Broth., Pfl., x, 471).
=*Bartramia *pallidifolia* C. M., in Hb. Macowan (Hedw., 1899, p. 91).
=*Breutelia *validinervis* Dixon, MSS., in Pillans 4286.
Cape, W. Table Mountain, Spielhaus, 1875 (as *B. Spielhausii* C. M.) ; Disa Stream, Table Mountain, Pillans 4286.
Cape, E. Boschberg, Macowan 1883.
Natal. Inanda, Rehm. 185.

477. *Breutelia angustifolia* (Rehm. 538) Sim.*

Plants ½–1½ inch high, tomentose below, simple or fasciculate from the base ; leaves shining, yellowish, erecto-patent, or more spreading when moist, lax on red stem, narrowly lanceolate, 3 mm. long, including slender acute arista ; sulcate far up, twisted and sinuate when dry, with expanded, almost entire margin, quadrangular cells, the basal marginal cells rather shorter than others but three to four times as long as wide, all almost smooth ; nerve narrow, coloured, extending through the arista ; whole leaf-base very fibrose, and often with a line of round red cells between the lamina and the fibres. Other characters not seen. Only known sterile from Rehm. 538, from Macmac ?
Transvaal, MacLea.

478. *Breutelia gnaphalea* (Palis ; Schimp. hb.), Broth., Pfl., x, 471.

=*Bartramia gnaphalea* C. M., Syn., i, 489.
=*Bartramia arcuata* Brid., Bry. Univ., ii, 35 p.p. (not W. P. Schimp.).
=*Bartramia tomentosa* Schw., Sp. M., p. 97.
=*Hypnum ? gnaphaleum* P. B., Prod., p. 64.

Tufts loose, yellowish ; stems 2–4 inches long, mostly erect, tomentose below, irregularly branched, 4–5 mm. diameter through the leaves being suberect, and often subfalcate ; leaves 3–4 mm. long, acuminate from a narrowed

---

* *Breutelia angustifolia* (Rehm. 538) Sim (sp. nov.).
Plantae 1–4 cmm. altae, plerumque simplices, infra tomentosae. Folia nitida, patula, laxa, anguste lanceolata, aristata, plicata, 3 mm. longa ; margo explanatus, paene integer ; cellae 4-gonae, paene leves.

base, often crisped, deeply plicate and strongly revolute the whole length, margin lightly denticulate below, dentate above ; upper cells linear, lower rectangular, mostly with low papillae in the lower angles ; basal marginal cells short

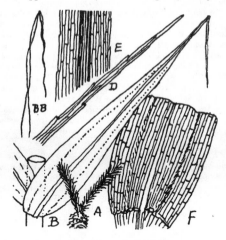

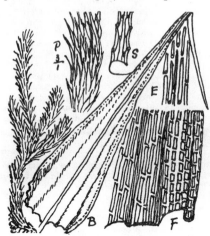

Breutelia angustifolia.   *Breutelia gnaphalea.*

and quadrate in five to eight rows ; nerve narrow, coloured, exserted in the subulate-denticulate apex.   Other parts not seen.

Natal.   Giant's Castle, 7000 feet, 1915, R. E. Symons (Sim 8668).

O.F.S.   Rensburg Kop, 1911, Wager (as *B. aristaria* C. M.).

Also tropical Africa and its islands.

Shaw includes *B. gnaphalea* from Stellenbosch and *B. arcuata* from Koudeveld, 5500 feet, near Graaff Reinet (1873, MacLea), but I have not seen his specimens.

479. *Breutelia subgnaphalea* (C. M.), Par., Ind., p. 154 (1894), and 2nd ed., p. 173 ; Broth., Pfl., x, 472.

=*Bartramia subgnaphalea* C. M., in Flora, 1890, p. 480.
=*Breutelia* *Macleana, Rehm. 539.

Loosely tufted, strong ; stems 2–3 inches long, tomentose in places, simple, irregularly branched or sub-fasciculately branched, red ; leaves lax, pellucid, 3–4 mm. long, yellowish, shining, squarrose, subcuneate and lightly revolute at the base, thence suddenly acuminate to the acute dentate apex, with about three deep plicae on each side for the whole length. Leaves not crisped even when dry ; margin usually expanded to near the base, dentate by exserted cells from the base to the apex, slightly below but very strongly upward.   Nerve narrow, coloured : lamina cells shortly linear, shortly mamillate at the base of each cell or almost smooth ; basal cells shorter, with one or two lines of wider, hyaline, alar cells at the margin.   Perichaetal leaves not different ; seta red, straight, 1–2 cm. long ; capsule oval (all capsules seen have been damaged).   A beautiful moss, very squarrose in its foliage, often densely tomentose, sometimes not so.

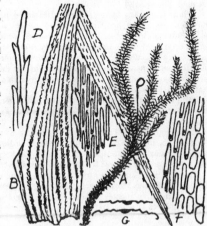

Dixon states (Trans. R Soc. S. Afr., viii, 3, 206), "*B. subgnaphalea* shows some variation in the alar cells, of which there are frequently two to three rows laxer and shorter than the median, but frequently these are wanting and there is only a single row of narrow cells almost similar to the rest except for being quite pellucid."   For the much more robust Table Mountain species, see *B. tabularis* (p. 449).

Cape, E.   Umtentu River, East Pondoland, July 1915, J. Burtt-Davy 15,353.

Natal.   Town Bush, Maritzburg, Sim 8702 (a very large form) ; Bushman's Pass, Giant's Castle, R. E. Symons (Sim 10,162) ; Koodoo Bush and Mont aux Sources, 7000 feet, July 1921, Sim 10,163 and 10,164, and H. Botha-Ried (Sim 8999) ; Little Berg, Cathkin, 6000 feet, July 1922, Miss Owen 34 and 39.

Transvaal.   Macmac, MacLea (Rehm. 939, as *B. Macleana* Rehm.).

Breutelia subgnaphalea.

Family 36. BRYACEAE.

Plants having a common aspect, but showing great variation in size, habit, and appearance ; usually caespitose, often densely so ; the stems usually erect, often branched at the base, and sometimes by innovation below the inflorescence ; the stems or branches usually club-shaped through the imbricate leaves increasing in size upward, but sometimes equally leaved throughout. Leaves lanceolate, ovate-lanceolate, or sometimes obovate-acuminate, usually from a rather wide base and sometimes twisted. Nerve present and usually excurrent as a hyaline point ; cells usually rhomboid or longer, large and lax, smooth, and with thin walls. Capsule cylindrical, clavate or pyriform, often with a tapering neck, suberect, inclined, cernuous or pendulous, on a long seta ; usually symmetrical, and without striae. Lid usually convex-apiculate, sometimes shortly conical or longer ; calyptra cucullate, smooth, cylindrical, caducous. Peristome usually double, sometimes outer only, sometimes inner only ; outer peristome of sixteen lanceolate teeth, horizontally striate within ; inner thin, usually of sixteen more or less perforated teeth, with or without cilia, and often with basal membrane. A natural group, often difficult to arrange on account of the peristome breaking off or partly doing so ; or through juvenile, senile, or imperfect condition being different from perfect. Brotherus separates into three groups on a habit character which is often unreliable in Bryoideae where basal inflorescence is frequent.

Synopsis of Bryaceae.

(Peristome very imperfect, consisting usually of basal membrane only . . . LEPTOSTOMUM.)
Peristome usually single ; inflorescence on short lateral branches near the root . . . 2.
　　Outer peristome usually absent . . . . . . 141. MIELICHHOFERIA.
　　Inner peristome absent . . . . . . 142. HAPLODONTIUM.
Peristome double ; male and female usually near together at the top of the main stem, or basal in some . 3.
　　No basal membrane present . . . . . . 143. ORTHODONTIUM.
　　Basal membrane present . . . . 4.
　　　　Outer peristome longer than the inner . . . . 144. BRACHYMENIUM.
　　　　Outer and inner peristomes about equal . . . . . 5.
　　　　　Upper leaf-cells narrow . . . . 6.
　　　　　Leaves narrow from a rather wider base, spreading . . . . 7.
　　　　　　Segments not ciliated . . . . . . 145. WEBERA.
　　　　　　Segments ciliated . . . . . . 146. LEPTOBRYUM.
　　　　　Leaves ovate, closely appressed and imbricated . . . 147. ANOMOBRYUM.
　　　　Leaf-cells all lax, rhomboid or hexagonal, longer than wide . . . 8.
　　　　　Leaves lanceolate-acute ; ring absent . . . . 148. MNIOBRYUM.
　　　　　Leaves ovate ; ring present . . . . . 149. BRYUM.
　　　　　　Plants stoloniferous . . . . . . 150. RHODOBRYUM.
　　　　　　Plants not stoloniferous . . . . . . 9.

(Genus LEPTOSTOMUM R. Br. Dr. Shaw, in South African list, describes in English a plant to which he gives the name *Leptostomum Gerrardii* Shaw (n. sp.) as follows : " Dioecious, stem leaves ovate, concave, becoming suddenly acuminated, ending with a long, crisped, toothed hair-point. Nerve excurrent. Perichaetal leaves ovate or lanceolate-subulate with very large ciliated hair-point. Capsule on a long pedicel, oblongo-ovate ; operculum minute, conical. This beautiful species turned up in a set of Gerrard's plants during my residence at Kew. The stems are striking in aspect, resembling those of *Myurum hebredarum*, but fluffy in appearance on account of the long hair-points of the leaves. It is allied to *Leptostomum splachnoides* Hk. and Arn. ; coll. on wood in Natal by Gerrard." He places it between *Mnium* and *Catharinea*.

There is nothing in the above indicating generic character except the fact that Shaw called it *Leptostomum*, while the reference to the prostrate *Myurum* (which resembles *Stereodon cupressiformis* and is nerveless) raises a doubt as to its being rightly placed in *Leptostomum*, which has usually erect, matted, forked, rhizoid-tomentose stems, the leaves appressed and somewhat spirally inclined, and more or less julaceous when dry, erect or erecto-patent when moist, usually rounded at the apex, with the nerve excurrent as a hair-point ; roundly hexagonal cells, a long seta, and an erect or nearly erect, regular, ovoid capsule, having a very short, more or less rudimentary peristome, without teeth, or with these only rudimentary, but with a complicated, basal membrane. Lid dome-shaped, calyptra evanescent, shortly cucullate, smooth.

Müller's description of the perianth of the genus, which is presumably what Shaw had to go by, is : " Per. simplex, internum ; membrana plus minus alta sedecies plicata albida tenerrima cellularis, apice nec in dentes nec in cilia producta."

The genus belongs almost entirely to the Southern Hemisphere ; it was placed by Müller in Mniaceae, by Mitten in Bryaceae, and by Brotherus in a separate order Leptostomaceae, who, following Fleischer, feels inclined, on account of the peculiar construction of the peristome, to connect the genus with Nematodontei (Polytrichales), but in Pfl., x, 404, he still retains it next to Bryaceae, and illustrates the S. American *L. splachnoides* Hk. mentioned above by Shaw.

I place no reliance on Shaw's identification, and am not aware if the specimen Shaw refers to is still available at Kew. Since even Hornschuch (Linnaea, xv (1841), 134) thought fit to comment on the relation of *Brachymenium julaceum* Hsch. to species of *Leptostomum*, I feel sure Shaw's plant is *Brachymenium pulchrum* with old capsules. His description fits exactly.)

Genus 141. MIELICHHOFERIA Horns., in Br. Germ.

Usually small caespitose plants, having the inflorescence on a lateral short stem at the base of the plant ; leaves lanceolate or ovate-lanceolate with hexagonal, rhomboid, or linear areolation ; nerve extending to near or sometimes beyond the leaf-apex. Capsule erect or cernuous, pear-shaped and regular, or clavate, or cylindrical and irregular, usually with a distinct neck, a convex-mucronate lid, small cucullate evanescent calyptra, the peristome usually single (outer absent and inner present); the processes narrow from a 16-plicate, wide or narrow membrane; ring wide, dehiscent; spores small, smooth. Many species, mostly American, but some from each continent.

Synopsis of Mielichhoferia.

480. *M. subnuda.* Leaves concave-amplexicaul.
481. *M. transvaaliensis.* Basal membrane very low; leaves entire, ovate-acuminate.
482, 483. *M. Eckloni.* Basal membrane high, leaves lanceolate, usually denticulate.

480. *Mielichhoferia subnuda* (Dixon, MSS.) Sim (new species).*

(= *M. *amplexicaulis* Sim, MSS.).

A minute gregarious moss, growing on mud, apparently dioecious; main stems very short, with rosule of imbricate shortly ovate or nearly semicircular, amplexicaul leaves 0·4 mm. long, among which stand the longer cylindrical paraphyses; subtending the rosule are one or more dense innovations 1–2 mm. long, usually with one or more comal rosules, from which the branch grows on again more lax-leaved. Leaves similar to the perichaetal leaves, subjulaceous, subacute, the strong nerve ceasing just below the apex, or excurrent as a mucro; margin erect and entire or almost so; cells very lax and pellucid, the basal cells quadrate, $15 \times 15\mu$, the upper cells irregularly hexagonal, $30 \times 15\mu$, the marginal cells hardly different. Seta 6–10 mm. long, erect, red, slender; capsule pyriform with little neck, microstomous, with flatly semiglobose lid; ring present; outer peristome absent, inner peristome membrane when present fairly deep, with occasional irregular processes, but often not present when it seems almost gymnostomous. Fruits freely and frequents mountain localities. Riverbank, Mooi River, Natal, 4000 feet, Aug. 1920, Sim 10,217 and 10,197, and also Dr. P. van der Bijl; Goodoo, Bergville and Ladysmith, Wager 738, 715, 686.

In the type the nerve is regularly shortly excurrent, ending in a sharp mucro; rosules very compact, as wide as long, or almost globose; very minute; brood-branchlets often present in the leaf-axils, but another form has shorter blunter leaves, and some run into each of these.

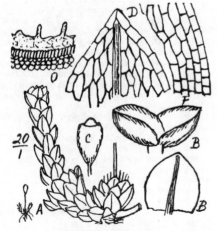

Mielichhoferia subnuda.

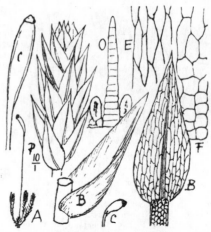

Mielichhoferia transvaaliensis.

481. *Mielichhoferia transvaaliensis* C. M., in Hedw., 1899, p. 64; Broth., Pfl., x, 351.

Densely caespitose, synoecious, pale green; fertile stems very short, sessile or stiped, subtended by numerous, erect, clavate, stiped, sterile stems up to 1 cm. long from a branched base or rhizomes, the stem itself stout and red. Branch leaves 1 mm. long, ovate-acuminate, imbricate, erecto-patent, widely clasping though narrowly attached, very concave at the base, very laxly areolated, entire or almost entire, acute, with spreading points; the nerve distinct but narrow; almost percurrent, the basal cells small and roundly quadrate, sub-basal cells widely and irregularly hexagonal, lamina-cells hexagonal, $80 \times 20\mu$. Lower leaves smaller, wider and scattered, the lowest very small and bract-like. Perichaetal leaves 0·5 mm. long, ovate-acute, very wide and lax, entire; nerve present. Seta 2–3 cm.

---

* *Mielichhoferia subnuda* (Dixon MSS.) Sim (sp. nov.).

Minuta, gregaria; caulis fertilis brevissimus, rosula foliorum imbricatorum alte convexorum breviter ovatorum amplexicaulium 0·4 mm. longorum ac latorum, nervo percurrente vel excurrente; caules steriles 2–3 mm. longi, singulis pluribusve rosulis adpressis foliisque laxis iulaceis similibus interpositis. Cellae laxissimae, 6-gonae, basales minores et quadratae. Seta 6–10 mm. longa; capsula erecta, pyriformis; operculum semi-globosum; interius peristom membranum, cum adest, satis altum, dentibus internis irregularibus.

long, erect, slender, red ; the capsule at first inclined, cylindric, and green, ultimately horizontal, pyriform and red ; lid hemispheric, shortly pointed, red ; ring shallow, dehiscent ; peristome of rather wide, yellowish, inner teeth almost without membrane and themselves easily dehiscent.

Cape, W. Claremont, Aug. 1922, M. Duncan.

Natal. Town Bush, Maritzburg, Apr. 1917, Sim 8650.

Transvaal. Duivels Krakler, near Lydenburg, April 1887, Dr. Wilms (*fide* C. M.).

482. *Mielichhoferia Eckloni* Horns., in Linn., xv (1841), 118 ; C. M., Syn., i, 230 ;
Broth., Pfl., x, 352, and fig. 302, p. 351.

=*Oreas *capensis* Bruch, in Coll. Zeyher.

=*Schizhymenium bryoides* Harv., in Hook., Ic. Pl. Rar., iii, 230.

=*Mielichhoferia Rehmanni* C. M., in Hedw., 1899, p. 64 ; Dixon, Trans. R. Soc. S. Afr., viii, 3, 199 ; Broth., Pfl., x, 353.

=*Leptochlaena Rehmanni* C. M., in Rehm. 215, 543, and 543*b*.

=*Leptochlaena Rehmanni*, var. *oranica*, Rehm. 216.

Mielichhoferia vallis-gratiae Hampe, Rehm. 542 (and *fide* Jaeg., Ad. *ined.*, see Kew Bull., 1923, No. 6, under No. 542).

M. squarrosula, C. M., Rehm. 210, and Hedw., 1899, p. 64.

Gregarious or caespitose, low, synoecious, with a silvery shine ; fertile stem basal, very short, almost bud-like, with comal leaves ovate-acuminate, acute, short, laxly areolated, almost entire, with nerve half-length. Sterile

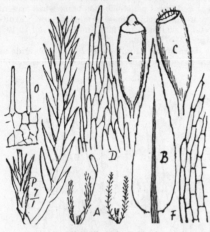

Mielichhoferia Eckloni.

branches simple on much-branched, stout, red stolons subtending the fertile stem, 2–10 mm. high, erect, often stipate and club-shaped with reduced lower leaves, but sometimes leaved to the base ; leaves 1–1·5 mm. long. erecto-patent, lanceolate-acute, slightly concave ; the margin subentire, denticulate or dentate almost the whole length but especially upward, the marginal cells being sometimes appressed, sometimes with erect or reflexed point. Nerve extending half-way or more, often brown, wide below, narrow above ; cells very lax, shortly hexagonal at the base, upper cells longer and narrower, and usually twisted-vermicular. Seta 1–3 cm. long, red, mostly straight, bent or twisted just below the capsule. Capsule suberect, slightly irregular, cylindrical and green with red ring-base till mature, afterwards wider and brown ; lid subhemispherical mamillate ; calyptra long, cucullate, caducous ; ring wide dehiscent ; peristome deeply set in the mouth, exceedingly variable ; usually inner only with deep membrane, slender and often imperfect pale teeth (sometimes irregularly branched and thereby connate) with or without cilia, but occasionally with perfect or imperfect outer teeth also.

Mountain specimens usually have seta shorter, capsule pyriform when mature, and branch foliage lax but are specifically the same.

A common moss on moist banks throughout South Africa ; occurs also in Australasia. The irregularity of its peristome, even in the same gathering, has caused much trouble, and the mature capsule is usually eperistomate.

Cape, W.   Window Gorge, Kirstenbosch, Sim 9484 ; Newlands, Rehm. 542 (as *M. vallis-gratiae* Hpe.); Claremont, Mrs. Duncan ; Paarl Rock, Sim 9646, 9650, 9644 ; Hex River, Miss Stephens ; Elgin, Sim 9593.

Cape, E.   Valley of Desolation, Graaff Reinet, MacLea, Rehm. 543*b* (as *M. Rehmanni* C. M.).

Natal. Van Reenen's Pass, Wager, 65 ; Eston, Sim 10,224 ; Karkloof, J. M. Sim ; Sweetwaters, Sim 10,226 ; Koodoo Bush, Sim 10,228 ; Inanda, J. M. Wood ; Cathkin, Miss Owen 37 ; Benvie, Sim 10,229.

Transvaal. Kaapsche Hoek, Wager 314, 322, 336 ; Macmac, MacLea, Rehm. 215, 543 ; Jessievale, Sim 10,023 ; Melville, Johannesburg, Sim 10,227 ; Witpoortje, Professor Moss.

O.F.S. The Kloof, Bethlehem, Sim 10,225 ; Leribe, Mme. Dieterlen 591.

S. Rhodesia. Zimbabwe, Sim 8770 ; Salisbury, Eyles 2984.

Genus 142. HAPLODONTIUM Hampe.

Slender little dioecious plants, growing in more or less dense green tufts ; stems erect, rhizoid-tomentose at the base, rosulate or club-shaped, with innovations from near the root only. Leaves suberect, densely imbricated when dry, concave, ovate-acute or ovate-lanceolate or lanceolate, indistinctly toothed or entire ; nerve strong to or near to the apex, seldom excurrent ; cells laxly rhomboid, with thin walls, those at the base shortly rectangular. Perichaetal leaves hardly different. Fertile stem at the crown, very short, subtended by longer sterile branches. Male bud thick, basal, many-leaved, with paraphyses. Seta long, curved or inclined at the top ; capsule cernuous, regular, cylindrical or ultimately pear-shaped, contracted toward the small mouth. Ring wide, dehiscent. Inner peristome absent, outer of sixteen equidistant equal teeth, outer membrane narrower than the inner, usually papillose. Lid small, concave-mamillate.

483. *Haplodontium reticulatum* (Hook.), Broth., Pfl., i, 3, 540, and x, 356.

= *Weisia reticulata* Hook., Bot. Misc., i (1830), 121, t. 29.
= *Mielichhoferia pellucida* Hampe, Ic. Pl. Dec., iii, t. 27.
= *Mielichhoferia Breutelii*, W P. S., in Breutel, Musc. Cap.
= *Bryum Breutelii* C. M., in Hedw., 1899, p. 65 (*fide* Broth., in Pfl., i, 3, 540, and x, 356).
= *Haplodontium Breutelii* (C. M., Broth. ; Par., Ind., 2nd ed., p. 301, in error).
= *Bryum *schizotrichum* C. M., in Rehm. 241.

Gregarious or caespitose, dioecious, green, growing on moist banks as a perennial but often reduced to a bare crown by drought or fire, growing again rapidly after rain. Stems rosulate to 5 mm. high ; leaves lax, spreading (incurved when dry), ovate-lanceolate, narrowed from the middle downward, and drawn to an acute point. Nerve strong, often brown, excurrent in a sharp point ; margin entire or almost so, slightly bordered by long narrow cells ; cells laxly hexagonal, basal cells shorter, wider and rectangular. Seta 2–3 cm. long, straight ; capsule inclined, at first cylindrical and green, afterwards pyriform and brown, regular. Ring dehiscent ; inner peristome absent, outer teeth equidistant, conical, closely trabeculate below, papillose.

Common on every moist bank in each province and Rhodesia.

C. M. (Syn., i, 232) wrongly connects our plant with *Miel. clavata* Br. and Sch., an Abyssinian plant, since separated.

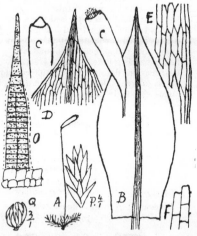

Haplodontium reticulatum.

Genus 143. ORTHODONTIUM Schw.

Slender, caespitose, light-coloured mosses with narrow subulate leaves, having long hexagonal-rhomboid areolation and autoecious axillary inflorescence. Capsule small, inclined, oval, with a tapering neck ; calyptra obliquely beaked ; lid shortly conical-pointed ; ring absent ; peristome double, inserted below the mouth ; outer peristome with lanceolate-cone-shaped trabeculate teeth ; inner almost without basal membrane and with thread-like processes as long as the outer teeth.

(*Orthodontium gracile* (Schw., MSS.) ; C. M., Syn., i, 238.

= *Bryum (Pohlia) gracile* Wils., in Gard., Musc. Brit., t. 34.
= *Stableria gracilis* (Wils.) Lindb. ; Broth., Pfl., x, 349, and fig. 299.

Is mentioned in Shaw's list as collected by Macowan and Shaw at Fern Kloof, Grahamstown, Sept. 1867, and occurs also in North Europe and North America, and the closely allied *Stableria aethiopica* (C. M.) Broth. in North Africa, but as there is nothing to support Shaw's record it must be considered a case of mistaken identity, probably relating to *O. lineare*. *O. gracile* has the processes much shorter than the peristome teeth, and has inclined cylindrical capsule with long neck.)

484. *Orthodontium lineare* Schw., Suppl., ii, 2, 124, t. 188 ; C. M., Syn., i, 238 ; Broth., x, 350 ;
Dixon, Trans. R. Soc. S. Afr., viii, 3, 199 ; (not *Aplodium lineare* Mitt.).

" Monoecious, small, laxly caespitose ; leaves very long and narrowly subulate, laxly areolate, flexuose, almost entire, nerve disappearing in the subula, subkeeled-concave ; capsule horizontal, oval, with subequal base, seta inclined at the top ; lid conic, suboblique, acute, short ; outer peristome teeth wide, densely trabeculate, inner equalling the outer, filiform, slender, articulate. Male flowers gemmiform, perigonial leaves small, obtuse, nerveless ; ring absent " (C. M., Syn., i, 239). Brotherus' characters are, " Capsule ribbed, when dry furrowed ; peristome teeth smooth, yellowish, short, usually 0·10 mm., seldom 0·20 mm. ; membrane not present, processes very narrow, as long as or longer than the peristome teeth."

Müller described from a specimen from the Cape (Menzies), and Shaw adds Graaff Reinet (MacLea). In relation to a specimen from George, C.P., 1916 (Wager 542), Mr. Dixon says, " By the kindness of the keeper of the Herb. Boissier I have been able to examine part of Schwaegrichen's type ; this agrees with the present plant exactly as far as it goes, and I have no hesitation in referring it there. Unfortunately the capsules of the type showed no complete peristome. Wager's plant has the peristome in good condition, showing short, almost smooth, cuneiform, outer teeth, rather distantly articulate, with very narrow-linear, smooth, articulate inner processes, markedly longer than the teeth. C. Müller's description of the teeth as densely barred, and the processes as equal to the teeth in length, does not agree, but, as he had not seen actual specimens, too much stress must not be laid on this discrepancy. The plication of the capsule is very variable, often wanting."

Genus 144. BRACHYMENIUM Schw.

Robust rhizoid-tomentose plants, ¼–1 inch high, growing in dense epiphytic tufts often high up on living trees, or else smaller soil-plants without tomentum, of varied habit and habitat. Stems more or less club-shaped, occasionally gemmiferous ; leaves julaceous or erecto-patent, entire in our species.　Nerve strong, often excurrent as an arista, sometimes ending in or below the apex ; cells laxly hexagonal or rhomboid, those at the base not larger but rectangular. Perichaetal leaves decreasing in size inward, lanceolate-ovate-acute.　Seta often long, erect or nearly so ; capsule erect, inclined or drooping, regular ; oval, pear-shaped, or cylindrical.　Ring dehiscent.　Outer peristome teeth longer than the inner, lanceolate-pointed, widest at the base, hyaline above ; the inner with very high, plaited, usually papillose basal membrane, bearing irregular, short, often rudimentary processes.　Cilia rudimentary or absent.　Lid small, shortly conical. or shortly beaked.

Synopsis of Brachymenium.

1. Leaves strongly aristate　.　.　.　.　.　.　.　.　.　.　.　2.
1. Leaves acute but not aristate　.　.　.　.　.　.　.　.　.　.　5.
2. Leaves straight and appressed when dry　.　.　.　.　.　.　.　3.
2. Leaves twisted when dry, 1·5–2 mm. long　.　.　.　.　.　.　4.
3. Leaves tumid-cochleate, gemmae absent　.　.　.　.　.　485. *B. pulchrum.*
3. Leaves not tumid, gemmae abundant　.　.　.　.　486. *B. campylotrichum.*
4. Leaf-margins reflexed, capsule erect　.　.　.　.　487. *B. rhodesiae.*
4. Leaf-margins flat or erect, capsule cernuous　.　.　.　488. *B. variabile.*
5. Stems julaceous, leaves many and dense, lanceolate-acuminate　.　489. *B. Borgenianum.*
5. Stems very laxly julaceous ; leaves lax, erecto-patent　.　.　490. *B. dicranoides.*

485. *Brachymenium pulchrum*, Hook., Bot. Misc., 1830, i, p. 136, t. 38 ; C. M., Syn., i, 324 ;
　　　Dixon, Trans. R. Soc. S. Afr., viii, 3, 200 ; Brunnthaler, 1913, i, 27.

　　　=*B. pulchrum* Hk. and *B. julaceum* Hsch. ; Broth., Pfl., x, 367, and fig. 321.
　　　=*B. korotranum* C. M., Syn., i, 324 (see Dixon, Trans. R. Soc. S. Afr., viii, 3, 200).
　　　=*B. julaceum* Hsch., in Linn., xv (1841), 133 (and *fide* C. M., Syn., i, 324).
　　　=*B. *eriophorum*, Rehm. 545.

　　(And see *Leptostomum Gerrardi* Shaw herein, p. 314.)

　　Dioecious ; growing in dense, tomentose, epiphytic tufts on branches of trees, usually on high branches in sunshine, occasionally on stones ; the tufts 2–4 inches diameter, 1 inch deep, and through brown tomentum exceedingly compact.

Brachymenium pulchrum.

Stems ½–1 inch long, closely placed, simple to much-branched, the lower branches short and old perichaetal, the upper branches 2–4 mm. long, all julaceous, but the perichaetal branches standing higher than the others and conspicuously white-awned.　Branch leaves 0·5–0·7 mm. long, densely imbricate, shortly ovate from a wide base, convex-cochleate so as to be subhemispherical, with erect, simple or dentate awn up to 0·7 mm. long ; nerve narrow but strong, weakening into the flat hyaline awn, which is often flexuose or even crisped.　Margin entire or almost so, usually bordered by one line of long, narrow, hyaline cells, which change to quadrate toward the base.　Basal cells small and quadrate, sometimes coloured ; sub-basal cells rectangular, $80 \times 20\mu$ soon passing into rhomboid sublinear cells $80 \times 10–5\mu$, the upper and basal cells hyaline, others bright green, all thin-walled.　On the fertile stems the upper leaves are longer and more hyaline and lax-celled, with wide concave nerveless hyaline white apex leading into the awn, and usually these stems stand higher than the others.　Vaginula short, brown, set with withered archegonia.　Seta straight, purple, erect, 1–2 mm. long ; capsule erect or almost so, at first clavate, later pyriform and brown, smooth, narrow-mouthed.　Peristome double, the outer of sixteen strong, rather blunt, lanceolate, trabeculate teeth, the inner membrane half as long or more, with irregularly undeveloped processes.　Ring dehiscent.　Lid shortly conical ; calyptra caducous.　A pretty moss of very distinct habit, present in every forest district in South Africa, including Rhodesia ; usually fertile, and extending from the coast to the mountains.　A form occurs having short hair-points and very dense concave leaves (Junod 4018, etc.).

　　B. eriophorum Rehm. is a vigorous specimen in which all the upper leaves are perichaetal and consequently with wide hyaline awn-base, but does not differ even as a variety.

486. *Brachymenium campylotrichum* (C. M.), Broth., Pfl., i, 3, and x, 367.
　　Dixon, S.A. Journ. Sc., xviii, 319, and Trans. R. Soc. S. Afr., viii, 3, 200.

　　　=*Bryum campylotrichum* C. M., in Hedw., 1899, p. 65.

　　Much like a small condition of *B. pulchrum*, growing on soil, but branching shows it is not juvenile ; stems 1–2 mm. high, julaceous, the leaves 0·7 mm. long, oblong-ovate (fig. *β*), shortly awned, very pellucid, with a white sheen ; closely imbricate, making the branches clavate ; the nerve strong and yellowish, ending in the short, often recurved and flexuose yellowish awn, usually not denticulate. Cells lax, the lower small and quadrate, the upper widely hexagonal, $80 \times 20\mu$, marginal cells not different (C. M. says margin pale ; I do not find it so). Müller does not mention gemmae, but I find many stems ending with a spike of reproductive dehiscent buds (250μ long), that spike subtended by a few short almost semicircular pellucid bracts, 0·4 mm. or less long, shortly aristate and very laxly areolated. Longer stems intermixed are not reproductive, so that condition may be the effect of snow, frost, or some similar cause not destructive to all. Sexual parts not known. Leaves concave, but not cochleariform as in *B. pulchrum.* Dixon says it is known at once from *B. pulchrum* in the scarcely bordered leaves.

Natal. Described by Müller from Helpmakaar, Dr. Wilms (1884) ; Giant's Castle, 7000 feet, R. E. Symons (Sim 10,198).

Transvaal. Rietfontein, Wager 253 p.p.

S. Rhodesia. Zimbabwe, 3000 feet, July 1920, Sim 8814.

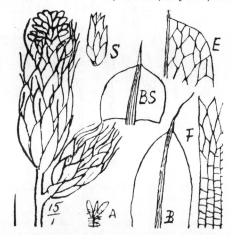

Brachymenium campylotrichum.

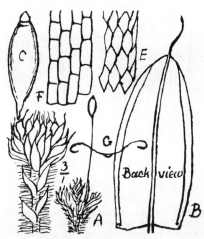

Brachymenium rhodesiae.

487. **Brachymenium rhodesiae** Dixon, S.A. Journ. Sc., xviii (June 1922), 319.

Autoecious ; a vigorous plant, growing on the ground in lax tufts ½–1 inch deep ; stems simple or branched, ½–1 inch high, the inflorescence at first terminal, soon lateral by innovation. Branch-leaves very lax except a comal green tuft, slightly twisted when dry, widely oblong-ovate, 2 mm. long, obtuse with excurrent, slender, acute, aristate nerve ; concave, with margin entire and recurved almost the whole length ; nerve strong, green, excurrent for one-fourth the length of the leaf, and usually reflexed above the lamina ; basal cells rectangular $80 \times 10\mu$; upper lamina cells hexagonal, $80 \times 10\mu$ or narrower, with usually one border-line of narrower hyaline cells along the margin. Per. leaves not different. Male inflorescence discoid, terminal on branches. Seta 2–3 cm. long, brown ; capsule erect, pyriform, microstomous, brown ; lid shortly conical, obtuse ; peristome teeth red below, narrowly lanceolate ; membrane adherent, processes not seen. Spores 20μ.

On Granite Hill, 6500 feet, Makoni, April 1918, Eyles 1317*a* ; on dead wood in bush, Umtali, 4000 feet, June 1919, Eyles 1730.

Brotherus omits this from Pfl., x. Why ?

488. **Brachymenium variabile** Dixon, in Uganda Mosses, Sm. Misc. Coll. 69, viii, 2 (1918) ; Broth., Pfl., x, 369.

Autoecious ; loosely caespitose on bark, very variable, fruiting freely ; fertile stems very short, with a comal tuft, and subtended by red-stemmed innovations almost leafless below, 2–3 mm. long, on which the leaves are erecto-patent and straight when moist, but more or less twisted round the plant when dry. Leaves 1·5 mm. long, widely ovate, carinate-keeled, not julaceous, closely imbricate in the coma, the nerve strong, usually yellowish, extending as a more or less elongated sharp yellowish arista ; margin erect, entire ; basal cells quadrate, 20μ long ; upper cells lax, pellucid, irregularly rhomboid, the marginal line somewhat narrower, sometimes forming an evident border, sometimes hardly different. Per. leaves ovate-lanceolate, more aristate, very laxly areolated ; vaginula short, stout, red ; seta 1·5–2 cm. long, red, cygneous at the top ; capsule 3 mm. long, horizontal or subpendulous ; at first clavate, finally pyriform, brown, with rather short peristome, " the outer teeth of which are very densely barred with highly projecting lamellae on the outer surface, deep orange in colour, strongly bordered, very finely and regularly papillose on the dorsal surface ; the basal membrane of the endostome is about half the height of the outer teeth, with well-developed, broad, obtuse segments, almost equal to the teeth, pale, and very delicately papillose " (Dixon). Lid very shortly conical. Uganda and South Africa.

Natal. Glyn Stream Falls near Hilton College, on Cycads, May 1921, Sim 10,194 ; Town Bush Road, Maritzburg, Sim 10,195 ; Zwaartkop Valley, Sim 10,196, 10,200.

S. Rhodesia.   Zimbabwe, 3000 feet, Sim 8800, 8823 ;   Rhodes Grave, Matopos, 5000 feet, Sim 8870, 8876, 8877. Wager lists *Brachym. capitulatum* (Mitt.) Paris, as *Bryum capitulatum* Mitt. (Journ. Linn. Soc., 1886, p. 306), which Brotherus places with *B. variabile* in the autoecious group from East Africa.   Nothing more is known of it here.

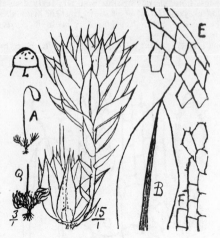

Brachymenium variabile.   *Brachymenium borgenianum.*

489. *Brachymenium Borgenianum* Hampe, in Linn., 1874, p. 211 ; Broth., Pfl., x, 367, and fig. 319A, E ; Dixon, S.A. Journ. Sc., xviii (1822), 319, and Trans. R. Soc. S. Afr., viii, 3, 200.

A small, dioecious, gregarious or caespitose soil-moss ; fertile stems very short, few-leaved, subtended by several slender, club-shaped, julaceous, shining branches 3–10 mm. high, less than 1 mm. diameter, on which the numerous leaves are closely appressed except the points ; leaves concave, lanceolate or ovate-lanceolate, 0·5–1 mm. long, acute ; the nerve strong, shortly excurrent ; the margin entire and not different ; basal cells small, quadrangular and often coloured ; lamina cells narrowly hexagonal or rhomboid, pellucid, thin-walled.   Perichaetal leaves in a comal tuft, rather larger but not different.   Seta 2 cm. long, red ; capsule erect, clavate with long neck, narrow-mouthed ; teeth widely conical-acute, papillose ; membrane high, without regular processes.

Widely distributed in South Africa, East Africa, and African islands.

Cape, W.   Ameib, Erongo Mountains, S.-W. Africa, 1916, Professor H. H. W. Pearson (P. Sladen, Mem. Exp., 9849).

Transvaal.   Pretoria, Sim 7781 ; Kaapsche Hoop, 1915, Wager 331.

S. Rhodesia.   Salisbury, Eyles 2765 ;   Umtali, Eyles 1737 ;   Matopos, Sim 8845, and Eyles 936 ;   Zimbabwe, Sim 8742.

490. *Brachymenium dicranoides* (Horns.), Jaeg., Ad., i, 575 ; Broth., Pfl., x, 366.

=*Bryum dicranoides* Horns., in Linnaea, 1841, p. 131 ; C. M., Syn., i, 309.
=*Brachymenium liliputanum* (C. M.), Broth., x, 366.
=*Bryum liliputanum* C. M., in Hedw., 1899, p. 66.
=*Brachymenium stenopyxis* C. M., in Rehm. 220 (not 241 as in Hedwigia).   Name only, but name preoccupied by the S. American *B. stenopyxis* C. M., in Linn., 1878, lxxix, 840.

Dioecious, very small ; main stem 1 mm. high, with a small comal tuft of perichaetal leaves and terminal seta, and subtended by several erect, laxly clavate branches 3–4 mm. long, having ovate-lanceolate, lax, pellucid, erecto-patent, keeled-concave leaves 0·4–0·7 mm. long, julaceous when dry ; margin entire, unbordered, erect ; nerve decurrent, strong, coloured, percurrent, or in perichaetal leaves sometimes, ending in a strong coloured arista.   Cells pellucid, hexagonal-rhomboid, 20μ long, the basal cells hardly different or more rectangular ; marginal cells little altered ; per. leaves more lanceolate and more aristate.   Seta red, slender, 1–2 cm. long, straight or cygneous at the top ; capsule oblong-cylindrical, horizontal or drooping, at first pale and nearly erect, afterwards brown ; ring present, somewhat microstomous ; lid shortly and acutely conical.   Outer per. teeth long, narrow, slender, subulate, brown below, rugulose ; membrane short, pale, with long, narrow, subulate, knotted, pale segments as long as the outer, with short hyaline cilia between.   Müller describes the perigonia as similar to the perichaetia, the antheridia as thick, large and numerous, and axillary gemmae at first green, afterwards brown, as being produced.

Cape, W.   Table Mountain and Camps Bay, Ecklon ; Swellendam, Dr. Pappe ; Rondebosch, Rehm. 219 ; Kloof Nek, Cape Town, Sim 9559 ; Cape Town, Rehm. 220 (as *B. stenopyris* C. M.).

Cape, E.   Dohne Hill, 4000 feet, 1897, Sim 7185.

Natal.   Home Rule, Polela, Sim 8666 p.p. ; Giant's Castle, 8000 feet, 1915, R. E. Symons (Sim 10,206 and 10,207) ; Cathkin, 6000 feet, Miss Owen 19 : Rosetta Farm, Sim 10,221, 10,230.

Transvaal.   Pretoria, Sim 7783.

In Rehm. 220 the capsules are all Tortula in what I have seen.

Probably frequent throughout South Africa, but minute, and looks juvenile. *Bryum pallido-julaceum* C. M., in Hedw., 1899, p. 66 (sterile), does not appear from description to differ, and type is not available in South Africa (Lake Chrissie, 1885, Dr. Wilms). Brunnthaler records it (p. 27) from Van Reenen's Pass as *Brachymenium pallido-julaceum* (C. M.) Par., and Dixon, writing *re Bryum subdecursivum* C. M. (in Rehm. 254 and in Hedw., 1899, p. 74), says, it "is a sterile *Brachymenium*, very near to *B. philonotula*, but nerve too weak. From description I should think it may well be *B. pallidojulaceum* C. M."

C. Müller, from the sterile specimens, placed *B. subdecursivum* C. M. in his *Bryum*, section *Erythrocarpidium*, and it has, he says, the habit of *B. decursivum* but differs in leaf-formation, and describes the cells as small and the leaf-margins erect. Brotherus (Pfl., x, 394) attributes this name to a Porto Rico *Bryum*, probably in error, for *Bryum decursivum* C. M., and my own reading of Müller's description, indicates what I have named *Bryum argenteum* var. *viride* in a sand-grown and red condition.

Bryum Neesii C. M., in Hedw., 1899, p. 66 (=*Brachymenium nutans* Nees, in Herb. Hampe), probably does not differ, though described as having leaves more imbricate, more boatshaped-oblong, shortly acuminate, with subrevolute margin and basal cells larger and more hexagonal than others (Gnadenthal, Breutel).

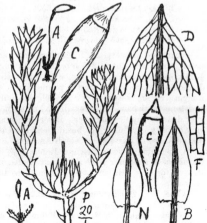

Brachymenium dicranoides.

Genus 145. WEBERA Hedw. (not Ehrh.).

Stems slender, simple or innovating, mostly from the base, ½–3 inches high, loosely tufted or gregarious on moist ground ; axis usually red ; leaves more or less lanceolate, decurrent, gradually longer and narrower up the stem, sometimes crowded and longer in a comal tuft ; the nerve usually reaching to or near to the apex ; the cells irregularly elongate-hexagonal, or almost linear ; capsule pyriform or clavate, usually decreasing rapidly into a narrower neck. Mouth small, regular. Peristome double, the outer and inner of more or less equal height ; the inner free, with a considerable or narrow basal membrane and long or irregular processes, with or without intermediate cilia, and these when present without appendages.

Closely allied to *Bryum*, the most evident differences being the narrower leaves, narrower cells, the cylindrical non-tapering neck, and the absence of cilia-appendages.

The genus has been divided into *Webera* and *Pohlia* ; in Pfl., i, 3, Brotherus put most of the species as herein understood into *Pohlia*, and used the name *Webera* (Ehrh.) and family Weberaceae for quite a different group of plants, of which *Diphyscium foliosum* Mohr. is typical ; but in Pfl., x, he reverts to *Webera* Hedw. for the present group, and the other genus *Webera* Ehrh. is not mentioned so far.

Synopsis of Webera.

1. Leaves 1–1·5 mm. long, or less, lanceolate-acute, nearly entire	.	.	.	491. *W. mielichhoferiacea.*
1. Leaves 2 mm. long or more	.	.	.	.	.	2.
2. Leaves not revolute, or revolute at the base only	.	.	.	.	.	3.
2. Leaves much revolute	.	.	.	.	.	4.
3. Leaves lanceolate from an ovate base, denticulate	.	.	.	492. *W. nutans.*
3. Leaves ovate-lanceolate, shortly pointed ; peristome reduced or single	.	.	493. *W. depauperata.*
3. Leaves ovate-lanceolate, shortly pointed ; peristome well developed	.	.	494. *W. MacLeai.*
4. Leaves strongly pointed or shortly awned	.	.	.	.	495. *W. revoluta.*
4. Upper leaves linear-lanceolate, denticulate ; nerve not excurrent	.	.	496. *W. leptoblepharon.*

Bryum (Dicranobryum) plumella C. M., in Hedw., 1899, p. 66 (*Pohlia*, Wager's list), is described (from immature fruits) without reference to the peristome, and Müller states that it is near *Bryum dicranoides* Horns. (now *Brachymenium*), but that species and *Bryum (Senodictyon nutantia) philonotula* C. M., in Hedw., 1899, p. 76 (from Kuilen, near Lydenburg, Dr. Wilms), and *Bryum (Senodictyon nutantia) pseudo-philonotula* C. M., in Hedw., 1899, p. 76 (from Lake Chrissie, Dr. Wilms), both *Pohlia* in Wager's list, both sterile and all three without gemmae, are now placed by Brotherus in *Webera*, § *Lamprophyllum*, where characters depend on peristome and the presence of gemmae. *B. plumella* (which is from Lydenburg, Dr. Wilms) agrees with *W. mielichhoferiacea* so far as description goes ; with the other two the descriptions are insufficient, and they cannot be retained, especially where Brotherus has them.

Webera annotina (Hedw.) Bruch., a Northern Hemisphere plant, is mentioned in Drege's list (from locality IV, A. 11), likely upon *W. nutans.*

Species known to me by name only, there being no named specimen in South African Herbaria : *Webera *glandiformis*, Rehm. 225 ; Cape Town (no specimens at Kew or British Museum). "*Webera Korbiana* C. M., Jaeg., Ad., i, 594 (1873–4)=*Bryum Korbianum* C. M., in Flora, 1874, No. 31, aquatic, Transvaal ? Prom. B. Sp.," *fide* Par., Ind., 2nd ed., v, 143. Mr. Dixon informs me this is a Libyan species, so presumably credited here in error.

491. *Webera mielichhoferiacea* (C. M.), Par. ; Broth., Pfl., x, 358.

=*Bryum mielichhoferiaceum*, C. M., in Hedw., 1899, p. 75.
=*Webera *gregaria*, Rehm. 549.
=*Webera *montis-cavi*, Rehm. 552.
=*Bryum *Haanii*, Rehm. 236 (*fide* Dixon).
=*Webera *Woodii*, Rehm. 547, 547*b*.
=*Pohlia mielichhoferiacea* (C. M.), Broth., i, 3, 547, Wager's list ; Dixon, Trans. R. Soc. S. Afr., viii, 3, 200.

Scattered or caespitose in flat low tufts, androgynous ; fertile stems very short, rosulate, subtended by one or more sterile red shoots, 2–5 mm. long, having lax, pellucid, erecto-patent, decurrent, lanceolate-acuminate leaves ½–1½ mm. long, keeled-concave, with thick yellowish nerve, slender subaristate apex, erect almost entire or closely denticulate margin, and shortly vermicular cells, the basal cells hardly different. Per. leaves wider, with revolute margin ; capsule rather large on ½ inch seta, horizontal, irregular, oval with thick neck, lid very shortly conical ; outer peristome teeth pale red, wide half-way up or more, then quickly narrowed ; rather closely barred, bordered irregularly, very densely papillose all over ; inner nearly as long but irregular ; linear, not appendiculate nor regularly nodose, very densely papillose ; the basal membrane as high as the ring. Small and scattered, probably often overlooked ; it occurs also in Europe and North America.

Concerning *Webera Enselini*, Rehm. 223, from mountains above Worcester, Mr. Dixon writes : " Quite indistinguishable from our *Pohlia elongata* Hedw. in fruit and vegetation. *P. mielichhoferiacea* (C. M.) may only be a starved form of this, but I should not like to be sure " (*Pohlia elongata* Hedw.=*Webera elongata* Schw. ; Dixon, 3rd ed., p. 332).

Cape, W. Devil's Peak, Rehm. 222 ; Paarl tram-line, Jan. 1919, Sim 9653.
Cape, E. Cave Mountain, above Graaff Reinet, MacLea (Rehm. 552, as *W. montis-cavi* Rehm.).
Natal. Inanda, J. M. Wood (Rehm. 547, as *W. Woodii* Rehm.) ; Van Reenen's Pass, 1915, Wager 285.
Transvaal. Pilgrim's Rest, MacLea (Rehm. 547β as *W. Woodii* Rehm., and Rehm. 549 as *W. gregaria* Rehm.).

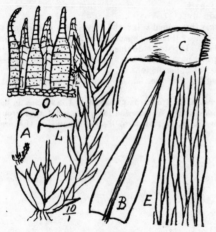

Webera mielichhoferiacea.

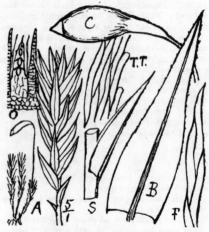

Webera nutans.

492. *Webera nutans* (Schr.), Hedw., Descr., i, 9, t. 4 ; W. P. Sch., Syn., 2nd ed., p. 396 ;
Dixon, 3rd ed., p. 333, Tab. 40 I ; Broth., Pfl., x, 360.

=*Bryum nutans* Schr., Spic. Fl. Lips., 81, No. 1043 ; C. M., Syn., i, 335.
=*Pohlia nutans* (Schr.), Lindb., M. Scan., xviii ; Braithw., M. Fl. ; Wager's list ; Broth., Pfl., i, 3, 548.
=*Webera *austro-nutans* C. M., in Rehm. 221 and 551 (*Pohlia*, in Wager's list).
=*Bryum austro-nutans* C. M. (from Kerguelen, *fide* Broth., and see Kew Bull., 1923, No. 6, under No. 221).
=*Webera afro-nutans* (C. M.), Par., Suppl. Ind., p. 327 (1900).
=*Bryum afro-nutans* C. M., in Hedw., 1899, p. 76.
=*Bryum Ecklonianum* C. M., in Bot. Zeit., 1855, p. 752.

Light green, usually laxly caespitose or gregarious on moist soil ; paroecious ; stems ½–2 inches long ; axis usually red ; lower leaves small, ovate-acute and keeled, larger but scattered upward, crowded and about 2 mm. long at the comal tuft, somewhat ovate at the base, narrowly lanceolate-acuminate, keeled, and abundantly denticulate by protruding teeth near the apex or in the upper half, but the margin is often revolute lower, and to the base. Nerve strong, percurrent, red with age. Lamina cells narrowly vermicular, $120 \times 10\mu$, basal cells rather wider ; perichaetal leaves similar, or longer and somewhat aristate. Seta 2–3 cm. long or more, pale ; capsule at first horizontal, later pendulous, elliptical, with distinct neck ; somewhat curved ; brown when mature ; lid shortly conical-acute ; calyptra white, caducous ; ring wide ; peristome teeth yellow, separate, closely trabeculate ; membrane fairly high with perforated processes and slender cilia. Antheridia in the axils of the upper leaves. Very variable, frequently sterile and scattered among grass in moist uplands ; fertile usually on claybanks. Much like a *Philonotis* in vegetative growth. Present in each province of the Union, and throughout the Northern Hemisphere.

Cape, W. Montagu Pass, Rehm. 221 ; Chaplin Point, Cape Peninsula, Sim 10,202 ; Intake, Woodhead Tunnel, Sim 9128 ; George, 1916, Wager 552.

Cape, E. Bailey's Grave, Perie, 4000 feet, March 1893, Sim 7172.

Natal. Misty Home, Hemuhemu, 5000 feet, Sim 10,201 ; Cathkin, 7000 feet, Sim 10,204 ; Van Reenen, Sim 7797 ; Giant's Castle, R. E. Symons ; Greenfields, Mooi River, Sim 10,231.

Transvaal. Pilgrim's Rest, MacLea (Rehm. 551) ; Piet Retief, Sim 10,203 ; Johannesburg, Miss Edwards.

O.F.S. Rydal Mount, Wager.

493. *Webera depauperata* Sim (new species).*

Laxly caespitose, shining, yellowish, the red axis showing through the yellowish leaves ; stems erect, 1–2 inches high, simple, nude below or with only a few small scales ; on upper two-thirds the leaves are laxly and equally placed, erecto-patent, 2–3 mm. long, straight or almost straight and hardly changed when dry, ovate-lanceolate from narrow base, acute, flat ; nerve fairly strong, pale when young, red when old, disappearing far below the leaf-apex ; margin expanded, or near the base slightly reflexed when dry ; entire, except a few denticulations near the apex ; cells narrowly twisted-vermicular, $200 \times 12\mu$; marginal cells not different ; basal cells rather shorter and wider. Old leaves mostly red ; young leaves quite green. Only three capsules found and these not in good condition, but from them it is evident that the capsule is pyriform, 2–3 mm. long, brown, suberect or nearly horizontal, with seta 3–4 cm. long ; the exothecium made of thick-walled quadrate cells, the only peristome with pale trabeculate teeth 250μ long ; basal membrane very short placed deep in the mouth. Ring absent.

Near *Webera cruda* Schwaeg., but distinct in the absence of inner peristome ; it may, however, be a reduced form (as to peristome) of it, or further specimens may show processes. *W. cruda* extends from Europe to New Zealand and South America, but is not recorded from South Africa.

W. depauperata is from the top of Giant's Castle, Natal, 8000 feet, R. E. Symons (Sim 10,208).

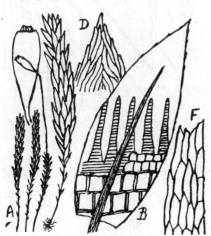

Webera depauperata.

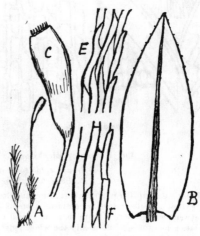

Webera Macleai.

494. *Webera Macleai* (Rehm. 548) Sim (new species).†

In yellowish tufts ; stems ½–1 inch high, red, unbranched ; lower leaves small, upper lax, 2 mm. long, subimbricate, erecto-patent, ovate-lanceolate, shortly pointed, lightly denticulate along the margin toward the apex : margin expanded ; nerve strong, percurrent, often red. Cells widely vermicular, $150 \times 10\mu$; basal cells as large but straighter and more rectangular. Leaves shrunk but hardly different when dry. Leaves closer on fertile stem ; seta erect, red, capsule 3 mm. long, nearly erect or suberect, cylindrical, wider above except at the mouth. Peristome teeth

---

* *Webera depauperata* Sim (sp. nov.).

Nitida, sub-flava ; caules nudi infra ; cetera folia 2-3 mm. longa, laxe pariterque disposita, ovato-lanceolata, patula, acuta ; margo explanatus, prope apicem denticulatus, nervus longe infra apicem desinens ; cellae 200×12 mm. torto-vermiculares ; folia haud multum mutata ubi sicca ; peristom. simplex ; dentes lati infra, arte trabeculati ; membranum humile.

† *Webera Macleai* (Rehm.), Sim (sp. nov.)=*Webera Macleai* Rehm. 548, ined.

Folia 2 mm. longa, sub-imbricata, erecto-patentia, ovato-lanceolata, breviter acuminata ; margo explanatus, leviter denticulatus ; nervus percurrens ; cellae grandes, late vermiculares. Capsula sub-erecta ; peristom. membranum humile, dentibus internis bene formatis.

strong, trabeculate, papillose ; membrane low, with well-developed processes. Mr. Dixon writes concerning the London specimens: "It is an unusually tall form of *P. elongata* Hedw., with the habit of *Mniobryum albicans.*"

Rhenosterberg Mountains, near Middelburg, MacLea (Rehm. 548).

495. *Webera revoluta* Sim (new species).*

Stems up to 3 inches long, mostly simple, straight, strong, usually red, equally and closely leaved except at the apex where leaves are more crowded, closely imbricated, erecto-patent or when dry suberect, and forming a bristly apex. Leaves 2–3 mm. long, nearly 1 mm. wide, ovate-acuminate, conduplicate, deeply and strongly keeled, and with the whole margin strongly revolute. Nerve strong, red, decurrent below, ending in a long, sharp, excurrent point or short awn. Cells long and narrow, the upper and inner rather rounded at the ends, others and especially the marginal cells narrower or vermicular ; basal cells irregularly quadrate or oblong, thick-walled.

Not unlike *W. nutans*, but stronger, closer-leaved, and with leaf-margins strongly revolute their whole length ; also like *Bryum aulacomnioides* but with narrow cells, revolute margins, and close growth. The habit is like that of a *Campylopus.*

Houtbosch, Lechlaba, Transvaal (as *Bryum*, No. 8), Rehm. 566.

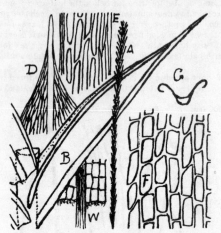

Webera revoluta.

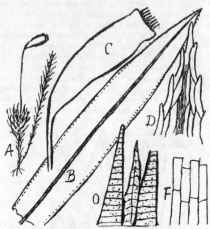

Webera leptoblepharon.

496. *Webera leptoblepharon* (C. M.), Jaeg., Ad., i, 600 ; Broth., Pfl., x, 359.

=*Bryum leptoblepharon* C. M., Syn., i, 337 ; Rehm. 550 ?
=*Pohlia leptoblepharon* (C. M.), Wager's list.

Müller's description is not satisfactory, and Rehmann's 550 is marked doubtful ; this description is from the latter, except as stated : loosely caespitose ; stems ½–1½ inch high, red, nude below, with scattered leaves above, and also between where there is more than one comal tuft ; stem leaves 2 mm. long ; comal leaves 3 mm. long, linear-lanceolate ; margin revolute nearly the whole length, but margin strongly denticulate where exposed ; nerve often red, percurrent, strong but narrow ; cells shortly vermicular, thick-walled ; basal cells rectangular ; capsule terminal on a strong stem, horizontal on a red seta, ultimately pendulous, brown, 3–3·5 mm. long, wide-mouthed ; teeth strong, red, papillose ; membrane low ; processes entire ; cilia not seen.

Müller says, "monoecious ; antheridia and archegonia thicker than long, numerous, without paraphyses ; inner floral leaves much smaller, acuminate." These I have not seen.

Cape, W. Zwellendam, and at Grootvadersbosch, Ecklon (1826).

Cape, E. The Wilderness, George, Apr. 1924, Miss A. Taylor.

Transvaal. Macmac, MacLea (Rehm. 550).

Genus 146. LEPTOBRYUM (Bry. Eur.) Wils.

Synoecious, or imperfectly dioecious, growing in lax tufts ; stems slender, branched at the base, or from bulbils on red protenema ; lower leaves lanceolate, upper leaves much longer in a crowded comal tuft, with long subulate leaves from wider sheathing base. Nerve extending to near the apex, margin denticulate, cells linear-rectangular. Capsule suberect or inclined, pear-shaped, with long neck. Outer peristome teeth, processes, and cilia of equal length and well developed ; processes perforated, cilia often appendaged. Very few species.

---

* *Webera revoluta* Sim (sp. nov.).

Similis *W. nutanti*, grandior tamen, foliis artioribus, 2–3 mm. longis, subula fortiori nec non marginibus revolutis.

497. *Leptobryum pyriforme* (Linn.), Wils., Bry. Brit., p. 219 ; Dixon, 3rd ed., p. 328, Tab. 40*c* ;
Broth., Pfl., x, 374, and fig. 326.

=*Mnium pyriforme* Linn., Sp. Pl., p. 1112.
=*Bryum pyriforme* Hedw. ; C. M., Syn., i, 330.

Soft, delicate, silky, and gregarious or laxly caespitose ; stems 1–2 cm. long, annual, very slender, unbranched, composed of delicate, thin-walled cells bearing chlorophyll granules in abundance ; the lower leaves short, distant and triangular, gradually longer and closer upward, with usually a many-leaved perichaetal comal tuft at the apex, the leaves of which have rather wide, sheathing, ascending base, above which the leaf spreads horizontally, tapering gradually to the long, subulate, channelled point 3–4 mm. long ; the nerve wide but rather indistinct, present nearly to the apex, the margin slightly denticulated by protruding cells, the lamina one cell thick, and of four to six rows of long linear cells, usually rectangular at their ends, and, while vigorous, containing numerous chloroplasts in one or

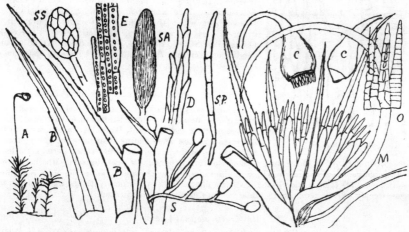

Leptobryum pyriforme.

two lines or adherent to the sides of the cell. In some conditions the lower leaves produce from their axils strong, papillose, red protenema, simple or forked, bearing large, oval, red, terminal or lateral gemmae, from which new stems arise, or these gemmae are sometimes axillary and shortly stalked or almost sessile, and probably these are resting tubers for winter or for dry times as they are not always present. " Barren plants with terminal flowers often occur, having the appearance of male plants, but they contain abortive archegonia mixed with the antheridia " (Dixon). Hence the plant is synoecious or imperfectly dioecious. Seta slender, 1–3 cm. long, often red ; capsule suberect, inclined or pendulous, pear-shaped with reduced neck, small, delicate. Per. teeth strong, red, trabeculate, and papillose, membrane fairly high, processes perforated, cilia slender, with or without appendages. A very delicate moss, growing on moist soil or brick-work, usually present in conservatories but not confined to these, though it is probable that its distribution is at least sometimes effected along with plants in pots. In some sites it is, however, certainly indigenous.

Cape, E. Sandili, Uitenhage, Sim 9028 ; George, 1916, H. A. Wager 570.

Natal. Maritzburg, April 1917, Sim 8676 and 10,205 ; Cannibal Caves, Oliver's Hoek Valley, 6000 feet, Aug. 1922, Miss Edwards.

Genus 147. ANOMOBRYUM Schimp.

Dioecious, in flat extensive carpets on banks, or hanging laxly in waterfalls ; stems slender, ½–2 inches long, usually simple ; leaves ovate, shining, closely appressed, julaceous ; nerve percurrent ; cells narrow, upper cells almost linear ; capsule horizontal or ultimately hanging ; peristome as in *Bryum* ; lid bell-shaped with a short beak ; ring wide, dehiscent.

498. *Anomobryum promontorii* (C. M.), Dixon, in Trans. R. Soc. S. Afr., viii, 201.

=*Bryum promontorii* C. M., in Hedw., 1899, p. 69.
=*Anomobryum procerrimum*, Rehm. 540 ; Dixon, Torr. Bot. Club, xliii, 68 ; Broth., Pfl., i, 3, 563 (as C. M., in error).
=*Bryum *procerramium* Par., Ind., p. 206 (1894).
=*Mielichhoferia *procerrima*, Rehm. 214 (not 219 as in Paris).
=*Mielichhoferia *procumbens* Rehm. (in error in Par., Ind., 2nd ed., under *B. promontorii*.)

Stems filiform, slender, julaceous, shining, pale green, ½–2 inches long, simple or branched below the inflorescence, growing in dense wet tufts, seldom fertile. Dioecious. Leaves closely imbricated whether dry or moist, ovate boat-shaped, slightly apiculate ; the strong yellowish nerve ending in the mucro ; margin almost entire or some cells projecting slightly. Cells very varied, the basal cells rectangular, $40 \times 15\mu$, sub-basal longer and irregularly hexagonal, upper cells linear-vermicular, $50–80 \times 8\mu$, all with thick walls.   Inflorescence terminal with longer, narrower, subulate, nerveless bracts, and soon lateral (and nearly basal) through one or two innovations just below it.   Male flowers gemmiform ; seta ½–1 inch long, pale red, cernuous ; capsule horizontal or subpendulous, brown, oblong with a short neck while the lid remains on, afterwards clavate and wide-mouthed ; lid shortly pointed.   Peristome strong, the outer teeth red below, hyaline above, erect, trabeculate, and when dry sometimes reflexed and pectinate inside ; the inner about equally long, very weak, on thin deep membrane, and cilia sometimes present.   The inner peristome and cilia are often difficult to find.   Dixon says (Torrey. Bot. Club, xliii, 68), " very near the northern *A. filiforme* (Dicks.) Husn., but which may perhaps be separable on the ground of its slightly pointed leaves and less highly appendiculate cilia."   Further comparison may prove these identical, in which case the latter name has precedence.   Very similar in general appearance to *Aongstroemia julacea* from which the entire leaves easily separate it.   A very common moss on wet banks, waterfalls, etc., in the eastern area, Orange Free State, Transvaal, and Rhodesia, extending under these conditions even into dry country, but I have no specimen of it from the Western Cape Province.

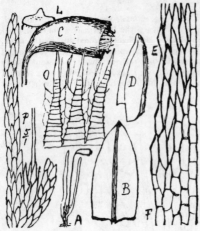

Anomobryum promontorii.

Genus 148. MNIOBRYUM W. P. Schimp.

Plants of various size, habit, and colour, usually in tufts or gregarious ; stem erect, rhizoid-tomentose, with erect basal shoots. Leaves erect or erecto-patent, lanceolate, acute, without border and not serrated ; the nerve ending at or near the point ; leaf-cells large and lax, rhomboid-hexagonal, thin ; inner perichaetal leaves small. Seta long, twisted when dry, curved at the top ; capsule more or less pendulous, pear-shaped or urn-shaped, with a wide mouth and rather large, rounded-acuminate lid ; ring absent, peristome crowded, double, the outer and inner equally long ; outer teeth lanceolate from a wide base, faintly papillose ; inner peristome with the basal membrane half its length ; processes deeply split or punctured ; ciliae two to three, somewhat jointed.

499. *Mniobryum albicans* Limpr.

Dr. Shaw lists *Bryum Wahlenbergii* Schw. (= *Webera albicans* Schimp. = *Bryum albicans* Wahl. = *Pohlia albicans* Lindb. = *Mniobryum albicans* Limpr.) from mountains near Middelburg, Cape Province, 7000 feet (MacLea), a pale glaucous-green, tufted moss resembling *Webera cruda* in habit but having short wide cells as compared with any *Webera* as here understood.   I have strong doubts as to the presence of this European species, but have lately heard from Mr. Dixon, who has examined *Bryum *amplirete*, Rehm. 249, in London, that it is a *Mniobryum*, sterile, very near to *M. albicans*, but with subobtuse leaves and entire margins, having oblong-lingulate leaves ; nerve stout at the base, narrowing upwards and ceasing just below the apex.   As Rehmann's species is undescribed and the name preoccupied both under *Bryum* and *Mniobryum*, I leave it under *albicans* for the present.

Concerning *Webera cruda* Schw., see under *Webera depauperata*.

Genus 149. BRYUM Dill.

Large or small, often densely tufted, and apt to become more so through the innovations being subterminal under the inflorescence.   Leaves numerous, ovate or ovate-acute, or boat-shaped, often concave and closely imbricated, sometimes blunt, but more usually with the nerve excurrent and sometimes with a long hyaline point.   Cells smooth, large and lax, more or less rhomboid-hexagonal, seldom linear, except a marginal line of narrow cells frequently present.   Dry leaves often twisted round the stem.   Capsule inclined or pendulous, pear-shaped or clavate, or sometimes oblong ; the calyptra cucullate, caducous ; the lid conical ; ring dehiscent ; peristome double, the outer teeth sixteen, undivided, lanceolate from a wide base, closely barred below, and also sometimes on the inner side ; inner peristome equally long, the basal membrane being about half that height, and plaited, adhering sometimes to the outer teeth ; the processes like the outer teeth in form, but keeled and often perforated along the keel, and often with two to four slender thread-like cilia of same length or shorter between them and opposite the outer teeth, these cilia bearing short cross-bar appendages.   A very large genus, containing many types, but all having a general resemblance. These are variously distributed into many subgenera or sections.

Synopsis of Bryum.

1. Leaves julaceous, upper half of leaf hyaline (§ *Argyrobryum*)   500. *B. argenteum.*
1. Leaves not julaceous nor hyaline   .   .   .   .   2.
2. Small ; capsules blood-red (§ *Erythrocarpidium*)   .   .   501. *B. erythrocarpum.*
2. Small or large, capsules green, brown, or red (subgenus *Eubryum*, which includes also the two previous species) 3.
3. Stems equally leaved, leaves decurrent   .   .   .   502. *B. decurrens.*
3. Stems equally leaved, leaves not decurrent   .   .   .   4.
3. Stems with comal tufts, or equally leaved, or both   .   .   9.
4. Leaves many, imbricate, suberect, acutely nerve-pointed (§ *Doliolidium*)   .   .   5.

4. Leaves scattered, equally distributed, or stem nude below.   Leaves straight when dry   .   .   .   6.
5. Nerve shortly excurrent, margin erect   .   .   .   .   .   503. *B. condensatum.*
5. Nerve far excurrent as a denticulate bristle ; margin often revolute   .   .   .   504. *B. rigidicuspis.*
6. Margin plane, more or less bordered (§ *Pseudotriquetra*)   .   .   .   .   .   .   7.
6. Margin revolute, not bordered (§ *Alpiniformia*)   .   .   .   .   .   .   8.
7. Nerve excurrent   .   .   .   .   505. *B. aulacomnioides.*
7. Nerve not excurrent   .   .   .   .   .   505β. *B. aulacomnoides,* var. *limbatum.*
8. Leaves widely ovate, acuminate, forming comal tuft, nerve slightly excurrent   .   506. *B. subcavifolium.*
8. Leaf-apex acute or nerve shortly excurrent, leaves not tufted   .   .   .   507. *B. alpinum.*
8. Leaf-apex rounded, cucullate, without mucro   .   .   .   .   508. *B. Muhlenbeckii.*
9. Leaves rigid, inflexed when dry, but not flexuose nor shrinking (? *Caespiti bryum*)   .   509. *B. radicale.*
9. Fertile stem shorter than the sterile ; leaves when dry sometimes twisted (§ *Trichophora*) .   .   10.
9. Larger ; leaves in terminal rosules, dry leaves usually twisted (§ *Rosulata*)   .   .   13.
10. Leaves entire, almost without border.   Nerve not excurrent   .   510. *B. brachymeniaceum.*
10. Leaves denticulate, bordered, and with long hair-point   .   .   .   .   11.
11. Leaves densely packed, straight when dry   .   .   .   .   511. *B. aterrimum.*
11. Leaves more or less twisted when dry   .   .   .   .   .   12.
12. Monoecious, leaves wide below   .   .   .   .   .   512. *B. torquescens.*
12. Dioecious ; leaves narrowed below and concave, border of several rows of cells   .   513. *B. capillare.*
12. Fertile stem sessile, rosulate ; leaves with narrow coloured border and incurved margins   514. *B. acuminatum.*
13. Rosule 5–6 mm. across ; border about four rows of cells   .   .   .   515. *B. pumilo-roseum.*
13. Rosule 6–8 mm. across, or more   .   .   .   .   .   14.
14. Hair-point short, reflexed ; leaves scarcely bordered, margin plane   .   .   516. *B. canariensiforme.*
14. Hair-point strong, long   .   .   .   .   .   .   15.
15. Leaves twisted when dry ; border about four rows of cells   .   .   .   517. *B. truncorum.*
15. Leaves erect when dry ; border of one to two rows of cells ; teeth spinose   .   518. *B. Mundii.*
15. Margin denticulate from one row of cells, hardly bordered, often strongly recurved   .   519. *B. canariense.*
15. Leaves entire, awned, almost without border   .   .   .   .   520. *B. polytrichoideum.*

Concerning the following names, the following notes are all that I know about them, viz. :—

Bryum **leptotrichaceum,* Rehm. 256, and *Bryum* **bartramioides,* Rehm. 257, are both represented in South African herbaria and in Kew Herbarium by *Dicranella Symonsii* Dixon, so that is evidently the plant he intended.

Bryum humidulum Sull., in Proc. Amer. Acad., 1859, p. 278, is there given a most meagre description, which might apply to half the species of *Eubryum.*

Bryum *capillarioides C. M., in Breutel Musc. Cap., is not described and is not represented in South African herbaria.

500. *Bryum argenteum* Linn., Sp. Pl., p. 1120 ; Brid., Bry. Univ., p. 657 ; C. M., Syn., i, 314 ; W. P. Sch.,
Syn., 2nd ed., p. 448 ; Dixon, 3rd ed., p. 375, and Tab. 45 L ; and S.A. Journ. Sc., xviii, 320.

=*Argyrobryum argenteum,* Kindb. Laubm. Schw. and Norw., p. 78.
=*Bryum capensi-argenteum* C. M., in Hedw., 1899, p. 67.

Dioecious, low, densely matted, often covering large areas, usually on soil or gravel, sometimes on bark, stones, or cement ; the upper half of each leaf without chlorophyll, and consequently the mat or cushion is silvery white. Stems

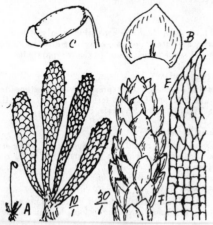

Bryum argenteum.

Bryum argenteum, var. *australe.*

2–5 mm. long, closely packed, shining, usually branched at the base below the nearly sessile inflorescence, julaceous, terete, usually club-shaped with scale-leaves below and larger cochleate leaves above ; leaves ½ mm. long, as wide or wider than long, clasping, very concave, imbricate, with usually slightly mucronate apex ; nerve very short, or up to half the leaf length, weak, often red ; margin entire, erect ; cells hyaline and very laxly hexagonal toward the leaf-apex ; much smaller, quadrate and often chlorophyllose or coloured toward the leaf-base ; seta almost basal, ¼ inch high, scarlet ; capsule very small, pendulous or nearly so, red, oval, almost without neck, with raised short lid ; when dry the capsule contracts below the mouth ; ring wide, dehiscent ; peristome with sharp outer teeth, red below ; inner teeth and cilia pale. Sessile axillary dehiscent buds sometimes present. Often not fertile, but capsules usually abundant where present. The type (as above and as per figure on page 327) is much like *Anomobryum promontorii*, but the position and form of the capsule differs.

Var. australe (Rehm.) Sim (new variety).

=B. oranicum C. M., in Hedw., 1899, p. 68.

Rather larger ; leaves shortly acuminate or pilose, the nerve usually coloured below and short, not extending into the leaf-point. Sessile axillary dehiscent buds often present. Kadziberg, O.F.S., etc. Dixon remarks (Kew Bull., 6, 1923) that the same form occurs in India.

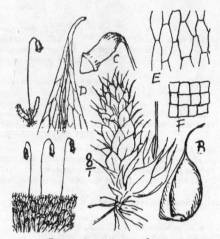

Bryum argenteum, var. lanatum.

Bryum argenteum, var. proliferum and var. viride.

Var. lanatum B. and S., Br. Eur. ; C. M., Syn., i, 314 ; W. P. Sch , Syn., 2nd ed., p. 448.

=Bryum argenteum, var. argyrotrichum Mitt., in Journ. Linn. Soc., 1886, p. 307.

=Bryum argyrotrichum (Mitt.), C. M., in Linn., 1875, p. 385.

=Bryum squarripilum, C. M., in Flora, 1886, p. 280.

=Bryum lanatum, Brid., Sp. M., iii, 20.

=Bryum argenteum Linn., var. incanum.

Stronger ; leaves regularly hair-pointed, the hair often as long as the leaf and consisting mostly of hyaline excurrent nerve ; perichaetal leaves longer, narrower, and more laxly areolated ; all leaves tapering into the hair-point. Not unlike *Brachymenium pulchrum*, but the pendulous capsule and acute peristome teeth easily separate it.

Var. rotundifolium Sim (new variety).

A very marked lax form, rather larger than the type, but having rounded, rather spreading leaves without mucro, but with strong percurrent red nerve, and with no quadrate basal cells ; not known fertile or proliferous. Giant's Castle, Natal, Sim 10,209, 10,210 ; Rosetta Farm, Sim 10,232 ; Victoria Falls, Rhodesia, Sim 8554.

Var. proliferum Sim (new variety).

Stems ½–1 inch high in close cushions, leaves closely julaceous, nearly round, hardly apiculate ; nerve ceasing below the apex ; stalked axillary buds abundant in summer, and deciduous, by which it is abundant in every cement water-way in every town ; perennial, and never fertile, Sim. (See also var. *viride*.)

Var. viride Sim (new variety).

Exactly like var. *proliferum* in habit and in propagation by deciduous axillary stalked buds, and always mixed among it as a green dimorphic form ; differs in having leaves ovate, more spreading, coloured green throughout, and with nerve percurrent ; not known fertile.

The cosmopolitan species *B. argenteum* is extremely variable, each variety named above is the limit of a South African variation type, but the intermediate grades connect them all up, and the forms cannot be separated specifically, though each form is permanent and they are extremely unlike in their extremes.

All these forms are common throughout South Africa, Bechuanaland, and S. Rhodesia except vars. *proliferum* and *viride*, which are mostly confined to cement-work in the towns, and var. *rotundifolium*, mostly on the mountains. The plant is much more common than its capsules, and some of its forms (e.g. *proliferum* and *riride*) have been watched for twenty-five years without fruiting.

Bryum *tenerrimum, Rehm. 251, is not represented in South African herbaria, but in the British Museum it "is a composite species; *Bryum argenteum* c. fr. mixed with a sterile, scrappy species, quite unworthy of notice" (Dixon).

501. *Bryum erythrocarpum* Schwaeg.

=*Bryum sanguineum* Brid.

Section *Erythrocarpidium* is a widely distributed group of small caespitose mosses with equally leaved innovations; the leaves usually decurrent, lanceolate, with incurved margins slightly denticulate above, and with red and pointed nerve. Capsule oblong-clavate, bright red, on a red seta. Somewhat similar to § *Doliolidium*, which, however, has acorn-like fruits and usually revolute leaf-margins.

In dealing with the three British species, Mr. Dixon says (2nd ed., p. 372) that they "form a distinct and natural group, characterised by their blood-red capsules and small leaves of fairly uniform type, acuminate, with the nerve shortly excurrent, and with small, rather narrow cells."

Brotherus (Pfl., x, 394) includes in this group *B. zuluense* Br. and Br., *B. Macleanum* C. M.. and *B. laxogemmaceum* C. M., as well as *B. subdecursivum* C. M. (in Rehm. 254 and Hedw., 1899, p. 74), which he wrongly ascribes to Porto Rico; the latter is only known sterile, and C. M. placed it in § *Erythrocarpidium* when describing it.

Shaw lists *B. atropurpureum* Wahl. from Graaff Reinet (MacLea), and *B. erythrocarpum* Schw., "On rocks, Bains Kloof; Graaff Reinet (MacLea)," which records are possibly of this species, or they may be *B. condensatum* C. M.; but there are no South African specimens named *B. erythrocarpum* at Kew, and neither that nor *atropurpureum* from South Africa in the British Museum. For Kew specimens of *B. atropurpureum*, see under *B. condensatum*.

Brotherus states that the Australian species are hardly separable specifically from *B. erythrocarpum*, and as I have seen only one unripe specimen of the group (Sim 10,220), and those described are mostly from few and imperfect specimens, I doubt whether all or any of them can be maintained even as subspecies to *B. erythrocarpum*. The species of this group described as South African are:

1. *B. zuluense* Br. and Bry. (Videns. Foch., 1911, iv, 14); heteroecious, stems 2 mm. high, with red papillose radicels, leafy, and with one to two innovations; leaves erecto-patent, appressed when dry, not decurrent, concave-keeled, lanceolate mucronate, without border. Comal leaves 2·5 mm. long, 0·6 mm. wide, with entire, narrowly recurved margins; lower leaves shorter and wider, from mid-leaf often serrulate or serrate. Nerve strong, yellowish, with a straight smooth mucro. Cells rhomboid or roundly hexagonal, gradually linear-rhomboid near the margin, basal cells roundly rectangular, red with age. Per. leaves smaller. Seta 10–25 mm. high, purple; capsule pendulous, 3 mm. long, red, pyriform-clavate with long curved neck, mouth of transverse cells. Teeth purple below, pale above, inner of shorter lanceolate processes, two intermediate ciliae and low membrane. Lid shortly conical-mamillate, shining; spores 0·2 mm., yellow, punctate. Eshowe, Zululand, Jan. 1909, H. Bryhn.

2. *B. laxogemmaceum* C. M., in Hedw., 1899, p. 75; dioecious, fertile stems very short and few-leaved, leaves oblong-acuminate from wide base, margin erect, nerve thick, yellow, ending suddenly at the leaf-apex; cells large, rhomboid, per. leaves larger, seta long, capsule reddish, minute, cernuous, oblong with short neck, lid conical. Esternek, Knysna, Rehm. 253; Hilton Road, Natal, Sim 10,220.

Rehmann's 253 is not represented in South African Herbaria, and at Kew and the British Museum only *Funaria hygrometrica* is present.

3. *B. Macleanum* C. M., in Hedw., 1899, p. 75; leaves narrowly oblong-ovate-acuminate, purplish; nerve narrow, purple, ending in or below the rather aristate apex; cells large, rhomboid; basal cells long-rectangular; leaves laxly and narrowly bordered, margin erect; seta purple; capsule purple, drooping, small, ovate-oblong from long neck; teeth red, processes pale, cilia in twos. Lydenburg, Transvaal, 1899, MacLea.

4. *B. subdecursivum* C. M., in Hedw., 1899, p. 74; sterile, buried in sand, with many globose, axillary innovations, and ovate-acuminate, concave, mucronate, entire leaves. Cape Town, Rehm. 254.

This has been dealt with under *Brachymenium dicranioides*, but seems to me to belong to *B. argenteum* L.

Wager 155 was distributed under the name *B. murale* Wils., but that species, which differs from *B. erythrocarpum* mostly in having its cells half as wide (say 5µ), is not known to be South African, and the specimen issued was widely different, viz. *B. canariensiforme* Dixon.

502. *Bryum decurrens* C. M., in Bot. Zeit., 1855, p. 751 (sterile, Cape), is unknown to me. Brotherus (Pfl., x, 386) includes it as *probably* belonging, as under, to his § *Eubryum*, *Leucodontium*, "stems equally leaved; leaves somewhat decurrent, almost entire; capsule pear-shaped, with neck," and his key gives, "leaves sharply pointed, bordered; nerve prominent or excurrent; dioecious; capsule not curved, regular, when dry almost top-shaped," and as the specimens are sterile this does not lead far. He does not seem to have seen it, and no specimens are available in London or in South African Herbaria.

503. *Bryum condensatum* (Hampe MSS.), C. M., in Bot. Zeit., 1858; Dixon, in S.A. Journ. Sc., xviii, 320; Broth., Pfl., x, 392 (and see C. M., in Linnaea, 1875, p. 388).

Synoecious; plant small, densely tufted on soil, green or grey; the fertile stem very short, red, subtended by several erect red stems 2–5 mm. high, more or less clavate, julaceous when dry; leaves suberect, closely imbricate, ovate-lanceolate, very concave, slightly decurrent; the nerve usually coloured, very strong and excurrent as a short or long subula; margin entire, erect or incurved, without margin; cells irregularly hexagonal, 30×20µ, firmly walled; the lowest cells quadrate, 25μ wide. Perichaetal leaves rather larger, more laxly reticulated, less pointed, strongly nerved. Seta 1–1·5 cm. high, red, with small, pendulous, oval, red capsule, wrinkled at the base by the neck which is as wide as the capsule and passes abruptly into the seta. Sometimes not unlike *B. argenteum* when in white condition, sometimes green and much more comose. See further under *B. rigidicuspis.* The Kew specimens named *B. atropurpureum* are all *B. condensatum* (Hpe.) in good condition and fruit, leg. Menzies and leg. Zeyher.

Cape, W. Sea Point, Table Mountain, Camps Bay, Kloof Nek, Sim 9315, 9298, 9321, 9558; Stellenbosch, Dr. Burtt-Davy, Miss Duthie; Kenilworth, M. Duncan; Valkens Ravine, Dr. Bews.

Transvaal. Müller's Farm, MacLea (Rehm. 563, as *Bryum*, No. 5).

S. Rhodesia. Zimbabwe, Sim 8763; Victoria Falls, Wager 909; Matopos, Eyles 938; Umtali, Eyles 1742β.

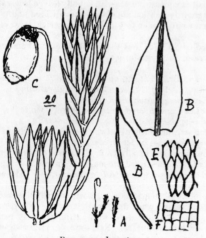

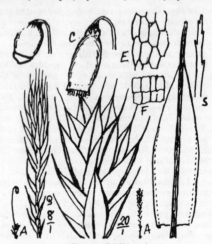

Bryum condensatum.

Bryum rigidicuspis.
S, apex of bristle.

504. *Bryum rigidicuspis* Dixon, S.A. Journ. Sc., xviii, 320; Broth., Pfl., xi, 530.

=*Bryum* *bulbilliferum, Rehm. 252 (=*B. bulbiferum* Rehm., *fide* Wager's list).

Closely packed, dioecious, dark green or brown; stems 2–15 mm. long, erect, rather rigid; the leaves numerous above, fewer below, erecto-patent when moist, when dry appressed above, incurved below; widely ovate, so concave as to appear lanceolate, somewhat decurrent; margin entire or almost entire, often revolute below; nerve very strong, not less above, usually coloured, excurrent as a strong yellowish bristle, denticulate at the apex; cells rhomboid-hexagonal, lax, 20μ long, the marginal cells narrower but hardly making a border; the basal cells tumid and quadrate. Perichaetal leaves hardly different, but fertile plant shorter. Seta ½ inch long, slender, red; capsule red, pendulous, 1 mm. long, oval, with short narrowed neck suddenly ceasing. Peristome yellowish, the base orange-red; lid conical, ring wide, of very narrow cells *Bryum coronatum* Schw., i, ii, 103 (C. M., Syn., i, 307), is the widely distributed tropical type of this group and comes very near *B. condensatum* C. M., from which the synoecious inflorescence and comose habit of that species distinguish it; *B. rigidicuspis* is known from both by the longer and denticulate bristle, hardly tapering to the point and usually coloured to the apex, and by the revolute margin and rather indistinct border and narrowed neck.

Brotherus (Pfl., x, 392) connects *B. horridulum* C. M. (Hedw., 1899, p. 67) from Mopea, Zambesia, with *B. coronatum*, though the leaves are described as long linear-acuminate, and as he omits both *B. rigidicuspis* and *B. condensatum* he probably includes both of these there, though they have differences. His illustration of *B. coronatum* Schw. (Pfl., x, 392) is very near to all, and all appear to be subspecies to it. Indeed Bryhn (p. 16) uses the name *B. coronatum* for specimens from Eshowe, Zululand.

Cape, W. Kloof Nek, Disa Ravine, Platteklip Stream, Sim 9565, 9147, 9392; Cape Town, Rehm. 252 (as *B.* *bulbilliferum Rehm.).

Cape, E. Driftsands, Port Elizabeth, Sim 9064.

Natal. Van Reenen Pass, Wager 74.

S. Rhodesia. Zimbabwe, Khami Ruins, Sim 8790, 8839, and 8867; Matopos, fertile, Eyles 938.

505. *Bryum aulacomnioides* C. M., in Hedw., 1899, p. 72.

=*B.* *afroturbinatum, Rehm. 247 and 247*b*.

Probably =*B. austroventricosum* C. M. (of Madagascar).

Dioecious; stems erect, often semi-aquatic, laxly tufted or gregarious, 2 inches high, simple or branched below an old inflorescence; the axis purplish red, equally furnished with leaves along the whole length; the leaves light green or yellowish, distant, spreading when moist, crisped when dry; the upper leaves meeting as an acute stem-point; leaves clasping with decurrent base, widely lanceolate or ovate-lanceolate, concave (or often flat when dry), with excurrent or much-excurrent point, which is slightly toothed at the apex, other parts of the margin not toothed, hardly bordered or narrowly bordered, and often more or less reflexed or revolute in the lower half ; nerve wide, deep red, and extending to the apex in the lower leaves, and forming the usually green excurrent point of the upper leaves. Cells hexagonal, lax, $80 \times 30\mu$, narrower toward the margin, where a line of long narrow cells $150 \times 10\mu$ form the border, or the border may be almost absent ; basal cells laxly quadrate or oblong, the outer narrower. Lower leaves and lowest leaves on innovations often wider and only percurrent, with wide, somewhat denticulate apex, and when old are persistent, scarious, and more flat and red. Per. leaves like stem leaves, with strongly excurrent nerve. Seta red, $1\frac{1}{2}$–2 inches long, with pendulous cylindric-pyriform capsule 3 mm. long, having curved narrower neck.

The scattered leaves, crisped when dry, on red axis, and long seta mark the plant, but it varies much in respect to wideness and position of margin, and also in respect to leaf-form, sometimes wide and deeply keeled, at other times lanceolate and nearly flat, while some forms are twisted or convolute and wide-bordered.

Usually in large open flat tufts on flat wet rocks ; sometimes more or less scattered on wet soil, in river beds which are irregular in flow, and hence they are sometimes aquatic for months and at other times dry for months, and this accounts for the variations recorded.

Müller mentions its close resemblance to the European and Australasian *B. bimum* Schreb., which again is a synoecious sub-species of the European dioecious *B. pseudotriquetrum* Schw. (*B. ventricosum* Dicks.), and Brotherus groups all these in his § *Pseudo-triquetra.* Further field work is necessary to decide whether *B. aulacomnioides* is distinct from these, or represents a local group of forms. Its habit and foliation are those of *Webera*, but the wide hexagonal cells are those of *Bryum.*

What was distributed as Rehm. 568, named "*Bryum* Sp., No. 10," from Lechlaba, Transvaal, was marked at Kew by Brotherus, "*B. *lechlabense*, n. sp.," but differs only in the nerve being percurrent or shortly excurrent, and is separable only as a form. I have that form also from Mont aux Sources and Giant's Castle, Sim 10,211 and 10,212.

Bryum *porphyroloma C. M., in Rehm. 250, from Devil's Peak, Cape Town, contains fragments mostly of *B. aulacomnioides* ; and Rehm. 250β from Kadziberg, O.F.S., is not represented in South African Herbaria. *B. aulacomnioides* is frequent on the high mountains, rare lower.

Cape, W. Devil's Peak (as above).

Cape, E. On shaded rocks, Kentani, 1200 feet, May 1907, Miss Alice Pegler ; Eastern Province, MacLea, Rehm. 567 ; Kokstad, in water, R. A. D. Magg.

Natal. Mont aux Sources, Sim 10,233, 10,234 ; Marshes, Hilton Road, Sim 8187 ; Giant's Castle, R. E. Symons.

Transvaal. Lechlaba (as above), Rehm. 568 ; Heidelberg, J. Gibbs.

O.F.S. Witteberg, above Kadziberg, Rehm. 247 ; Bethlehem, Rehm. 247 (both as *B. afroturbinatum* Rehm.).

Bryum aulacomnioides.

Var. *limbatum* Sim (new var.).*

Growing in fairly compact, saturated tufts $\frac{1}{2}$–1 inch high, brown below, green above ; stems erect, the axis firm, often red ; leaves scattered, ovate-lanceolate, 2 mm. long, 0·6 mm. wide, from a narrower, clasping, but not decurrent base ; subacute, concave below ; nerve strong, green or coloured, uniting in the point with the wide firm border of four or more rows of narrow cells which extend round the leaf ; margin entire (or at the apex almost so), often shortly reflexed toward the base. Cells $50 \times 20\mu$; basal cells rather larger and rectangular, but all enclosed in the border of long narrow cells. When dry the leaves are more conduplicate, much undulated, and apparently more acute. A short dense form of *B. aulacomnioides.* Other parts not seen.

Cape, W. Cape Town, Sim 9534.

Natal. Rosetta Farm, 4000 feet ; Allerthorpe, Dargle Road, 3500 feet ; Edgehill, Mooi River, 4000 feet ; Sim 10,235, 10,236, 10,237 ; Little Berg, Cathkin, 6000 feet, Miss Owen 40 ; Giant's Castle, 8000 feet, R. E. Symons ; Impolweni Bridge, Giant's

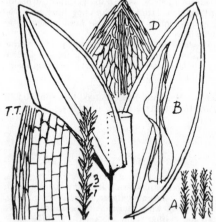

Bryum aulacomnioides, var. *limbatum.*

Castle, Mont aux Sources, Sim 10,247, 10,245, 10,246.

Transvaal. Rosehaugh, 3000 feet, Dec. 1914, Sim 8031.

---

* *Bryum aulacomnioides* var. *limbatum* Sim (var. nov.).

Forma brevis densa. Folia sparsa, ovato-lanceolata, 2 mm. longa, connexa, haud decurrentia, fortiter marginata, sub-acuta, nervo forti margineque in apicem concurrentibus.

506. *Bryum* (§ *Alpiniformia*) *subcavifolium* (C. M.; Par.) Dixon (new species).*

=*Bryum* *cavifolium C. M., in Rehm. 255, altered by Paris to *B.* *subcavifolium in view of the American *B. cavifolium* Schimp.

No specimen exists in South African Herbaria, but Mr. Dixon has kindly sent the following particulars from London : " Sterile, in dense tufts, prettily variegated with red (§ *Alpiniformia*, I think, certainly), but leaves widely ovate, acuminate, very little altered when dry, only appressed in a dense comal tuft. Leaves very concave, margin recurved, quite entire ; nerve rather strong, terete, often red, *slightly excurrent*. Cells small, with firm walls, basal, all rather small, hexagonal-rectangular, and short or subquadrate at angles, upper narrowly hexagonal-rhomboid, and often subsinuose, *gradually* narrowed to margin and there linear but not incrassate ; hence no differentiated border ; all with the primordial utricle very distinct."

Camps Bay, Cape Peninsula, Rehm. 255, as *B. cavifolium* Rehm.

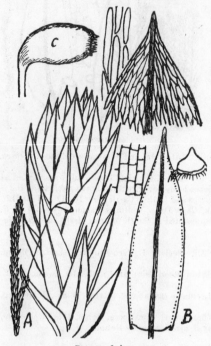

Bryum alpinum.

507. *Bryum alpinum* Huds., Fl. Angl., p. 415 ; C. M., Syn., i, 285; Dixon, 3rd ed., p. 372, and Tab. 45G, and in Smithson. Misc. Coll., 69, 2, 2 ; Mitt., Linn. Soc. Journ., xx, 146, 307.

=*Bryum Wilmsii* C. M., in Hedw., 1899, p. 74.
=*Bryum afro-alpinum* C. M., in Hedw., 1899, p. 73, sterile as described, not as distributed by Rehmann in Rehm. 248.

A mountain moss, growing usually on moist flat rocks ; dioecious ; metallic crimson-red, shining ; tufts fairly compact, 1–2 inches deep, flat-cushioned ; stems 1–2 inches long, erect, mostly simple, with the seta terminal on a short lower branch. Leaves oblong-lanceolate, 2 mm. long, usually bright red, occasionally olive green ; when dry erecto-patent, straight, imbricated, and hardly shrunk ; when moist spreading more, concave, with entire margin reflexed from base to arista. Nerve strong, decurrent, excurrent (in the South African specimens), deep crimson, ending in a short stiff arista ; cells incrassate, coloured, linear-rhomboid, $50 \times 8\mu$, the outer longer up to $120 \times 8\mu$ but not forming a border ; basal cells quadrate, $12 \times 12\mu$, gradually longer upward, the outer line usually smaller so far as basal. Seta crimson, 1 inch high ; capsule crimson, horizontal or nearly so, 5 mm. long, oval, pyriform ; lid convex-mamillate. Peristome teeth short, the outer pale, separate, erect, or with incurved pectinate apex ; membrane fairly high ; processes and cilia stand above the incurved teeth when dry but are often absent.

The African plant with strong arista is *B. alpinum*, var. *viride* Husn., and was separated by C. M. as *B. Wilmsii* C. M.; it occurs at high altitudes throughout Africa, and does not differ from the *B. alpinum* of Europe, though a more obtuse form with shorter nerve is more common there.

Bryum transvaalo-alpinum C. M., in Hedw., 1899, p. 73 (=*Philonotis*, Broth., Pfl., i, 3, 598), from between Middelburg and Lydenburg, Transvaal, may belong here, with short nerve ; without seeing the type I see no reason to think it a *Philonotis* or a distinct *Bryum* (see under *Philonotis*).

Cape. E. Dohne Hill, 1898, 5000 feet, Sim 7212.

Natal. Mistyhome, Hemuhemu, 5000 feet ; Zwart Kop, 4000 feet ; Mont aux Sources, 7000 feet, Sim 10,238, 10,239, 10,240. (Rehm. 248c is from Inchanga. I have not seen it, and the locality is at low altitude.)

O.F.S. Kadziberg, Rehm. 248β.

Transvaal. Spitzberg, 1887, Dr. Wilms (as *B. Wilmsii* C. M.).

Also from Mount Elgin, Uganda.

508. *Bryum Muhlenbeckii* Br. and Sch. ; Dixon, 3rd ed., p. 373, and Tab. 45 I ; C. M., Syn., i, 285.

A common, dioecious, mountain moss, growing on flat moist rocks ; similar to *B. alpinum* but more vigorous, usually olive-green coloured or with the tips purplish ; leaves more drawn together in a terminal point ; stems more or less tomentose and seldom fertile ; lower leaves somewhat lax, short, and ovate-acute ; upper leaves 2–3 mm. long, imbricate, concave-cucullate, when dry imbricate, straight, and not shrivelled, when moist more spreading. Nerve strong, coloured, incurved near the point (making the leaf cucullate), and ceasing below the apex. Tufts often in

---

* *Bryum subcavifolium* (C. M. ; Par.) Dixon (sp. nov.).

(§ *Alpiniformia*) Folia late ovata, acuminata, minimum mutata ubi sicca, tantum adpressa in cumulo denso comali. Folia admodum concava, margo integer, recurvatus ; nervus fortis, teres, saepe ruber, tenuiter excurrens ; cellae parvae, anguste hexagono-rhomboideae, saepe sub-sinuosae, sensim attenuatae ad marginem ibique lineares at non incrassatae, inde marginem distinctum haud efficientes ; cellae basales parvae, hexagono-rectangulares, breves vel sub-quadratae ad angulos.

horizontal layers ; up to a certain level stems many-leaved in many rows, above that level stems are more lax, leaves fewer in about five rows, drawn to a point at the apex. Leaves oblong or elliptic, rounded at the almost entire and often slightly mucronate apex ; margin revolute to the shoulder ; cells incrassate ; basal cells quadrate, 50×12 ; lamina cells rhomboid, 50×12 ; marginal cells rather longer and narrower, 70×10, and the apex ending in short, variable, and irregular incrassate cells. Capsule oblong-pyriform, horizontal, brown. Common on the Drakensberg.

Bryum *prionotes Shaw (on rocks on the highest point of the Katberg) probably belongs here.

Natal. Giant's Castle, 8000 feet, R. E. Symons (Sim 8656) ; Upper Bushman's River, 7000–8000 feet, Sim 8655 ; Tugela Valley and Gorge, Mont aux Sources, 6000 feet, Professor Bews (Sim 8657), and 7000 feet, Sim 10,241 ; Miss Edwards ; Oliver's Hoek, Sim 10,242.

O.F.S. Kadziberg, Rehm. 248 (as distributed at least in some of the collections).

Concerning *Bryum *laetum* Mitt., Mr. Dixon writes that there are good specimens of this at Kew, from Basutoland (Cooper), Orange Free State (Cooper), and Cape of Good Hope (Miss Bowker). They are green slender forms of *B. Muhlenbeckii*, rather off type, and no two of them exactly alike.

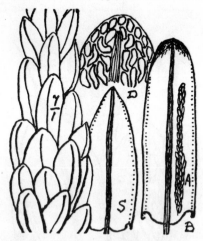

Bryum Muhlenbeckii.
S, subacute lower leaf.

Bryum brachymeniaceum.

509. *Bryum radicale* (Rehm. 245, ined.), Dixon (new species).*

Not represented in South African herbaria, but Mr. Dixon, who has seen it in London, favours with the following particulars : " Tufts dense, reddish brown, 2–3 cm. high. Old capsules brown, subcylindric, with *very slightly* tapering neck, pendulous. Spores small. Outer teeth deep orange, rather narrow, and irregular, highly trabeculate within. Leaves ovate, very acuminate, red at base, not decurrent, *rigid*, inflexed when dry, but not flexuose or shrinking ; setae long ; alar cells and those of insertion *very lax and large*, upper rather firm-walled, rather wide and short, but variable ; marginal in two to three rows, linear, thin-walled, forming a *quite distinct but very narrow border* ; margin quite plane ; nerve *stout, orange-red*, rather longly excurrent. Leaves entire. Not synoecious, probably dioecious." Fruit seen was in poor condition, and peristome only seen in one capsule and very worn (§ *Caespitibryum* ?).

510. *Bryum brachymeniaceum* C. M., in Hedw., 1899, p. 71 ; Broth., Pfl., x, 398.

= *Webera brachymeniacea* C. M., in Rehm. 224 (mixed, see Kew Bull., 1923, p. 6, No. 542, and No. 553).

Dioecious ; caespitose on wet soil ; green to red ; fertile stems very short, subtended by $\frac{1}{4}$–$\frac{1}{2}$-inch sterile stems, often on red axes, with scattered leaves below and somewhat clavate coma in which the leaves are erecto-patent when dry and more spreading when moist. Comal leaves 1·5–2 mm. long, concave, densely imbricate, oblong-acuminate, widest at the base, rounded or subacute, firm, entire ; nerve strong, usually purplish, percurrent or nearly so ; margin entire, erect, or lower half narrowly revolute, one line of narrow cells giving slight border ; lamina cells long-hexagonal, $40–50 \times 15\mu$, the lower larger ; basal cells quadrate, $40 \times 15–20\mu$, per. leaves rather larger, ovate-lanceolate, strongly nerved, almost entire. Seta $\frac{1}{2}$–1 inch long, red, terminal on a short basal branch ; capsule horizontal or pendulous, 3 mm. long including wide tapering neck ; lid convex-mamillate ; ring wide, dehiscent ; peristome yellowish, membrane high ; only imperfect processes seen.

---

* *Bryum radicale* (Rehm. 245) Dixon (sp. nov.).

(§ *Caespitibryum* ?) Cumuli densi, 2–3 cmm. alti ; folia ovata, admodum acuminata, haud decurrentia, rigida, inflexa si sicca, non flexuosa tamen nec crispata ; cellae alares laxissimae ac maximae, cellae superiores latae ac breves, marginales 2–3 ordines lineares, marginem distinctum angustum efficientes ; margo integer, planus ; nervus robustus, excurrens, ruber ; seta longa, capsula sub-cylindrica collo tenuiter decrescente, pendula. Peristom. dentes angusti. Dioecum ?

Cape, W. Rondebosch, Rehm. 224 (Coll., No. 318) ; Newlands, Rehm. 542 and 553.

Natal. Gordon Falls, near Edendale, 3000 feet, Sim 10,243 ; Camperdown, 1500 feet, Sim 10,244 ; Acutts' stream, Rosetta, 4500 feet, Sim 10,213 ; top of Giant's Castle, 8000 feet, R. E. Symons.

511. *Bryum aterrimum* (C. M., in Rehm. 235), Sim (new species).*

Near *B. capillare*, but with short, dense, many-leaved stems 2–4 mm. long, the fertile branch basal and with similar leaves except that the inner are shorter, pointless, and without nerve in the upper half. Stem leaves densely imbricated, somewhat concave, straight when dry, erecto-patent, ovate-lanceolate, firmly made, with reflexed yellowish bristle ; nerve strong, coloured, excurrent in the bristle ; border yellowish, of one or two rows of narrow long cells $(100 \times 20\mu)$ and denticulate near the apex ; lamina cells hexagonal, $60 \times 40\mu$; basal cells quadrate, $35 \times 35\mu$; margin seldom revolute toward the base. Seta 1–2 cm. long, red, with pendulous brown capsule 3 mm. long with short tapering neck. Peristome teeth red, the membrane high, and inner processes and cilia standing above the inflexed teeth when dry.

Cape, W. Between Knysna and Belvedere, on clay, Rehm. 235.

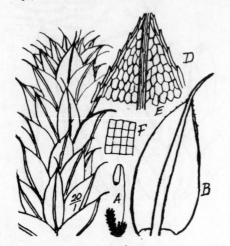

Bryum aterrimum.

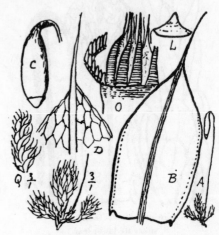

Bryum torquescens.

512. *Bryum torquescens* Br. and Sch. ; Bry. Eur., iv, 49, t. 358 ; C. M., Syn., i, 277 ; W. P. Sch., Syn., 2nd ed., p. 431.

=*B. capillare* L., var. *torquescens* Husn., Dixon, 3rd ed., p. 367.

A small, gregarious or loosely caespitose, yellowish-green, synoecious moss ; stems 2–5 mm. high, the fertile short and the innovations longer and more laxly leaved, with the leaves suberect, and often twisted round the stem when dry ; leaves ovate-acute from a wide base, tapering at apex into an acute bristle one-third to one-half the length of` the leaf ; margin revolute or reflexed from the base to above the middle, the margin above that indistinctly denticulate or subentire, the subula flat at the base, entire and acute ; nerve wide and coloured at the base, narrower and paler upward ; cells lax but thick-walled, hexagonal, $70 \times 35\mu$, and often with the walls coloured, marginal line narrow, others all alike, and basal cells only rather longer. Seta red, 1 inch long ; capsule pendulous, red, cylindrical, 3 mm. long, tapering slightly to the curved neck ; lid convex-apiculate, red ; ring very deep, dehiscent ; peristome pale or white under a lens and forming a pointed dome, but the per. teeth yellowish in the lower half, closely trabeculate (and pectinate inside when dry), the membrane high with long processes and intermediate, long, slender cilia.

This, which is a European species, differs little from the northern *B. capillare* L., except that this is synoecious while *B. capillare* is dioecious, and on this W. P. Schimper (Synopsis, 2nd ed., p. 432) says *re B. torquescens*, "raro flores masculi archegoniis destituti reperiuntur."

Brotherus (Pfl., x, 397) includes in the synoecious group *B. torquescentulum* C. M. (Hedw., 1899, p. 71), and Bryhn records the name from Eshowe ; Brotherus remarks it is not dioecious as stated and emphasised by C. M. in describing it, but if it is not dioecious it is *B. torquescens*, and if dioecious it is *B. capillare* L., as the more or less twisted condition is common to both. The var. *nutans* described by C. M., in Hedw., 1899, p. 71, has no separate character. *B. porphyreothrix* C. M., in Hedw., 1899, p. 70 (=*B. porphyrotrix* C. M., in Rehm. 242) is described as monoecious, with the

---

* *Bryum aterrimum* (C.M. ap. Rehm. 235) Sim (sp. nov.).

Simile *B. capillari*, ramis autem brevibus dense vestitis foliorum arte imbricatorum ovato-lanceolatorum, rectorum ubi sicca ; nervus robustus, ruber, exsertus ut seta sub-flava ; margo supra denticulatus ; margo sub-flavus, 1–2 ordinum cellarum angustarum. Capsula pendula collo brevi decrescente. Peristom. bene formatum.

remark, " Planta mascula feminea simillima capitulifera, ad basin ejusdem, foliis similibus," and in view of Schimper's remark quoted above this seems an insufficient distinction, though Brotherus (Pfl., x, 397) classes it separately as autoecious, and as a species unknown to him, placed so on Müller's description. He also mentions two lines of narrow marginal cells. It seems unsatisfactory (at least meantime) to separate this group too far on these characters only, and Müller did so only at the stage of his life when anything (or almost nothing) made a species.

Size also is no character, for large and small plants, simple or tufted, and with long or short setae and capsules, all occur in one tuft, intermixed. *Re* my 10,218, Dixon calls it from description *B. porphyreothrix*, but says it is very near, indeed, to *B. obconicum* B. and S., but may be distinct in the shorter peristome, the very narrow processes, and short seta.

B. *pseudodecursivum Par., Suppl. Ind., p. 70 (1900), changed by Paris from *B. decursivum* C. M., in Hedw., 1899, p. 70 (on account of the S. American *B. decursivum* C. M., in Hedw., 1898, p. 224), is a small much-branched condition from Rondehous, near Cape Town, Rehm. 246, and is synoecious.

Specimens issued as Wager 153, from Rydal Mount, O.F.S., under the name *B. inclinatum* B. and S., belong here, and that European species which belongs to a different group is not known to be South African.

Cape, W. Tokai ; Camps Bay ; Bishop's Court ; Platteklip ; Disa Ravine ; Sim 9506, 9310, 9457, 9391, 9149 ; Constantia Nek, Sim 9342, 9349, 9357, 9360 ; Slongoli Stair, Sim 9174, 9196, 9213, 9214 ; Mowbray, Pillans 3983 ; Table Mountain, Pillans 4003 ; Stinkwater, Rehm. 561 (as *Bryum*, No. 3).

Natal. Hilton College, Sim 10,218.

O.F.S. Rydal Mount, Wager.

Transvaal. Houtbosch. Rehm. 560 ; Heidelberg, Wager.

S. Rhodesia. Umtali, 3700 feet, F. Eyles.

513. *Bryum capillare* Linn., Suppl., 1856, No. 30 ; C. M., Syn., i, 281 ; W. P. Sch., Syn., 2nd ed., p. 449, a very variable species, almost cosmopolitan in some form, and includes *B. capillare* L., var. *capense* C. M. = *B. capense* (Hampe MSS.), C. M., Syn., i, 281.

= *Bryum* *pycnoloma C. M., in Rehm. 239, *fide* Dixon, of which there is no specimen in South African herbaria.

A small plant with nearly sessile fertile branch, subtended by several short sterile branches on which the concave, ovate-spathulate leaves, from narrower base and gradually tapering into the long yellowish hair-point, have margin almost entire above, reflexed below, and bordered by three to four lines of usually yellowish narrow cells. Lamina cells hexagonal, basal cells smaller and quadrate, but bordered by longer narrower cells. Capsule cylindrical, horizontal or pendulous, wide-mouthed, with persistent convex-conical lid ; dioecious ; male flowers terminal, bud-like, or sub-discoid. Hardly differs in any essential particular from *B. torquescens* B. and S., except that it is dioecious, and even that is a doubtful difference (see *B. torquescens*), while its negative characters vary much.

I do not doubt the European bryologists who treat these two as distinct, but in South Africa much has to be learned before the others reputedly South African can be separated, and I can only place the so-called dioecious species as belonging here or to wrong identifications, and the so-called synoecious and autoecious species as belonging to *B. torquescens*.

South African specimens named *B. obconicum* Hsch. (a European species) belong here.

B. caespiticium Linn., recorded from the Cape by Shaw (Macowan), is probably this misidentified.

B. **syntrichiaefolium* C. M., in Rehm. 231, from Rondebosch, is not represented in South African herbaria, but Mr. Dixon writes that at Kew it is clearly either *B. capillare* L. or *B. torquescens* B. and S., according to inflorescence which he has not been able to dissect.

B. lonchopyxis C. M. (Hedw., 1899, p. 72) is a small and branched

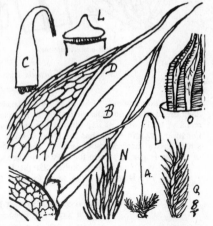

Bryum capillare, var. *capense.*

condition, somewhat yellowish-margined, with long seta and large capsules, and said by C. M. to have three intermediate cilia in the peristome. Why Müller placed *B. lonchopyxis* in § *Bimum* is difficult to understand ; Brotherus places it in § *Trichophora*.

Wager distributed *B. capillare* under the name *B. Pappeanum* C. M., which plant I know by name only, viz. *B. Pappeanum* C. M., in Bot. Zeit., 1855, p. 752, and 1859, p, 206, and its var. *humile* C. M., in Bot. Zeit., 1859, p. 206. Wager's specimen does not prove that Müller's plant is synonymous. Brunnthaler records it (1913, i, 28) from Genadendal.

B. pallens Sw. (on mountains, 7000 feet, near Graaff Reinet, MacLea, *fide* Shaw's list) has been recorded probably in error for *B. capillare*, and so also *B. turbinatum* (Hedw.) Schw. (in bogs on mountains near Middelburg, C.P., MacLea, *fide* Shaw's list).

B. capillare Linn., var. *capense*, is described by C. M. as having " ramis elongatis laxissimis foliis remotissimis flaccidis viridissimis, thecis laevissimis," but as compact forms also occur, it seems to me that *B. capillare* is a common South African species and var. *capense* only a condition. It is abundant in every part of South Africa from the coast to the mountains, including S. Rhodesia, and is usually fertile ; also present at Lourenzo Marques, P.E. Africa.

514. *Bryum* (§ *Trichophora*) *acuminatum* Sim (new species).*

A small plant, the fertile stem almost a sessile rosette subtended by one or more innovations, leafless below, laxly leaved above ; leaves 1·5 mm. long, 0·5 mm. wide, ovate-acuminate, concave, twisted and incurved toward the base when dry ; the per. leaves narrower and more awned than the innovation leaves ; all strongly nerved, with the coloured nerve ending in an acute bristle ; margin entire, not reflexed, but with about one line of coloured cells making a narrow border, inside of which all cells are very lax and pellucid-hexagonal, $80 \times 40\mu$; the basal cells rectangular, $100 \times 50\mu$; seta red, $\frac{1}{2}$–1 inch long, hooked at the top, the capsule pendulous, 2 mm. long, pyriform, brown ; the lid shortly conical ; peristome strong, the outer teeth red, closely trabeculate ; the membrane of the inner high ; processes as long as the outer teeth, and with two intermediate cilia.

Like *B. capillare*, var. *capense*, but smaller, different in habit, with narrow coloured border, and incurved margins in lower part.

Natal.   Little Berg, Cathkin, 6000 feet, July 1922, Miss Owen 15.

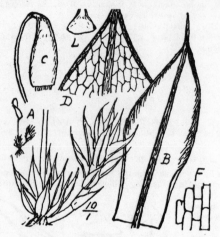

Bryum acuminatum.

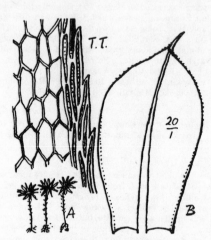

Bryum pumili-roseum (after Dixon).

515. *Bryum pumili-roseum* Dixon, in Trans. R. Soc. S. Afr., viii, 3, 203, and Pfl., xi, 9.

= *Bryum *chrysoloma* C. M., in Rehm. 229.

Not known to me.   Described by Mr. Dixon as being like a small form of *R. roseum*.   A small, densely caespitose, deep green moss, not known fertile, but occasionally gemmiferous ; stems simple, mostly nude or scale-clad, with coma of horizontal, widely obovate or spathulate, subacute leaves, 2–3 mm. long, and not decurrent ; base rather narrower, margin somewhat recurved toward the base on one or both sides, expanded and dentate above ; nerve strong below, narrow above, with a short reflexed exserted point.   Cells small, 10–16μ wide, rhomboid ; marginal cells long, linear, in two to four rows, incrassate, forming a strong rufescent border.

Cape, W.   Camps Bay, Rehm. 229 (as *B. chrysoloma* C. M.), *fide* Dixon : not in South African herbaria.
Cape, E.   Hogsback, 4000 feet, Tyumie, 1917, D. Henderson 326, 327, 337.

516. *Bryum canariensiforme* Dixon, Bull. Torr. Bot. Club, xliii, 68 (March 1916) ;
and Trans. R. Soc. S. Afr., viii, 3, 202, 203 ; Broth., Pfl., x, 400.

= *B. *afrobillardieri* Dixon, in hb. and letters.

Dioecious ; densely caespitose, light green ; stems $\frac{1}{4}$–1 inch high ; simple or once or several times forked at successive comal tufts ; when simple the stems are usually clad in leaves throughout, when branched the branches are usually leafless and with red axis below, gradually or suddenly increasing into a coma 5–6 mm. across when moist, club-shaped and 2–3 mm. across when dry.   Leaves ovate-acuminate, 2–4 mm. long, concave, especially toward the point, spreading widely, the nerve strong and coloured, and reflexed toward the apex, ending in a short acute entire bristle ; margin expanded (not revolute), entire below, and finely denticulate in the upper half ; the protruding cells $85 \times 10\mu$, but not forming a border ; inner cells $80 \times 20\mu$, those near the nerve larger ; basal cells rectangular, $100 \times 20\mu$, with longer and narrower cells along the margin.   When dry the leaves often gather together in an imbricate ball and are almost straight or hardly twisted.   Per. leaves shorter, but with longer denticulate awn.   Seta red, erect, 1–1$\frac{1}{2}$ inch high ; the capsule pendulous, 3–4 mm. long, obconic, tapering to the neck, brown ; lid convex-apiculate.

---

* *Bryum* (*Trichophora*) *acuminatum* Sim (sp. nov.).

Planta minuta, rosula sessili fertili innovationibusque clavatis.   Similis *B. capillari* var. *capensis*, minor autem, discrepans habitu, margine angusto colorato et marginibus incurvatis in parte inferiore.

Closely related to the Australasian *B. billardieri* Schw. (C. M., Syn., i, 253), which has the leaves more suberect when moist, and was formerly known as *B. afrobillardieri* Dixon, ined.

In describing it, Mr. Dixon says : " Very near *B. canariense* Brid., but differs in stature a little slighter, leaves very concave, margin erect, not recurved, apex shortly mucronate ; theca lid shortly conical, subotuse, inflorescence autoecious ? fig. 1."

What was distributed as Wager 155 had no resemblance to *B. murale* Wils., and belongs here.

A common species, exceedingly variable in the arrangement of its leaves, but constant in its margin and lid, common in the forestal or scrub parts of each province and S. Rhodesia, including Kosi Bay, Zululand, Aitken and Gale 10. Mr. Dixon states he has it also from Mexico.

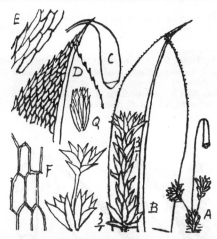

Bryum canariensiforme.

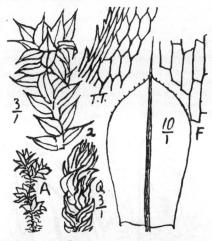

Bryum truncorum.
2. var. *pycnophyllum*, wet and dry.

517. *Bryum truncorum* Bory, in Brid., Bryol. Univ., i, 699 ; C. M., Syn., i, 254 ; Besch., Fl. Reunion, p. 238 ; Dixon, S.A. Journ. Sc., xviii, 321.

=*B. erythrocaulon* Schw., Suppl. I, ii, p. 127, t. 80 (from Bourbon) ; see Kew Bull., 1923, 6, No. 557.
=*B. *syntrichioides* C. M., in Rehm. 228 and 557, and forms.
=*B. ramosum* (Hk.), Mitt., M. Ind. Or., p. 75 ; and see Dixon, Trans. R. Soc. S. Afr., viii, 3, 203.

A comparatively small dioecious species, deep green, laxly caespitose, tomentose below, up to 1 inch high including several successive stellate comal tufts, between which the leaves, though more sparse, are almost continuous, or they may be continuous all up the stem, with evidence of occasional comal crowding, and when dry all are more or less twisted round the stem. Comal leaves 3–4 mm. long, erecto-patent, widely obovate-acuminate from comparatively wide base, the nerve strong, usually red, and extending as a long, slender, pale arista ; the margin reflexed in the lower half, closely spinulose-serrate in the upper half, the yellowish border consisting of about four rows of narrow cells $(100 \times 10\mu)$ the whole length of the leaf, behind which the lamina cells are large, lax, and mostly pellucid, $60 \times 30\mu$, while the basal cells are rectangular, $120 \times 40\mu$, the walls often red. Per. leaves shorter and narrower. Capsule on a long red seta, oblong-cylindrical, inclined ; lid convex-mamillate, ring dehiscent. Bescherelle says : " Peristomii dentes generis, ciliis binis raro tribus brevioribus appendiculatis."

B. truncorum Bory, var. *pycnophyllum* Dixon, S.A. Sc. Journ., xviii, 321, has stem an inch or more long, and the leaves are more equally distributed in this than in the type, and this seems to be the condition common in South Africa and is *B. erythrocaulon* Schw., though both occur on one plant and they cannot be separated. Only known sterile in South Africa.

Common in each Province, and Rhodesia, on stones or tree-trunks, usually on the mountains, though down to the coast at Knysna.

518. *Bryum Mundii* C. M., in Bot. Zeit., 1859, p. 206 (as *Mundtii* in error).

=*B. canariense* (Schw.), C. M., Syn., i, 253 p.p., *fide* Paris.
=*B. Mundii* C. M., var. *flaccidum* C. M., Bot. Zeit., 1859, p. 206.
=*B. pervirens* C. M., in Rabenh. Bryoth. Europ., p. 1399 (from Boschberg, Som. E.), *fide* Dixon, from British Museum specimen. There is no specimen at Kew, or in South African herbaria.
=*B. *subtorquatum* Horns., in Hb. Kunz, as var. of *B. canariense*.

Dioecious, growing in large lax tufts ; stems red-tomentose where not leafy, innovated at ½-inch distances, yellowish, and with the leaves in rosettes, each terminal to its branch. Leaves 7–8 mm. long, when moist erecto-patent, when dry more erect, imbricate, and not twisted ; leaves oblong-acuminate, concave, yellowish, the nerve strong, yellow or yellowish red, extending far beyond the leaf-apex in a slightly denticulate, acute, yellowish bristle, which is usually reflexed and sometimes somewhat twisted. Margin entire, and slightly revolute in the lower half, closely spinulose or serrate toward the apex, a thin yellowish border of two to three lines of narrow cells enclosing the much wider hexagonal lamina cells, $70 \times 17\mu$, the basal cells being larger, more lax and pellucid, and more rectangular, up to $120 \times 35\mu$. Setae one to three in a rosule, erect, strong, red ; the capsule orange-brown, inclined or subpendulous, rather asymmetrical and slightly gibbous, long-pyriform or obconic, 8–10 mm. long, including the tapering neck ; lid convex-apiculate ; peristome strong, the teeth erect and orange-red, with incurved slender apex, the membrane high and pale, nearly white, and the processes standing erect and widely perforated, with slender, nodose, intermediate cilia not longly appendiculate.

Müller's description in Synopsis Musc. of *B. canariense* Schw. agrees entirely with this, but the North African plant on which it was founded is said to have differences ; in any case the name is preoccupied (see further, Dixon, Trans. R. Soc. S. Afr., viii, 3, 203).

Frequent in each province and Rhodesia, from the coast to the mountains.

Bryum ochropyxis C. M., in Rehm. 233 (not 283 as shown by Paris), is not represented in South African herbaria, but Mr. Dixon informs me the British Museum specimen shows it to be *B. Mundii*, and the Kew specimen is a weak form of same.

Mr. Dixon also informs me that *B. *Rehmanni* C. M. (Rehm. 234) is a robust form of *C. Mundii* C. M., and that *B. *monilicaule* C. M. (Rehm. 244) is a sterile, short, dense, rather strict form of *B. Mundii* C. M. Neither are represented in South African herbaria.

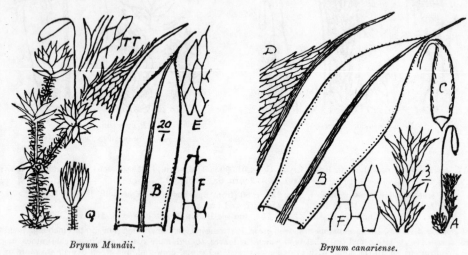

Bryum Mundii.   *Bryum canariense.*

519. *Bryum canariense* Brid., Mant. M., p. 118 ; and Bry. Univ., i, 672 (not *B. canariense* of various other authors).

Dioecious ; stems erect, laxly caespitose, deep green, up to ½ inch high, the fruiting stem usually simple and low, with one or more innovations which are either shortly stipate or equally leafy up to the next coma, the leaves suberect when moist ; erect, imbricate, and not twisted when dry ; the lower leaves small and scattered, the upper ovate-acuminate from a long narrow base, concave ; the nerve wide and coloured below, narrow above, and extending as a long firm bristle from the acute-denticulate leaf-apex ; margin revolute from the base to near the apex ; cells openly hexagonal, $80 \times 15\mu$ toward the nerve, rather narrower toward the margin, but not bordered ; basal cells hexagonal, $80 \times 30\mu$. Seta red, 1–1½ inch high ; capsule pendulous, cylindrical, tapering at the neck, brown ; lid acutely conical. (See Dixon, Torr. Bot. Club, xliii, 69, and Trans. R. Soc. S. Afr., viii, 3, 202, 203.)

South Europe, N.-E. Africa and islands. Not well understood here ; Witpoortje, Transvaal, Professor Moss.

520. *Bryum polytrichoideum* C. M., in Bot. Zeit., 1859, p. 206 ; Dixon, Torr. Bot. Club., xliii, 69, and Trans. R. Soc. S. Afr., viii, 3, 203.

Plants laxly caespitose ; stems erect, mostly simple, ½–1 inch high, nude below, many-leaved upward ; leaves appressed and forming a clavate stem when dry, with erecto-patent leaves when moist, or occasionally several successive comal tufts are produced. Leaves closely imbricated, 2–3 mm. long, 1 mm. wide, the upper the largest, usually straight when dry, sometimes slightly twisted ; oblong-acuminate or wider in the upper half, the nerve very strong throughout, green or yellowish, and ending in an acute bristle ; margin entire, expanded, or at the base slightly revolute ; cells hexagonal, $50 \times 25\mu$, those nearest the nerve largest, the border-line of cells shorter, rhomboid, and hyaline ; this extends to the base but does not form a border ; basal cells hexagonal and larger near the nerve, gradually smaller outward. Capsule pendulous, red, 3 mm. long, elliptical, on a terminal red seta 2 cm. high. Other parts not seen by me. A strong polytrichoid-looking moss, distinct in its entire unbordered leaves.

Transvaal. Pretoria, March 1918, G. H. Gray (Sim 10,214) ; Macmac, MacLea (Rehm. 558, as *Bryum*, No. 1), and Lechlaba, Rehm. 558β, as *Bryum*, No. 1β.

Genus 150. RHODOBRYUM W. P. Sch.

Strong, vigorous, dioecious mosses, one to several inches high, growing gregariously in very lax open cushions and extending by underground stolons or by prostrate old stems, from which upright stems rise, having the leaves mostly small and scale-like up the stem, but very large in a spreading rosulate crown, having the inflorescence central and terminal, frequently with several capsules from one rosette, and usually one or more innovations from inside the rosette which repeat the process. Leaves widely elliptical-lanceolate or spathulate, acute, usually serrated, with nerve to or near to the apex, and sometimes with bordered margin. Cells large,, thin-walled, laxly rounded hexagonal, toward the base larger, and longer rectangular. Seta long, curved above ; capsule horizontal or pendulous, clavate-cylindrical with a short neck, slightly incurved ; lid conical ; peristome large, as in *Bryum*. The male inflorescence is subdiscoid, with small leaves among the antheridia.

Very large and pretty mosses, often having the habit and appearance of *Mnium*, but the peristome and areolation of *Bryum*.

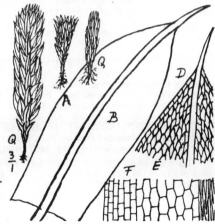

Bryum polytrichoideum.

Frequent in forests or smaller bushes, on the ground or on decayed stumps.

Synopsis of Rhodobryum.

1. Border of several lines of cells	**521.** *R. umbraculum.*
1. Border of one line of cells	2.
2. Leaf much narrowed below ; capsule tapering to the base ; dry leaves horizontal .	**522.** *R. Commersonii.*
2. Leaf not much narrowed below ; capsule not tapering at the base ; dry leaves suberect .	**523.** *R. roseum.*

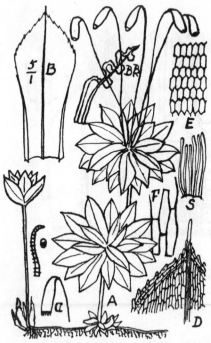

Rhodobryum umbraculum.

521. *Rhodobryum umbraculum* (Burch. Hook.), Par., Ind., 2nd ed., p. 202 ; Broth., Pfl., x, 401.

=*Bryum umbraculum* Burch. MSS., in Hook., Musc. Exot., t. 133 ; Mitt., Journ. Linn. Soc., xxii, 146, 307 ; C. M., Syn., i, 248.
=*Polla umbraculum*, Brid., Bry. Univ., i, 698.
=*Mnium spathulatum* Horns., in Linn., 1841, p. 135.
=*Mnium umbraculum* Schw., Suppl. II, ii, 1, 22, t. 157.

Dioecious, deep green, loosely gregarious over yards of surface, sometimes crowded into tufts, usually on the ground or on humus-clad stones or tree-stumps. Plant 3–6 cm. high ; stems 2–3 cm. high, leafless below except for a few scales, then bearing a rosette of ten to twenty-five leaves. From among the leaves of old rosettes one to three innovations arise, ultimately themselves forming rosettes, and this goes on indefinitely, the older ones ultimately becoming prostrate and tomentose or covered with humus, innovations from them then having the appearance of suckers. Leaves very large, horizontal when moist, tending upward when dry, 6–12 mm. long, 3 mm. wide, spathulate-acute ; margin entire and reflexed in the lower half of the leaf (and more reflexed when dry), serrate or strongly serrate in the upper portion, and bordered with several rows of narrower and often purple cells. Nerve strong, extending to or beyond the apex, usually as a long, strong, hyaline point. Upper cells large, $100 \times 35\mu$, hexagonal, twice as long as wide, in very regular lines, very chlorophyllose, basal cells larger and rectangular, $200–250 \times 50\mu$. Inflorescence forming the centre of the rosette, consisting of a few archegonia surrounded by numerous linear, red paraphyses. Seta 3 cm. long, bright red, three to five rising together from the centre of a rosette ; capsules grey, pendulous, oblong-cylindrical ; lid shortly conical-apiculate. Peristome prominent, light red, the processes when dry longer than the teeth, perforated, and with several interposed cilia.

Cape, W. Outeniqualand, Burchell.

Cape, E.　Fern Kloof, Grahamstown, 1919, Miss Farquhar ; Philippstown, Ecklon ; Perie Forest, 4000 feet, 1892, Sim 7130, 7226 ; Umzimkulu, Gr. E., Tyson 2553.

Natal.　Umlaas, Drège ; Giant's Castle, 7000 feet, 1915, R. E. Symons ; Oliver's Hoek, Miss Edwards ; Van Reenen ; Mont aux Sources ; Giant's Castle ; Nels Rust, Sim 7792, 10,248, 10,250, 10,251.

Transvaal.　The Downs, Pietersburg, 4000 feet, Rev. H. A. Junod 4012.

S. Rhodesia.　Zimbabwe, Sim 10,249.

522. *Rhodobryum Commersonii* (Schw.), Paris, Ind., 2nd ed., p. 198 ; Dixon, S.A. Journ. Sc., xviii, 321 ; Broth., Pfl., x, 404.

= *Bryum Commersonii* Brid., Mant. M., p. 119 ; C. M., Syn., i, 249 ; Besch., Fl. Reunion, p. 239.
= *Bryum (Polla) Commersonii* Brid., Bry. Univ., i, 700.
= *Mnium Commersonii* Schw., Suppl., I, ii, 134, t. 80.
= *Bryum *argutidens*, Rehm. 556, which he says = 230 forma *prolifera* (not 330 as written) = *R. spinidens* (Ren. and Card.), *fide* Thériot, in letter to Dixon, see Kew Bull., 1923, 6, No. 556.
= *B. *argutidens* Rehm., forma *prolifera*, Rehm. 555 and 555β (both = Rehm., No. 230, not 330 as written) ; 550 β being = *R. laxiroseum* (C. M.), *fide* Thériot, in letter to Dixon, see Kew Bull., 1923, 6, No. 555β.

Plant dioecious, robust, 1–1½ inch high, the fertile stem usually only ½ inch high from the ground, but terminal on a prostrate old stem, and with innovations rising from it ; scale-leaved below, but the leaves gradually larger upward, rather lax, and with an open stellate disc and almost flat leaves at the apex, these leaves up to 10 mm. long, 5 mm. wide, flatly concave ; base spathulate, narrow ; lamina widely ovate-acute above, the nerve strong and yellowish, percurrent in a sharp bristle, the margin hardly revolute or only at the very base, elsewhere expanded and denticulate from below the middle, closely and sharply spinulose toward the apex, the teeth single or double, in one row of narrow cells, behind which the lamina has large lax cells $100 \times 50\mu$; basal cells more rectangular, $120 \times 40\mu$ with one or two rows of long marginal cells ; per. leaves smaller, the outer ovate-acuminate, subspinose, the inner smaller and narrower.　Seta 1½–1¾ inch high, red, erect to near the apex, capsule horizontal, oblong-cylindric from tapering slightly bent neck, brown, 5–10 mm. long ; teeth brown, closely trabeculate when dry, pectinate on the inside ; membrane high, its processes about six, perforated, lid convex with rather long conical beak.　Peristome as in *R. roseum* from which it differs in the hardly noticeable border and the sharply spinulose teeth.　When dry the leaves are very undulate but widely spreading, the lower separate.

On bark or humus, tropical Africa, Australia, and South Africa.

Cape, E.　Kentani, 1907, Miss Pegler 1448.

Natal.　New Hanover and Mont aux Sources, Sim 10,252, 10,254 ; Giant's Castle, A. E. Symons.

Transvaal.　The Downs, Pietersburg, Junod 4019 ; Lechlaba, Rehm. 555, 556 ; Pilgrims Rest, Rehm. 555β.

S. Rhodesia.　Mountain bush, Umtali, 4500 feet, Eyles 1736.

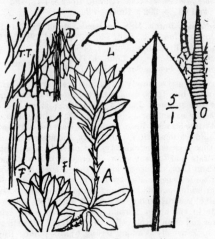

Rhodobryum Commersonii.

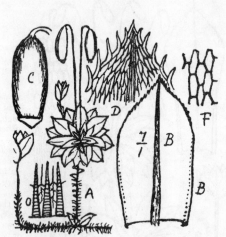

Rhodobryum roseum.

523. *Rhodobryum roseum* (Weis.) W. P. Sch. ; Limpr.; Broth., Pfl., x, 404.

= *Bryum roseum* Sch., Spic. Fl. Lips., p. 34, No. 1848 ; Dixon, 3rd ed., p. 376, Tab. 45 M ; C. M., Syn., i, 247 ; W. P. Sch., Syn., 2nd ed., p. 463.
= *Mnium roseum* Weis., Pl. Crypt. Fl. Gott., p. 157.
= *Bryum proliferum* Sibth., Fl. Oxon., p. 292 ; Braithw., Br. M. Fl.
= *Rhodobryum leucothrix* (C. M.), Broth., x, 404.
= *Bryum leucothrix* C. M., in Hedw., 1899, p. 70, and in Rehm. 554 and 554β, and in Musc. Macowan, No. 20.

=*Rhodobryum capense* Par., Suppl. Ind., p. 229 (1900).

=*Rhodobryum leptothrix* (C. M., as *Bryum*), Par., 2nd ed., p. 200, in error, as no such name was used by C. M.

=*Bryum *integrifolium*, Rehm. 226 (=*B. leucothrix* C. M., *fide* C. M.).

Laxly caespitose over considerable areas, growing from prostrate old stems like stolons ; dioecious ; stems 1–2 inches high, tomentose below, with few, small, scale-like leaves among the tomentum in the lower portion, then a flat comal rosette ¾–1 inch across, in which the outer leaves are up to 10 mm. long, the inner gradually much smaller, with central fertile inflorescence having numerous, slender, red paraphyses and a few (one to six) seta, as well as usually one or more stalked rosetted innovations ; leaves 4–6 mm. long, 3 mm. wide, ovate-acute, entire and slightly revolute toward the base, sharply and closely spinulose toward the apex, the teeth-cells forming a narrow hyaline border, usually only one cell deep. Nerve strong, often red, percurrent or ending in or just below the apex ; lamina cells laxly hexagonal, 120×40, chlorophyllose, thin-walled ; basal cells larger and more rectangular-hexagonal, 160× 60μ, hyaline. Seta purple, 1–1½ inch long ; capsule pendulous, oblong-cylindrical, somewhat high-backed, grey or ultimately brown ; lid convex-mamillate ; peristome strong, red ; membrane high ; processes standing higher than the teeth when dry. Leaves suberect and much undulated, and more evidently spinose when dry.

Occurs also in Britain, Northern Europe, Northern Asia, Uganda, etc.

Cape, E. Kakazella Hill, Dohne, 1898, Sim 7215 ; Longbushkop, Alexandra Forest, 1920, J. W. Sim.

Natal. Cathkin, 7000 feet, Sim 10,255 ; Little Berg, Cathkin, 6000 feet, Miss Owen 67 and 68 ; Polela, Haygarth, No. 1, and Sim 8075 ; Lynedoch, N.R., 1917, Dr. Bews ; Giant's Castle, 8000 feet, R. E. Symons ; Zwart Kop, 1917, Sim 10,256 ; Koodoobush, Mont aux Sources, Sim 10,257 ; Majuba, Rehm. 554 ; Van Reenen, Rehm. 226.

Transvaal. Lemona Wood, Spelonken, 1918, Rev. H. A. Junod ; Pilgrims Rest, Rehm. 554β.

Family 37. MNIACEAE.

Strong plants, having erect, more or less rosulate, fertile stems, and long prostrate runners with scattered distichous alternate leaves, much crisped when dry. In the South African species leaves oval, strongly bordered, spinulose ; cells large, roundly hexagonal, pellucid ; nerve strong, percurrent ; seta long, terminal, often several in a coma ; capsule cylindrical, without neck ; peristome as in *Bryum*, but cilia not or hardly appendiculate. Only one South African genus.

Genus 151. MNIUM (Dill. p.p.) Linn. emend.

Tall plants, with erect fertile stems ending in a terminal rosette of large leaves, the lower leaves lax and smaller, and below these rise long, prostrate, distichous-leaved, rooting, simple or branched, barren shoots, from which further erect fertile stems rise, usually where they have rooted. Leaves large, pellucid, rounded, elliptical, oval, or spathulate, strongly nerved, with a distinct border of narrower cells, sometimes toothed, with very lax, large, smooth, roundedhexagonal areolation ; cells nearly as broad as long, except the basal cells. South African species synoecious ; inflorescence terminal, often producing several capsules ; seta long ; capsule oblong-cylindrical, without neck, usually horizontal or pendulous ; calyptra narrow, fugacious ; lid beaked. Formerly included in Bryaceae, but easily distinguished by the more rounded leaf-cells.

524. *Mnium rostratum*, Schrad., in Linn. (Gmel.), Syst. Nat., xiii ed., II, ii, 1330, n. 28 ; C. M., Syn., i, 158, and ii, 554 ; W. P. Sch., Syn., 2nd ed., p. 480 ; Dixon, 3rd ed., p. 380, Tab. 46β ; Broth., Pfl., x, 415.

=*Mnium Eckloni* C. M., in Bot. Zeit., 1855, p. 749 ;

and many old, but no recent, synonyms.

In loose cushions in streams or wet ground. Fertile stems 2–3 cm. high, erect, woody, usually simple above, with primarily sterile branches at the base, which are 2–3 inches long, simple or branched, prostrate or arching, permanently sterile or rooting and sending up another fertile stem. Leaves on the prostrate stems complanate, distant, oblong-elliptical, 2–4 mm. long, 1·5–2 mm. wide ; those on the erect fertile stems similar but larger and more acuminate,

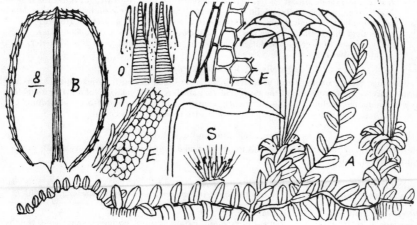

Mnium rostratum.

set all round the stem, scattered and small below but rosulate at the disc, and 5–6 mm. long, 2–3 mm. wide. **Leaves** shortly decurrent, rounded and apiculate at the apex, the nerve very strong on the under-side and shortly and usually obliquely excurrent in the mucro ; margin very pronounced for the whole length of the leaf, of about four rows of linear cells, the marginal cells partly free and often forming one line of short hyaline teeth from base to apex, in other cases they are more appressed and the leaf is almost entire ; other cells regularly hexagonal and in lines, and about 25μ diameter, except those at the base which are 25 × 35μ ; inflorescence discoid, synoecious ; archegonia and androecia very numerous, mixed in many groups, the archegonia longer than the androecia, and the paraphyses still longer. From one to ten capsules are produced per disc. Seta strong, erect, about 3 cm. long, with a membranaceous sheath at the base. Capsule erect when quite young, afterwards horizontal or pendulous, widely cylindrical, with a long and somewhat curved persistent beak ; calyptra fugacious. Peristome yellow, the outer teeth closely barred. This species varies much in the leaf-form, the wideness of the border, and the presence of teeth, sometimes being almost entire. When dry the leaves are exceedingly crisped and contorted, and they moisten very slowly.

Var. *Reidii* Sim (new variety).*

Plant grows in a perennial perpendicular cushion several yards long and wide, on the face of a dripping waterfall, where there is constant saturation and drip throughout the whole mass, but where the occasional larger flow of flood-water pours beyond the moss. Plant pendent, 6–18 inches long, branched, ending in one or more unbranched stolons 6–12 inches long, which overlap in mass and form the cushion referred to above. Floriferous stems several, situated near the base, apparently lateral, 1–3 inches long, simple or producing another floriferous branch or a stolon below the rosule ; leaves on the floriferous stems scattered, irregularly placed round the stem and rosulate at the apex ; leaves on the barren stems regularly alternate and distichous, elliptical, emarginate, up to ¼ inch long, one to two lines wide, simply toothed, thinly pellucid, somewhat undulated, strongly nerved, distinctly margined, not decurrent ; cells large, six-sided ; the thickened margin composed of three to four lines of long narrow cells, the outer line of cells along the whole margin protruding as teeth ; nerve shortly excurrent. Leaves on the floriferous stems and especially in the rosule more obovate and much more decidedly toothed, but the teeth simple and in a single line. Inside the rosule are several small lanceolate leaves and the abundant synoecious inflorescence. Capsules not seen. Stems rooting at the base only. A most beautiful moss, unique in its mass-production, and resembling a filmy fern in its habit.

Natal. Waterfall in Town Bush, Maritzburg, first shown to me by Professor Reid, after whom it is named (Sim 7552). Also, though less elongated : Overwood, Polela, 4000 feet, 1915, Sim 8552 ; and Buccleuch Forest, 4000 feet, 1914, W. Leighton (Sim 8551).

Mnium rostratum is almost cosmopolitan and is widely distributed in South Africa, usually in upland forest streams or marshes, or in wet grasslands, and fruit is rare.

Cape, W. Window Gorge Waterfall, Kirstenbosch, January 1919, Sim 9475 ; Devils Peak, Rehm. 261*c*.

Cape, E. Eveleyn Valley, Perie Forest, 1892, 4000 feet, Sim 7001, 7091 ; Dohne Hill, 4500 feet, 1898, Sim 7201 ; Paradise Kloof, Grahamstown, 1921, Miss L. L. Britten ; Boschberg, James Sim.

Natal. Little Berg, Cathkin, Miss Owen 57 and 69, and Sim 10,259 ; Giant's Castle, R. E. Symons ; Mont aux Sources, Sim 10,258 ; Oliver's Hoek Valley, 6000 feet, August 1922, Miss Edwards ; Van Reenen Pass, Rehm. 261 (as *M. Eckloni* C. M.).

O.F.S. Kadziberg, Rehm. 261*β* (as *M. Eckloni* C. M.) ; Rydal Mount, December 1910, Wager 436 (distributed as *M. subglobosum* B. and S.).

Transvaal. Pilgrims Rest, MacLea (Rehm. 570).

Besides *M. rostratum* Shaw lists *M. affine* Bland. (Boschberg, Macowan), *M. cuspidatum* Hedw. (in Hb. Hook, from the Cape), *M. undulatum* (Linn.), Weis (Cape), *M. punctatum* Hedw. (Cape) ; C. M., Syn., i, 161, includes *M. undulatum*, and Wager adds *M. subglobosum* B. and S. (as above). I have no reason to think either of these is South African, but lest others may differ, I subjoin synopsis of these species as found in Europe.

Synoecious, leaf entire, border 1-stratose, capsule subglobose, lid conical-rostrate, *M. subglobosum* B. and S.
   ,,   leaf denticulate or almost entire, border thickened, lid long, rostrate, *M. rostratum* Schrad.
   ,,   leaf toothed, border thickened, lid convex-mamillate, membrane perforated, *M. cuspidatum* Hedw.
Dioecious, leaf entire, lid long-conical, obtuse, leaf-margin 2–4-stratose, *M. punctatum* Hedw.
   ,,   leaf-teeth of three cells, in single row, leaf oval-acute, lid mamillate, *M. affine* Bland.
   ,,   leaves ligulate with cross undulations, border narrow, thickened, cells 15μ diameter, *M. undulatum* (Linn.), Weis.

Family 38. RHIZOGONIACEAE Broth.

Erect, rather rigid plants, having basal or lateral inflorescence, and in South African species strongly bordered, spinose-dentate, lanceolate leaves all round the stem, and very small round areolation (about 10μ diameter). Capsule inclined ; peristome double, included inside the mouth, consisting of sixteen outer teeth which infold singly or in pairs over the membrane half that height, which is crowned with lanceolate processes.

Genus 152. RHIZOGONIUM Brid.

Characters as above, in South African species leaves all round the stem, leaf-cells very small, peristome double, seta long, capsule not ribbed.

Synopsis.

Dioecious, leaves lanceolate, teeth in one line, fertile inflorescence lateral . . 525. *R. vallis-gratiae.*
Synoecious, leaves linear-lanceolate, teeth often in two lines, inflorescence basal, seta very long, 526. *R. spiniforme.*

---

* *Mnium rostratum* Schrad. var. *Reidii* Sim (var. nov.).
    Planta pendens, 6–18″ longa, toris proveniens in aspergine cataractarum ; grandis laxaque tota, caulibus brevibus basalibus floriferis, sterilibus longis.

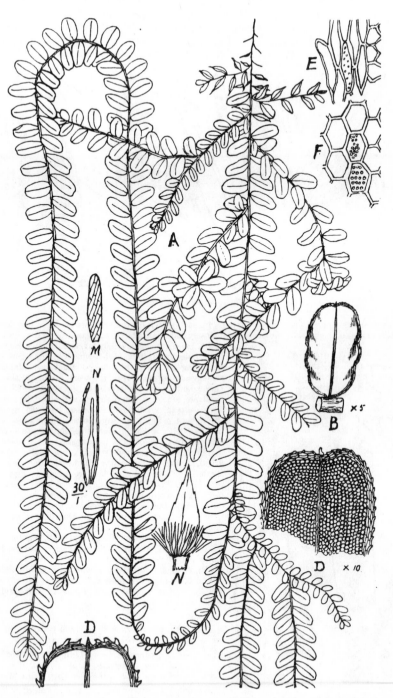

Mnium rostratum, var. Reidii.

525. *Rhizogonium vallis-gratiae* (Hpe.), Jaeg. Ad., i, 687.

=*Mnium vallis-gratiae* Hpe., in Bot. Zeit., 1859, p. 205.

A rigid, erect, gregarious or caespitose, dioecious moss 2–3 cm. high ; stems rising a few from one base, then simple or branched half up, and bearing the capsules lateral, about half up, from a perichaetium of a few leaves shorter than

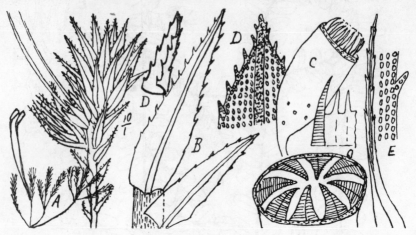

Rhizogonium vallis-gratiae.

the stem leaves, enclosing one or several strong red seta, each with large red vaginula nearly as high as the per. leaves. Stem leaves sparse and small below, numerous and closely imbricate above, appressed when dry, erecto-patent when moist, 1·5–2 mm. long, widely lanceolate-acute from a wide clasping but not decurrent base, keeled-concave ; the nerve very strong and extending to the point ; border thickened, almost entire below, strongly dentate upward ; the teeth usually of three cells, and the nerve dentate on the keel near the apex. Cells minute, roundly quadrate, separate, 10μ diameter and about 10μ apart, in straight lines, pellucid, not different at the leaf-base ; the margin 2–3 stratose throughout. Below the inflorescence innovations when young are bud-like with almost linear, strongly serrate leaves from wider clasping base, the subula almost entirely nerve except the serrate margin. Seta erect, red, strong, 2 cm. high ; capsule oval-cylindrical, inclined, wide-mouthed, and with a few sunk stomata toward the base. Capsule wall of large round cells ; mouth of about four rows of small round red cells ; ring not seen ; peristome inflexed in eight pairs of smooth yellowish teeth 400μ long, closely trabeculate outside and inside ; membrane pale and thin, about half as high, with short lanceolate processes. Only known on and near Table Mountain and Hex River, Cape, but frequent there.

Cape, W. Window Gorge Waterfall, Kirstenbosch, Sim 9471, 9477 ; and 1917, Miss Michell ; Blinkwater Ravine, Professor Bews ; Newlands, Rehm. 263*c* and Sim 9451 ; Disa Ravine, Sim 9124 and 9151 ; Slongoli, Sim 9192 ; Devils Peak, Rehm. 263*β* ; Hout Bay, Rehm. 263 ; above Worcester, Rehm. 263*d* and *c*.

Dixon points out that 263*c* *Fissidens glaucescens* on some labels is an error for 283*c*.

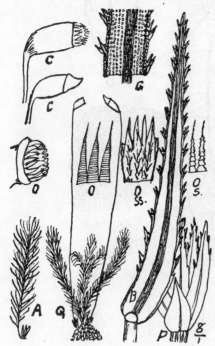

Rhizogonium spiniforme.

OS, outer teeth, dry ; OSS, part of inner peristome.

526. *Rhizogonium spiniforme* (Linn.), Bruch., in Flora, 1846, p. 134 ; Bryol. Jav., II, i, 131.

=*Hypnum spiniforme*, Linn., Sp. Pl., 1587.

=*Mnium spiniforme*, C. M., Syn., i, 175.

=*Pyrrhobryum spiniforme* Mitt., Sam. M., p. 174.

Synoecious, caespitose in widely spread mats or smaller tufts ; stems tufted, erect, 2–3 cm. long, mostly simple, but a few with two to three branches ; stems wiry, erect, rigid, leafless below ; the lower leaves small, ovate-acute, entire ; the upper leaves

5 mm. long, 0·5 mm. wide, placed all round the stem, linear-lanceolate, acute or pointed, clasping at the base, strongly nerved throughout ; border thickened, coarsely serrate above, sparsely denticulate or sometimes serrate below, the teeth usually one large, acutely triangular, hyaline cell, and usually in two rows (*i.e.* two are opposite together) though sometimes single ; cells equal, minute, running in lines, rounded, separate, pellucid ; nerve firm, striated, formed of long slender cells and toothed on the back near the apex. Perichaetal leaves in a cluster of six to eight at the base of other stems, 1–2 mm. long, the outer short, ovate-acuminate, and entire, the inner longer and somewhat toothed, all shorter than the other leaves. Seta red, 5–6 cm. long, erect, firm ; capsule green, horizontal or inclined, oblong, 2 mm. long ; beak rostrate, 1 mm. long ; peristome double, outer of sixteen simple, firm, conical, red teeth, showing bars when dry, and tipped by slender pale ends which incurve when dry ; basal membrane of inner peristome very high, much plaited, with thin processes. Occurs in the mountain forests of all tropics, and also in coast forests in South Africa.

Cape, W. Hermanus, 1920, Miss Stephens ; Window Gorge Waterfall, 1919, Sim 10,260.

Cape, E. Pirie Forest and Eveleyn Valley, 1893, Sim 7236, 7256, 7146, 7066 ; Knysna Woods, Rehm. 262, and Miss Duthie 39 ; Kentani, Miss Pegler 1270 ; Ntsubane Forest, Lusikisiki, E. Pondoland, 1917, G. H. B. Fraser (S.A. Mus., 12,762) ; Storms River, Zitzikama, 1917, J. Burtt-Davy ; Deepwaal, Knysna, Dr. J. Phillips 14 and Rehm. 262 ; Oudtshoorn, Miss Taylor.

Natal. Town Bush, Maritzburg ; Buccleuch ; Zwart Kop ; Ngomi Forest ; Ngoya Forest, Sim 10,261, 8564, 8565, 8566, 10,262.

S. Rhodesia. Inyanga Forest, 6000 feet, J. S. Henkel (Eyles 2632).

Shaw catalogues " *R. pennatum* " (misprinted *permatum*) " Hook. f. and Mitt. var., Katberg Forests, 1869," but that is a New Zealand species having its leaves in two rows, and is evidently a mistake.

Subtribe B. Diplolepideae Pleurocarpae.

Usually more or less prostrate or arching perennial plants continuing to grow from the apex of the stem and branches, often matted, the branches often pinnately or subpinnately arranged, seldom dichotomous, but in some cases the branches are erect from a prostrate, nearly leafless rhizome or stolon. Fructification always lateral, *i.e.* the inflorescence from which it grows is not terminal, but produced as a lateral bud, though in some cases this bud develops so that the seta appears to be terminal on a short lateral branch. Very seldom do the male and female organs occur in the same inflorescence. Peristome usually double, or more or less perfect. A few genera hold a somewhat indefinite position, or depart from their type, and have been placed alternately in Acrocarpae and Pleurocarpae according to the views of the various authors. Thus Hedwigiaceae has terminal fructification, and is included by some in Grimmiaceae, while *Anoectangium* in Orthotrichaceae has lateral inflorescence, and *Macromitrium* and various other related genera have pleurocarpous habit of growth.

Synopsis of Families in Diplolepideae Pleurocarpae.

Peristome absent, leaves without nerve, capsule immersed or almost so.
  Small prostrate bark epiphytes . . . . 39. ERPODIACEAE.
  Robust, suberect, straggling, or hanging plants . . . . . 40. HEDWIGIACEAE.
Peristome (in the South African genera) very rudimentary. Vigorous aquatic plants with linear twisted cells and
  nerveless leaves . . . . 41. FONTINALACEAE.
Outer peristome present, inner usually present.
  Leaves all round the stem, not complanate.
    Capsule immersed ; calyptra laciniate at the base, rough above ; cells parenchymatous ; nerve single ; cells sometimes papillose. Small bark epiphytes . . . . . 42. CRYPHAEAE.
    Capsule just emergent, calyptra cucullate ; cells parenchymatous, papillose ; marginal cells at leaf-base modified ; stems stoloniferous or rhizome-like . . . . . 43. PRIONODONTACEAE.
    Capsule exserted ; calyptra cucullate ; cells smooth.
      Cells parenchymatous ; marginal cells at leaf-base modified ; stems stoloniferous 44. LEUCODONTACEAE.
      Cells prosenchymatous, usually smooth ; outer cells of stem not thickened ; basal leaf-cells quadrate but no differentiated alar cells ; inner peristome absent, or thread-like processes without basal membrane. Nerve single, seldom absent ; capsule erect and regular ; small bark parasites or epiphytes 45. FABRONIACEAE.
      Outer cells of stem thickened ; nerve single ; capsule inclined, horizontal, or pendulous,
         46. BRACHYTHECIACEAE.
  Leaves either round the stem or in flattened arrangement. Dorsal leaves not different from the others.
    Cells parenchymatous, papillose ; no differentiated alar cells. Nerve single in South African species. Paraphyllia usually numerous. Capsule exserted, erect or inclined. Calyptra cucullate . 47. LESKEACEAE.
    Cells small, prosenchymatous or parenchymatous, usually smooth ; alar cells not much modified. Capsule usually immersed, erect, symmetrical, occasionally inclined. Leaves thin and scarious, usually complanate,
       48. NECKERACEAE.
  Cells prosenchymatous.
    Capsule erect and regular ; leaf-cells sometimes papillose ; alar cells differentiated, numerous, square or broader than long. Nerve usually single and long, sometimes short and double or absent 49. ENTODONTACEAE.
    Capsule usually cernuous or horizontal, not pendulous ; leaves often falcate ; nerve very short and weak, single, double, or absent. Cells usually smooth ; alar cells all alike . . 50. HYPNACEAE.
    Capsule inclined, irregular ; lid sharply beaked ; cells smooth or papillose, one row of the alar cells inflated, elongated, thin-walled ; midrib short and double or absent . . . 51. SEMATOPHYLLACEAE.

Leaves (in South African genera) distinctly complanate, delicate ; cells large, parenchymatous, smooth (except *Callicostella*).

    Sparsely branched, prostrate, with spreading side-leaves and appressed, often different, back and front leaves. Nerve double, single, or absent ; no differentiated alar cells. Cells of stem not thickened. Peristome double .   .   .   .   52. HOOKERIACEAE.

    Rhizomatous, with erect frondose branches. Leaves distichous, with a row of small leaves on the under-surface of stem. Marginal cells of stem not thickened. Peristome double in South African genera. No differentiated alar cells. Capsule pendulous .   .   .   .   53. HYPOPTERYGIACEAE.

    Pinnately branched ; leaves distichous, with two rows of small leaves on under-surface of stem. Marginal cells of stem thickened, lumen small .   .   .   .   .   54. RHACOPILACEAE.

Family 39. ERPODIACEAE.

    Slender little plants, more or less matted on tree-bark, with prostrate, simple or branched extension stems from which short, erect, somewhat club-shaped, fertile branchlets rise, as also the small, thick, bud-like, axillary male inflorescence, the patches flat, shining, and sometimes extensive. Leaves in four or more rows all round the stem, imbricate, concave-pointed, without nerve or distinct border. Cells lax, thin-walled, usually papillose, roundish quadrate, those in the point more elongated ; the alar cells little different and not coloured. Perichaetal leaves longer, more clasping, often more or less including the nearly sessile, erect little capsule ; lid more or less dome-shaped, mamillate ; calyptra usually bell-shaped or cylindrical, striate, and lobed at the base ; occasionally cucullate and not striate. Ring usually present, peristome absent in South African genera ; when present elsewhere consisting of outer peristome only, having sixteen lanceolate teeth. A tropical and subtropical family of minute bark-mosses.

Synopsis.

153. AULACOPILUM. Calyptra cylindrical-cucullate, twisted, covering the whole capsule ; prostrate stems somewhat flattened, with leaves dimorphous.

154. ERPODIUM. Calyptra bell-shaped, lobed, striate, not twisted, covering only the upper half of the capsule ; leaves all round the stem, not dimorphous.

Genus 153. AULACOPILUM Wils.

    Very slender little prostrate bark-mosses, irregularly pinnately branched, and having short, upright, fertile branches on which the setae are terminal. Leaves when moist erect or suberect, when dry closely appressed, concave, often hyaline-pointed or with a long hair-point, usually dimorphous on the prostrate stems, those below ovate-unsymmetrical, those on the upper surface lanceolate-symmetrical, seldom all symmetrical. Cells lax, rounded or roundly hexagonal and papillose, or in some cases oval and smooth, with a marginal line of smaller quadrate cells, or among the alar cells more than one line. Perichaetal leaves larger and erect, clasping ; capsule oval or cylindrical, more or less exserted, on a very short seta ; ring and peristome absent ; lid usually conical ; calyptra large, covering the whole capsule, striate, twisted, split on one side, deeply lobed, rough on the ridges above. Autoecious.

527. *Aulacopilum trichophyllum* (Aongst. MSS.), C. M., in Bot. Zeit., 1862, p. 393 ; Dixon, S.A. Sc. Journ., xviii, 324 ; Broth., Pfl., i, 3, 711, and xi, 5.

    A small, autoecious epiphyte on bark or stones ; stems prostrate, matted, $\frac{1}{2}$ inch long, irregularly pinnately many-branched ; leaves imbricate, mostly in two rows of suberect, ovate-concave, clasping, irregular, hyaline-pointed leaves 1 mm. long, but with some smaller, lanceolate, regular leaves on the lower surface, both kinds nerveless with rounded, densely but minutely papillose cells, 15μ diameter, which give the margins a crenate and densely papillose appearance ; the hair-point is hyaline and smooth, or its lower cells papillose ; alar cells not different. When moist the leaves are expanded, but when dry closely appressed. Fertile stems lateral, very short, with few but larger similar perichaetal leaves, and shortly suberect, erect, oblong capsule with short conical lid and larger white calyptra, open on one side but ridged, and with the ridges twisted and somewhat papillose. Male flower bud-like, small, axillary.

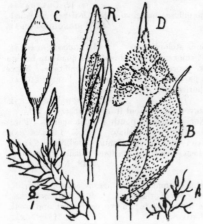

Aulacopilum trichophyllum.

    Natal. Eshowe (*fide* Bryhn, 1911, p. 17).

    Transvaal. Lemona Wood, Spelonken, in mats on bark, March 1918, Rev. H. A. Junod 12 ; Houtbush, Rehm. 599.

    S. Rhodesia. Acropolis, Zimbabwe, 3500 feet, in little tufts on rocks, Sim 8816.

    A. glaucum Wils., in Lond. Journ. Bot., 1848, p. 91 ; C. M., Syn., ii, 186, is credited to New Zealand and South Africa by Paris, the latter locality *fide* Mitten, but I presume in error, as Brotherus confines it to Australasia ; C. M. describes it as glaucous, with smaller perichaetal leaves and smooth calyptra.

    A. incanum Mitt., Journ. Linn. Soc., 1872, p. 308 (Broth., Pfl., i, 3. 711, and xi, 5), credited to South Africa only, is not known to me, and I have not seen the description, but Dixon places it in *A. trichophyllum*, and Brotherus places these two together.

Genus 154. ERPODIUM Brid.

Slender and usually small, prostrate, rooting bark-epiphytes, irregularly and often closely set with more or less erect little club-shaped, fertile branches, on which almost sessile capsules are terminal, and often almost included among the rather long perichaetal leaves. Branches flattened when horizontal, round when erect, the leaves closely imbricated when dry, sometimes spreading when moist, more or less ovate in form, concave, obtuse or pointed or often with a hyaline point or a long hair-point; margin flat, entire; nerve absent; cells smooth or papillose, oval or hexagonal, those at the margin smaller, quadrate, the several rows of marginal basal cells quadrate or oblique. Perichaetal leaves erect, long, clasping the capsule; seta very short; capsule erect, oval; ring wide, long-adherent; peristome absent. Lid shortly conical or domeshaped-mucronate. Calyptra shortly bell-shaped, ridged, rough above on the ridges, lobed at the base, not twisted, and covering only the upper part of the capsule. Tropical and subtropical.

Müller places this genus as a section of Pilotrichum.

Synopsis.

Leaves hair-pointed.
Cells 20μ diameter, papillose . . . . 528. *E. Hanningtoni.*
Cells 40μ diameter, smooth . . . . 529. *E. transvaaliense.*
Leaf not hair-pointed . . . . 530. *E. grossirete.*

528. *Erpodium Hanningtoni* Mitt., in Journ. Linn. Soc., 1886, p. 313, and pl. 16, figs. 4–7; Dixon, S.A. Sc. Journ., xviii, 324; Dixon, Torrey Bot. Club Bull., xliii, 69; Broth., Pfl., xi, 3.

= *E. Menyhardtii* C. M., in Verh. d. K.K. Zool. Bot. Gesells. in Wien, xliii, 13–14; Broth., Pfl., xi, 3.
= *E. Joannes-Meyeri* C. M., in Flora, 1890, p. 486; Broth., Pfl., xi, 3, and fig. 421.

A minute bark-epiphyte, with prostrate, rooting, pinnate stems 1–2 cm. long, from which short fertile stems stand erect, having a few ordinary stem leaves below, then six to eight much larger perichaetal leaves rather longer than the capsule. Leaves all round the stem, but secund where it is rooting, 1 mm. long, with hyaline point often 0·5 mm. long, widely ovate, or sometimes narrowed above the middle, deeply concave, imbricate-conchiform, nerveless; the hair flat, tapering from a wide base below the leaf-apex, often denticulate, and with longish hyaline cells; lamina cells round, 20μ wide, turgid, densely but minutely papillose, and not different at the leaf-base. Perichaetal leaves 1·5–2 mm. long, 1 mm. wide, roundly ovate, deeply concave, clasping the capsule and rather longer than it, and having a flat hyaline point nearly equally long, tapering from wide base, and usually denticulate. Cells of per. leaves rhomboid, 80–100μ long, slightly papillose, and usually bordered by a line of shorter obliquely quadrate cells. Capsule almost sessile, immersed, oval, large, with pale, nearly flat, long-mamillate lid. Ring wide. Calyptra short, white, bell-shaped, only covering the lid, furrowed above and lobed at the base. Spores 40μ wide, very green.

O.R.C. Kroonstad, Sim 7703.

Transvaal. Kaapmuiden, 1910, Wager 258 and 440 (wrongly distributed as *Hedwegia Macowani* C. M.).

S. Rhodesia. Zimbabwe, 3000 feet, Sim 8801, on bark; Matopos, on rock, Wager 899; Victoria Falls, Wager 907; Baroma, Rev. Menyhardt. Found also in East Africa.

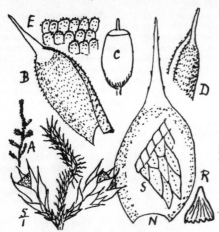

Erpodium Hanningtoni.
S, cells of perichaetal leaves.

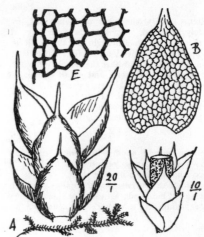

Erpodium transvaaliense.

529. *Erpodium transvaaliense* Broth. and Wag., Trans. R. Soc. S. Afr., viii, 3, 208, and pl. xii, fig. 13; Broth., Pfl., xi, 3.

Vigorous, green, prostrate on bark, 1–2 inches long, almost hoary, with leaf-points and white perichaetal leaves; irregularly pinnated or branched; the fertile branchlets short; and erect leaves 1–1·5 mm. long, with hyaline points

0·5 mm. long on upper leaves; the apex rounded, without hair on lower leaves, widely ovate at the base, somewhat reduced above mid-leaf, very concave, imbricate-conchiform, placed all round the stem; cells very large, smooth, 40μ wide, roundly hexagonal except the smaller marginal line of quadrate cells all round, basal and alar cells not different. Per. leaves white, widely clasping, as long as the capsule, hyaline-pointed from below the leaf-tip, very concave, the cells long-hexagonal. Capsule almost immersed, shortly oblong, lid and calyptra not seen.

Transvaal. Wolhuter's Kop, April 1912, H. A. Wager.

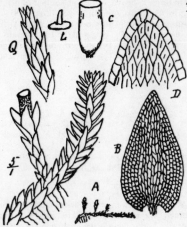

Erpodium grossirete.

530. *Erpodium grossirete* C. M., in Verh. d. K.K. zool.-bot. Gesells., in Vien, xliii, 13–14; Broth., Pfl., xi, 3.

=*E. distichum* Wag. and Dix., Trans. R. Soc. S. Afr., viii, 3, 208, and pl. xii, fig. 14; Dixon, S.A. Sc. Journ., xviii, 324 (1922); Wager's list; Broth., Pfl., xi, 3.

Small, green, prostrate on tree-trunks, rooting, 1–2 cm. long, somewhat pinnate, with short, erect, lateral, fertile stems. Leaves more or less complanate (not distichous) when moist, often with two lateral rows of clasping and spreading leaves, and an upper and under more flat row, but also often irregular, or where rooting frequently secund, and when dry the leaves are appressed or almost julaceous. Leaves 1 mm. long, ovate or ovate-acuminate, blunt, without hair-point or only apiculate, rounded in to a narrow base, and attached to the stem by central prosenchymatous cells. Nerve absent. Leaf mostly composed of quadrate cells 20μ wide, of which at least one row goes round the point, but the inner portion of the leaf has larger hexagonal cells 30μ long; cells all smooth. Inner per. leaves larger, with rounded apex, sheathing the sessile capsule but not covering it, and composed mostly of lax, long-hexagonal cells. Capsule oblong, white, half-exposed; lid flat, mamillate; calyptra short, bell-shaped, ridged, and lobed. Where described as *E. distichum* reasons are given why *E. distichum* should be separated from *E. grossirete*, but my further specimens where both conditions occur on the same plant remove that contention and sink the new name.

Natal. Maritzburg, Wager 226.

Transvaal. Barberton, Wager 279.

Portuguese East Africa. Hellett's Concession, Magude, 500 feet, March 1921, Sim 8988.

S. Rhodesia. Lo Mogundi, Eyles 2702.

Family 40. HEDWIGIACEAE.

Stems usually one to several inches long, erect, suberect, prostrate, or hanging; dichotomously, irregularly, or pinnately branched, rooting at the base only. Leaves surrounding the stem in eight rows, firm, nerveless, wide, pointed, with or without margin, usually papillose; the cells often small, usually roundish, in regular lines, the central basal portion often with longer cells; alar cells differentiated in *Rhacocarpus*. Inflorescence autoecious or dioecious, seldom synoecious, bud-like, the female inflorescence terminal either on the main or secondary branches. Calyptra either small and mitriform and naked or hairy, or several-lobed, or else large, cucullate, long-beaked, and naked. Capsule either immersed, shortly exserted, or on a long seta, urn-shaped, pear-shaped, or elliptic, smooth; lid widely convex or shortly conical. Peristome absent in South African genera. This small family has been variously placed. Schimper and also Dixon place it in Grimmiaceae, but I follow Brotherus in placing it among the pleurocarpous mosses, even though it has terminal fructification.

Synopsis of Hedwigiaceae.

Capsule immersed; papillae 2- or more-pointed.
 Leaves with a spinosely denticulate hyaline point . . . . 155. HEDWIGIA.
 Leaves not having a hyaline point . . . . 156. HEDWIGIDIUM.
Capsule more or less emergent; papillae mostly 1-pointed; sometimes smooth or almost so.
 Leaves not bordered; seta long or short; alar cells not differentiated . . . 157. BRAUNIA.
 Leaves bordered; seta long; alar cells prominent . . . . . . 158. RHACOCARPUS.

Genus 155. HEDWIGIA Ehrh.

Stems at first erect, afterwards straggling, one to several inches long, not stoloniferous, very loosely caespitose, slightly forked, and with shorter lateral branches, more or less hoary and when dry rigid. Leaves widely oval, nerveless, sometimes furrowed, without hyaline margin, without differentiated alar cells, but with a spinosely denticulate hyaline point; margin recurved below; cells roundish except the central at the base which are longer; upper cells with sinuose walls and strongly papillose, the papillae several per cell. Capsule immersed, subsessile, shortly urn-shaped, gymnostomous, not furrowed; lid widely convex, shortly pointed; calyptra conical, covering the point of the lid only, caducous; perichaetal leaves with long, diaphanous, long-ciliated points. Autoecious, seldom synoecious.

531. *Hedwigia albicans* (Web.), Lindb., in Hartm. Skand. Fl., 9th ed., ii, 54 ; Broth., Pfl., i, 3, 715, and fig. 535.

=*Fontinalis albicans* Web., Spic. Fl. Goett., p. 38, No. 115.
=*Hedwigia ciliata* (Ehrh. MSS.), Hedw., Descr., i, 107 ; Dixon, 3rd ed., p. 171, Tab. 22A.
=*Hedwigia Macowaniana* C. M., in Flora, 1888, p. 415 (nom.), and in Hedw., 1899, p. 122, and Rehm. 596β and c.
=*Hedwigia Joannis Meyeri* C. M. (*fide* Broth.).
And many other European synonyms.

Autoecious, laxly caespitose, 1–3 feet high, usually suberect or ascending on basaltic rock, rigid when dry, whitish or hoary or very pale green, irregularly branched, often dichotomously branched below an old capsule, sometimes much-branched ; leaves round the stem in about eight rows, spreading when moist, but appressed or semijulaceous and closely imbricated when dry, except that the upper part or the hyaline points often spread outward, giving the plant a hoary appearance. Leaves 2 mm. long, widely ovate from rounded short base, very concave, nerveless, without striations, point acuminate, margins often narrowly revolute in the lower half, the upper half crenate and densely papillose ; above that denticulated by long hyaline cells, and the apex itself hyaline. Cells 20μ long, in straight separate lines, turgid, with several strong, pellucid papillae on each, the cells oblong or quadrate, with sinuose thick walls, but the central area toward the leaf-base has pellucid, somewhat porose, narrowly rectangular cells up to 70μ long ; the quite basal cells are shorter and often coloured, and the leaf-apex has a few elliptical cells up to 40μ long. The leaves and per. leaves are everywhere tumid and papillose on the back on lamina cells, and the leaves on prostrate lower branches are sometimes squarrose. Inflorescence at first terminal, soon lateral by innovation. Per. leaves like others or rather larger, but with four to six long, hyaline-jointed, smooth cilia on each side near the apex, each terminating one or several rows of papillose cells. Capsule red, shortly cup-shaped, wide-mouthed, almost sessile, immersed among the perichaetal leaves, composed of large square cells, and without ring or peristome ; lid convex-pointed ; calyptra small, fugacious, yellow, conical, only covering the lid. When prostrate or suberect the leaves (or some of them) are often secund. Frequent on mountain rocks. I have no specimen from Western Cape Province.

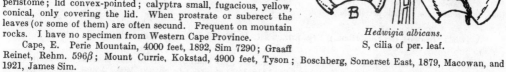

Hedwigia albicans.
S, cilia of per. leaf.

Cape, E. Perie Mountain, 4000 feet, 1892, Sim 7290 ; Graaff Reinet, Rehm. 596β ; Mount Currie, Kokstad, 4900 feet, Tyson ; Boschberg, Somerset East, 1879, Macowan, and 1921, James Sim.

Natal. Hemuhemu, Sim 10,265 ; Karkloof, Rehm. 496c ; Drakensberg, Wylie ; Van Reenen, Wager ; Mont aux Sources, 7000 feet, July 1921, Sim 10,264.

Transvaal. Houtbush, Rehm. 596.

Genus 156. HEDWIGIDIUM Br. Eur.

Stems straggling, often stoloniferous or growing from prostrate old stems, loosely caespitose, rigid when dry, not hoary ; slightly branched, the branches sometimes flagellate ; leaves firm, nearly smooth, ovate-lanceolate, longitudinally plicate, nerveless, acute but the point not diaphanous ; margin revolute, the cells small, rectangular, erose, with sinuose walls, and several minute papillae ; the central cells at the base are linear. Capsule obovate-globose, with a narrow neck and short seta but hardly emergent, when dry wrinkled. Lid conical-rostellate ; calyptra small, cucullate. Inflorescence autoecious, terminal.

532. *Hedwigidium imberbe* (Sm.), Bryol. Eur., vol. iii, Mon., p. 3, t. 1 ; Schimp., Syn., 2nd ed., p. 281 ; Broth., Pfl., i, 3, 717.

=*Gymnostomum imberbe* Sm., Eng. Bot., t. 2237.
=*Hedwigia imberbis* Spruce, Musc. Pyr., p. 338 ; Dixon, 2nd ed., p. 175, and pl. 24, fig. J.
=*Neckera imberbis* C. M., Syn., ii, 105.
=*Braunia maritima* C. M., in Hedw., 1899, p. 124, and Rehm. 306, 307.
=*Hedwigidium maritimum* C. M., Par. Suppl. Ind., p. 179 (1900), and 2nd ed., p. 305 ; Broth., Pfl., xi, 70.
=*Braunia Macowaniana* C. M., in Hedw., 1899, p. 123, and in Rehm. 597 ; Par., Ind., 2nd ed., p. 167.
=*Braunia erosa* C. M., in Hedw., 1899, and Rehm. 305.

Autoecious, fairly vigorous ; stems 1–2 inches high, simple or irregularly branched, often rising from stolons or from prostrate old stems. Leaves densely crowded, in many rows, 1–1·5 mm. long, spreading when moist, widely ovate, shortly pointed, deeply concave-cochleate with revolute margins up to the flattened point, thence denticulate along the short acute apex. Leaves often closely appressed and somewhat 2–3-ridged when dry, especially on young stems, nerveless ; cells in separate straight lines, oblong, 15μ long, with sinuose walls and several minute papillae per cell on the outer side, but the central leaf-area at the base has long, narrow, pellucid, rectangular cells 70–100μ long. Leaf-apex usually chlorophyllose. Perichaetium and capsule not seen in South Africa, but the capsule is described in Europe as scarcely emergent with unciliated per. leaves, conical-rostellate lid, and small cucullate calyptra.

The Table Mountain plant distributed in Rehm. 305 and 306 has been considered identical with the nearly cosmopolitan *H. imberbe*, but has not yet been found fertile in South Africa. I have collected the same thing on the top of the Zwart Kop Mountain, Natal, 5000 feet altitude (Sim 8674, sterile), and have doubts that when capsules are found it may prove to be a form of *Braunia secunda* (which has long seta). My figure is from Rehm. 306 (Table Mountain, C.P.), in describing which as *B. maritima* C. M. cites also Lydenburg, Transvaal (Dr. Wilms), a far inland locality, which specimen I have not seen. C. M., in Hedw., 1899, p. 123, gives Boschberg, C.P., and Jammerlapen, Natal, as localities for *B. Macowaniana* C. M. (both sterile), and Rondebosch for *B. erosa* C. M. (sterile). Rehm. 307, from Table Mountain, distributed as *B. maritima* C. M. var. *rufescens*, agrees except that it is of brown colour and less erose.

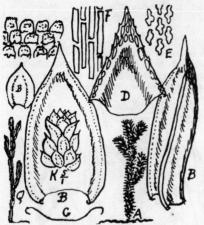

Hedwigidium imberbe.

Dixon, in Trans. R. Soc. S. Afr., viii, 3, 209, inclines to maintain *H. erosum* (C. M.), Par., but his description agrees with the above, and he adds : "The perichaetal leaves are erose above and subundulate at the margin, distantly ciliolate-toothed half-way down."

Genus 157. Braunia Br. Eur.

Stems irregularly branched, widely caespitose, with small-leaved stolons or flagellae ; leaves ovate-oblong-acuminate, striate or furrowed, nerveless, the apex erosely denticulate, often not hyaline, the margin slightly revolute ; cells mostly oblong-sinuose, the central basal cells vermicular, all minutely papillose or almost smooth, except toward the leaf-apex where large papillae occur in some species ; capsule usually on a longish seta, lateral, erect, usually gymnostomous, microstomous, with long clasping, perichaetal leaves ; lid conical ; calyptra cucullate.

Synopsis.

533. *B. secunda*. Without peristome.
(534. *B. peristomata*. With peristome.)

533. *Braunia secunda* (Hook.), Bry. Eur., Fasc. 29–30, vol. iii ; Dixon, Sm. Misc. Coll., lxxii, 3, 10 ; Broth., Pfl., xi, 71.

Dixon critically examined this species and gives the synonymy bearing on the African plant in above paper as :
" =*Hedwigia secunda* Hook., Musc. Exot., pl. 46, 1818–1820 (t. 99, *fide* Mitt.) ; Mitt., Journ. Linn. Soc., 1886, pp. 146, 310.
=*Neckera macropelma* C. M., Syn., ii, 104 (1851).
=*Braunia macropelma* Jaeg., Adumb., ii, 87 (1869–1870).
=*Hedwigia indica* Mitt., Journ. Linn. Soc. Bot., p. 3, Suppl., p. 123 (1859).
=*Braunia indica* Par., Ind., p. 149 (1894) ; Mitt., Journ. Linn. Soc., 1886, pp. 146, 310.
=*Neckera diaphana* C. M., Syn., ii, 105 (1851).
=*Braunia diaphana* Jaeg., Adumb., ii, 87 (1869–70)."
And he states that as Mitten includes the American plant also, that must be placed here.

Shaw cites as South African *B. Schimperiana* C. M. (Syn., ii, 104), and *B. julacea* Schw., which specimens may both belong here, or the latter to *Leucodon*.

Exceedingly variable, sometimes on the same plant, more frequently in different mats, or above and below in the same one. Autoecious, loosely caespitose, usually suberect, somewhat stoloniferous or flagellately rhizomatous, more or less pinnately or irregularly branched, the older brown branches and often all branches with leaves squarrose when moist and sometimes when dry also, the younger glaucous or light green branches usually terete and somewhat julaceous when dry, with appressed leaves which are suberect when moist ; flagellate branches are also sometimes formed, which, when young, have narrower leaves with long, hyaline, crispate points forming a woolly tuft, and on these flagellate branches the further leaves are reduced till the stem is almost nude, ending in a tuft of brown rhizoids ; these rooting flagellae may occur on any part of a stem from the base to near the apex. Leaves widely oval, deeply concave, with a variable point, usually short, green, scabrid or jagged, sometimes thin, hyaline, smooth or denticulate, and more or less long. Leaves often surrounding the stem regularly, and with the point slightly reflexed all round, but often on the younger branches the leaves are secund or falcate secund, or all the leaves may be so. When moist the leaves are deeply cochleate-concave and more or less striate, when dry the older ones (and sometimes all) are flatter and distinctly 3-ribbed ; nerve absent, margin usually revolute and entire, apex erosely scabrid ; cells mostly oblong-sinuose, the central basal cells vermicular, all cells minutely papillose except those near the leaf-apex, which are strongly papillose, with one to three high white papillae per cell on the inner cells on both sides, as also on the marginal cells. Perichaetium lateral, its inner leaves long and closely sheathing the seta, some of the outer leaves often narrow and long-pointed. Seta ½ inch long ; capsule erect, cylindrical or narrowly elliptical with tapering neck, gymnostomous, red, 2–3 mm. long, without ring ; lid sharply and obliquely beaked ; calyptra small, stramineous, cucullate, glabrous, evanescent.

Dixon places his *B. brachytheca* (in Sm. Misc. Coll., lxxii, 3, 10) from Central Africa as a subspecies of this, differing only in having the capsule subglobose or shortly elliptical, without neck, subplicate when old, which has not as yet been found in South Africa.

The European *B. alopecura* (Brid.) Limpr., which also has been credited to South Africa, differs, according to Brotherus, in having a short lid-point, and has not been seen here; and Brotherus, in Pfl., i, 3, 718, mentions a *B. squarrosula* (Hamp., as *Harrisoma*), having globose capsule and beaked flat lid, as South African, concerning which I know no more except that Paris credits it to Central America only, under the name *Harrisoma squarrosula* (Hampe), Br. Eur.

Brotherus also (in Pfl., i, 3, 1) placed *B. diaphana* (C. M.), Jaeg., in § *Macromidium*, which has almost immersed fruits; but Müller's description was from sterile plants, and no almost immersed fruits of this genus have been seen here. In Pfl., xi, 71, Brotherus brings it into *B. secunda*, and in Pfl., xi, 72, he also, doubtfully, places *B. Elliotii* Broth., for which see *Leucodon assimilis* herein.

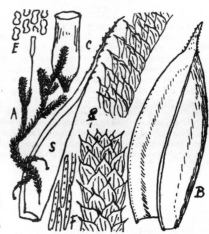
Braunia secunda.

Forma *longipila* has all the leaves on some branches long and dentate, but other branches on the same plant are on the usual plan. Other specimens have the leaf-point long and very scabrid, the longish cells there having three to four long papillae, while the margin has them irregularly large or small, and single or in pairs.

B. secunda is widely distributed through Africa, India, and Central America, usually at fairly high altitudes.

Cape, W. Slongoli, Sim 9207; Paarl Rock, Sim 9627.

Cape, E. Kakazela Mountain, Dohne, 1898, Sim 7206, 7216; Boschberg, Somerset East, Rehm. 598*d*; Cave Mountain, Graaff Reinet, Rehm. 598*c*; Enyembe, East Griqualand, 6500 feet, Tyson.

Natal. Mont aux Sources, 7000 feet, Sim; Giant's Castle, 8000 feet, R. E. Symons; Van Reenen, Wager 163; Bulwer, Haygarth 20.

Transvaal. Houtbush, Rehm. 598; Snellskop, Lechlaba, Rehm. 598*β*.

S. Rhodesia. Salisbury, Eyles 1572; Zimbabwe, Sim 8828, 8738, 8745, 8811, 8779; Odanzi River, Manica, A. J. Teague 166; Makoni and Macheka, Eyles 835 and 1993; Matopos, Mme. Borle 56, Eyles 1026, and Sim 8871 and 8849; Marandellas, Eyles 3884, 3887.

Var. *pinnata* Sim (new variety).* Stems 3 inches long, erect, unforked, but pinnately many-branched; the branches mostly 1–2 cm. long, drooping, flagellate, and gradually reduced, with nude ends bearing terminal rhizoid tufts. Leaves less concave than in the type, from 2 mm. length downward, the margin of normal leaves is recurved for the whole length, the point chlorophyllose and scabrid, and the cells there with several large, scattered, single or clustered papillae; but the depauperate leaves are long and narrow, and often more or less hyaline. Perichaetium

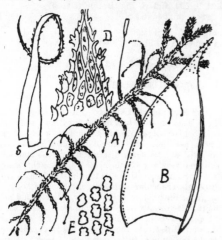
Braunia secunda, var. *pinnata.*

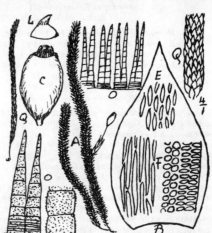
Braunia peristomata.
O, peristome teeth and enlargements of same.

---

* *Braunia secunda* Bry. Eur., var. *pinnata* Sim (var. nov.).
Caules 3″ longi, pinnatim pluri-ramosi, rami plerumque breves, flagellati, radicantes. Perich. in caule principe. Cetera ut in typo.

lateral on the main stem, short, its inner leaves long and closely convolute, shortly pointed ; seta ½-inch long or more ; capsule cylindrical, red. Giant's Castle, Natal, 8000 feet altitude, 1915, R. E. Symons (Sim 10,266). Dixon has it also from Angola. What Wager 163, from Van Reenen, wrongly distributed as *Leucodon assimilis* belongs here.

534. *Braunia peristomata* Dixon, in S.A. Sc. Journ., xviii, 324 (1922).

Autoecious ; laxly caespitose on trees and stones, 1–2½ inches long, vigorous, simple or few-branched or sometimes many-branched, suberect ; older parts brown, with permanently squarrose leaves, younger parts shining and silvery, julaceous, with leaves closely appressed when dry ; leaves 2 mm. long, densely imbricate, ovate-acuminate, with erect entire margins, no nerve, and the lamina cells elliptical-acute, 25μ long, smooth, separate ; the cells toward the leaf-base centre much longer and narrowly vermicular, up to 100μ long, while the outward cells there grade through elliptical-acute and rhomboid to nearly round, 15μ wide, with one or two lines narrower and set transversely. Seta 1 cm. long, the lower half closely sheathed by a few long perichaetal leaves ; capsule oval, or shortly elliptical, microstomous, without ring, but with well-developed, single peristome set low in the capsule-mouth, and inflexed into it when dry. Teeth in pairs, 150μ long, pale, slender at the points, closely trabeculate below, laxly trabeculate above, and densely papillose ; capsule-wall red, of large quadrate cells, smaller at the mouth. Lid convex with short oblique beak. The smooth elliptical areolation and general habit resemble *Leucodon assimilis*, and Dixon agrees in letter of 25th April 1925 that these may prove to be identical, in which case *Braunia* as a genus is eperistomate and *Leucodon* is peristomate.

Zimbabwe Ruins, S. Rhodesia, 3000 feet, common, Sim 8750, 8778, 8793, 8809, 8743 ; Fort Victoria, Sim 8843.

Genus 158. RHACOCARPUS Lindb.

Stems several inches long, usually fairly robust, irregularly or almost pinnately branched, growing in wide compact mats ; dioecious, both sexes with inflorescence terminal on short branches. Leaves deeply concave, spreading, nerveless, oblong-acute, often with apical hair or bristle, the margin erect, slightly denticulate near the apex, elsewhere having a smooth, pellucid, or yellowish border of long cells not or only sparingly papillose. Leaf finely and minutely papillose-granulate on both sides ; cells linear or vermicular, those at the base often yellowish brown, smooth. Alar auricles pronounced, concave, with numerous quadrate cells, sometimes thick-walled and discoloured. Perichaetal leaves longer, erect, convolute, without border, the alar cells not differentiated. Seta ½ inch long, straight ; capsule erect, globose, urn-shaped, furrowed when dry ; lid convex-rostrate ; calyptra cucullate, nude, evanescent.

Synopsis.

535. *R. Ecklonianus.* Leaves sharply acute or hair-pointed, branches pointed.
536. *R. Rehmanni.* Leaves subacute, branches blunt.

535. *Rhacocarpus Ecklonianus* (C. M.), Broth., Pfl., i, 3, 722, and xi, 75 ; Brunnthaler 29 ; Broth., in Marloth's Flora of South Africa, p. 254, and pl. 7.

=*Harrisonia Eckloniana* C. M., in Oesterr. Bot. Zeitschr., 1897, xlvii, 398.
=*Rhacocarpus *piliferus* (Rehm.), Par., Ind., 1897, and 2nd ed., iv, 146.
=*Harrisonia *pilifera*, Rehm. 314 (not 197 as cited by Paris).
=*Rhacocarpus cuspidatus** (Rehm.), Par., Suppl. Ind., p. 291 (1900), and 2nd ed., iv., 144 (sterile).
=*Harrisonia cuspidata**, Rehm. 312.
=*Harrisonia gracillima* C. M., in Oesterr. Bot. Zeitschr., 1897, xlvii, 391 (sterile).
=*Rhacocarpus gracillimus* (C. M.), Broth., Pfl., xi, 74.

Dioecious, caespitose or scattered in wet places, naturally in closely pinnated stems 2–3 inches long, often in detached pieces or fragments ; upper pinnae crowded, 3–6 mm. long, each tapering to a rigid point from a wide base ; lower branches less pointed. Leaves 2 mm. long, closely imbricated, oblong-acute or -acuminate, very concave-clasping, imbricate and often cochleate, with incurved red alar lobes ; nerveless ; the margin erect or often convolute upward, slightly revolute near the base, often indurated and coloured red for some distance above the alar lobes, thence slightly denticulate and usually bordered by long hyaline cells in several rows to the apex, which is either sharply pointed (form *cuspidatus*) or extending as a long, thin, hyaline hair-point, sometimes as long as the lamina (form *piliferus*), but these forms intergrade ; lamina cells irregularly elliptical-vermicular, either in contact or adjacent and with translucent holes between, densely and minutely granular-papillose ; central lower cells longer, pellucid, and somewhat connected by pores. Alar cells quadrate, indurated, red, smooth, with usually long narrow lumen. Perichaetium terminal on short upper branches, often on the terminal branch ; per. leaves hardly different but longer, more convolute round the seta, shortly hair-pointed and without border or alar groups ; seta about ½ inch long ; capsule erect, roundly oblong, red, when dry wrinkled ; peristome and ring absent ; lid convex, with sharp oblique beak ; calyptra cucullate, with somewhat lobed base.

Abundant on Table Mountain in every stream and sphagnum patch, absent elsewhere so far as is known, seldom fertile, and the leaf very variable as to the extent of hair-point, and as to its coloration.

R. Humboldtii (Hook.), Lindb., though recorded as a variety from Madagascar has not yet been detected in South Africa, and the whole genus requires collation with many specimens from all known habitats, especially in view of the curious geographical distribution in South Africa.

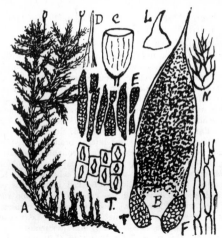

Rhacocarpus Ecklonianus.

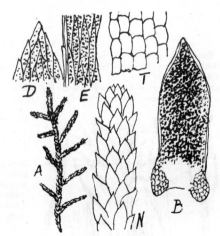

Rhacocarpus Rehmanni.

536. *Rhacocarpus Rehmanni* (C. M.), Par., Ind., p. 1070, and 2nd ed., iv, 146; Broth., Pfl., i, 3, 720, and xi, 74.

=*Harrisonia Rehmanni* C. M., in Rehm. 311 (not 270 as cited by Paris), and in Oesterr. Bot. Zeitschr., xlvii, 398 (1897).
=**Anoectangium Humboldtiana* Breut., in Sched.
=*Rhacocarpus *cucullatus* (Rehm. 313, as *Harrisonia*), Par., Ind., 1897, p. 1069.
=*Rhacocarpus Breutelianus* (C. M.), Broth., Pfl., i, 3, 720, and xi, 74.
=*Harrisonia Breuteliana* C. M., in Oesterr. Bot. Zeitschr., xlvii, 398 (1897).
=*Rhacocarpus *capensis* (W. P. Sch.), Paris, Suppl. Ind., 1900, p. 291.
=*Harrisonia *capensis* W. P. Sch., in Breutel, Musc. Cap.

Dioecious; laxly caespitose in very wet places or among sphagnum, etc.; 1–2 inches long, pinnately or irregularly branched; branches not tapering, blunt; leaves 1·5 mm. long, oblong, shortly pointed, with blunt, slightly denticulate apex; deeply concave, with erect, slightly bordered margin; the central basal cells pellucid, long and narrow; alar cells roundly quadrate with thin walls; and the lamina cells widely vermicular and densely granular-papillose. Other characters not seen. Very near *R. Ecklonianus.*

Cape, W. Table Mountain, Sim 9191, 9119, 9163, 9272, 9295; Montagu Pass, Rehm. 311.

Family 41. FONTINALACEAE.

Aquatic plants, in many cases constantly submerged in flowing streams, the rooting base attached to stones, and the stems several or many inches long, irregularly or somewhat pinnately much-branched, pliant, floating out in the stream and not again rooting, but in *Wardia* the stems are shorter and rigid. Leaves usually in three or five rows, smooth, more or less oval-acuminate, often without or with narrow nerve, and with the cells all long and narrow except the alar cells which are sometimes much wider. Inflorescence bud-like, terminal or lateral on short lateral branchlets from the main stems; perichaetal leaves imbricated and sheathing; capsules usually immersed, but in *Wardia* exserted on a ½-inch seta, erect, regular, without throat; ring absent; peristome double, single, rudimentary or absent, when present the outer teeth sixteen, linear, barred; the inner either of free cilia or more or less connected in a 16-keeled membrane. Abundant in the streams of the Northern Hemisphere, represented in South Africa by only two genera, one of which is endemic and monotypic, the other having two species here, is probably introduced.

Synopsis.

159. WARDIA. Capsule on a long seta.
160. FONTINALIS. Capsule immersed.

Genus 159. WARDIA Harv.

Dioecious aquatic plants found in the mountain streams of Western Cape Province, with fairly strong, rigid, much-branched stems up to 2 inches long, the leaves in five rows, widely lanceolate, without nerve, and with linear-acuminate, somewhat twisted cells, those at the base little different except the alar cells, ten to twelve on each side, which are as long as, but much wider than the others, being roundish oblong. Perichaetium lateral as a short branchlet, many-leaved, the upper large, wide, and rosulate; seta ½ inch long, twisted; capsule erect, oval, wide-mouthed, afterwards urn-shaped, without ring, and with a rudimentary peristome consisting of a short, erect, irregular membrane, longitudinally and transversely striate, papillose, but not distinctly toothed. Lid shortly conical, obliquely beaked, dehiscent, but remaining attached to the top of the columella. Calyptra smooth, regular, gradually narrowed from base to apex, and split on one side the whole distance. Male inflorescence unknown.

23

537. *Wardia hygrometrica* Harv., in Hook., Comp. of the Bot. Mag., ii, 183, t. 15 ; Card., Monogr. Fontin., p. 129 ; Broth., Pfl., i, 3, 723, fig. 541, and xi, 55, fig. 471.

=*Neckera* (§ *Harrisonia*) *hygrometrica* C. M., Syn., ii, 667.

Dioecious ; stems 1–3 inches long, attached to stones in running streams, the older stems black and wiry with all the leaves broken off, simple below but repeatedly branched above, the branches more or less clothed in erecto-patent leaves, except that toward the point usually the leaves are closely imbricate-convolute, forming a solid tapering point ; on simple basal regrowths these points are less evident.   Leaves 1–1·5 mm. long, widely oval, nerveless, deeply concave, with erect or incurved, entire, unbordered margins, the point shortly acuminate and acute, and the base wide, with

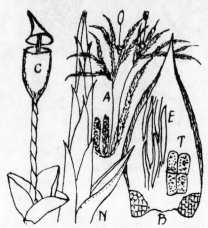

Wardia hygrometrica.

two more or less evident alar areas having large, oblong, granular, colourless cells, while the lamina cells are thick-walled, vermicular and hardly granular, almost pellucid when old, deep green when young.   Perichaetium terminal on short lateral branch, with short, rounded, concave, perichaetal leaves having vermicular areolation. Seta ½ inch long, thick, exceedingly hygrometric ; capsule oval-oblong, grey or dark, rather wide-mouthed, with central columella to which the convex, straight, or slightly obliquely beaked lid remains permanently attached, though rising off the capsule ; peristome rudimentary, consisting of an irregular, roughly papillose, yellowish membrane.   Ring absent ; calyptra cucullate, acute.   Leaves and branches sometimes subfalcate, or affected by stream current.

Frequent in every stream on Table Mountain, C.P., over about 2000 feet altitude, also present at Tokai and Montagu Pass, H. A. Wager 580, and Stellenbosch Mountain, Garside 69 ; but absent from all eastern mountains.   A wonderful moss, with a wonderful endemic distribution.

Genus 160. FONTINALIS Dill.

Aquatic dioecious mosses, usually floating in running water, rooting at the base only, with stems simple and rigid below, much-branched above, either leaved all round or in three rows, and usually pointed at the apex.   Leaves lanceolate or wider, 2–4 mm. long, keeled or plane, entire, nerveless, pointed, and with rather distinct alar groups of quadrate cells.   Cells vermicular. Per. leaves shorter and rounded ; fructification not found so far in South Africa, but in Europe and elsewhere the capsule is lateral, almost immersed, with double peristome of red lanceolate teeth, and inner dome of sixteen cilia united by transverse processes.

Synopsis.

538. *F. antipyretica.*   Stems many inches long, leaves folded or keeled.
539. *F. Duthieae.*   Stems 2–3 inches long, leaves not folded.

538. *Fontinalis antipyretica* Linn., Sp. Pl., p. 1571 ; Broth., Pfl., xi, 58, pl. 474 ; Dixon, 3rd ed., p. 390, Tab. 48c, and many other authors.

=*Pilotrichum antipyreticum* C. M., Syn., ii, 148.

Aquatic, the stems attached to stones at the stem-base only, pliable, and floating out freely, sometimes 12 inches in length, clustered, leafless and rigid at the base, abundantly pinnately branched upward ; leaves on the main stem 3–4 mm. in length, 1–1·5 mm. wide, nerveless but usually somewhat folded or keeled, oblong-lanceolate, acute, or sometimes long-pointed, with erect entire margins and often one side revolute near the base ; the apical leaves convolute and forming a solid tapering stem-point.   Lamina cells shortly vermicular or sometimes long-vermicular, the lower cells usually wider and grading into a rather clearly defined alar group in which the cells are wide, shortly hexagonal, thin-walled, and colourless.   Leaves on branches much smaller, the lower triangular, but with alar cells along the base (fig. S, page 355).

Found in some abundance in the Mill Stream, Stellenbosch, by Miss Duthie, June 1919, a foot long, but without capsules, also by Garside, at Jonker's Hoek.   The leaves are often shredded, apparently by stream-current.   It is a common moss, in several varieties in European streams, and is likely introduced here, probably with trout ova, but it may be that only one sex is present.   The common form in Europe has its leaves in three rows, so the stems are more or less trigonous, but there are forms there in which that does not show, and in the Stellenbosch plant it is not apparent.   The capsule shown in fig. on page 355 is from a European specimen.

539. *Fontinalis Duthieae* (Dixon MSS.) Sim (new species).*

Aquatic or semiaquatic, submerged or suberect, matted ; stems 2–3 inches long, the lower part black and rigid, simple and nude, or having only scattered, short, ovate-lanceolate, acute leaves or the remains of damaged leaves, the

---

* *Fontinalis Duthieae* (Dixon MSS.) Sim (sp. nov.).
Aquatica vel semi-aquatica ; caules 2–3″ longi, atri, infra rigidi ac simplices, supra libere ramosi.   Folia crebra, patula, late lanceolata, nervo carentia, 2 mm. longa, a tergo plana, ad basin connexa et concava, marginibus explanatis integris, cellis vermicularibus, cellis alaribus oblongo-quadratis.

upper part freely branched, set with numerous spreading, widely lanceolate, nerveless leaves 2 mm. long, flat on the back, bristling all round, and at the branch-tips closely imbricated and convolute, forming a solid pointed tip from a

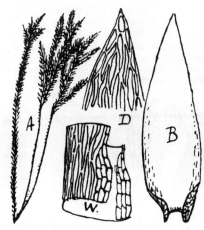

Fontinalis Duthieae.

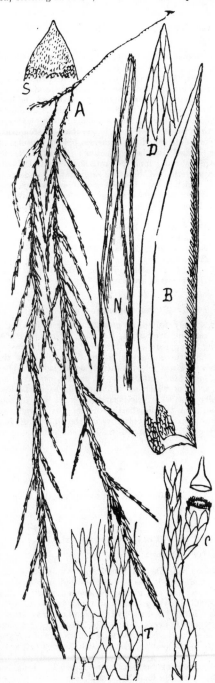

Fontinalis antipyretica.
S, lower leaf of branch.

wider base. Leaves clasping and concave at the base, nearly flat above, with expanded or rising entire margins, vermicular lamina cells, and oblong-quadrate, usually colourless or rusty alar cells. Other features not seen.

Platteklip Rock, Cape Town, C.P., 500 feet, June 1919, Sim 9389 ; The Saddle, Table Mountain, C.P., 2200 feet, Sim 9385.

Family 42. CRYPHAEAE.

Primary stems or rhizomes procumbent, leafless ; main stems erect or hanging, distichously but irregularly pinnately branched ; leaves set equally round the stem, those on the main stem larger than those on the branches, usually more or less ovate-lanceolate, with nerve to near the apex; the cells elliptical, smooth or papillose; the outer lower cells smaller and quadrate. Perichaetia lateral, sessile, or forming a short branchlet, with long hair-pointed perichaetal leaves among which the sessile capsule is immersed. Calyptra small, conical, rough ; peristome usually double.

Genus 161. CRYPHAEA Brid.

Stems irregularly subpinnately or sometimes 2-pinnately branched, usually epiphytic on bark ; leaves erecto-patent or spreading when moist, imbricate when dry, ovate-lanceolate, the nerve single and strong, not reaching the apex; cells oval or elliptical, nearly smooth, the outer lower cells smaller and quadrate. Inflorescence autoecious ; flowers numerous ; perichaetia sessile or on short branches, with many large, convolute, hair-pointed leaves, among which the capsule is immersed. Calyptra small, conical, split at the base ; ring wide ; peristome double, the outer of sixteen linear-lanceolate, articulated granular teeth confluent at the base, the inner of sixteen processes, filiform above, keeled below, and united in a narrow basal membrane. A tropical and subtropical genus, rare in South Africa.

540. *Cryphaea exigua* (C. M.), Jaeg., Ad., ii, 95 ; Broth., Pfl., i, 3, 741, and xi, 79.

=*Pilotrichum exiguum* C. M., Syn., p. 166.
=*Cryphaea dentata* Mitt., in Linn. Soc. Journ., 1886, pp. 146, 311 ; Broth., Pfl., i, 3, 741, and xi, 79.
=*Cryphaea *Breuteliana* W. P. Sch., in Breutel's Musc. Cap.

Autoecious ; plants erect, 1–3 inches high, rigid, pinnately few-branched, the branches fairly long and distichous, and perichaetia sessile laterally on the upper part of the main stem. Leaves imbricated, erecto-patent, closely appressed when dry, 2 mm. long, widely lanceolate with acuminate apex, the nerve strong, ending some distance below the apex ; the margins narrowly revolute in the lower half and coarsely few-toothed or serrate toward the apex. Cells

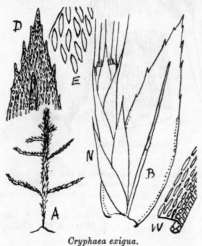

Cryphaea exigua.

all elliptical, almost smooth, 20μ long, except those toward and at the margin near the base, which are quadrate and only 10μ long, the transition being gradual. Branch leaves similar to stem-leaves but smaller. Perichaetium single, few-leaved, longer than the stem-leaves, the leaves closely convolute round the immersed oblong capsule, with long, slender, few-toothed subulae projecting ; lid short, conical, acuminate ; calyptra glabrous ; inner perianth shorter than the outer.

Mitten must have overlooked Müller's *P. exiguum* when he described *Cryphaea dentata*, in which he calls the leaves ovate-acuminate, which hardly differs, and other characters as similar. *C. laxifolia* Mitt., from Usagara, seems to differ mostly in its leaves being more spreading when moist.

Cape, W.   Portland, Knysna, Rehm. 315.
Cape, E.   Phillipstown (Hb. Göttsch., *fide* C. M.).
Natal.   Umgoya Mountains, Zululand, Mr. Plant, and also Mr. Keit (*fide* Mitt., as *C. dentata*).
Transvaal.   Lechlaba, Houtbush, Rehm. 600. (I have doubts about this specimen ; it is mixed and sterile.)

Family 43. PRIONODONTACEAE.

Very vigorous, loosely tufted, or gregarious mosses hanging on bark or standing erect on tree-humus, the long, ultimately leafless, rooting stems having the appearance of rhizomes but actually prostrate old stems, from which the numerous vigorous young stems rise ; these are erect, simple or irregularly pinnately branched, 1–6 inches long, not root-producing ; scale clad at the base, thence with numerous leaves all round the stem ; leaves 2–4 mm. long, lanceolate, furrowed, serrate, nerved to near the apex, and with numerous lax, roundish-oval, elliptical cells, most of which have one papilla each, the lower central cells long-hexagonal and the lower marginal cells gradually shorter. Fructification not yet seen in South Africa, but elsewhere it is known to be lateral, the seta very short, the capsule shortly emergent, oval, erect, with obliquely short-beaked, conical lid, and a small, conical, cucullate, entire calyptra only covering the lid. Ring broad, dehiscent. Peristome double, teeth and processes long, strongly papillose, arranged in a high and a low cone, and variously connate or free at the apex. Cilia absent.

Genus 162. PRIONODON C. M.

The only genus, growing on trees or decayed stumps in tropical and subtropical wet forest, mostly American, but found also in Central Africa, though not found fertile as yet in Africa.

541. *Prionodon Rehmanni* Mitt., in Journ. Linn. Soc., 1886, pp. 146, 311 ; Broth., Pfl., xi, 114.

=*Prionodon *africanus* Rehm. 606 and 606β.

Prostrate stems 1–6 inches long, black, wiry, with a few old leaves still adherent, producing erect or ascending or drooping stems 3–8 inches high and ¼–1 inch apart, which have only small scale-leaves near the base but are abundantly leafy upward, irregularly branched ; branches in contact with soil, rooting and giving out slender, nearly leafless, rhizomatous branches, which on rooting produce an erect leafy stem. Leaves produced all round the stem, erecto-patent, often more or less secund, 3–4 mm. long or more, flat, from a broad base, widely lanceolate, often tapering to a long point, exceedingly brittle, few leaves retaining their points, when dry deeply 2-ribbed (4-furrowed). On young active shoots the leaves are longer, and with longer and more slender point than those grown more slowly. Margin coarsely toothed, the teeth irregular as to size, some at right angle to the margin, others much inclined forward or hooked, usually consisting of one large hyaline cell and several smaller cells. Apex acute, usually strongly toothed ; nerve strong, brownish, extending almost to the apex ; cells small, quadrangular, with round lumen 15μ wide, each having one mamillate papilla, those at the base of the leaf longer, up to 30μ long, hexagonal, and mostly with one papilla, but gradually toward the margin they are reduced to small quadrate cells 8μ wide in many rows. Inflorescence and fruit not seen. Mitten's description *folia subcompressa* hardly applies in the forest, though the plant appears so after drying under pressure. Abundant in the very wet parts of forests in Cape Province, Natal, and Transvaal, and also in East Africa.

Cape, E. Perie Forest, 1892, Hogsback, 1898, etc., Sim 7126, 7087 ; Perie, B. H. Dodd 1270.

Natal. Townbush, Maritzburg ; Overwood, Polela ; Zwart Kop ; Karkloof, Sim 8570, 9365, 9367, 8569 ; bush below Cathkin Peak, Sim.

Transvaal. Lydenburg, MacLea (Rehm. 606) ; Houtbush, Rehm. 606β.

Known also in Central Africa and Mauritius.

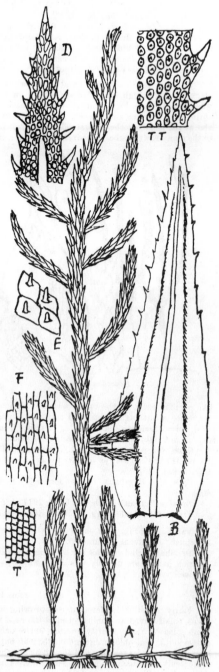

Prionodon Rehmanni.

Family 44. Leucodontaceae.

Mostly vigorous, usually dioecious, rather firm plants, growing loosely on bark or stumps, having prostrate, stoloniferous, main stem, and numerous erect or pendulous, simple or branched, much-branched or subpinnately branched, non-rooting, secondary stems. Leaves equally placed all round the stem, symmetrical, imbricate, spreading or secund, more or less ovate-acute, often somewhat decurrent, firm, shining, sometimes furrowed, 1-layered, without border ; the nerve slender, simple, double, or absent ; the cells rhomboid or linear, usually smooth, those toward the base vermicular ; the marginal basal cells numerous, quadrate or rounded ; no distinct alar cells in South African genera. Paraphyllia absent in South African genera. Inner perichaetal leaves long and sheathing ; vaginula long, nude or nearly so ; lid obliquely beaked from a conical base ; calyptra large, dimidiate, cucullate or extending below the capsule, smooth or somewhat hairy. Capsule on a long or short seta, hardly emergent or well emergent, erect or cernuous, symmetrical. Peristome simple or double ; outer teeth well developed, usually papillose and occasionally striate ; inner with low basal membrane, and usually the processes rudimentary or absent, though occasionally well developed. Cilia absent. In my check-list I included *Porotrichum, Porothamnium,* and *Thamnium* in this family, but they are now divided up and removed to Neckeraceae.

Synopsis of Leucodontaceae.

Nerve absent ; leaf-cells smooth . . 163. Leucodon.
Nerve single ; leaf-cells smooth . 164. Forsstroemia.
(Nerve single, or with one strong nerve and one
 or two small side-nerves at its base . Antitrichia.)
Nerve double ; leaf-cells papillose . . 165. Pterogonium.

Genus 163. Leucodon Schw.

Fairly robust mosses growing on stones or bark, with prostrate, wiry, main stems, and numerous erect, simple or forked, leafy stems, often stoloniferous, rooting at the base only, and closely leaved all round the stem. Leaves symmetrical, ovate or ovate-lanceolate, entire or toothed at the apex, sometimes longitudinally furrowed, nerveless, imbricated when dry ; cells smooth, oval to vermicular ; the central basal cells long and narrow, gradually merging through obliquely oval to quadrate, and small at the margin ; lower cells sometimes rusty brown. Perichaetium lateral, high on the stem, its leaves long and closely sheathing. Calyptra cucullate, enclosing the whole capsule and drawn close below it. Lid beaked, straight, from a conical base. Seta fairly long ; capsule oval, smooth, small-mouthed ; ring dehiscent. Peristome normally double ; the inner sometimes absent or rudimentary ; outer teeth lanceolate, slender, papillose, split or forked.

542. *Leucodon assimilis* (C. M.), Jaeg., Ad., ii, 124 ; Broth., Pfl., i, 3, 749, and xi, 92.

=*Neckera assimilis* C. M., Syn., ii, 92.
=*Leucodon capensis** W. P. Sch., in Breutel, Musc. Cap. ; Renauld Prodr. ; Broth., Pfl., i, 3, 749, and xi, 92.
=*Braunia Elliotii* Broth., in Engl. Bot. Jahrb., 1897, p. 253, sterile, Zambesia (*fide* Dixon) ; Dixon, Trans. R. Soc. S. Afr., viii, 3, 209 ; Broth., Pfl., i, 3, 718 ; Dixon, Torrey Bot. Club., xliii, 70.
(See also under *Braunia peristomata*, which may be identical.)
(Concerning *Leskea maritima* Hook., see note under *Rhaphidorrhynchium* ; it evidently belongs here.)

Primary stem black, wiry, and rhizomatous, sending up at close intervals erect, leafy, terete, slightly branched stems 1–3 inches high, scale-clad at the base, on which upward the many leaves are erecto-patent when moist, julaceous when dry, 1·5 mm. long, roundly ovate-concave and shortly acuminate, closely imbricated, clasping, nerveless, with entire erect margins and the cells smooth, the upper cells with oval-elliptical lumen, the mid-cells longer, elliptical, and at the leaf-base the central cells are vermicular, gradually changing outward to elliptical-subtransverse, thence to the many rows of small, quadrate, marginal cells, but the leaf-point cells are often all tumid, making an appearance

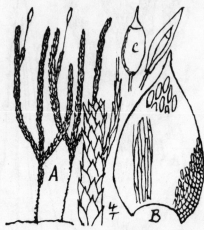

Leucodon assimilis.

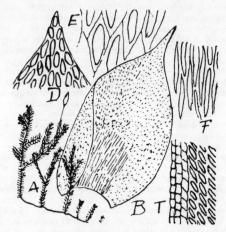

Leucodon assimilis, var. *humilis.*

like papillae there on the upper surface. Perichaetium lateral near the stem-apex, its leaves 4 mm. long, closely convolute, sheathing the seta, and with short subulate point ; seta 1 cm. to ½ inch long ; capsule brown, oval or elliptical, microstomous ; lid rostrate, straight, from conical base ; calyptra white, cylindrical, acute, enclosing the capsule but tight below it, split high on one side. Peristome not seen in good condition. Found also in Madagascar.

Cape, W. Grootvadersbosch, Swellendam, Dr. Pappe ; The Wilderness, George, Miss Taylor ; Blanco, Knysna, Rehm. 319 and 320 (as var. *gracilis* Rehm., which is much mixed with *Squamidium* and fertile *Macromitrium*) ; Camps Bay, Rehm. 319β.

Cape, E. Hogsback, D. Henderson 334 ; Addo Bush, Ecklon ; Perie Forest, Sim 7270 ; Hankey, 1922, Jas. Sim.

Natal. Burdon's Bush, Karkloof, 1917, J. M. Sim ; Town Bush, Maritzburg, Sim.

Transvaal. Lydenburg, MacLea (Rehm. 605β).

Var. *humilis* Sim (new variety).* Only 1 inch high, more bushy, with leaves permanently spreading. Lechlaba, Houtbush, Transvaal, Rehm. 605.

Dr. Shaw included in his list *Braunia julaceus* Schw., as collected by Shaw on Katberg and found in India and America, which presumably was *Leucodon assimilis* mistaken for *L. julaceus* (Linn.), Sull., a North American near relative.

Genus 164. FORSSTROEMIA Lindb.

Fairly robust, stoloniferous mosses growing on stones or bark, with numerous erect, irregularly pinnately branched stems, small-leaved and unbranched at the base and closely leaved above, with erecto-patent leaves set all round the stems ; leaves ovate, shortly pointed, entire or almost so, and with the margin somewhat reflexed. Nerve single ; cells smooth, oval-oblong throughout the leaf, but smaller and quadrate toward the margin at the base. Perichaetium lateral, its inner leaves long and sheathing the short seta ; capsule emergent, oval to cylindrical ; lid shortly conical, acute ; calyptra cucullate, usually set with erect hairs. Peristome teeth lanceolate, faintly papillose above, sometimes more or less split ; inner peristome rudimentary or absent. Brotherus (Pfl., xi, 87) includes this in Cryphaeaceae.

---

* *Leucodon assimilis* (C. M.), Jaeg., var. *humilis* Sim (var. nov.).
Tantum 1″ altus, caespitosior quam typus, foliis constanter patulis.

543. *Forsstroemia producta* (Horns.), Par., Suppl. Ind., p. 167 ; Broth., Pfl., i, 3, 1, 759, and xi, 88.

=*Neckera producta* Horns., in Mund and Maire's Musc. Cap. ; C. M., Syn., ii, 94.
=*Pterogonium productum* Horns., in Linnaea, 1841, p. 138 (except errors).
=*Braunia producta* (Horns.), Shaw's Catalogue, p. 381.
=*Dusenia producta* C. M., in Hedw., 1899, p. 128.
=*Lasia producta* Jaeg., ii, 108 ; C. M., in Rehm. 317.

A fairly vigorous, erect bark-moss, 1½–2 inches high, rising in somewhat dendroid manner from prostrate, black, wiry rhizomes or fallen stems (as leaves sometimes remain on them) ; the stems firm, irregularly pinnate, or sometimes slightly 2-pinnate, with numerous leaves, horizontal when moist, subjulaceous when dry ; the leaves 1–1½ mm. long, shortly pointed, widely ovate, concave, and clasping, not plicate, entire or almost entire, strongly nerved more than

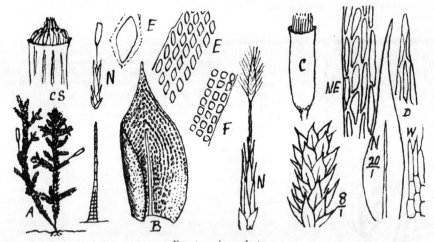

Forsstroemia producta.
CS, dry condition of peristome and capsule.

halfway, the margin often reflexed but not revolute ; cells oval at the apex and margin, elsewhere rhomboid, 20μ long, and these extend to the base in the leaf-centre, but toward the basal margin they are gradually smaller (to 10μ diameter) in many rows.   Perichaetium lateral, sessile, its inner leaves lanceolate, long-pointed, 4 mm. long, sheathing the seta, and including also numerous very long-jointed capillary paraphyses ; per. leaves nerved half-way, the cells elliptical vermicular, up to 100μ long at the base, except the lowest which are shorter and oblong, 50μ long. Seta 1 cm. long ; capsule erect, cylindrical, with obliquely rostrate lid and split calyptra enclosing the capsule, and set with erect hairs. Capsule somewhat wrinkled when old.   Peristome teeth erect, inflexed with erect points when dry, narrowly lanceolate, smooth. Autoecious, male flower bud-like, sessile, lateral, outer leaves small, inner longer ; anthers four to five, paraphyses none.   Hornschuch wrongly describes the seta as one line long, and the calyptra as not hairy.

Cape, W. Zwellendam, Mund ; Knysna Woods, Rehm. 317 ; Montagu Pass, Miss Taylor.

Cape, E. Hangklip, Mund and Maire, 1821 ; Alexandria Forest, Jas. Sim ; Albany, Dregé, 1831 ; Krakakama, Ecklon ; Dohne Hill, 5000 feet, 1898, Sim 7351 ; Boschberg, Somerset East, Macowan ; Hamilton Reservoir, Grahamstown, Miss Farquhar 89 ; Alexandria Forest, Longbush Kop (6 miles from the sea), James Sim ; King William's Town, 1892, Sim 7106.

Also from Mount Meru, East Africa, 4000 feet (W. Leighton).

(Genus ANTITRICHIA Brid. Harvey, in his Genera of South African Plants, 1st ed. (1838), under *Anomodon* Hook., says, "*A. curtipendula* is found in this country" ; and in Pfl., i, 3, 755, and xi, 98, Brotherus, who figures it there, includes South Africa among its many habitats, but he omits it from his key in Marloth's Flora of South Africa, and no other collector records its presence, nor have I found it, and I presume it was a case of mistaken identity in early collecting.   The plant is a very vigorous one, 4–12 inches long, usually pendulous on tree-bark, frequent in the Northern Hemisphere (and occurs also in Uganda), and having pinnate stems, numerous ovate-acuminate leaves 1–2 mm. long, somewhat plicate with revolute margin sharply dentate near the apex, and one wide nerve reaching three-fourths the length of the leaf, which has one or several weak nerves branching off near its base.   Seta ½ inch long ; capsule large, elliptical ; calyptra smooth ; peristome double, the outer teeth strong, the inner processes slender, without basal membrane.   Per. leaves long, sheathing.   It must be collected again, or specimens produced, before it can be accepted as South African.)

Genus 165. PTEROGONIUM Sw.

Vigorous but slender dioecious mosses, forming masses on stones and stumps, with wiry, rhizomatous, leafless, prostrate stems and erect, much-branched or dendroid, but slender leafy stems, small-leaved at the base and closely leaved upward, with spreading, concave, ovate-acute leaves, toothed toward the apex. Nerve rather short, double from the base or forked, not always present ; cells rhomboid or oblong except those toward the basal margin. Seta long, enclosed in long per. leaves ; capsule erect, cylindrical, with short thick neck ; ring dehiscent ; lid bluntly conical ; calyptra cucullate, reaching below the middle of the capsule, slightly hairy ; peristome double ; outer teeth lanceolate, papillose above, papillose-striate below ; inner peristome with narrow basal membrane and short subulate keeled processes, without cilia.

544. *Pterogonium ornithopodioides* (Huds.), Lindb., in Oefv. af K. Vet.-Akad. Förk., 1863, p. 411 ; Braithw., M. Fl. ; Broth., Pfl., i, 3, 757, and xi, 99.

=*Hypnum ornithopodioides* Huds., Fl. Angl., 1st ed., p. 430.
=*Pterogonium gracile* Sw., Disp. M. Suec., p. 26 ; W. P. Schimp., Syn., 2nd ed., p. 575 ; Dixon, 3rd ed., p. 404, Tab. 49 L.
=*Pterogonium gracile* Sw., var. **capense* Rehm.
=*Forsstroemia *dendroides* Dixon, in letter.
=*Neckera gracilis* C. M., Syn., ii, 97.

Pterogonium ornithopodioides.

Slender and julaceous when dry, more or less expanded when wet, growing in loose tufts on stones or soil, the rhizomatous stems prostrate, black, and wiry, the leafy stems erect or suberect, simple and small-leaved at the base, irregularly fastigiately branched or bipinnately branched, the branches slender and obtuse, and all bending to one side, often flagellate, and those which touch the ground produce roots and fresh plants from the ends ; long, slender, flagellate branches may start anywhere, even from the main stem, scale-clad or leafless their whole length. Leaves closely imbricate all round the stem, 1 mm. long, ovate-acuminate with rather cordate base ; often homomallous ; concave, not plicate, the base-centre decurrent on the stem ; nerve weak and short, often double, and often absent ; margin erect, sharply dentate or denticulated by single exserted cells in the upper half, and by smaller teeth to near the base. Apical and lamina cells oval to elliptical, 30μ long, gradually longer and vermicular toward the centre of the base, whence they are reduced through obliquely oval to small quadrate or transversely oval, and 10μ wide ; the branch leaves gradually reduced to small lanceolate form with oval cells on the flagellate branches. Leaves are often very scabrous on the back ; some of Rehmann's are so, mine are not. Perichaetal leaves convolute ; apex acuminate, entire ; cells long and lax. Fruit not yet found in South Africa, but described in Europe, where it is widely distributed, as : "Capsule on a long seta, erect or slightly curved, sub-cylindric ; lid conical, rather obtuse ; peristome pale yellow, inner of short processes, without cilia, on a narrow basal membrane. Dioecious." Calyptra slightly hairy.

Rehmann called all his specimens var. *capense*, but I see no reason for a varietal name.

Cape, W. Disa Ravine, Table Mountain, Sim 9113, 9134, 9138 ; Platteklip Ravine, 3000 feet, Sim 9218 ; Window Gorge, Kirstenbosch, and Devils Peak, Sim 9241, 9472 ; Lower Plateau, Table Mountain, Pillans ; Kloof, near Chaplin Point, Sim 9367 ; Montagu Pass, Rehm. 318 ; rocks at Stinkwater, Rehm. 318β.

Transvaal. Lechlaba, Houtbosch, Rehm. 603.

Wager's list includes *Pterogonium abruptum* Wright, and *P. decipiens* Wright, dealt with herein.

Family 45. FABRONIACEAE.

Slender little bark-epiphytes or parasites, as always found on living bark. Plants closely appressed to the bark and rooting into it, the vegetative extension stems simple or forked, flattened against the bark, or more or less regularly pinnate ; the fertile branchlets very short, erect or suberect from the older portions of the other ; the whole forming a very closely appressed mat, covering sometimes large areas of bark so thinly as to be only visible as a shining coating with imbricate julaceous habit when dry, but usually quite evident and bright green with spread-out leaves when moist. Stem without central strand ; outer cells of stem not indurated ; paraphyllia not present. Leaves crowded, placed round the stem, seldom secund except on extension shoots, ovate-lanceolate, concave, not bordered, not plicate ; nerve single, short, occasionally absent ; cells usually prosenchymatous, laxly rhomboid, smooth, thick-walled ; basal cells quadrate, but there are no differentiated alar cells. Perichaetal branches rooting. Vaginula usually naked. Capsule very small, erect, regular, oval, well exserted on a short seta, contracted below the mouth when dry ; the lid large, convex-apiculate or rostrate ; the calyptra small, cucullate, smooth, naked, caducous ; peristome single or double or absent ; in sixteen pairs, or more usually in eight double pairs, or separate ; the inner peristome absent or having filiform or subulate processes, without basal membrane, or in *Helicodontium* having plaited membrane and keeled processes. Autoecious or dioecious ; male and female plants similar. Spores small. Seldom absent from old trees, especially scattered trees exposed to light, and often found from base to top of the tree, but selecting certain species of trees in preference to others.

Synopsis of Fabroniaceae.

Inner peristome absent, outer present.
  Silky shining plants.
    Slender little plants ; teeth of the outer peristome wide and blunt, or in one section absent   166. FABRONIA.
    Stronger ; teeth of the outer peristome lanceolate, narrow, in pairs   .   .   .   167. ISCHYRODON.
  Slender plants, not silky shining ; teeth of the outer peristome widely lanceolate, in pairs 168. DIMERODONTIUM.
  Peristome double. Teeth of the outer peristome separate, lanceolate, not striate ; basal membrane of inner peristome
    very narrow or absent   .   .   .   .   .   .   169. SCHWETSCHKEA.

Genus 166. FABRONIA Raddi (in Atti dell Acad. de Sc. di Siena, ix, 230 (1808)).

Slender little silky epiphytes, closely matted on bark, with short, erect or ascending stems in the mat, and with closely appressed, pinnate or single extension shoots, the former closely leaved all round, the latter more lax and often with the leaves turned upward. Leaves spreading when moist, imbricate in most species or sometimes somewhat secund when dry, concave, ovate or ovate-lanceolate, usually subulate or hair-pointed ; the margin flat, entire or toothed or ciliated ; nerve single or almost invisible, usually weak and short ; cells prosenchymatous, long-rhomboid or -hexagonal ; outer basal cells quadrate. Inner perichaetal leaves sheathing, subulate, toothed or ciliated, without nerve. Seta smooth, short, twisted when dry. Capsule erect, regular, urn-shaped ; the teeth broad and short, without border, finely papillose in longitudinal lines, in eight bundles, inflexed when moist, reflexed when dry, sometimes split along or in parts of the mid-line. One section is gymnostomous. Lid convex or conical, pointed or beaked. Autoecious or dioecious.

Synopsis of Fabronia.

Without peristome   .   .   .   .   545. *F. Wageri.*
With peristome.
  Leaves entire.
    Leaves lanceolate   .   .   .   .   .   .   546. *F. Gueinzii.*
    Leaves ovate-acuminate, hardly concave   .   .   .   547. *F. Rehmanni.*
    Leaves ovate-pointed, very concave ; cells large and short   .   548. *F. leikipiae.*
  Leaves more or less dentate   .   .   .   .   549. *F. pilifera.*
  Leaves ciliate-dentate, with long hair-point   .   .   550. *F. abyssinica.*
  Leaves shorter, ciliated, without long hair-point   .   .   .   551. *F. victoriae.*

F. Eckloniana (Hpe., MSS.), C. M., in Bot. Zeit., 1859, p. 247, is known to me by name only.

545. *Fabronia Wageri* Dixon, in Bull. Torrey Bot. Club., xliii, 73, and fig. 4 ; Broth., Pfl., xi, 285.

Fairly strong for this genus ; extension branches prostrate and ½–2 inches long, with mostly homomallous leaves, pointing upward ; erect branches very numerous, subpinnately arranged, 3–6 mm. long, densely leafy ; the leaves closely imbricate, expanded when moist, appressed when dry, ovate-lanceolate, 1 mm. long or more ; the nerve strong, extending half-way ; the margin expanded, entire ; the apex entire, chlorophyllose when young, the cells rhomboid, $50 \times 10\mu$, smooth, at first green ; the basal cells numerous, quadrate, $10 \times 10\mu$, chlorophyllose when young. Perichaetium lateral on stem and branches ; its leaves concave, the inner sheathing, shortly pointed, hardly entire, the margin being irregular ; cells all long, pellucid ; seta 4–6 mm. long, pale ; capsule erect, oval or oval-oblong, 1·5 mm. long, with short tapering neck, no peristome, and distinct mouth-band of five to six rows of small transverse cells ; other capsule cells red, large, with sinuose walls. Lid convex, hardly apiculate. Dioecious, male flowers not seen. Dixon made a new section *Gymnoischyrodon* for this, similar to § *Pseudoischyrodon*, except that the peristome is absent, the lid not apiculate, and the vigour greater.

Ischyrodon *leptocladus, Rehm. 633, hardly differs except in having the leaf-point longer and may be a form of this, but in the absence of fruit cannot be definitely placed. Rehmann's specimens of it are marked " c. fr.," but no fruit is present in the specimens in the British Museum, Kew, or the South African Herbaria ; it is from the oaks in the Cape Town avenue, where also *F. Wageri* is found. The only other locality known to me for *F. Wageri* is on the old oaks in the Stellenbosch streets, where it grows abundantly and fruits freely ; Sim 4573, 9611, 9612, 9579, 9582, etc.

Fabronia Wageri.

546. *Fabronia Gueinzii* (Hpe., MSS.), C. M., Syn., ii, 37 ; Broth., Pfl., i, 3, 905.
               = *Fabronia *hypnoides*, Rehm. 629.

Dioecious, small, prostrate ; stems slender, subpinnate, 1–2 inches long, the upright stems densely tufted, 3–4 mm. long, closely leaved ; the leaves 1 mm. long, lanceolate with acuminate point or hair-point, entire ; nerve wide and faint, extending half-way; cells vermicular, 70μ long, except toward the leaf-base where they pass through rhomboid to the quadrate alar cells 10μ wide, which also are chlorophyllose when young. Per. leaves few, erect, sheathing, pointed, almost entire, shorter than the stem-leaves. Seta 3 mm. long, red; capsule erect, oval, without neck; teeth short, in eight wide bundles. "Lid and peristome not seen; male flowers numerous, turgid, gemmaceous" (C. M.).

Brotherus (Pfl., i, 3, 205) brings *F. Breutelii* (Hpe., MSS.), C. M., in Bot. Zeit., 1859, p. 247, alongside of this, having upper leaf-cells oblong-rhomboid, whereas in *F. Gueinzii* they are linear.

Natal. Mooi River, 4000 feet, T. R. Sim. Collected also by Dr. Gueinzius long ago.

Transvaal. On Faurea at Houtbush, Rehm. 629 (as *F. hypnoides*).

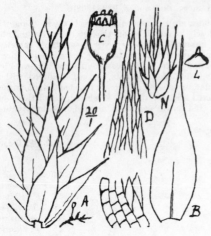

Fabronia Gueinzii.

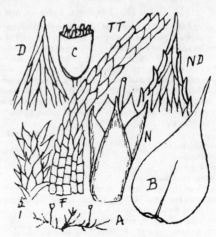

Fabronia Rehmanni.

547. *Fabronia Rehmanni* C. M., in Hedw., 1899, p. 131; Broth., Pfl., i, 3, 904, and xi, 283; Dixon, Trans. R. Soc. S. Afr., viii, 3, 213.

Small, autoecious; stems up to 2 cm. long, prostrate, pinnate, with short erect branches; leaves numerous, closely appressed when dry, spreading when moist, chlorophyllose when young, 1 mm. long, ovate-acuminate, entire, faintly nerved half-way; cells shortly rhomboid, 50μ long, shorter and wider toward the leaf-base, the alar cells numerous and quadrate, 10μ wide, the marginal cells at mid-leaf slightly overlapping but hardly denticulate. Per. leaves short, wide, nerveless, with short dentate point. Seta 3–5 mm. long; capsule erect, oval-oblong, verruculose, cup-shaped when dry, with eight bundles of short teeth inserted low in the mouth. Lid deeply convex. Leaves on prostrate stems narrower and homomallous.

Var. **julacea* Rehm. has very compact, catkin-formed branches, from which only the numerous hair-points show out (Rehm. 627 on *Eugenia cordata*, near Maritzburg).

F. Macowaniana C. M., in Hedw., 1899, p. 132 (Broth., Pfl., xi, 283), from Somerset East, seems to be the same thing, "without nerve." The nerve is often faint and of very similar cells to the others, and of the same colour, but I have not yet found a specimen in which it is not visible at least in some leaves.

Cape, W. Bishops Court, Cape Town avenue, and Kloofnek, Sim 9523, 9524, and 9560; George, Wager 561; Cape Town, Rehm. 347; Claremont, W. Duncan.

Cape, E. Perie, 1892, and Toise River on Cycads, Sim 7241 and 7286; Port Elizabeth, Wager 172; Kokstad, Dr. Burtt-Davy, 1918.

Natal. Maritzburg, many places; Josephine Bridge, January 1920; Cathkin, 1919; Blinkwater, Durban, Sim; Polela, Haygarth; Little Berg, Miss Owen.

Transvaal. Zelikats Nek, Magaliesberg, Sim; Spelonken, Rev. H. A. Junod 10.

S. Rhodesia. Zimbabwe Acropolis, Sim 9524, 9523, 9560.

548. *Fabronia leikipiae* C. M., in Flora, 1890, p. 487; Broth., Pfl., xi, 283.

Densely matted on bark; extension stems slender and straggling, with scattered homomallous leaves and brown rhizoid tomentum; upright stems often club-shaped, with scattered leaves below and many closely imbricated leaves above, all leaves very concave, ovate-acuminate in a hyaline hair-point, suberect when moist, julaceous when dry, 0·75 mm. long; nerve extending half-way; cells shortly rhomboid, 30×12μ; quadrate basal cells very numerous, 12×12, extending more than half-way up the leaf; sometimes almost all the marginal cells are of this kind; margin mostly quite entire, but occasionally a cell projects or is exserted, and on extension stems sometimes these are frequent. Per. leaves erect, short, similar, but smaller and more toothed. Seta 3–5 mm. long; capsule pear-shaped, small-mouthed when fresh, verruculose, the cell-walls exceedingly crisped. Teeth in eight bundles, inflexed when moist, wide and short, but 2-pointed, lightly striate.

S. Rhodesia. Zimbabwe, 3000 feet, frequent on trees, Sim 8783, 8767, 8821 ; Salisbury, Eyles 1447. The original description was from Aberdare, and I have it also from Mount Meru, 6000 feet, East Africa (W. Leighton).

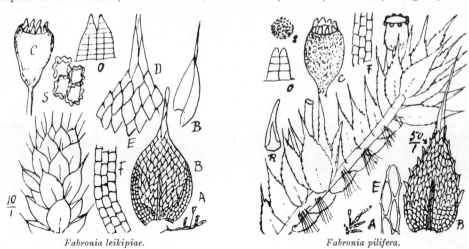

Fabronia leikipiae.   *Fabronia pilifera.*

549. *Fabronia pilifera* Horns., in Linnaea, xv, 136 (1841) ; C. M., Syn., ii, 35 ;
Broth., Pfl., i, 3, 904, and xi, 284.

=*F. angolensis* Welw. and Duby, in Mem. Genève, 1871, p. 1, t. 1, f. 2 (see Dixon, Trans. R. Soc. S. Afr., viii, 3, 213) ;
Broth., Pfl., xi, 284.

=*F. transvaalensis* C. M., in Hedw., 1899, p. 131 (see Dixon, Trans. R. Soc. S. Afr., viii, 3, 213) ; Broth., Pfl., xi, 284.

Autoecious ; deep green, widely caespitose or densely matted on bark ; the prostrate pinnate stems 1–2 cm. long, rooting freely by brown rhizoids, and with lax homomallous leaves pointing upward ; the matted portion set with erect, closely leaved stems 3–5 mm. long, its leaves imbricate, often julaceous when dry, spreading widely when moist ; leaves ovate-acuminate, somewhat concave, the hyaline point half as long as the leaf, or often more ; nerve wide, faint, extending half-way, often almost indistinguishable ; the margin more or less toothed by single exserted cells except near the base ; the lamina cells lax, shortly and widely oblong-rhomboidal, 50μ long, those forming the point longer ; alar cells very numerous, quadrate, $20 \times 15\mu$; per. leaves erect, pellucid, long-sheathing, ovate-acuminate, nerveless, ciliate-dentate. Seta 3–5 mm. long, twisted when dry ; capsule 500μ long, obovate, with short tapering neck and verruculose surface ; when dry shortly cup-shaped and wide-mouthed, with reflexed teeth ; when moist the mouth narrow, with eight bundles of short, blunt, erect teeth set deeply in the mouth. Lid shortly rostrate or rostellate ; calyptra conical, split on one side. Spores 25μ, tuberculate.

F. pilifera is described as having only one or two teeth close to the base of the hair-point, while *F. angolensis* has more numerous teeth, but both occur on one plant and the numerous small teeth is the more common condition ; a shorter capsule has, however, often been noticed on Rhodesian plants.

Cape, E. Kentani, 1917, Miss Pegler ; Rhenosterberg, MacLea (Rehm. 630) ; Windvogelberg, 5000 feet, 1832, Dregé ; Sterkspruit, Herschel, Hepburn.

Natal. Maritzburg ; Rosetta Farm ; Blinkwater ; Karkloof ; Sweetwaters, Howick ; Hilton College (T. R. Sim) ; Nottingham Road, Dr. v. d. Bijl 9.

O.F.S. Eagles Nest, Professor Potts ; Kroonstad, Wager 170.

Transvaal. Melville, Johannesburg, Sim ; Elim, Louis Trichard, Rev. H. A. Junod ; Pretoria and Rietfontein, Wager.

S. Rhodesia. Salisbury, Eyles 1573, 1447 ; Zimbabwe, Sim 8815 ; Marandillas, Eyles 3883 ; Victoria Falls, Eyles 1306 ; Mazoe, Ironmask, Eyles 616.

Re Eyles 1575 (Salisbury), Dixon points out it has short points and may be referred to var. *acuminata* Gepp.

550. *Fabronia abyssinica*, C. M., Syn., ii, 35 ; Broth., Pfl., i, 3, 904, and xi, 284 ;
Dixon, Trans. R. Soc. S. Afr., viii, 3, 212, and 213.

=*F. *macroblepharis* Br. and Sch., in Schimp., Musc. Abyss., Nos. 466 and 477.

=*F. *aureonitens*, Rehm. 628.

=*F. perciliata* C. M., in Hedw., 1899, p. 131 ; Dixon, in Trans. R. Soc. S. Afr., viii, 3, 213 ; Broth., Pfl., xi, 285.

=*F. *densifolia*, Rehm. 348.

=*F. Schweinfurthii* C. M. (*fide* Dixon).

=*F. vallis-gratiae* (Hpe., MSS.), C. M., Bot. Zeit., 1859, p. 247 ; Bryhn, p. 19 ; Broth., Pfl., i, 3, 904 (see Dixon, Trans. R. Soc. S. Afr., viii, 3, 213).

Autoecious ; prostrate stems long and straggling, up to $1\frac{1}{2}$ inch long, often growing on soil but usually on bark, little-branched, pale, with homomallous leaves ; erect stems 3–5 mm. long, closely leaved, often densely catkin-like, with hair-points only showing out ; leaves 1–1·5 mm. long, ovate-acuminate, hair-pointed, concave, nerved half-way, with rather long, twisted, linear areolation ; basal cells few and quadrate ; margin more or less strongly ciliated from the base to near the apex by single hyaline cells ; per. leaves short, nerveless, ciliated ; capsule 4–6 mm. long, erect, pear-shaped when fresh, shortly and widely cup-shaped when dry, but with eight bundles of rather long peristome teeth sometimes one-third as long as the dry capsule ; lid obliquely rostellate ; beak as long as the capsule is wide.

Shaw's record of *F. pusilla* Raddi (a European ciliated species), from near Graaff Reinet, Stellenbosch, and Bedford, likely belongs here.

Cape, W. Devils Peak and Cape Town, Sim 9234 ; Table Mountain, Rehm. 348 (as *F. densifolia* Rehm.).

Cape, E. Rhenosterberg, MacLea (Rehm. 628).

Natal. On Sugarbush, Mont aux Sources and Gordon Falls, Edendale, T. R. Sim ; Giant's Castle, 1915, R. E. Symons (Sim 8694) ; Arcadia, Scheeper's Nek, Sim.

O.F.S. Eagles Nest and Wilde Als Kloof, Professor Potts.

Transvaal. Noorddrift, Waterberg, Pretoria, and Rietfontein, Wager ; Woodbush, Jenkins.

S. Rhodesia. Zimbabwe, Sim 8804, 8808, 8769, 8762 ; Matopos, Sim 8455 ; Victoria Falls, Wager 905 ; Salisbury, Rua River, and Bulawayo, Eyles 684, 1323, and 1053.

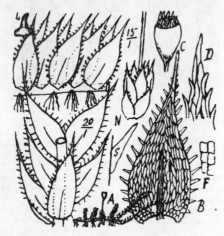

Fabronia abyssinica.

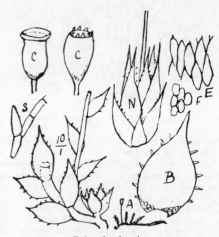

Fabronia victoriae.

C, capsules, one fresh and one dry ; S, hyaline
marginal cilia.

551. *Fabronia victoriae* Dixon, S.A. Journ. Sc., xviii, 328 ; Broth., Pfl., xi, 285.

Autoecious ; densely matted in rather large patches on bark, but the stems slender and the leaves spreading or suberect even when dry, and small-leaved branches are frequent. Leaves small, 0·75 mm. long, widely ovate with short acuminate point or sometimes shortly hair-pointed ; nerve absent or very indistinct, its cells almost conforming with other leaf-cells ; margin strongly ciliated ; branch leaves more acuminate and less regularly ciliated ; cells pellucid, widely rhomboid, $40 \times 15\mu$; alar cells few, quadrate, $10 \times 10\mu$, marginal cilia hyaline, of one conical, hyaline cell each ; per. leaves erect, somewhat sheathing, long-pointed, somewhat ciliated ; capsule erect, ovate, with short neck, when dry contracted below the mouth. Male flowers bud-like, basal.

Natal. Blinkwater, 3000 feet. 1918, Sim 10,280 (slightly drawn-out form).

S. Rhodesia. Victoria Falls, 3000 feet, Sim 8943 ; Salisbury, March 1919, Eyles 1573.

Genus 167. ISCHYRODON C. M., in Linnaea, xxxix, 443 (1875), as section, and in Hedw., xxxviii, 132 (1899), as genus.

Fairly robust, silky, yellowish matted, autoecious or dioecious plants, with closely branched, prostrate, rooting stems, these branches erect, simple, blunt or pointed ; leaves numerous, spreading when moist, compact when dry, imbricate, concave, ovate-lanceolate, subulate-pointed or aristate, with entire margin, somewhat reflexed at the base ; nerve wide, ending about mid-leaf ; cells linear-rhomboid, alar cells numerous, large, and quadrate ; seta short, straight, smooth ; capsule erect, regular, shortly oval. Teeth of the outer peristome in pairs, widely lanceolate, longer and narrower than on *Fabronia*, narrowly pointed, closely papillose ; inner peristome absent ; lid obliquely beaked from a conical base ; monotypic and South African only, growing on bark. *Ischyrodon leptocladus* Rehm. probably belongs to *Fabronia*.

552. *Ischyrodon seriolus* C. M., in Hedw., 1899, p. 132 ; Broth., i, 3, 902, fig. 661, and xi, 288.

=*Fabronia seriola* C. M., Bot. Zeit., 1864, p. 367, and in Linnaea, 1875, p. 443.
=*Hypnum seriolum*, Hpe., MSS., in Musc. Eckl.
=*Ischyrodon* *Rehmanni C. M., in Rehm. 353 (not 179, as given by C. M.).
=*Brachythecium afroalbicans* Dixon, in Trans. R. Soc. S. Afr., viii, 3, 221 (1920), and pl. xii, fig. 21.
=*Lepyrodon capensis*, Rehm. 634.

Shining, yellow, exceedingly variable in general appearance, normally with prostrate extension stems 1–2 inches long, having homomallous leaves and closely set with ovoid erect stems, often tapering to the point, closely leaved all round ; the leaves numerous, imbricate, appressed when dry, erecto-patent when moist, 1·5 mm. long, ovate-acuminate, concave, somewhat cordate at the base ; the apex either shortly pointed or shortly hair-pointed ; the nerve wide, extending about half-way or more, narrow above, wide below ; the margin entire and erect, especially inflexed about mid-leaf, the lower part of the margin on one side sometimes reflexed ; the cells widely linear, $100 \times 10\mu$; the basal cells very numerous, lax, pellucid, quadrate, $20 \times 20\mu$, extending one-third of the leaf-length along the sides and for a few cells up at the nerve, but suddenly differentiated ; all cells sometimes green and chlorophyllose, more frequently empty and pellucid ; but variations occur having these same general characters, but with pendulous, yellowish, slender, silky, extension stems 2–3 inches long, with few, scattered, yellowish, short, acute branches near the base, lightly clad in aristate leaves all round the stem, and not unlike a *Brachythecium*. Apparently dioecious. The perichaetia are very small, bud-like, and set on the main stem ; the leaves shorter and wider than other leaves, sheathing, shortly brought to a point, entire or with a few big teeth ; cells linear-rhomboid, and nerve absent ; archegonia about six ; seta ½ inch long ; capsule oval ; teeth in pairs, widely lanceolate, erect, laxly trabeculate, and closely papillose.
The compact upright branches resemble *Fabronia Wageri*.

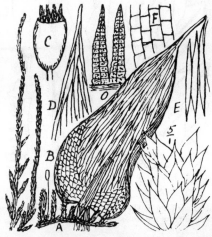

Ischyrodon seriolus.

C. M., in Hedw., xxxviii, 132, mentions a " forma *minor* " as *I. Rehmanni* C. M., Rehm. 353. Rehmann also issued Rehm. 351, *I. seriotus* C. M. ; Rehm. 351β, do, forma *tenella* C. M. ; and Rehm. 352, *I. seriotus* var. *albicans* Rehm. (in each case named *seriotus* in error instead of *seriolus*), but Rehmann's specimens show no good characters for these names, *tenella*, *albicans*, and *minor* or *Rehmanni* ; they are simply or hardly local or individual variations without stability.
My many specimens are all from the Cape Peninsula except one from Paarl (Sim 9620) and one from Kimberley (Miss Wilman), and it does not seem to have been collected elsewhere. Only one specimen is fertile.

Genus 168. DIMERODONTIUM Mitt. (Musc. Austr. Amer., p. 540 (1869)).

Slender and stiff, matted, little, dark green, but not silky, autoecious bark-epiphytes (often mixed among *Fabroniae*), with prostrate, rooting, irregularly many-branched stems ; the branches short, erect or suberect, usually simple, with numerous leaves all round, which are imbricate when dry, but when moist spreading, very concave, cordate or widely ovate, blunt ; the margin entire and flat, or toward the base somewhat incurved. Nerve strong, ending just below the leaf-apex ; cells laxly rhomboid or oval, the alar cells quadrate. Inner perichaetal leaves erect, pale, broadly ovate-lanceolate, shortly pointed, with large cells and very weak nerve. Vaginula slightly hairy. Seta ½ inch or less, smooth, twisted when dry. Capsule erect, regular, oblong, when dry smooth. Outer peristome teeth in pairs, broadly lanceolate, forked nearly to the base, distantly jointed, papillose ; inner peristome absent. Spores small. Lid conical, blunt. Mostly South American.

553. *Dimerodontium africanum* C. M., in Hedw., 1899, p. 134, and in Rehm. 354 ; Broth., Pfl., x, 912, and xi, 295.

=*Leskea* *Breutelii W. P. Sch., in Breutel, Musc. Cap. (*fide* C. M., in Hedw., 1899, p. 134) ; Par., Ind., p. 740 (1896).
=*Dimerodontium* *Breutelii (W. P. Sch.), Par., Suppl. Ind., p. 128 (1900), and 2nd ed., ii, 78.
=*Dimerodontium carnifolium* C. M., in Hedw., 1899, p. 134 ; Par., Ind., 2nd ed., ii, 79 ; Broth., Pfl., xi, 295.
=*Leskea carnifolia* C. M., in Rehm. 358 (*fide* C. M., in Hedw., 1899, p. 135).

Monoecious, closely matted on bark, deep green ; stems prostrate, ½–1 inch long, rooting, with leaves often homomallous, branched ; branches simple or shortly branched or subpinnate ; the leaves appressed during dry weather, spreading when moist, densely imbricate, 0·5 mm. long, firm, widely ovate, shortly and bluntly pointed, concave, entire, with wide nerve to near the leaf-apex ; cells rather incrassate, those near the leaf-apex obliquely elliptical and 15μ long, oval and rather shorter lower ; the basal cells very numerous, the outer transverse, the inner quadrate, $12 \times 12\mu$; per. leaves up to 1 mm. long, convolute, erect, long-acuminate, with slight nerve and large lax cells $50 \times 10\mu$. Seta ½ inch long, red ; capsule erect, cylindrical, red, with short, bluntly conical lid. Peristome short ; outer peristome only is present, set deep in the capsule-mouth, with narrow, few-barred papillose teeth in pairs ; ring small-celled, permanent ; capsule-wall of large lax cells ; spores very small, green. Like *Lindbergia* in everything except the peristome.

    D. carnifolium is placed by Brotherus (Pfl., x, 912) in a group having larger size, but neither the description nor Rehmann's specimens show that or any other difference except yellow colour, which is purely conditional.

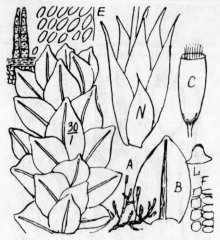

Dimerodontium africanum.

    Cape, W. Claremont and Cape Town, Rehm. 354 and 357 ; Brack River and Montagu Pass, Breutel ; Bishops Court, Sim 9459, 9460 ; Rondebush, Rehm. 358 (*D. carnifolium* (C. M.)) ; George, Wager 346, 561 p.p.

    Cape, E. Gwacwaba River, King William's Town, 2500 feet, Sim 7116 ; Perie Forest, Sim ; Gobogobo, F. Leighton.

    Natal. Mont aux Sources, Sim.

Genus 169. SCHWETSCHKEA, C. M., in Linnaea, xxxix, 429 (1875).

    Slender, little, matted, shining, autoecious bark-epiphytes, with prostrate, more or less regularly pinnately branched, rooting stems, the branches short, erect or spreading, mostly simple, closely leaved. Leaves when dry erect or sometimes homomallous, when moist spreading, lanceolate, pointed, with the margin flat, minutely denticulated by protruding cells. Nerve thin, reaching the middle of the leaf or more ; cells usually oblong-hexagonal, the alar cells numerous and quadrate. Inner perichaetal leaves ovate-lanceolate, subulate-pointed. Seta short, smooth or rough above. Vaginula with numerous long paraphyses. Capsule erect or nearly so, regular or almost so, oval or cylindrical ; ring narrow, dehiscent. Peristome double ; teeth of the outer peristome lanceolate, smooth below, papillose above, distantly barred ; inner peristome basal membrane narrow, or almost absent ; processes about as long as the outer teeth, narrowly lanceolate, keeled, papillose. Cilia absent ; spores small ; lid obliquely beaked from a dome-shaped base.

    554. *Schwetschkea Rehmanni* C. M., in Hedw., 1899, p. 133, and Rehm. 346 ; Broth., Pfl., xi, 293.

=*Helicodontium lanceolatum* (Hpe. and C. M.), Jaeg., Ad., ii, 292 ; Dixon, in Trans. R. Soc. S. Afr., viii, 3, 214 ; Broth., Pfl., xi, 292.
=*Hypnum lanceolatum* Hpe. and C. M., in C. M., Syn., ii, 411.
=*Leskea lanceolata* Hpe. and C. M., in Linnaea, xviii, 702 (1844).

    A shining, yellowish-green, little bark-epiphyte, with prostrate stems ½–1½ inch long, from which rise many short blunt stems 2–5 mm. long on which the lanceolate-acute leaves 2 mm. long are laxly arranged, and are spreading when moist, with margin erect, entire or minutely denticulated : nerve weak and narrow, disappearing about mid-leaf ; cells 40μ long, rhomboid or often twisted, with narrow lumen, and basal cells quadrate toward the margin. Perichaetal leaves erect, similar to stem-leaves, sheathing and longer pointed. Seta 1 cm. long ; capsule oval-oblong with short, obliquely rostellate lid, and calyptra small, split, glabrous ; peristome teeth short, narrowly lanceolate, barred and papillose ; inner as long, hyaline and keeled, without evident membrane. Several other species occur in Africa.

Schwetschkea Rehmanni.

    Cape, W. Blanco, Knysna, Rehm. 346 ; Tow River, on trees, 1875, Rehmann ; Bishops Court, Sim.

    Cape, E. Krakakamma, Ecklon, and George, Wager 541, both as *Helicodontium lanceolatum* Jaeg.

Family 46. BRACHYTHECIACEAE.

    Very diverse as to habit and size, but usually somewhat silky and shining ; the stems prostrate, recurved or spreading, seldom erect, often stoloniform in places, rooting freely, and often irregularly pinnate ; branches usually pointed, often flagellate and rooting at the point ; usually without paraphyllia ; the stem with central strand and with indurated cells. Leaves numerous, placed all round the stem, spreading or erecto-patent, seldom homomallous, usually lanceolate, ovate-lanceolate or cordate, usually long- and slender-pointed, seldom obtuse ; when stoloniferous the leaves there different from the others. Nerve single, usually ending below the apex ; cells parenchymatous, usually elongated rhomboid, linear or vermicular, smooth or sometimes papillose upward ; those of the leaf-base lax and coloured ; the alar cells usually differentiated, quadrate, chlorophyllose or empty, not swollen, forming a concave angle. Seta rather long, often rough. Capsule usually inclined or horizontal, short, smooth, oval or oblong and high-backed, seldom erect and regular, never hanging. Peristome double, the two of equal length. Teeth of the outer peristome lanceolate-subulate, usually striate; lamillæ numerous, well developed. Inner peristome free, basal membrane wide; processes keeled, lanceolate-subulate; cilia usually developed, seldom rudimentary or absent. Lid conical, obtuse or acute or long-beaked. Calyptra cucullate, caducous, usually naked.

Synopsis of Brachytheciaceae.

Capsule erect and regular; basal membrane very narrow; leaves with many deep furrows.
   (Inner peristome somewhat attached to the outer . . . . . . HOMALOTHECIUM.)
   Inner peristome free . . . . 170. PLEUROPUS.
Capsule inclined or horizontal, irregular; basal membrane of inner peristome wide; leaves smooth or slightly furrowed.
   Lid conical, shortly pointed; leaves somewhat furrowed; alar cells differentiated   171. BRACHYTHECIUM.
   Lid long-beaked; alar cells not numerous or not much differentiated, not forming a group at the leaf-margin.
      Stem- and branch-leaves little different; cells long and smooth, not thickened; leaves not furrowed.
         Leaves widely ovate; nerve often protruding as a mucro on the back of the leaf; branches mostly flattened; seta usually rough . . . . 172. OXYRRHYNCHIUM.
         Leaves narrowly ovate to lanceolate; nerve not protruding; branches somewhat flattened; seta smooth; plants robust . . . . 173. RHYNCHOSTEGIUM.
         Leaves narrowly lanceolate; branches sometimes pinnately arranged; seta rough; plants slender .
                                     174. RHYNCHOSTEGIELLA.
   (Stem- and branch-leaves often unlike, evidently furrowed; leaves widely ovate; nerve often protruding as a mucro; cells long, smooth, not thickened; seta rough . . . . EURHYNCHIUM.)

Genus 170. PLEUROPUS Griff., Not., p. 468, and Icon. Pl. Asiat., ii, t. 90 (1849).

Robust, dioecious, matted mosses growing on tree-stems, with long, prostrate or stolon-like stems, often abundantly rhizoid on the under-surface, with numerous short and erect, or longer and pinnate, or bushy-branched, spreading branches, without paraphyllia, making a loose or curled tuft. Leaves erecto-patent or when moist spreading, large, often somewhat secund, not decurrent, rather concave, with several deep furrows very evident when dry; widely lanceolate from a widely subcordate base, sharply serrate, somewhat reflexed at the basal angles; nerve single, rather wide, ending below the leaf-apex. Cells rhomboid-vermicular, more or less twisted, smooth; angles usually slightly auricled but not concave-auricled; the lower and auricle cells shorter than others, or oval and quadrate. Perichaetium not rooting, inner perichaetal leaves with long narrow points. Seta long, red, smooth; capsule erect, ovate-cylindrical, regular, not contracted below the mouth when dry. Ring distinct. Peristome double, the inner much the shorter, and free from the outer. Teeth of the outer peristome linear or linear-lanceolate, in our plant striate, papillose upward, with well-developed lamella.

555. *Pleuropus sericeus* (Horns.), Broth., Pfl., i, 3, 1138, and xi, 357.

=*Leucodon sericeus* Horns., MSS., in Musc. Cap., Mund and Maire, in Hb. Reg. Berol; Rehm. 320, 321, and 321β.
=*Leucodon sericeus* Horns., var. *afrostriatus* (C. M.), Par.
=*Neckera sericea* C. M., Syn., ii, 113.
=*Palamocladium sericeum*, var. *afrostriatum* C. M., in Hedw., 1899, p. 135.
=*Astrodontium sericeum* (Horns.), Jaeg., Rehm. 604 and 604β.
=*Eurhynchium *afrostriatum* C. M., in hb. G. Winter, 1884 (see Hedw., 1899, p. 135).
=*Pleuropus sericeus*, var. **afrostriatus* (C. M.), Dixon, in Trans. R. Soc. S. Afr., viii, 3, 221.

A robust dioecious moss, common in every forest, usually epiphytal on trees but also often on stones or soil, the main stem prostrate and rooting freely, several inches long, sending out numerous erect, stiff, or curved branches ½–1 inch long, and also occasional erect or pendulous branches several inches long, pinnately or fastigiately many-branched, and on which usually the perichaetia are lateral and sessile. Exceedingly irregular in its habit-form, but distinct from other South African mosses in its vigour together with its furrowed leaves; leaves 3–4 mm. long, nearly 1 mm. wide at the base, subcordate at the base, tapering regularly from base to acute apex, deeply several-furrowed, nearly flat, laxly appressed but not twisted when dry, fully expanded when moist, often the margin strongly reflexed when dry; nerve strong, wide below, extending to near the leaf-apex; margin strongly denticulate or dentate the whole length, except at the very base where the margin is often somewhat reflexed and the hardly evident auricle not toothed; perichaetium sessile and lateral on main branches, its lower leaves wide and short, the upper leaves 2–3 mm. long, ovate-acuminate, with long, thin, pellucid subula, nerved below or nerveless, lightly toothed and laxly reticulated. The vaginula set with many abortive archegonia. Cells elliptical or shortly vermicular, 50µ long, the lower shorter and the outer basal cells quadrate, 10µ diameter. Occasionally long, slender, flagellate, terminal branches are produced, with small, scattered, nerveless leaves and terminal rhizoid tufts. Seta 2 cm. long, red, smooth; capsule erect, regular, ovate-cylindrical, tapering through a graded neck to the seta, brown; when old microstomous, cylindrical, with short neck and the mouth suddenly inflexed at a right angle all round like a reducing veil. I have not seen a peristome on any of my many capsules, as all are slightly over-ripe or else young, but it is described as double, the outer with lanceolate, barred, papillose teeth, the inner with shorter or rudimentary processes, fairly deep membrane and no cilia. Male plants more slender, the perigonial bud lateral, its leaves short and subulate.

There has been evident confusion between this and *Homalothecium sericeum* B. and S. (Broth., Pfl., xi, 355=

Hypnum sericeum C. M., Syn., p. 356); the latter is a well-known European moss and may be the same as this, at least its general appearance is much the same, but it may differ in regard to its inner peristome and perhaps elsewhere; further collation of fresh, good material is necessary. If it differs, Paris loses this; if not, he omits its separate synonymy (see *Homalothecium sericeum* (Linn.)). If they prove to be identical, *Hypnum sericeum* Linn., Sp. Pl., 1st ed., p. 1129, is the oldest name. *Leskea sericea* Hedw., Descr., iv., fasc. 11, p. 43, t. 17 (1794), is another old name. In any case Paris credits South Africa with *Palamocladium Bonplandii* (Hk.) Broth. (=*Hypnum Bonplandii* C. M., Syn., ii, 463), which may be distinct as a South American plant, but I believe the South African record is on *P. sericeus.* Brotherus (Pfl., i, 3, 1138) thus separates the two :

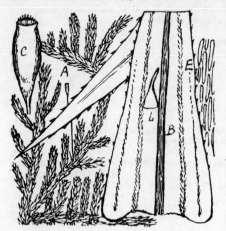

Pleuropus sericeus.

> *P. sericeus* (Horns.). Inner peristome irregular ; cilia absent.
> *P. Bonplandii* (Hk.). Inner peristome normal : processes much shorter than the teeth ; cilia short.

Among many other localities may be mentioned :

Cape, W. Knysna woods, Miss Duthie, and Rehm. 321 ; Clermont, Rehm. 321*β* ; George, Knysna, and Olifant's Hoek (Ecklon, *fide* C. M.) ; Montagu Pass, Miss Taylor.

Cape, E. Perie, 1892, Sim 7154, 7138 ; Eveleyn Valley, Sim 7043, 7019 ; Hangklip, Queenstown (Mund and Maire, *fide* C. M.); Malowe, Gr. East, Tyson ; Alexandria Forest, Jas. Sim ; Uitenhage, M. Duncan ; Toise River, 1919, Dr. Brownlee ; Kokstad, W. Tyson 1236 ; Boschberg, Jas. Sim ; Zitzikamma, 1917, J. Burtt-Davy ; Boschberg, Jas. Sim, etc., and Macowan.

Natal. Mont aux Sources, 6000 feet, common ; Zwart Kop, Giant's Castle (Sim).

Transvaal. Lydenburg, MacLea (Rehm. 604) ; Lechlaba, Houtbush (Rehm. 604*β*). Also Uganda, 12,000 feet (Dummer), and certainly it will be found in S. Rhodesia.

Genus 171. BRACHYTHECIUM Br. Eur.

Slender or fairly robust mosses of varied size and aspect, yellowish, shining, growing on earth, stones, or trees, with stems prostrate, decurved, spreading or suberect, bushy rooted, closely leaved, irregularly forked or pinnate, and abundantly stoloniferous at the point; paraphyllia only present where branches rise ; stem- and branch-leaves evidently different in size, stem-leaves erecto-patent or spreading, somewhat concave, somewhat furrowed, lanceolate or ovate from a wider, slightly decurrent base ; the margins flat or revolute on one or both sides, denticulate or serrate all round or at the apex only, seldom quite entire. Nerve single, rather wide ; cells long-rhomboid or linear, smooth ; those at the base shorter and more lax ; alar cells quadrate, rectangular and hexagonal, forming a distinct group in the often somewhat concave angles. Branch leaves shorter and narrower, with shorter and weaker nerve. Perichaetium rooting ; inner perichaetal leaves long and finely pointed ; seta long, smooth or somewhat rough ; capsule inclined or horizontal, seldom erect, usually shortly ovate and high-backed, seldom oblong-cylindric. Peristome double, inner and outer equally long ; teeth of the outer strong, striate, papillose upward, with close lamellae ; inner peristome free, yellow or orange, with wide basal membrane ; processes widely lanceolate, long-pointed, keeled, split in places or forked ; cilia present, knotted or appendiculate, seldom rudimentary or absent. Lid shortly conical, blunt or shortly pointed ; calyptra naked, cylindrical.

An extensive genus, in which several species vary much in outward appearance, which has led to much synonymy ; the genus is abundant almost everywhere in South Africa, on soil, among grass, as also in the forests, but the distribution of some species is imperfectly known.

Synopsis of Brachythecium.

Dioecious.

Perichaetal leaves subfimbriate at the base	556. *B. pinnatum.*
Branch leaves narrowly ovate-lanceolate ; leaves denticulate .	557. *B. implicatum.*
Branch leaves widely ovate, narrowly acuminate	558. *B. subrutabulum.*

Autoecious.

Stem-leaves 1 mm. wide or more, very large, with long dentate point .	559. *B. salebrosum.*
Leaves smaller ; point often twisted	560. *B. pseudoplumosum.*

556. *Brachythecium pinnatum* Dixon, in Trans. R. Soc. S. Afr., viii, 3, 222 (1920), and pl. xii, fig. 22.

Somewhat matted, light green, shining, dioecious ; stems 4–6 cm. long, prostrate, closely pinnate, with slender branches 5–6 mm. long ; stem-leaves 2 mm. long, cordate, the upper half much narrower, ending in an acute, flat subula ; nerve narrow, evident in lower half only ; margin almost entire ; cells linear-vermicular, up to $100 \times 5\mu$, lower cells shorter, alar cells segregated, numerous, small, transversely oblong ; branch leaves smaller, much narrower, plicate, oblong-lanceolate, the upper cells exserted at their tips ; perichaetal leaves compactly bundled, the inner with long narrow point, almost entire, but with subfimbriate teeth near the base. Seta 1–1·5 cm. long, smooth, red ; capsule nearly horizontal, irregular, high-backed ; lid shortly rostellate. Male flowers numerous with ovate-acuminate, spreading bracts.

Cape, E. Knysna, Wager 520.

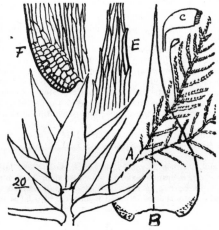

Brachythecium pinnatum.

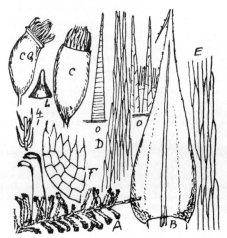

Brachythecium implicatum.

557. *Brachythecium implicatum* (Horns.), Jaeg., Ad., ii, 388 ; Broth., Pfl., i, 3, 1143, and xi, 361.

=*Hypnum implicatum* Horns., in Mund and Maire's Musc. Cap. ; C. M., Syn., ii, 362.
=*Brachythecium *inandae*, Rehm. 381.
=*Brachythecium afrosalebrosum* C. M., in Rehm. 385, and in Hedw., 1899, p. 136 ; *fide* Dixon, Kew Bull., 6 (1923).
=*Brachythecium *adscendens* Hampe, Rehm. 378, *fide* Dixon, Kew Bull., 6 (1923).
=*Brachythecium *knysnae* C. M., in Rehm. 387, and in Hedw., 1899, p. 137 ; *fide* Dixon, Kew Bull., 6 (1923).
=*Brachythecium afrovelutinum* C. M., Rehm. 379, and in Hedw., 1899, p. 135 ; *fide* Dixon, Kew Bull., 6 (1923).
=*Brachythecium *Macowani*, Rehm. 380, *fide* Dixon, Kew Bull., 6 (1923).
=*Brachythecium *Dicksoni*, Rehm. 383β, *fide* C. M., and also Dixon, Kew Bull., 6 (1923).
=*Brachythecium *sulfurescens* C. M., in Sched., *fide* C. M.
=*Brachythecium erythropyxis*, Rehm. 382 and 382β and c, and in Hedw., 1899, p. 137 ; Broth., Pfl., xi, 361.
I have seen all these Rehmann numbers except 378, and agree that they belong here.

Dioecious ; stems 1–3 inches long, shining, yellowish green, much intricate, slender, pinnate or irregularly pinnate or sometimes fastigiate ; the branches many, usually 1 cm. long or less, curved or straight ; stem-leaves to 2 mm. long, lightly furrowed, not crowded, the branch leaves rather smaller and seldom furrowed, all leaves flat above from a clasping, very concave base with somewhat decurrent fibres ; lamina narrowly ovate-lanceolate, with often long, slender, flat or cirrhate, pellucid, almost entire, nerveless subula ; nerve thin, wide below, ending near the subula, sometimes inconspicuous ; margin varying on one plant from almost entire to strongly denticulate, especially on the branch leaves, expanded or often reflexed or sometimes revolute on one or both sides ; cells linear, 75–100μ long, smaller near the subula, pellucid, thick-walled ; alar cells few, quadrate, not sunk, pellucid, gradually lengthening upward. Perichaetal leaves sheathing, nearly like others, thin, pellucid, almost entire, almost nerveless, arching outward, and with long slender point. Seta 2 cm. long, smooth, red ; capsule somewhat inclined, irregular, brown, often somewhat strumose, oval-elliptical, when dry contracted below the mouth and more irregular. Peristome very large, outer teeth yellowish, incurved round the inner erect cone, but when moist and pressed they are extended, regularly lanceolate, externally many-barred below, few-barred and slender above, the bars forming a pectinate row on the inner side when dry ; inner teeth pale, as high as the outer, with high plicate membrane ; short slender cilia, 1–2–3 present. Lid shortly conical ; calyptra cylindrical, narrow. The leaves extend outward from the stem but bend upward, especially when dry, forming the concave leaf-base.

This moss (and all the genus) has the same colour as *Pleuropus sericeus*, but is much smaller and more slender, less furrowed, and with regularly inclined capsule and large permanent peristome. Very common ; among other localities are :

Cape, W. Newlands Avenue and Ravine, and in grass, Cape Town, Sim 9444, 9446, 9542 ; Knysna, J. Burtt-Davy 17,019, and Wordsell, No. 8.

Cape, E. Keurbooms River, J. Burtt-Davy 17,039 ; Uitenhage, Sim 9004 ; Toise River, etc., Lovedale, Rev. J. Henderson.

Natal. Mont aux Sources, Mogg 4213 and 4225 ; Little Berg, Cathkin, Miss Owen 55 and 56 ; York, J. M. Sim ; Gordon Falls, Scheeper's Nek, Ihluku. Josephine Bridge on Umkomaas, Inanda, Van Reenen, Mooi River, Bulwer, Inchanga, Cathkin Bush, etc. (Sim).

Transvaal. Johannesburg, Miss Edwards ; Pilgrims Rest, MacLea ; Lechlaba, Houtbush, Rehm. 653 ; Kaap-schehoek, etc

O.F.S. Molmonspruit, Rehm. 380 ; Kadziberg, Rehm. 383β. Also Mount Meru, East Africa, 1916, W. Leighton.
S. Rhodesia. Matopoas, 5000 feet, Sim 8454 ; Zimbabwe, Sim.

24

558. *Brachythecium subrutabulum* (C. M.), Jaeg., Ad., ii, 398 ; Broth., Pfl., i, 3, 1143, and xi, 362.

=*Hypnum subrutabulum* C. M., in Bot. Zeit., 1856, p. 455.

Vigorous, dioecious, yellowish plants in extensive prostrate mats ; stems 1–3 inches long, usually arching and rooting again at the tip, secundly pinnate ; the branches slender and simple ; stem-leaves 2 mm. long, widely and shortly ovate-acuminate, the point long and narrow, almost entire ; leaf deeply concave, 2-plicate, decurrent, nerved about half-way ; margin expanded, denticulate or subentire ; cells linear, 70μ long ; alar cells forming a rather distinct but not sunk group and crossing the leaf-base ; branch leaves as long, narrower below, wider above, nearly lanceolate-acuminate, cochleate-concave, nerved half-way, and denticulate along the margin. Perichaetal leaves short and small, clasping at the base, with narrow subula, spreading or reflexed. Seta 2 cm. long, red, almost smooth ; capsule bent ; mouth oblique ; peristome rather short ; outer teeth widely lanceolate, closely barred, the bars showing as a pectinate row on the inside ; inner as high on high membrane ; cilia present ; ring absent ; mouth not differentiated ; lid shortly conical.

Natal. Ihluku, Harding, 3000 feet, Jan. 1920, Sim 10,267, under wattles ; Rosetta Farm, 1918 ; Benvie, York, 4000 feet, Sim ; Koodoo bush, Mont aux Sources, Sim ; Giant's Castle, 8000 feet, Symons.

O.F.S. The Kloof, Bethlehem, Jan. 1920, Sim.

Concerning *B. *orthopyxis*, Rehm. 377, Dixon writes : " Habit between *B. rutabulum* and *B. salebrosum*, fairly robust, silky. Leaves ovate, shortly acuminate, denticulate above (*i.e.* branch leaves) ; seta less than 1 cm. long, smooth ; capsule short, very little curved, inclined or suberect. Cells narrow, a few enlarged and subquadrate all along line of insertion, not forming marked auricles. I do not know any species it could belong to. No lid seen." Rehm. 377 is from Cape Town.

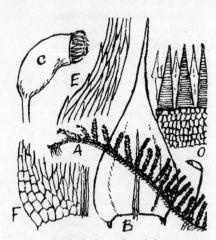

Brachythecium subrutabulum.

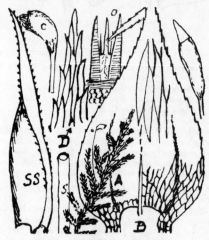

Brachythecium salebrosum.
S, smooth seta ; SS, branch leaf.

559. *Brachythecium salebrosum* Br. and Sch. ; Dixon, 3rd ed., p. 442 ; Broth., Pfl., xi, 362.

=*Hypnum salebrosum* Hoffm.

Autoecious, fairly strong, in yellowish tufts ; stems 1–2 inches long, suberect, pinnately or somewhat irregularly branched ; the branches short and mostly simple ; leaves 2–3 mm. long; those on the main stem wide ; shortly ovate-acuminate with long, narrow, dentate point, usually 2-plicate, concave, pellucid, strongly decurrent, with almost invisible nerve reaching about half-way ; margin expanded, toothed from near the base or sometimes only denticulate ; areolation exceedingly lax ; lamina cells widely linear, 100×12μ, gradually reduced at the base to quadrate or oblong cells 35–40×20μ, these cells connecting with the stem at the corners. Branch leaves shorter, narrower, concave-cochleate, not plicate, more strongly toothed. Perichaetium sessile, small-leaved, the leaves ovate-acuminate, spreading widely or reflexed, nerveless, laxly areolated, margin denticulate. Seta 2 cm. long, red, smooth ; capsule brown, thick-skinned, much inclined, very irregular ; outer teeth shortly lanceolate, closely barred, smooth ; inner teeth as high, on high membrane, with short slender cilia. Larger in all parts than *B. subrutabulum* but closely related. Some specimens have the leaves as narrow as in *B. implicatum*, but toothed throughout.

Cape, E. Hell's Gate, Uitenhage, 200 feet, Jan. 1922, Sim 9039.

Natal. Rosetta Farm, 4000 feet, and Lynedoch, Nottingham Road, T. R. Sim ; Drakensberg, 1822, H. Rolfes 181.

Transvaal. Lechlaba, Houtbush, Rehm. 652 (as *Brachythecium*, No. 2).

S. Rhodesia. Umtali, 1919, Eyles 1722.

Occurs also in Europe, where the leaves are usually narrower and more like *B. implicatum*.

560. *Brachythecium pseudoplumosum* Brid., Mant. M., p. 170, and Bry. Univ., ii, 472 ; C. M., Syn., ii, 350.

Almost or quite = *B. plumosum* (Sw.), Bry. Eur., vi, Mon., pt. 4, p. 597 ; Broth., Pfl., xi, 365.
= *Hypnum plumosum* Sw., in Act. Holm., 1795, p. 256 (not C. M., Syn. ?).

Fairly strong, autoecious, yellowish, shining, stramineous, usually on stones, not much-branched ; the prostrate branches pinnate, with rather small-leaved branches ; the erect branches dichotomously or irregularly few-branched, strong, terete through imbricate leaves when dry, but with leaves spreading when moist ; larger leaves 1·5 mm. long, narrowly ovate-lanceolate, deeply concave to near the rather reflexed, usually oblique point, erect or subsecund when dry, imbricate-cochleate, faintly plicate or not plicate, all the leaves on some branches sometimes serrulate. Nerve wide below, faint above, often extending to the subula ; margin erect or slightly reflexed, almost entire ; cells shortly linear, 70µ long, pellucid, often twisted, seldom pointed, the apical cells rather shorter, the basal cells shorter, and the pellucid alar cells quadrate, forming a cavity. Per. leaves wider, sheathing, shorter, more ovate, with long narrow subula, arched backward, laxly reticulated, almost entire, and almost nerveless. Seta 2–3 cm. long, red, the upper half rough ; capsule inclined, brown, oval-oblong, with high back and short neck ; ring narrow ; lid shortly conical; calyptra cylindrical ; peristome as in *B. implicatum.*

This is in Europe included in *B. plumosum* B. and S., which has the seta *nearly smooth* and the nerve ending at *mid-leaf,* and which is found in Europe and N. America ; and Brotherus (Pfl., i. 3, 1147, and xi, 365, 366) places the following South African names, as well as several mid-African names, as being specifically inseparable from *B. plumosum,* viz. :

B. plumosiforme (W. P. Sch., MSS.), Broth. = *Hypnum plumosiforme* C. M., in Bot. Zeit., 1858, p. 171. (Dixon, Trans. R. Soc. S. Afr., viii, 3, 223 (1920), records it from Hout Bay and Hogsback, and says it varies in form and direction of capsule and also in the per. leaves. He had compared the type plant in Schimper's collection.)

Brachythecium pseudoplumosum.

B. minutirete C. M., in Hedw., 1899, p. 136.

B. pulchrirete C. M., in Rehm. 143 and in Hedw., 1899, p. 138.

And he also places as almost inseparable from *B. populeum* (Hedw.), Br. Eur., which he says has nerve as here described, viz. :

B. *afrovelutinum C. M., in Rehm. 379 (seta smooth), but see herein under *B. implicatum.* It is a sterile scrap of a young plant ; Dixon states, no specimens are in Kew Herbarium or in the British Museum.

B. strictopatens C. M., in Hedw., 1899, p. 137 (maintained in Broth., Pfl., xi, 365).

To me these groups appear inseparable into fixed species though variations occur, covering several even in one patch.

Cape, W. Rondebosch, Rehm. 379 ; Genadendal, R. Schlechter ; Hout Bay, H. A. Wager 601 (*fide* Dixon as *plumosiforme*).

Cape, E. Grahamstown, Webster 773, and D. Henderson 360, both *fide* Dixon, as *stricto-patens* and *plumosiforme.* Natal. Weenen County, 4000 feet, March 1898, J. M. Wood ; on stones in bed of a stream, 2000–3000 feet, Maritzburg, H. A. Wager (as *B. *Wageri,* Paris, MS.).

Brotherus also places as unknown the South African :

B. pseudovelutinum (Hpe.), Jaeg., Ad., ii, 391 = *Hypnum,* C. M., in Bot. Zeit., 1858, p. 171 = *Leptorhynchostegium pseudovelutinum* Hampe, MSS.

B. pseudopopuleum W. P. Sch., MSS. = *Hypnum,* C. M., in Bot. Zeit., 1858, p. 170.

Concerning these I have no further information than is given above, and they are not represented in South African herbaria, but Dixon writes concerning *B. *truncorum,* Rehm. 386 (Belvedere): " I think quite certainly identical with *B. pseudopopuleum* Sch." ; and concerning *B. pseudovelutinum* (Hpe.), Jaeg., he says : " Small, autoecious, fruiting well ; seta 1 cm. or less, smooth ; capsule minute, 1 mm. long, dark, suberect, curved. Leaves cordate at the base, ovate-lanceolate ; differs from *B. velutinum* in smooth seta, wider leaves, and more pellucid cells." The type is from Genadendal.

Genus 172. OXYRRHYNCHIUM (Br. Eur.) Warnst.

Tufted or matted, moist-earth or -stone mosses, sometimes inundated, of varied size and aspect, either shining or not, with elongated, prostrate, rooted, often stoloniferous, irregularly pinnated stems ; branches often laxly leaved, light in colour, and with the leaves more or less complanate ; paraphyllia present only where the branches rise ; stem- and branch-leaves not evidently different, or different in size only ; old leaves striate but not furrowed, all leaves somewhat concave ; stem-leaves laxly erecto-patent, somewhat decurrent, ovate or roundish-ovate, and drawn to a short point ; denticulate or serrate with flat margin. Nerve single, reaching half-way or more, sometimes ending suddenly or in a protruding mucro on the back ; cells long and smooth, the basal cells shorter and more thick-walled. Seta long, rough or smooth ; capsule inclined or horizontal, oval or high-backed ; peristome as in *Brachythecium* ;

lid long-beaked from a dome-shaped base ; calyptra naked.   Usually dioecious.   Formerly included in *Eurhynchium*, and separated on very slight grounds.

Synopsis of Oxyrrhynchium.

561. *O. Macowanianum.*   Fairly strong ; seta smooth ; leaves denticulate.
562. *O. subasperum.*   Fairly strong ; seta rough ; leaves dentate.
563. *O. confervoidum.*   Very slender and confervoid ; leaves denticulate.

561. *Oxyrrhynchium Macowanianum* (Paris), Broth., Pfl., i, 3, 1155.

= *Rhynchostegium Macowanianum* Paris, Suppl. Ind., p. 302 (1900).
= *R. afrorusciforme* C. M., in Hedw., 1899, p. 139 (changed by Paris to *O. Macowanianum*, in view of the West African *R. afrorusciforme* C. M., in Dusen, M. Cameroons, p. 664 = *Hypnum* Broth., in Engl. Bot. Jahrb., 1897, p. 280, and = *Oxyrrhynchium afrorusciforme* (C. M.), Broth., in Pfl., i, 3, 1155).

Autoecious ; prostrate on moist stones or soil, yellowish green, shining, subpinnately branched, with erect, red, smooth seta an inch long or more, and inclined, slightly strumose, ovate-oblong, irregular capsule, with prominent permanent peristome.   Leaves crowded, in about eight rows, 1·5 mm. long, imbricated when dry, spreading when moist, rather keeled than furrowed, narrowly ovate-acuminate, slightly homomallous, subconcave ; the acute point subtriangular, not drawn out ; nerve ending above mid-leaf ; marginal cells elliptical, 40μ long, pellucid, overlapping and so more or less denticulate, making a pellucid border ; inner cells widely vermicular, often twisted, 70μ long, those near the leaf-apex shorter and more elliptical ; basal cells quadrate-oblong, 15μ long, filling the whole base but going highest at the margin.   Branches somewhat complanate, with more denticulate leaves.   Inflorescence sessile on main stem ; per. leaves shorter, spreading, more subulate-pointed, more laxly areolated, and less nerved than others. Seta 2–3 mm. long, red, smooth ; capsule inclined, oval with short neck ; peristome as in *O. subasperum*.   The fairly wide, shortly pointed, denticulate leaves without furrows, the numerous basal cells, and smooth seta indicate the plant.

Cape, E.   Boschberg, Somerset East, on trunks of trees, Macowan.
Natal.   Zwart Kop, Sim.
S. Rhodesia.   Floating encrusted in water in Sinoia Cave, Lo Mogundi, 3900 feet, Aug. 1920, Dr. Nobbs.

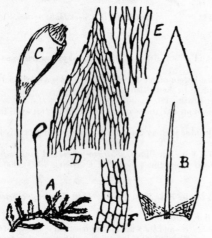

Oxyrrhynchium Macowanianum.

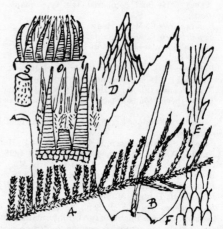

Oxyrrhynchium subasperum.
S, rough seta.

562. *Oxyrrhynchium subasperum* (Dixon, in litt.) Sim (new species).*

Stem 2–4 inches long, dark green, prostrate on damp banks and rocks, rooting occasionally, and bearing the sessile autoecious inflorescence and numerous, mostly simple, shining, yellowish branches 1–2 cm. long, arranged pectinately ; leaves somewhat laxly complanate in about eight rows, 1·5 mm. long, widely ovate-acuminate, acute but not subulate, clasping, decurrent, striated when old ; nerve fairly strong, extending to near the apex, ending suddenly or in a slight bristle ; margin expanded, dentate from near the base, each tooth one rather wide, exserted cell ; other lamina cells

---

* *Oxyrrhynchium subasperum* (Dixon MSS.) Sim (sp. nov.).
Caules 2–4″ longi, pinnatim ramosi, rami 1 cm. longi ; folia 1·5 mm. longa, late ovato-acuminata, connexa, dentata, nervata prope ad apicem ; cellae marginales pellucidae, exsertae : interiores vermiculares, sensim breviores latioresque ad basin ; seta aspera, 2–3 cm. longa ; capsula horizontalis ; peristom. dentes pariter alti, interiores perforati ad carinam ; membranum altum.

much narrower, vermicular, $80 \times 10\mu$, gradually shorter and wider near the base where they are oblong, $30 \times 20\mu$, and thick-walled. Autoecious; per. leaves numerous, wide, short, clasping, shortly pointed, dentate; seta 2 cm. long or more, red, *rough*; capsule dark brown, shortly oblong, horizontal, somewhat irregular, contracted below the mouth when dry, without ring or mouth-band; peristome red, permanent; outer inflexed toward the inner cone, but when moist and pressed all equally high; outer closely barred, inner perforated along the keel and on very high membrane, with two slender, short cilia between each. Lid beaked.

Cape, W. Knysna, J. Burtt-Davy.

Natal. Adamshurst, Merrivale, Sept. 1919, Sim 8699; Mtunzini, Zululand, Miss Edwards.

563. *Oxyrrhynchium confervoidum* Sim (new species).*

Confervoid and exceedingly slender in detail, still making a bright green, prostrate mat, rooting freely, with yellow rhizoids. Stems up to 2 inches long, simple or somewhat pinnately branched, often freely branched, delicate in texture as if etiolated. Leaves very lax or distant, placed round the stem, spreading, 0·5–0·75 mm. long, ovate-acuminate, subulate at the apex, concave and keeled below, clasping, pellucid; those on the ultimate branches smaller; nerve thin, difficult to see, extending half-way or more, and forming the keel; margin expanded, denticulate from near the base; cells pellucid, smooth, linear-oblong, those toward the base shorter, and those in the clasping corners almost quadrate. Other parts not seen.

Natal. Alexandra Park, Maritzburg, 3000 feet, July 1921, Sim 10,278β.

Dixon says, very near *O. pumilum* (Wils.), for which see Bry. Eur., Tab. 525β.

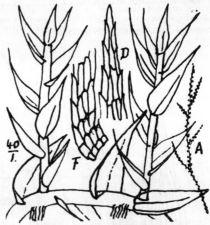

Oxyrrhynchium confervoidum.

Genus 173. RHYNCHOSTEGIUM Br. Eur.

Usually fairly robust, tufted, autoecious mosses, somewhat shining when dry, growing on soil or stones, with prostrate, freely rooted, irregularly or pinnately branched stems, sometimes stoloniferous; branches closely leaved, often flattened more or less; paraphyllia absent. Leaves spreading on all sides, seldom imbricate, slightly or not decurrent, usually somewhat concave, not furrowed, ovate or ovate-lanceolate from a wide base, more or less pointed or long-pointed; margin often denticulate-serrulate; nerve single, smooth, ending above the middle of the leaf, seldom forked at the apex; cells long, prosenchymatous, smooth, those at the base and the alar cells shorter and wider. Perichaetium rooting; inner perichaetal leaves sheathing, subulate, and recurved above. Seta long, smooth; capsule inclined or horizontal, oval and rather high-backed, or longer and almost regular, when dry contracted below the mouth. Ring distinct; peristome as in *Brachythecium*; lid conical, long-beaked; calyptra smooth. A very unsatisfactory group, in which the wider leaf separates it from *Rhynchostegiella*, and the beaked lid and subcomplanate arrangement separates it from *Brachythecium*, but many species have been placed here only to be moved elsewhere later.

After removing various names to other genera the following first four are stated by Brotherus to be unknown to him, and all this list are unknown to me and not represented in South African herbaria, viz.:

R. subconfertum (C. M.), Jaeg., Ad., ii, 431 = *Hypnum subconfertum* C. M., in Bot. Zeit., 1856, p. 438 (not W. P. Sch.).

R. subenerve (Hpe. and C. M.), Jaeg., Ad., ii, 430 = *Hypnum subenerve* Hpe. and C. M., in Bot. Zeit., 1855, p. 785.

R. subserrulatum (C. M.), Jaeg., Ad., ii, 437 = *Hypnum subserrulatum* C. M., in Bot. Zeit., 1856, p. 439.

R. chrysophylloides (Hpe.), Jaeg.

R. Teesdalioides W. P. Sch., in Breutel, Musc. Cap. = *Eurynchium Teesdalioides*, Jaeg., Ad., ii, 425 = *R. pseudoconfertum*, fide Hpe. (see under *Eurhyncostegiella* herein).

Brotherus allows to stand, *R. senodictyon* (C. M.), Jaeg.; *R. membranosum* (C. M.); *R. brachypterum* (Horns.), Jaeg., and *R. rhaphidorrhynchum* (C. M.), Jaeg., but I cannot separate the first three.

Synopsis.

564. *R. brachypterum.* Branches and leaves subcomplanate, ovate-lanceolate, basal cells few.

565. *R. rhaphidorrhynchum.* Branches subterete, leaves ovate-acuminate, basal cells many.

---

* *Oxyrrhynchium confervoideum* Sim (sp. nov.).

Confervoideum. Caules ad 2″ longi, graciles, ramosi, radicantes. Folia laxissima vel distantia, circum caulem posita, patula, 0·5 ad 0·75 mm. longa, ovato-acuminata, ad apicem subulata, concava ac carinata, connexa, pellucida. Nervus tenuis, ad dimidium longiusve pertinens, margo explanatus, denticulatus; cellae leves, lineo-oblongae, breviores infra, infimae paene quadratae.

564. *Rhynchostegium brachypterum* (Horns.), Jaeg., Ad., ii, 438 ; Broth., Pfl., i, 3, 1163, and xi, 374;
C. M., in Hedw., 1899, p. 152 ; Dixon, Trans. R. Soc. S. Afr., viii, 3, 223 (1920).

=*Hypnum brachypterum* Horns., in Linnaea, 1841, p. 142 ; C. M., Syn., ii, 248.
=*Hypnum *arboreum* W. P. Sch., in Breutel, Musc. Cap. (*fide* Hampe ; C. M.).
=*Brachythecium rhynchostegioides* C. M., in Rehm. 375 (Table Mountain).
=*Rhynchostegium subbrachypterum* Broth. and Bryhn ; Bryhn, p. 26 ; Broth., Pfl., xi, 374.
=*Hypnum *odontophyllum*, Rehm. 395 (*fide* Dixon).

Plant autoecious, prostrate, slender, shining green, rooting only occasionally, growing in flat mats, usually near moisture ; stems 1–2 inches long, laxly complanate, with spreading leaves 2 mm. long ; leaves variable, lanceolate or ovate-lanceolate from wide base ; main-stem leaves with long subulate point, branch leaves with acute, somewhat oblique apex, somewhat concave below, flat above, with denticulate-serrulate or sometimes almost dentate expanded margin, and weak narrow nerve extending to beyond mid-leaf ; cells linear, $100 \times 7\mu$, gradually changed at the base to irregularly rectangular, $40 \times 17\mu$ or less, of which there are comparatively few. When dry the leaves are sometimes unchanged, sometimes much crisped (possibly an effect of grass-fire). Perichaetal leaves forming a thickened bulb-like seta-base, the outer ovate-lanceolate, entire, nerveless, the inner longer, closely sheathing, quickly acuminate,

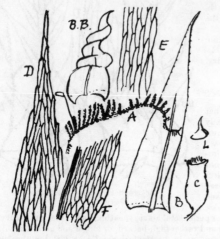

Rhynchostegium brachypterum.

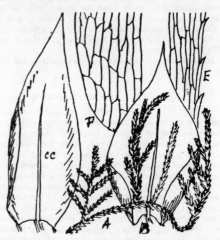

Rhynchostegium brachypterum, var. *nitens.*

denticulate, nerveless. Seta 1 cm. long, red, smooth ; capsule horizontal, ovate-oblong or oblong, brown, with short, obliquely beaked lid ; peristome teeth reddish below, pale above ; inner yellowish, perforated, with two nodose cilia between. Müller described *Hypnum brachypterum* in Syn., ii, 248, from Hornschuch's description and locality ; then in Hedw., 1899, p. 152, he described as *Rhynchostegium brachypterum* (Hedw., in Sched., as *Hypnum*) the same species from Genadendal (Breutel), apparently from specimen, forgetting his earlier publication.

R. *brachypterum* is frequent in forest streams throughout South Africa.

Var. *nitens* Sim* is a strong form with laxly pinnate, erect, non-complanate branches ; stem-leaves cordate-ovate, acute, 1·5 mm. long, 1 mm. wide ; branch leaves rather longer and narrower, stramineous and shining, slightly complanate, regular, concave, strongly decurrent, spreading whether dry or wet, those toward the branch-point appressed and imbricate ; cells about $50 \times 10\mu$ lower cells $40 \times 20\mu$, but not segregated as alar groups. Giant's Castle, Natal, 8000 feet, 1915, R. E. Symons ; Sim 10,276.

R. *senodictyon* (C. M.), Jaeg., Ad., ii, 438 ; Broth., Pfl., i, 3, 1163, and xi, 374=*Hypnum senodictyon* C. M., Syn., ii, 247, is an incompletely described, prostrate, dormant stream condition, and R. *membranosum* C. M., in Rehm. 376 (as *Brachythecium*) and in Hedw., 1899, p. 135 (mis-spelled R. *membranaceum*, in Broth., Pfl., xi, 374), a vigorous, thin, membranaceous condition of the same plant, but no specimens are available in the South African herbaria.

---

* *Rhynchostegium brachypterum*, var. *nitens* Sim (var. nov.).
Caules principes prostrati foliis cordato-ovatis acutis 1·5 × 1 mm., caules plurimi sub-erecti irregulariter pinnati 1–2″ alti, ramis ad ½″ longis, folia gerentibus tenuiter complanata, concava, decurrentia, patula, nitida, sub-flava, longiora atque angustiora cauli-foliis itemque leviter denticulata, nervo simplici ⅔ folii longo, levia, circiter 50 × 10 mm., rotundiora magisque rectangularia basin versus, ad 40 × 20 mm., non tamen segregata ut cellae alares.

565. *Rhynchostegium rhaphidorrhynchum* (C. M.), Jaeg., Ad., ii, 432 ; C. M., in Hedw., 1899, p. 152 ;
Broth., Pfl., i, 3, 1163, and xi, 375 ; Bryhn, p. 27 ; Rehm. 370.

=*Hypnum rhaphidorrhynchum* C. M., Syn., p. 354.
=*Rhynchostegium *subrhaphidorrhynchum* Wag. and Br., Wager's list.

Autoecious ; plant slender, prostrate, rooting, dark green or yellowish green ; stems 1–1½ inch long, irregularly pinnate ; pinnae ½ inch long, simple, crowded, with leaves all round ; leaves only subcomplanate, ovate-acuminate, acute, not cordate, 1½–2 mm. long, or the stem-leaves rather longer with long subulate points ; leaves not furrowed, concave, decurrent, denticulate-serrulate ; margins often narrowly reflexed at the base, expanded elsewhere ; toothed by protruding cell-ends from the base upward, but mostly upward, the lower teeth often obscured by the reflexed margin ; nerve weak, extending to or near mid-leaf ; cells linear, 100μ long, quickly reduced at the leaf-base to many small quadrate or oblong cells extending across the leaf, of which the lower row are the larger, the alar cells not different from these and hardly sunk ; per. leaves sheathing, short, quickly narrowed to a long, spreading or reflexed subula, nerveless. Seta 1–2 cm. long, smooth, red ; capsule oblong, horizontal, arched, with wide ring and long-beaked lid. Peristome teeth red, inner teeth perforated with two cilia between (much appendiculate, C. M.). Variable as to the length of the leaves and of the subula points, branch leaves often smaller than those of the main stem, more distinctly toothed and less regularly nerved. Not uncommon in forest or scrub country, extending also into tropical East Africa.

Rhynchostegium rhaphidorrhynchum.

Genus 174. RHYNCHOSTEGIELLA (Br. Eur.), Limpr.,
Laubm., iii, 207 (1896).

Usually small, slender, densely matted, autoecious stone- or bark-mosses ; silky yellowish when dry, occasionally more lax and several inches long, pinnate or bipinnate, with complanate branches and laxly complanate leaves ; paraphyllia almost scale-like, present where branches rise ; stem- and branch-leaves little different, not decurrent, spreading all round or squarrose on smaller branches, hardly concave, without furrows, usually narrowly lanceolate, the angles not concave. Margin expanded, entire, denticulate or toothed ; nerve single, long, smooth on the back, but sometimes ending in a mucro ; cells long, linear-oblong, chlorophyllose, smooth ; the basal and alar cells little different, or smaller and more quadrate. Seta long, rough or smooth ; capsule inclined or horizontal, oval or oblong, almost regular ; cells of the exothecium collenchymatous. Ring distinct. Peristome as in *Brachystegium.* Lid long-beaked from a dome-shaped base. Calyptra naked.

The genus is divided into :

§ *Eurhynchostegiella* Broth. Leaves long and slender-pointed, entire or toothed, the nerve not ending in a protruding mucro ; cells linear ; seta rough.

§§ *Leptorhynchostegium* (C. M.), Broth. Leaves less long-pointed ; nerve ending in a protruding mucro below the leaf-apex ; cells shortly linear ; seta smooth. This section, now named *Eurhynchiella* by Brotherus, is represented in South Africa by :

566. *Rhynchostegiella Zeyheri* (Spreng.), Broth., Pfl., i, 3, 1163 ; Brunnthaler, p. 33.

=*Hypnum Zeyheri* Spreng., MSS. ; C. M., in Bot. Zeit., 1855, p. 785.
=*Eurhynchiella Zeyheri* (Schimp.), Fleisch. ; Broth., Pfl., xi, 380.
=*Rhynchostegium Zeyheri* (Spreng.), Jaeg., Ad., ii, 430 ; C. M., in Hedw., 1899, p. 140 (=*Hypnum vagans*, Hpe., not Hook).
=*Rhynchostegium Zeyheri* (Spr.), Jaeg., var. *condensatum* C. M., in Hedw., 1899, p. 140 = Rehm. 361, *Eurhynchium *faucium* C. M. (*fide* Dixon, Kew Bull., 6, 1923, at No. 361).
=*Hypnum Ecklonianum* Sch., in herb. (*fide* Hpe.) ; C. M., in Hedw., 1899, p. 140.
=*Rhynchostegium natali-strigosum* C. M., in Hedw., 1899, p. 139.
=*Rhynchostegiella natali-strigosum* Broth., Pfl., i, 3, 1162 (*fide* Dixon, in Kew Bull., 6, 1923), and Pfl., xi, 380, as *Eurhynchiella* Fleisch.
=*Rhynchostegium afrostrigosum* C. M., in Hedw., 1899, p. 140 ; Rehm. 364 (as *Eurhynchium*), and Rehm. 363 (as *Eurhynchium Muelleri* Rehm.) ; Broth., Pfl., i, 3, 1162 (as *Rhynchostegiella*), *fide* Dixon, in Kew Bull., 6, 1923, and as *Eurhynchiella* Fleisch. ; Broth., Pfl., xi, 380.
=*Rhynchostegium afrostrigosum* C. M., in Hedw., 1899, p. 140, var. *thuiopsis* C. M., in Hedw., 1899, p. 140.
=*Eurhynchium thuiopsis* C. M., in Rehm. 362 (*fide* Dixon, in Kew Bull., 6, 1923).
=*Rhynchostegium lepto-eurhynchium* C. M., in Hedw., 1899, p. 140 ; Broth., Pfl., i, 3, 1162 (as *Rhynchostegiella*), and xi, 380 (as *Eurhynchiella* Fleisch.).
=*Rhynchostegium aristoreptans* C. M., in Hedw., 1899, p. 141 ; Broth., Pfl., i, 3, 1162 (as *Rhynchostegiella*), and xi, 380, as *Eurhynchium* Fleisch.
=*Rhynchostegium pseudoconfertum* (C. M.), Jaeg., Ad., ii, 431 =*Hypnum pseudoconfertum* C. M., in Bot. Zeit., 1858, p. 70 (*fide* Dixon).

Autoecious, very pale yellow, lax, slender, dwarf, in close tufts with thin spreading leaves ; stems 1–2 cm. long, prostrate, 2-pinnate, the pinnae about 5-branched, all branches rising erect to the same level if crowded, or if lax spreading, the final branches 3–5 mm. long, slightly complanate, often with leaves in four rows ; leaves erecto-patent or spreading, 0·5–1·25μ long, lanceolate-acute but not subulate, dentate or strongly denticulate, with fairly strong

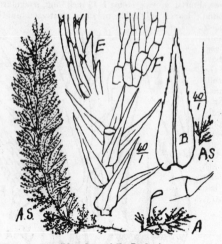

Rhynchostegiella Zeyheri.
AS, var. *frondosa*, natural size.

Rhynchostegiella Zeyheri.

nerve often present to near the apex ; cells rhomboid, 50 × 20μ, the marginal teeth often wider and oblong, the basal cells gradually shorter and wider. Seta 1 cm. long, red. smooth ; capsule small, inclined, brown, with obliquely beaked lid, wide ring, and long, narrow, white calyptra. Per. teeth red, narrow ; processes narrow not perforated ; cilia two, small.

Var. *frondosa* Sim (new variety).* Fronds 2 inches long or more, ½ inch wide, from a nearly stipate base, 2-pinnate ; pinnae and branches alternate, laxly distichous, subcomplanate, with leaves to 1·5 mm. long ; end of

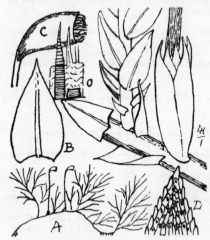

Rhynchostegiella Zeyheri.

the frond sometimes extending as a simple scale-leaved innovation up to 1 inch long, from which leaved branches are produced above that. Disa Ravine, Table Mountain. Sim 9370 (Pillans 4309, as *Amblystegium capense* Dixon, MSS.) comes exceedingly close to this, but Dixon still feels inclined to maintain the latter, and says : "The one thing stable about *R. Zeyheri* is the narrow upper cells with the distinct small group of *minute opaque alar cells*, which is indeed a generic character. Here the basal cells are all enlarged and lax, which, I think, is never the case with *R. Zeyheri*, and the angles are *decurrent*, with quite lax cells." This requires more investigation. Wide variations occur in the leafage of *R. Zeyheri* ; some plants have the leaves all small, others all large, and wherever branchless parts occur, reduced scale-like leaves with subulate dentate apex from wide almost cordate base occur. The smallest forms are often submerged in mud, but merge gradually into the long lax var. *frondosa*, which occurs only, so far as I know, under the daily mists and showers of Disa Ravine. The type is abundant everywhere on and round Table Mountain, and occurs also at Grahamstown, Miss Farquhar ; Perie Forest, Cape, E., and Blinkwater Forest, Natal, J. M. Sim. *R. pseudo-confertum* is from Genadendal.

Mr. Dixon named my 9445 var. *obtusata* in Hb. ; the branch leaves are very small and obtuse or subobtuse, but are not very stable. This is very variable, and includes *Rigodium dentatum* Dixon, in Bull. Torrey Bot. Club, xliii, 79, and fig. 9 (=*Isothecium afromyosuroides* C. M., in Hedw., 1899, p. 147 ; Broth., Pfl., i, 3, 1230, and still retained by Brotherus in Pfl., xi, 211 and 212=*Eurhynchium afromyosuroides* C. M., in Rehm. 369 ; Paris, 2nd ed., p. 163) (upper figure on this page), and also *Thuidium *natalense* Sim (in letters) (lower figure on this page). The former is

---

* *Rhynchostegiella Zeyheri* Broth., var. *frondosa* Sim (var. nov.).
Frondes 2″ plusve longae, ½″ latae, a basi prope stipata, 2-pinnatae ; pinnae ramique alternantes, laxe distichi, sub-complanati.

very different in habit, and the latter has the leaves on the main stem scoop-like, 0·75 mm. long, denticulate ; cells oval-elliptic, drawn to the upper end, but not clearly papillose.

Cape, W. Cape Town, Rehm. 369 ; var. *minor* from Rondebosch, and var. *major* from Claremont (*fide* Müller), Cape Town, Rev. H. Friend.

Natal. Buccleuch, Sim 10,268.

Rhynchostegiella sublaevipes Broth. and Bryhn ; Bryhn, p. 26 ; Broth., Pfl., xi, 377 (on bark, Eshowe, Zululand, January 1909, scarce, H. Bryhn), has no difference described except " basal cells uniform, quadrate or shortly rectangular, others upward longer," and " seta subsmooth, upper part only, and slightly rough" ; it may or may not be different.

Fleischer has placed our *Rhynchostegiella* in *Eurhynchiella* (Laubm., Java, iv, 1566 (1922)), including names as indicated above, together with :

E. leskeifolia (C. M.), Fleisch.

E. modesta (Rehm.), Fleisch. (=*Eurhynchium modestum*, Rehm. 365, Stinkwater, C.P.).

E. Teesdalioides (Schimp.), Fleisch. (see under *Rhynchostegium* herein).

And stated that all are very closely allied.

(Genus *Eurhynchium* held a good many South African species fifty years ago, but these have since been reallocated, till now the genus which still exists holds no known South African species.)

Family 47. LESKEACEAE.

Mosses of varied habit, usually prostrate and much-branched, rooting from the under-surface, and often with distantly leaved stolons starting from various parts. Secondary branches usually erect, simple or irregularly forked or somewhat pinnately branched, or in *Thuidium* 1–3-pinnate in one plane ; paraphyllia usually numerous, seldom absent. Leaves often dimorphous, those on the main stems differing from those on the more ultimate divisions. Lower leaves distant, small, smooth, and usually nerveless ; stem-leaves numerous, usually spreading all round the stem, seldom in flattened arrangement, symmetrical, pointed, seldom blunt or rounded, concave, sometimes shortly 2-furrowed at the base, formed of one layer of cells, one or both sides rough from papillae or mamillae, seldom smooth. Nerve single and strong in South African species, sometimes short and weak, and then double or forked elsewhere. Cells usually papillose, lax, quadrate or rhomboidal, or at the central portion of the leaf-base longer and coloured. Branch leaves shorter and narrower than the stem-leaves. Seta long and straight ; capsule erect and regular, or inclined and somewhat curved ; ring usually distinct ; lid conical or beaked from convex base. Calyptra cucullate, usually naked. Peristome double ; inner peristome with high, 16-plicate, basal membrane and keeled processes, often with long or rudimentary intermediate cilia.

This family differs from Leucodontaceae in having papillose cells, and from Hypnaceae in its short, rhomboidal or rounded cells.

Synopsis of Leskeaceae.

§ ANOMODONTEAE. Main stems stoloniform ; seta rising from the branches ; paraphyllia absent ; leaves uniform ; processes of inner peristome thread-like, rudimentary or absent.

Slender plants, the nerve not passing the middle of the leaf, often shorter, inner peristome processes absent,
175. HAPLOHYMENIUM.

Robust plants, the nerve ending at or near the leaf-apex. Nerve twisted above ; cells very small, quadrate, smooth ; processes rudimentary . . . . . .176. HERPETINEURON.

Seta rising from the main stem, which is not stoloniform.

§ LESKEAE. Leaves uniform.

Capsule erect, regular, usually straight ; basal membrane of inner peristome narrow ; processes linear or absent.

Dioecious ; teeth without lamellae ; processes thread-like . . . 177. LESKEELLA.

(Teeth of outer peristome with well-developed lamellae ; processes narrowly linear. Autoecious LESKEA.)

Teeth of outer peristome with narrow lamellae ; processes absent. Autoecious . 178. LINDBERGIA.

Capsule inclined or horizontal, more or less irregular ; basal membrane of inner peristome wide ; processes wide . . . . 179. PSEUDOLESKEA.

§ THUIDEAE. Leaves not uniform (except *Haplocladium*).

Autoecious slender plants ; stems pinnate ; leaves dimorphous ; cells with numerous papillae on both sides ; end-cell pointed or blunt ; lid shortly beaked from a conical base . . . 180. RAUIA.

Stems pinnate, leaves uniform, almost smooth or with one papilla in the lumen, or with projecting papillose angles ; end-cell pointed ; lid shortly conical . . . . 181. HAPLOCLADIUM.

Dioecious ; stems 1–2–3-pinnate ; leaves heteromorphous ; cells with numerous low papillae ; end-cell 2–4-pointed . . . . 182. THUIDIUM.

Including :

Stems very slender, 1–2-pinnate ; end-cell blunt, 2-pointed ; lid with long narrow beak,
(*a*) THUIDIELLA.

Stems strong, 2–3-pinnate ; end-cell of the branch leaves (except in *T. tamariscinum*) blunt, 2–4-pointed ; lid longer, with thicker beak . . . . . . (*β*) EUTHUIDIUM.

Genus 175. HAPLOHYMENIUM, Doz. and Molk. (in Ann. Sc. Nat., 1844, i, 310).

Slender, firm, dioecious, matted mosses growing on bark or rocks, the stems stoloniform, wiry, forked, prostrate, widely spreading, rooting occasionally, and with the branches irregularly pinnate, the branchlets short and blunt ;

the leaves usually spreading all round, but occasionally somewhat flattened ; paraphyllia absent. Leaves of the main stem fewer and smaller than those on the branches, which are numerous or crowded, spreading when moist, imbricate when dry, not auricled, more or less lingulate from a concave oval base, blunt or shortly pointed, flat and entire. Nerve smooth, usually weak, reaching not more than half-way ; cells turgid, roundly hexagonal, with usually several papillae per cell ; the papillae sometimes 2-pointed ; the marginal cells smaller, quadrate or rhomboid ; the central basal cells oblong, pellucid. Vaginula with many paraphyses. Seta short, straight, smooth, rising from the branches ; capsule erect, oval, smooth ; ring wide, dehiscent in fragments ; lid shortly beaked from a conical base ; calyptra cucullate, rough at the apex, set with a few, long, erect hairs, and having the margin irregularly lobed. Peristome double ; the teeth with some irregular appendages ; the inner peristome a very narrow, smooth, basal membrane, without processes and without cilia.

567. *Haplohymenium pseudotriste* (C. M.), Broth., Pfl., i, 3, 986, and xi, 313.

=*Hypnum pseudotriste* C. M., in Bot. Zeit., 1855, p. 786.
=*Anomodon pseudotristis* C. M., in Rehm. 343.
=*Haplohymenium exile* (Mitt.), Broth., Pfl., i, 3, 1236, and xi, 313.
=*Anomodon exilis* Mitt., Journ. Linn. Soc., 1872, p. 309 (sterile).
=**Lindbergia natalensis* Sim, in letters.

Stem prostrate, very slender, rooting occasionally, forked, mostly simply and irregularly pinnate, 1–2 inches long ; the pinnae and pinnules numerous, each smaller-leaved, 5–8 mm. long ; stem-leaves 0·75 mm. long, widely ligulate from a wider clasping base, rounded or blunt at the apex, concave at the leaf-base, nearly flat above ; nerve very short and weak, single ; margin crenulate all round through rounded papillose cells ; cells nearly round, tumid, with several papillae on each, perichaetia lateral on branches, sessile, few-leaved, these leaves lorate from wider base ; other features not seen.

Cape, W.　Prom. Bon. Sp., Ecklon (*fide* C. M.) ; Portland, Knysna, Rehm. 343.

Cape, E.　Eveleyn Valley, Perie, 4500 feet, 1888, Sim 7012 ; Paradise Kloof, Grahamstown, Miss Farquhar, 90*β* ; East Pondoland, 1899, Sim 10,270, and 1915, Burtt-Davy 15,351 ; Dohne Hill, 1888, Sim 7210 ; Alexandria Forest, Jas. Sim.

Natal.　Ngoya Forest, Zululand, 1000 feet, Sim 10,271 ; Eshowe, Zululand, Bryhn (p. 21) ; Natal, Wager 6.

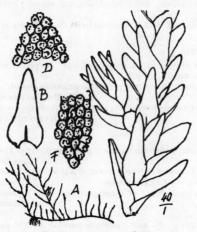

Haplohymenium pseudotriste.

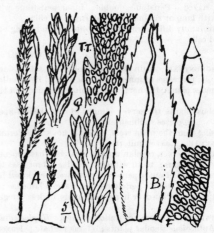

Herpetineuron Toccoae.

Genus 176. HERPETINEURON (C. M.), Card., in Beihefte z. Bot. Centralbl., xix, Abt. ii, p. 127 (1905).

Fairly strong, stiff, laxly tufted, dioecious mosses, having long, prostrate, wiry, stoloniform, small-leaved, rooting main stem, and many erect or spreading, simple, forked or branched, closely leaved, secondary stems, and sometimes flagellate ends or branches. Paraphyllia absent. Leaves laxly imbricate and somewhat twisted when dry, erecto-patent when moist, keeled-concave, oblong-lanceolate, sharply pointed, 2-furrowed at the base ; the margin flat, sharply and irregularly toothed upward ; nerve strong, much twisted upward, smooth, disappearing just below the apex. Cells very small, not thickened, quadrate, smooth ; the basal cells not lengthened but transverse. Fruit (not seen on South African specimens) described as having : Vaginula with numerous paraphyses ; seta long, straight, smooth ; capsule erect, regular, cylindrical, with small mouth ; lid shortly beaked from a conical base ; calyptra cucullate, naked ; peristome double, outer teeth narrowly lanceolate, papillose, with low lamellae, inner peristome finely papillose, with narrow basal membrane.

568. *Herpetineuron Toccoae* (Sull. and Lesq.), Card. ; Broth., i, 3, 990, and xi, 315.

= *Anomodon Toccoae* Sull., M. of U.S. ; Sull. and Lesq., Icon. M., p. 121, t. 76 ; Lesq. and Jam., Man., p. 306 ; Paris, 2nd ed., i, 58.
= *Anomodon *robustus*, Rehm. 639 and 639*β* ; Par., Ind., 1st ed., p. 45.

Loosely tufted ; main root prostrate, black, wiry, with only fragmentary leaves ; erect stems 2 inches long' closely leaved above, closely enclosed in shorter, ovate-lanceolate, subulate, entire but nerved leaves below ; branches sometimes flagellate at the apex ; branch leaves set all round the stem, widely lanceolate-ligulate, 2 mm. long, half-twisted when dry, erecto-patent when moist, ending rather quickly in an acute point ; somewhat concave ; nerve strong, extending to the apex, much twisted above ; margin expanded throughout, roughly serrate in the upper half, dentate in the lower half to near the base ; larger teeth simple or with small mid-teeth, up to 100μ long and 30μ wide, and containing up to 100 cells ; cells minute, smooth, obliquely quadrate, 10μ long, in rows, those toward the leaf-base not larger or smaller, but transverse. Capsule not seen in South Africa, but included in my figure from foreign specimens.

Laingsnek, Natal, Rehm. 639 ; in shade on the mountains above Lydenburg, Transvaal, Rehm. 639*β*.

Genus 177. LESKEELLA (Limpr.), Loesk, Moosfl. d. Harz., p. 255 ; Broth., Pfl., xi, 302.

Dioecious, prostrate, rooting occasionally, closely set with erect short branches, closely leaved and without paraphyllia. Leaves uniform, imbricated and 2-plicate when dry, erecto-patent or secund when moist ; cordate at the base, soon long-pointed with narrow, reflexed, entire margin ; nerve strong, percurrent in the subula ; cells smooth, oval, inner basal cells rectangular, marginal basal cells quadrate. Branch leaves smaller, with short weak nerve. Per. leaves erect, white, sheathing, with long narrow point, long weak nerve, and longish cells. Capsule erect, regular ; teeth lanceolate, striate ; processes thread-like, on plaited membrane ; cilia usually absent. Lid beaked.

569. *Leskeella zuluensis* Broth. and Bryhn, in Bryhn, p. 21.

Not known to me, but described as dioecious, slender, deep green, prostrate, 2 cm. long, with few short, erect, obtuse, irregularly branched, densely foliose branches. Stem-leaves crowded, horizontal, imbricated when dry, ovate-cordate base quickly lanceolate, concave, 2-sulcate at the base, and with recurved margin, wide nerve for one-third the length ; basal cells rounded-quadrate or shortly transversely rectangular, other cells shortly oval, smooth. Branch leaves rather smaller, with flat margins. Inner per. leaves hyaline, sheathing, subulate, with very weak nerve ; seta 8–10 mm. long ; mature theca not seen, nor gonidia. Entumeni, Zululand, 1908, H. Bryhn.

(Genus LESKEA, though formerly regarded as South African, has now had its South African species moved into *Lindbergia*, because no processes are present in its inner peristome.)

Genus 178. LINDBERGIA Kindb., Sp. Eur. and Northam. Bryin., p. 13 (1896).

Slender, small, laxly tufted, brownish-green, opaque, autoecious mosses, epiphytic on tree-trunks ; stems long, prostrate, tufted-rooted, closely leaved all round, forked, the forks usually irregularly pinnated ; branches short or unequal, blunt ; paraphyllia absent. Leaves spreading or suberect, imbricated when dry, rather concave, more or less decurrent, lanceolate or ovate-lanceolate, shortly subulate-pointed, without furrows ; margin flat and entire, often somewhat reflexed toward the base. Nerve fairly strong, more or less wide, smooth, ceasing below the apex. Cells lax, roundish oval or rhomboid hexagonal, not thickened, smooth or with one papilla per cell ; the marginal row often smaller, quadrate or rhomboid, as also the basal cells toward the margin, in many rows, but with no special alar cells. Perichaetium not rooting ; inner perichaetal leaves larger than the stem leaves, pale, erect, lanceolate or subulate-pointed from a sheathing base ; margin entire or denticulate ; nerve short ; cells elongated, smooth. Seta 5–10 mm. long, straight, thin, red, smooth. Capsule erect, regular, oval-oblong, thin-walled, brown, short-necked ; stomata few on the neck ; ring differentiated or absent ; peristome inserted far below the mouth ; teeth of the outer peristome lanceolate, blunt, pale or yellow, not striate, densely papillose, with zigzag median line and narrow lamellae below ; inner peristome finely papillose ; basal membrane small or high, processes and cilia absent. Spores 25–30μ ; lid conical, blunt ; calyptra cucullate, naked.

Synopsis.

Leaves not appressed when dry ; leaves with long acumen	570. *L. haplocladioides.*
Branch leaves julaceous when dry.	
Branches few, curved when dry ; leaves bluntly acuminate . . .	571. *L. pseudoleskeoides.*
Pinnate ; leaves cordate-acuminate, spreading widely when moist . .	572. *L. patentifolia.*
Pinnate, slender ; leaves widely lanceolate-acute, erecto-patent when moist . .	573. *L. viridis.*

570. *Lindbergia haplocladioides* Dixon, in Bull. Torrey Bot. Club, xliii, 75, and in Trans. R. Soc. S. Afr., viii, 3, 215 ; Broth., Pfl., xi, 301.

A slender, little, dark green moss, epiphytic on bark, rooting freely, and sometimes homomallous opposite roots. Autoecious. Stems 6–12 mm. long, very slender, laxly pinnate or 2-pinnate, all laxly leaved, and some branches flagellate or smaller and smaller-leaved ; the leaves all round the stem, spreading when moist, turned upward or with the point outward but not appressed when dry ; variable as to size on different branches from 0·3–1·25 mm. long, ovate-acuminate, with long, narrow, acute point on all leaves, entire, or the cells at the apex slightly overlapping ; nerve narrow, extending to the point of the acumen ; lamina cells smooth, oval or rhomboid, 12μ long, the marginal cells often projecting so that the leaf is slightly crenulate ; acumen cells longer and narrower ; alar cells many, transverse, gradually merging into the lamina cell-form ; main stem-leaves larger and with longer points than others. Perichaetium sideways on the main stem ; inner leaves 1·5 mm. long, convolute, nerveless, shortly pointed, with long cells. Seta short, red, smooth ; capsule widely cylindrical ; teeth pale, shortly lanceolate, densely papillose ; membrane low, papillose.

Cape, E. Boschberg, Somerset East, Jas. Sim.

Natal. Zwartkop Valley, Maritzburg, and Donnybrook, Sim.

Transvaal. Pretoria (among *Fabronia*), Sim 7702 ; Rydal Mount, 1914, Wager 29.

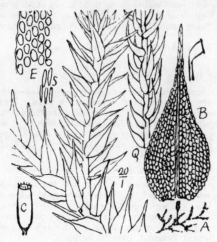

Lindbergia haplocladioides.

S, cells in acumen.

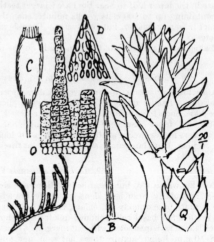

Lindbergia pseudoleskeoides.

571. *Lindbergia pseudoleskeoides* Dixon, in Bull. Torrey Bot. Club, xliii, 75, and fig. 5,
and in Trans. R. Soc. S. Afr., viii, 3, 215 (1920) ; Broth., Pfl., xi, 301.

Caespitose in low mats on bark ; stems 2–3 cm. long, irregularly few-branched, some of the branches strong, 10–12 mm. long, others weak and small, 2–3 mm. long ; the branches obtuse and terete through appressed leaves and usually curved when dry ; the main stem and older branches often bristling with erecto-patent leaves even when dry ; all leaves erecto-patent when moist, or some almost horizontal ; branch leaves very numerous, 0·5–1 mm. lcng on different branches, closely imbricated when dry, widely ovate-acuminate, subobtuse, somewhat concave, with expanded entire margins, or often reflexed toward the base. Cells roundly oval, incrassate, subopaque, smooth, gradually oblique downward, and at the base transverse and crowded at the margin, not much different nearer the nerve, but when empty, as sometimes happens, the cells are then shortly 6-sided, with thin walls, and appear larger. Main-stem leaves laxly placed, longer and more pointed ; perichaetium sessile on side of main stem, erect ; the outer leaves like branch leaves, the inner much longer, convolute, shortly pointed, with no nerve, entire margins, and large, bluntly vermicular, often twisted, pellucid cells. Seta 1 cm. high, red, smooth ; capsule erect, cylindrical, with slight throat, pale red ; lid conical, obtuse ; peristome fairly strong, the teeth pale red, bluntly lanceolate, densely papillose, and the membrane pale, one-third as high, of short, densely papillose squares, and without processes or cilia. Abundant in scrub districts in each province, including Rhodesia, Portuguese East Africa, and Orange Free State. Zimbabwe, Sim 8746, 8824, and 8744, etc. ; also in British East Africa.

572. *Lindbergia patentifolia* Dixon, in Smithson. Misc. Coll., lxix, 8, 7, fig. 6,
and in Trans. R. Soc. S. Afr., viii, 3, 215 ; Broth., Pfl., xi, 301.

=*Leskea *fallax*, Rehm. 635.
=*Leskea *fallax*, var. *robusta*, Rehm. 636.

Prostrate in masses on tree-bark ; autoecious ; stems 1–2 inches long, procumbent, rooting occasionally, simply pinnate ; the branches simple, straight, 3–8 mm. long, spreading or suberect, densely foliose ; the perichaetium sessile and erect on the main stem. Leaves very numerous, closely imbricate when dry, spreading horizontally when moist or with up-curved tips, 1 mm. long, widely ovate-acuminate from wide cordate base, acute, entire or slightly crenulate, keeled-convex, but with expanded or often lightly reflexed margins ; nerve wide, extending to near the point ; cells oval or rhomboid in the upper half of the leaf, 15μ long ; often quadrate and tumid in the centre, so as to appear lightly papillose on both sides and rough on the back of the nerve, the central cells at leaf-base obliquely elliptical, 30μ long, gradually reduced and more oblique and elliptical (15μ long) toward the margin where they are transversely quadrate, 8μ long, but not forming any separate group. Leaves on the main stem often rather larger, more subulate, and homomallous if the stem is rooting, except the lowest leaves which are shortly triangular. Perichaetal outer leaves small, spreading, and deltoid, the inner leaves 1 mm. long, closely clasping, slightly denticulate, and with short subula, nearly nerveless, and with vermicular cells up to 100μ long, the apical cells shorter. Seta erect, red, smooth; capsule erect, red, cylindrical or oval-cylindrical, 2 mm. long, with slight neck and somewhat contracted mouth; teeth pale, narrowly conical, finely striate, terminal joints papillose, lower bordered; membrane nearly half as high, firm, like three transverse rows of bricks, densely papillose, without processes or cilia. Ring a row of oblong long cells one cell deep; capsule-wall of long, lax, oblong cells, except near the mouth where they are small, round, and deep red. Lid deeply convex-apiculate, too large for the young capsule, and made of large transversely hexagonal cells 25μ across.

Very like *Pseudoleskea*, but with the capsule erect and with no processes or cilia; also like *Dimerodontium*, which, however, has no membrane.

Cape, E. Cave Mountain, Graaff Reinet, Rehm. 636; Gwacwaba River, King William's Town, 1898, Sim 7104.

Natal. Van Reenen, Sim; Bulwer, 1915, W. J. Haygarth (Wood 31); Cathkin, 6000 feet, Sim; Zwartkop Valley, Maritzburg, Sim.

Transvaal. Lechlaba, Rehm. 635; Waterval Onder, Wager 165.

S. Rhodesia. Zimbabwe, Sim 8774. Also found in East Africa.

Lindbergia patentifolia.

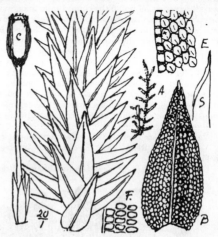

Lindbergia viridis.
S, point of perichaetal leaf.

573. *Lindbergia viridis* Dixon, in Trans. R. Soc. S. Afr., viii, 3, 214, and 215 (1920), and pl. xii, 15; Broth., Pfl., xi, 300.

Very slender, little, dark green bark-epiphyte, with pinnate stems 5–15 mm. long; the branches spreading or suberect, 3–4 mm. long, julaceous when dry; branch leaves numerous, imbricate, erecto-patent when moist, 0·75 mm. long, 0·25 mm. wide, all alike, somewhat keeled, clasping at the base, widely lanceolate, acute, with distinct, thin, narrow nerve to near the apex; cells pellucid, roundly quadrate, 15μ wide, marginal often transverse, toward the leaf-base gradually longer and more oblique or transverse; seta short, red, 7–8 mm. high; perichaetal leaves sheathing, pellucid, nerveless, shortly pointed (fig. S). Capsule cylindrical; teeth pale, columnar, blunt, papillose; membrane one-third as long, densely papillose.

Kentani, Transkei, Miss Pegler (Sim 8322); Kroonstad, O.F.S., Wager 170; Kaapschehoek, Transvaal, Wager.

The Abyssinian *L. abbreviata* (W. P. Sch.), Broth., Pfl., xi, 300, differs from South African species as *per* Dixon, in Bull. Torrey Bot. Club, xliii, 75, in having points longer and reflexed, branches seldom curved when dry, capsule ovate-cylindrical, darker, and of stouter texture, and the lid conical-rostellate; and Brotherus describes it as having an evident papilla on the lumen.

Genus 179. PSEUDOLESKEA Br. Eur.

Slender matted mosses, growing on stones, tree-stumps, or moist soil; autoecious and without paraphyllia in South African species; the main stem prostrate, rooting and slender, irregularly or somewhat pinnately branched; the branches irregular as to size but not flagellate. Leaves uniform, placed all round the stem, spreading when moist, imbricate when dry, often secund but not regularly so; ovate or ovate-lanceolate, concave, the margin flat or reflexed, sometimes denticulate. Nerve strong, ending near the leaf-apex; cells small, oval or elliptical, smooth or 1-papillate per cell. Branch leaves smaller. Perichaetium on the main stem or its forks, forming a cylinder; inner perichaetal leaves sheathing and long-pointed; capsule inclined or horizontal, short, turgid, bent, irregular, thick-skinned; lid shortly pointed ; calyptra cucullate, naked ; peristome double ; teeth lanceolate, bordered, striate ; inner peristome processes as high or higher, lanceolate, plaited or perforated, papillose ; basal membrane wide, papillose ; cilia present in South African species.

Synopsis.

574. *P. claviramea.* Lower cells quadrate ; processes higher than the teeth, keeled or perforated.
575. *P. Macowaniana.* Inner lower cells elliptical ; processes as high as the teeth.

574. *Pseudoleskea claviramea* (C. M.), C. M., in Rehm. 355 ; Bryhn, p. 21 ; Brunnthaler, p. 32 ;
Brotherus, Pfl., i, 3, 1000, and xi, 306.

=*Hypnum clavirameum* C. M., in Bot. Zeit., 1855, p. 787.
=*Anomadon clavirameus* (C. M.), Jaeg., ii, 305.

Dark green, autoecious, caespitose in extensive patches on stones, soil, or sometimes bark ; stems 1–2 inches long, rooting below, suberect above, frequently forked, the forks pinnate or innovated ; the ultimate branches 5–15 mm. long, often curved, often or usually starting from an innovation in which the lower leaves are small and the upper much larger, giving the branch a club-shaped form.   No paraphyllia are present ; and leaves are usually gone where rooting takes place, so none are seen to be secund ; but the erect forks carry main-stem leaves, and also perichaetia set sideways.   Leaves set all round the stem, 1 mm. long, ovate and shortly acute, imbricate or julaceous when dry, spreading nearly or quite horizontally when moist, very concave, with high rising sides which reflex widely along the margin. Nerve wide, sometimes weak, sometimes coloured, extending to near the apex, covered with lamina cells ; margin entire ; cells oval or elliptical toward the leaf-apex, 17–20μ long, tumid but not papillose ; lower cells quadrate, the inner 12μ long, the marginal transverse and 8μ across ; no alar cells and no longer basal cells.   Main-stem leaves rather longer and longer pointed.   Inner per. leaves sheathing, long-pointed, slightly denticulate, with rhomboid cells 30–50μ long.   Seta red, ½ inch long, smooth ; capsule nearly horizontal, oval-elliptical with slight neck, bent, and when dry contracted below the large mouth.   Peristome strong, outer teeth red, closely trabeculate ; inner higher, on high membrane, nearly as wide, keeled and sometimes perforated, acute ; cilia short, slender.   Young calyptra conical, white, shining, covering the capsule, and with long ridge-like cells.   Common everywhere on stones, stumps, or humus, including Orange Free State, Rhodesia, Portuguese East Africa, and East Africa ; usually in scrub bush.

Rehmann distributed two named varieties, not worth varietal names or to be more than conditional forms, viz. :
Var. *acuminata*, Rehm. 356.   More acuminate, slightly denticulate.
Var. *aquatica*, Rehm. 642.   Leaves short, impregnated with iron, sometimes very slightly denticulate at the apex (Lechlaba).
C. M. described a var. *denudata* in Abhandl. Brem., vii, 213, from Madagascar.

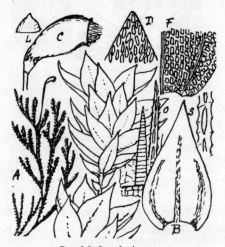

Pseudoleskea claviramea.
S, perforated part of a process.

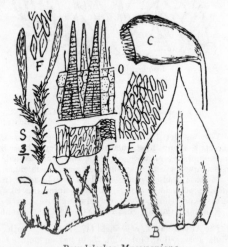

Pseudoleskea Macowaniana.
S, lower stem with spreading leaves and
innovation with julaceous leaves.

575. *Pseudoleskea Macowaniana* C. M., in Hedw., 1899, p. 148, and in Rehm. 643 ;
Dixon, in Trans. R. Soc. S. Afr., viii, 3, 215.

=*Amblystegium filiforme* Wager and Wright (in Trans. R. Soc. S. Afr., iv, 1, 2, and pl. 1β).

Fairly robust, yellowish, autoecious ; the main stem 1–2 inches long, prostrate, with spreading or broken leaves, and rooting occasionally, irregularly branched, the branch leaves when dry either suberect or imbricate, and innovations from broken branch-tips club-shaped, julaceous, with few leaves below and numerous leaves above.   Paraphyllia absent ; stem-leaves 1·5 mm. long, branch leaves 1 mm. long, all ovate-acuminate with rather long acute point ; deeply concave with erect sides, margin flat above and then reflexed, especially in the lower half ; nerve wide, disappearing below the point, projecting along the lower surface, and covered on each surface with long lamina cells ; cells oval, smooth but tumid, 20μ long, the outer rendering the margin sinuose-denticulate ; mid-cells at the base also oval or rhomboid, 30μ long, gradually transverse toward the margin where they are about 10μ across ; the cells on the nerve about 25μ long. Stem-leaves rather longer and more subulate. Inner per. leaves 1 mm. long, sheathing, lanceolate-subulate, with fairly long, slender, subdenticulate point ; nerve very weak below, but increasing upward and forming the point ; cells rhomboid, 50–70μ long ; seta $\frac{1}{2}$ inch long, red, smooth ; capsule horizontal, elliptical, very irregular, slightly bent ; lid convex-mucronate ; peristome fairly strong ; teeth pale, bordered, closely trabeculate, faintly striate, and upper sections papillose ; membrane very high, pellucid, closely punctate ; processes as high as the teeth, papillose ; cilia slender.

P. capilliramea C. M., in Hedw., 1899, p. 148 ; Broth., Pfl., xi, 306 = *Hypnum leskeifolium*, Par., 2nd ed., iii, 54, and iv, 102 = *Brachythecium leskeifolium* C. M., in Rehm. 373 = *Rhynchostegium leskeifolium* Hedw., 1899, p. 139 (see Kew Bull., No. 6, 1923, No. 373), is a small, somewhat etiolated, slender form of this, and from it *P. *laxifolia*, Rehm. 640 (Lechlaba), does not differ, and as described by C. M., in Hedw., 1899, p. 149, *P. *leskeoides* (Schimp., ined.), Broth., ined., seems to be the same. Dixon (in Trans. R. Soc. S. Afr., viii, 3, 215) states there are no specimens of it in Schimper's herbarium.

The habit and vigour vary in accordance with whether it is prostrate and sunk in mud, or erect, or pendent from humus or bark.

Cape, W. Tows River, Rehm. 373 (as *B. leskeifolium* C. M.), and probably elsewhere in West, East, and Natal.

Cape, E. Hell's Gate, Uitenhage, Sim ; Boschberg, Macowan 3, Rehm. 643.

Natal. Van Reenen, Wager 1911 (as *Amblystegium filiforme* W. and W.).

Transvaal. Dixon cites for *P. Macowaniana*, Macomo's Hoek, 1897, Mrs. Williams ; Noorddrift, Waterberg, Wager 403 ; and for *P. leskeoides* (Sch.), Broth., *fide* Dixon, Grahamstown, S. Burton, and Woodbush, T. Jenkins.

Brotherus places in the same group, described as " Branches sometimes flagellate, leaves more or less evidently secund," *P. pseudo-attenuata* (C. M., as *Hypnum*, in Bot. Zeit., 1855, p. 786), Broth. = *Anomodon*, Par., 2nd ed., iii, 57 ; no specimen of that or of *P. capensis* W. P. Sch., in Breutel, Musc. Cap. (ined.), is available in South African herbaria.

Genus 180. Rauia Austin, in Bull. Torrey Bot. Club., vii, 16 (1880).

Slender, autoecious, matted, forest mosses, having prostrate, simple or forked main stems, each fork regularly or irregularly pinnate ; the pinnae spreading, short, blunt, with leaves all round, and numerous simple or forked papillose paraphyllia ; leaves dimorphous, spreading when moist, imbricate when dry. Stem-leaves lanceolate or subulate from a wider subbasal portion, very concave, with margin reflexed and untoothed but sometimes papillose-crenulate. Nerve strong, ending below the apex ; cells uniform, roundish, with one central papilla each ; papillae numerous on both surfaces and at each margin ; lower cells more lax and pellucid. Branch leaves ovate-lanceolate, shortly pointed, with the nerve rough on the ridge. Leaves when dry inflexed and incurved. Seta long, smooth ; capsule horizontal in South African species, cylindrical, arcuate ; ring dehiscent ; lid shortly beaked from a conical base ; calyptra cucullate. Peristome double ; teeth bordered and striate ; processes as long, keeled or split ; cilia two or three.

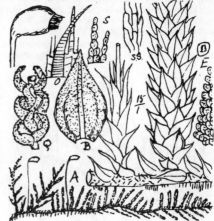

Rauia abbreviata.
S, paraphyllia ; SS, cells of per. leaves.

576. Rauia abbreviata (Broth.), Broth., Pfl., i, 3, 1005, and xi, 322.*

= *Pseudoleskea abbreviata* Broth., in Engl. Bot. Jahr., 1897, p. 282.

A laxly caespitose, autoecious moss ; the main-stem 2–3 inches long, procumbent and rooting, slightly forked ; each fork pinnately many-branched above ; these branches 2–6 mm. long, simple, spreading or suberect, and the perichaetium sessile on the main stem. Branch leaves very numerous, inrolled and incurved when dry, widely expanded when moist, set all round the stem, 0·5 mm. long, widely ovate-acute, deeply concave, nerved to the apex ; the margins expanded and regularly slightly reflexed nearly the whole length, papillose-crenate ; the cells minute, 8μ diameter, roundish quadrate, dense and crowded, mostly 1-papillose ; the central lower cells rather larger and less dense ; the nerve covered with papillose cells in the upper part of the leaf and so is rough on the back there. When dry these leaves are inrolled and incurved, chain-like, and very papillose. Leaves of the main stem much more lax, longer, ovate-acuminate, usually on the upper side and pointing upward ; interstices on the stem between leaves occupied by many minute, few-celled, linear, green paraphyllia of various sizes and shapes (fig. S). Sometimes slender, flagellate, terminal branches are produced, with scattered, small, densely papillose leaves. Perichaetium

---

* *Rauia abbreviata* (Broth.), Broth. (Addendum de capsula).

Capsula 1·5 mm. longa, ovoidea, horizontalis, valde irregularis, in seta levi rubra 2 mm. longa ; capsula turgida ubi humida, multum constricta infra os ubi sicca ; peristomium robustum, dentes sub-flavi, marginati, arte trabeculati infra, graciles supra ; membranum altissimum, dense papillosum, dentibus internis gracilibus ad altitudinem dentium externorum, nec non ciliis binis brevibus gracilibus inter utrumque par.

sessile on or alongside the main stem, conspicuously stramineous in a green tuft ; the lower leaves short and convex, the upper leaves 2 mm. long, closely sheathing, quickly brought to a long subulate point, nerveless, almost entire, and with the cells laxly and widely twisted vermicular, 20–50μ long (fig. SS). Perigonia small, bud-like, abundant on the main stem, with small ovate acute leaves, the inner the longer and more laxly areolated. Seta 2 cm. long, erect, red, twisted when dry ; capsule 1·5 mm. long, ovoid, horizontal, very irregular, turgid when moist, much constricted below the mouth when dry ; ring dehiscent ; peristome strongly developed ; the teeth yellowish and erect, many-barred below and bordered, slender and paler above ; membrane wide, closely papillose, with slender processes as high as the teeth, and shorter slender cilia between each pair.

S. Rhodesia. Zimbabwe, 3000 feet, July 1920, Sim 8774β ; Khami Ruins, 5000 feet, Sim 8863 ; also Mount Meru, E. Africa, 6000 feet, W. Leighton ; and elsewhere in tropical East Africa.

Genus 181. HAPLOCLADIUM (C. M.), C. M., in Nuov. Giorn. Bot. Ital., iii, 116 (1896).

Slender, matted, autoecious mosses, with prostrate, forked, leafy and rooting, regularly or irregularly pinnate stems ; the pinnae close, spreading, short, blunt and simple, sometimes longer and again pinnate ; paraphyllia in South African species few on the main stems, absent on the branchlets, which vary much in size and development. Leaves more or less uniform except in size, or on the smallest branchlets ; when dry incurved, when moist erecto-patent or subhorizontal ; stem-leaves lanceolate from a wider base, 2-furrowed ; margin reflexed toward the base, entire or almost entire ; nerve strong, extending to the leaf-apex or excurrent, usually smooth ; cells pellucid, not incrassate, oval or oblong-hexagonal, usually with one papilla per cell, or drawn upward ; basal marginal cells quadrate, inner oblong. Leaves of the branchlets narrower, shorter pointed, and with nerve shorter. Perichaetium rooting ; seta long, smooth ; capsule oblong, subhorizontal ; ring present ; lid arched mucronate ; calyptra cucullate, naked. Peristome double ; teeth lanceolate, bordered, with numerous lamellae ; inner peristome with wide and plaited basal membrane ; processes as high as the teeth, keeled, sometimes split ; cilia two to three, knotted.

577. *Haplocladium angustifolium* (Hpe. and C. M.), Broth., Pfl., i, 3, 1108, and xi, 320 ;
Dixon, in Bull. Torrey Bot. Club, xliii, 76.

=*Hypnum angustifolium* Hpe. and C. M., in Bot. Zeit., 1855, p. 788.
=*Thuidium angustifolium* (Hpe. and C. M.), Jaeg., Ad., ii, 318.
=*Pseudoleskea angustifolia* Par., Ind., p. 1033 (1897).
=*Thuidium *pinnatulum*, Rehm. 360 ; *fide* C. M., Hedw., 1899, p. 149 (not Lindb.).
=*Thuidium Rehmanni* Sb., in Jaeg., Ad., ii, 741.
=*Thuidium amplexicaule* (Rehm.), C. M., in Hedw., 1899, p. 149.
=*Hypnum amplexicaule*, Rehm. 392 and 680 ; Par., Ind., p. 611 (1895).
=*Haplocladium amplexicaule* (C. M.), Broth., Pfl., i, 3, 1108, and xi, 320 ; Dixon, in Bull. Torrey Bot. Club., xliii, 76.
=*Haplocladium transvaaliense* C. M., in Hedw., 1899, p. 149 ; Broth., Pfl., xi, 320.
=*Thuidium transvaaliense* (C. M.), Par., Ind. Suppl., p. 321 (1900), and 2nd ed., v, 23.
=*Haplocladium angustifolium*, var. *viride*, Broth. and Bryhn ; Bryhn, p. 21.
=*Thuidium *Cooperi* Mitt., MSS. (*fide* Dixon).

Very small in its parts and still forming large, dark green or yellowish, flat masses on soil ; autoecious ; stems 1–3 inches long, slender, prostrate, rooting occasionally, pinnate or occasionally 2-pinnate, and bearing the inflorescence on the main stem ; when pinnate the branches 2–4 mm. long ; when bipinnate the branch leading on to form a new stem and the first branchlets very slender and with different leaves ; paraphyllia present but not abundant on the main stem though abundant on its youngest parts ; absent from other stems ; paraphyllia simple or slightly branched, erect, consisting usually of four to eight cells on end. Leaves scattered, incurved when dry ; main-stem leaves 1·5–1·75 mm. long including point, spreading, ovate-acuminate in the lower half, thence a long narrow acumen is spreading or sometimes reflexed ; nerve narrow but strong below, extending to the point of the acumen ; leaves somewhat concave, denticulate and long-pointed ; margin often slightly reflexed, entire or lightly denticulate ; cells oblong, 20μ long, tumid but almost smooth or slightly drawn to the upper end ; acumen cells rather longer ; nerve cells rather longer and sometimes giving a rough surface on the back ; lower cells rather smaller, especially the marginal ones, but no separate basal or alar cells. Main-stem leaves furrowed when dry, the point leaves of extension stems more denticulate and longer pointed than others and sometimes homomallous, other leaves not furrowed ; branch leaves similar to stem-leaves but smaller ; branchlet leaves only about 0·3–0·5 mm. long, scattered, lanceolate without subula ; the branchlet sometimes reduced upward as if flagellate. Perichaetium sessile sideways on the main stem, the outer leaves small and reflexed, the inner lanceolate-convolute, long, changed quickly into a longer flat hair-point occupied by the nerve, sometimes all brown and scarious, with white calyptra (Sim 10.284). Seta ½ inch long ; capsule small, nearly horizontal, high-backed, very irregular, often contracted below the

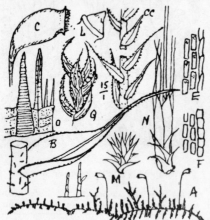

Haplocladium angustifolium.

mouth; peristome fairly strong; outer teeth yellowish red, closely trabeculate, bordered; membrane one-third as high, plaited; the processes as high as the teeth, pale, slender, keeled or split; cilia two, slender, half as long. Lid convex-pointed, short; calyptra white, long, closing below the capsule when young. Male flowers in small open buds on the main stem.

An exceedingly etiolated and almost smooth condition is *Hypnum flaccidulum* MacLea (Rehm. 676), from Rhenosterberg, C.P.

Cape, W. Kirstenbosch, Pillans 3330 and 3352.

Cape, E. Toise River, 3000 feet, 1919, Dr. Brownlee; Boschberg, Somerset East, Jas. Sim; Mount Currie, 5500 feet, Tyson.

Natal. Ihluku, Harding, 3000 feet, June 1920, Sim; Adamshurst, Merrivale, Sim 8697β; Mooi River; Cathkin, 7000 feet; Rosetta Farm; Zwart Kop; Mont aux Sources, Sim; Maritzburg, 1911, J. B. Pole-Evans, p. 75; Van Reenen, Wager; also Rehmann; Donnybrook, Sim 8560 p.p.; Knoll, Hilton Road, Sim; Blinkwater, Sim 10,280β; Chard, Karkoof, 5000 feet, Sim; Ekombe, Zululand, L. M. Titlestad; near Eshowe, Zululand, Haakon Bryhn; York, Sim; Van Reenen Pass, Rehm. 360 and 392 (*H. amplexicaule*).

Transvaal. Macomo's Hoek, Mrs. Clarke Williams; near Pretoria, Wager 186.

Genus 182. THUIDIUM Br. Eur.

Loosely tufted or cushioned, autoecious or dioecious mosses, growing on humus, stones, or decayed stumps; some very slender, others very vigorous and more or less rigid; some prostrate and rooting, others spreading or suberect, and rooted mostly at the base; stems little forked but regularly and complanately 1–2–3-pinnate, and often beautifully frondose; paraphyllia numerous on the main stems, simple, forked or foliose, papillose, often absent from the branches and from the ultimate pinnae. Leaves dimorphous, erecto-patent, never secund; stem-leaves furrowed, lanceolate-subulate from a cordate base; margin often reflexed below, entire or toothed above, usually papillose-crenate; nerve strong, reaching near to the apex, seldom excurrent; the ridge sometimes rough with papillae; cells uniform, roundish or oval-hexagonal, both sides bearing numerous papillae or only on the keel, or both sides with only one papilla per cell; the branch leaves often different in arrangement from those on the stem, the ultimate classes much smaller, usually ovate-lanceolate, with shorter and weaker nerve. Axis in every case very stout as compared with the size of leaf. Perichaetium usually rooting. Seta long, smooth or rough; capsule inclined or horizontal, oval-oblong or cylindrical, usually narrowed below the mouth when dry, and often with a short tapering neck to the seta. Lid obliquely beaked from an arched conical base; calyptra cucullate, usually naked, seldom set with one-celled hairs. Peristome double; teeth lanceolate-subulate, bordered, striate, with close lamellae; inner peristome smooth or finely papillose, with wide and plaited basal membrane; processes as high as the teeth, lanceolate-subulate, keeled, sometimes perforated; cilia two to four, strong and knotted or shortly appendaged, seldom rudimentary or absent. A genus of very pretty and attractive mosses, represented in all countries.

Synopsis of Thuidium.

Very slender confervoid species (*Thuidiella*).

Pinnate	578. *T. laevipes.*
Mostly pinnate	579. *T. sublaevipes.*
Young parts pinnate, old parts 2-pinnate.	
Very slender. Stem leaves minute, deltoid; branch leaves larger, blunt	580. *T. torrentium.*
Larger. Leaves larger, deeply concave, more acute	581. *T. versicolor.*
Stems 2-pinnate; seta rough in upper part	582. *T. borbonicum.*
Much larger; frondose; 2–3-pinnate (*Euthuidium*).	
2-pinnate; papillose; seta smooth	583. *T. promontorii.*
3-pinnate; deltoid; dendroid-complanate; nearly smooth	584. *T. thamniopse.*

578. *Thuidium laevipes* Mitt., Journ. Linn. Soc., cxlvi, 22, 318 (1866); Broth., Pfl., xi, 325.

Laxly matted, prostrate, small, autoecious, usually brown; stem 4–6 cm. long, simple or forked, simply pinnate, densely clad with paraphyllia; leaves cordate-amplexicaul, 1 mm. long, rounded at the apex, incurved, very concave, nerved to near the apex; branches simple, scattered, laxly set with leaves 0·5–0·7 mm. long, oval-oblong with rounded subacute apex, spreading, with incurved point even when dry, concave, and somewhat twisted. Margin papillose-crenulate; nerve extending to near the apex; cells 8μ diameter, rounded, dense, papillose. Perichaetal leaves 2 mm. long, ovate below, subulate above, the inner often with two to four long simple cilia on each side at the shoulder, each consisting of one line of cells on end; seta pale red, smooth; capsule inclined, brown, oval when moist, constricted below the mouth when dry; peristome as in *T. versicolor*.

Transvaal. Kaapsche Hoek, Wager (as a reduced form of *T. sublaevipes*).

Occurs also in East Africa and Uganda, from whence above description is drawn.

579. *Thuidium sublaevipes* Dixon, Trans. R. Soc. S. Afr., viii, 3, 216, and pl. xii, fig. 16; Broth., Pfl., xi, 325.

Dixon describes this as "near *T. laevipes*, but differs in habit stronger, branches less complanate, leaves when dry more incurved-crisped; seta longer, up to 2·5 cm. long (the other 1–1·5 cm.); capsule almost twice larger, longer, 2 mm. long, curved, when dry suberect.

"Kaapsche Hoek, Transvaal, Wager 295.

"Tyumie, C.P., 4000–6000 feet, D. Henderson 358."

I find *T. sublaevipes* often 2-pinnate and almost like *T. versicolor* except in the ciliated perichaetal leaves, with which it agrees with *T. laevipes* from Central Africa, but though specimens with pinnate stem, short seta, and small capsule occur at Kaapsche Hoek, as well as others answering *T. sublaevipes*, I doubt if two species are present, and consider one variable species includes all here.

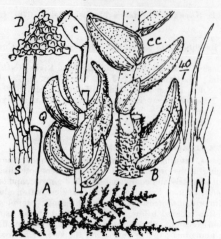

Thuidium sublaevipes.

S, shoulder of perichaetal leaf, with cilia.

Thuidium torrentium.

S, pectinate inner surface of outer teeth.

580. *Thuidium torrentium* C. M., in Hedw., 1899, p. 149 ; Rehm. 644, and Musc. Macowan, No. 44 ; Broth., Pfl., i, 3, 1012, and xi, 324.

Very slender, autoecious, straggling in low masses on damp stones in or near streams in shade, light green, rooting occasionally ; stem 1–3 inches long and more or less 2-pinnate, or extending again as a 1-pinnate stem ; paraphyllia abundant, very short, simple ; leaves bract-like, deltoid, acute, 0·3 mm. long, spreading or reflexed ; branches 3–8 mm. long, the larger pinnate, the smaller simple, with few or no paraphyllia ; leaves of the larger branches ovate below, bluntly pointed, nerved to near the apex, very papillose ; leaves of the smaller branches very small, rather distant, bluntly elliptical, suberect when moist, incurved chain-like when dry ; nerve percurrent, pale, narrow ; cells roundish, 10μ wide ; basal cells up to 15μ long, all dense and very bristly with papillae when dry.   Perichaetal leaves 1·5 mm. long, sheathing, subulate, subentire, with cells 25μ long, short hair-point, weak nerve, and without cilia ; seta 2 cm. long, red, smooth ; capsule brown, horizontal or hanging, ovoid with wide mouth and very large peristome ; teeth red, very closely trabeculate outside and pectinate inside ; processes as high, pale, plicate, on very high membrane ; cilia two, rather shorter.

Cape, E.   Near Somerset East, Macowan ; Hogsback, D. Henderson 325 ; Perie Mountain, 4000 feet, 1892, Sim 7167 and 7239.

Transvaal. Rosehaugh, Sim 8556 ; Barberton and Kaapsche Hoek, Wager 325 ; The Downs, Pietersburg. Junod 4005*a* ; Houtbush, Rehm. 644 ; Lydenburg, MacLea (Rehm. 644β).

581. *Thuidium versicolor* (C. M.), W. P. Sch., in Breutel, Musc. Cap. ; Broth., Pfl., i, 3, 1013, and xi, 324 ; Bryhn, p. 22 ; Brunnthaler, p. 32 ; Dixon, in S.A. Journ. Sc., xviii, 329.

=*Hypnum versicolor* C. M., Syn., ii, 494.
=*Hypnum *minutulum* Horns., in Dregé, Musc. Cap. (not of Hedw. or of Sull.).

Slender and small, but larger than *T. torrentium*, autoecious, growing in extensive, bright green, flat, low mats on wet stones, tree stumps, or bark, and often intermixed among other mosses ; stems very slender, 1–3 inches long, often rooting ; older parts 2-pinnate, with abundant paraphyllia up to 0·4 mm. long, younger parts 1-pinnate, without paraphyllia except on young main stems ; main-stem leaves amplexicaul, plicate when dry, widely cordate-ovate, quickly narrowed or conduplicate in upper half to a long narrow point (often reflexed), very concave and scoop-shaped, 1 mm. long, few or numerous, spreading, those on leader-tips often smaller, crowded, and homomallous ; the margin often reflexed in the lower half (fig. SS), nerved to the apex, dense ; cells oval, 1-papillate, 0·8μ long ; the nerve having longer cells acutely pointed on the ridge at the upper end (fig. SSS) ; branch and branchlet leaves 0·4 mm. and 0·2 mm. respectively, lanceolate-concave, more or less acute, nerved to the apex, incurved when dry, and then bristling with papillae. The paraphyllia are simple or branched, not foliose. Per. leaves widely lanceolate, convolute, shortly hair-pointed, thinly stramineous, with longer cells, thin nerve throughout, and denticulate margin but no cilia. Seta 2 cm. long, smooth, red ; capsule brown, horizontal-oblong with short neck, very irregular, with very large, red, well-developed peristome, dehiscent ring, and conical-beaked, rather long, oblique lid.

Cape, W. Montagu Pass and Portland, Rehm. 359 and 359β ; Blanco, Rehmann ; Oudtshoorn, Miss Taylor.

Cape, E. Alexandria Forest, Jas. Sim ; Keurbooms River, J. Burtt-Davy 17,038 ; Perie, Kologha, Katberg, Kakazela, Griqualand East, etc., Sim.

Natal. Donnybrook, Maritzburg, Ngoya, Polela, Sim 8560, 8562, 8563, 8559 ; Centocow, Mont aux Sources, Zwart Kop, Elandskop, Buccleuch, etc., Sim ; Giant's Castle, Symons ; Van Reenen, Rehm. 359c.

Transvaal. Macmac, MacLea (Rehm. 465) ; The Downs, Pietersburg, Junod 4005.

S. Rhodesia. Victoria Falls, Brunnthaler.

Also occurs in East Africa.

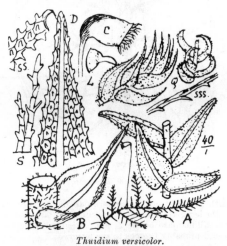

Thuidium versicolor.

S, paraphyllia ; SS, margin and back of leaf where reflexed ; SSS, back of nerve.

Thuidium borbonicum.

S, marginal cells.

582. *Thuidium borbonicum* Besch., Fl. Bryol. Reunion, etc., p. 289 ; Par., Ind., v, 5 ; Broth., Pfl., i, 3, 1012, and xi, 325 ; Dixon, S.A. Journ. Sc., xviii, 329.

Small, confervoid, autoecious, deep green, in dense low mats ; stem slender, 1–2 inches long, regularly 2-pinnate, densely but shortly paraphylliate ; pinnae 3–5 mm. long, without paraphyllia ; pinnules 2–3 mm. long, with numerous small, ovate-acuminate, concave leaves 0·25 mm. long, laxly inflexed and crisped when dry ; stem-leaves 0·5 mm. long, widely cordate-hastate with reflexed mucro, nerved to near the apex ; margin flat, cells strongly papillose. Branch leaves intermediate in size and form. Perichaetal leaves erect, 1·5 mm. long, lanceolate-subulate, suberect, dentate, with excurrent nerve and longish cells but no cilia. Seta 2 cm. long, pale red, smooth below, rough above ; capsule small, inclined, short, very irregular ; peristome very large, red ; lid obliquely rostrate. Bescherelle does not mention the roughness on upper part of seta ; he says seta smooth, but Brotherus makes that a character, and I find it is so. Abundant in the rain forest at Victoria Falls ; known also in East Africa and East African islands.

S. Rhodesia. Rhodesia Falls, Sim 8893, and Miss Farquhar 23, Wager 908 ; Cheeseman, Sept. 1905, etc.

583. *Thuidium promontorii* (C. M.), Par., Ind., 2nd ed., v, 16 ; Broth., Pfl., i, 3, 1016, and xi, 326.

=*Tamariscella promontorii* C. M., in Hedw., 1899, p. 150.

=*Thuidium tamariscinum* C. M., Syn., ii, 483 p.p., and of all South African records prior to 1899 ; also of Wager's check-list, p. 15.

Dioecious, procumbent, in large lax mats several feet across and inches deep ; stems suberect or erect, 1–5 inches high, ½–1½ inch wide, rising from prostrate old stems either leafy or leafless, frondose, closely and complanately 2-pinnate, with convex, appressed, widely cordate-amplexicaul, 2–4-plicate leaves 1·5–2 mm. long, quickly narrowed into a lanceolate-acute apex ; these leaves subtending each pinna or often more abundant, the stem densely clad in paraphyllia, especially the oldest and youngest parts, the youngest parts also sometimes closely clad in imbricate, alternate, cochleate stem-leaves before branches appear ; when dry these leaves have inflexed margins, and the apical leaves are often falcate-homomallous. Pinnae 0·5–1·5 cm. long, with numerous small, close, complanate pinnules 2–4 mm. long, without paraphyllia, on which the small convex leaves are ovate-acute, more or less incurved, and very rough with papillae, especially when dry ; the terminal cell, like others, being made up of papillae. Perichaetal leaves numerous, 2 mm. long, sheathing, stramineous, toothed or serrated, and sometimes with a few long cilia on each side at the shoulder. Seta 1½–2 inches long, red, smooth ; capsule shortly cylindrical, inclined, high-backed ; lid acutely beaked. Formerly, on account of superficial resemblance, this was taken to be *Thuidium tamariscinum* C. M. (of Europe), which, however, is 2–3-pinnate and has the terminal cell of the branch leaves pointed and simple, and does not occur among my South African specimens.

Present or abundant in every mountain bush in Eastern Cape Province, Natal, Transvaal, and Orange Free State, and occasionally in the coastward bushes, including Ngoya Forest, Lower Zululand, and Knysna (Miss Duthie), but I

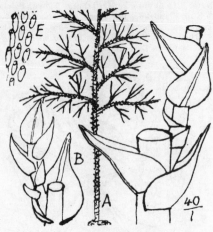

Thuidium thamniopse.

Thidiuum promontorii.

S, teeth on per. leaves ; CC, incurved papillose branch leaf.

have no further record from the western part of Cape Province nor from Rhodesia, though Mitten recorded *T. tamariscinum* Hedw. from Kilimanjaro before *T. promontorii* was separated and the record probably belongs to the latter.

584. *Thuidium thamniopse* (Rehm.), Sim (new species).*

Stems erect, dendroid, complanate-frondose, deltoid, 3-pinnate, 2 inches high or more, without paraphyllia, and with rather scattered, widely ovate-acute, amplexicaul, scoop-shaped, erecto-patent leaves 0·75 mm. long, nerved to the apex, with denticulate margins, and cells 20μ long, each drawn to the upper end but hardly papillose ; the marginal cells over-lapping upward, other parts not seen ; possibly not a *Thuidium*. Macmac, MacLea (Rehm. 647).

Family 48. NECKERACEAE.

Slender or robust, usually somewhat stiff, loosely tufted, or hanging mosses, having the main stem procumbent and the secondary fertile stems erect or pendulous, not rooting, often pinnately branched, seldom simple ; terete or flattened, growth being continued by elongation, by lateral branches, or by stolons. Leaves in many rows, spreading all round, or more often complanate ; symmetrical or unsymmetrical ; of various forms, usually wide ; smooth or almost so, occasionally transversely undulated. Nerve usually weak, single or double, often absent. Areolation minute ; the upper cells rhomboid or shortly linear, smooth or papillose ; the lower linear or linear elongate, angular, sometimes coloured ; alar cells sometimes differentiated. Perichaetal branches not rooting ; vaginula naked or hairy. Capsule usually immersed among the long perichaetal leaves, sometimes emergent or on a long seta, erect, symmetrical, rarely curved. Lid pointed or shortly beaked, from a conical base ; calyptra conical or cucullate, naked or pilose. Peristome usually double ; inner peristome rarely absent, usually without intermediate cilia. Dioecious, rarely autoecious or synoecious ; the male plant resembling the female.

Synopsis of Neckeraceae.

§ PTEROBRYEAE. Leaves not flattened, closely symmetrical ; basal cells more or less differentiated. Peristome usually double, with properistome ; outer teeth smooth, irregularly tumid ; inner peristome with very rudimentary processes and without cilia, or absent.

Nerve absent (or occasionally very short and double) .   .   .   .   . 183. RENAULDIA.

* *Thuidium thamniopse* (Rehm. ined.) Sim (sp. nov.).
    Caules erecti, dendroidei, complanato-frondosi, deltoidei, 3-pinnati, 2″ plusve alti, sine paraphylliis ; folia sub-sparsa, late ovato-acuta, amplexicaulia, erecto-patentia, 0·75 mm. longa ; nervus ad apicem pertinens ; margines denticulati ; cellae 20 mm. longae, contractae quaeque ad finem superiorem, vix autem papillosae.

Nerve single (in the South African species).
    Alar cells differentiated   .   .   .   .   184. PTEROBRYOPSIS.
    (Alar cells not differentiated   .   .   .   .   .   .   ORTHOSTICHOPSIS.)
§ METEOREAE (Meteoriaceae of Brotherus). Usually epiphytal and hanging. Leaves not or hardly complanate, symmetrical, often auricled; cells longish; alar cells usually lax or distinct; peristome double; outer teeth not tumid, inner well developed; seta smooth.
    Nerve single.
        Leaves not at all complanate.
            Cells smooth; alar cells quadrate, distinct   .   .   .   .   185. SQUAMIDIUM.
            Cells with several papillae on each; alar cells hardly differentiated   .   .   186. PAPILLARIA.
        Leaves, or some of them, somewhat complanate.
            Cells linear, several papillae on each; seta short and smooth; outer peristome striate, papillose only toward the apex   .   .   .   .   .   .   187. FLORIBUNDARIA.
            Cells oblong, one papilla on each; seta long and rather rough; outer peristome papillose, not striate,
                                  188. AEROBRYOPSIS.
    Nerve absent; leaves in rows; cells smooth; alar cells distinct .   .   .   189. PILOTRICHELLA.
§ TRACHYPODEAE. As in Meteoreae, but seta papillose   .   .   .   190. TRACHYPODOPSIS.
§ NECKEREAE. Secondary stems distinctly complanate, usually pinnate, seldom dendroid. Leaves unsymmetrical, in eight rows, the side leaves spreading, the back and front leaves appressed and arranged alternately right and left. Cells smooth, the upper rhomboid, the lower linear. Peristome double; outer teeth not tumid; inner peristome well developed.
Branches closely inrolled when dry, paraphylliate; nerve single; properistome not present   191. LEPTODON.
Branches not closely inrolled when dry, usually not paraphylliate.
    Nerve single; leaves auricled; properistome present .   .   .   192. CALYPTOTHECIUM.
    Nerve very short and weak, forked or absent; leaves not auricled; properistome not present   193. NECKERA.
§ THAMNIEAE. Secondary stems erect and dendroid or frondose, usually more or less complanate. Leaves symmetrical or almost so, serrate, with single nerve not reaching the apex. Cells smooth or nearly so. Peristome double; the outer teeth not tumid; the inner well developed.
(Leaves small, roundish, with one small papilla on the lumen   .   .   .   PINNATELLA.)
Leaves smooth or somewhat papillose at the upper end.
    (Outer peristome teeth papillose above, striate at the base only   .   .   POROTRICHUM.)
    Outer peristome teeth striate .   .   .   .   .   194. POROTHAMNIUM.

Genus 183. RENAULDIA C. M.

Robust plants growing in large cushions among other mosses on bark, with rhizomatous, rooting main stems and erect, closely leaved secondary branches several inches long, irregularly or subpinnately many-branched, the secondary stems subcomplanate with small leaves toward the base; the branchlets spreading, not closely flattened, ½ inch long, seldom flagellate. Leaves erecto-patent when moist, closer when dry, very concave, cordate-ovate-acute; side leaves spreading, others laxly suberect, often absent on branches; nerve absent or very short and double; cells widely linear, blunt, smooth, with intercellular pores. Basal cells shorter, rounder, and discoloured. Perichaetal leaves longer than stem leaves, tapering from a sheathing base, extending beyond the almost sessile oval capsule. Seta smooth. Ring absent; lid shortly conical and sharp-pointed; calyptra covering the lid only, naked. Peristome normally double, with properistome; teeth smooth or nearly so, irregularly tumid, its parts hardly connate at the base; inner peristome very rudimentary, absent in the South African species.

585. *Renauldia Hoehnelii* (C. M.), Broth., Pfl., i, 3, 792 (1906), and xi, 146, 533; Dixon, in Journ. of Bot., 1915, p. 22 (Synonymy, p. 23).

=*Neckera Hoehnelii* C. M., in Flora, 1890, p. 489 (not *N. Hoehneliana* C. M., in Flora, 1890, p. 490).
=*Trachyloma africanum*, Rehm. 332 and 619.
=*Calyptothecium africanum* Mitt., Journ. Linn. Soc., 1896, p. 312; Broth., in Engl. Bot. Jahrb., 1894, p. 128.
=*Renauldia africana* Broth., Pfl., i, 3, 792.
    Probably also *Porotrichum rostrifolium* C. M., in Hedw., 1899, p. 128, but see Kew Bull., No. 6 (1923), under No. 333c, where *P. pennaeforme* is suggested, but Müller's description is insufficient and incomplete.

    Laxly caespitose, hanging on tree-stems or suberect on humus. Dioecious. Rhizomes wiry, slender, at first scale-clad, producing several stems together or from that up to 2 inches apart. Stems ascending or drooping, firm, nearly bare at the base, abundantly leafy above, 2–6 inches long, irregularly pinnately branched, the branches up to 1 inch long; leaves produced all round the stem, imbricate, loosely complanate, 3 mm. long from a wide clasping base, oblong-acute, very concave, the sides infolded, making the leaf boat-shaped without keel, and hooded but not united at the apex where the leaf is abruptly reduced to a rather firm, long, erect point; the sides of the complanate side leaves meeting upwards (*i.e.* in the direction of the stem or branches); margin entire, nerve absent; cells bluntly vermicular, 80–100μ long in two layers; connected by pores, alike on all parts of the leaf except the very base where they are shorter and yellowish. Leaves not wrinkled when dry. Perichaetium sessile on the main stem; per. leaves longer, more lanceolate and longer pointed than others, somewhat infolded; capsule almost sessile, oval with narrowed but still rather wide mouth, immersed, with sixteen conical-subulate incurved teeth often bridged at the base, and with many irregular but tumid properistome cells; inner peristome very rudimentary or absent (I do not find any). Occasional flagellate rooting branches are produced, either separately from the stem or in continuation of a branch, on which the leaves or scales are few and small, and slightly denticulate. Lower part of main stem dendroid, with small, more open, and often longer pointed leaves. Male plant more slender than the fertile plant. Found only in moist forest regions, but occurs also in East Africa. *Calyptothecium* differs mostly in being more definitely complanate throughout; this is only laxly flattened. These genera have been defined differently by different authors.

Cape, W. Knysna, Rehm. 332; The Wilderness, George, Miss Taylor.

Cape, E. Perie Forest (fertile), Sim 7062; Eveleyn Valley and Hogsback, etc., Sim 7125, 7014; Malowe, Griqualand East, Tyson 2574.

Natal. Drakensberg bushes, Weenen County, Dr. J. M. Wood; Zwart Kop, Sim, Taylors, Dr. Bews.

Transvaal. Lydenburg, MacLea (Rehm. 619); Donkerhoek, Piet Retief, 1916, J. S. Henkel.

Renauldia Hoehnelii.
S, leaf from base of stem.

Genus 184. PTEROBRYOPSIS Fleischer.

Slender or robust, brownish, shining, dioecious mosses growing on trees in tropical and subtropical forest regions; main stems prostrate on bark, rooting, with or without small dry perished leaves; secondary stems horizontal or hanging, small-leaved at the base, gradually larger leaved upward; the leaves set all round the stem; stems irregularly branched or dendroid-pinnate, seldom simple, sometimes flagellate or foxtail-form alternately; branchlets usually short and thick, or sometimes flagellate. Leaves very concave; when dry laxly imbricate, smooth, usually oval and short-pointed, often wide-based, symmetrical, entire or almost so, and without border. Nerve single (in the one South African species), or sometimes double or absent. Cells linear or narrowly rhomboidal, smooth, thick-walled, those at the leaf-base roundly quadrate, lax, coloured; alar cells usually more or less quadrate, sometimes hardly differentiated. Perichaetal leaves long and narrowly pointed, from a wide sheathing base. Seta smooth, sometimes longish, sometimes very short; capsule well exposed in the South African species, in some other cases hardly emergent, oval-oblong; calyptra cucullate, pointed, naked or hairy. Lid sharp-beaked from a widely concave base, sometimes oblique; ring distinct, dehiscent or permanent. Peristome double; outer peristome with lanceolate, irregularly tumid teeth, somewhat connate at the base; inner peristome usually rudimentary and without cilia. Gemmae are sometimes present.

586. Brotherus places herein *P. *natalensis* (Rehm.), Broth., xi, 141 (=*Neckera *natalensis*, Rehm. 615, Natal, Buchanan), which is undescribed and unknown to me, and not represented in South African herbaria nor published. He includes it (Pfl., i, 3, 803) in § IV *Pterobryodendron* Fleisch.. which has "capsule exserted; calyptra cucullate; secondary stems usually pinnate or dendroid, seldom simple, without thread-like branches. Leaves usually gradually short-pointed; nerve single," and in the group which has secondary stems closely and more or less regularly pinnate.

Dixon, who has examined Rehm. 615, has favoured me with a drawing of the leaf, and remarks, "Near to some of the Indian species, but not *very* near to any. Stem robust, pinnate, 8 cm. long; branches up to 1 cm. Leaves very concave; alar cells very numerous, large, pellucid; leaves yellow at mid-base."

Other species in the genus belong to tropical Africa, India, and elsewhere.

(Genus ORTHOSTICHOPSIS Broth., which resembles a lax condition of hanging *Pilotrichella*. and has similarly exserted capsule and very similar cochleate-concave leaves arranged in straight lines, comes in here because it has a properistome and only rudimentary inner peristome, a long, single, narrow nerve, and alar cells not differentiated. It has not yet been found in South Africa, but occurs further north and in the east African islands and is likely to be found here.)

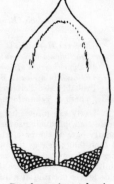

Pterobyropsis natalensis,
by Dixon.

Genus 185. SQUAMIDIUM (C. M.), Broth.

Slender, pendent, light green or dark green, shining mosses growing on tree stems and branches; main stem prostrate and rooting; secondary stems long and hanging, irregularly pinnate (or occasionally somewhat bipinnate) or more frequently placed four pinnae together almost in a whorl, with a ½-inch space or thereby between them and the next whorl; these pinnae usually simple, ½ inch long. and blunt; leaves of two forms, closely placed round the stem, erecto-patent when moist, loosely imbricate-cochleate when dry, oval-acute from a somewhat decurrent base; smooth, pointed or hair-pointed, with the margin entire; the stem-leaves or leaves of extension branches ovate-lanceolate, flat, large, and open; the branch leaves shorter and concave-cochleate. Nerve narrow, not reaching the apex and not always present in cochleate leaves; cells linear, twisted, smooth, those at the base shorter and more lax; the alar cells quite distinct, numerous, quadrate-hexagonal. Inner perichaetal leaves long, narrow-pointed from a wider sheathing-base; seta short or very short; capsule large, oval, smooth, immersed or shortly exserted; ring differentiated, permanent; lid sharply beaked from a wider convex base; calyptra conical, several-lobed, only covering the top of the capsule, more or less hairy. Peristome double, well developed, the outer teeth long, tapering from a wide base, somewhat papillose; inner processes equally high, subulate, and keeled, with narrow basal membrane and no cilia. Spores small, papillose.

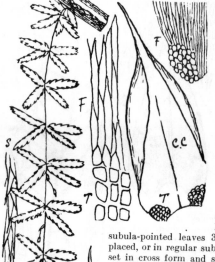

587. *Squamidium Rehmanni* (C. M.), Broth., Pfl., i, 3, 809, and xi, 157; Dixon, Trans. R. Soc. S. Afr., viii, 3, 209.

=*Meteorium Rehmanni* C. M., in Hedw., 1899, p. 127.
=*Pilotrichella Rehmanni* C. M., in Rehm. 323 and 610; Par., Ind., p. 949 (1896).

Dioecious, shining, brown or dark, with light green new shoots; stems and leading shoots mostly simple, pendent, slender, 3–8 inches long, with cordate-ovate-acute, open, appressed, rather sparse, subula-pointed leaves 3 mm. long, and with branches either numerous and irregularly placed, or in regular subwhorls about ½ inch apart, usually four in a whorl, these branches set in cross form and spreading at right angles, ½–1 inch long, closely set with cochleate-imbricate, shorter, more oblong leaves 2 mm. long, deeply concave, with inflexed margins and acute subula; nerve clear, very narrow, needle-like, extending about half-way but often almost invisible; margins entire; cells linear, 100–150μ long, 10μ wide; basal cells shorter and wider; alar cells quadrate, tumid, pellucid, forming a large and distinct group just below where the leaf widens. Very seldom fertile in South Africa, but with exserted, small, erect capsule and beaked lid; other species in the Mascarenes, East Africa, South America, etc., belonging to the same group have immersed capsules on the branches. Branches, after a period of maturity, occasionally change into long extension branches with open leaves.

Very like *Pilotrichella*, but the whorled arrangement and clear nerve are distinctive when present, but that is not always. The form of main-stem leaf is different and it is usually present and appressed, sometimes in quantity; when so, the appearance is much changed and more flattened.

Cape, W. Oudtshoorn, Miss Taylor; Diepwall, Knysna, Phillips; Montagu Pass, Rehm. 323; Knysna, Rehm. 320 p.p. (mixed with *Leucodon*).

Cape, E. Toise River, Dr. Brownlee; Katberg, Mrs. C. Williams; Hogsback, D. Henderson 346β.

Natal. Entumeni, Zululand, Bryhn, p. 17; Cathkin Bush, 7000 feet, 1918, Sim.

Transvaal. Lechlaba, Houtbush, Rehm. 610.

Genus 186. PAPILLARIA C. M.

Dioecious. Main stem creeping and rooting, usually on tree-branches; secondary stems numerous, usually long and hanging, sometimes very long, regularly or irregularly pinnated; the branches, branchlets, and leaves not complanate; leaves erecto-patent when moist, imbricated when dry, ovate-acute with a wider or auricled lower portion, rounded in to the base and slightly decurrent; the apex usually subulate or hair-pointed; margin entire or nearly so, without border; nerve single, wide, not reaching the apex. Cells elliptical-hexagonal, opaque, with several papillae on each, those at the central basal portion of leaf pellucid and smooth or papillose, as also the alar cells which are quadrate or not much differentiated in form, but arranged in oblique rows. Perichaetia sessile on the branches, inner perichaetal leaves long, erect, subulate-pointed. Seta short or more often up to ½ inch long, straight, smooth. Vaginula hairy; capsule emergent in South African species, more or less oval, regular, smooth. Ring hardly evident. Peristome double; outer teeth lanceolate, papillose, not striate; inner peristome with hyaline papillose linear processes as high as the teeth, and very narrow basal membrane. Cilia absent. Lid beaked from a conical base. Calyptra hairy.

Papillaria flexicaulis (Tayl.), Jaeg., Ad., ii, 175=*Pilotrichum* (Mitt.), which is Australasian, and *P. fuscescens* (Hk.), Jaeg., which belongs to East Asia and East Indies, are both wrongly credited to South Africa in Shaw's catalogue.

Squamidium Rehmanni.
S, part of an extension stem.

Synopsis.

588. *P. africana.* Stems long, pendent, usually pinnate, slender.

589. *P. natalensis.* Stems irregularly few-branched, not slender.

588. *Papillaria africana* (C. M.), Jaeg., Ad., ii, 176; Brunn-thaler, p. 30; Broth., Pfl., i, 3, 816, and xi, 163; Dixon, S.A. Journ. Sc., xviii, 325; Rehm. 324.

= *Neckera africana* C. M., Syn., ii, 137.
= *Dendropogon africanum* W. P. Sch., in Breutel, Musc. Cap.
= *Meteorium africanum* (C. M.), Mitt., Journ. Linn. Soc. (1886), xxii, 146, 314.
= *P. africana*, var. *natalensis, Rehm. 325 and 614, and var. *rupestris, Rehm. 324β, and var. *tenuis C. M., Rehm. 326.

Rhizoid stems rooting along branches; stems pendent, brownish, 4–18 inches long, slender, simple or forked, pinnate or slightly 2-pinnate; the pinnae up to 2 cm. long, horizontal or usually reflexed on the stem, not complanate, tapering from base to the smaller-leaved acute point, with occasionally one to two pinnules. Leaves 1–3 mm. long, alike except as to size, very numerous, closely imbricated, ovate-acuminate with long subulate point, deeply concave as far as the nerve goes, with erect, wrinkled-up sides and reflexed alar areas at the base, and the sides usually undulate when dry above that; nerve wide and strong, often pellucid or rusty, extending beyond mid-leaf; margins entire, or denticulate at the base; lamina opaque, dense, usually blue green, its cells mostly with separate oval lumen 20μ long and containing about three low papillae each; cells of the leaf-point considerably longer; at the leaf-base lines of pellucid or hardly papillose vermicular cells about 100μ long radiate from the nerve-base, gradually merging into lamina cells, but on the reflexed basal margin small quadrate cells 10μ wide occur in many rows, not forming a separate alar group. When dry the leaves are often closely appressed, especially on the more slender branches, and on account of undulation old dry leaves often appear plicate; when moist the leaves are erecto-patent. New branches begin with dense tufts of short, almost triangular leaves. Perichaetia sessile on the branches, with few lanceolate-acute leaves enclosing numerous linear-jointed paraphyses as long as the seta and capsule. Capsule exserted; seta 3–4 mm. long, smooth; capsule erect, oval; peristome strong, teeth nearly white, inner with very low membrane. Lid obliquely beaked. Mitten mentions that fruit had not been seen, and Paris repeats that; but I find it occurs in our forests, though not abundantly. The folded-in leaf-base, acuminate flat apex, and dull blue-green colour under the microscope, together with the habit of growth, are very distinctive. Vigorous apical groups of leaves are often longer pointed than usual, and are slightly denticulate and have twisted point.

Abundant in every moist forest from Knysna eastward, including Rhodesia and East tropical Africa; usually hanging as festoons on tree branches.

589. *Papillaria natalensis* Sim (new species).*

Main stem prostrate, black, wiry, clad with leaf-remains; secondary and further stems erect, up to 2 inches high, simple, forked, or lightly branched; the leaves laxly appressed, apparently furrowed through reflexed margins, ovate-acuminate-acute, somewhat concave at the rather cordate base; the nerve wide, yellowish, covered with lamina cells, and extending to near the apex. Apical cells showing elliptical lumen; lamina cells with oval lumen or with hexagonal form where that can be seen; lower central cells much longer, up to 100μ, and auricle cells laxly elliptical, the outer projecting as denticulations on the reflexed border from base to near apex; all cells tumid and having three to ten low, rounded papillae, giving a minutely crenate margin where visible; other parts not seen. Very different from *P. africana* in general appearance and also in formation.

Town Bush, Maritzburg, Natal, 3000 feet, July 1916, Sim 8703, rare.

Papillaria africana.

---

* *Papillaria natalensis* Sim (sp. nov.).

Caules frondosi erecti, ad 2″ alti, haud multum ramosi; folia laxe adpressa, ovato-acuminata, acuta, sub-concava ad basin paulo cordatam; nervus latus, prope ad apicem pertinens; margines reflexi a basi prope ad apicem; cellae lumine ovato; cellae mediae basales multo longiores; cellae auriculares laxe ellipticae, cellae universae 3–10 papillas humiles rotundatas gerentes.

Genus 187. FLORIBUNDARIA C. M. (in Linnaea, 1876, p. 267, as § of *Papillaria* ; afterwards as genus) ;
Fleischer, in Hedw., xliv, 301 (1905).

Growing on bark or stones in very wet places or round waterfalls, sometimes submerged and floating out as flagellate or long, very much-branched plants from the fixed base. Main stem often very long, rooting ; secondary stems usually irregularly pinnately branched or sometimes 2-pinnate. Leaves laxly arranged, those of the stem- and branch-bases small, subulate from a wide base, with evident alar cells ; those in the branches and branchlets often somewhat complanate, spreading, lanceolate, with subulate or hyaline point, the margin flat, not bordered, somewhat denticulate, and the pellucid nerve reaching to or beyond the middle. Cells long and narrow, papillose, often several papillae together, those at the leaf-base more lax and smooth, those in the angles more quadrate, smooth. Inflorescence dioecious, placed on the branches and branchlets ; the perichaetal leaves not spreading and without nerve ; the seta short and smooth ; the capsule ovoid, erect, regular, straight or slightly bent, well exserted ; ring not evident ; lid short, obliquely beaked from a conical base ; calyptra conical, small, hairy. Peristome double ; the outer of sixteen lanceolate, striate, almost free teeth, papillose at the point only ; the inner equally high, faintly papillose, with plaited basal membrane and keeled or sometimes split processes ; cilia absent or rudimentary. The genus contains several African and other species besides the South African species ; it was formerly included in *Papillaria*, from which genus this is distinguished by the peristome being striate and being papillose only toward the apex of the teeth, and by the somewhat flattened arrangement of the leaves.

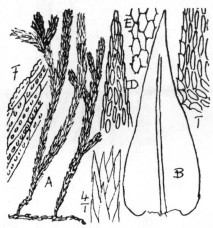

Papillaria natalensis.

590. *Floribundaria floribunda* (D. and M.), Fleischer, in Hedw., 1905, p. 301, with fig. ;
Broth., Pfl., i, 3, 822, and xi, 170.

=*Papillaria floribunda* (Dz. and Mk.), C. M., in Linn., 1876, p. 267.
=*Hypnum floribundum* C. M., Syn., ii, 265.
=*Meteorium floribundum*, Bryol. Jav., ii, 91.
=*Neckera floribunda* C. M., in Linn., 1868–1869, p. 9.
=*Papillaria fulvastra* Besch., Fl. Bryol. Reunion, p. 124.
=*Floribundaria Morokae* C. M., in Herb. of New Guinea.
=*Papillaria Robellardi* (C. M., in herb.), Besch., Fl. Reunion.

Dioecious ; very tender ; in pale yellowish or light green masses, usually hanging from or attached to wet rocks, sometimes on living tree-branches, always in an intricate and extensive mat ; main stem long, prostrate, rooting, often leaf-clad and emitting numerous branches 1–3 inches long, usually irregularly pinnate with long flagellate branchlets, sometimes closely pinnate, with shorter but stronger branches. Main-stem and base leaves shortly lanceolate from a wide base ; all other leaves lanceolate-subulate from a clasping or subcordate base, and arranged usually in four rows, but spreading to two sides so as to appear distichous and pectinate even when dry. Leaves vary from 2×0·5 mm. on leading branches down to 0·75 mm.×50μ on flagellate shoots, and are slightly convex, nearly horizontal, lax, pellucidly 1-nerved half-way, minutely crenulate through low papillae, and slightly denticulate from base to apex, especially denticulate toward the apex ; the cells very long and narrow, containing six or more low papillae in line, which form the slight crenulations on the border and surface ; the lowest cells shorter and pellucid, and a few small, quadrate cells occupy the basal margin, which is usually reflexed on larger leaves ; smaller leaves are sometimes without nerve. Perichaetia sessile on the branches and branchlets ; per. leaves erect, short, pointed ; male flowers smaller and like open buds ; seta 3–4 mm. long ; capsule small, ovoid, regular, straight ; the peristome teeth faintly transversely striate, the processes as high as the teeth, cilia absent. Ring absent ; lid shortly obliquely beaked ; calyptra slightly hairy ; fruit rare in South Africa.

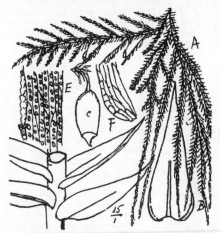

Floribundaria floribunda.

Rehm. 611, from Woodbush, Transvaal, named *Papillaria floribundula* C. M., var. *MacLeana* Rehm., agrees with the other South African specimens, but Fleischer keeps up *F. floribundula* from the Comores. *F. floribunda* occurs also in East Asia, East Indies, and East Africa.

Cape, E. Perie Forest, Eveleyn Valley, and Hogsback, Sim.

Natal. Overwood, Polela ; Town Bush, Maritzburg ; Zwart Kop and Mont aux Sources, Sim 8547, 8546, 8545.

· Transvaal. Rosehaugh, Sim 7429; Lydenburg, MacLea (Rehm. 611); Lechlaba, Houtbush, Rehm. 612 (as *P. nitidula* Rehm.).

Genus 188. AEROBRYOPSIS Fleis. (in Hedw., xliv, 304 (1905)).

Rather vigorous, dioecious, light green, loosely cushioned or often long and hanging plants ; main-stem prostrate, closely pinnate ; branches suberect, horizontal or hanging irregularly or nearly regularly, pinnate, and with rather widely spreading, laxly placed leaves, set all round and not closely appressed, though somewhat flattened. Leaves lightly concave, ovate-lanceolate, tapering to a long point, not bordered, finely serrated, nerved to or beyond the middle ; cells oblong or elliptic, often with one papilla per cell ; those at the base shorter and more lax, but there are no differenti- ated alar cells. Leaves erecto-patent and somewhat crisped or undulated when dry. Perichaetal leaves long and tapering ; archegonia several, with long slender paraphyses. Seta long, somewhat rough ; capsule erect, straight, shortly cylindrical ; ring not evident ; peristome double ; teeth of the outer lanceolate, papillose, shortly connate below, not striate ; inner hyaline, papil- lose, with short membrane and linear, somewhat split processes. Lid long and obliquely beaked from a wider convex base, calyptra cucullate, somewhat hairy. Usually hanging on trees or among other mosses on moss-clad stumps. This was separated out of *Aerobryum* on account of having the peristome teeth not striate and the seta rather rough, but I find that in our plant the peristome teeth are striate.

591. *Aerobryopsis capensis* (C. M.), Fleisch. in Hedw., 1905, p. 305 ; Broth., Pfl., i, 3, 821, and xi, 165 ; Bryhn, p. 18 ; Brunn- thaler, p. 30.

=*Aerobryum capense* (W. P. Sch.), C. M., in Linn., 1876, p. 262.
=*Meteorium capensum* W. P. Schimp., in Breutel, Musc. Cap.
=*Neckera capensis* C. M., in Bot. Zeit., 1858, p. 165.
=*Aerobryum capense* C. M., var. *rupestris* C. M., in Rehm. 322*b*.

Dioecious ; main stems prostrate, long, producing numerous, sometimes crowded, short, simple, densely leaved branches 1–2 cm. long, some or many of which grow into long, weak, straggling, laxly leaved, yellowish, secondary stems 2–12 inches long, which are irregularly branched ; the branches spreading, laxly leaved, and mostly 1–1·5 cm. long. When the secondary stems are mature and prostrate, these in turn become main stems and continue the exten- sion. Sometimes it straggles loosely among other mosses, more usually it forms a soft silky cushion of this species alone, but with secondary branches straggling over and through whatever is near. Usually it grows on humus in forest, on decayed tree-stumps, or on mossy tree-stumps, and is exceedingly polymorphic. Leaves set all round the stem, patent ; those on the dense branches numerous, 2 mm. long, cordate-lanceolate, long-pointed, plane, slightly denti- culate, firmly nerved at the base and with nerve extending half- way ; those on the secondary stems and on their branches stramineous, spreading at right angles to the stem, much longer, from a wide ovate-lanceolate base gradually tapering to a very long filiform point ; the lower half somewhat concave, the upper half undulate when dry ; the margin minutely denticulate ; the nerve narrow and disappearing about half-way up. Leaves of the dense branches sometimes plicate, others usually plane. Cells elongated, narrow, vermicular, those toward the leaf-point and those toward the leaf-base rather shorter. Inflorescence sessile on the secondary stems and· on the branches ; perichaetal leaves narrower than branch leaves, with long filiform point and entire margin. Seta 2 cm. long, slender, red, erect, smooth or rough only in upper part ; capsule inclined, oval-oblong, with wide mouth ; peristome erect, the red teeth striate below, pectinate inside, papillose and very slender and knotted above ; inner peristome higher, on very high

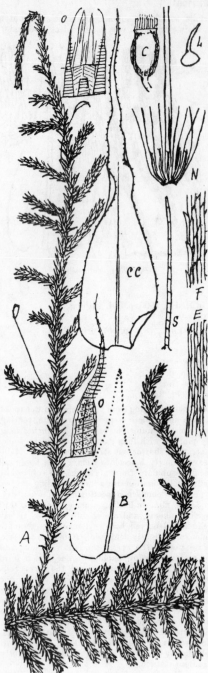

Aerobryopsis capensis.

membrane ; process punctate, sometimes split, with two slender cilia ; lid short with a long beak ; calyptra slightly hairy ; paraphyses very long, usually of a slender hanging or trailing form, and of yellowish colour when in forest, but when exposed to sunshine it assumes more of the dense, shorter-leaved character, and is of a dirty brown colour under, with the exposed parts pale green, and is less rambling than the forest form. This condition is *A. capense* C. M., var. *rupestris* Rehm. (Rehm. 608), but is not a fixed distinct variety. *A. capense* has the leaf-form, areolation, and often nearly the habit of the European *Brachythecium glareosum*, B. and S., which, however, has the leaves erecto-patent instead of spreading, the filiform point twisted, and the lid shortly pointed, not beaked, and is included in *Brachythecium* by Dixon.

Abundant wherever high forests are fairly open and moist, seldom found outside the forest or scrub forest ; occurs also in East Africa.

Cape, W. Knysna, Rehm. 322 and Burtt-Davy 17,015 ; Oudtshoorn and Montagu Pass, Miss Taylor, and Rehm. 322*β*.

Cape, E. Perie Forest, Sim 7015, 7032, 7056 ; Grahamstown, Miss Britten ; Van Staadens Pass, Miss Farquhar ; Hankey and Alexandria Forests, Jas. Sim ; E. Pondoland, E. Griqualand, Tembuland, etc., Sim.

Natal. Zwart Kop, Sim 7441 ; Mistyhome, Buccleuch, Karkloof, etc., Sim ; Ekombe, Zululand, Bryhn.

Transvaal. Lechlaba, Houtbush, Rehm. 608.

Genus 189. PILOTRICHELLA (C. M.) Besch.

Long, slender, brownish mosses usually hanging epiphytically on tree-branches in tropical and subtropical forests ; abundant in South African forests. Main stem long, not flattened ; secondary stems numerous, not flattened and not in one plane, often very long, more or less pinnate or 2-pinnate. Leaves concave, more or less boat-shaped, spreading when moist, loosely imbricated when dry, smooth, more or less distinctly auricled at the base, and abruptly cucullate-mucronate at the apex. Margin entire or almost so ; nerve absent. Cells narrowly linear, twisted, smooth ; at the base shorter and discoloured ; the alar cells forming a distinct small group of large, roundish, sometimes discoloured cells. Perichaetia sessile on stem or branches ; inner perichaetal leaves long and narrowly pointed, from a sheathing base, and sometimes enclosing many long paraphyllia. Seta short, erect or suberect ; capsule more or less oval ; calyptra cucullate, covering half the capsule, somewhat hairy. Ring absent. Peristome double ; the outer teeth lanceolate from a wide base, papillose, and in some species striate at the base ; inner per. processes shorter, linear, not keeled, papillose, sometimes perforated.

There are many species from Africa and America, none from Asia or Australasia. The South African species as well as most of the African species belong to the § *Orthostichella* C. M., having leaves more or less in rows, often spiral ; alar cells few and not discoloured ; seta short ; outer peristome teeth not or lightly striate, and spores small.

Synopsis.

592. *P. panduraefolia.* Branch leaves short, involute, cochleate ; point long, acute.
593. *P. chrysoneura.* Branch leaves longer, wider, flatter ; margins erect but not involute ; alar group small.

592. *Pilotrichella panduraefolia* (C. M.), Jaeg., Ad., ii, 159 ; Broth., Pfl., xi, 158.

=*Neckera panduraefolia*, in Bot. Zeit., 1855, p. 767 (not *P. panduraefolia* C. M., in K. Sw. Vet. Akad. Handl., 1895).
=*P. Kuntzei* C. M., in Hedw., 1899, p. 127 ; sunk by Dixon, Trans. R. Soc. S. Afr., viii, 3, 209.

Dioecious ; shining, yellowish brown, or sometimes green, epiphytic on trees, the primary or continuation stems rooting occasionally but set with large ordinary leaves ; secondary branches at first 2–3 inches long, and dendroid, pinnate or somewhat 2-pinnate, occasionally bearing capsules thereon, but eventually extending into long, slender stems 6–18 inches long, clad in cochleate leaves and bearing subpinnately many branchlets irregularly 1–4 cm. long, either blunt or tapering and placed all round the stem, which branchlets also bear capsules and are clad in closely imbricate cochleate leaves 1–1·5 mm. long, often arranged more or less regularly in five straight or spiral lines, but sometimes apparently irregular. Leaves cochleate from a wide cordate but not auriculate base, twice as long as the width of the base, but themselves much wider than the base since the margins are far incurved-involute ; this condition suddenly ceases at the leaf-apex, and a short or long subula, often reflexed, tapers to its point. The lowest leaves in a new branch are nearly convex-semicircular, nerveless, and either without point or shortly pointed, and with larger rhomboid cells than other leaves. All leaves entire, with small, sunk, alar group ; leaves nerveless and very tumid ; cells smooth, vermicular, up to $100 \times 10\mu$, those in the subula and those at the leaf-base rather shorter and wider ; alar cells few, roundly quadrate, usually pellucid, sunk. Perichaetium sessile sideways on the branchlets, few-leaved ; the leaves lanceolate-acuminate, erect, nerveless, entire, and slightly sheathing, and often enclosing many longer linear paraphyses nearly as long as the seta, and consisting of two to three lines of cells at the base. Seta 3–4 mm. long, pale, smooth ; capsule erect, oval-oblong, red, small, with obliquely rostellate lid and somewhat hairy, cucullate calyptra. Peristome double ; the teeth yellowish, bordered, papillose, faintly striate ; inner membrane low, papillose, with slender papillose processes.

A condition occurs having occasional bunches of closely fastigiate branches, no doubt due to insect agency (Sim 8571). Sometimes very slender ; sometimes fairly strong (branches 2 mm. wide). Common in very wet forests, rare elsewhere. Used by the ton in Natal for nursery packing.

Cape, E. Eveleyn Valley, Sim 7051, 7006, 7018 ; Fort Grey, Sim 8572 ; Uitenhage, Hankey, and Van Staadens, James Sim.

Natal. Cathkin Bush, Zwart Kop, Town Bush, Karkloof, Upper Tugela, Umkomaas, etc., Sim ; Ekombe, Zululand, Bryhn, p. 17.

Occurs also in East Africa.

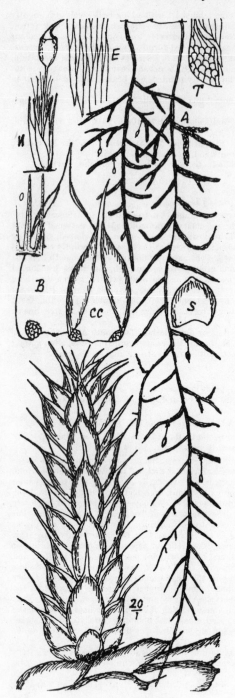

Pilotrichella panduraefolia.
S, lower leaf of new branch.

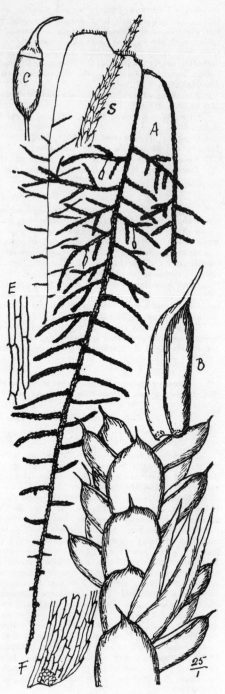

Pilotrichella chrysoneura.
S, cuspidate condition.

593. *Pilotrichella chrysoneura* (Hpe.), Jaeg., Ad., ii, 100.

=*Neckera chrysoneura* Hpe., in Linn., 1876, p. 263.

Very similar to *P. panduraefolia* in its epiphytic growth and in both short dendroid and long stems, but rather more 2-pinnate and with larger leaves, up to 1·5 mm. long, with shorter points. The leaves are usually in fine regular rows, deeply cochleate, with the margins erect toward the stem but not more inflexed, consequently they appear much larger and more flattened than those of *P. panduraefolia*, and the branches appear more heavily leaf-clad. Old stems sometimes stretch and root at the apex, starting out again as vigorous long stems, and the apical leaders have cochleate leaves, not ovate-lanceolate leaves as in *Squamidium*. Leaves entire, nerveless, with blunt cells up to 100 × 10μ ; the basal cells rather less, and the alar group small and of very small rounded cells. Perichaetium axillary, sessile, few-leaved, the leaves stramineous, longer than stem leaves, lanceolate-acute, sheathing from a wide base. Capsule and lid similar to those of *P. panduraefolia* or rather longer.

A condition occurs (fig. S) in which the branchlets taper to a fine point. This is presumably *P. cuspidata* Broth., in Engler's Bot. Jahrb., 1897, p. 255, and Pfl., xi, 158 (sterile from Pondoland), but these branchlets develop later into mature obtuse branchlets and it is not different.

Concerning *P. conferta* Ren. and Card. (in Bull. Soc. Roy. Bot. Belg., 1899, p. 233, sterile, from Australia), which is claimed to be South African in Wager's 257, from Barberton, and in Broth., Pfl., xi, 158, Dixon's remarks (Torrey Bot. Club, xliii, 71) on this species indicate that the transition from " short, subdendroid, rather robust and rigid branching " to ordinary long stems is not fully appreciated, and I am satisfied that no South African species can be separated on these grounds.

Cape, E. Hankey, Jas. Sim ; Gobogobo, King William's Town. F. Leighton ; Queenstown, Miss E. A. Graham. Natal. Cathkin Bush, 1920, T. R. Sim.

Transvaal. Lechlaba and Houtbush, Rehm. 609 and 609β ; The Downs, Pietersburg, Junod 4004a ; Modjadjes, Rogers.

Genus 190. TRACHYPODOPSIS Fleisch.

Vigorous, dioecious, laxly pendent on bark, yellowish ; the main stem prostrate, more or less leafy, the secondary stems several inches long, irregularly pinnate ; the pinnules simple and like the stems slightly flattened, closely clad in long, lanceolate, acute, furrowed, serrate or serrulate leaves nerved to near the apex, and with elliptic cells, 1-papillate at the middle of the lumen, and somewhat sunk alar groups containing small, smooth, roundish cells. Capsule not seen in South African specimens, but from elsewhere it is described as regular and shortly exserted on a rough seta ; ring absent ; peristome double ; teeth papillose below, hyaline above ; membrane low, smooth ; processes slender, split, papillose ; cilia absent ; lid small, sharply beaked. Perichaetal leaves long and narrowly pointed.

594. *Trachypodopsis serrulata* (Palis.) Fleisch.; Broth., Pfl., i, 3, 831, and xi, 122 ; Dixon, in S.A. Journ. Sc., xviii, 325.

=*Trachypus serrulatus* (P. B.), Besch., Fl. Bryol. Reunion, etc., p. 128 ; Wager's list 13 ; Mitt., Journ. Linn. Soc., xxii, 146, 314.

=*Pilotrichum serrulatum* P. B., Prodr., p. 83 (1805).

=*Neckera serrulata* Brid., Bry. Univ., ii, 238 ; C. M., Syn., ii, 140.

=*Papillaria serrulata* Jaeg., Ad., ii, 178.

=*Meteorium serrulatum* Mitt., in Linn. Soc. Journ., 1863, p. 156.

=*Hypnum ericitorum* Brid., Sp. Musc., ii, 97.

=*Neckera Macleana*, Rehm. 616.

And Brotherus (Pfl., i, 3, 832) mentions as hardly specifically distinct—

Trachypus nudicaulis (C. M.) Besch., from Comora.

Trachypus Rutenbergii C. M., from Madagascar.

Trachypus Quintasianus C. M., from the Cameroons and St. Thomé.

Stems suberect or hanging, ultimately prostrate and rooting, and sending up fresh stems 4–6 inches high, irregularly pinnate, yellowish brown, with branches and leaves subcomplanate ; the branches simple, ½–1 inch long, patent ; leaves numerous, imbricate, widely lanceolate from rather a wide symmetrical base, produced all round the stem, 3–4 mm. long, lightly 4-furrowed longitudinally when dry, transversely undulated, the points long and persistent, and much crisped and twisted when dry, and remaining so even when wet. Margin toothed, the teeth large, varying in size but each consisting of a single cell ; teeth extending to the leaf-base, though smaller, and produced by almost every marginal cell ; cells long and narrow, those at the leaf-apex elliptical-vermicular, others vermicular and 1-papillose ; basal cells rather wider and shorter ; alar groups distinct, with small rounded cells. Nerve strong, disappearing some distance below the apex. Rare. Included in Wager's list without locality.

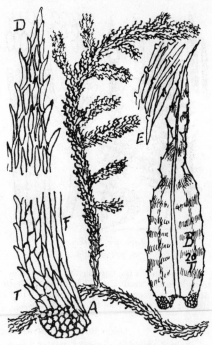

Trachypodopsis serrulata.

Transvaal. Mountains above Lydenburg, MacLea (Rehm. 616, as *Neckera Macleana* R.).
S. Rhodesia. Wet forests, Inyanga, 6000 feet, Aug. 1920, J. S. Henkel (F. Eyles 2630).
Known also from Central and West Africa and East African islands.

Genus 191. LEPTODON Mohr.

Fairly strong, dioecious mosses; the main stem straggling, rhizomatous, and wiry, rooting among other mosses on stones or bark; the secondary stems 1–2 inches long, erect, pinnately or 2-pinnately branched; the lower branches short, the upper longer, placed in complanate arrangement, and when dry the stem and branches are all spirally inrolled into a rather compact ball. Branches closely leaved, with numerous paraphyllia, sometimes flagellate, with small leaves; leaves in eight rows, the side leaves complanate and spreading, the dorsal and ventral leaves appressed and alternately set to right and left; all the leaves somewhat decurrent, unsymmetrical, roundish or shortly oval, with rounded apex, concave, entire, incurved at the base; nerve single, going half-way; cells small, roundish or oval, smooth or almost so, longer along the nerve in the middle of the leaf: seta short, almost straight; capsule exserted, oval, erect, smooth, symmetrical; vaginula hairy, the hairs reaching the base of the capsule; ring absent; peristome double, the outer of sixteen small, lanceolate, slightly papillose teeth hardly connate below, the inner a basal membrane only, or with rudimentary processes. Lid obliquely rostrate from a conical base; calyptra cucullate, set with long erect hairs.

595. **Leptodon Smithii** (Dicks.), Mohr. Obs., p. 27; Brid., Bry. Univ., ii, 197; Bryol. Eur., v, t. 439; Schimp., Syn., 2nd ed., p. 562; Broth., Pfl., i, 3, 835, fig. 621, and xi, 180; Dixon, 3rd ed., p. 416, Tab. 50K.

=*Hookeria convoluta* Spreng., Syst. Veg., iv, 2, 324.
=*Hypnum circinatum* Santi Viagg., p. 209.
=*Lasia Smithii* Brid., Mant. M., p. 133.
=*Maschalocarpus Smithii* Spreng., Syst. Veg., iv, i, 159.
=*Neckera Smithii* C. M., Syn., ii, 118 and 669.
=*Orthotrichum Smithii* Brid., Musc. rec., ii, 2, 33.
=*Pilotrichum Smithii* P. B., Prodr., p. 83.
=*Polytrichum Smithii* Brid., Sp. M., i, 140.
=*Pterogonium Smithii* Sw., in Schrad. Journ., ii, 173.
=*Leptodon* *mollis, Rehm. 602 and 602β.

Dioecious; laxly intermixed among other mosses or hanging from bark, expanded when moist, closely inrolled as to leaves and stems when dry; the main stem prostrate, very laxly clad in minute, acute, nerveless leaves; secondary stems rising in subdendroid fashion, branchless below, complanate-frondose above; paraphylliate; 2–3 inches long, ½–1 inch wide; closely pinnate or usually 2-pinnate, or with irregular fresh fronds rising anywhere instead of branchlets; pinnules often reduced toward the point or flagellate, and always with the leaves smaller than the stem-leaves. Stem-leaves crowded, 1 mm. long, slightly irregular, concave, sometimes plicate, oval-oblong, with rounded or sometimes obtuse apex, slightly decurrent, with entire margins, reflexed toward the base; nerve very wide below, weak, tapering out above mid-leaf, made of lamina cells; cells small, smooth, 20µ long, shortly hexagonal, but tumid and showing only oval lumen, those toward the leaf-base longer and twisted, except the reflexed marginal portion where the cells are about 10µ long, quadrate or shortly oval. Perichaetia mostly on the inner surface, sessile on the stem or branches, few-leaved; outer leaves short and spreading, the inner narrow above, convolute, and consequently apparently very narrow or acuminate; paraphyses numerous and longer than these leaves. Seta 5 mm. long, smooth, nearly straight; capsule well-exserted, oval, brown; lid beaked; calyptra somewhat hairy. Ring absent; outer peristome of sixteen short, white, linear-lanceolate, perforated, papillose teeth; inner consisting of a low membrane only. On branchlets and young shoots (which latter sometimes produce rhizoids) the reduced leaves may be in fewer than eight rows, and not compressed, and the leaf-point is semicircular. The plant is very rigid when dry, and incurved. This species is exceedingly variable in its form and division, and yet exceedingly simple in all other characters.

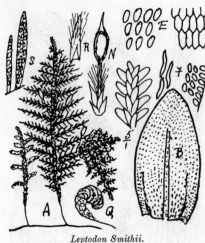

Leptodon Smithii.
S, paraphyllia.

Abundant on the Cape Peninsula and in every forest in the Cape Province, Natal, and Transvaal, from the coast to high on the mountains. I have no specimen from Rhodesia, but expect it occurs there and in Portuguese East Africa, as it occurs elsewhere in Africa and its islands, mid-Europe, West Asia, South America, and Australasia. The other species of *Leptodon* belong to Africa.

Genus 192. CALYPTOTHECIUM Mitt. (in Journ. Linn. Soc. Bot., x, 190 (1868)).

Robust, yellowish, dioecious, shining mosses, growing on the ground in the tropics and subtropics, having creeping, small-leaved, rooting main stems, sometimes small-leaved stolons, and usually long, straggling, pinnately branched, secondary stems without paraphyllia ; the branchlets simple, blunt, rather short, and sometimes more or less flagellate. Leaves more or less in eight rows, the side leaves complanate and spreading, the back and front leaves laxly appressed in two rows on each side and placed right and left alternately, ovate acuminate, unsymmetrical from a usually auricled oblique base, concave, entire or almost so, frequently transversely undulated, with single nerve usually present and extending to mid-leaf. Cells linear, smooth ; those at the apex rhomboid, those at the base lax and brownish ; alar cells not differentiated. Perichaetal leaves sheathing. subulate ; seta very short ; capsule quite immersed, oval or urn-shaped ; ring absent ; lid shortly straight-beaked from a convex base ; calyptra small, lobed or split, naked ; peristome double, teeth lanceolate, smooth, almost free, with properistome ; inner peristome of longer, linear, almost smooth, keeled processes, without basal membrane and without cilia. *Renauldia Hoehnelii* (as *R. africana*) used to be included here. but is less flattened.

Dixon (S.A. Sc. Journ., xviii, 325) suggests that the two following species may yet have to be united under *C. acutifolium*.

Synopsis.

596. *C. Brotheri.* Leaves very concave, with acute reflexed point.
597. *C. acutifolium.* Leaves bluntly pointed, undulated.

596. *Calyptothecium Brotheri* (Par.), Dixon, S.A. Journ. Sc., xviii, 325.

=*Neckera Brotheri* Par., Ind., Ed. 1, Suppl., p. 254.
=*Calyptothecium subacutifolium* Broth., in Engl. Bot. Jahrb., xxiv, 254 (1897) (not *C. subacutifolium* (Geh. and Hampe), Broth. =*Neckera subacutifolia* G. and H., in Flora, p. 379 (1881), which is S. American).
=*Calyptothecium Beyrichii* Broth., in Eng. Pfl. Musci., ii, 839 (=i, 3, 839).

Main stem basal, prostrate, eventually wiry, at first green, small-leaved, and branched ; erect stems seen only 1–2 inches high and sterile, but simple or pinnate, complanate, closely leaved except at the base ; the leaves complanate, in two rows of lateral spreading leaves at each side and two rows each back and front of appressed leaves, all slightly unequal at the subcordate base, more or less oblong, boat-shaped or deeply concave to near the apex, which is triangular, flat, reflexed, sharply acute and subdenticulate ; margins elsewhere erect. Nerve extending as far as the leaf is concave ; cells about 70μ long, showing narrowly elliptical lumen and three to six granules or low papillae in each. Basal cells rather shorter, and a few, representing an alar group, are oval and about 10μ long. Other parts not seen.

Dixon, in S.A. Journ. Sc., xviii, 325 (1922), explains the above synonymy.

Cape, E. Pondoland, Beyrich.

S. Rhodesia. Victoria Falls, 3000 feet, Sim 8910, 8922, Wager 914 ; Palm Grove, 2500 feet, Sim 8913, and Brunnthaler (all sterile).

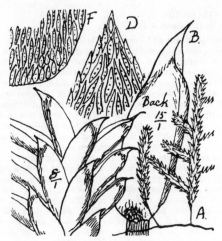

Calyptothecium Brotheri.
B, back view.

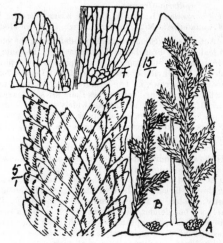

Calyptothecium acutifolium.
B, leaf.

597. *Calyptothecium acutifolium* (Brid.), Broth., in litt. ; Broth., Pfl., i, 3, 839 ;
Dixon, in S.A. Journ. Sc., xviii, 325 (1922).

=*Neckera acutifolia* C. M., Syn., ii, 48 ; Par., Ind., p. 842 (1896).
=*Neckera (Distichia) acutifolia* Brid., Bryol. Univ., ii, 48.
=*Calyptothecium pseudocrispum* (C. M.), Broth. (=*Neckera pseudocrispa*, Rehm. 328).

Main stem basal, prostrate, wiry, small-leaved ; erect stems only seen sterile here, 2 inches high, irregularly pinnate, complanate, with two pairs of rows of spreading side leaves and a pair of appressed leaves each back and front; leaves imbricate, shining, oblong from cordate base, with short, triangular, blunt, and suboblique point; transversely undulate and somewhat crisped when dry, nearly flat, but usually with one margin more or less folded under; nerve strong below, extending to near the apex; margin entire; lamina cells rhomboid or widely rhomboid, 50μ long, not showing lumen; apical cells more so; basal cells shorter and wider, enclosing a small group on each side of small, rounded, alar cells. Other parts not seen. Van Reenen's Pass, Natal, Rehm. 328, as *Neckera pseudocrispa* Rehm. I am not aware that this has yet been seen anywhere fertile, but C. M. (Syn., ii, 48) cannot separate the foliage from a South American species which has immersed oval capsules and obliquely rostrate lid. Bescherelle (Florule Bry. de la Reunion, etc., p. 273) describes it as dioecious, the sterile Reunion plant growing stems 15–20 cm. long, laxly pinnate or 2-pinnate; leaves ovate-acuminate, subnavicular; the perigynia lax, their leaves sheathing, lanceolate-acuminate, subentire, obsoletely nerved, and the archegonia few and long. Dixon (S.A. Journ. Sc.) states that the Van Reenen plant is certainly identical with *C. acutifolium*, which I have not seen otherwise.

Genus 193. NECKERA Hedw.

Vigorous, shining, light-coloured mosses, usually hanging in masses from tree-trunks or decayed stumps in moist forests; main stem prostrate, rooting, and sometimes stoloniferous; secondary stems up to several inches long, usually hanging, complanately and pinnately or 2-pinnately branched; the branches of equal width throughout, usually without paraphyllia in South African species. Leaves complanate in eight rows or fewer; the side leaves widely spreading, and the back and front leaves appressed and alternately pointing right and left; flat, usually obliquely ovate and unsymmetrical, with obtuse or mucronate apex, sometimes transverse shining undulations, and the margin somewhat incurved toward the base. Nerve (in the South African species) exceedingly short and weak, forked or absent. Cells rather small, elliptical or rhomboid; those toward the base linear; the alar cells usually different, small and quadrate. Seta usually very short. Capsule immersed in South African species, erect, symmetrical; lid shortly and obliquely rostrate from a wide base; calyptra cucullate, naked; peristome double; teeth lanceolate, slightly connate at the base, smooth or papillose, often perforated; inner of fragile linear processes, often perforated, almost without basal membrane and without cilia.

Neckera was formerly a very large genus, but as many other genera have been extracted from it, its characters have been modified, and the name frequently appears now as a synonym.

598. *Neckera Valentiniana* Besch., Florule Bryol. de la Reunion, p. 273 (132); Broth., Pfl., i, 3, 844, and xi, 185 Dixon, Trans. R. Soc. S. Afr., viii, 3, 210, and Smithson. Misc. Coll., lxxii, 3, 13.

=*N.* *capensis* W. P. S., in Breutel, Musc. Cap. (not C. M., which is *Aerobryopsis*); Broth., Pfl., i, 3, 844; Brunnthaler, p. 30; Bryhn, p. 18.

=*N.* *africana* W. P. Schimp. (Herb. Mont. coll. Zeyher, as *N. intermedia* Brid.), not C. M., Syn., p. 137, which is *Papillaria africana*.

=*N.* *Borgeniana* Kiaer=*N. Borgeni* Kiaer; Broth., Pfl., i, 3, 844 (*fide* Dixon).

=*N.* *undulatifolia* Mitt., in Herb. Kew; C. M., in Hedw., 1899, p. 126 (name only).

Autoecious. Stems extending indefinitely, the older portion prostrate and eventually leafless, and bearing old and leafless side-branches and perichaetia; the younger portion erect or ascending, or from tree-trunks horizontal or drooping, 4–9 cm. long, irregularly pinnate; the pinnae simple or slightly branched unless they start away as new leaders. Paraphyllia absent or very few. Leaves 2–2·5 mm. long, complanate, apparently distichous but actually in four to eight rows, *i.e.* two lateral on each side and usually two on the upper face and two on the lower face, though sometimes on the smaller branches there are side leaves only, of which the bases overlap over the axis. Leaves asymmetrical, transversely undulated, widely oblong, widest below, auriculate on the lower side, shortly and obliquely triangular-pointed from its full upper width, somewhat convex, with the margin often inflexed in the lower half on one side; apex bluntly mucronate or subacute, varying in this immensely; margin minutely but closely serrulate; nerve either absent or barely visible as one or two short unequal nerves; cells all pellucid, elliptical toward the leaf-apex and margin, linear up to $100\times10\mu$ elsewhere, shorter at the base, and with a small alar group on one side of quadrate cells showing circular lumen. Perichaetia abundant on the main stem, sessile, scarious, few-leaved; the leaves imbricate, ovate with long subulate point, longer than the capsule, with linear cells, no nerve, and slightly denticulate or almost entire margin. Capsule brown, immersed, nearly sessile, oval, often contracted below the mouth when dry; ring absent; lid conical, shortly rostrate; calyptra cucullate, glabrous; peristome double, the outer teeth red, lanceolate, faintly striolate at the base, smooth or faintly papillose above; processes rather shorter, often perforated, well developed but slender, without membrane; cilia absent. Sometimes with flagellate branches; always so undulated as to give a crisped appearance over all, even when moist, from which the points sometimes incurve making the apex appear rounded. Very closely allied to *Neckera pennata* Hedw. of Europe, which, however, has the inner peristome less developed or rudimentary, and all early collections in South Africa were made under that name before the two were separated, hence in C. M., Synopsis, it is credited also to South Africa. The differences between this and *N. capensis* given by Brotherus are unstable, and Dixon mentions (Smithson. Misc. Coll., lxxii, 3, 13) that all the African species are inseparable from one another by vegetative characters alone, but in a letter (25th April 1917) he mentions that my 7498 has the teeth very finely papillose and a *very* slight tendency to striolation at the base, and the processes are narrow, irregular, from one-third to two-thirds the length of the teeth, and· so is *N. capensis* Schp.; while in my 8543 the teeth are almost smooth and pellucid throughout, and the processes almost equal in length, and so is *N. Valentiniana* Besch. But I do not look on these differences as stable or constant. *N. capensis* B. and S., var. *laevigata*, Rehm. 618, is only a young condition.

Abundant in every moist forest from Table Mountain to Transvaal. There is an Ecklon and Zeyher specimen in the Albany Museum Herbarium named *N. remota* Br. and Sch. (an Abyssinian species) which is simply *N. Valentiniana*.

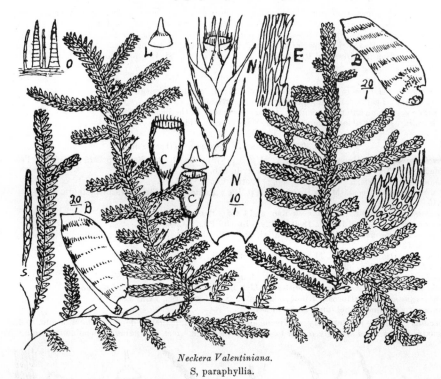

Neckera Valentiniana.
S, paraphyllia.

(Genus HOMALIA Brid. was included in my check-list on account of specimens from Barberton distributed under the name *Homalia trichomanoides* Brid. by Wager, but my examination of these specimens, and comparison with European specimens, convince me that the plants distributed, so far as I have seen them, are not that species but *Porothamnium comorense*, an exceedingly variable plant and sometimes like *Homalia* in leafage, but always more dendroid in its normal forms, while its flagellate and subaquatic forms are less like *Homalia*. So far I see no reason to consider *Homalia* South African; besides the non-dendroid habit, it is autoecious and has rhomboid leaf-cells, while *Porothamnium* is dioecious and has the cells more linear.)

Genus 194. POROTHAMNIUM Fleisch., Laubm. Java, 1908, p. 925.

Dioecious, robust mosses, with long, stoloniferous, rooting main-stems, and with complanately branched, pinnate or often 2-pinnate, more or less dendroid or frondose secondary stems, branchless and small-leaved at the base and with spreading complanate branches above, usually erect or arching except when flagellate or semiaquatic, when the long slender branchlets are usually pendent or floating out from a rigid, wiry, secondary stem. Leaves often irregularly in four rows, or more regular and complanate, spreading alternately right and left on each surface, somewhat concave, shovel-form, smooth or furrowed, somewhat oblique and unsymmetrical, not decurrent, oval-elliptic, acute serrate or serrulate, with nerve half-way or more, and elliptic-linear, sometimes papillose cells. Perichaetal leaves long, lanceolate or acuminate from a sheathing base. Seta about an inch long, straight or nearly so, regular or nearly so, with dehiscent ring, shortly beaked lid from a wide base, and smooth calyptra. Peristome double; the outer teeth striate, shortly connate below; the inner as long, nearly smooth, with wide basal membrane and split subulate processes and with several cilia, sometimes having appendages. Spores small.

This exceedingly variable group has caused much confusion; the genus *Thamnium* was cut out of bryological literature because preoccupied (see Dixon, 2nd ed., p. 409, and 3rd ed., p. 408); many species were then placed in *Porotrichum*, from which Fleischer separated those having striate outer teeth as *Porothamnium*, but Brotherus (Pfl., i, 3, 855, and xi, 198) places *P. Chauveti* (from Mascarenes and South Africa) in *Porotrichum*, and *P. pennaeforme* C. M. in *Porothamnium* (p. 862), section *Euthamnium*, and then at p. 1229 (and **xi**, 198) moves it to section *Pseudoporotrichum*, while Paris places *P. Chauveti* as a variety of *P. pennaeforme*. C. Müller, in Hedw., 1899, on Rehmann's barren specimens, evidently failed to realise the extreme variability of the African species, which quite corresponds with similar variability in their British close allies, *P. alopecurum* Mitt., *Neckera complanata* Hübn., and *Homalia trichomanoides*, all of which closely resemble our plants in leaf and habit. This is a group, like *Stereodon cupressiformis*, in which no one character is quite stable, and in which an office botanist, dealing with small specimens, however numerous, is likely to make several species; while the field botanist, familiar with the variability not as fixed varieties,

26

but as conditional changes, even on the same plant, can only group all together, without even varietal names. *Porothamnium* and *Pilotrichella* are used by the ton in Natal for nursery packing (as sphagnum or oak leaves are in England), and I have had constant opportunity of seeing that our first two species are hardly distinct, and each varies immensely, and also agrees with other African species so-called (except *P. Woodii*), to the extent that I believe one good species covers all that section in Africa and its islands. Brotherus, in Pfl., xi, 196, maintains *Porotrichum*, *Thamnium*, and *Porothamnium*.

Synopsis.

599. *P. comorense.* Leaves in two regular rows on each side, strongly dentate.
600. *P. natalense.* Leaves rather irregular, smaller, nearly entire, often reduced.
601. *P. capense.* Stems filiform, procumbent ; nerve dentate at the back above.
602. *P. Woodii.* Leaves oblong-lingulate, entire ; nerve very wide.

599. Porothamnium comorense (Hpe.) Sim.

=*Porotrichum comorense* Hampe, in Linnaea, xl, 270 (1876) ; Besch., Fl. Reun., p. 275 ; Renauld, Prod. Fl. Bry. Madag., p. 206 ; Dixon, in S.A. Journ. Sc., xviii, 326.
=*Porothamnium pennaeforme* (C. M.), Fleisch. ; Broth., Pfl., xi, 199.
=*Porotrichum pennaeforme* C. M., in Hedw., 1899, p. 128 ; Rehm. 333, 333β, 620 ; Mitt., Journ. Linn. Soc., xxii, 146, 315.
=*Porotrichum pennaeforme*, var. *Chauveti* Ren. and Card ; Ren., Prodr., p. 207.
=*Thamnium afrum* C. M., in Hedw., 1899, p. 129 ; Broth., Pfl., xi, 200.
=*Thamnium Hildebrandtii* C. M., in Linnaea, xl, 287 ; Besch., Fl. Reun., p. 276 ; Ren., Prodr., p. 210 ; Wager's list, p. 13.
=*Porothamnium Hildebrandtii* (C. M.), Fleisch. ; Broth., Pfl., xi, 200.
=*Neckera *pterops*, Rehm. 329.
=*Porotrichum *pterops*, Rehm. 621, 621*b* ; Mitt., Journ. Linn. Soc., xxii, 146, 315.
=*Porotrichum madagassum* Kiaer (*fide* Mitten), which Brotherus still upholds (Pfl., xi, 198).
=*Thamnium pennaeforme* (Horns.), Kindb. ; Dixon, in S.A. Journ. Sc., xviii, 326.
=*Thamnium *complanatum* Schimp., MSS., in herb.

Dioecious ; rhizome slender, wiry, producing erect, usually complanately dendroid stems, 5 mm. to 4 cm. apart ; stems woody, firm, from a strongly rooted base, 2–6 inches high, erect, ascending or pendulous, shining, irregularly

Porothamnium comorense.

pinnate or bipinnate ; the branches and leaves complanate, forming a flat frond 2–4 inches across each way, except in the modified branches or forms. Leaves very small on the lower part of the stem, those on the older branches much larger, 3–4 mm. long, larger and more imbricate than those on the younger. Leaves overlapping, spreading more or less regularly in four lateral rows, two on the upper surface closely pressed against two on the under-surface, but this is less marked and the leaves are more scattered and smaller on the younger branches. Leaves oblong-acuminate from a clasping base, shortly pointed, regular or slightly oblique, somewhat concave through the lower margin being inflexed or conduplicate. Nerve single, yellowish, fairly strong, usually reaching half-way or more, occasionally unequally forked far up ; margin dentate from base to apex but most strongly above ; cells small, rhomboid or elliptical, longer, up to 50µ long at mid-leaf, and again reduced toward the base and corners. Fructification usually on the main stems ; perichaetal leaves high-sheathing, with long subulate point, nerveless or almost so, entire, and with long, narrow, pellucid cells. Seta 2–3 cm. long, cygneous at the top, brown ; capsule somewhat inclined, oval to shortly cylindrical, with a rostrate lid. Ring present ; peristome strong, outer teeth red, striate ; inner higher, erect, split, slender. Whole frond inclined to roll backward when dry ; branches long, often flagellate at the apex.

Present in every forest and scrub from Cape Town to the Zambesi, and this or very closely allied forms occur throughout Africa and its islands. Sometimes the leaves are wonderfully like those of *Levierella*.

600. *Porothamnium natalense* (C. M.), Lindb., in Hedw., xli, 209 (1902) ; Broth., Pfl., i, 3, 861, and xi, 199.

=*Porotrichum natalense* C. M., in Hedw., 1899, p. 120.
=*Porotrichum pennaeforme*, var. *brachyphyllum, Rehm. 334 and 334β (sterile).
(On some labels 334c should be 434c *Sphagnum mollissimum*.)

Very similar to *P. comorense* but usually smaller, with stems narrower and not so definitely regular in its leafage, while many of its stems are weak and trailing and frequently unbranched, and others are 2-pinnately flagellate throughout. with numerous or scattered small, yellowish-green, lanceolate leaves, with or without nerve, and with the margin denticulate, subentire or entire, as also occurs on normal plants. Leaves 1–1·5 mm. long, acute, the cells

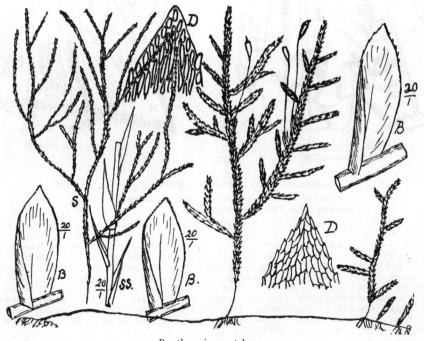

Porothamnium natalense.
S, flagellate frond ; SS, part of same.

rhomboid-hexagonal, or when tumid showing oval lumen on apical cells, elliptical lumen on lamina cells, and oval lumen gradually reached in basal cells. Nerve extending half-way, or more or less, not toothed on the back. Seta slightly inclined, 1–1½ inch long, brown ; capsule as in *P. comorense*. The two species often grow intermixed and are difficult to separate.

Abundant throughout the forestal regions of South Africa, often on or hanging from rocks, stumps, etc., sometimes growing erect on humus. Exceedingly variable within itself, even on the same plant.

601. *Porothamnium capense* (B. and D.) Sim. (New combination.)

=*Thamnium capense* Broth. and Dixon, Bull. Torrey Bot. Club., xliii, 71 (1916), and fig. 2 ; Broth., Pfl., xi, 201.

A weak straggling moss 2–4 inches long, with main stem often black and leafless, or with only fragmentary remains of leaves, but the branches, branchlets, and leaves complanate and dark green. Stems sometimes forked ; forks irregularly subpinnate or somewhat 2-pinnate ; the ultimate branches 5–10 mm. long, spreading ; leaves 2 mm. long, widely oblong-acute, concave and keeled ; side leaves spreading or slightly incurved, others appressed (hence foliage complanate). Keels very evident, spinulose-dentate outside in the upper half, formed of the strong nerve which extends, somewhat flexuose above, to near the apex. Margin expanded, denticulate below, strongly serrate above, with irregularly double unequal teeth ; cells pellucid, irregularly deltoid or angular ; the apical cells larger and more lax ; lamina cells 10μ wide, closely packed, the lowest rather larger and longer. The nerve dentate on the back, which shows easily on almost every leaf, is a character not seen among my many specimens of *Porothamnium* from Cape Town and elsewhere, and even the serrated leaf is distinctive. Fruit not seen.

Received in packing from Cape Town (G. Webster 710) by Mr. Dixon, who has kindly lent me specimens.

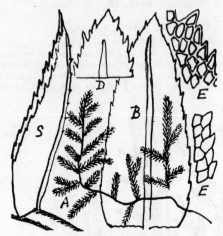
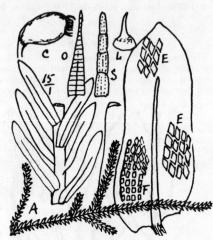

Porothamnium capense.
S, side view of leaf.

Porothamnium Woodii.

602. *Porothamnium Woodii* (Rehm. and Macowan) Sim (new species).*

=*Hookeria* Woodii*, Rehmann and Macowan, in Albany Museum.
=*Mniadelphus Woodii* Sim (in letters).

Stems 2–3 inches long, prostrate, with few irregular branches ; leaves lax, unsymmetrical at the base, lying complanately when moist, laxly appressed when dry, 2 mm. long, oblong-lingulate, shortly rounded to an obtuse submucronate point, base rather wider ; nerve very wide, extending to near the apex and with fibrous connection to the stem ; margin entire, not bordered ; cells rhomboid-elliptical, small, 15μ long, the lower lines on the larger leaf-base smaller and quadrate. Seta 1 inch long, rising from lower part of the branches, smooth ; perichaetal leaves like other leaves or with less nerve ; capsule horizontal oval, with shortly rostrate lid from convex base. Peristome teeth furrowed along the mid-line, somewhat bordered and closely cross-striate in the lower half, nearly white and papillose in perpendicular lines in the upper half ; inner peristome not seen. The leaves are lax anywhere, but the young branches on specimens seen are very lax indeed, as if recently exposed to dense shade, and drawn thereby.

J. M. Wood 285, in Albany Museum Herbarium. Although locality is not stated, all Wood's specimens at that time were from Inanda, Natal, where he was farming.

Family 49. ENTODONTACEAE.

Slender or somewhat robust, autoecious or dioecious, usually loosely matted, shining mosses, growing on stumps, rocks, or occasionally on the ground, with rooting, prostrate, spreading or suberect, often stoloniform, pinnate or irregularly branched stems, with homogeneous leaves all round the stem or in flattened arrangement. Leaves of one layer of cells, somewhat unsymmetrical, of various forms ; nerve usually single, more or less long, but never excurrent,

---

* *Porothamnium Woodii* (R. et M.) Sim (sp. nov.).
Rami pauci, irregulares ; folia oblongo-lingulata, breviter rotundata ad apicem obtusum sub-mucronatum ; basis paulo latior. Nervus latissimus, prope ad apicem pertinens. Margo integer, non marginatus.

or occasionally double, very short or absent. Cells usually long, sometimes papillose, very seldom quadrate or rhomboid, and smooth or with one papilla per cell; the alar cells differentiated, usually quadrate or rhomboid; vaginula naked; seta usually long; capsule erect and regular, not wrinkled; lid long-beaked or short-beaked from a conical base; calyptra cucullate, naked; peristome usually double or the inner occasionally absent; outer peristome usually normal, papillose, and occasionally striate, sometimes perforated, seldom smooth, and also seldom irregularly thickened; inner peristome with usually low, plicate basal membrane, and narrow, linear or lanceolate, keeled, and frequently more or less split processes, and with the cilia either rudimentary or absent. Most frequent in warm countries.

Synopsis of Entodontaceae.

Cells, or some of them, papillose; nerve double, short or absent . . . 195. Trachyphyllum.
Cells all smooth.
Processes much shorter than the teeth, thread-like; nerve double, often short or absent 196. Erythrodontium.
Processes as long as the teeth, or nearly so (absent in *Levierella*).
Nerve absent; inner peristome without basal membrane; branches not flattened; leaves not narrowed to the base . . . . 197. Platygyrium.
Nerve single; inner peristome absent . . . . 198. Levierella.
Nerve single, usually present; leaves narrowed to the base; peristome double . 199. Stereophyllum.
Nerve double, very short or absent.
Alar cells segregated; branch leaves usually flattened; inner peristome without basal membrane,
200. Entodon.
(Alar cells not sharply segregated; branches not flattened, often secund; basal membrane wide Pylaisia.)

Genus 195. Trachyphyllum Gepp, in Pl. Welw., ii, 2, 298 (1901).

Slender, firm, matted, yellowish- or brownish-green, shining, dioecious plants; stem very long, creeping, fixed down by numerous rhizoids, closely leaved, usually regularly and closely branched; branches short, spreading or suberect, when dry twisted. Stem leaves when dry closely imbricate, when moist slightly spreading, somewhat concave, not furrowed nor decurrent, broadly ovate, narrowly pointed, flat margined, slightly toothed upward through protruding papillose cells. Nerve double, short, strong at the base; cells elliptical, at the upper end of the back strongly papillose; at the leaf-base near the nerve the cells are in oblique rows and rhomboid hexagonal; the alar cells in many rows, quadrate or rhomboid, chlorophyllose; perichaetium wide; inner perichaetal leaves lanceolate-subulate from an oblong base, sharply toothed toward the apex. Fruit not yet seen in South Africa. Seta ½ inch long, twisted, thin, yellowish red; capsule horizontal, irregular, high-backed, oval, with short neck, when dry somewhat contracted below the mouth, and with stomata on the neck. Peristome double, inserted near the mouth, the two equally long. Teeth of the outer peristome linear-lanceolate, yellow, closely striated and papillose, with closely placed lamellae; basal membrane of the inner peristome wide, hyaline, smooth; processes lanceolate-subulate, perforated, finely papillose; cilia three, well developed, papillose, knotted; spores about 15μ diameter; lid shortly beaked.

Mr. Dixon has twice stated, that when the scattered African species (for they are all African) come to be collated, all may be united under one, in which case *T. fabronioides* (C. M.), Gepp; Broth., Pfl., xi, 384, fig. 702, is the name that has precedence, but meantime our two species, as extreme forms, are well separated. *T. fabronioides* is illustrated by Brotherus in Pfl., i, 3, 889, fig. 651, and Mr. Dixon writes that *Pterogonium decipiens* Wright, Journ. of Bot., 1892, p. 264, is *T. fabronioides*.

Synopsis.

603. *T. gastrodes.* Leaves 0·5 mm. long; lower cells all round.
604. *T. maximum.* Leaves 1 mm. long; central lower cells laxly rhomboid.

603. **Trachyphyllum gastrodes** (Welw. and Duby), Gepp; Dixon, in Bull. Torrey Bot. Club, xliii, 72, and in S.A. Journ. Sc., xviii, 326; Broth., Pfl., xi, 384.

=*Heterocladium gastrodes* (W. and D.), Jaeg., Ad., ii, 312.
=*Hypnum gastrodes* Welw. and Duby, in Mem. Geneve, 1871, p. 15, t. 5, fig. 1.

Stems 1–2 inches long, prostrate, rooting, making an intricate low mat; when dry the branches are terete and julaceous; when moist the leaves are spreading widely, widely cordate-oval, 0·5 mm. long, concave; the stem leaves shortly narrowed to a subulate point, the branch leaves shortly acute, or often the branches are shortly flagellate, with reduced leaves. Nerve double, very short and hardly visible, or absent; margin slightly denticulate near the apex; upper cells long-hexagonal, 30μ long, pellucid or slightly papillose; basal (alar) cells round, densely papillose, occupying the whole leaf-base and more than half-way up the leaf, giving it a roughly crenulate margin, and making the back of the leaf very rough. Male inflorescence discoid on the stem and branches. Fruit not seen.

Transvaal. Barberton, Wager.
S. Rhodesia. Abundant at Khami ruins, 5000 feet altitude, July 1920, Sim 8840; Zimbabwe, 3000 feet, Sim 8760 and 8756β; Rhodes Grave, Matopos, 5000 feet, Sim 8473, 8876β, 8872β.
Originally found in Angola.

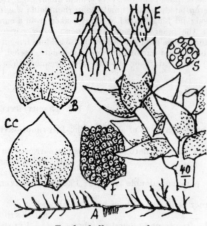
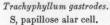

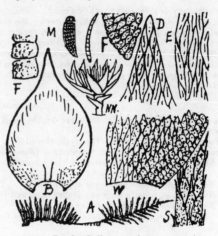

Trachyphyllum gastrodes.

S, papillose alar cell.

Trachyphyllum maximum.

S, terete dry branch.

604. *Trachyphyllum maximum* Dixon, in S.A. Journ. Sc., xviii, 326 ; Broth., Pfl., xi, 384.

Stems strong for this genus, 1–2 inches long, laxly prostrate, closely pinnate, with branches spreading or erect, about 5 mm. long, when dry turgidly terete through appressed leaves, or the leaves, especially on older growth, are often suberect or even erecto-patent ; when moist the leaves are erecto-patent or more spreading, 0·8–1 mm. long, ovate-acuminate, rather long-pointed, very concave, entire, with erect margins and almost invisible, short, double nerve. Apical cells rhomboid, 25µ long ; mid-leaf cells rather longer and somewhat twisted ; lower central cells laxly rhomboid, gradually changed to crowded, transversely elliptical, and densely papillose at the margin. The rhomboid cells are laxly papillose, or often have a twisted line in the lumen. Male flowers abundant on the branches, shortly stalked, discoid, with about six antheridia, about six longer linear paraphyses (fig. M), and short ovate-subulate bracts.

Transvaal. Witpoortje Kloof, Professor Moss.

S. Rhodesia. Makoni, Timasu, on a granite hill, 6500 feet, April 1918, Dr. E. A. Nobbs (F. Eyles 1317β, in Herb. Sim).

Genus 196. ERYTHRODONTIUM Hampe, Symb., viii, 274 (1870).

Slender, autoecious, somewhat firm, shining mosses, having long, prostrate, forked main stems, forming flat mats on bark or stones. Branches long, closely leaved, nearly regularly pinnate, with short, terete, spreading branchlets. Leaves imbricate when dry, concave, oval or ovate from a decurrent base, shortly and narrowly pointed, with margin flat or reflexed at the base, and entire or finely toothed at the apex. Nerve double, often short or absent ; cells oblong-elliptic, smooth ; the alar cells in rows, roundish quadrate or rhomboid. Fruit not seen in South Africa. Seta long ; capsule erect, straight, cylindrical, short necked ; ring absent ; lid long beaked from a conical base ; peristome double ; teeth of the outer peristome lanceolate, flat, bordered ; inner peristome of short, thread-like, caducous processes, without basal membrane and without cilia.

605. *Erythrodontium abruptum* (C. H. W.), Broth., Pfl., i, 3, 888, and xi, 382.

=*Pterogonium abruptum* C. H. Wright, London Journ. of Bot., 1892, p. 264.

Stems 1–2 inches long, prostrate, rooting, shining green, closely pinnate ; branches 5–10 mm. long, simple, slender, turgidly but closely julaceous when dry, with the leaf-tips reflexed ; when moist the branch leaves are crowded, imbricate, set all round the stem except where rooting, erecto-patent, very concave, chlorophyllose, ovate, with recurved triangular flat point ; the margins erect and entire ; the nerve very weak and short and widely double when visible, which is seldom ; upper cells twisted and narrowly rhomboid, 20µ long ; central basal cells laxly hexagonal and pellucid, gradually changing outward to numerous, rounded or quadrate, pellucid alar cells 10µ wide.

Much like *Trachyphyllum* in general appearance, but the cells are smooth and the leaf not rough on the back.

Fruit has not been seen on this plant, but capsules of *E. subjulaceum* (C. M.), Par., from Central Africa, are shown in fig. S, p. 407 ; the seta there is smooth, 2·5 cm. long ; capsule large and oval, with bright red papillose teeth, and short, sharp, obliquely beaked lid, but the plant has longer and fewer branches, and is more dendroid in habit. There are several other African species.

E. abruptum was described from Zambesia, and I found it at Zimbabwe, S. Rhodesia (sterile), Sim 8776. Mr. Dixon, who has examined the type, states that it is=*E. rotundifolium* (C. M.), which itself is very doubtfully distinct from *E. subjulaceum* (C. M.), both Broth., Pfl., xi, 382, as mid-African.

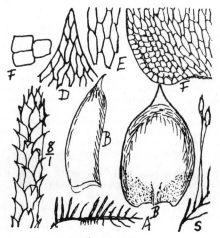

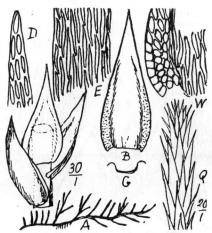

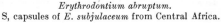

Erythrodontium abruptum.
S, capsules of *E. subjulaceum* from Central Africa.

Platygyrium afrum.

Genus 197. PLATYGYRIUM Br. Eur.

Rather vigorous, dioecious, matted, shining mosses, having long, prostrate, closely leaved, irregularly pinnate stems rooting freely from the under-surface; the branches usually short and simple, seldom longer and branched. Leaves uniform; when moist spreading all round the stem, not plicate, somewhat decurrent, ovate-lanceolate, sharply pointed; when dry imbricate; the margin reflexed up the leaf, smooth; nerve absent; cells smooth; apical cells rhomboid, downward longer; the alar cells numerous, quadrate, and rather large. Seta long, smooth; capsule erect, regular or almost so, narrowly oval, with a short neck; ring wide, dehiscent; lid shortly and obliquely beaked from a conical base; calyptra cucullate, covering half the capsule or more. Peristome double, the two equally long, teeth of the outer lanceolate, widely bordered, not striate; inner peristome with narrowly linear, somewhat split processes, without basal membrane; cilia absent; gemmae sometimes present on the ends of the branches.

606. *Platygyrium afrum* C. M., in Hedw., 1899, p. 133 (see Kew Bull., No. 6, 1823, No. 345) = *Pylaisia africana*, in Rehm. 345. (Brotherus includes *Platygyrium* and *Pylaisia* in Hypnaceae, but *P. afrum* is not shown in either.)

A small, trailing, intricately matted moss, seen only sterile, consequently its genus is uncertain since that depends largely on the presence or absence of membrane to inner peristome. Müller at first named it *Pylaisia* in Rehmann's Exsiccatae, but twenty-five years later wrote, " On account of its julaceous branches it belongs rather to *Platygyrium* than to *Pylaisia*." Stem 1–2 inches long, pinnate or somewhat 2-pinnate, slender, julaceous when dry; leaves all alike except as to size, apparently lanceolate because exceedingly concave, but ovate-acute if spread out, 1 mm. long, with flattened margins above the rising sides, that margin occupied for some distance with small, roundish or quadrate, smooth cells, while the rest of the leaf is occupied by very close and narrow, elliptic-vermicular, smooth cells, shorter toward the leaf-apex. Nerve often absent, occasionally there is a faint trace of short double nerve. Margin entire; leaf usually straight, occasionally oblique or falcate where the stem is rooting.

Cape, W. On rocks, Camps Bay, Rehm.

Genus 198. LEVIERELLA C. M., in Nuov. Giorn. bot. ital., 1897, p. 73.

Autoecious, fairly strong, green or yellowish, matted mosses; main stem stoloniform, rooting occasionally, and with scattered ovate-lanceolate leaves with short nerve, toothed apex, and no differentiated alar cells; secondary stems erect, 1–2 cm. high, dendroid, and few-branched; leaves laxly imbricate when dry, spreading when moist, pellucid or chlorophyllose, somewhat concave, smooth, ovate-acute, narrowly reflexed at the base, toothed from there to the apex; nerve single, ending at or above mid-leaf; cells elliptic, twisted, smooth; the basal marginal cells quadrate, in many rows. Seta 1 cm. high, straight; capsule erect, straight, with dehiscent ring, smooth teeth, no inner peristome, and shortly beaked lid.

607. *Levierella fabroniacea*, var. *abyssinica* (Broth.), Dixon, in Trans. R. Soc. S. Afr., viii, 3, 211; Broth., Pfl., xi, 394.

      = *Levierella abyssinica* Broth., in Pfl., i, 3, 894.
      = *Rozea abyssinica* Broth., in Herbaria (*Rozea* has inner peristome).
      = *Cylindrothecium abbreviatum* Schimp., MS., in herb.
      = *Cylindrothecium* *cyrtocladon Besch., in Herb. Mus. Paris.

Growing in lax, light green, shining mats on bark or humus; main stem prostrate, black, wiry; erect stems 1–3 cm. long, dendroid in habit, subpinnately or fastigiately few-branched; leaves 1–1·5 mm. long, ovate-acute, set all round the stem, or sometimes secund, laxly appressed, and almost julaceous when dry, not plicate, erecto-patent when moist, somewhat concave, with the margin reflexed near the base. Nerve very wide at the base, tapering out beyond mid-leaf, smooth on the back; apex acute; margin sharply dentate from near the base of the leaf; cells with oval to elliptical twisted lumen 30μ long, with twisted line along its middle (SS), chlorophyllose when green, dotted and drawn

Levierella fabroniacea, var. *abyssinica.*

S, point of perichaetal leaf; SS, twisted line in cell lumen.

to one end when dry; marginal cells pellucid, exserted as teeth; lower cells rather longer, except the marginal in many lines which are quadrate and small on the margin, extending shortly up the margin and larger and oblique toward the nerve. Perichaetal leaves high-sheathing; the inner lanceolate-pointed from sheathing base, nearly entire, and almost nerveless. Seta smooth, red, straight, 1 cm. long; capsule erect, oval, with shortly conical-rostrate lid, sometimes oblique. Teeth red, smooth, of about ten joints without mid-line; inner peristome absent; ring deep, enclosing the teeth, dehiscent. Forms of *Porothamnium* sometimes closely resemble this in leaf-form, etc.

Natal. Polela, 4000 feet, 1916, Sim 10,272; Little Berg, Cathkin, 6000 feet, July 1922, Miss Owen 59; Bulwer, 5000 feet, 1919, Sim 10,273.

Transvaal. Kaapsche Hoek, 6000 feet, 1915, Wager 301, 305.

I have it also from Mount Meru, East Africa, and Uganda, and it occurs in Abyssinia. Dixon states that the African variety is slightly more acutely toothed than the Indian, and that he can find no other difference.

Genus 199. STEREOPHYLLUM Mitt., Musc. Ind. Or., 1859, p. 117.

Medium-sized mosses, having prostrate rooting stems, forming mats on moist bark or stones in the tropics and subtropics; the stems pinnate or forked, with blunt branches, and the leaves usually dimorphous, lax, and unsymmetrical, in loosely flattened arrangement; when moist often directed upward; the front leaves absent, the side leaves large and spreading or incurved upward, somewhat concave, ovate-lanceolate or elliptic, blunt or pointed, usually with flat, entire margin, sometimes faintly toothed toward the apex. Nerve single, fairly long, seldom absent; cells laxly rhomboid or linear, usually smooth; the alar cells numerous, quadrate or rhomboid, hyaline or opaque. Seta short or long; capsule inclined or horizontal, irregular, oval, seldom regular, erect or pendulous. Lid shortly pointed or obliquely beaked from a conical base; ring wide and dehiscent, seldom absent. Peristome double; outer teeth lanceolate-subulate, bordered, hyaline above, closely striate, often obliquely striate upward, with zigzag mid-line. Inner peristome hyaline, faintly papillose, with wide basal membrane and keeled lanceolate processes as high as the teeth, split in places.

Stereophylloideae, including *Stereophyllum*, is placed in *Plagiotheciaceae* by Brotherus, Pfl., xi, 396.

Synopsis of Stereophyllum.

Cilia single or absent.
Capsule horizontal; leaves not auriculate; apex rounded; cells not alike . . . 608. *S. odontocalyx.*
Capsule inclined; leaves not auriculate; apex acute; cells alike except at the leaf-base . 609. *S. natalense.*
Capsule erect, regular; leaves auriculate . . . . 610. *S. zuluense.*

608. *Stereophyllum odontocalyx* (C. M.), Jaeg., Ad., i, 541; Dixon, in Bull. Torrey Bot. Club., xliii, 73, 899, and in Trans. R. Soc. S. Afr., viii, 3, 212, and in S.A. Journ. Sc., xviii, 327; Broth., Pfl., i, 3, 899, and xi, 397.

=*Hypnum odontocalyx* C. M., Syn., ii, 232.
=*Hookeria *radiculosa*, in Dregé, Musc. Cap.

Autoecious; stems mostly prostrate, rooting occasionally, 2–3 inches long, very sparsely branched, the branch and stem leaves alike; leaves complanate, usually imbricated, the side leaves distichous and spreading widely in two rows face to face on each side, the upper surface leaves appressed, tending more up the stem, and placed alternately right and left, and the under-surface leaves usually absent, but on young or tender stems the leaves are often distant and the lines difficult to trace, or the leaves are apparently irregular. Side leaves about 2 mm. long, obliquely inserted, each leaf half-clasping the stem, and consequently the lower margin is inflexed (this makes the nerve appear excentric though not so), ovate-oblong, concave below, nearly flat above, shortly and obtusely pointed, or sometimes with a short, obtuse, wide mucro; surface leaves 1·5 mm. long, ovate, obtuse, somewhat concave. Nerve very wide at the base, tapering out above mid-leaf or near the apex, usually coloured; margins entire or minutely denticulate; apical cells with elliptical lumen 25μ long; mid-leaf cells with longer twisted lumen $35–40\mu$ long; basal cells again shorter, and toward the margin gradually reduced to crowded quadrate cells 10μ wide, but nowhere segregated as an alar group; all cells when young chlorophyllose in many bundles but not papillose; when old these disappear, or show as dry dots, or are drawn to the cell-apex; auricle not present; perichaetal leaves long, clasping below, with subulate, almost entire apex, and hardly nerved at the base. Seta 1–2 cm. long, smooth, red, straight; capsule small, brown, high-backed, inclined or horizontal, very unequal, when dry contracted below the mouth; if erect then high-backed and irregular; ring early dehiscent; peristome teeth erect at the base, then strongly incurved and pectinate on the inner surface, red, striate, slightly papillose toward the point, closely barred; the inner higher, wide, folded, thin, pellucid, slightly perforated, on a high membrane and with shorter cilia. Lid shortly beaked; calyptra white, erect, straight, and narrow. C. M. states, "leaf-apex distinctly serrulate," but I do not find it so on any of my many specimens

Long ago Brotherus named, in schedule for the late Dr. J. M. Wood, a Natal specimen, collected in 1898, *S. natalense* Broth., but that name has not been published, and having seen the duplicate retained, I find no difference; and in Trans. R. Soc. S. Afr., viii, 3, 212, Dixon says the same in regard to *S. * Wageri* Broth. (ined.).

S. odontocalyx grows usually in or near water, sometimes on moist soil or bark; when in water it develops longer stems, more lax, scattered, and larger leaves, longer seta, and usually has fewer capsules and fewer branches, but this grades into the other at the water's edge. It occurs at Mount Meru in East Africa, and in Uganda.

Cape, E. Collected by Dregé. Hankey, 2000 feet, 1916, Jas. Sim.

Natal. Weenen County, 3000–4000 feet, March 1898, J. M. Wood.

Transvaal. Elim, Louis Trichardt, 1918, Rev. H. A. Junod; Wolhuters Kop, April 1913, Wager 198; Rosehaugh, Sim; Spelonken, 1918, Junod.

S. Rhodesia. Muzoe, Umtali, Great Zimbabwe Temple Ruins, Victoria Falls, frequent and mostly fertile (Sim, Eyles, Wager), as cited by Dixon; Inyanga, Umtali, Henkel.

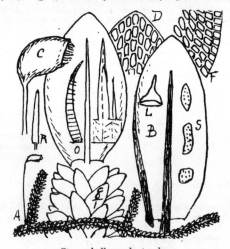

Stereophyllum odontocalyx.
S, apical, mid-leaf, and basal cells.

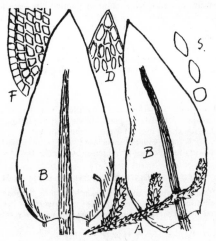

Stereophyllum natalense.
S, apical, mid-leaf, and basal cells.

609. *Stereophyllum natalense* Sim (new species), not of Brotherus, ined.*

Autoecious; stem prostrate, rooting, few-branched, 1–2 inches long, semi-aquatic; leaves complanate, 2 mm. or over in length, less regular in arrangement than in *S. odontocalyx*, which it resembles; upper surface leaves usually largest, widely ovate at the base, then oblong-acuminate; side leaves oblique, the lower side much larger than the other; nerve and leaf bent downward above mid-leaf; leaf almost flat when moist, without folded margin, somewhat crisped when dry. Apex triangular and almost acute; nerve decurrent, very strong at the base, and strong as far as it goes, which is three-fourths leaf-length; margin entire; cells smooth, showing rhomboid-elliptical lumen about 25µ long, all alike except that toward the basal margin they gradually merge into shorter, more rhomboid form, which in the marginal few lines becomes transversely quadrate about 10µ wide. Perichaetal leaves shorter than others, few, lanceolate-acuminate, with denticulate margins, rather longer cells, and no nerve. Seta ½ inch long, red, smooth; capsule small, inclined, brown; peristome and other parts not seen.

Ngoya Forest, Zululand, 1000 feet, 1915, Sim 10,275.

610. *Stereophyllum zuluense* Broth. and Bryhn; Bryhn, p. 18; Broth., Pfl., xi, 399.

Unknown to me, and the description of erect regular capsule and auriculate leaves does not indicate *Stereophyllum*, especially of § *Eustereophyllum*, of which the inclined or horizontal capsule is a character. Bryhn's description includes: Autoecious; male flowers axillary, bud-like; sericeous; stem complanate, to 20 mm. long, prostrate,

---

* *Stereophyllum natalense* Sim (sp. nov.).

Autoecum, semi-aquaticum, prostratum, pauci-ramosum; folia sub-irregularia, oblongo-acuminata, acuta, a basi latiore late-ovata; nervus fortissimus; cellae universae similes praeterquam ad folio-basin ubi sensim minores et transverse quadratae.

rhizoid, laxly leaved. Side leaves distichous, spreading, concave, not sulcate; base obliquely ovate; leaf obliquely elongate-lingulate; apex rounded, triangular, recurved, about 2·2 mm. long and 0·8 mm. wide; margins plane, entire, except apex, where erose-dentate by protruding teeth. Nerve thin, two-thirds length of leaf; leaf-base asymmetric, on one side auriculate. Cells of auricle quadrate, numerous, reaching to near the nerve; other basal cells elongate-rectangular; the upper cells elongate-hexagonal and at length gradually linear; cells all chlorophyllose and smooth. Surface leaves about 2 mm. long and 0·7 mm. wide; base oblique, leaf ovate-lanceolate-lingulate, erecto-patent; apex rounded; nerve rather shorter than the leaf; auricle cells few. Per. leaves three or four, lanceolate-subulate, laxly areolated, the inner without nerve. Seta 10 mm. long, slender, flexuous, reddish. Capsule regular, erect, cylindrical, about 2 mm. long, pale, brown with age. Peristome of the group. Eshowe, Zululand, on tree-trunks, Nov. 1908, scarce, Haakon Bryhn.

Genus 200. ENTODON C. M., in Bot. Zeit., 1844, p. 740.

More or less robust, somewhat rigid mosses, matted on tree-stems or on humus or rocks, with prostrate and spreading, seldom suberect or incurved or reflexed, somewhat rooting stems, with or without stolons and with the leaves numerous and homogeneous, but usually in more or less flattened arrangement, and the stems forked or irregularly pinnately much-branched, the branches mostly short and either blunt or pointed, occasionally some of them longer and pinnately branched. Leaves often crowded, somewhat decurrent at mid-base, concave; side leaves flatly arranged, the back and front leaves often closely imbricate, either ovate or ovate-lanceolate from a narrow base, or oval from an ovate base, blunt or pointed; margin flat or reflexed at the base only, entire or serrulate toward the apex. Nerve double, very short or absent; cells linear, smooth, at the leaf-base laxer and more thickened; the alar cells hyaline, forming a lax quadrate group. Seta long; capsule erect, straight, or nearly so. Lid conical or shortly and obliquely beaked. Calyptra cucullate, beaked, covering half of the capsule. Peristome double; outer teeth lanceolate, not bordered, orange or purple, sometimes perforated; inner peristome without basal membrane, and with rudimentary or linear processes, short or about as high as the teeth; cilia absent; columella usually far exserted and permanent. When dry the point of each leaf is often decurved, giving a narrow branch outline.

Synopsis.

§ *Erythropus.* Peristome teeth papillose, not striate . . . . 611. *E. natalensis.*
§ *Xanthopus.* Peristome teeth striate, not papillose.
    Leaves short, cymbiform, not complanate . . . . . . 612. *E. cymbifolius.*
    Leaves long, widely lanceolate; alar cells numerous; seta 1–2 cm. . . . 613. *E. Dregeanus.*
    Leaves long, ovate-acuminate; alar cells few; seta 2–3 cm. . . . 614. *E. brevirameus.*

611. *Entodon natalensis* (Rehm. 331 and 650); C. M., in Hedw., 1899, p. 133; Broth., Pfl., i, 3, 880, and xi, 389.

Similar in appearance and branching to *E. Dregeanus*, but more compact; leaves shorter, wider, more boat-shaped, and all more appressed, either closely imbricate or laxly placed, usually directed up the stem; margin slightly denticulated near the apex, the basal margin narrowly reflexed, and the base attached up the stem from the decurrent

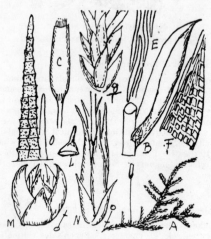

Entodon natalensis.

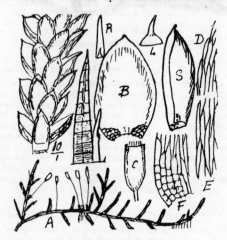

Entodon cymbifolius.
S, side view of leaf.

leaf-centre. Nerve obsoletely 2-nerved; cells vermicular, 50–100μ long, apical and basal cells shorter, all chlorophyllose when green; basal cells rather segregated, quadrate. Perichaetal leaves all erect and sheathing, up to 1 mm. high, with lanceolate-acuminate, almost entire apex. Seta straight, red, smooth, ½ inch high; capsule erect, straight, regular, cylindrical, but widest near the base ; ring not seen (said to be present and deciduous) ; outer per. teeth erect, red below, papillose the whole length, not striate ; inner processes very rudimentary. Columella far exserted. Male flowers small, globose, with few ovate-acute bracts.

Wager lists a var. **longisetum* W. and D. (ined.) from Kaapsche Hoek, which I have not seen ; the long-seta species here is *E. brevirameus.*

Van Reenens, Natal, Rehm. 331 (sterile) ; and Lydenburg, Transvaal, MacLea (fertile), Rehm. 650 ; Kaapsche Hoek, Transvaal (*fide* Wager).

612. *Entodon cymbifolius* Wager and Dixon, in Trans. R. Soc. S. Afr., viii, 3, 211 ; Broth., Pfl., xi, 391.

Autoecious ; growing prostrate in extensive, flat, shining, sericeous, intricate patches ; stems 1–3 inches long, forking occasionally, almost regularly pinnately branched, and rooting where in contact with soil. Branches 3–5 mm. long, simple, turgidly julaceous and crowding together when dry, spreading widely when moist, with the capsules standing erect at right angles, and with numerous short, imbricate, boat-shaped leaves 1 mm. long or less ; branch leaves smaller ; leaves widely ovate, with apiculate reflexed apex, deeply concave, cochleate, placed all round the stem and branches, and not flattened ; margin erect, minutely denticulate near the apex ; nerve usually invisible, occasionally faintly shown as very short and double ; cells narrowly linear, about $50 \times 5\mu$; apical and basal cells rather shorter ; alar cells a small segregated group of quadrate cells about 10μ diameter. Outer perichaetal leaves short and spreading, inner erect, sheathing, nerveless, 2 mm. long, including subulate, almost entire points. Seta single, 1 cm. long, yellow, straight, smooth ; capsule small, erect, cylindric, regular, red, with red teeth widely barred, horizontally striate in the lower part, then slightly papillose, then hyaline, often split or perforated in the upper half. Inner peristome very rudimentary, somewhat attached to the teeth ; columella exserted further than the teeth. Lid shortly and obliquely beaked ; calyptra cucullate, straight, white, dehiscing early.

Natal. Mont aux Sources, 7000 feet, and Blinkwater Forest, 4000 feet, T. R. Sim.

Transvaal. Moorddrift, Waterberg, 1916, Wager 400.

S. Rhodesia. Zimbabwe, 3000 feet, July 1920, Sim 8786, abundantly fertile.

613. *Entodon Dregeanus* (Horns.), C. M., in Linn., xviii, 706 (1844), and in Syn. Musc., ii, 63 ; Broth., Pfl., i, 3, 881, and xi, 391.

= *Neckera Dregeana* Horns., xv, 139 (1841).

Autoecious, shining green, matted and prostrate or suberect ; stems dichotomously forked, 1–2 inches long, rooting occasionally, the forks subpinnately or irregularly and not complanately few-branched ; leaves complanate, the side leaves spreading widely, surface leaves closely appressed, directed more up the stem ; side leaves lanceolate from narrowed base, concave, with acute, often reflexed point ; surface leaves rather wider, not striate ; nerve usually not visible, but when seen it is very faint and double, very short or unequal. Margin slightly denticulate in the upper half, sometimes clearly denticulate ; cells with linear-vermicular lumen, 100μ long or more at mid-leaf, about 50μ long at apex and base ; alar cells segregated, quadrate, fairly large. Perichaetia often many in proximity ; the bracts erect, inner bracts subulate-acuminate, nerveless, almost entire. Seta smooth, yellowish, 1–2 cm. long ; capsule erect, cylindrical, brown, with red mouth-band and no ring ; teeth red, smooth and horizontally striate in the lower half, narrower, incurved, yellow and obliquely striate above, not papillose, and hyaline at the point, usually perforated in the upper half ; inner peristome without membrane, exceedingly variable in length, usually much shorter and rudimentary, but sometimes fairly well developed and longitudinally striate. Calyptra white, naked.

Brotherus includes it in his group with several seta from a flower ; I find only one in every case, and Hornschuch's description does not indicate more though he mentions pistils six to eight with paraphyses, and in the small globose male flower antheridia five to six without paraphyses.

E. Dregeanus Horns., var. *imbricatus,* Rehm. 649 = *E. enervis,* Rehm. 330 (*fide* Rehm.), does not differ, and according to C. M. (Hedw., 1899, p. 133), *E. perpinnatus* C. M. (in Sched., Musc. Macowan, p. 45) and *E. enervis,* Rehm. 330 (and Rev. Bryol., 1878, p. 71, name only), belong here, though, as that was before *E. brevirameus* was separated, it seems more likely that *E. perpinnatus* belongs to it.

Frequent in every moist forest in eastern Cape Province and eastward to Free State and Zambesi ; also recorded from East Africa, Mascarenes, and Belgian Congo.

614. *Entodon brevirameus* Dixon, in Bull. Torrey Bot. Club, xliii, 73, and in Trans. R. Soc. S. Afr., viii, 3, 211 ; Broth., Pfl., xi, 391.

Autoecious ; widely spreading, flat on wet stones, light green, shining, rooting freely, and forking occasionally ; stems 2–3 inches long, almost regularly pinnate ; the branches complanate. 5–8 mm. long ; leaves complanate, 2 mm. long ; side leaves decurrent at mid-base, not striate, spreading nearly to right angles ; others appressed ; all ovate-acuminate, regular, somewhat concave, with subulate apex, expanded margin denticulate at the top of each long marginal cell, fairly closely denticulate near the apex where the cells are shorter. Nerve usually absent ; occasionally there is a trace of a faint, short, double nerve ; cells smooth, narrowly linear up to $100 \times 5\mu$, when young containing two to three close rows of small, round, chlorophyll grains ; alar group small, its cells quadrate, chlorophyllose when young, rather rectangular, but sometimes also spreading across the base. Seta 2–3 cm. long, straight, pale red, smooth ; perichaetal leaves few, erect or spreading, lanceolate from sheathing base ; capsule erect, brown, 3 mm. long, cylindric ; ring absent ; teeth red, sometimes split, obliquely striate all the length ; inner peristome without membrane, processes shorter than the teeth, much more slender, pale, longitudinally striate. Spores smooth. Columella exserted further than the teeth and remaining so permanently. Plant semi-aquatic in habit.

Natal. Ihluku, Harding, 3000 feet, Jan. 1920, and Edendale Falls, 3000 feet, Sim ; Lynedoch, Professor Bews ; Maritzburg, Sim.

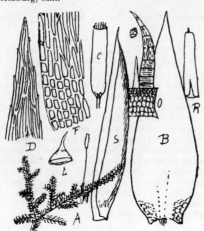

Entodon Dregeanus.

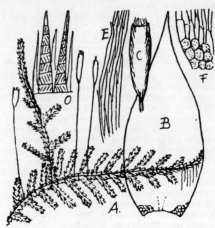

Entodon brevirameus.

Transvaal. Lemona Wood, Spelonken, and the Downs, Pietersburg, Rev. H. A. Junod ; Kaapsche Hoek (*fide* Wager).

S. Rhodesia. Zimbabwe, Sim.

Wager's Barberton specimen (254), recorded in Trans. R. Soc. S. Afr., viii, 3, 210, as *E. geminidens* (Besch.), Broth. (=*Cylindrothecium geminidens* Besch., in Flor. Bry. Reunion, p. 293 or p. 152), probably belongs here, at least there is nothing in Bescherelle's description to prevent this belonging to his plant, but specimens may show a difference.

Family 50. Hypnaceae.

Exceedingly varied in habit ; plants large or small, abundantly rooted, subsimple or much branched, the ramification scattered or pinnate, dendroid, creeping, prostrate, procumbent, suberect or erect ; stems wiry, often stoloniform ; paraphyllia seldom present. Leaves in many rows, spreading all round, squarrose or complanate, often secund or falcate-secund, of many forms, never round, sometimes unsymmetrical, of one layer of cells, with or without nerve, the nerve when present weak and short, simple, double or forked, seldom excurrent ; leaves often subscarious, usually smooth and shining, with long, narrow, linear or vermicular, and sometimes papillate cells ; the basal cells more lax and discoloured, and the alar cells quadrate, usually hyaline, forming a distinct group. Autoecious or dioecious, rarely synoecious. Calyptra cucullate, smooth, or rarely with a few hairs. Vaginula usually naked. Capsule on a long, usually smooth seta, usually cernuous or horizontal, never pendulous, more or less incurved, rarely erect and regular, usually smooth. Peristome double, the two of equal height ; the outer of sixteen teeth, united at the base, usually bordered and striated, with zigzag mid-line ; the inner with a narrow or wide basal membrane, 16-plicate, and the keels continued through the processes, and cilia are also often present. Spores small, smooth.

Hypnaceae has from time to time varied immensely as to its definition, and even now Dixon and Jameson (2nd and 3rd ed.) include thereunder all the British species of pleurocarpous mosses with elongated smooth areolation and well-developed bryoid peristome, except such as for other reasons come under Neckeraceae. By doing so, Brachytheciaceae and Sematophyllaceae as used here are included, and they further remark : " Several of the genera here included with erect symmetrical capsules are frequently separated under the title of Orthotheciaceae, but the distinction cannot be held of real importance since it leads to the separation of genera obviously closely allied."

The classification used here is that adopted by Brotherus (Nat. Pfl., 1st ed.), and used by him again in Marloth's Flora of South Africa, but it is widely departed from in Pfl., xi, 445, without making classification clearer.

Synopsis of the Sections of Hypnaceae.

Leaves usually dimorphous, symmetrical, transversely inserted ; nerve double or absent ; stem leaves usually from
 a wide base, drawn to a long or short point . . . . . 2. Hylocomieae.
Stem leaves and branch leaves little different.
 Leaves transversely inserted, symmetrical ; nerve single, more or less long, seldom double ; lid never beaked,
                 1. Amblystegieae.
 Leaves transversely inserted and symmetrical, or somewhat obliquely inserted and more or less unsymmetrical ;
  nerve double or absent ; lid sometimes beaked . . . . 3. Stereodonteae.
 Branches mostly flattened-leaved ; side leaves obliquely inserted, spreading complanately, usually unsymmetrical ;
  nerve double or absent ; lid conical or short, seldom long-beaked . . 4. Plagiothecieae.

Subfamily 1. AMBLYSTEGIEAE (now separated as AMBLYSTEGIACEAE, in Broth., Pfl., xi, 332).

Stems irregularly or 1-pinnately branched, without stolons; the branches with leaves set all round. Leaves transversely inserted, symmetrical; stem and branch leaves almost alike, the latter usually only smaller and with weaker nerve. Nerve single, more or less long, seldom double, short or absent. Cells smooth, very seldom papillose; alar cells differentiated. Seta smooth. Capsule usually inclined or horizontal, cylindrical, curved, often 2-coloured. Teeth of the outer peristome often toothed like steps toward the apex. Lid shortly conical and mucronate from an arched base.

Synopsis of Genera of Amblystegieae.

Leaves bordered   .   .   .   .   204. SCIAROMIUM.
Leaves not bordered.
   Cells hexagonal, 2–4–6 times as long as wide (in the South African species); leaves secund,
                                                       201. HYGROAMBLYSTEGIUM.
   Cells laxly quadrate or oblong-hexagonal, seldom linear; leaves spreading widely   .   202. AMBLYSTEGIUM.
   Cells long and linear, usually very narrow; leaves usually falcate-secund   .   .   203. DREPANOCLADUS.
   (Cells long and linear, usually very narrow; leaves erecto-patent or imbricate, ovate-oblong or rounded, or mucronate; alar cells lax and numerous; nerve double, very short or absent   .   .   ACROCLADIUM.)

Genus 201. HYGROAMBLYSTEGIUM Loesk., Moosfl. d. Harz., 1903, p. 298.

Autoecious or dioecious, slender or fairly strong, usually stiff mosses, forming tufts, or floating or pendent. Stem long, rooting, usually inundated, and often with the leaves skeletonised, seldom prostrate or erect, usually regularly pinnate; the pinnae simple; paraphyllia usually absent; leaves fairly close, spreading on all sides or secund, concave, not furrowed, seldom but sometimes long-decurrent, usually ovate or oblong-lanceolate, long-pointed; margin flat and entire or distantly serrulate; nerve straight, seldom reaching the apex, but sometimes excurrent as a point; cells hexagonal, 2–4–6 times as long as wide; the alar cells differentiated and forming segregated corners. Perichaetium rooting; seta long or very long; capsule inclined or horizontal, oblong-cylindrical, curved; lid arched and mucronate; ring present. Peristome double; teeth widely bordered, striate, narrowed by steps upward, pale and papillose at the point, with numerous wide lamellae; inner peristome finely papillose, with wide basal membrane; processes keeled, usually split in places. Cilia strong, knotted, or shortly appendaged. Spores small.

Synopsis of Hygroamblystegium.

Stems mostly simple and slender; alar cells indistinct   .   .   .   .   615. *H. caudicaule.*
Stems pinnate, dwarf; alar cells few and pellucid   .   .   .   .   616. *H. filicinum.*

615. *Hygroamblystegium caudicaule* (C. M.), Broth., Pfl., i, 3, 1028, and xi, 337.

        = *Drepanophyllaria caudicaulis* C. M., in Hedw., 1899, p. 150.
        = *Hypnum *pendulum*, Rehm. 404 (not C. M.).

Hanging in the drip of waterfalls in masses; stems slender, mostly simple, 4–6 inches long, or only slightly branched with slender branches; not rhizoid, without paraphyllia, pendulous, but with the branch-tips and often the leaf-tips reflexed or falcate; all leaves appressed when dry, spreading widely when wet; leaves 1·5–2 mm. long, the stem leaves short and wide (up to 1 mm. wide), ovate-acuminate, acute, concave, with strong yellowish nerve to the usually reflexed apex; margin entire, and there is usually a small indefinite group of small, quadrate, pellucid alar cells; branch leaves longer and narrower, with long narrow subula filled with nerve, ovate-lanceolate but apparently lanceolate because concave; nerve yellowish, strong to the apex, usually curved backward, sometimes falcate, especially among apical leaves; alar cells very indistinct, all cells rhomboid-hexagonal, about $30 \times 10\mu$; alar cells about $15 \times 15\mu$. Inflorescence and fruit not seen.

   Müller says " without alar cells," but I do not find Rehmann's plants so.

   Cape, W. Cataract at Devils Peak, 1875, Rehm. 404, and again 1919, Sim 7962; Kloof near Chaplin Point, Cape Peninsula, Sim 9374.

616. *Hygroamblystegium filicinum* (Linn.), Loesk.; Broth., Pfl., i, 3, 1028, and fig. 740 (but not in Pfl., xi, 337).

        = *Amblystegium filicinum* De Not.; Dixon, 3rd ed., p. 497, Tab. 57c.
        = *Hypnum filicinum* Linn.; Schimp., Syn., etc.

As known here, this is a dwarf, terrestrial, bright yellow species, with stems 2–3 cm. long, simple or pinnately branched, rooting at the base only, scrambling among other mosses (e.g. *Dicranella Symonsii* Dixon), and destitute of paraphyllia, but in Europe it sometimes floats out and is much larger and more subject to produce rhizoids and paraphyllia along the stem. Stem and leaves not complanate, the larger stems with decurrent, straight or falcate leaves 1·75 mm. long, or shorter and wider, narrowly attached, very concave at the base, spreading widely when moist, appressed and almost falcate when dry, with yellowish nerve to near the apex, and minutely denticulate or subentire, with rhomboid-hexagonal cells about $50 \times 8\mu$, the lower wider, and the alar cells few, up to $20 \times 12\mu$, with a few small, roundish cells between them and the lamina cells; all cells thin-walled and usually pellucid; the smaller branches with leaves 0·5–1 mm. long, scattered, spreading outward and then upward, somewhat concave below, flat above, subulate, pellucid, often entire, and usually without distinct alar cells and often somewhat falcate. Dioecious; not seen fertile here as yet, but said to have seta 1¼–2 inches long; capsule subcylindrical, rather turgid, arcuate, and lid conical apiculate.

Natal. Upper Bushman's River, 7000 feet, Sept. 1905, Sim 8670β; Mooi River at Rosetta Farm, 4000 feet, Aug. 1918, T. R. Sim.

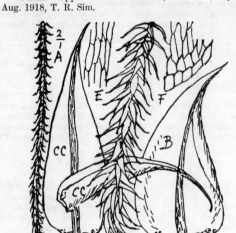

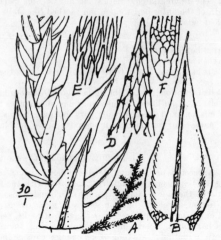

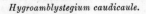

Hygroamblystegium caudicaule.　　　　　　*Hygroamblystegium filicinum.*

Genus 202. AMBLYSTEGIUM Br. Eur.

Usually small or very small, but occasionally more or less robust mosses, growing in lax mats on the ground, on stones, or on tree-stems, never stoloniferous, but extending from the apex of the stem and branches. Branches prostrate, irregularly forked or pinnated. Paraphyllia absent. Leaves in five to eight rows, usually set all round, seldom complanate, patent and subsecund, cordate, oval or oblong-lanceolate, long-pointed, not bordered, concave, not plaited, entire or toothed. Stem- and branch-leaves almost alike. Nerve thin, single, ending at or beyond mid-leaf. Areolation lax; cells loosely quadrate or oblong-hexagonal, never linear or vermicular, smooth; alar cells quadrate or rectangular, forming distinct, almost concave angles. Perichaetal branch rooting, very short; perichaetal leaves similar to stem leaves, serrated at the apex. Autoecious; antheridia oval, small, without paraphyses. Seta long, erect, suberect or incurved, smooth. Capsule oblong or cylindrical, incurved, smooth; lid large, tumid-conical, obtuse but mucronate. Ring present; peristome double; teeth lanceolate-subulate, closely jointed, bordered, striate; upward papillose and toothed; processes usually keeled, not split; basal membrane wide; cilia usually present, knotted or appendaged.

617. *Amblystegium riparium* (Linn.), B. and S., Bryol. Eur.; Broth., Pfl., i, 3, 1024 (not xi, 339); Schimp., Syn., ii, 717; Braithw., etc.

=*Hypnum riparium* Linn., Sp. Pl., p. 1129; C. M., Syn., ii, 321; Dixon, 3rd ed., p. 503, Tab. 57e, etc.
=*Hypnum *fluitans* Rehm. (not Linn.), and its vars. *rivulare* and *longifolium*, Rehm. 399, 400.
=*Amblystegium riparium*, var. *rivulare*, Rehm. 668, 669 (the longer floating-form).

Autoecious, dark or light green, very variable, prostrate on mud, 1–2 inches long and irregularly pinnate, or submerged, longer and unbranched, less fertile, and with more lax pectinate leaf-arrangement. Leaves all round the stem, erecto-patent or spreading whether dry or wet, 2–2·5 mm. long, ovate-lanceolate in the lower half, lanceolate-acuminate in the upper half, the basal brown leaves shorter and wider than those on young shoots, but branches begin growth with a few small rounded leaves. Leaves decurrent, shortly connected, entire or, in the lower half, minutely denticulate, somewhat concave below, flat above, usually pellucid or lightly chlorophyllose unless old and crusted, in which case they are brown and opaque; nerve extending about three-fourths of leaf-length; cells variable, the mid-leaf cells linear, 50–100×8–10μ, decreasing in length to $50 \times 10\mu$ downward, and gradually wider and shorter to $20 \times 15\mu$ at the base, where they are lax and rounded and not segregated as alar cells. Perichaetium rooting. Perichaetal leaves similar to stem leaves, erect; perigonial buds axillary, subglobose, and short-bracted; seta ½ inch long, red, straight, smooth; capsule horizontal, curved, irregular; teeth red, bordered, inflexed, the lower part closely lamellate and distinctly cross-striate; above the middle pale and punctate, with bars extending out; inner peristome with high lamellae, pale, plaited, and minutely papillose.

Cape, W. Woodbourne, Knysna, Miss Duthie 50.
Natal. Kosi Bay, Zululand, July 1920, Aitken and Gale; Giant's Castle, 7000 feet, R. E. Symons; Greenfield, Mooi River, floating in a very cold spring pool, 3500 feet, Jan. 1918, Sim 9664.
O.F.S. Kadziberg, Rehm. 399, 399β, 400.

Transvaal. Macmac, MacLea (Rehm. 668, 669).

Amblystegium filiforme Wager and Wright (Broth., Pfl., xi, 340), as described and figured in Trans. R. Soc. S. Afr., iv, 1, 2 (1914), has no relationship with *Amblystegium*, as is shown by the round or roundish cells.   One named packet

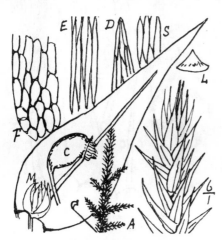

Amblystegium riparium.
S, supra-basal cells.

Amblystegium riparium, var. *rivulare.*
S, point of branch leaf ; F, one alar cell.

of this supplied by Wager contains *Haplocladium angustifolium* and *Pseudoleskea Macowaniana* mixed ; probably the latter was the plant.

The authority on which *Amblystegium varium* Lindb. was included in my check-list was a former identification of Eyles 402 as that species, but it has since been described as distinct and as *Isopterygium aquaticum* Dixon.

Genus 203. DREPANOCLADUS (C. M.), Roth., in Hedw., xxxviii, Beibl., 1899, p. 6.

Dioecious, seldom autoecious, fairly strong, shining, swamp or peat mosses, with prostrate, procumbent, suberect, pendent in drip or floating stems, usually without rhizoids ; irregularly or regularly, closely or distantly pinnated, seldom almost single, usually with falcate branch-points ; pseudoparaphyllia scarce only near where the branches rise. Leaves usually falcate-secund or hooked, seldom erect or spreading, more or less concave, with or without furrows, ovate or cordate-lanceolate from a narrowed and somewhat decurrent base, short-pointed or subulate-pointed, entire or slightly denticulate. Nerve single, narrow, and reaching to mid-leaf, or stronger and ending in the subula, seldom excurrent. Cells usually long-linear, smooth, the alar cells laxly hexagonal, forming distinct, hardly concave groups. Seta long or very long ; lid convex-conical, with small point. Capsule inclined or horizontal, cylindrical, curved, smooth ; when dry, contracted below the mouth. Ring usually differentiated.

Synopsis of Drepanocladus.

Every leaf strongly uncinate .	.	.	.	.	.	.	.	618. *D. uncinatus.*
Leaves straight or oblique, or apical leaves homomallous-falcate.								
Aquatic, strong ; leaves ovate-acuminate ; cells very narrow		.	.	.	619. *D. sparsus.*			
Erect, slender, lanceolate-acuminate from wide base ; cells wider	.	.	.	.	. 620. *D. Hallii.*			

618. *Drepanocladus uncinatus* (Hedw.), Warnst., in Beitr. = Bot. Centralb., xiii, 397 (1903) ;
Broth., Pfl., i, 3, 1033, and xi, 343.

=*Hypnum uncinatum* Hedw., Descr., iv, 65, t. 25 ; C. M., Syn., ii, 322 ; W. P. Sch., Syn., ii, 738, etc.

Laxly caespitose, yellowish, autoecious ; stems 1–2 inches high, suberect or erect, pinnate or 2-pinnate ; branches short ; branch-points and every leaf strongly uncinate ; leaves 3 mm. long, concave throughout, keeled, decurrent, tapering from a wide, somewhat auricled base to the point, and deeply furrowed whether wet or dry almost the whole length. Nerve narrow but distinct and extending well into the subula. Margin slightly denticulate, often appearing entire ; cells linear-vermicular, pointed, thin-walled, rather wider and shorter toward the leaf-base, and with quadrate or rhomboid hyaline cells of various sizes forming the indistinct auricle. Perichaetium sessile on the stem, erect, nearly white ; its leaves very long, sheathing, acuminate, plicate ; seta 1½ inch high, red ; capsule subcylindric, inclined, bent, red ; ring present ; lid shortly rostrate ; peristome complete.

O.F.S. Rydal Mountain, Dec. 1910, Wager 183.

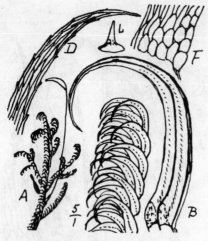

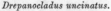

Drepanocladus uncinatus.

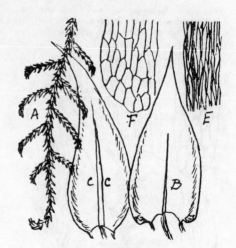

Drepanocladus sparsus.

619. *Drepanocladus sparsus* C. M., in Hedw., 1899, p. 151 ; Broth., Pfl., xi, 344.

=*Hypnum sparsifolium*, Rehm. (not Hampe).
=*Hypnum exannulatum*, Rehm. 397 (*fide* Dixon).

A submerged or drip aquatic several inches long, growing in masses, and either simple or more or less pinnated, with branches about ½ inch long, spreading widely ; stem leaves 1·5 mm. long, widely ovate-acuminate, deeply concave, not furrowed, spreading, decurrent, with entire margins, slender nerve half leaf-length, and long, narrow, rather obscure lamina cells up to $100 \times 8\mu$, those nearer the base shorter and wider, and those in the alar groups shortly hexagonal, $20 \times 15\mu$ and less. Branch leaves scattered, spreading, mostly pellucid, deeply concave, often falcate and long-pointed, nerved three-fourths leaf-length, entire, and with lamina and basal cells similar to the stem leaves, but often no alar cells. The lamina cells are much more narrow than those of *Amblystegium riparium*. It is not known fertile.

Brotherus (Pfl., xi, 344) places *D. sparsus* C. M. and *D. afrofluitans* (C. M.) as South African representatives of his group *Drepanocladus* (and also of his group *Warnstorfia*), but to my mind the shorter wider cells of the latter remove it to *Amblystegium*.

Cape, E.   Kokstad, May 1917, B. O. D. Mogg.

Natal.   Fields Hill, Rehm. 397 (as *Hypnum afro-exannulatum* Rehm.).

O.F.S.   Kadziberg, Rehm. 398.

Rhodesia.   Makoni, 4700 feet, June 1917, Eyles 787.

Concerning *Hypnum *Worthii*, Rehm. 393 (Inchanga, Natal), Dixon, after seeing the specimen, writes : " It is a *Campylium* I should say, with colour and habit of the larger forms of *C. polygamum* ; very glossy leaves, usually very finely longitudinally striate ; cells *very* narrow, basal fewer than in *C. polygamum*, but well defined. The only doubt as to its validity as a species would be its resemblance to *Drepanocladus sparsus*, non-falcate-leaved form, but I do not think it is that." His sketch agrees with mine of *D. sparsus.*

Campylium and *Drepanocladus* are very closely related, and the above remarks are insufficient to do anything further with *Hypnum Worthii* Rehm.

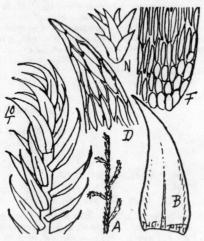

Drepanocladus Hallii.

620. *Drepanocladus Hallii*, Dixon, in Bull. Torrey Bot. Club, xliii, 77 (1916).

A small, yellowish, dioecious mud-moss 1–2 inches long, with few-branched suberect stems about 5 mm. long, spreading and somewhat incurved at the point, or larger and floating ; leaves about 1–2 mm. long, lanceolate-acuminate from wide base, concave, homomallous-falcate at the branch-points ; lower the leaf is either oblique or straight, widely attached to a thick axis, spreading, entire, with nerve often faint and usually ending at or about mid-leaf. Apical cells tumid, elliptical, chlorophyllose ; lamina cells longer and narrower ($50–100 \times 8\mu$), reduced again near the base to rhomboid, and with the alar cells numerous, rectangular or laxly ovoid, grouped in the corners or extending nearly across the base, pellucid but thick-walled. Branch leaves narrower, more pellucid and more acuminate. Female flowers abundant, small, with short, nerveless, spreading, pellucid leaves. Male flowers and fruit not seen.

Cape of Good Hope, 1912, S. W. Hall.

Natal. Maritzburg, Rehm. 675 ; Laing's Nek, Rehm. 675β, both as *Hypnum *brevifolium* Rehm. Paris, on account of the earlier and different *Hypnum brevifolium* Lindb., changed this to *H. *natalense*, overlooking that Rehmann had already used that name for his 674 (*Rhaphidostegium Dregei*), but all these African names *brevifolium* and *natalense* are unpublished.

Genus 204. Sciaromium Mitt., Musc. Austro. Amer., p. 571.

Fairly strong, simple or slightly branched aquatics, having oval leaves set all round the stem, imbricate, and strongly bordered, the borders and strong nerve uniting to form a firm oblique point ; paraphyllia absent. Fruit not seen yet in South Africa, but from South America it is described as having seta 1–3 cm. long ; capsule inclined, irregular ; ring present ; peristome teeth yellow, bordered, striate, at the point pale and papillose, and the joints extruding ; inner peristome with high membrane and narrow, keeled, perforated processes, and cilia one to three, short, knotted. Lid convex-mucronate. Our plant belongs to § *Limbidium*, which is mostly South American, and has the leaves entire and border strong.

621. *Sciaromium capense* Dixon, in Trans. R. Soc. S. Afr.,
viii, 3, 216, and pl. vii, fig. 17 ; Broth., Pfl., xi, 339.

A black aquatic with simple or shortly few-pinnated stems several inches long, densely set all round with erect or almost erect imbricated leaves, sometimes homomallous at the point, especially toward the branch-point. Leaves nearly oval, with long, usually oblique point, 1·75 mm. long, 0·5 mm. wide, entire, flat or somewhat keeled, more or less dense throughout, but especially the very wide nerve and margins which are thick and solid and of several strata, and at the top of the lamina these unite to form the dense points. Cells with elliptic incrassate lumen $15 \times 6\mu$, rather longer toward the leaf-base ; those on the nerve and borders similar. Other characters not seen.

Cape, E. Near Hogsback, Tyumie, 4000–6000 feet, 1916, D. B. and M. Henderson 213.

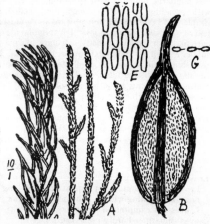

Sciaromium capense.
G, section across leaf.

Subfamily 2. Hylocomieae.

Stem regularly 1–2–3-pinnate, always without stolons ; branches with leaves set all round. Leaves transversely inserted, symmetrical, usually dimorphous, sometimes somewhat furrowed. Stem leaves usually long-pointed from a widely oval or cordate base. Nerve double, short or absent. Cells long, prosenchymatous, often papillose at the upper end. Capsule usually inclined or horizontal (seldom hanging, or erect and regular), oval or oblong, not curved, usually firm-skinned. Peristome normal. Lid usually conical, seldom beaked.

The South African species all have paraphyllia scarce or absent, but the genus *Hylocomium*, which extends south as far as Northern Africa, has the stem surface densely coated with them.

Hylocomiaceae is now (Broth., Pfl., xi, 483) separated as a family including only *Leptohymenium* of those mentioned below.

Synopsis of Subfamily Hylocomieae.

Leaves more or less spreading, pointed ; seta smooth ; alar cells differentiated ; lid conical or shortly beaked.

(Capsule erect, regular ; paraphyllia absent . . . . . Leptohymenium.)

Capsule inclined or hanging, more or less irregular ; leaves dimorphous or heteromorphous ; stem leaves lanceolate-subulate . . . . 205. Microthamnium.

Stem leaves imbricate, rounded or with short bluntish point ; alar cells numerous, forming a concave group above, but the *Hypnum* restricted definition is not now adhered to by Brotherus, who now includes in *Hypnum* what are here separated as *Stereodon*.

(Genus Leptohymenium Schwaegr., Suppl., iii, 1, 2, t. 246.

Although Paris gives the same source as the authority for this name as Brotherus does, it is evident that the two adopted different definitions, for while Brotherus (Pfl., i, 3, 1052) only mentions two species of this genus—both Asiatic—Paris places here twenty species, several of which are African, including two said to be South African. These are :

(1) *Leptohymenium Breutelii* (W. P. Sch., MSS.), Jaeg., Ad., ii, 346 = *Neckera Breutelii* C. M., in Bot. Zeit., 1858, p. 165 (Prom. B. Sp.), concerning which I know nothing further, except that it is not represented in London ; and

(2) *Leptohymenium dentatum* Schw., Suppl., iv, 332 = *Neckera dentata* C. M., Syn., ii, 88 = *Pleuropus dentatus* Griff. in Schw., Suppl., iv, 332.

27

Müller's description of the latter, in his section " Leaves 2-nerved " is : " Monoecious ; prostrate, compact and intricate, dark green, pinnate with many short, slender, densely leafy branches ; stem leaves erecto-patent, lanceolate-acuminate, entire, and obsolete-nerved ; cells narrow, alar cells laxly quadrate, pellucid ; perichaetal leaves sheathing, long acuminate, the outer smaller ; seta long, flexuous, rusty brown, rigid ; capsule erect, short, subturgid, oblong-oval, brown, with the mouth constricted ; lid slightly conical, oblique ; ring absent ; peristome as in *N. abyssinica* " (=*Entodon Schimperi* Hpe.).

This has not been placed and is not represented in London or South African herbaria. The locality is " Prom. B. Spei."

Genus 205. MICROTHAMNIUM Mitt., Musc. Austr. Amer., 1869, p. 21.

Rather slender, but firm, matted, autoecious or seldom dioecious mosses having prostrate stems, either more or less regularly pinnate, with spreading, short, straight, closely leaved, blunt branches (§ *Pseudomicrothamnium*), or with spreading, more or less long and crooked shoots, sometimes rooting at the tip, which at the base are subdendroid, simple and laxly leaved, but upward have clustered pointed branches which are 1–2-pinnate and frequently closely leaved (§ *Eumicrothamnium*). Paraphyllia absent ; leaves dimorphous, sometimes furrowed, usually shortly decurrent, with flat and usually serrulate margin. Stem leaves more or less spreading, cordate-ovate or lanceolate-subulate from a widely ovate or cordate base. Nerve double, very short or reaching to or near mid-leaf, or absent ; cells shortly linear, usually papillate in the upper angle ; those at the leaf-base shorter and more lax, not discoloured ; the alar cells differentiated, small, oval or rectangular. Branch leaves smaller, shorter pointed, serrate. Perichaetium rooting ; seta long, smooth ; capsule inclined or hanging, oboval, oval, oblong, or cylindrical, more or less irregular, smooth. Ring differentiated ; lid more or less arched or conical and shortly pointed or short-beaked, seldom long-beaked. *Microthamnium* Mitt., *Stereohypnum* Hampe, *Mittenothamnium* Henn., and *Rhizohypnum* Hampe, are all synonyms, but the first is established by use and familiarity, as well as by priority, and is retained meantime (see footnote, Dixon, Journ. of Bot., 1915, p. 20, and more finally in Journ. of Bot., lx, 281, Oct. 1922).

Microthamnium and *Acanthocladiella* are now put in § Ctenidioideae of Hypnaceae, in Broth., Pfl., xi, 466.

Synopsis of Microthamnium.

Leaves strongly papillose.
Leaves ovate-acute, very concave ; beak long	622. *M. cavifolium.*
Leaves spreading widely, subcompressed, lax ; calyptra hairy ; beak short . . .	623. *M. patens.*
Leaves not compressed ; calyptra hairy	624. *M. horridulum.*

Leaves smooth or hardly papillose.
Leaves ovate-acute, very concave ; seta 1·5 cm. long; beak short . . .	625. *M. pseudoreptans.*
Leaves ovate-acute ; seta very short	626. *M. squarrosulum.*
2-pinnate ; leaves ovate-lanceolate, denticulate	627. *M. cygnicollum.*
Mostly 1-pinnate ; leaves ovate-lanceolate, entire	628. *M. ctenidioides.*

622. *Microthamnium cavifolium* (Rehm.), Dixon, in Journ. of Bot., 1915, p. 21.

=*Eurhynchium cavifolium*, Rehm. 368 (not *Microthamnium cygnicollum* (Hpe.), as implied by C. M., in Hedw., 1899, p. 144, *re* Rehm. 368).
=*Microthamnium *lutescens* Mitt., MSS., in Hb. Mitt. (*fide* Dixon).

Prostrate, scattered among other mosses, or laxly matted, not rooting, pale green, shining, irregularly or more or less regularly pinnate ; the pinnae distichously arranged, subcompressed, up to 1 cm. long, and occasionally again subpinnate. Leaves 1 mm. long, wide and deeply concave, cordate-ovate, acuminate, from rather a wide base, erecto-patent and cochleate, with the margin often flattened out, and the point flattened or reflexed, or longer and flexuose. Nerve double, fairly strong in some leaves, more often absent. Margin denticulate the whole length, sharply so in the upper half. Cells shortly and widely linear throughout, all drawn to the upper end of the cell, which overlaps and projects, making the leaf on both surfaces densely scaberulous. Alar cells small, quadrate, making a distinct group. Perichaetal leaves erect, sheathing, lanceolate-acuminate, slightly denticulate. Seta red, 1·5–2 cm. long, cygneous at the top making the capsule horizontal. Capsule brown, oval with slight neck, often contracted below the mouth ; lid long-beaked, as long as the capsule ; calyptra nude, white ; peristome normal.

Natal. Buccleuch, Cramond, April 1918, W. Leighton ; Inanda, Rehm. 368, and Inanda, J. M. Wood (Rehm. 654) ; Maritzburg, Sim, and also Wager.

Transvaal. Lechlaba, Houtbush, Rehm. 654β and 655 ; Macmac, MacLea (Rehm. 655β).

623. *Microthamnium patens* (Hpe.), Jaeg., Ad., ii, 472 ; Dixon, Journ. of Bot., 1915, p. 21 ; Broth., Pfl., xi, 471.

=*Chrysohypnum patens* Hampe.
=*Eurhynchium *brevirostre*, Rehm. 366 ; Par., Ind., 1895, p. 441, *fide* C. M.

Plant small, mostly prostrate, 2-pinnate ; leaves of both stem and branches lax, spreading at right angles to the axis and all round it, ovate-acuminate, subcompressed, 1–1·5 mm. long, with expanded, slightly denticulate margin, the double nerve only sometimes visible and very short. Cells shortly oblong-linear, drawn to the upper end, and there projecting on the lower surface which is very rough thereby. Alar cells rather wider but hardly different. Perichaetal leaves few, small, spreading and narrow, surrounding long paraphyses. Seta red, smooth, 1 cm. long, erect to near the top; capsule small, horizontal, brown; beak shortly rostellate; calyptra hairy; peristome normal; the teeth strongly pectinate on the inner surface where they bend inward.

Cape, E. Pondoland, July 1899, T. R. Sim.

Natal. Inanda, Rehm. 366; Ekombe, Entumeni, Zululand, Bryhn 22; Transvaal, Pilgrims Rest, MacLea; Zwart Kop, 4000 feet; Giant's Castle, 8000 feet, Sim.

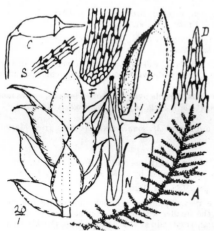

Microthamnium cavifolium.

S, projecting cell-ends.

Microthamnium patens.

624. *Microthamnium horridulum* Broth., in Eng. Bot. Jahrb., 1897, p. 263, and in Pfl., i, 3, 1049, and xi, 471.

Unknown to me, but in describing *M. ctenidioides* (Trans. R. Soc. S. Afr., xviii, 217) Dixon says, " This is a slender species with stellately spreading leaves, in habit resembling some *Ctenidia*; *M. horridulum* Broth., from Pondoland, is near it, but has more shortly pointed leaves, wider at the base. (*M. Shawii* Rehm., ined., is very similar to *M. horridulum*, if indeed distinct.)" He indicates also that it is autoecious. In Pfl., i, 3, 1049, Brotherus adds : " Calyptra hairy; leaves spreading, widely oval, lanceolate or subulate-pointed; cells papillose, alar cells small; seta 1 cm. long; capsule oval; lid conical; branches hardly flattened; dry capsule contracted under the mouth."

M. Shawii mentioned above is *Eurhynchium Shawii*, Rehm. 367 and 656β, both from Inanda, Natal (Coll., J. M. Wood), and disposed of above.

The only difference shown by Brotherus (Pfl., i, 3, 1049) from *M. horridulum* by *M. afro-elegantulum* C. M. (in Dusen, M. Cameroons, No. 283; Broth., in Engler's Bot. Jahrb., 1897, p. 263), which is also included in Bryhn's list from Ekombe, Zululand, is that the dry capsule is not contracted below the mouth. I doubt the stability of this character.

625. *Microthamnium pseudoreptans* (C. M.), Jaeg., Ad., ii, 494; Dixon, in S.A. Journ. Sc., xviii, 329; Broth., Pfl., i, 3, 1051; Dixon, Trans. R. Soc. S. Afr., viii, 3, 218.

=*Hypnum pseudoreptans* C. M., in Bot. Zeit., 1859, p. 439, and Rehm. 418 and β.

=*Hypnum glabrifolium* C. M., in Flora, 1890, p. 496, *fide* Dixon.

=*Microthamnium glabrifolium* Par., Ind., p. 809.

=*M. reptans* (Sw.) Mitt., var. *pseudoreptans* (C. M.) Dix.; Broth., Pfl., xi, 472.

Autoecious, mostly prostrate, forming considerable mats on soil or humus, in which the 2-pinnate stems 2–3 inches long bend over and root at the point, while occasional new stems 2–3 inches long rise dendroid-fashion from these, branchless below, closely and not or hardly complanately branched, 2-pinnate above, clad with small concave or sometimes almost cochleate leaves which are not or hardly compressed and not closely imbricate, and in the more lax conditions are more or less spreading when moist and laxly suberect when dry. Stem leaves ovate-acute and about 1 mm. long on the prostrate and erect stems, very concave with rather long point on the main stems; margins erect, denticulate in the upper half; nerve usually absent; cells vermicular, 50–70μ long, with very few indistinct papillae, and hardly changed to top or base, but with three or four large alar cells about 35×12μ at each corner, with a few

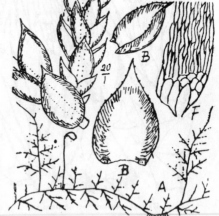

Microthamnium pseudoreptans.

much smaller quadrate cells above them. Perichaetal leaves longer, narrower, and more subulate and denticulate, erect or often somewhat homomallous. Seta red, smooth, 1·5 cm. long, curved at the top ; capsule small, inclined ; lid shortly beaked ; calyptra smooth, white ; peristome normal.

Common in every forest or shrubland and in most grass-lands throughout South Africa and South Rhodesia

626. *Microthamnium squarrosulum* (C. M.), Jaeg., Ad., p. 490 ; Broth., Pfl., i, 3, 1051 ;
Dixon, Journ. of Bot., 1915, p. 21 ; Broth., Pfl., xi, 472.

=*M. squamosulum* (C. M.), Jaeg., Ad., p. 490=*Hypnum squamosulum* C. M., in Bot. Zeit., 1856, p. 440
(both as printed in Par., Ind., 2nd ed.).

Dixon, describing *M. cavifolium*, says, " *M. squarrosulum* (C. M.), Jaeg., *e descr.*, seems to be very near in vegetative characters, but the fruiting characters do not at all agree, ' ped. perbrevi, flavo ; operculo conico-acuminata recto,' etc."

Brotherus places it in section *Eumicrothamnium*, B., " autoecious, capsule oval, lid shortly beaked, leaves lanceolate-subulate from a cordate base," along with *M. pseudoreptans* (C. M.), Jaeg., and *M. cygnicollum* (Hampe), C. M., but adds for *M. squarrosulum* " seta 1 cm."

Brotherus confuses matters by including also (Pfl., xi, 471) a different *M. squarrosulum* (Card.), from Mexico.

627. *Microthamnium cygnicollum* (Hpe.), C. M., in Hedw., 1899, p. 144 (name only) ; Broth., Pfl., xi, 472 ;
Dixon, in Journ. of Bot., 1915, p. 21 (not *Eur. cavifolium* R., as implied by C. M. and stated by Paris).

=*Chrysohypnum cygnicollum* Hpe., MSS., in Herb.

Stems dendroid at the base, arching over and rooting at the apex, 2-pinnate, 1-3 inches long ; the branches up to ½ inch long, the larger with leaves imbricate, erecto-patent ; the smaller with leaves often as long but much more scattered and more spreading ; the stem leaves hardly 1 mm. long, very wide at the clasping base, quickly narrowed to a long flat subula ; the larger branches with leaves 1-1·5 mm. long, ovate-lanceolate, concave ; nerve not visible ; margin erect, denticulate or strongly denticulate ; back of the leaf hardly or only slightly papillose, the cells being shortly and widely vermicular, rather drawn to the apex, but hardly papillose ; alar cells numerous on stem leaves, few on branch leaves. Perichaetal leaves numerous, erect or homomallous, lanceolate-acuminate, denticulate ; perichaetium axillary. Seta red, erect to near the top, then cygneous, so the small, brown, oval, subsymmetrical capsule is subpendent ; the lid is shortly beaked, and the stout, red, and bordered outer teeth shortly connate at the base, closely barred, densely striate, with firm mid-line to far up, and bent inward and pectinate on the inner surface, then erect, slender, and hyaline at the top. Inner peristome and ring not seen. Male flower globose, axillary, with short ovate-acuminate, nearly entire leaves. I have seen no nerve on leaves of this species.

Cape, W. Kooksbosch, Breutel.
Natal. Umpumulo, 1867, Rev. Borgen.
Transvaal. The Downs, Pietersburg, 4000 feet, Rev. H. A. Junod 4008.
S. Rhodesia. Inyanga, 6000 feet, J. S. Henkel.

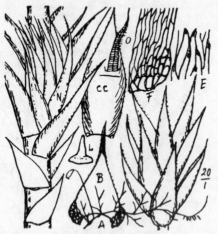

Microthamnium cygnicollum.

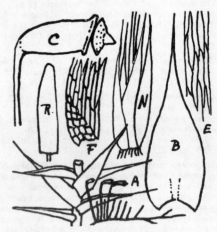

Microthamnium ctenidioides.

628. *Microthamnium ctenidioides*, Dixon, in Trans. R. Soc. S. Afr., viii, 3, 217 and plate xii, fig. 18.

A slender, little, prostrate, yellowish, autoecious moss ; stems rooting occasionally, 1-2 cm. long, with numerous short, irregular, or pinnately arranged branches ; leaves numerous, 1·5-2 mm. long, placed all round the stem, spreading, ovate-cordate, and clasping at the base, lanceolate-subulate upward, often with long subula, entire or almost so ; nerve short, weak, and double, or absent ; cells pellucid, smooth, linear, twisted ; lower and outer rhomboid, alar shorter and quadrate. Perichaetia erect, rooting, longer than the stem leaves, cylindrical, several in proximity, with leaves long-subulate. Seta 1 cm. long, smooth ; capsule inclined or horizontal, irregular, with shortly conical lid and cylindrical nude calyptra. Hogsback, Cape, E., 4000–6000 feet, 1916, D. C. and M. Henderson 220.

Dixon, in describing it, adds the note quoted here under *M. horridulum*, to which is added : "*Ctenidium squarrifolium* (C. M.) from the Cameroons is much like it, but is dioecious. The autoecious inflorescence is the principal argument, perhaps, for placing the present species in *Microthamnium*, some other species of which, also, are very closely related to species of *Ctenidium*." For *C. squarrifolium* (C. M.), Broth., see Broth., Pfl., xi, 468.

(Genus Hypnum Dill.

This was formerly the largest genus in the Bryophyta, and from it, time after time, sections have been separated as new genera. So far has this been carried, that Brotherus (Pfl., i, 3, 1060) reduces *Hypnum* to one species for the whole world—*H. Schreberi* Willd., and that is not recorded from South Africa. Brotherus was not closely followed in this, as Dixon (2nd and 3rd ed.) adopts Schimper's arrangement, and includes forty (forty-one) British species in *Hypnum*. In dealing with the South African mosses my check-list showed thirty-three species under *Hypnum*, most of which were placed thereunder while a wide definition was fully accepted, but at present, though some of these have not as yet been definitely placed elsewhere, there is no reason to believe that any South African species belongs to this genus in its restricted sense.

Brotherus (in Pfl., xi, 451) now brings in an under-family Hypnoideae of Hypnaceae in which he confines *Stereodon* to a few American and Chinese species having erect, regular, cylindrical capsules with double peristome without striae, and thus restores all the species herein included in *Stereodon* to *Hypnum*. His new arrangement includes *Hypnum*, *Ectropothecium*, *Isopterygium*, and *Vesicularia* as South African genera.)

Subfamily 3. Stereodonteae.

Stem more or less regularly once- seldom twice-pinnate, without stolons. Paraphyllia unusual. Leaves either horizontally inserted and symmetrical, or more or less obliquely inserted and unsymmetrical. Stem- and branch-leaves hardly different, the latter usually only smaller. Nerve double and short or absent. Cells narrowly prosenchymatous, smooth, seldom papillose toward the leaf-apex. Alar cells distinct ; capsule seldom erect and regular, usually inclined or horizontal and high-backed, oblong or cylindrical, often somewhat curved. Peristome normal ; lid usually conical, seldom beaked.

The recorded South African genera have the capsule inclined or horizontal, high-backed, and the inner peristome normal. Stereodonteae is now merged in Hypnoideae by Brotherus (Pfl., xi, 451), except *Acanthocladiella* put into Ctenidioideae.

Synopsis of the Stereodonteae.

Alar cells evidently differentiated in a large group. Leaves entire	.	.	.	.	206. Stereodon.
Alar cells in a small group ; leaves denticulate	.	.	.	.	207. Acanthocladiella.
Basal alar cells large, long ; others small, parenchymatous	.	.	.	.	208. Acanthocladium.
Alar cells not evidently differentiated, or few and small	.	.	.	.	209. Ectropothecium.

Genus 206. Stereodon (Brid.), Mitt., Musc. Austr. Amer., 1869, p. 22 ; now separated, and these species are included in *Hypnum* by Brotherus, in Pfl., xi, 451, 452.

Fairly strong, tufted, yellowish, shining, dioecious or seldom autoecious mosses, with rather long decumbent or suberect stems (erect only when closely packed), without stolons, and with or without rhizoids and paraphyllia, or the latter present only where the branches rise. Stems single or forked and usually irregularly, or more seldom regularly, pinnated, with often hooked or falcate branch-tips. Leaves set all round the stem but somewhat 2-ranked, usually falcate-secund, sometimes quite uncinate ; not or hardly decurrent, more or less concave, ovate, or cordate-lanceolate ; either shortly pointed or subulate pointed ; those on the back, sides, and front somewhat different. Nerve short and double if present, more usually absent ; cells narrowly prosenchymatous, both sides smooth, those at the leaf-base usually thick-walled and discoloured ; the alar cells parenchymatous, usually forming a hardly auricled concave group. Inner perichaetal leaves usually furrowed, long and finely pointed ; seta long ; capsule inclined or horizontal (in South African species), oblong or cylindrical, somewhat curved ; lid shortly conical-acute, or dome-shaped and short-pointed. There are many species or forms, widely distributed. The South African species belongs to subgenus *Drepanium* (Schimp.), Mitt., which has no paraphyllia ; leaves falcate or secund ; capsule inclined or horizontal, irregular, when dry more or less curved and wrinkled ; teeth of the outer peristome closely striate ; inner peristome normal ; spores small.

Synopsis.

629. *S. cupressiformis.* Alar cells numerous ; paraphyllia absent.
630. *S. alboalaris.* Paraphyllia numerous.
631. *S. aduncoides.* Cells at leaf-insertion few, large ; alar cells smaller.

629. *Stereodon cupressiformis* (Linn.), Brid., Bryol. Univ., ii, 605.

=*Hypnum cupressiforme* Linn., Sp. Pl., p. 1126; Hedw., iv, 59; C. M., Syn., ii, 289; Dixon, 3rd ed., p. 528, and Tab. 59β; Broth., Pfl., xi, 454.
=*Hypnum dicladum*, Rehm. 416, which contains this and *Brachythecium implicatum* mixed.
=*Hypnum *semirevolutum*, Rehm. 412.

Exceedingly variable even in one tuft, but much more so from different sites; this could easily make ten or more herbarium species by selecting types, but all these have features in common, and the field bryologist finds they all grade into one another, so that specific separation is impossible without having connecting links all through. It is dioecious and seldom fertile; it is usually pinnate, seldom 2-pinnate, and in the stronger forms the branch-tips are uncinate.

The most common cosmopolitan form is that shown as in fig. A, a shining, greenish-yellow plant growing in dense intricate mats usually upon stones or soil, varying from a close prostrate habit to an inch depth and fairly lax, and though pinnate on every fork it often has so many forks as lower branches as to appear 2-pinnate. Branches sometimes blunt, sometimes tapering to a point. Leaves oblong-lanceolate-falcate, 2 mm. long, smooth, concave, with erect margins, not plicate, often strongly uncinate, and though produced all round the stem they often part in the middle on the upper side so as to appear distichous in two rows on that surface, while the under-surface is a confused mass of leaf-points. Nerve absent or usually so (double and short when visible); margin erect, practically

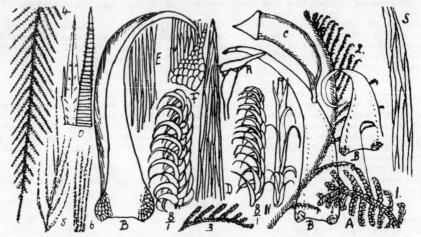

Stereodon cupressiformis and some of its forms.
S, leaf margin.

entire; cells vermicular, and shining above; alar cells abundant and mostly small, in flattened out but hardly auriculate angles; leaf-base wide. Perichaetal leaves sheathing and acuminate, the inner erect and clasping, the lower spreading. Seta 1 inch long, red, smooth; capsule inclined and curved, red, cylindrical, with short, acute, conical beak, white smooth calyptra, and complete peristome. Common everywhere.

From among many forms I select the few most distinct in South Africa, and of these I only know the type fertile, and that seldom so.

1. Type as above described.

2. Form *julaceus* (fig. 2). Golden yellow, closely pinnate, vigorous, larger than the type, but leaves short, more straight and appressed, consequently almost julaceous above, with extruded points on the lower side. A mountain form.

3. Form *minus* (fig. 3). Smaller than the type, shining, yellow, with abundant lax falcate-uncinate leaves which bristle all over it. Known only from the mountain-tops of Natal.

4. Form *afrocupressiformis* (fig. 4). Pale yellowish green, closely pinnate, pendent from branches and very regular, or on the ground straggling and less regular, much more slender than the type and very unlike it, the leaves being lax, homomallous or subfalcate, and nearly straight, but with the leaf-base construction and alar cells similar to the type, and leaves not complanate. Table Mountain, Eastern Province, and elsewhere in moist surroundings. =*Cupressina afrocupressiformis* C. M., in Hedw., 1899, p. 144.

5. Form *filiformis* (fig. 5). Pale green, very slender, laxly pinnated with long, slender, filiform pinnae and small, nearly straight leaves. A more slender condition of No. 4.

6. Form *resupinatus* (fig. 6). Still more slender, silky, erect or pendent, simple or almost so, with almost erect narrow leaves all round the stem, the points mostly on one side, and the margin often slightly reflexed near the apex. Table Mountain, etc.

Concerning forms 4 and 5, see Dixon, in Smithson. Misc. Coll., lxxii, 3, 14, where he calls a similar Central African form *S. cupressiformis* var. *Hoehnelii* (C. M.), Dixon.

Forms with erect or almost erect regular capsule occur in Europe and also in Central Africa and may occur here, but fruit is rare.

Flagellate forms on robust plants and aquatic forms also occur but are not common.

Cupressina basaltica C. M., in Hedw., 1899, p. 145 ; *Cupressina afrocupressiformis* C. M., in Borgen M. Mad., p. 67 ; *Hypnum dicladum* C. M., in Bot. Zeit., 1855, p. 784 (sterile), and *H. semirevolutum* C. M., in Bot. Zeit., 1855, p. 784 (=*H. drepanophyllum* C. M., Syn., ii, 319=*Leskea uncinata* Brid., ii, 323=*Hookeria uncinata* Smith, Trans. Linn. Soc., ix, 281, t. 33 (P.B.S., Menzies)), belong here, *fide* Brotherus ; also *Hypnum crassicaule*, Rehm. 414=*Cupressina crassicaulis* C. M., in Hedw , 1899, p. 145 (sterile, Natal), is this, *fide* Brotherus, but C. M., in Rehm. 678, made it =*Hypnum anotis* C. M., as a robust swamp form, while I find that Rehmann's specimen of it, 414β, from Fields Hill, Natal, is that also, *i.e. Ectropothecium regulare.* No. 414 is not represented in South African herbaria, but Müller's description of *H. crassicaule* makes it without alar cells, and rightly says Rehmann's name is unsuitable, though he uses it.

S. cupressiformis belongs to his subgenus *Drepanium* in which paraphyllia are absent ; and paraphyllia have not been seen on any of my specimens.

630. *Stereodon alboalaris* Broth., in Engler's Bot. Jahrb., 1897, p. 261 (sterile, from Zambesia) ; Broth., Pfl., i, 3, 1073.

=*Hypnum alboalare* (Broth.), Par., Suppl. Ind., 1900, p. 194.

I do not have specimen or full description of this, but Brotherus (Pfl., i, 3, 1073) puts it in his subgenus 3, *Heterophyllium* Sch., having many paraphyllia, leaves on all sides, erecto-patent, not furrowed ; lower cells yellowish ; alar cells lax, usually pale golden or brown, some special, making a sharply defined group ; and inside that subgenus he places it in group β, *b*, *a*, described as " Autoecious ; stem with central string ; leaves where the stem is rhizoid, erect and unsymmetrical, elsewhere all round the stem and symmetrical, erecto-patent, concave ; inner perichaetal leaves not plicate ; capsule small ; branches curved ; leaves subulate-pointed with recurved margins and slightly toothed upward ; lid conical."

631. *Stereodon aduncoides* (C. M.), Broth., in Pfl., i, 3, 1071 (§ G, β).

=*Hypnum aduncoides* C. M., Syn., ii, 295 ; Besch., Flor. Bryol. de la Reunion, p. 323 ; Renauld, Fl. Bry. de Mad., p. 259 ; Broth., Pfl., xi, 454.
=*Hypnum stereodon cupressiforme*, var. *aduncoides* Brid., Bryol. Univ., ii, 612.
=*Hypnum *afropurum*, Rehm. 679 and 679β.
=*Hypnum *basalticum* C. M., in Musc. Macow.. No. 37.
=*Cupressina basaltica* C. M., in Hedw., 1899, p. 145.
=*Stereodon zickendrathii* Broth.

A very vigorous, laxly caespitose, shining, yellowish, dioecious moss, remarkably uncinate in every leaf and bud ; stems 3–4 inches high, suberect, regularly pinnate ; branches 1 cm. long spreading, hooked ; leaves 2–2·5 mm. long, pellucid, circinate-falcate, very concave, ovate-acuminate, usually nerveless, sometimes with short, indistinct, double nerve ; margin expanded, slightly denticulate (or more so on slender adventitious growths). Cells pellucid vermicular, $100 \times 5\mu$; lower cells shorter and wider ; alar cells few, large, roundly quadrate, pellucid or yellowish. Fruit unknown.

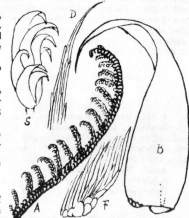

Stereodon aduncoides.

S, bud.

The general appearance is much like that of *Drepanocladus uncinatus*, but that has a long single nerve and more alar cells. Dixon, in Trans. R. Soc. S. Afr., viii, 3, 218, remarks that the single row of basal marginal cells is characteristic. but I find forms in which the alar cells are more numerous.

Frequent in high mountain localities, scarce elsewhere ; known also from Bourbon, Madagascar, and India.

Cape, E. Ingeli Forest, Tyson 1443 ; Knysna Forest, W. C. Wordsell, No. 7 (*fide* Dixon) ; Boschberg, Somerset East, Macowan ; Cave Mountain, Oudeberg, Rehm. 679β (as *H. afropurum* Rehm.) ; Perie, Sim 7131 ; Kakazella, Sim.

Natal. Giant's Castle, 8000 feet, 1915, R. E. Symons ; The Gorge, Mont aux Sources, 7000 feet, Sim ; Sentinel Peak, 10,000 feet, H. G. Botha-Reid.

Transvaal. Lechlaba, Houtbush, Rehm. 679 (as *H. afropurum* Rehm.).

Genus 207. ACANTHOCLADIELLA Fleisch., lately separated out of *Acanthocladium*, is similar to it in most respects. I have not seen description of genus or species, but Mr. Dixon has favoured me with specimen of *A. transvaaliensis* Thér. and Dixon, from which description and drawing are made. The Madagascar *Microthamnium flexile* R. and C., in Bull. Soc. Bot. Belg., t. xxix, 1890=*Sematophyllum flexile* (R. and C.), Renauld, p. 235 (see Broth., Pfl., i, 3, 1076, and xi, 473), which is the type of the genus, to which have been added *A. congoana* Thér. and Dixon, and *A. transvaaliensis* Thér. and Dixon, as African representatives, does not indicate a generic character. The genus is now moved with *Microthamnium* into Ctenidioideae by Broth., Pfl., xi, 466.

632. *Acanthocladiella transvaaliensis* Thér. and Dixon, in herb. and letter.

Paraphyllia absent ; stem suberect, simple or laxly pinnate ; leaves 1·5 mm. long, placed all round the stem, straight, imbricate, erecto-patent or closer, concave, not furrowed, lanceolate-acute from narrower base ; the margin erect toward the base, more or less recurved below the apex, often on one side only, slightly toothed ; apex rather sharply toothed ; nerve absent ; cells linear, smooth, thin-walled, and mostly pellucid, 70–100μ long, rather shorter and wider toward the leaf-base, with small distinct alar groups of small, swollen, quadrate cells, sometimes yellowish. Branches similar, sometimes tapering to a small-leaved point.   Fruit not seen.   In this species, Mr. Dixon states, the recurved margin is less regularly evident than in the other species ; in *A. flexile* it is the lower part of the leaf that has revolute margin, while the upper half is serrate, and an unequal double nerve goes one-third to one-half leaf-length.

Woodbush, Transvaal, H. A. Wager, 1922–23 (Herb., H. N. Dixon, p. 979).

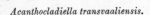

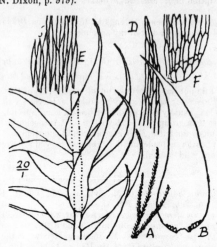

Acanthocladiella transvaaliensis.   *Acanthocladium trichocolea.*

Genus 208. Acanthocladium Mitt. ; Broth., Pfl., i, 3, 1075.

Dioecious ; pale yellowish-green, straggling, rooting, singly or doubly pinnate mosses, without bark or central string and with very few or no paraphyllia.   Leaves ovate-lanceolate, erecto-patent, concave, not furrowed ; margin erect or recurved toward the base, incurved upward, entire or slightly denticulate ; nerve absent, or if present very short and double ; cells linear, smooth, or on the back papillose ; alar group of one row fairly large and roundish, and another row of smaller round cells.   Branch leaves smaller, narrower, sharper toothed.   Inner per. leaves erect, not furrowed, sheathing, with subulate, sharply toothed point.   Seta purple.   Capsule horizontal, oblong, high-backed, smooth, when dry and old usually contracted below the mouth.   Ring present.   The genus is now included in Sematophyllaceae in Broth., Pfl., xi, 412.

633. *Acanthocladium trichocolea* (C. M.), Broth., in litt. ; Dixon, in S.A. Journ. Sc., xviii, 330.

=*Hypnum trichocolea* C. M., in Flora, 1888, p. 417.
=*Sematophyllum trichocolea* Par., Ind., 1897, p. 1171.

A small moss, only seen 2 cm. long, straggling among other mosses, and sparsely irregularly branched.   Paraphyllia absent ; leaves 1·75 mm. long, lanceolate-subulate from narrow base, either straight or slightly falcate-homomallous, or both on one plant, spreading, pellucid, slightly concave ; margin entire or waved, erect toward the base, often reflexed where the leaf bends ; apex long, subulate ; cells linear, up to 70μ long, wider and shorter toward the base, and a few hyaline cells larger and rounded or quadrate at the alar corners.   Other features not seen.

Natal.   Ngoya Forest, Zululand, Sim.
S. Rhodesia.   In wet forest, Inyanga, S. Rhodesia, 6000 feet, Aug. 1920, J. S. Henkel.
Occurs also in tropical East Africa.

Genus 209. Ectropothecium Mitt. (in Journ. Linn. Soc., 1868, pages 22 and 180, p.p.).

From strong to very slender, stiff, yellowish, loosely matted, shining, autoecious or dioecious plants, with fairly long, creeping or decumbent, seldom hanging or epiphytic stems (only erect when in close tufts), rooting occasionally ; single or forked, usually regularly pinnate or pectinate ; branches spreading, usually short and simple, often with the leaves somewhat flattened ; paraphyllia either small and subulate or absent ; leaves closely placed, not or shortly decurrent, somewhat unsymmetrical, more or less secund or falcate, the side, back, and front leaves different.   Stem leaves ovate, oval or ovate-lanceolate, shortly or subulately pointed.   Nerve short and double, or absent.   Cells narrowly prosenchymatous ; toward the leaf-apex papillose ; at the leaf-base shorter and wider. Alar cells few and small, rectangular or quadrate. Perichaetium usually rooting ; inner perichaetal leaves widely lanceolate, long pointed and finely pointed ; seta long and smooth ; capsule horizontal or hanging, oval or urn-shaped, or oblong-cylindrical ; when dry straight ; often contracted below the wide mouth, not furrowed ; the wall often rather rough from protruding cells. Ring distinct ; lid large, flatly arched or conical, with short point or short beak. Calyptra naked, or seldom set with 1-celled hairs. This is placed in § Hypnoideae, in Broth., Pfl., xi, 451.

Many species on rocks and earth in the subtropics.

Synopsis of Ectropothecium.

Regularly pinnate, branches many, short.
  Paraphyllia absent .   .   .   .   634. *E. regulare.*
  Paraphyllia present .   .   .   .   635. *E. Perrotii.*
Irregularly few-branched, epiphytal .   .   .   .   .   636. *E. brevisetum.*

634. *Ectropothecium regulare* (Brid.), Jaeg., Ad., ii, 531 ; Broth., Pfl., i, 3, 1064, and xi, 475.

  =*Hypnum stereodon cupressiforme*, var. *regulare*, Brid., Bry. Univ., ii, 609.
  =*Hypnum regulare*, C. M., Syn., ii, 307.
  =*Hypnum porrectirameum* C. M., in Dusen, Musc. Cameroons, No. 3 ; Par., Ind., 1896, p. 670.
  =*Hypnum *crassicaule*, Rehm. 414 and 414β.
  =*Hypnum anotis*, Rehm. 413 and 413β ; Par., Ind., p. 612.
  =*Rhaphidostegium anotis* (Rehm.), Par., Suppl. Ind., 1900, p. 295, and 2nd ed., iv, 166.
  =*Cupressina anotis* C. M., in Hedw., 1899, p. 146.

Dioecious ; growing in large lax mats ; common in eastern forest localities ; light green, suberect, densely pinnate, without paraphyllia, and not known fertile except one capsule mentioned hereunder. Branches 5–10 mm. long, complanately arranged, spreading at right angles to the stem, and uncinate at the apex. Leaves not compressed, numerous, falcate, and folding under ; stem leaves 1 mm. long, branch leaves 1·5–1·75 mm. long, both from wide clasping base ; nerve absent ; margin slightly denticulate to near the base, strongly denticulate near the apex ; cells narrowly vermicular, 70–100μ long, the lower shorter and wider, without distinct alar cells, but with the outer row sometimes shorter. Perichaetium of many narrow, spreading, or reflexed lanceolate leaves enclosing the red inflorescence. Not known fertile except that C. M., in describing his *Cupressina anotis* from Esternek, Knysna, on which was one capsule, says, " Seta short, slender, purple, slightly curved ; capsule oval, with a short neck, and slightly constricted below the mouth." Several Mascarene species are hardly different.

  Cape, W. Belvedere and Knysna, Rehm. 413 and β ; Diepwall, Knysna, J. Phillips ; The Wilderness, George, Miss Taylor.

  Cape, E. Perie Forest and Eveleyn Valley, East Griqualand, etc., Sim.

  Natal. Mont aux Sources, Sim 8997 ; Ngoya, Inanda, Giant's Castle, Zwart Kop, Cathkin Peak, Karkloof, etc.

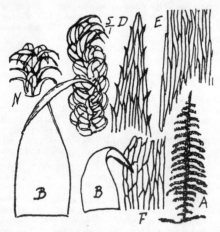

Ectropothecium regulare.

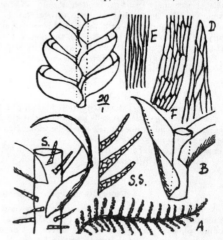

Ectropothecium Perrotii.
S, stem with paraphyllia ; SS, paraphyllia.

635. *Ectropothecium Perrotii* Ren. and Card., in Bull. Soc. roy. bot. Belg., 1896, p. 332 ;
Dixon, S.A. Journ. Sc., xviii, 330 ; Broth., Pfl., xi, 457.

Procumbent but closely matted, dark green, closely pinnate stems, rooting occasionally, the larger stems sprinkled with paraphyllia ; branches numerous, short, distichously arranged ; stem leaves subcompressed, shortly ovate-acuminate from a wide clasping base, 1 mm. long, somewhat concave below, nerveless or almost so ; branch leaves about as long but narrower and falcate, clearly in two series on the upper surface and folding under, with longer subula, and making branch-tips uncinate ; margin almost entire ; cells narrowly vermicular, smooth, about $70 \times 7\mu$, lower cells shorter and blunter, the lowest still shorter and wider, but not making an alar group, sometimes with one to two large lax cells below this, perhaps belonging to the stem.   Flowers and fruit not known.

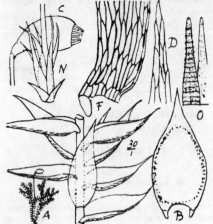

Dixon says (S.A. Journ. Sc., xviii, 330), " Very near *E. regulare* and perhaps only a form or variety of that species," but to me it appears to be different.

S. Rhodesia.   Inyanga, 6000 feet, J. S. Henkel (Herb., Eyles 2623) : Victoria Falls, 3000 feet, Sim 8900 feet, Wager 903, and Miss Farquhar 23.

636. *Ectropothecium brevisetum* Dixon, S.A. Journ. Sc., xviii, 334 ; Broth., Pfl., xi, 458.

Small, prostrate, yellowish green, complanate on bark, ½ inch long, irregularly few-branched, closely matted, the leaves lying almost flat on the bark and apparently in two rows though actually in several ; leaves 1 mm. long, ovate-acuminate, straight or almost so, spreading, rather deeply concave, pellucid green, practically entire, nerveless ; the cells linear, 80μ long, smooth, the lowest reduced in length but not forming an alar group, a few at the angles subquadrate, lax, chlorophyllose, and usually one large oval, hyaline cell connects with the stem.   Autoecious ; perichaetal leaves erect or slightly homomallous, ovate below, lanceolate-subulate above, entire ; seta slender, pale, 4–6 mm. long ; capsule horizontal or subpendulous, small, 0·75 mm. long, with slight neck ; lid pale, conical-rostellate, elliptical ; peristome erect, dense, yellowish and closely cross-striate below ; inner and outer of equal height, paler, and densely papillose ; calyptra white, smooth.

Portuguese East Africa.   Sheringden, near the Limpopo, March 1919, Rev. H. A. Junod.

Ectropothecium brevisetum.

Subfamily 4. PLAGIOTHECIEAE.

Stems irregularly branched, or more or less irregularly, seldom regularly, pinnated, sometimes having stolons. Branches with leaves in flattened arrangement ; paraphyllia absent ; side leaves obliquely inserted, spreading in two ranks, sometimes secundly curved, usually unsymmetrical. Stem and branch leaves little different. Nerve short, double or unequally two-legged, often absent. Cells elongated, rhomboid, linear, or vermicular, seldom parenchymatous ; in the angles not or little differentiated. Seta usually smooth. Capsule seldom erect and regular, usually inclined, horizontal or hanging, and irregular, oblong or cylindrical, when dry often wrinkled. Lid pointed or bluntly conical, or shortly beaked. All the South African genera have the leaves long-pointed and the cells smooth on the lumen, but in some cases they are slightly papillose at the upper angle.

Plagiothecium and *Catagonium* are now placed in Plagiotheciaceae ; *Isopterygium* and *Vesicularia* in Hypnoideae, by Brotherus, in Pfl., xi, 452.

Synopsis of the Plagiothecieae.

Stem with indurated outer cells.

Genus 210. ISOPTERYGIUM Mitt. (Musc. Austr. Amer., 1869, p. 21).

Autoecious or dioecious, slender, weak, silky matted mosses, with prostrate rooting stems, freely stoloniferous, and scattered irregular branches, the branches similar to the main stem, and usually the leaves somewhat flattened and dimorphous ; paraphyllia absent.   Stem and branch leaves uniform, obliquely inserted, smooth ; the front and back leaves alternately pointing right and left, and usually symmetrical ; the side leaves spreading in two rows, symmetrical or not quite, with the wings alternately twisted right and left ; oval-oblong or oblong from a wide, hardly decurrent base, and shortly pointed, or ovate or oblong-lanceolate, shortly or subulately or hair-pointed ; margin flat and entire or serrulate, seldom sharply toothed.   Nerve double and very short, or absent.   Cells narrowly prosenchymatous, smooth, or at the upper corner somewhat drawn to a papilla, at the leaf-base shorter and more thick-walled ; the alar cells not different.   Perichaetium rooting ; inner perichaetal leaves erect, partly sheathing, subulate-pointed.   Seta long, smooth.   Capsule almost erect, or inclined, or horizontal, with distinct throat ; oval, oblong, or cylindrical, almost regular, or somewhat high-backed ; smooth or occasionally furrowed.   Ring differentiated or not. Teeth of outer peristome subulate, free, yellowish, usually hyaline-bordered, striate, hyaline and papillose at the point; mid-line zigzag and lamellae numerous; cilia one to two knotted, seldom three, with short appendages. Lid arched-conical, sometimes beaked.

Mr. Dixon informs me *Hypnum *lechlabense*, Rehm. 671, is a small sterile *Isopterygium*, unfit for separation.

Synopsis of Isopterygium.

Leaves 1 mm. long, or less.
  Inner perichaetal leaves toothed at base of subula . . . . 637. *I. brachycarpum.*
  Inner per. leaves not toothed.
    Leaves with acute points . . . . 638. *I. leucophanes.*
    Leaves with rather blunt points . . . . . . 639. *I. leucopsis.*
Leaves over 1 mm. long.
  Leaves very lax, straight, concave . . . . . 640. *I. aquaticum.*
  Leaves ovate-acuminate, 2 mm. long; side leaves oblique.
    Cells pellucid . . . . 641. *I. strangulatum.*
    Cells with a line of chloroplasts . . . . . 642. *I. punctulatum.*
  Leaf with long twisted subulate apex.
    Leaves sometimes 2-nerved, often denticulate; plant large; stems simple . . 643. *I. Taylori.*
    Plant small; stems pinnate; leaves nerveless, entire . . . 644. *I. taxithellioides.*

637. *Isopterygium brachycarpum* Dixon, Trans. R. Soc. S Afr., viii, 3, 218, and pl. xii, 19; Broth., Pfl., xi, 462.

    = *Hypnum *argenteum*, Rehm. 673.

Closely matted on damp soil; stems ½–1 inch long, deep green, shining; branches few, subpinnately and complanately arranged, often sub-falcate at the tip, 3–5 mm. long; leaves complanate, often subfalcate, 1 mm. long, shortly ovate-acuminate; side leaves spreading and somewhat folded, all concave; margin entire or almost entire; nerve when present double and very short and weak, usually absent; cells linear, up to 100×5–8μ, those at the apex and base rather smaller; basal cells roundly ventricose, connecting with larger hyaline stem-cells; branch leaves are more acuminate and some are slightly denticulate; autoecious; male flowers small, abundant. Perichaetium with many leaves, the outer small and spreading, the inner erect, sheathing widely and quickly reduced to a subula, almost entire except that it often has one or two large teeth at the base of the subula. Seta slender, pale, 8 mm. high; capsule small, horizontal, oval, without neck, and with short, obliquely rostellate lid.

  Cape, W. Knysna, Wager 512.
  Natal. Inanda, J. M. Wood (Rehm. 673).
  Transvaal. Rietfontein, Wager 234 (sent to Mr. Dixon for publication as "*Plagiothecium subrhynchostegioides* Broth. and Wager, n. sp. Broth., in litt. ad. H. A. Wager," but not different from this and so not published) and 429.

Isopterygium brachycarpum.

638. *Isopterygium leucophanes* (Hpe.), Jaeg., Ad., ii, 505; Broth., Pfl., xi, 462.

    = *Hypnum leucophanes* Hampe, in Bot. Zeit., 1858, p. 169.
    = *Hypnum *Knuthei*, Rehm. 670.
    = *Hypnum *gracillimum*, Rehm. 672.

Flatly matted on bark and humus, dark green; stems ½–1 inch long, slender, rooting occasionally, simple or pinnate, with few short complanate branches, often flagellate; leaves 0·5–0·75 mm. long, ovate-acuminate, more or less complanate, laxly set round the stem, on flagellae quite lax, concave, clasping at the base; the side leaves unequally folded and with the point more or less regularly oblique or falcate; smaller branches and flagellae distichous, with leaves in two ranks only; nerve usually absent, sometimes faintly and shortly 2-nerved; margin entire or minutely denticulate; cells widely linear and thick-walled; cells at mid-leaf up to $75 \times 8\mu$, upper and lower cells shorter, down to $20 \times 8\mu$, lowest cells deltoid but not alar. Perichaetal leaves small, hyaline, sheathing below, lanceolate-acuminate, enclosing the succulent vaginula and, at the right stage, numerous longer paraphyses. Seta 6–8 mm. long; capsule inclined, 1 mm. long, with tumid large cells when dry. Peristome teeth long and slender; inner processes pellucid on high membrane. Lid mamillate or shortly beaked. *I. subleucopsis* Broth. and Bryhn (Bryhn, p. 22; Broth., Pfl., xi, 462) seems from description to belong here, though the stem leaves—not the branch leaves—are said to have piliferous apex.

*Isopterygium *argillicollum* Wag. and Broth. (Barberton) of Wager's list, which is a small compact condition which I cannot distinguish from *I. leucophanes*, but the name is preoccupied by (Ren. and Card., as *Microthamnion*) Broth., Pfl., i, 3, 1081, and xi, 461.

Cape, W.   Kloof near Chaplin Point, Cape Peninsula, Sim 9738 ;   Blinkwater ravine, Sim 8614.
Cape, E.   Kentani, Miss Pegler 2331.
Natal.   Inanda, J. M. Wood (Rehm. 672) ;   Maritzburg, Sim 10,278a ;   Zwart Kop, Sim.
Transvaal.   Wolhuter's Kop, April 1913, Wager 180 ;   Lechlaba, Houtbush, Rehm. 670.

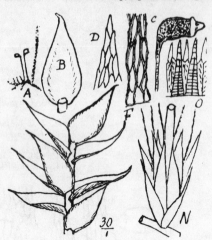

Isopterygium leucophanes.

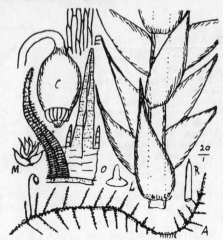

Isopterygium leucophanes.

639. *Isopterygium leucopsis* (C. M.), Paris ;   Broth., Pfl., xi, 462.
=*Taxicaulis leucopsis* C. M., in Hedw., 1899, p. 142.
(Both changed in error in Paris' Index to **leucopyxis.*)

    Caespitose ; stems prostrate, 2–3 cm. long, rooting, irregularly pinnate ; leaves irregularly complanate, spreading, 0·5 mm. long, oblong from a narrow base, shortly acuminate, straight, acute or bluntly pointed, spreading, clasping somewhat, slightly concave ; margin flat, entire or slightly denticulate ; nerveless. Cells linear-oblong, 70–100 × 8μ all alike ; involucral leaves rather long-pointed, entire ; seta 1–1·5 cm. long, slender ; capsule minute, inclined, 1 mm. long, ovate-oblong, with wide mouth and very large peristome ; lid small, conical-acute.
    Cape, W.   Blanco, Rehmann.
    Natal.   Maritzburg, Sim ;   Natal, Wager 179, 180β.

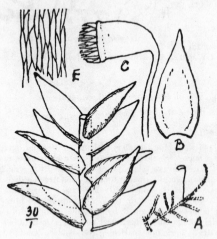

Isopterygium leucopsis.

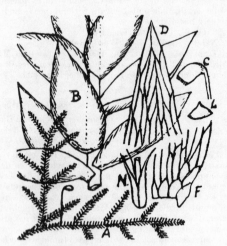

Isopterygium aquaticum.

640. *Isopterygium aquaticum* Dixon, S.A. Journ. Sc., xviii, 3, 330.
    Submerged and floating out, shining and yellow above, dark green below, several inches long and sparsely pinnated, or amphibious with smaller land condition, ½–1 inch long, more closely and freely pinnated and more fertile ;

in each case branches and leaves flatly complanate; the leaves spreading widely both when dry and wet, ovate, shortly acute, or sometimes obtuse or subacute when coated with slime, concave, entire below, entire or slightly denticulate toward the apex; cells linear, up to 100μ long, smooth, hardly different to the base, where the lowest cells are short and irregular with about one larger, ventricose, connecting cell, but no alar group. Autoecious. Perichaetal leaves few, sheathing below; lax-celled at the base, subulate above, slightly denticulate or occasionally with a large tooth at the base of the subula. Seta 1 cm. long, strong, dark; capsule horizontal or hanging, small, brown, elliptical; lid convex-mamillate; peristome normal; paraphyses short; wasted archegonia several.

S. Rhodesia. Under running water, Makoni, 4700 feet, Eyles 780; on stones by stream, Mazoe, 4600 feet, Eyles 402 (named at first in error *Amblystegium varium*).

641. *Isopterygium strangulatum* (Hampe), Broth., Pfl., i, 3, 1082, and xi, 462.

=*Hypnum strangulatum* Hampe, in Bot. Zeit., 1858, p. 169.
=*Rhaphidostegium strangulatum* (Hpe.), Jaeg., Ad., ii, 474; and Par., Ind., 1896, p. 689.
=*Hypnum *subincurvans* W. P. Schimp., in Breutel, M. Cap.

Straggling, subaquatic or amphibious, 2–3 inches long, sparingly branched; branches and leaves distichous whether wet or dry; leaves 2 mm. long, lax, ovate, with long narrow point, often twisted when dry, spreading or erecto-patent, set all round the stem, but side leaves folded and hence apparently oblique and reflexed at the point, other leaves straight, concave, entire, shortly or unequally 2-nerved, or nerve absent; cells smooth, pellucid, up to 100×8μ at mid-leaf, shorter to apex and base; the lower two lines irregularly oblong, but hardly forming an alar group. Dioecious (?), male flowers abundant, small, axillary to every leaf on some stems; female inflorescence and capsule not seen.

Cape, W. Kloof near Chaplin Point, Jan. 1919, Sim 9379; on wet shaded banks of stream on west slopes of the Muizenberg, Cape Peninsula, Pillans.

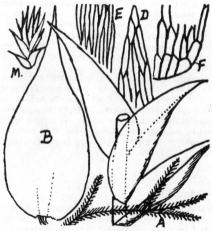

Isopterygium strangulatum.

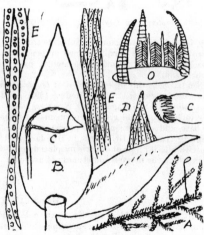

Isopterygium punctulatum.

642. *Isopterygium punctulatum* Broth. and Wager, Trans. R. Soc. S. Afr., iv, 1, 5, and pl. iic.

Stems prostrate and straggling on bark or moist ground, 2–3 inches long, rooting freely with red rhizoids, usually pinnate, occasionally 2-pinnate, the branches 1–1·5 cm. long, spreading distichously. Stem leaves 1·5–1·75 mm. long, irregularly complanate, ovate-acuminate, bluntly pointed, twisted when dry, concave below, lax, spreading; the side leaves slightly asymmetric-falcate from a narrow clasping base; margin faintly denticulate or almost entire, often somewhat reflexed below the apex; cells narrowly vermicular, 100–150×6–8μ each when fresh, containing ten to twenty chloroplasts, the lower cells rather wider and shorter, but not forming an alar group. Perichaetal leaves ovate-clasping, with long, narrow, almost entire, pellucid subula. Seta 10–15 mm. long, red; capsule cylindric, irregular, inclined or horizontal, brown; peristome normal; teeth very large; lid conical.

Cape, W. Under water, Kasteels Poort, Table Mountain, March 1919, N. S. Pillans.
Transvaal. Wolhuter's Kop, April 1913, Wager 181; near Rustenburg, Wager.

"This moss is characterised by its leaves having no midrib and no auricles, the length of the cells, and the smooth capsule" (Wager), which characters apply almost as well to any South African species in the genus.

643. *Isopterygium Taylori* Sim (new species).*

A densely matted, shining moss; stems 1½–2 inches long, mostly simple, closely packed, and parallel, brown below, yellowish green wherever exposed; leaves 1·5–2 mm. long, rising all round the stem, but the upper and lower surface leaves appressed and the side leaves erecto-patent, so the foliage is complanate. Leaves pellucid, smooth, narrowly lanceolate from a slightly cordate base, somewhat concave; apex subulate; nerve absent or often weakly and unequally 2-nerved; cells pellucid, very long and narrow (to $120 \times 5\mu$) from apex to base, cells of the connecting line small and shortly quadrate; margin subentire or slightly denticulate through the long marginal cells being often exserted at the tip. A very beautiful forest species.

Cape, E. The Wilderness, George, April 1924, Miss A. Taylor (Sim 10,287).

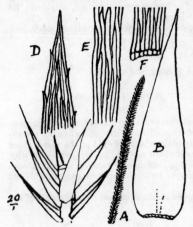

Isopterygium Taylori.

644. *Isopterygium taxithellioides* Broth. and Bryhn; Bryhn, p. 23; Broth., Pfl., xi, 462.

Not known to me. It is described as "autoecious; male flowers near the perichaetium; soft, caespitose, tufts densely intricate, shining, light green above; stems elongate, prostrate, and depressed; rooting, densely leaved, complanate, pinnately branched; branches horizontal, up to 8 mm. long, densely leaved, complanate; leaves dense, spreading, complanate; apex recurved, flatly concave; base ovate; lamellae narrowly lanceolate, gradually contracted to an oblique acute subula; margin erect, entire, ecostate; one to two rows basal cells inflated, rectangular, hyaline; others linear, chlorophyllose; all smooth. Stem leaves about 1·5 mm. long, 0·5 mm. wide; branch leaves 1·2 mm. long. Perichaetium rooting, inner bracts erect, quickly subulate from semi-sheathing base. Seta 10–15 mm. high, thin, flexuous, smooth, pale brown. Capsule small, inclined or pendulous, asymmetric from a tapering neck, ovate-cylindrical, arched, high-backed, contracted under the mouth, pale, smooth; lid convex-conical, acutely rostrate. Eshowe, Zululand, Haakon Bryhn."

Then he spoils this by saying, "near *Isopterygium complanatulum* (C. M.), Broth.," which I understand is *Brachythecium complanatulum* C. M., in Hedw., 1899, p. 138, apparently an immature plant, which has been *Isopterygium complanatulum* C. M. (Wager's list), *Sematophyllum complanatulum* (C. M.), Jaeg. (*fide* Paris = *Acanthocladium*, where it is not included). Müller's description answers *Isopterygium*, all except that the plant is minute, the margin revolute at the base, and a narrow nerve is percurrent in the long, straight, slightly denticulate acumen. It is from Spitzkop, Lydenburg, Transvaal (Dr. Wilms, in Hb. Jack, in 1899), and must be seen again to place it. But why compare this new species with it? one or other is misplaced or misdescribed. Possibly both are *Brachythecium implicatum* (Horns.), Jaeg., in which the nerve is sometimes not evident.

Genus 211. PLAGIOTHECIUM Bry. Eur.

Fairly vigorous, seldom slender, weak, pale yellowish, silky, shining, large-leaved, loosely matted, autoecious or dioecious mosses, with prostrate spreading stems (erect when crowded), moderately rooted, and producing abundant erect stolons, small-leaved at the base; the branches scattered and irregular and similar to the main-stem, usually with the leaves quite flattened, the branches narrowed to a point or flagellate; stem without paraphyllia; stem and branch leaves uniform, obliquely inserted, not furrowed, seldom transversely undulate, the back and front leaves alternately pointing right and left, usually symmetrical and slightly concave; the side leaves spreading in two ranks, usually unsymmetrical with alternately right and left twisted wings, widely lanceolate, ovate or ovate-oblong from a somewhat decurrent base, long-pointed or sometimes hair-pointed; margin usually flat and entire. Nerve short, usually double or unequally 2-legged, sometimes absent. Cells numerous, chlorophyllose, elongated, rhomboid or linear, thin-walled; those toward the leaf-base shorter and wider, the alar cells more lax, hyaline, and thin-walled. Perichaetium rooting; inner per. leaves erect, high-sheathing, without furrows, and without nerve. Seta long, red; capsule almost erect or inclined, with evident throat, oblong or cylindrical, regular or slightly high-backed, when dry furrowed lengthways or smooth; when dry and empty often horizontal and wrinkled. Ring dehiscent in fragments. Teeth of the inner peristome free, subulate, yellowish, with hyaline border, usually striate, and with zig-zag mid-line and lamellae numerous. Lid arched-conical, pointed, seldom beaked.

Plagiothecium *arboreum* W. P. Sch., in Breutel, Musc. Cap., is known to me by name only.

---

* *Isopterygium Taylori* Sim (sp. nov.).

Caespitosum, flavo-viride supra, caulibus simplicibus ad 2″ longis. Folia 1·5″ ad 2″ longa, undique circum caulem surgentia, superiora autem inferioraque adpressa, lateralia erecto-patentia, ita ut ordo complanatus sit. Folia pellucida, levia, anguste lanceolata a basi sub-cordata, sub-concava; apex subulatus; nervus absens vel languide 2-nervatus; cellae universae ad 120×5 mm.; cellae connexae breves ac quadratae; margo paene integer leviterve denticulatus.

Synopsis of Plagiothecium.

Leaf-cells very narrow.
   Dioecious.   Alar cells very few .   .   .   .   646. *P. rhynchostegioides.*
   Autoecious.   Alar cells numerous .   .   .   .   .   .   647. *P. Hendersonii.*
Leaf-cells very wide .   .   .   .   648. *P. membranosulum.*

645. *Plagiothecium rhynchostegioides* C. M., in Hedw., 1899, p. 143 ; Broth., Pfl., i, 3, 1086, and xi, 403.

        =*Hypnum *Moorei*, Rehm. 396 ; Par., Ind., 1896, p. 658.
        =*Plagiothecium *Moorei* (Rehm.), Par., Suppl. Ind., 1900, p. 275.

Dioecious, laxly tufted or matted, in extensive yellowish-green patches (brown underneath) on moist soil ; stems 2–4 inches long, prostrate, weak, seldom rooting, irregularly or fastigiately branched above by branches up to 1 cm. long, simple or forked, lower branches usually dominated and crowded off, branches often tapering to a point and small-leaved at the apex. Leaves ovate-acuminate, falcate, point subulate ; surface leaves appressed, symmetrical at base and oblique at apex, the side leaves spreading, especially when moist, narrow-based, somewhat concave, unequally folded in the lower half and irregular at the base ; margin entire, often narrowly revolute ; pellucid, decurrent, without nerve or with very faint, short, double nerve ; cells widely linear, up to $70 \times 10\mu$, shorter toward the base, where three to four ventricose hyaline cells occupy the alar corners ; perichaetal leaves suberect, entire, with long, narrow, subulate tips. Seta red, 2 cm. long ; capsule inclined or drooping, 1·5 mm. long, oblong, without neck or with very short neck, irregular, when dry contracted below the mouth ; peristome large, red, normal ; lid convex-mamillate.

Cape, W. Platteklip Ravine, Sim 9286 ; Blinkwater Ravine, Sim 8613 ; on wet shaded rocks in bed of stream, west slopes of Muizenberg, Pillans ; Table Mountain, Rehmann, Hex River above Axell's Farm, Rehm. 396 (as *Hypnum Moorei*).

Natal. Top of Giant's Castle, 8000 feet, Symons ; Rosetta Farm, 4000 feet, Sim ; Groenkopje, York, 3000 feet, 1918, J. M. Sim.

Transvaal. Rietfontein, Wager 422 ; Witpoortjes, 5000 feet, Dr. Moss.

S. Rhodesia. On wet soil by stream, Matopos, July 1920, F. Eyles 2542.

Plagiothecium sphagnadelphus C. M., in Hedw., 1899, p. 143 (from Hout Bay), seems from description to belong here, and is not represented as Rehm. 388 in South African herbaria, and Brotherus places it among the unknown to him.

Plagiothecium rhynchostegioides.
S, leaves on branch-point.

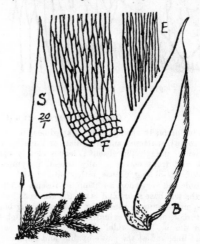

Plagiothecium Hendersonii.
S, perichaetal leaf, $\frac{20}{1}$.

646. *Plagiothecium Hendersonii* Dixon, in Trans. R. Soc. S. Afr., viii, 3, 219, and pl. xii, 20 ; Broth., Pfl., xi, 403.

Shining, light green, prostrate, autoecious ; stems 1–2 inches high, irregularly branched or pinnate, with branches about 6 mm. long, 2 mm. wide, spreading ; leaves complanate, side leaves erecto-patent or spreading, surface leaves appressed, all very concave from slightly above the base, ovate-acuminate, almost entire ; nerve absent, or short, weak, and double ; upper cells $100 \times 5\mu$, lower cells shorter and wider with incrassate porous walls ; alar cells numerous, suddenly wider, quadrate, pellucid. Perichaetia on the main stem, 3 mm. long ; their outer leaves spreading, inner suberect, lanceolate-acute but not hair-pointed, subentire. Seta red, 1 cm. long. No mature capsule seen.

Cape, E. Hogsback, Tyumie, 4000–6000 feet, 1917, D. Henderson 365 and 366 on bark.

The upper cells are as in *Isopterygium*, but the lower cells and alar cells clearly place it in *Plagiothecium*.

647. *Plagiothecium membranosulum* C. M., in Rehm. 389, and in Hedw., 1899, p. 144 ; Broth., Pfl., xi, 403.

= *P.* *selaginelloides, Rehm. 390 (Cape Town).

Autoecious ; shining, yellowish green, delicate, laxly caespitose on wet soil, or straggling or pendent in sphagnoid drip ; in the former case up to 1 inch long, pinnately few-branched ; in the latter case 2–3 inches long, simple or 1–2-branched or somewhat flagellate ; the leaves lax, complanate and almost in two series, spreading widely or recurved when moist, more erecto-patent and somewhat crisped when dry, 2 mm. long, clasping and asymmetric at the decurrent base, the margin on the lower side narrowly folding under ; lamina obliquely ovate-acuminate, the subulate point long and entire ; margin entire ; nerve absent or obsolete ; texture very pellucid ; cells widely linear, up to $100 \times 20\mu$, hardly different at the base ; per. leaves short, narrowly acuminate, closely appressed, few ; male flowers small, open, few-bracted ; seta 2 cm. long, red, smooth ; capsule almost erect but very irregular at the mouth, tapering to the base ; peristome red, normal ; lid shortly conical, mamillate. Seldom fertile, especially the semi-aquatic form.

Cape, W. Window Gorge, Sim 9463, 9481, 9489, 9494 ; Newlands Ravine, Sim 9448, 9424 ; Kirstenbosch, Pillans 3332 ; Table Mountain, Rehm. 389 ; Montagu Pass, Rehm. ; Rondebosch, Rehm. 390 (as *P. selaginelloides*).
Natal. Inanda, Rehm. 667 ; Mooi River, 4000 feet, T. R. Sim.

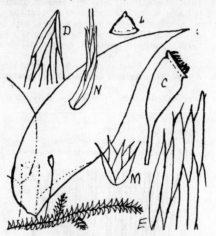

Plagiothecium membranosulum.

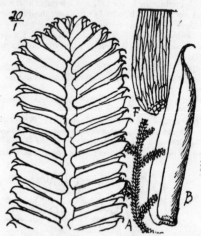

Catagonium mucronatum.

Genus 212. CATAGONIUM (C. M.), C. M., in Flora, 1896, p. 468.

Slender or fairly strong, stiff, silky, dioecious, matted mosses, with rather elongated, decumbent or suberect, sparsely rooted, scattered and irregular, or sometimes clustered obtuse branches, with the leaves in quite flattened arrangement ; paraphyllia absent ; leaves closely placed, obliquely inserted, smooth, arranged in two side-ranks, the back and front leaves usually absent, unsymmetrical, very concave, alternately pointing right and left, not decurrent, ovate-oblong, obtuse, either with or without a hair-point or short point ; margin erect and usually entire ; nerve double and very short and thin, or absent. Cells very long and narrow, smooth or drawn up to the upper end ; those at the leaf-base short and lax ; the alar cells not differentiated. Perichaetium rooting little ; inner per. leaves subulate-pointed from a sheathing, oval or oblong base, nerveless. Seta long, red, smooth ; capsule inclined, with distinct throat, oblong or cylindrical, somewhat high-backed ; when dry smooth, not wrinkled, and not contracted below the mouth. Ring wide. Teeth of the outer peristome free, subulate, yellow, narrowly bordered, striate, with zigzag mid-line, and at the point hyaline and papillose ; lamellae numerous. Inner peristome hyaline, finely papillose, with wide basal membrane, processes as high as the teeth, split in places. Cilia well developed, knotted ; spores small ; lid beaked from a shortly conical base.

648. *Catagonium mucronatum* (C. M.), Broth., Pfl., i, 3, 1088.

= *Acrocladium mucronatum* (C. M.), Jaeg., Ad., ii, 509 ; Par., Ind., 2nd ed., i, 5.
= *Hypnum mucronatum* C. M., Syn., ii, 262, and Rehm. 405.

Laxly caespitose, dioecious, pale yellowish green, prostrate on mud, laxly pinnate, the branches few and distichous, spreading, with the leaves numerous, close, and quite flatly and complanately arranged in a double row on each side of the stem, with few or no surface leaves ; the leaves erecto-patent, concave, boat-shaped, with straight rounded boat-base, and erect sides meeting quickly in a short, reflexed, hooked subula. Nerve when present very short, weak, and double, often absent ; margin entire ; cells smooth, linear, $70–100 \times 5–8\mu$, little changed to the base except one to two lines of small, incrassate, rounded cells where the leaf gathers in to the connection all round. No alar cells are present ; inflorescence and fruit not seen by me, but Müller (in Synopsis) describes it as : "Seta rigid, thick, rather long, red ; capsule inclined or horizontal, unequally obovate-cylindrical, thin-skinned ; lid conical-obtuse ; ring simple ; peristome teeth lanceolate, pale, with faint mid-line, the inner hyaline ; two slender cilia interposed hardly shorter.''

C. politum (Hk.f. and W.), Broth., Pfl., i, 3, 1088, and fig. 772 and p. 1237=*Acrocladium politum* (Hk.f. and W.), Mitt.=*Hypnum politum* Hk.f. and W., Lond. Journ. of Bot., 1844, p. 553 ; C. M., Syn., ii, 263, is credited to South Africa by Paris, Ind., 2nd ed., i, 7, but the description indicates a more attenuated plant than ours, with long acumen. Brotherus keeps them separate, and shows *C. politum* from South America and Australasia only, so for the present I keep the South African plant separate, but if identical *C. politum* is the earlier name and has priority.

Cape, W.   Disa Ravine, Sim 9115, 9121 ; Kloof near Chaplin Point, Devil's Peak, Blinkwater Ravine, and Camps Bay, Sim 9368, 9227, 9255, 9323 ; Cape Town and Table Mountain, Rehm. 405 and 405*b* ; Montagu Pass, Rehm. 405*c*.

Cape, E.   Hangklip, Queenstown, Maire and Mund (C. M.'s type).

Natal.   Inanda, Rehm. 405*d*.

Genus 213. VESICULARIA (C. M.), C. M., in Flora, 1896, p. 407.

(At first in C. M., Syn., ii, 233, as a subgenus of *Hypnum*, representing *Ectropothecium*.)

Autoecious, slender or fairly large, weak, shining or opaque mosses, growing in wide flat mats ; the outer cells of the stem not indurated ; the stem elongated, prostrate, sparsely rooted, single or forked, usually pinnate or pectinate ; branches spreading, short and simple ; the leaf arrangement flattened ; paraphyllia absent.   Leaves closely placed, not decurrent ; the back, front, and side leaves different.   Side leaves spreading or somewhat secund, seldom almost falcate-secund, either widely ovate or ovate-oval, short-, or long-, or hair-pointed, flat and entire at the margin, or somewhat toothed toward the apex.   Nerve double and very short, or absent.   Cells more or less lax, oval or oblong or elongated-rhomboid-hexagonal, smooth ; all round or toward the leaf-base only one marginal row elongated and narrow, making an evident border ; the alar cells not differentiated.   Back and front leaves much smaller.   Inner perichaetal leaves erect, slender, pointed, from an ovate or oblong base.   Seta fairly long, smooth ; capsule horizontal or hanging, oval or urn-shaped, seldom cylindrical, usually straight, often contracted under the mouth, not furrowed. Ring distinct.   Peristome normal.   Lid flat, arched or shortly conical, with point, or seldom short-beaked.

649.   *Vesicularia sphaerocarpa* (C. M.), Broth., Pfl., i, 3, 1094, and xi, 464.

=*Hypnum sphaerocarpum* C. M., Syn., ii, 238.
=*Leucomium sphaerocarpum* (C. M.), Jaeg., Ad., ii, 340.
=*Pterygophyllum *Montagnei* Belang., in Sched.
=*Ectropothecium sphaerocarpon*, Besch., Flor. Reunion, p. 321 (or p. 180).
=*Hypnum *Woodii*, Rehm. 393 (*fide* Dixon).

Autoecious, growing in extensive, nearly flat, intricate patches, shining green or yellowish green ; stems 2–4 inches long, rather weak, prostrate or subprostrate, regularly and complanately branched near the apex, often more irregularly branched below ; branches 1–1·5 cm. long ; leaves laxly complanate, somewhat concave ; side leaves spreading horizontally, 1·5–2 mm. long, ovate, rounded to the base, variable from blunt to subulate at the apex, sometimes somewhat falcate, especially at the branch apex ; subula sometimes reflexed ; the surface leaves smaller and often narrower, usually in two rows, acute or cuspidate ; margin entire or slightly denticulate ; nerve usually absent, when present very short and weak, sometimes double, or occasionally single, or a forked one which may be the coalescence of two short nerves ; cells laxly hexagonal, $30 \times 15\mu$–$45 \times 12\mu$, rather thick-walled, beautifully and regularly reticulated, and with the outer row or two rows narrower at the leaf-base, shorter upward, more pellucid, and almost forming a border.   Alar cells not different.   Leaves somewhat crisped when dry.   Perichaetal leaves shortly ovate and sheathing at the base, the points soon narrow, very long, almost entire, and usually homomallous or subfalcate.   Male inflorescence small, subglobose, near the perichaetium ; seta up to 2 cm. long or more, smooth, curved at the top ; capsule horizontal or hanging, roundly oblong-spherical ; lid convex-mamillate.   Peristome teeth short ; bars numerous and narrow ; processes and high membrane pale ; processes often split ; cilia two to three, shorter, pale ; calyptra small, smooth.

The leaf-apex varies immensely from branch to branch, but the extremes often occur on the same plant, a vigorous young shoot having long denticulate homomallous apex, while the slower growth has the more obtuse, entire, erect apex ; also the variation in cell-wideness to some extent corresponds, the long and falcate leaves having usually longer cells.

V. *transvaalensis Wager and Broth. was proposed for a green vigorous condition and is included in Wager's list, but was not published and does not differ.

Vesicularia sphaerocarpa.
S, apex of vigorous leaf ;
SS, apex of per. leaf.

28

Cape, W. Knysna, Burtt-Davy 17,001 ; Mossel River, Jan. 1913, Professor Potts.

Natal. New Germany, Dr. v. d. Bijl, April 1917 ; Zwart Kop and Karkloof, Sim ; Inanda, Rehm. 393, as *Hypnum Woodii* Rehm.

Transvaal. Barberton, 1914, Wager 259, and W. Hendry, June 1917.

Family 51. SEMATOPHYLLACEAE.

Slender or robust, weak or firm, prostrate or suberect, rooting, irregularly branched or sometimes pinnate plants, with the leaves in many rows, either standing round the stem or in more or less flattened arrangement ; paraphyllia absent ; leaves usually of one layer of cells, usually homogeneous and symmetrical, of various forms ; nerve double and very short, or often absent ; cells usually oblong or linear, smooth or papillose ; the basal and alar cells differentiated and usually oval. Vaginula naked. Capsule usually inclined or hanging, more or less oval, unsymmetrical, not striate, on a fairly long seta ; lid sharply beaked from a wide or conical base ; calyptra cucullate, naked ; peristome usually double, the two equally high ; the outer teeth separate to the base, usually striate, seldom smooth ; the inner sometimes absent, when present it has wide basal membrane and keeled subulate or linear processes, without cilia.

Synopsis of the Sematophyllaceae.

Inner peristome absent ; leaf-cells smooth	214. MEIOTHECIUM.
Peristome double ; leaf-cells smooth or ends projecting.	
Mid-line of teeth furrowed. Leaves usually erect	215. SEMATOPHYLLUM.
Mid-line of teeth zigzag. Leaves often secund	216. RHAPHIDORRHYNCHIUM.
Peristome double ; leaf-cells with papillae on the lumen . .	217. TRICHOSTELIUM.

Genus 214. MEIOTHECIUM Mitt.

Fairly robust mosses, with long, prostrate, rooting, irregularly or pinnately branched stems ; the branches either simple or again branched somewhat, thickly set with leaves, either all round the stem or in somewhat flattened arrangement. Leaves often somewhat secund, not furrowed, or with a faint central furrow, without nerve, oval or ovate, shortly or lanceolately pointed, entire, often having a wide incurved margin, often somewhat dimorphous, with smaller, longer-pointed front leaves. Cells smooth, usually rhomboid toward the leaf-apex ; marginal cells differentiated ; the cells toward the leaf-base longer and at the very base yellow ; the alar cells oval, small, hyaline or yellow, forming a distinct alar group. Seta short, smooth, or in the South African species set with mamillae upward. Capsule erect or inclined, oval or elliptic, smooth ; inner peristome absent ; teeth of the outer peristome papillose, not striate, alternate ones absent in the South African species. Lid shortly beaked from a conical base ; calyptra small, cucullate, more or less rough upward.

650. *Meiothecium fuscescens* (Horns.), Broth., Pfl., i, 3, 1103, and xi, 420.

=*Neckera fuscescens* C. M., Syn., ii, 77 (not 114 as put by Par., Ind., 2nd ed.).
=*Pterogonium fuscescens* Horns., in Sched.
=*Pterogoniella fuscescens* (Horns.), Jaeg., Ad., ii, 114.

Not known by me. C. M., in Synopsis, includes it in § III, *Pterogonium*, of *Neckera*, defined, " Plants pleurocarpous, small, tender ; stem decumbent, subcompressed, more or less pinnate ; leaves homomallous, cochleate ; cells oblong ; alar cells vesicular, yellowish ; nerve double, obsolete, often shown as a short yellow line. Capsule exserted, small, when overmature hanging ; peristome often single " ; and in that section he places it among those with leaves shortly acuminate, compressed ; margin revolute ; and the species is described : " Monoecious ; dense, prostrate, appressed ; branches leafy, grey-green ; apex somewhat falcate, short, flat ; leaves lax and homomallous ; stem leaves with wide ovate base, obtuse or subacuminate, deeply concave, subplicate, entire ; margin everywhere revolute ; cells minute, rhomboid, pellucid ; lower alar cells small, vesiculose, pale yellow ; upper where the margin is revolute similar ; other cells longer and narrower ; perichaetal leaves lanceolate-acuminate, acute ; base thin and lax, yellowish. Capsule on very short erect seta, subinclined, oblong, everywhere equal ; lid conical, obliquely acuminate ; calyptra slightly scaberulous at the apex. Cape, Bergius."

Brotherus includes it in *Meiothecium*, § III *Eumeiothecium* : " Seta more or less mamillate upward ; alternate peristome teeth absent ; teeth papillose ; spores 25–30μ, sometimes to 40μ ; lid short and not subulate-beaked ; calyptra more or less rough," and therein in the group " Leaves usually more or less secund, somewhat concave, usually shortly pointed ; cells not incrassate ; upper cells rhomboid, the marginal cells different ; seta 2–3 mm., seldom 5 mm. ; capsule inclined, somewhat irregular, small, oval or ovate, sometimes slightly bent ; when dry and developed usually contracted under the mouth ; leaf-cells smooth. Autoecious."

I am not aware that it has been seen except in Bergius original gathering. It seems to differ from everything else we have here.

Genus 215. SEMATOPHYLLUM (Mitt.), Jaeg. (=*Rhaphidostegium*, § *Aptychus*).

Slender or somewhat robust, autoecious mosses, matted on tree stems, with rather elongated, prostrate and tomentose or occasionally hanging, closely branched stems ; the branches spreading or erect, short or long, seldom simple, usually pinnately branched, with numerous leaves usually placed all round the stem, but sometimes in flattened arrangement ; the branch-point often acute through long leaf-tips closely placed together. Leaves homogeneous, erect or erecto-patent, sometimes somewhat secund, channelled-concave, somewhat auricled or almost cordate at the base, oblong-elliptic or oval, blunt or with short or subulate point, entire, or at the apex serrulate, sometimes drawn suddenly into a long subulate point. Nerve absent or faintly 2-nerved. Cells linear or oblong, usually incrassate and discoloured, with long lumen, smooth or almost smooth; the alar cells elliptical, hyaline or pale, forming a small concave group. Seta long or fairly long, smooth or papillose upward; capsule almost erect, inclined or horizontal, seldom hanging, oval to cylindrical. Lid sharply beaked from a conical base. Calyptra cucullate, smooth. Peristome double; teeth lanceolate or subulate, striate, bordered, with furrowed mid-line, and with well-developed lamellae; inner peristome as high, with keeled processes and with wide basal membrane; the cilia one or two, or rudimentary.

Synopsis of Sematophyllum.

Leaves oblong-lanceolate, straight; point long, flat . . . . . . 651. *S. Wageri.*
Leaves ovate-lanceolate, secund; point long, flat . . . . . 652. *S. brachycarpum.*
Leaves ovate, with long, flat, or twisted or recurved point; cells twisted . . . 653. *S. sphaeropyxis.*
Leaves ovate, very concave, with short flat point; cells rhomboid-hexagonal . . . 654. *S. caespitosum.*

651. *Sematophyllum Wageri* Wright, in Trans. R. Soc. S. Afr., iv, 1, 5, and pl. ii*b*.

A flatly caespitose, shining, yellowish-green bark-epiphyte; stems ½ inch long, simple or few-branched, with leaves closely imbricated all round, forming a terete growth except where rooting, which occurs freely and where leaves are subsecund; leaves oblong-lanceolate, usually straight, 1·5–1·75 mm. long, very concave, tapering to an acute flat apex, and somewhat contracted at the base, where there is a fibrous decurrent attachment; margin expanded flatly, entire; nerve absent; cells oblong-hexagonal in the upper half, about 45μ long; toward the base longer and narrower, all green, but the alar corners occupied by three to four large, tumid, elliptical, hyaline cells, and above these a few small subquadrate cells, with a line of narrow marginal cells. Perichaetium rooting, shorter than the leaves, and of similar oblong-lanceolate, shortly pointed, but more appressed leaves; seta 8–12 mm. long, straight, red; capsule small, oblong, inclined or horizontal, 1–1·5 mm. long, irregular, when fully developed contracted under the mouth. Peristome double, the teeth short, inflexed, yellowish, closely barred, hardly papillose, furrowed down the mid-line, pectinate on the inner side where bent, hyaline at the apex; processes pale, narrow, smooth; lid obliquely rostrate; spores smooth, 12–15μ diameter. Van Reenen, Natal, Jan. 1911, Wager 197.

Probably too near to *Sematophyllum brachycarpum.*

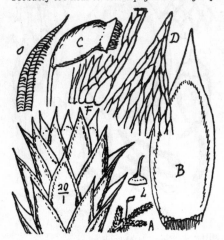

Sematophyllum Wageri.

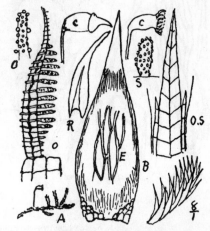

Sematophyllum brachycarpum.
O, tooth, and apex of same; OS, process;
S, mamilla on inner side of tooth.

652. *Sematophyllum brachycarpum* (Hampe), Broth., Pfl., xi, 431 = *Rhaphidostegium brachycarpum* (Hampe), Jaeg., Ad., ii, 462; Dixon, in Trans. R. Soc. S. Afr., viii, 3, 221; Bryhn, p. 25; Brunnthaler, p. 33.

= *Rhynchostegium brachycarpum* (Hampe), Rehm. 420.
= *Hypnum brachycarpum* Hampe, Icon. Musc., t. 11; C. M., Syn., ii, 328.
= *Rhaphidostegium Reichardtii* (Rehm.), Par., Suppl. Ind., p. 297 = *Sematophyllum*, Broth., Pfl., xi, 4?2.
= *Aptychus Reichardtii* C. M., in Hedw., 1899, p. 141.
= *Hypnum Reichardtii*, Rehm. 402.

A close bark-epiphyte, especially on decayed stumps, forming compact, dark green to yellowish, shining mats, usually abundantly fertile. Prostrate stems 1–4 cm. long, sparingly or pinnately branched; the branches ascending; the leaves often regular round the ascending stems, but at the apex, especially of horizontal stems and branches, more or less secund upward ; leaves spreading when moist, ascending when dry, lanceolate-subulate, entire, very concave except toward the apex, which is flat ; the margin ascending or somewhat flattened outward or reflexed toward the base ; nerve absent ; alar group of about three large cells, with a few small quadrate cells in three or four rows above them up the margin ; other cells linear-rhomboid, pellucid, smooth. Some leaves narrower than others. Perichaetium rooting ; per. leaves few, like the stem leaves, not closely sheathing. Seta red, smooth, slender, 5–7 mm. long ; capsule small, horizontal or somewhat hanging, oval ; the lid wider, rostrate from a convex base ; calyptra cucullate, rostrate, long, covering the capsule at first. Capsule on maturity constricted below the very wide mouth. Outer perichaetal teeth usually inflexed, thick, and having a line of about twelve prominent papillose mamillae along the central line on the inner surface ; processes as high as the teeth, keeled, papillose below, with high basal membrane and single cilia between. Spores large. Very abundant on wattle stumps in Natal, and present in all the forestal and scrub regions of South Africa from Table Mountain to the Zambesi.

653. *Sematophyllum sphaeropyxis* (Rehm), Broth., Pfl., xi, 432 = *Rhaphidostegium sphaeropyxis* (C. M.), Par., Suppl. Ind., 1900, p. 297 ; Dixon, Trans. R. Soc. S. Afr., viii, 3, 221.

= *Rhynchostegium sphaeropyxis*, Rehm. 372 ; Par., Ind., 1897, p. 1137.
= *Aptychus sphaeropyxis* C. M., in Hedw., 1899, p. 141.

Fairly robust, prostrate or suberect, light green, 1–4 cm. long, slightly branched ; branches very short ; leaves ascending, clasping, intricate, numerous, regular ; the tightly appressed apical leaves making the branch-apex pointed ; the branch either erect or slightly curved upward. Leaves very concave, conchoid, ovate-elliptical so far as concave, then rapidly narrowed into a flat or twisted or reflexed subula ; lower leaves spreading widely when moist ; leaf-margin narrowly reflexed, entire ; nerve absent ; cells linear or twisted vermicular ; lower cells yellowish, the alar group of about three large tumid cells and a few quadrate cells above them. Perichaetal leaves few, similar to the others, or slightly longer and more sheathing and with fairly wide subula. Seta 1–1·5 cm. long, erect, red, smooth ; capsule very small, 1 mm. long, nearly horizontal, brown ; peristome inflexed, red, pectinate inside, with zigzag mid-line and scarious border.

Natal and Transvaal.

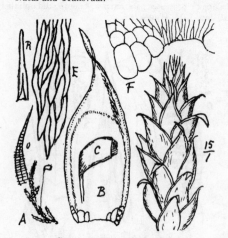

Sematophyllum sphaeropyxis.

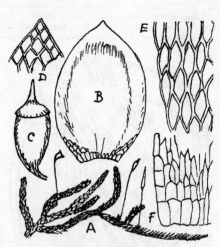

Sematophyllum caespitosum.

654. *Sematophyllum caespitosum* (Sw.), Mitt. ; Broth., Pfl., xi, 432.

= *Rhaphidostegium caespitosum* (Sw.), Jaeg., Ad., ii, 454 ; Dixon, in Journ. of Bot., 1920, p. 81, and in Kew Bull., No. 6, 1823, No. 421.
= *Hypnum caespitosum* Sw., Prod., p. 142.
= *Rhaphidostegium sphaerotheca* (C. M.), Jaeg., Ad., i, 468 ; Dixon, in Trans. R. Soc. S. Afr., viii, 3, 221.
= *Rhynchostegium sphaerotheca* C. M., in Rehm. 421 and *b, c.*
= *Hypnum sphaerotheca* C. M., Syn., ii, 333.
= *Hypnum *lithophilum* Horns., in Musc., Mund and Maire, (not Fl. Bras.).
= *Achrirhynchum *cupressinum* Hpe., in Sched.
= *Rhynchostegium *inclinatum* W. P. S., in Breutel, M. Cap. (not *R. inclinatum* (C. M., as *Taxicaulis*) Broth.).
= *Pterogoniella Stuhlmanni* Broth., in Engler's Bot. Jahrb., 1894, p. 208.
= *Rhynchostegium *julaceum*, Rehm. 371 (not 375 as given by Paris).

(Probably not *Leskea caespitosa* Hedw.)

Fairly robust, light green or yellowish green ; stems mostly prostrate and rooting, 1–3 inches long, irregularly branched ; the branches spreading or ascending, densely leaved below, sparsely leaved above ; the leaves all round the stem, laxly ascending or sometimes subjulaceous or julaceous and regular, except on prostrate rooting parts, where the leaves are ascending and secund. Leaves broadly ovate, obtusely pointed, very concave; the margin and apex flat or somewhat revolute; nerve occasionally present, very faint, short and double, usually absent. Alar group of four to six rather large swollen cells and sometimes rather numerous smaller quadrate cells up the margin; other cells hexagonal-elliptical, large, lax, pellucid or granular, smooth; the mid-basal cells narrower and longer; the apical cells shorter and rhomboid. Perichaetal leaves few, clasping, not much differentiated. Seta smooth, red, 1·5 cm. long, straight; capsule inclined or horizontal, oval-globose, with a slight neck, ultimately wide-mouthed and contracted below the mouth. Lid convex with slender beak, or sometimes conical-rostrate. Calyptra covering the erect young capsule, white, cucullate, smooth. Peristome normal for the genus; teeth inflexed with ascending points, much crested in front, papillose; mid-line somewhat zigzag; processes plaited, papillose; the basal membrane deep, and the cilia single, small, and inflexed, or absent. Spores large.

Dixon, in Journ. of Bot., March 1920, p. 81, shows what an exceedingly variable, widely distributed plant this is, and gives a long list of synonyms, mostly exotic, which hardly concerns us; but we are interested in how he limits the species, and this is mostly by the perichaetal leaves; and he says, " these leaves, erect, not differing greatly in size but usually narrower than the stem and branch leaves, with rather broadly tapering, not very finely acuminate points, entire or nearly so," are fairly constant.

It occurs throughout Africa and its islands as well as in tropical America, Australia, etc., and throughout the forest parts of South Africa from Table Mountain (abundant) to the Zambesi.

Pterogoniella Stuhlmanni Broth. has the leaves rather closely julaceous, subcircinate when dry (Salisbury, Eyles 683), but is the same species.

R. sphaerotheca var. *aquaticum*, Rehm. 423 (Stinkwater), shows nothing except aquatic conditions.

R. sphaerotheca C. M., var. *gracilis*. More slender, less branched, mostly prostrate and smaller in all its parts; leaves oval and more pointed; nerve absent; cells elliptical, smooth; capsule more distinctly necked; seta 1 cm. long. Van Reenen, Natal.

R. sphaerotheca C. M., var. *procerum*, Rehm. 422. Vigorous, 1–2 inches long; leaves lax, spreading, larger, rather narrower and rather more pointed than the type, deeply concave, with flat sides and apex, and recurved margin

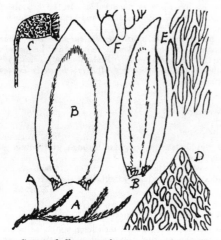

Sematophyllum caespitosum, var. *procerum.*

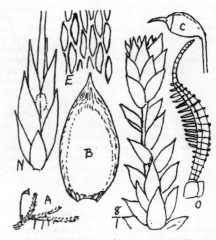

Sematophyllum caespitosum, var. *gracile.*

in the lower part. Nerve absent; cells longer than in type and more vermicular twisted. Inanda and Karkloof (Natal), and Lydenburg, Transvaal (Rehm. 660).

Brotherus (Pfl., xi, 433) cites Cameroons and South Africa for *S. Danckelmanii* (C. M., as *Hypnum*) Broth., but I know of nothing to support the South African record. It is very near *S. caespitosum.*

Genus 216. RHAPHIDORRHYNCHIUM (Besch.); Fleisch., Laubm. Java, iv, 1245.

(Formerly part of *Rhaphidostegium* Bry. Eur.)

Slender or robust autoecious mosses forming flat mats, usually on decayed tree-stumps, on charcoal from forest or wattle fires, or on stones, with prostrate, rooting, long, irregularly or pinnately branched stems; the branches erect or spreading horizontally, short and simple or longer and branched, often tomentose but not fixed by decurved leaves. Leaves uniform, placed all round the stem or in somewhat flattened arrangement, sometimes more or less secund upward, sometimes secund and falcate downward where rooting, concave, not furrowed, lanceolate, with blunt, shortly pointed or subulate apex. Nerve absent; cells linear or oblong, smooth or papillose, those at the base narrowly rectangular, pale or discoloured; the alar cells large, oval, hyaline or coloured, forming a distinct short line. Seta long, red, usually smooth; capsule inclined or horizontal, more or less oval, smooth, its cells roundish, collenchymatous.   Peristome double, the two of equal height ; teeth subulate, striate below, with zigzag, flat mid-line and hyaline border ; basal membrane wide and processes usually split in places ; cilia one or two or rudimentary. Lid sharply beaked from a wider base ; calyptra smooth.   The cells are not papillose, but occasionally the upper end of the cell is slightly raised in the upper part of the leaf.   The alar group consists of a few large, oval, swollen cells forming a distinct basal line together with a few quadrate small cells above them toward the leaf-margin.   The peri- stome teeth are usually strongly inflexed, with ascending points ; they are slightly grooved zigzag along the mid-line on the outer surface, and are remarkably crested on the inner surface by a row of about twelve terete papillose mamillae along the mid-line.   The inner peristome has a high, plaited, basal membrane ; the processes stand erect through between the outer teeth, and there is often one rudimentary inflexed cilia between the processes.   The seta is smooth and the capsule is very small, but it varies in form with the stage of development ; immediately before maturity it is usually oval or shortly cylindrical, but after it opens it becomes very wide-mouthed and constricted below the mouth. It is the most abundant genus in Natal, having taken possession of burned wattle stumps as peculiarly its own by the second or third year after burning.

The species closely resemble one another, and if dealt with as *Stereodon cupressiforme* is, there would only be two species.

Regarding *Rhaphidostegium maritimum* (Hook.), Jaeg., Ad., ii, 454 ; Par., Ind., iv, 181, which Brotherus states (Pfl., i, 3, 1114) he had not seen, Müller's description of *Hypnum maritimum* C. M. (Syn., ii, 328), from Burchell's speci- men, having only immature fruit, differs so considerably from Hooker's description of *Leskea maritima* Hook., Musc. Exot., t. 166 (from sandy scrub at the shore at Plettenberg Bay, Burchell, which also Müller cites), that one is con- strained to think they were dealing with different plants, though Müller quotes Hooker as synonymous (and Brid., ii, 294), and both appear to have seen only Burchell's specimens, indeed no other are known, and these I have not seen. But Hooker's description and figure represent an erect, slightly branched or simple plant 1½ inch high, with im- bricate, appressed, concave, ovate-acuminulate, nerveless leaves, forming a julaceous stem (evidently *Leucodon assimilis*), with erect lateral seta 1 inch long, suberect cylindrical capsule with sixteen free lanceolate teeth, and an inner peristome having deep basal membrane and short triangular processes without cilia.   But he describes the perichaetal leaves as minute and the perichaetium as *hairy* (which is also illustrated) ; these two, as well as other characters, indicate that the plant is not a *Rhaphidostegium* (though they do not help to place it elsewhere).   Hooker remarks, " The very erect and thickly tufted mode of growth of this moss is very remarkable.   In the disposition of the leaves it resembles *Hypnum moniliforme*, *H. trifarium*, *H. flexile*, and *Leskea ericoides*, and near the latter it will naturally enough rank.   The whole plant is very brittle, but particularly the fruit stalk."

Müller, on the contrary, describes it as *widely prostrate* ; leaves densely imbricate, *somewhat secund* ; margin *reflexed* ; nerve *subdistinct*, *double* ; cells large, elliptical ; perichaetal leaves *ovate, long-acuminate*, and he includes it in his *Aptychus* section of *Hypnum*, which corresponds with part of *Rhaphidostegium*.   Probably the specimens con- tained two different julaceous plants and each dealt with a different one of these, but Hooker's figure is *Leucodon*. Visitors to Plettenberg Bay might please look for some further material.

Synopsis of Rhaphidorrhynchium.

§ *Microcalpe.*   Leaves complanate, rigidly straight, erecto-patent.
   Leaves numerous ; side leaves distichous   .   .   .   .   .   . 655. *R. zuluense.*
§ *Cupressinopsis.*   Leaves lanceolate, subulate-pointed ; apex denticulate or irregular through protruding cells ;
      upper cells narrow ; upper cells with upper end sometimes raised.
   Stem apex blunt ; leaves falcate-secund upward, evidently toothed at the apex   .   . 656. *R. Dregei.*
   Stem apex pointed, straight or curved ; leaves mostly regular, crenate or hardly dentate at the apex,
                                                                                                   657. *R. Gueinzii.*

655. *Rhaphidorrhynchium (Microcalpe) zuluense* Sim (new species).[*]

Very flatly appressed on bark, yellowish pale green, autoecious (?).   Stems 1–1½ inch long, simple or pinnately branched ; branches up to 1 cm. long, with perichaetia from near their base ; leaves numerous, very flatly complanate, erecto-patent, 1·5–2 mm. long, pellucid, lanceolate from a cordate base which is much narrowed below to a narrow reflexed attachment made of a few large vesicular orange cells ; two lines of concave leaves folding together at each side, and there is a central straight line on the upper surface of rather smaller appressed leaves.   Side leaves very concave toward the base, often compressed and split, nerveless, drawn out to a long entire point, and very rigidly straight.   Margin erect, entire ; cells long and narrow ($100 \times 5\mu$) ; apical cells shorter and wider ; basal cells long and narrow till in contact with the inflated golden connecting cells.   Perichaetium near base of a branch, its leaves like stem leaves ; seta about 10 mm. long, red, smooth ; capsule inclined or horizontal, ovate-oblong, dense, with obliquely beaked red lid nearly as long ; when mature the capsule is slightly constricted below the mouth and peristome rises widely.   Only outer peristome seen ; teeth sixteen, pale, short, closely trabeculate, furrowed on mid-line, bordered, faintly striate.   No ring seen.
Ngoya Forest, Zululand, 1000 feet, 1915, Sim 10,285.

---

[*] *Rhaphidorrhynchium (Microcalpe) zuluense* Sim (sp. nov.).
   Folia crebra, erecto-patentia, rigide recta, disticha duabus lineis rectis utrinque, linea una media adpressa in superficie superiore.   Folia a basi cordata lanceolato-cuspidata, integra, nervo carentia.   Cellae longae atque angustae, praeter lineam inferiorem cellarum basalium inflatarum flavarum.   Capsula horizontalis ; operculum rostro longo angusto.

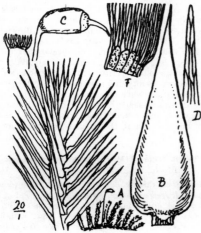

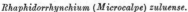

Rhaphidorrhynchium (Microcalpe) zuluense.

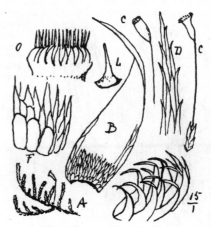

Rhaphidorrhynchium Dregei.

656. *Rhaphidorrhynchium Dregei* (C. M.), Broth., Pfl., xi, 427.

= *Hypnum Dregei* C. M., Syn., ii, 311, and Rehm. 411 and *b*.
= *Rhaphidostegium Dregei* (C. M.), Jaeg., Ad., xi, 473 ; Dixon, Trans. R. Soc. S. Afr., viii, 3, 220 ; Dixon, in Kew Bull., No. 6, 1923, in No. 401.
= *Cupressina Dregei* C. M., in Hedw., 1899, p. 146.
= *Rhaphidostegium krakakammae* (C. M.), Jaeg., Ad., ii, 476 (*fide* Dixon) (= *Rhaphidorrhynchium* Broth., Pfl., xi, 427).
= *Hypnum krakakammae* C. M., Syn., ii, 285.
= *Rhaphidostegium tapeinophyllum* (C. M.), Par., Suppl. Ind., 1900, p. 298.
= *Hypnum tapeinophyllum* C. M., in Rehm. 410 (= *Rhaphidorrhynchium* Broth., Pfl., xi, 427).
= *Cupressina tapeinophylla* C. M., in Hedw., 1899, p. 146.
= *Rhynchostegium* *Breutelianum W. P. S., in Breutel, Musc. Cap.
= *Rhaphidostegium Rehmanni* (C. M.), Par., Suppl. Ind., 1900, p. 297 (= *Rhaphidorrhynchium* Broth., Pfl., xi, 427).
= *Aptychus Rehmanni* C. M., in Hedw., 1899, p. 142 (*Hypnum*, Rehm. 401).
= *Rhaphidostegium dentigerum* (C. M.), Par., Suppl. Ind., 1900, p. 29 (= *Rhaphidorrhynchium* Broth., Pfl., xi, 427).
= *Rhynchostegium dentigerum* C. M., in Rehm. 419 (not *Hypnum*, C. M., in Rehm. 331, as put in Paris' Index).
= *Cupressina dentigera* C. M., in Hedw., 1899, p. 146.
= *Hypnum natalense*, Rehm. 674 (not Paris).

Stems prostrate, epiphytic, ascending or suberect, yellowish and shining, much branched, 2–3 cm. long ; the branches ascending, 3–10 mm. long, rather slender, densely leafy ; the leaves lanceolate-subulate, falcate, very concave throughout, the apex of each leaf twisted over in one direction making the leaf-arrangement regularly and distinctly secund and blunt at the top ; leaf-margin usually ascending, seldom flattened, entire except in the subula where it is often more or less toothed by protruding cells. Nerve absent ; alar group of three to four large swollen cells, above which and in the leaf-centre are short cells gradually extending into the ordinary linear or rhomboid cells of the leaf, which are smooth or almost so, except that the upper cells are papillose at the back. Perichaetal leaves few, concave and regular, longer and less subulate. Seta 1–1·2 cm. long, slender, red, straight, smooth. Capsule nearly erect or somewhat inclined, oval, when dry constricted below the wide mouth. Peristome teeth inflexed, lanceolate-subulate, rather thin, not furrowed down the mid-line, but very pronouncedly mamillate along the inner surface mid-line, these mamillae papillose, and because the tooth is thinner than the mamillae, are even more prominent than in *R. brachy-carpum.* Inner teeth standing erect between the outer teeth, plaited, with high basal membrane, and the cilia rudimentary or absent. Lid rostrate ; beak long and slender.

Abundant in forest and scrub regions of South Africa from Table Mountain to the Zambesi.

657. *Rhaphidorrhynchium Gueinzii* (Hampe), Broth., Pfl., xi, 428 = *Rhaphidostegium Gueinzii* (Hampe), Jaeg., Ad., ii, 473 ; Dixon, in Trans. R. Soc. S. Afr., viii, 3, 320 (Rehmann's labels read *Quinzii*).

= *Hypnum Gueinzii* Hampe, Ic. Musc., t. 12 ; C. M., Syn., ii, 311 ; Rehm. 409, and *b, c*.
= *Rhynchostegium Gueinzii* C. M., Rehm. 659, and var. *julascens.*
= *Hypnum cupressiforme*, var., Horns., in Dregé, Musc. Cap.
= *Hypnum Gueinzii* Hpe., var. *prostrata* C. M., Syn., ii, 311.
= *Rhaphidostegium hyalotis* (C. M.), Par., Suppl. Ind., 1900, p. 296 = *Rhaphidorrhynchium hyalotis* (C. M.), Broth., Pfl., xi, 428.
= *Cupressina hyalotis* C. M., in Hedw., 1899, p. 147.
= *Hypnum hyalotis* C. M., in Rehm. 408.

Moderately robust, dark green below with yellow tips ; stems up to 2 inches long, irregularly branched ; the branches spreading or suberect, 3–10 mm. long, the tip either erect or somewhat curved, but in either case sharp-pointed by congested young leaf-points. Leaves lanceolate-subulate, regular or slightly falcate, erecto-patent, concave ; the margin erect except the flat apical portion, almost entire but crenulated by protruding cells on the subula ; nerve

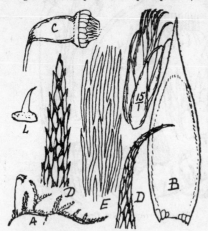

Rhaphidorrhynchium Gueinzii.

absent ; cells in two layers linear-vermicular, smooth or drawn to a raised upper end ; the cells toward the apex linear-rhomboid, those toward the base often discoloured, shorter and wider, but passing suddenly to the swollen cells without intermediate quadrate cells. Swollen cells sometimes two to three, large, sometimes four to six and smaller. Perichaetal leaves hardly differentiated or shortly pointed or subobtuse, very closely denticulated in the upper part. Seta 1 inch long, red, smooth ; capsule nearly horizontal, oval, wide-mouthed when dry. Abundant on and round Table Mountain, and present in Knysna, Eastern Province, Natal, O.F.S., and Transvaal forests.

Genus 217. Trichostelium (Mitt.), Jaeg., Ad., ii, 447 ; Broth., Pfl., i, 3, 1116.

Deep green, autoecious, shining, epiphytic mosses, with pinnate, short, rooting stems ; leaves all alike, spreading, lanceolate, with reflexed tips, erect almost entire margin, no nerve, linear-elliptical cells with one or more papillae on the lumen, and large, oval, basal cells. Perichaetal leaves erect, long-pointed ; seta exserted ; capsule inclined or hanging, small, oval-cylindrical, with short throat. Peristome double ; teeth not furrowed, striated ; basal membrane high ; processes as high as the teeth ; cilia one to two ; spores small ; lid acutely rostrate ; calyptra smooth.

658. *Trichostelium perchlorosum* Broth. and Bryhn, in Bryhn, p. 24 ; Broth., Pfl., xi, 439.

Unknown to me, described in Videnskap. Forhand., 1911, No. 4, p. 24, in section *Papillidium*, thus : " Autoecious, male flowers near the perichaetium, slight, caespitose, in dense, dark green, flat, shining mats. Stem prostrate, 20–30 mm. long, tufted-tomentose, rigid, closely pinnate ; branches spreading, up to 10 mm. long, and with the leaves about 2 mm. wide, hardly attenuated, flat or rising. Leaves densely crowded, soft, concave, channelled at the apex, complanate, horizontally spreading, with recurved apex, lanceolate-subulate from ovate base, 1·25–1·5 mm. long, 0·3–0·4 mm. wide, little changed when dry ; margin erect ; apex minutely or sharply serrulated by prominent cells ; nerve absent. Alar cells uniseriate, on each side three, large, oval, inflated, hyaline ; inner basal cells elongate-rectangular, yellow, slightly thickened ; suprabasal cells shortly rectangular, others gradually linear, not thickened ; at mid-leaf one central papilla on the back often present ; very chlorophyllose. Perichaetium rooting. Inner per. leaves erect, narrowly lanceolate, gradually filiform, without nerve ; cells smooth ; margin entire. Seta 5–8 mm. high, slender, flexuous, red, rough or roughish, smooth below. Capsule small, inclined or pendulous, cylindrical-arched, tapering by a short throat ; asymmetric ; subtumid on the back ; not constricted under the mouth when dry, brown ; exothecium especially on the side of the capsule next the seta slightly mamillate. Peristome as in the genus. Lid long-conical, acutely rostrate.

Ntingwe Wood, Ekombe, Zululand, on the bark of trees, L. M. Titlestad."

Family 52. Hookeriaceae.

Delicate, somewhat succulent, more or less pellucid, sparingly branched, and usually prostrate plants, without stolons and without paraphyllia, growing on moist or wet ground, or on wet stumps or stones, with the branches and leaves usually complanate ; the leaves consisting of one layer of cells, pellucid, usually unsymmetrical ; the side leaves spreading and the back and front leaves appressed, and usually alternately and obliquely pointing right and left. Nerve single, double, forked, or absent. Cells smooth in South African genera (except *Callicostella*), usually large, lax, roundish or hexagonal ; the basal cells sometimes coloured ; alar cells not differentiated, but marginal cells sometimes different. Perichaetal branch very small, few-leaved, sometimes rooting. Seta well developed, rough or smooth ; capsule erect or cernuous, usually regular ; lid rostrate from a conical base ; calyptra mitriform, lobed or ciliated. Peristome double, well developed ; the processes about as long as the teeth ; cilia absent or rudimentary. The species of Hookeriaceae have been frequently moved from one genus to another, the generic characters used varying with the author, and usually founded on vegetative organs, so that now the synonymy is very involved.

Dr. Harvey, in first edition of his Genera of South African Plants, includes Hookeria, and mentions four species which, " *with others*," he states, " are found in our moist mountain woods." These are as under, and I have noted opposite to each what I know about it, viz. :

Hookeria lucens .	Sm. A British species, without nerve, not known from South African localities.
Hookeria laetevirens	Hk. and Tay. See under *Cyclodictyon.*
Hookeria rotulata .	Sm. An Australasian *Hypopterygium* [(Hedw.), Brid.], not known here though still maintained in Wager's list.
Hookeria laricina .	See under *Hypopterygium.*

These early records can only be looked on as approximate, and the genus *Hookeria* (as here understood), having no nerve, is not now known to be South African.

Synopsis of Hookeriaceae.

Nerve single, undivided	218. DISTICHOPHYLLUM.
Nerve single at the base, usually forked above	219. ERIOPUS.
Nerve double from the base.	
Leaves margined	220. CYCLODICTYON.
Leaves not margined.	
Cells usually smooth	221. HOOKERIOPSIS.
Cells usually with one papilla per cell	222. CALLICOSTELLA.

Genus 218. DISTICHOPHYLLUM Doz. and Molk.

Prostrate or compactly matted, light green, monoecious mosses, somewhat crisped when dry. Stem short, rooting, simple or branched ; leaves all alike except as to size, ovate, obovate or lanceolate, 1-nerved to near the apex. usually bordered by long narrow cells, the lamina cells being roundly hexagonal, the lower cells larger. Seta 6–10 mm. long, red ; capsule erect or slightly inclined, oval with short throat ; ring absent ; peristome double, equally high. Teeth lanceolate-subulate, papillose above, with zigzag mid-line ; processes pale, perforated, papillose, with low membrane ; lid shortly rostrate ; calyptra erect, campanulate, fringed at the base.

Synopsis of Distichophyllum.

659. *D. Taylori.* Cells 13–16μ wide.
660. *D. mniifolium.* Cells 6–8μ wide.

659. *Distichophyllum Taylori* Sim (new species).*

Closely tufted, on humus, light green, somewhat crisped when dry. Stem simple or forked, 1 cm. long, suberect, tomentose. Leaves placed round the stem, equal, not complanate, suberect, 1·5 mm. long, or some shorter intermixed, obovate-acute, with rather a wide base ; nerve strong and usually yellowish, extending to near the leaf-apex ; margin entire, strongly bordered all round with long narrow cells ($100 \times 7\mu$), often yellowish ; apex acute or pointed, consisting of border cells ; cells roundly hexagonal, 15μ diameter, thick-walled, smooth, pellucid ; those near the leaf-base gradually longer to $35 \times 20\mu$, the lowest line very lax and usually yellowish. Other parts not seen. The Wilderness, George, Cape, W., July 1922, Miss Taylor (Sim 10,281).

Near *D. imbricatum* Mitt. (Marion Island), but has much weaker leaf-border and rather narrower cells.

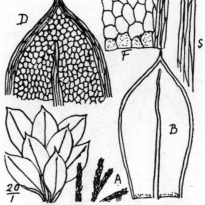

Distichophyllum Taylori.

660. *Distichophyllum mniifolium* (Horns.) Sim (new combination).

=*Hookeria mniifolia* Horns., in Linn., 1841, p. 141 (not *Hookeria mniifolia* Mont., 1842).
=*Mniadelphus Hornschuchii* C. M., in Syn. ii, 22, and in Hedw., 1899, p. 130 ; Par., Ind., 2nd ed., iii, 257.

Unknown to me, but described by Hornschuch as the smallest then known of the genus ; the plant minute ; stems two to three lines long, usually simple or slightly branched, with red radicels. Leaves distichous, spathulate, mucronulate, bordered, entire, 1-nerved, pellucid ; nerve obscure, ending above mid-leaf ; basal cells lax, sublinear or subquadrate, hyaline ; upper cells smaller, subcircular. Outer per. leaves similar, inner smaller, erect, oblong, shortly acuminate ; seta short, erect, smooth, purplish. Capsule suberect, oblong, narrow, subapophysate, when dry contracted at the base and below the mouth, smooth, brown ; the peristome of sixteen pale lanceolate teeth and sixteen equally high lanceolate-subulate processes with plicate hyaline membrane ; lid conical-acuminate with curved apex, shorter by half than the capsule. Koratra, Dregé, 1831.

C. M., in Syn., ii, 22, took over most of this description as *Mniadelphus Hornschuchii* C. M., because two species had been described as *Hookeria mniifolia*, but Hornschuch's name had priority. However, in Hedw., 1899, p. 130, C. M. again described *Mniadelphus Hornschuchii* as a new species ; apparently the same plant, but from Genadendal (Breutel), sterile, and described as having the nerve ending near the acumen. I have no other localities, though Shaw states that he found *Mniadelphus pulchellus*, Hampe (an Australasian species), on the Katberg, which most likely was the same plant.

---

* *Distichophyllum Taylori* Sim (sp. nov.).

Caespitosum ; caules 1 cm. alti, plerumque simplices ; folia aequa, undique circum caulem disposita, non complanata, 1·5 mm. longa, obovato-acuta ; nervus simplex, fortis prope ad folii apicem ; margo integer, cellis longis angustis marginatus ; cellae superiores rotunde hexagonae, 15 mm. diam., leves, pellucidae ; cellae basales longiores ; cetera parum cognovi.

Dixon, who has seen both the above described, finds them different in the size of the cells, and also that *D. Taylori* has slightly larger, more acute leaves, with longer cuspidate points, but that they are much alike.

I think one of these species of *Distichophyllum* may also be what Shaw recorded from the Katberg as *Cyathophorum bulbosum* C. M. (another Australian species), which has a single or forked nerve, though one species of the genus, *Cyathophorum africanum* Dixon, in Sm. Misc. Coll., lxix, 8, 5, has been found in Uganda, and it is just possible that it may be on the Katberg. My Uganda specimens have longer, narrower, serrate side leaves than this, smaller lanceolate surface leaves, and a gemmiparous bud in every axil, but that bud is not always present.

Genus 219. ERIOPUS Brid., Bry. Univ., ii, 339.

Fairly tender, vigorous plants, spreading on the ground and with erect, dendroid, branched stems; the side leaves complanate, spreading widely, very large, unsymmetrical, standing alternately right and left, more or less oval, shortly pointed, toothed in upper half; surface leaves much smaller, appressed, toothed; nerve single at the base, forked upward, with unequal arms; cells very lax, roundly hexagonal, smooth, those at the leaf-base longer and looser, and at the margins there are several rows of longer and narrower cells; seta short or long, papillose or bristly, not twisted; capsule inclined or hanging, small, oval, with a short thickened neck. Lid straight, acute from a wider base; calyptra mitriform, smooth or rough, with laciniated lobes at the base.

661. *Eriopus mniaceus* (C. M.), Broth., Pfl., i, 3, 931, and xi, 233.

= *Hookeria mniacea* C. M., in Bot. Zeit., 1859, p. 247, and in Rehm. 335, and in Hedw., 1899, p. 130 (where Rehm. 338 should be 337).
= *Pterygophyllum mniaceum* (C. M.), Jaeg., Ad., ii, 146.
= *Pterygophyllum* *Rehmanni C. M., in Rehm. 337 (*fide* C. M., who cites it R. 338); Par., Ind., 1897, p. 1054.
= *Pterygophyllum* *sublucens C. M., in Rehm. 336 (*fide* C. M., in Hedw., who calls it Coll., No. 89); Par., Ind., 1897, p. 1054.

Plants erect, dendroid; stems simple or irregularly branched, 1–2 inches high, stout, with small scattered leaves toward the base but regularly distichous distant leaves higher up, along with a right and left row of appressed smaller leaves on the upper surface.

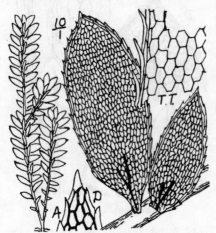

Eriopus mniaceus.

Leaves elliptical, sometimes shorter; the side leaves elliptical, 4–5 mm. long, spreading widely, nearly regular at the shortly acute point, and at the base quickly narrowed to the narrow connection, the lower side usually folded at the base. Surface leaves more regular, flatly appressed, 2 mm. long, with lower margin not reflexed; both kinds strongly bordered by about two lines of long, pellucid or yellowish cells, of which each on the outer line protrudes at the upper end and gives either small or large teeth; other cells hexagonal, with roundish lumen $100 \times 50\mu$, the apical cells shorter and the basal cells longer, all thick-walled; remarkably like *Mnium* areolation; nerve usually single at the base, branched into two unequal legs, yellowish, and not extending half-way. Leaves rather crowded at the branch-tops, as if comal for inflorescence, but no inflorescence or capsule seen.

Müller's note in Hedw., 1899, p. 130, indicates that he accounts for his three names by variation in size, etc., so it would appear that by that time he had realised that variation did occur. But it is a rare moss, except in the Knysna Wilderness region.

Cape, W. The Wilderness, George, Miss A. Taylor (in several packets); Blanco, Knysna, Rehm. 335; Montagu Pass, Rehm. 336; Devil's Peak at the cataract, Rehm. 337; Kumakala, Breutel, 1862.

Genus 220. CYCLODICTYON Mitt.

Stem long, creeping, weak, rooting, irregularly or pinnately branched, and with the leaves in a laxly flattened arrangement. Leaves in five to eight rows, heteromorphous, not symmetrical, the side leaves the larger, spreading, long-ovate, pointed, bordered, toothed toward the apex. Nerves two, divergent, not reaching the apex, smooth except near the points; cells smooth, lax, roundly hexagonal, those at the leaf-base longer; the marginal cells narrow and hyaline. Capsule arcuate or horizontal, irregular, smooth, with a thick neck. Lid long-beaked from a shortly conical base. Calyptra conical, lobed at the base, naked. Peristome double; cilia absent.

Synopsis.

Border of two, three, or more lines of narrowly hexagonal cells	.	.	.	.	662. *C. laetevirens.*
Border one to two lines of hyaline, narrowly elliptical cells	.	.	.	.	663. *C. vallisgratiae.*

662. *Cyclodictyon laetevirens* Mitt., in Journ. of the Linn. Soc., 1863, p. 136 ; Broth., Pfl., xi, 237 = *Hookeria laetevirens* Hk. and Gr., Musc. Brit., 1st ed., p. 89, t. 127 ; C. M., Syn., ii, 187 ; Bry. Eur., v, t. 447 ; W. P. Sch., Syn., 2nd ed., p. 282 ; Dixon, 3rd ed., p. 401, Tab. 49i.

= *Pterygophyllum laetevirens* Brid., Bryol. Univ., ii, 350.

Autoecious ; prostrate on mud or wet wood ; stems 1–2 inches long, sparingly and irregularly branched ; the branches flat and sometimes rooting ; leaves usually complanate, heteromorphous, the side leaves spreading widely, 2–2·5 mm. long, asymmetrical, oblong, somewhat concave, abruptly rounded to a short acute point ; 2-nerved, the base with one nerve nearly central and the other on the lower side near the margin, nerves equal or unequal, extending beyond mid-leaf or near to the apex ; the lower side usually reflexed at the base as far as the nearest nerve, making the leaf appear more unequal ; margin entire in the lower half, strongly dentate by triangular hyaline cells in the upper half ; lamina cells roundly hexagonal or with nearly circular lumen, deep-walled, pellucid, $35 \times 15\mu$, except two, three, or more lines along the margin, which also are hexagonal, but longer and narrower, forming a rather indistinct border ; those in the lower half of the leaf are up to $50 \times 12\mu$. Surface leaves few, about 1 mm. long, lanceolate, 2-nerved, bordered. Perichaetal leaves similar, smaller. Seta 2 cm. long, erect, red, smooth ; capsule inclined, turgidly oblong, with short neck ; teeth lanceolate, bordered, deeply furrowed along the mid-line, trabeculate, closely cross-striate and red in the lower half, paler and papillose above ; processes grey, equally high ; lid acutely rostrate. Found in Europe, North and South Africa.

Cape, E.   Perie Forest, 3500 feet, 1892, Sim 7232.
Natal.   Sweetwaters, 2500 feet, June 1916, Sim.

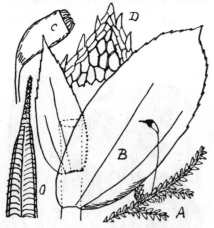

Cyclodictyon laetevirens.

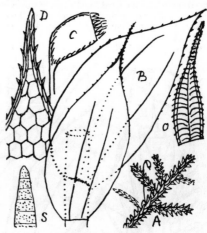

Cyclodictyon vallisgratiae.

663. *Cyclodictyon vallisgratiae* (Hampe), Broth., Pfl., i, 3, 935, and xi, 237.

= *Hookeria vallisgratiae* Hampe, MSS. ; C. M., in Bot. Zeit., 1858, p. 169.
= *Hookeria *subcordata* Hampe, MSS., in Sched.
= *Hookeria Breuteliana* Hpe., in Sched. ; C. M., Bot. Zeit., 1859, p. 247 (see C. M., in Hedw., 1899, p. 152).
= *Hookeria natalensis*, Rehm. 622 and 622b = *Pterygophyllum natalense*, Rehm. 338 (*fide* Rehm. 622).

Autoecious ; prostrate, sometimes matted, on wet mud ; stems 1–2 inches long, with leaves 3–4 mm. wide, irregularly branched ; the branches more or less complanate, up to 1·5 cm. long, rooting occasionally ; leaves complanate, pellucid, membranaceous, heteromorphous, the side leaves 2 mm. long, asymmetrical, spreading, oblong-ovate, acuminate, acute ; the surface leaves 1·5 mm. long, the same form, appressed, less spreading, and often right and left alternately ; both slightly concave, 2-nerved ; the nerves strong and extending to near the leaf-apex, smooth on the back or with the end protruding ; margin expanded, distinctly bordered by one row of hyaline elliptical cells with protruding points, making it denticulate in the upper half, and usually another row of narrow cells, both widely different from the roundly hexagonal cells, $35 \times 20\mu$ inside, which are smooth, chlorophyllose when vigorous, pellucid when old, laxly deep-walled ; the basal cells rather longer, hexagonal ; surface leaves usually present on the upper surface, fewer on the under-surface ; perichaetal leaves similar, but smaller, lax, hardly nerved, without margin, and entire. Seta 1–1·5 cm. long, red, smooth ; capsule brown, 2 mm. long, oval, horizontal or pendulous, with rather long, lanceolate, inflexed, deeply furrowed, deep brown, bordered and cross-striated teeth, papillose and pale above, and equally high, pale, faintly punctulate, blunt processes on short membrane.

Cape, W.   Devil's Peak, Rehm. 340, 341, and Sim 9233 ; Constantia Nek, Sim 9338, 9351 ; The Wilderness, George, Miss Taylor.
Cape, E.   Ugie, Sim 8576 ; Perie Forest, 1892, Sim 7227 ; Grahamstown, Miss Farquhar 73.
Natal.   Inanda, Wood (Rehm. 622b) ; Umpomulu, Buchanan, (Rehm. 622) ; Natal, Wager 160.
Transvaal.   Witpoortje Kloof, Professor Moss ; Rosehaugh, 3000 feet, 1914, Sim.

Genus 221. HOOKERIOPSIS (Besch.), Jaeg., Ad., ii, 262.

Fairly strong, prostrate, branched, flatly leaved mosses, the branches irregular, or pinnate, or subdendroid. Leaves in eight rows, heteromorphous, somewhat unsymmetrical, shortly pointed, serrate, not bordered, and with cells laxly hexagonal, usually smooth; nerve double from the base, strong, the two nerves equal. Seta long, smooth. Capsule arcuate or horizontal, with a thickened neck; lid subulate from a wide base; calyptra conical, shortly lobed or nearly entire, naked, covering half the capsule. Peristome double; cilia absent.

664. *Hookeriopsis Pappeana* (Hampe), Jaeg., Ad., ii, 264; Broth., Pfl., i, 3, 942, and xi, 243; Bryhn, p. 20.

= *Hookeria Pappeana* Hampe, Ic. Musc., t. 2; C. M., Syn., ii, 194, and Hedw., 1899, p. 151.
= *Hookeria macropyxis*, Rehm. 339 (wrongly printed *megalopyxis* in Par. Ind.).

Caespitose in mud or on wet alpine stones, usually shining brownish red; stems ½–1½ inch long, slightly branched; leaves nearly equal in size, numerous, set all round the stem, subcomplanate, 2–2·5 mm. long, apparently widely lanceolate, actually much wider because deeply concave, rounded to a narrow base, acuminate to acute apex, strongly 2-nerved to above mid-leaf; nerve smooth, often ending in a spicule on the back of the leaf; margin strongly dentate upward by large single, hyaline, extruded, acute cells; other cells all equal, without any border, except that the lower cells, 50–70 × 15μ, are longer than the upper ones; margin erect, partly lost to view by the concavity of the leaf; perichaetal leaves much shorter, lanceolate-acuminate, nerved, nearly entire; seta red, 1 cm. long, smooth, with subulate apex, suberect to cygneous; capsule small, 1·5 mm. long, oval with slight neck, brown; lid sharply rostrate; peristome dark red; teeth deeply furrowed, red, barred, bordered, striate below, paler and papillose above; processes equally high on narrow membrane. Found in East and South Africa.

Cape, E. Ugie, Transkei, 5000 feet, June 1899, Sim 8575; Paradise Kloof, Coldspring, Grahamstown, May 1921, Miss Britten 2758.

Natal. Inanda, J. M. Wood (Rehm. 339 and 623, as *Hookeria *macropyxis* Rehm.).

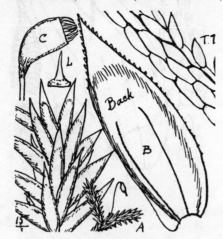

Hookeriopsis Pappeana.

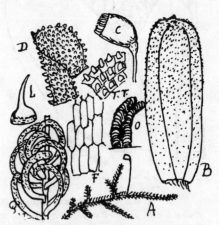

Callicostella tristis.

Genus 222. CALLICOSTELLA (C. M.), Jaeg.

Rather long, prostrate, rooting, flatly leaved mosses, more or less irregularly or pinnately branched; leaves set round the stem, hardly complanate, somewhat unsymmetrical; the back and front leaves obliquely placed; the side leaves hardly larger; all spreading and then erect; when dry all crumpled up and inward, when moist flat; long-ovate or elliptical, with or without point and not margined; nerve double from the base, strong, toothed along the ridge. Cells oval-hexagonal, often with one papilla per cell, those at the leaf-base much longer and smooth. Seta fairly long, smooth or rough; capsule horizontal, somewhat irregular, with double peristome and no cilia. Lid subulate from a wide base; calyptra conical, shortly lobed at the base, rough above.

Synopsis.

Leaves lingulate . . . . 665. *C. tristis.*
Leaves ovate-lingulate . . . . 666. *C. applanata.*

665. *Callicostella tristis* (Rehm. 342), Broth., Pfl., i, 3, 938, and xi, 239.

= *Hookeria tristis* (Rehm.), Par., Ind., 1895, p. 587; C. M., in Hedw., 1900, p. 130.

Flatly caespitose on soil, dark green, autoecious; stems 1–2 inches long, irregularly branched; the branches ½ inch long or less, spreading or suberect; leaves lax, lingulate, 1·5 mm. long, 0·5 mm. wide, somewhat irregular through the nerves starting on one side ; base with wide fibrous attachment ; leaves involute and inflexed when dry, spreading and then erect when moist ; apex suddenly rounded to a depressed mucro, one side often somewhat higher than the other ; nerve double from the base, each strong and ascending to near the leaf-apex, and spiculate on the back ; margin expanded when moist, nearly entire toward the base, strongly papillate-erose upward. Upper cells dense and opaque, about 20μ long, each bearing a large central hyaline papilla ; lower cells gradually longer, more pellucid, and less papillate ; lower cells rectangular-hexagonal, thin-walled, pellucid, and smooth, about $70 \times 15\mu$. Perichaetal leaves similar or shorter ; seta 1–1·5 cm. high, red ; capsule nearly horizontal, oval, brown, with short tapering neck ; peristome very red and solid ; teeth widely and deeply furrowed, closely barred, cross-striate ; membrane wide ; processes lanceolate-subulate, keeled, finely papillose ; cilia absent and ring absent. Lid obliquely rostrate from a convex base.

Müller's description in Hedw., 1899, p. 130, from a sterile plant and with no mention of papillae is exceedingly incomplete and misleading.

Natal. Inanda, J. M. Wood (Rehm. 624 and 342).

Transvaal. Lydenburg, MacLea (Rehm. 626) ; Wolhuter's Kop, 1912, Wager 159 ; Mahalisberg, MacLea (Rehm. 625).

666. *Callicostella applanata* Broth. and Bryhn ; Bryhn, p. 19 ; Broth., Pfl., xi, 239.

= *Hypnum *Macleanum*, Rehm. 677.

Stems prostrate or when crowded erect or suberect, rooting, 2–3 inches long, glaucous green, branching ; branches simple, pinnate or somewhat 2-pinnate, irregular ; leaves numerous, complanate, asymmetric, densely chlorophyllose-granular, 1·5–1·75 mm. long, 0·75–1 mm. wide, concave, widely ovate-lingulate, shortly mucronate, widest above the base, narrower toward the rounded or almost truncate apex, 2-nerved from the lower corner at the base ; nerves strong to near the apex, smooth on the back, but often spurred on the back at the apex, forming two keels ; margin crenulate in the upper half of the leaf through protruding cells, smooth and straight toward the base ; upper cells roundly hexagonal, 10μ wide, indistinctly 1-papillate, gradually larger but granular to near the leaf-base ; lowest cells laxly rectangular, to $40 \times 15\mu$, hyaline ; fruit not seen. Side leaves spreading widely ; surface leaves appressed, smaller and more symmetrical at the base ; when dry, all recurved from the keels to form a triangular cylinder, or somewhat crisped.

Natal. Wet ground, Eshowe, Zululand, 1908, Haakon Bryhn.

Transvaal. Mahaliesberg (?), MacLea (Rehm. 677, as *H. Macleanum* R.).

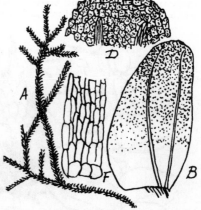

Callicostella applanata.

Family 53. HYPOPTERYGIACEAE.

Beautiful, delicate, light green mosses, having prostrate, tomentose-rhizomatous main stems, from which rise erect, simple, pinnate, flabellate or dendroid secondary branches having two rows of distichous larger leaves in one plane, and one or two rows of smaller appressed leaves on the under-surface, resembling the amphigastria of Hepaticae. Leaves ovate-acute to widely lanceolate, usually obliquely inserted, pellucid, unsymmetrical, 1-nerved in the lower half, with lax, rounded, rhomboid, smooth areolation, often bordered by longer cells, usually toothed, and with no differentiated alar cells ; capsule on a long or short seta, usually inclined or hanging, regular ; peristome double (in South African genera) ; outer peristome of sixteen lanceolate teeth, connate at the base, closely barred transversely, and with zigzag mid-line ; inner peristome with plaited basal membrane and keeled processes, and often having intermediate cilia ; lid beaked ; calyptra conical-beaked, naked.

Synopsis of the Hypopterygiaceae.

(Secondary stem simple ; under-leaves in one line ; seta very short, axillary . . . CYATHOPHORUM.)
Secondary stem pinnate, fan-like or dendroid ; under-leaves in two lines ; seta long . 223. HYPOPTERYGIUM.

(Genus CYATHOPHORUM Palis. Secondary stems erect, simple or almost so ; the lower portion leafless, the upper set with two distichous rows of obliquely lanceolate-ovate, bordered, toothed leaves in one plane, and a single line of smaller leaves on the under-surface. Capsule erect, oval, regular, on a very short axillary seta. Inner peristome with three cilia between the processes. Mostly an Australasian genus, of which Dr. Shaw records that *C. bulbosum* Hedw. was collected by him on the Katberg. It has not been recorded since, and I doubt Shaw's identification. See further under *Distichophyllum mniifolium* herein.)

Genus 223. HYPOPTERYGIUM Brid.

Main stem tomentose-rhizomatous, from which erect secondary stems ¼–1 inch apart rise, which are simple below but pinnate or flabellate above, and dendroid in habit, ½–2 inches high, with branches and leaves on one plane, and usually with a double row of small leaves on the under-surface ; the whole frond-like stem crisped and closely inrolled when dry, flat and beautifully pellucid green when moist. Leaves distichous, obliquely ovate-acute, bordered, usually serrate or ciliate, with nerve half-way or more, or sometimes excurrent ; under-leaves similar but smaller, more pointed or hair-pointed and appressed. Cells roundly hexagonal or rhomboid, very lax and pellucid, smooth ; the border cells longer. Capsules numerous, from near the furcation of the erect stems, with fairly long, erect, or often cygneous seta, oval or oblong, with wide dehiscent ring : lid beaked from a conical base ; calyptra conical or cucullate, naked.

A genus of very beautiful mosses, usually growing among other mosses on moist stones or bark in forest ; frequent but not common.

Synopsis of Hypopterygium.

Fronds flabellate	. . . .	667. *H. laricinum.*
Fronds pinnate	. . . .	668. *H. pennaeforme.*
Stems 2-pinnate ; cells narrowly linear	. . . . . .	669. *H. lutescens.*

667. *Hypopterygium laricinum* (Hook.), Brid., Bry. Univ., ii, 714, p.p. ; C. M., Syn., ii, 7, p.p. ; Marloth's Flora, p. 62, and pl. 7*f* ; Bryhn, p. 20 ; Brunnthaler, p. 32 ; Broth., Pfl., i, 3, 971, and xi, 275.

=*Hypnum laricinum* Hook., Musc. Exot. (excluding Andine specimens).
=*Hookeria laricina*, Harv. Gen., 1st ed.
=*Hypopterygium* *capense W. P. Sch., in Breutel, Musc. Cap.

(*H. mauritianum* (Hpe.), Besch., from Mauritius, which has only one line of border cells on the leaves and practically no nerve in the stipular leaves, was mixed with this in Bridel's and in Müller's descriptions.)

Main stem rhizomatous, black, tomentose, leafless, giving off erect stems at ½-inch distances or thereby ; erect stems dendroid, scale-clad and branchless below, 2-pinnately flabellate above (or branches are often simple), ½–1½ inch high ; flabellate frond ½–1 inch diameter, subcircular ; leaves distichous, alternate, spreading widely, obliquely

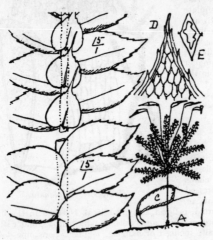

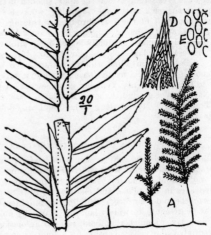

Hypopterygium laricinum.   Hypopterygium pennaeforme.

ovate-acuminate, acute or ovate-acute, 1½ mm. long, pellucid, very light green, dentate along the upper margin and only near the apex on the lower margin ; apex often obliquely tending upward ; nerve starting from lower corner of the base, strong, extending to near the apex ; cells rhomboid, $40 \times 20\mu$; lower cells little different ; border cells long and narrow in two to three lines, the outer extruded as teeth ; stipular leaves in one line, much smaller, appressed, cordate, with long, slender, almost entire subula, toothed above, the nerve extending half-way or more ; perichaetal leaves rather large, ovate-acuminate, acute, sheathing, hardly bordered, not toothed and not or only shortly nerved ; seta 1·5–2 cm. long, red ; capsule inclined or horizontal or eventually hanging, ovate-oblong, with straight subulate-rostrate beak as long as the capsule. Calyptra white, cucullate, not fringed at the base, rather persistent, and sometimes set on sideways.

Usually present in forest streams, often very scattered, seldom but sometimes caespitose. Peristome teeth as in the genus.

Cape, W. Kirstenbosch, Sim 4462 and 9478, Pillans 3353 ; Table Mountain, Sim 9112, and Miss Michell 3 ; Newlands Ravine, Sim 9430 ; Clermont and Rondebosch, Rehm. 298 and 298*c* ; Montagu Pass, 298*b*.

Cape, E. Eveleyn Valley, Perie, 4000 feet, 1892, Sim 7238 ; Grahamstown, Miss Farquhar 21 ; Longbushkop, Alexandria Forest, and Hankey, Jas. Sim ; Diepwall, J. C. Phillips ; Wilderness, George, Miss Taylor ; Knysna and Zitzikamma, Burtt-Davy.

Natal. Buccleuch, Sim 8550 ; Mont aux Sources, 7000 feet, Maritzburg, Van Reenen, and Zwart Kop, Sim.

Transvaal. Rosehaugh, Sim 8549 ; Barberton, Wager ; The Downs, Pietersburg, Junod.

Rhodesia. Inyanga, J. S. Henkel (Eyles 2634).

It occurs also in tropical East Africa and in Fernando Po.

668. *Hypopterygium pennaeforme* (Thun.), Horns., in Linn., 1841, p. 143 ; Brid., Bry. Univ., ii, 717 ;
C. M., Syn., ii, 10 ; C. M., in Hedw., 1899, pp. 124–129 ; Bryhn, p. 40.

= *Hypnum pennaeforme*, Thun., Prodr. Fl. Cap., ii, 175 (?).
= *Lopidium pennaeforme* (Thun.), Fleisch. ; Broth., Pfl., xi, 271.

Light green ; laxly tufted or mixed among other mosses ; stems rhizomatous, tomentose, leafless, sending up erect stems at about ½-inch distances ; erect stems dendroid, 1–3 inches high, up to ½ inch wide, scale-clad and branchless below, complanately pinnate above, or sometimes simple branches spreading, with lower leaves small and deltoid, other leaves distichous, erecto-patent, 1·50–1·75 mm. long, concave, obliquely lanceolate-acute or sometimes lingulate, with quickly subulate apex, clasping at the base, which is rounded on the upper side and often inflexed on the lower side ; the nerve strong, starting at the lower basal corner, central above, extending almost to the apex. Margin bordered by about two rows of long cells, the outer line protruding somewhat as teeth on both sides toward the apex, elsewhere giving a waved margin ; lamina cells with roundish or oval lumen, 12–15μ long, hardly different to the leaf-base ; stipular leaves appressed in one row along the stem, much smaller, 1 mm. long, lanceolate from a rounded clasping base, more toothed, and nerved almost into the apex. Other parts not seen by me, but Hornschuch gives full particulars, though a doubt may be cast on his identification since he describes quite wrongly where he says, "retis areolis linearibus angustis basi folii laxioribus " ; at least my plant is not so. Hornschuch himself doubts the identity of his plant with that of Thunberg, whose description was incomplete but, as Hornschuch puts it, was wide enough to cover this. His description includes the following : Flowers monoecious, male not seen, but jointed yellow paraphyses are often present in the axils ; per. leaves erect, imbricate, widely oblong, sheathing, with spreading, long, slender points, entire, nerveless ; the inner large and much narrower ; seta ½ inch, erect, cygneous, smooth ; peristome double, the teeth incurved when dry, lanceolate-subulate, solid ; reddish, pale above ; processes as high, keeled, with two rather shorter cilia between.

In forest streams and among wet epiphytes, rare ; South African only.
Cape, W. Outeniqua, Thunberg.
Cape, E. The Wilderness, George, Miss Taylor ; Storm's River, Burtt-Davy ; Blanco, Rehm. 301 ; Eveleyn Valley, 4000 feet, 1892, and Perie, 3000 feet, 1892, Sim 7027, 7064, and 7054 ; Pondoland, 1832, Dregé.
A specimen named *Hypopterygium struthiopteris* Brid. (C. M., Syn., p. 4), from Kentani, Transkei (Miss Pegler 25), in the Albany Museum, contains both *H. laricinum* and *H. pennaeforme*, but not *H. struthiopteris*, which is a Mascarene species described as having stems simple ; leaves lanceolate-acuminate ; nerve excurrent ; leaves narrowly bordered, distantly denticulate ; cells round ; stipuliform leaves much less, narrowly lanceolate, long-pointed, with subcordate base ; but it has no claim on South Africa.

669. *Hypopterygium lutescens* Horns., in Linn., xv, 144 ; C. M., Syn., ii, 11.

Described as : Stem 3 inches high, dendroid, 2-pinnate ; lower leaves scale-like, ovate-acuminate, entire, almost without nerve ; branches complanate ; leaves erecto-patent, laxly imbricate, distichous, obliquely inserted, oblong-ligulate, obliquely mucronulate, concave, serrated at the apex, pellucid ; nerve obscure, disappearing near the apex ; cells narrowly linear ; stipules distant, subappressed, ovate-acute, deeply concave, serrulate. Cape, Dregé.
(Narrow cells are not *Hypopterygium* ; perhaps *Porothamnium*, T.R.S.)

Family 54. RHACOPILACEAE.

Fairly robust mosses, forming wide, flat, dark green mats, often submerged or almost so ; stems very long, often closely rhizoid-tomentose, more or less regularly pinnate ; the leaves numerous and complanate, dimorphous ; the branches when dry either straight or curled up ; side leaves usually distichous, obliquely inserted, often secund opposite roots ; when dry either flatly spread or reflexed or involute and crisped, when moist spreading, unsymmetrical ; ovate, oblong or oval, obtuse and hair-pointed, without border, or border serrulate or serrate ; nerve strong, single, more or less excurrent, plano-convex in section, with 2-rowed lax front cells. Cells rounded or oval-hexagonal, chlorophyllose, smooth (in our species), those toward the base more lax and oblong, or little differentiated. Front leaves very small, distantly placed, in two rows, imbricate, more or less long-pointed from an oval or cordate base, entire or more or less toothed or serrated. Nerve far excurrent. Inner per. leaves erect, long-pointed from an ovate base, with excurrent nerve. Vaginula closely set with paraphyses in season. Seta elongated, red, smooth ; capsule oblong or cylindrical, when dry deeply furrowed, some of the furrows forked, erect and regular or nearly so, or inclined or horizontal and more or less bent, with short throat. Ring wide, dehiscent ; peristome double ; teeth lanceolate, closely striate and finely papillose, with zigzag mid-line and numerous normal lamellae ; processes equally high, wide, deeply forked, on high basal membrane ; cilia three, well developed, knotted or shortly appendiculate. Spores small. Lid long-beaked from a convex base. Calyptra cucullate, set with a few long, erect, jointed hairs. One genus only.

Genus 224. RHACOPILUM Palis, Prod., 1905, p. 36.

Characters as above. Grows on moist tree stems or on more or less submerged rocks in the tropics and subtropics.

670. *Rhacopilum capense* C. M., in Rehm. 297 and *b*, and in Hedw., 1899, p. 124 ; Bryhn, p. 20 ; Brunnthaler, p. 33 Broth., Pfl., i, 3, 977, and xi, 53.

Stems prostrate, often rooting freely, forming large mats on semi-submerged stones, or straggling among other mosses, autoecious, frequently fertile. Stems 2–4 inches long, pinnate ; the branches lax, irregular and unequal, 1–4 cm. long ; leaves lingulate or ovate-lanceolate or lanceolate, shortly rounded or tapering to the point, with strong central, far-excurrent nerve ; when dry involute, forming a cylindrical tube. Cells with oval lumen, 15–20μ long, the lower cells hardly larger, marginal cells hyaline, and toward the leaf-apex each second cell protrudes with sharp point, making the leaf irregularly serrate in the upper half. Stipular leaves on upper surface much smaller, in two rows, cordate,

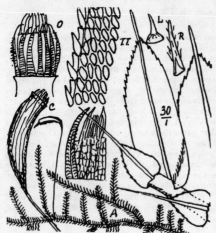

Rhacopilum capense.

with long hair-point, slightly toothed; per. leaves sheathing, lanceolate-subulate, erect, entire, mostly with excurrent nerve and larger rectangular lower cells; enclosed among them are, in season, numerous longer capillary, jointed, hyaline paraphyses; seta 2 cm. long, red; capsule 3·5–4 mm. long, nearly horizontal, cylindrical, somewhat arched, when dry furrowed with forked furrows; peristome normal; membrane very high. Lid rather obliquely beaked from convex base; calyptra cucullate, somewhat hairy. In the Mascarenes there are many species—some only known sterile; and Renauld (pp. 263–270) deals with them, and mentions that Wright (Journ. of Bot., 1888) identified some as *R. tomentosum* Hedw., a South American species which has also been credited to South Africa, but claims that a wider interpretation of *R. tomentosum* than that now in use is necessary for its inclusion; the same applies in continental and South Africa; we have much variation in leaf-form, even on the same plant, and also in length of seta, but we have only one species, whether the name applied be *R. tomentosum* Hedw., in a wide sense, or *R. capense* C. M., in the more restricted sense. But I doubt if the Mascarene species all differ from ours; Renauld gives a short diagnosis of other African species to assist him locally, but that of *R. capense* C. M. (p. 269) does not well describe the South and East African plants; Müller's description of *R. tomentosum* C. M. (Syn., ii, 12) describes it much better, but I have had no South American specimen of it for comparison. Brotherus' grouping is not easily understood to separate these species. Abundant in every forest stream from the Fish River to the Zambesi and beyond; I have no specimen or record from Cape, W., beyond Knysna; in the Transvaal low veld it is the most common stream-moss, even where there is no forest.

From Rehmann's Exsiccata Musci Austro-Africani (numbering 680 numbers besides *b, c, d,* etc., packets, so far as known) the following numbers were not accounted for in Kew Bull., No. 6, 1923, and have not yet been found, and may never have been issued, viz.: 1, 2, 3, 4, 5, 6, 77, 78, 79, 80, 81, 92, 93, 112, 116, 134, 136, 180, 183, 211, 213, 227, 237, 238, 240, 258, 271, 272, 292, 300, 303, 304, 310, 314*b*, 334*c*, 338, 384, 406, 407, 424, 452, 656. The following of his numbers and names are *nomina nuda* and not represented in South African herbaria, and concerning which I have no further information than that given hereunder, but as C. Müller, in 1899, described all that he considered fit of Rehmann's plants, it can be assumed that these required no further attention and can be sunk, unless specimens in Europe prove description necessary, in which case Rehmann's name has no claim, and they become new species, even if these are used, viz.:

11. *Sphagnum coronatum* C. M., var. *fluctuans,* Axelsfarm, Hex River Mountains.
23. *Trematodon tortilis* Rehm., Oakford, Natal (Wood 272, Albany Museum).
35. *Campylopus *tenellus* Rehm., Cape Town, C.P.
36. *C. *mollis,* Rehm., Inanda, Natal (*Palinocraspis Rigida,* Pfl., x, 188).
38. *C. *Sudabyanus* Rehm., Kadziberg, O.F.S.
39. *C. *Schunkei* Rehm., Montagu Pass, C.P.
48. *C. chlorophyllosus* C. M., var. *rivularis* Rehm., Table Mountain, C.P.
49. *C. chlorophyllosus,* var. *compactus* Rehm., Montagu Pass, C.P.
50. *C. *silvaticus* Rehm., Blanco, C.P. (Pfl., x, 186).
57. *C. *turgidus* Rehm., Kadziberg, O.F.S.
65. *C. *Marillacii* Rehm., Montagu Pass, C.P.
68. *C. *echinatus* Rehm., var. *brevipilus* Rehm., Cape Town, C.P.
103. *Barbula *plicata* Rehm., Caledon River, O.F.S. (see under *Weisiopsis*).
105. (*B. *torquescens* Schp., Cape Town, but not so at Kew.)
143. *Amphoridium *africanum* Rehm., Witteberg, Kadziberg, O.F.S.
148. *Zygodon *per-reflexus* C. M., Rondebosch, C.P.
176. *Entostodon *crassipes* Rehm., Gamkoo River, C.P. (Paris omits it and enters *Barbula syrrhopodontioides* as 176 instead of 476).
178. *E. *Kadzianus* Rehm., Kadziberg, O.F.S.
190. *Glyphocarpus *pseudocomosus* Rehm., Kadziberg, O.F.S.
195. *Philonotis *hyophila* Rehm., Kadziberg, O.F.S.
196. *P. *molmonica* Rehm., Molmonspruit, O.F.S.
212 and 541. *Mielichhoferia *elongata* Rehm., Van Reenen, Natal. (This name is preoccupied.)
225. *Webera *glandiformis* Rehm., Cape Town, C.P.
243. *Bryum *dimorphum* C. M., altered by Paris to *B. *afrodimorphum* as other name was preoccupied by *Mnium dimorphum* C. M.=*Bryum,* Broth. (Not in Kew Herb. nor British Museum.)
264. *Atrichum *nudifolium* C. M., Blanco, C.P.
268. *Pogonatum *tortifolium* Rehm., Umgeni, Natal.

288. *Fissidens *glaucescens* Hsch., var. *remotifolius* Rehm., Van Reenen.
299. *Hypopterygium *laricinum* Br., var. *compactum*, Montagu Pass, C.P.
308. *Braunia *flagellaris* Rehm., Worcester, C.P.
309. *B. *pseudodiaphana* Rehm., Bethlehem, O.R.C.
349. *Fabronia *oraniae* Rehm., Bloemfontein, O.R.C.
350. *F. *capensis* Rehm., Rondebosch.
436. *Gymnostomum *obtusatum* Rehm., Lechlaba, Transvaal.
457. *Phascum *assimile* C. M., Cape Town.
460. *Pottia *verrucosa* Rehm., Graaff Reinet, C.P.
524. *Funaria *serrata* Rehm., Cape Town (name preoccupied).
589. *Fissidens *pseudorufescens* Rehm. (mixed here).
612. *Papillaria *nitidula* Rehm., Lechlaba, Natal.
631. *Schwetschkea *aristaria* Rehm., Lechlaba, Transvaal.
657. *Eurhynchium *Macleanus* Rehm., Macmac, Transvaal.
658. *E. *ovatifolium* Rehm., Macmac, Transvaal.

There are also a few names which are represented in Rehmann's collection by other numbers under the same name and are dealt with herein, though not from the following numbers, which have not been seen, viz. :

14. *Sphagnum oligodon* Rehm., Inanda, Natal.
17b. *S. mollissimum* C. M., Montagu Pass, C.P.
100. *Barbula *basutensis* Rehm., Caledon River, O.F.S. (placed in *Hyophila* in R. 458).
248b and c. *Bryum afro-alpinum* Rehm., Kadziberg, O.F.S.
250b. *B. *porphyroloma* C. M., Kadziberg, O.F.S.

Also a few more or less described species have not been seen, viz. :

95. *Barbula Laureri* Ltz., Cape Town (evidently *B. Laureriana* Ltz., see from index).
388. *Plagiothecium sphagnadelphus* C. M. (see *P. rhynchostegioides*).

Additional Species. (See page 313.)
671. *Breutelia tabularis* (Dixon, MSS.) Sim (new species).*

Lax, yellowish, shining ; stems 3–6 inches high, simple or with few irregularly placed branches, densely leafy, with leaves the stem is 10–12 mm. wide ; leaves erecto-patent or more spreading whether dry or wet, not sheathing, 3–6 mm. long, roundly and widely ovate-lanceolate at the base, lanceolate at the apex, plicate all along when dry,

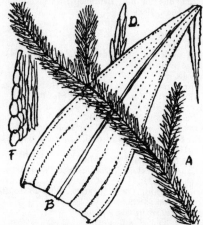

Breutelia tabularis.

at the base only when moist ; nerve narrow, extending to the denticulate apex ; margin denticulate but strongly revolute nearly its whole length ; cells very long and narrow, chlorophyllose ; lower cells rather shorter, interporous, all nearly smooth, but drawn to or shortly 1-papillose at the base ; alar cells few, large, rounded, lax, pellucid ; leaf-base often yellowish ; not seen fertile or floriferous.

Resembles *B. subgnaphalea*, " but is much larger, with different habit, scarcely at all tomentose, the leaves much less finely acuminate, and the margin strongly recurved " (Dixon).

Cape, W.   Platteklip Ravine, Table Mountain, Sim 9277 (where it is mixed with an exceedingly similar *Leucoloma*) ; apparently rare, in partly shaded spots among large boulders at base of southern slopes of hills north of Woodhead Reservoir, Table Mountain, Jan. 1919, N. S. Pillans 3335 and 4899.

---

* *Breutelia tabularis* (Dixon, MSS.) Sim (sp. nov.).
   Major, flavus, nitens, ramispaucis, non-tomentosis ; folia plurima, non-amplexicaulis, 3–6 mm. longa, ovata-lanceolata basi, lanceolata apice, sicca multiplicata ; margine ubique valde revoluto ; cellulis elongatis, alaribus paucis, rotundatis, pellucidis.

Bibliography of the most necessary works on Bryophyta and contractions used herein.

Bescherelle, E., Florule Bryologique de la Reunion, 1879.
„   " Flora de l'Ile de la Reunion, Hepatiques " (in Journ. de Bot., 1895).
Bescherelle and Spruce, Bull de la Soc. Bot. de France, 1889 (*re* Madagascar).
Bridel's Muscologia Recentiorum, 1797–1801.
„   Muscologia Supplementum, 1806–1817.
„   Bryologia Universa, 1826–1827.

Pfl.   Brotherus, in Engler and Prantl's Die Naturlichen Pflanzenfamilien, Mosses, 1893, and in vols. x, 1924,
                and xi, 1925.
„   in Engler's Botanische Jahrbucher, 1895.
„   V. F., "Ergebn. siner Botanisch. Forschungsreise nach Deutsch. Ostafrika and Sud-Afrika
                (Kapland, Natal, and Rhodesia) Musci," Denkschr. der Math. Nat. K.K. Akad. Wien.,
                Bd. 88, 1913.
Bryhn, N., "Bryophyta nonnulla in Zululand collecta," Videnskapsselskapets Forh., 1911, No. 4,
                Kristiana, 1911.
Campbell, Mosses and Ferns, 1895.
Cardot, The Mosses of the Azores, 1897.
Carrington, B., British Hepaticae, 1874–1876, and Exsiccatae of same, 290 numbers, 1872–1890.
Cavers, The Interrelationships of the Bryophyta.
Dixon, H. N., Student's Handbook of the British Mosses.
„   "New and Interesting S.A. Mosses," Trans. R. Soc. S. Afr., 1920.
„   "Bryophyta of S. Rhodesia," S.A. Journ. Sc., June 1922 (Mosses, Dixon ; Hepaticae,
                Sim).
„   "Studies in the Bryology of New Zealand," N.Z. Inst. Bull. 3 (1913–1914).
„   "New and Interesting South African Mosses," Trans. R. Soc. S. Afr., vol. viii (1920).
„   Many other bryological papers.
Dregé, Zwei-pflanzen Geographische Documente, 1843.
Dum.   Dumortier, Sylloge Jungermannidearum indegenarum Europae, 1831.
„   "Hepaticae Europeae" (in Bull. Soc. R. bot. Belg., 1874–1875).
Dusen, New and little-known Mosses from the West Coast of Africa, 1895–1896.
Evans, A. W., Revised List of New England Hepaticae, 1913.
Fleischer, in Hedwigia, 1890 onward, on "Laubmoosflora von Java."
Goebel, Organography of Plants, 1905.
Gottsche, C. M. (see Nees).
Gray, S. F., Nat. Arr. Brit. Plants.
Hampe, Icones Muscorum, 1844.
„   "Ennumeratio Hepaticarum, etc.," in Linn., xi, 552 (1854).
Harvey, Genera of South African Plants, 1st ed., 1838.
„   Thesaurus Capensis.
Hedweg, Species Muscorum Frondosorum, edited after Hedweg's death by Schwaegrichen, 1801–1856.
„   Descriptio et adumbratio Muscorum, etc., 1787–1797.
Hooker, W. J., Musci Exotici, 1817–1820 (176 plates).
„   British Jungermanniae, 1813–1817.
Hooker's Handbook of the New Zealand Flora, vol. ii, 1867.
Hornschuch, "Muscorum Frondosorum Novorum," Linnaea, xv, 1841.
Howe, A. M., Californian Hepaticae.
Jaeger, Genera and Species Muscorum, 1871–1879.
Lehmann, "Hepaticae Capensis Ecklonii," in Linnaea, 1829–1831.
„   Hepaticae Capensis, 1834.
Lehmann and Lindenberg, in Lehm., Pugill. pl. stirp., 1828–1831, afterwards by Lehmann to 1857.
Lindenberg, Species Hepaticarum, 1829.
Linnaea, Papers by various writers in many parts.
Linneus, Systema Naturae, 1735 ; ed. xiii, 1793.
Sp. Pl. „   Species Plantarum, 1753, 1764.
„   Genera Plantarum, 1737 ; ed. iii, 1752.
Macvicar, Student's Handbook of the British Hepaticae, 1912.
Marloth's Flora of South Africa (Mosses by Brotherus, Hepaticae by Diels).
Mitten, in Hooker's Handbook of the N.Z. Flora, 1867.
„   in Journ. Linn. Soc., No. 91 (1878), and No. 146 (1887).
Montagne, in Webb and Berthlot's Hist. Nat. des Isles Canar.
C. M. ⎫ Müller, C., Synopsis Muscorum Frondosorum.
Syn. Musc. ⎭ „   in Hedwigia, vol. xxxviii (1899).
„   in Botanische Zeitung, 1855–1864.
„   in Botanisches Centralblad, 1881.
„   in Flora, 1887–1888.
„   in Oesterreichische Botanische Zeitung, 1897.
„   Musci Schweinfurthiana.
Nees von Esenbeck, in Linnaea, 1831, etc.

N. G. L. Nees, Gottsche, and Lehmann, Synopsis Hepaticarum, 1844–1847.

Paris. Paris, Index Bryologicus, 2nd ed., 1903.

Hep. Nat. Pearson, W. H., "Hepaticae Natalensis," Christ. Viden. Sels. Forh., 1886, No. 3.

Hep. Knys.   ,, "Hepaticae Knysnanae," Christ. Viden. Sels. Forh., 1887, No. 9.

    ,, "Hepaticae Madagascar," Christ. Viden. Sels. Forh., 1891, No. 2, and 1892, Nos. 8 and 14.

    ,, The Hepaticae of the British Isles, 1899–1902.

Rehm. Rehmann, Dr. A., Exsicc. Musci Austro-Africani, 1875–1877.

    ,, List of same in Kew Bulletin, No. 6, 1923, Dixon and Gepp.

    Renauld, Prodr. de la Flore Bryologique de Madag., etc., 1897.

    Renauld and Cardot, Histoire des Mousses de Madag., 1899–1914.

Pfl. or }
Nat. Pfl. } Schiffner, in Engler and Prantl's Die Naturlichen Pflanzenfamilien Hepaticae, 1893.

    Schimper, Synopsis Muscorum Europaeorum, 1876.

    Schwaegrichen, Prodr. Hist. Hep., 1814.

    Shaw, in Cape Monthly Magazine, 1878.

    Sim, T. R., "South African Hepaticae," S.A. Journ. Sc., May 1916.

    ,, "The Geographical Distribution of the S.A. Bryophyta," S.A. Journ. Sc., April 1918.

    ,, Check-list of the Bryophyta of South Africa, 1915.

    ,, Handbook of the Bryophyta of South Africa, pts. 1–2, 1916.

    Sprengel, Syst. Veget., 1827.

Spruce. Spruce, "Hepaticae Amazonicae et Andinae," Trans. Bot. Soc. Edin., xv, 1885.

    ,, On Cephalozia, 1882.

St., Steph. Stephani, F., "Species Hepaticarum," in Bull. de l'Hb. Boissier, and in part atterwards issued separately.

    ,, in Hedwigia, etc., and in Engler's Bot. Jahrb. (1895).

Syn. Hep. Synopsis Hepaticarum, Gottsche, Lindenberg, and Nees, 1845–1847.

Thun. Thunberg, Prod. Fl. Cap., 1794–1800.

    ,, Flora Capensis, 1813.

    Warnstorf, in Engler and Prantl's Die Naturlichen Pflanzenfamilien Sphagnaceae, 1893.

    ,, in Hedwigia, 1890.

    ,, in Engler's Pflanzenreich.

    Weber, Prod. Hep., 1816.

INDEX

PRINTED IN GREAT BRITAIN BY NEILL AND CO., LTD., EDINBURGH.